# Op Amps and Linear Integrated Circuits for Technicians

# Op Amps and Linear Integrated Circuits for Technicians

## Second Edition

Frank R. Dungan

President, FRADUN Enterprises

## NOTICE TO THE READER

Publisher does not warrant or guarantee any of the products described herein or perform any independent analysis in connection with any of the product information contained herein. Publisher does not assume, and expressly disclaims, any obligation to obtain and include information other than that provided to it by the manufacturer.

The reader is expressly warned to consider and adopt all safety precautions that might be indicated by the activities described herein and to avoid all potential hazards. By following the instructions contained herein, the reader willingly assumes all risks in connection with such instructions.

The publisher makes no representations or warranties of any kind, including but not limited to, the warranties of fitness for particular purpose or merchantability, nor are any such representations implied with respect to the material set forth herein, and the publisher takes no responsibility with respect to such material. The publisher shall not be liable for any special, consequential or exemplary damages resulting, in whole or in part, from the readers' use of, or reliance upon, this material.

Cover photo courtesy of Signetics Company

Delmar Staff
Administrative Editor: Wendy Welch
Project Editor: Laura Gulotty
Senior Production Supervisor: Larry Main

Art and Design Manager: Rita C. Stevens
Art Coordinator: Michael Nelson
Senior Design Supervisor: Susan C. Mathews

For more information address Delmar Publishers Inc.,
2 Computer Drive West, Box 15-015
Albany, New York 12212

Copyright © 1992
By Delmar Publishers Inc.

All rights reserved. Certain portions of this work copyright © 1984 under the title *Linear Integrated Circuits for Technicians.* No part of this work covered by the copyright hereon may be reproduced or used in any form or by any means—graphic, electronic, or mechanical, including photocopying, recording, taping, or information storage and retrieval systems—without written permission of the publisher.

Printed in the United States of America
Published simultaneously in Canada
by Nelson Canada,
A division of The Thomson Corporation

10 9 8 7 6 5 4 3 2 1

**Library of Congress Cataloging-in-Publication Data**

Dungan, Frank R., 1924–
Op amps and linear integrated circuits for technicians / Frank R. Dungan.—2nd ed.
p. cm.
Rev. ed. of: Linear integrated circuits for technicians.
Includes index.
ISBN 0-8273-5086-4 (textbook).—ISBN 0-8273-5087-2 (instructor guide)
1. Linear integrated circuits.
I. Dungan, Frank R., 1924–
Linear integrated circuits for technicians.
II. Title.
TK7874.D87 1991
621.3815—dc20

91–29271
CIP

# Contents





920796

# Preface

Industry demands that electronic technicians possess a thorough knowledge of modern linear electronics. Unfortunately, however, available text material on linear integrated circuits has been unequal to the task. Existing books have ranged from highly theoretical, design-based presentations to project-oriented books addressed to the electronics hobbyist. Integrated circuit "handbooks" are available, but these are more useful to the technician working in industry than they are to the student who is encountering the subject for the first time. This book has been written to remedy the situation and provide future technicians with the knowledge of linear integrated circuits that is needed on the job or that is prerequisite for other electronics courses.

*Op Amps & Linear Integrated Circuits for Technicians* is a descriptive approach to the subject that minimizes mathematical analysis and design considerations while emphasizing applications and troubleshooting techniques. It is intended for use in the first course in linear integrated circuits in an electronics curriculum. Prerequisites include a course in dc/ac fundamentals and a basic understanding of semiconductor devices. The mathematics in the presentation is limited to algebra and trigonometry. To ensure student understanding of the material, a full range of learning aids is presented in the book including chapter objectives, chapter introductions, frequent examples, and extensive end-of-chapter questions and problems. For reference, appendixes containing extensive series of data sheets are included.

Revisions in this edition include expanded examples and end-of-chapter questions and problems, new devices, more data sheets, and application notes for the devices discussed in the text. A breakdown of the chapters is as follows:

**Chapter 1**—Linear and digital signals and devices are compared, an overview of linear applications is given, and a simple explanation of fabrication and packaging is provided.

**Chapter 2**—The simple operational amplifier is introduced as the building block for linear integrated circuits. The 741 is used as a model because it is the most available device that is easily used and extremely versatile. For clarification, this chapter provides a review of the internal stages of the op amp, its operating characteristics, and some safety precautions for its use.

**Chapter 3**—The basic op amp circuits, many of which are used as the basis for further examination in later chapters, are introduced in Chapter 3. A thorough knowledge of the circuits in this chapter is essential for the reader to easily gain an understanding of the circuits to follow.

**Chapter 4**—An in-depth study of comparators and their many uses are covered. The comparator as a sensing device is important to linear circuit applications.

**Chapter 5**—For devices to operate within specified limits, voltage and current regulation is essential. This vital subject is covered in Chapter 5. The additional topics of heat sinking and regulator protection circuits are also included.

**Chapter 6**—The three basic types of oscillators and the basic wave-shaping circuits are covered in this chapter. These basic circuits are used as a basis for more complicated circuits.

**Chapter 7**—Active filters, important in all analog systems, are covered at length in Chapter 7. Simple design techniques are demonstrated, and the reader is introduced to the concept of higher-order filters.

**Chapter 8**—The major signal processing circuits, important circuits for industrial operations, are introduced. Instrumentation devices and amplifiers are examined.

**Chapter 9**—Important subjects for both digital and analog device study, the digital-to-analog converter and the analog-to-digital converter, are discussed in this chapter. Several applications for these devices are demonstrated.

**Chapter 10**—An introduction to communications systems is provided in Chapter 10. This chapter covers the basic concepts of modulation and demodulation in AM and FM systems in a straightforward, easily understood manner. Compandors and phase-locked loop devices are also covered.

**Chapter 11**—One of the more versatile devices, the timer, is covered in Chapter 11. The well-known 555 timer is used as the model, but other devices used in more specialized systems are also covered.

**Chapter 12**—Many special circuits and devices are introduced and examined. Circuits used in consumer products and other practical circuits are discussed.

**Chapter 13**—Troubleshooting, always an important subject for the electronics technician, is discussed in Chapter 13. Several practical troubleshooting methods are discussed, as are typical integrated circuit failures.

Appendixes at the end of the text include manufacturers' data sheets and application notes that provide additional circuits and applications. Also provided at the end of the text are a glossary of terms and answers to selected end-of-chapter questions and problems. The accompanying laboratory manual, *Experiments in Integrated Circuits for Technicians,* provides correlated lab work and is available from Delmar Publishers. An instructor's manual, also available from the publisher, provides answers to the text problems and solutions to lab experiments.

For their valuable suggestions during the preparation of this new edition, the author extends sincere thanks to Richard G. Anthony, Cuyahoga Community College; John W. Berry, DeVry Institute of Technology; Jim Hallam, Community College of Spokane; Seyed A. Hosseini, Savannah Technical School; George J. Lynn, East Central Community College; Joe McLaughlin, Northern Maine Technical College; Gary Mullett, Springfield Technical Community College; and Bill Robertson, ITT Technical Institute.

Those who participated in the first edition continue to impact the second and their contributions are greatly appreciated: George Bruce, Spartanburg Technical College; Lou Gross, Columbus Technical Institute; Gerald E. Jensen, Western Iowa Technical Community College; and Jerry D. Mullen, Tarrant County Junior College—South Campus.

Also, thanks are extended to the editorial and production staff of Delmar Publishers for their invaluable assistance in the publication of this revised text.

# Manufacturer Acknowledgments

**Chapter 1:** Figures 1-3, 1-4, 1-6, 1-7, 1-8, 1-9, 1-10, Signetics Company, copyright 1980; Figures 1-5, 1-11, Part A: Signetics Company, copyright 1989; Part B: National Semiconductor Corporation, copyright 1980; Figure 1-13, Fairchild Camera & Instrument Corporation, copyright 1982. **Chapter 2:** Figures 2-2, 2-12, Signetics Company, Linear Division, copyright 1989. **Chapter 3:** Figure 3-9, Signetics Company, copyright 1976; Figures 3-20, 3-21, 3-22, 3-23, National Semiconductor Corporation, copyright 1980; Figure 3-24, Harris Semiconductor Products Division, copyright 1980; Figure 3-25, National Semiconductor Corporation, copyright 1980. **Chapter 4:** Figure 4-2, Signetics Company, copyright 1989; Figures 4-6, 4-7, 4-8, 4-10, National Semiconductor Corporation, copyright 1980; Figure 4-12, Signetics Company, copyright 1989; Figures 4-14, 4-15, National Semiconductor Corporation, copyright 1980. **Chapter 5:** Figures 5-1, 5-2, RCA Corporation, copyright 1981; Figure 5-4, Table 5-1, Fairchild Camera & Instrument Corporation, Linear Division, copyright 1978; Figures 5-7, 5-8, John Wiley & Sons, Inc., copyright 1981; Figures 5-9, 5-10, National Semiconductor

Corporation, copyright 1980; Figure 5-11, National Semiconductor Corporation, copyright 1973; Figures 5-12, 5-13B, National Semiconductor Corporation, copyright 1980; Figure 5-13A, National Semiconductor Corporation, copyright 1973; Figure 5-14, Fairchild Camera & Instrument Corporation, copyright 1978; Figure 5-15, Raytheon Company, copyright 1978; Figures 5-16, 5-17, Signetics Company, copyright 1989; Figures 5-18, 5-19, National Semiconductor Corporation, copyright 1980; Figures 5-20, 5-21, 5-22, 5-23, Fairchild Camera & Instrument Corporation, copyright 1978; Figure 5-24, John Wiley & Sons, copyright 1981; Figures 5-25, 5-30, 5-31, Fairchild Camera & Instrument Corporation, copyright 1978. **Chapter 6:** Figures 6-13, 6-15, 6-30, National Semiconductor Corporation, copyright 1980. **Chapter 7:** Figure 7-30, National Semiconductor Corporation, copyright 1980. **Chapter 8:** Figures 8-4, 8-5, 8-6, 8-7, 8-8, reprinted with permission from book #2000 *Encyclopedia of Electronics,* by Stan Gibilisco and Neil Sclater, copyright © 1985 by **TAB BOOKS,** a division of McGraw-Hill, Blue Ridge Summit, PA 17294 (1-800-233-1128 or 1-717-794-2191); Figures 8-9, 8-10, 8-11, 8-12, 8-13, 8-14, 8-15, 8-17, 8-18, 8-19, 8-20, National Semiconductor Corporation, copyright 1989. **Chapter 9:** Figures 9-1, 9-2, National Semiconductor Corporation, copyright 1980; Figures 9-10, 9-11, National Semiconductor Corporation, copyright 1980; Figures 9-3, 9-8, 9-12, 9-14, 9-17, 9-18, 9-19, 9-20, 9-21, 9-22, Signetics Company, copyright 1989; Figure 9-27, National Semiconductor Corporation, copyright 1980; Figures 9-30, 9-31, 9-32, 9-33, 9-34, Signetics Company, copyright 1989. **Chapter 10:** Figures 10-8, 10-9, 10-10, 10-11, National Semiconductor Corporation, copyright 1980; Figure 10-15, RCA Corporation, copyright 1981; Figure 10-16, National Semiconductor Corporation, copyright 1980; Figures 10-17, 10-18, 10-19, Signetics Company, copyright 1989; Figure 10-21, National Semiconductor Corporation, copyright 1980; Figure 10-22, Raytheon Company, Semiconductor Division, copyright 1978; Figure 10-23, Breton Publishers, copyright 1982; Figure 10-25, Raytheon Company, Semiconductor Division, copyright 1978; Figures 10-26, 10-27, Signetics Company, copyright 1989. **Chapter 11:** Figure 11-1, Fairchild Camera & Instrument Corporation, Linear Division, copyright 1982; Figure 11-2, Part A: Fairchild Camera & Instrument Corporation, Linear Division, copyright 1982; Part B: National Semiconductor Corporation, copyright 1980; Figure 11-4, Fairchild Camera & Instrument Corporation, Linear Division, copyright 1982; Figure 11-6, National Semiconductor Corporation, copyright 1980; Figure 11-9, Part B, Signetics Company, copyright 1989; Figures 11-10, 11-11, 11-12, 11-13, 11-14, Signetics Company, copyright 1989; Figures 11-15, 11-16, 11-17, 11-18, 11-19, 11-20, 11-21, Fairchild Camera & Instrument Corporation, Linear Division, copyright 1982; Figures 11-22, 11-23, 11-24, 11-25, 11-26, 11-27, 11-28, National Semiconductor Corporation, copyright 1980; Figure 11-29, Fairchild Camera & Instrument Corporation, Linear Division, copyright 1982; Figures 11-30, 11-31, National Semiconductor Corporation, copyright 1980; Figures 11-33, 11-34, Fairchild Camera & Instrument Corporation, Linear Division, copyright 1982; **Chapter 12:** Figures 12-1, 12-2, 12-3, 12-4, 12-5, 12-6, 12-7, 12-8, National Semiconductor Corporation, copyright 1980; Figures 12-9, 12-10, Signetics Company, copyright

1989; Figures 12-11, 12-12, 12-13, 12-14, 12-15, 12-16, 12-17, 12-18, Motorola Semiconductor Products, Inc., copyright 1981; Figures 12-19, 12-20, 12-21, 12-22, 12-23, Table 12-1, 12-25 (Part B), National Semiconductor Corporation, copyright 1980; Figures 12-26, 12-27, Fairchild Camera & Instrument Corporation, Linear Division, copyright 1982; Figure 12-28, Signetics Company, copyright 1989; Figures 12-32, 12-33, 12-34, 12-35, 12-36, 12-37, 12-38, 12-39, National Semiconductor Corporation, copyright 1980. **Chapter 13:** Figure 13-8, Signetics Company, copyright 1989; Figures 13-9, 13-10, 13-11, 13-12, National Semiconductor Corporation, copyright 1980. **Appendix A:** Fairchild Camera & Instrument Corporation, copyright 1982. **Appendix B:** National Semiconductor Corporation, copyright 1980. **Appendix C:** Fairchild Camera & Instrument Corporation, copyright 1982. **Appendix D:** Fairchild Camera & Instrument Corporation, Linear Division, copyright 1982. **Appendix E:** Signetics Company, copyright 1989. **Appendix F:** Raytheon Company, Semiconductor Division, copyright 1978. **Appendix G:** Signetics Company, copyright 1982. **Appendix H:** Signetics Company, copyright 1982. **Appendix I:** Signetics Company, copyright 1989. **Appendix J:** Signetics Company, copyright 1989. **Appendix K:** Signetics Company, copyright 1989. **Appendix L:** Reprinted with permission from Raytheon Company, Semiconductor Division, copyright 1989, Linear Integrated Circuits, pp. 11-9–11-30. **Appendix M:** National Semiconductor Corporation, copyright 1980.

# Chapter 1

# Linear Integrated Circuits

## 1.1 Introduction

This chapter provides an overview of the subject of linear integrated circuits (LICs) and introduces basic concepts. You will learn what linear integrated circuits are, how they are used, and why they are used. You will also see how linear circuits are implemented with hardware. This chapter provides the background and base upon which you will build your knowledge of LIC principles and applications. This foundation will put the concept of LICs into perspective so you can relate them directly to the field of electronics. The information presented here will give you a clear understanding of the need for and uses of LICs.

LICs are so widely used today that it is almost impossible to think of electronic equipment without them. LICs are used in virtually every area of electronics, even in digital systems. They have greatly improved the reliability and capability of electronic equipment. At the same time, they have greatly reduced the cost of building and repairing electronic equipment. There is also strong potential for further advances and improvements.

# 1.2 Objectives

When you complete this chapter, you should be able to:

- ☐ Distinguish between analog and digital signals.
- ☐ Define the characteristics of analog signals.
- ☐ Distinguish between linear and digital applications.
- ☐ Define the purposes of linear systems.
- ☐ Describe the fabrication, packaging, and numbering of LICs.

# 1.3 Comparing Linear and Digital Signals and Devices

There are two basic types of electronic signals and techniques used to generate, transfer, and process information. These are digital and analog signals.

## Digital Signals

*Digital* signals are essentially a series of pulses or rapidly changing voltage levels (usually two) that vary in discrete (individual or single) steps or increments. This two-level, rapid-switching characteristic is fundamental to digital signals.

Figure 1–1 shows several examples of digital signals. Notice how these signals switch sharply between two distinct levels. In Figure 1–1A, the signal switches between 0 V and +5 V, the ideal values for the standard logic signals. A more practical example is shown in Figure 1–1B, in which the transition occurs between some level above 0 V (0.8 V) and some level less than the ideal +5 V (2.8 V). The signal in Figure 1–1C alternates between equal positive (high) and negative (low) values. Finally, Figure 1–1D represents a signal in which the highest value is 0 V and the lowest value is –5 V, ideal values for a negative-going signal. The sharpness of the rise and fall of the signals shown in Figure 1–1 would be ideal, but, practically, the ideal condition cannot be met because of the finite delay time in transition from one level to another inherent in all electronic devices and circuits, although it can be approached. Electronic circuits that generate and process on-off, true-false, and up-down signals are called digital, logic, or pulse circuits.

## Analog Signals

The second form of signal is the *analog* signal. We will concentrate on analog signals because they are used in linear systems. An analog signal is an alternating

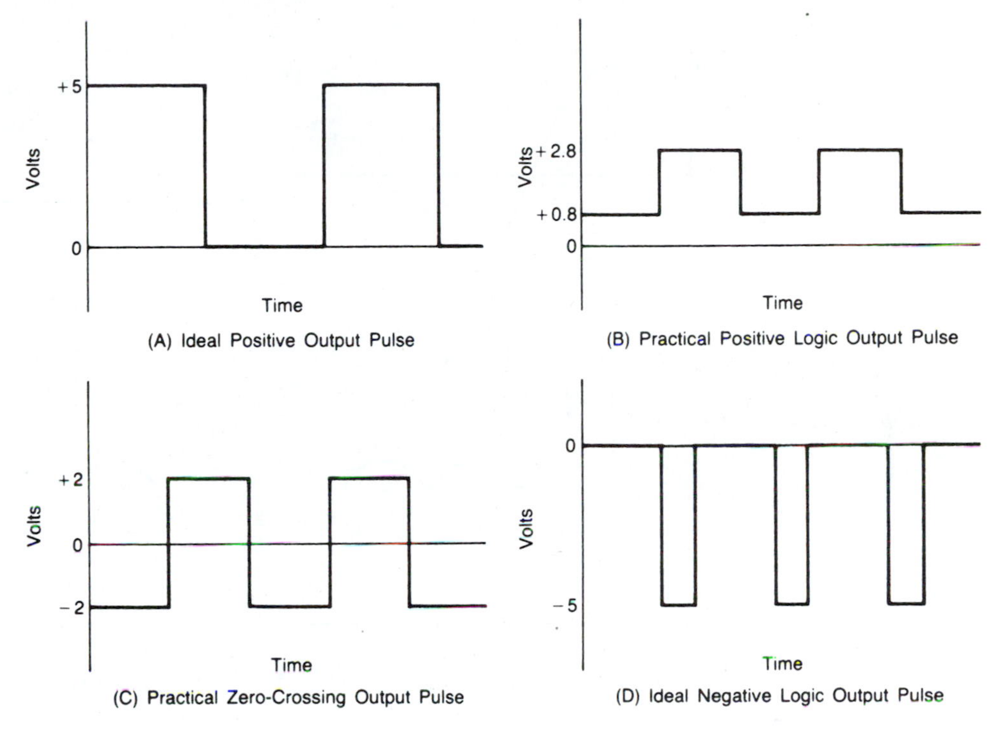


**Figure 1-1.** Digital Signal Waveforms

current (ac) or direct current (dc) voltage or current that varies smoothly or continuously. It is a signal that does not change abruptly or in steps.

Analog signals can exist in a wide variety of forms. Figure 1-2 shows six types of analog signals. The sine wave, shown in Figure 1-2A, is the most common analog signal. Some examples of *sinusoidal* (sine wave) applications include the audio tones and radio waves. A fixed dc voltage, as shown in Figure 1-2B, is also an analog signal, even though it may appear to be a digital signal. Consider the fact that a battery has a fixed dc output. The output value will gradually decrease over time, even if the battery is unused. Other examples of analog signals are a pulsating dc voltage, shown in Figure 1-2C, and a varying dc voltage or current. Note that these varying signals can assume either positive values, as in Figure 1-2D, or negative values, as in Figure 1-2E. Finally, Figure 1-2F shows a random but continuously varying waveform. There are an infinite number of such analog signals.

## Some Comparisons

Now we will examine analog and digital methods by using devices that are familiar to you. Table 1-1 lists a number of such items. The light bulb is an interesting

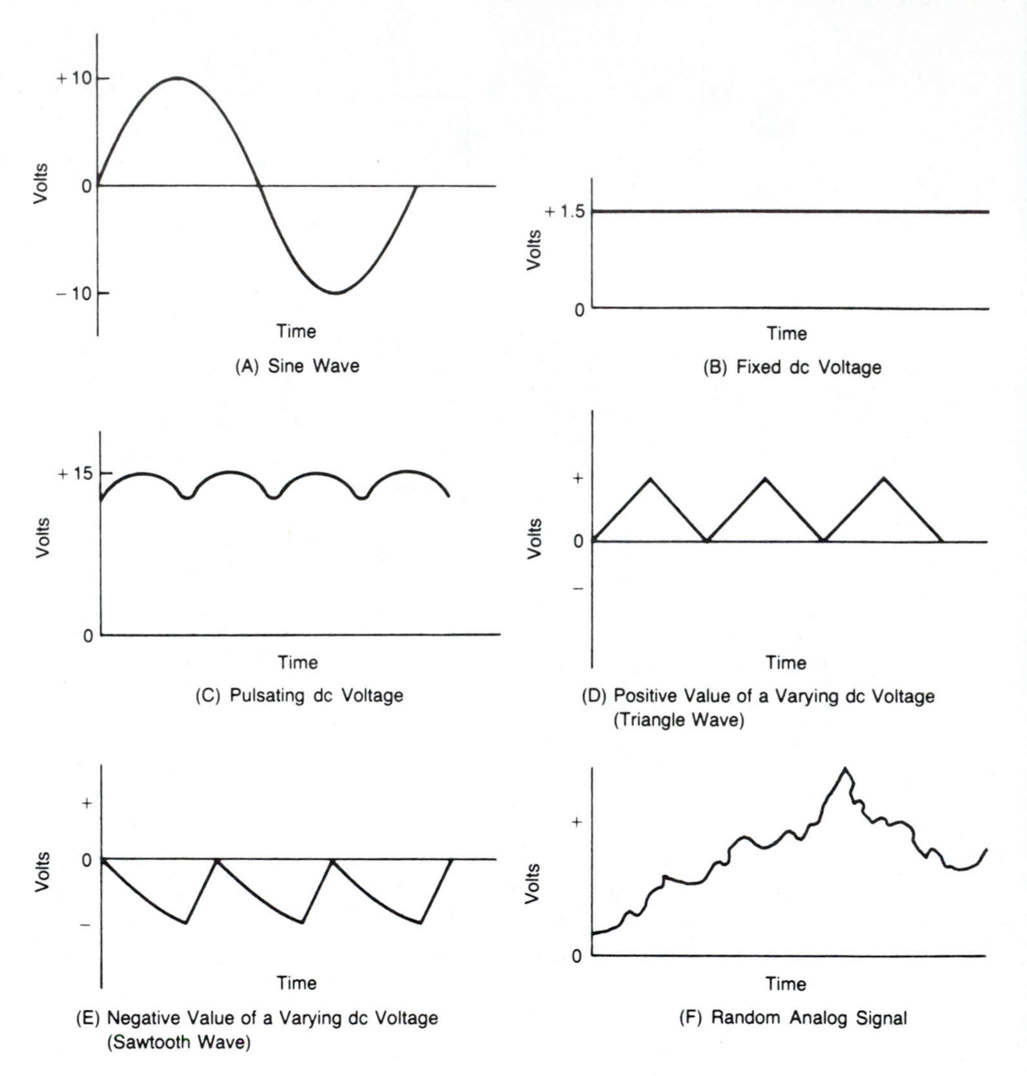


**Figure 1–2.** Analog Signal Waveforms

example to note, because it can be either a digital or an analog device, depending on how it is used. By using a dimmer in the circuit, the amount of current through the light bulb can be set to any level we choose, up to its maximum rated value. We can vary the current through the bulb continuously and thus vary its brightness over an infinite number of levels. The light bulb is an analog device when used in this manner.

The light bulb can also be used as a digital device, for example, when the current through the bulb and the brightness of the bulb vary in discrete steps. Without the dimmer in the circuit, a switch turns the lamp on or off at a set brightness level. Another example of the light bulb used as a digital device is

**Table 1–1 Comparison of Analog and Digital Devices**

| Analog | Digital |
|---|---|
| Electric light bulb | Electric light bulb |
| Volume control on a radio or TV set | TV channel selector |
| Variac or dimmer control | Any type on-off switch or switch with positive detents |
| Automobile speedometer | Automobile odometer |
| Typical wall clocks and wristwatches | Digital clocks and wristwatches |
| Meters: standard voltmeter, ohmmeter, and ammeter | Meters: digital readout multimeters |

the three-way lamp, which has three different light levels that are obtained by discrete steps of a switch.

Many of the other devices listed in Table 1–1 that we think of in analog terms also appear in the digital column. For example, the speedometer on your car is an analog device, but newer models of cars offer the speed indicator in digital form. For comparison, consider the analog speedometer and the odometer portion of that speedometer. While the speed is indicated by the needle in a continuously varying and smooth manner, the odometer indicates the miles traveled in a digital manner. Since the odometer records mileage in increments of one mile or one-tenth mile, it is a digital device. You should be able to list many more comparisons of analog and digital devices on your own.

## 1.4 Linear Systems Defined

A *linear system* can be defined as any unit or assembly of components that generates or processes the analog signals we have discussed. Such systems are everywhere, performing a multitude of tasks. We have but to look in our homes to see examples of their uses. Radio and television receivers are prime examples of applications of these systems. The power supplies of the receivers are regulated with analog devices; the filtering action of the television tuner and radio station selector chooses only specific frequencies, rejecting all others; and the various amplifiers in both radio and television receivers are linear devices.

Today, almost all the units and assemblies mentioned in the preceding paragraph are packaged as LICs, the subject of this book.

# 1.5 An Overview of Linear Applications

The broad capabilities of LICs make them an important and vital part of electronic circuitry. Table 1–2 lists some applications of LICs. The first column identifies the unit or system that the integrated circuit (IC) is used in, and the second column identifies the purpose or function of the IC. These applications will be discussed in detail in later chapters.

Table 1–2 lists only a few of the many uses of LICs. Some additional applications are timers, phase-locked loops, logarithmic amplifiers, and multipliers.

**Table 1–2 Applications of Linear Integrated Circuits**

| System | Purpose |
|---|---|
| Power supplies | Regulate voltage or current |
| Radio and TV receivers | Amplify audio, RF, and IF signals |
| Signal generators | Generate and shape desired waveforms such as square, triangle, and sawtooth waves |
| Instrumentation amplifiers | Process output signals from instruments and transducers, such as strain gauges and thermocouples |
| Communications systems | Generate, modulate, and detect desired frequencies as well as filter and amplify modulated signals |
| Converters | Convert voltage to current and current to voltage (Important interface devices are analog-to-digital (ADC) and digital-to-analog (DAC) converters) |
| Oscillators | Provide necessary amplification and phase inversion |

# 1.6 Fabrication of Linear Integrated Circuits

You do not need to know how a device is constructed in order to use it, but there are advantages to understanding the construction. By knowing how the IC is fabricated, you will be better prepared to follow handling precautions and to select the right device for the job at hand.

There are four basic ways to manufacture integrated circuits:

1. Monolithic (the most widely used method)
2. Thin-film
3. Thick-film
4. Hybrid

Hundreds of integrated circuits are fabricated side-by-side on a single silicon wafer that is approximately 0.25 mm thick and two or more inches in diameter. Each of the ICs may contain thousands of elements. The circuitry is built up layer-by-layer.

## Monolithic Construction

A *monolithic* IC is constructed entirely on a single silicon chip. The various junctions that make up components such as diodes, transistors, resistors, and capacitors are formed by diffusing semiconductor materials into the basic semiconductor substrate. The materials to be diffused into the substrate are in gaseous form, deposited on the substrate through masking operations, under high temperature. With this method, the entire circuit is formed on a single base, providing the name monolithic IC.

## Thin-Film and Thick-Film Construction

*Thin-film* and *thick-film* techniques are used primarily for constructing passive networks such as resistor ladders, filters, attenuators, and phase-shift networks. Such ICs are manufactured by depositing resistive and conductive materials, through a series of masking steps, on a nonconducting base such as ceramic. The networks formed can be made extremely small, and component tolerances can be made closer than equivalent components made by the monolithic method. For this reason, thin-film and thick-film techniques are preferred for high precision circuits.

### Thin-Film

In the thin-film technique, resistors, capacitors, and conductors are formed from thin layers of metals and oxides that are deposited on an insulating substrate such as glass or ceramic. The resistors are usually formed of either tantalum or nichrome, the resistive value being determined by the length, width, and thickness of the material. The capacitors are constructed by depositing a thin metal film on the substrate, followed by an oxide coating for a dielectric, and topped by another thin metal film. The capacitor value is determined by the area of the plates and the thickness and type of the dielectric material used. The interconnecting conductors are extremely thin strips of gold, aluminum, or platinum deposited on the substrate. To insure that the desired films are deposited in the desired locations on the substrate, a series of masking steps is used, as in the monolithic process.

### Thick-Film

The thick-film technique uses a silk-screen process to form resistors, capacitors, and conductors on an insulating substrate. In this process, a very fine wire screen is placed over the substrate and a metalized ink is forced through the screen with a squeegee. The painted surfaces are submitted to temperatures higher than 600 °C and harden to form the interconnecting conductors. Resistors and capacitors are formed in the silk-screen process by forcing materials in paste form through a wire screen onto the substrate, and then submitting the substrate to extremely high temperatures. Resistors with tolerances as low as $\pm 0.5\%$ can be obtained by carefully trimming the resistive material after construction, using a sandblasting or laser trimming technique. Thick-film capacitors are limited to relatively low values.

## Hybrid Construction

*Hybrid* ICs are combinations of monolithic, thin-film, or thick-film circuits. A hybrid IC may consist of two or more monolithic chips interconnected in a single package; a monolithic circuit combined with a thin-film or thick-film passive network; or monolithic circuits, thin-film or thick-film circuits, and individual semiconductor component chips combined in a single package. Because of the different techniques required to construct the various circuits, hybrid ICs are more complex and, therefore, more expensive than other types.

## A Fabrication Procedure

A procedure used by Signetics Corporation will serve to demonstrate a typical fabrication of LICs. There are three major steps in the Signetics procedure—masking, etching, and diffusion.

Fabrication of a typical IC begins with a polished silicon wafer. The silicon is treated with a *dopant,* (an impurity) to give the wafer a positive (P-type) electrical charge. During this initial *oxidation,* a thin layer of silicon dioxide, ($SiO_2$) a very pure form of glass, is grown on the surface, as shown in Figure 1–3A. The oxidized wafer then is subjected to masking.

## Masking

*Masking*, sometimes called photomasking, is the name given to a series of steps that selectively cuts openings or windows into the oxide, metal, or glass surfaces of a wafer. The masking operation is a photographic process that uses a *photoresist* (film) that is sensitive to ultraviolet (UV) light. When the UV light strikes the photoresist, it hardens. Then, when the photoresist is chemically developed, the hardened area of the resist struck by UV light (clear areas of the mask) will not develop away; the soft areas not struck by the UV light (dark areas of the mask) will develop away. These steps are shown in Figure 1–3B through 1–3E. You should remember that the same pattern is being reproduced on every die (each tiny segment) of the wafer at the same time, as shown in Figure 1–3F.

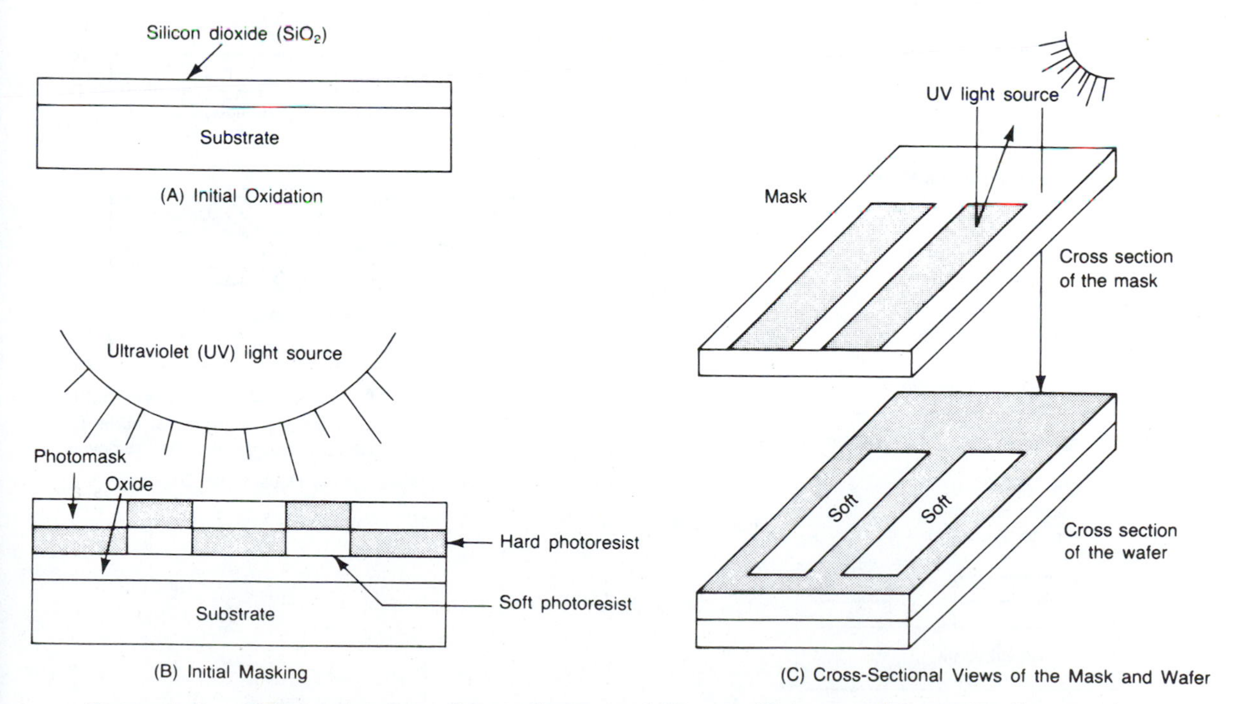


**Figure 1–3.** Fabrication Steps for an Integrated Circuit (Courtesy of Signetics Company) (Continued on p. 10)

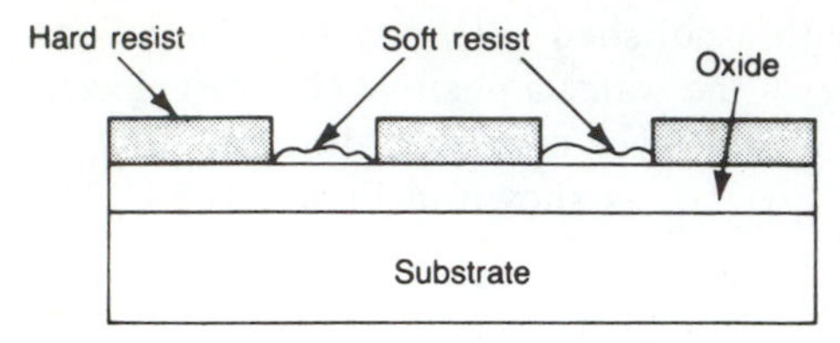


(D) Wafer after First Masking, before Developing

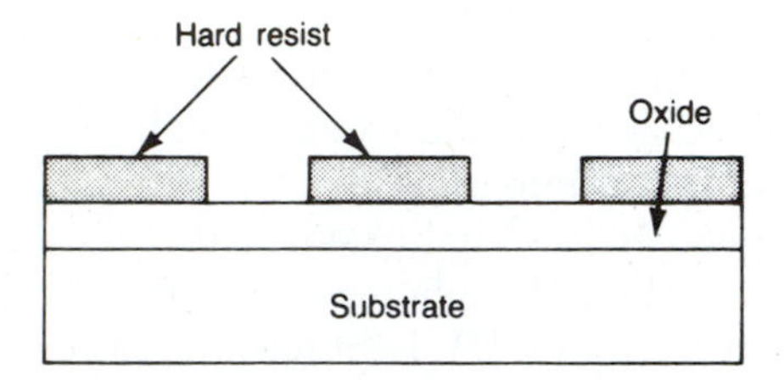


(E) Wafer after Soft Areas (Not Subjected to UV Light) Have Been Developed Away by Chemical Action

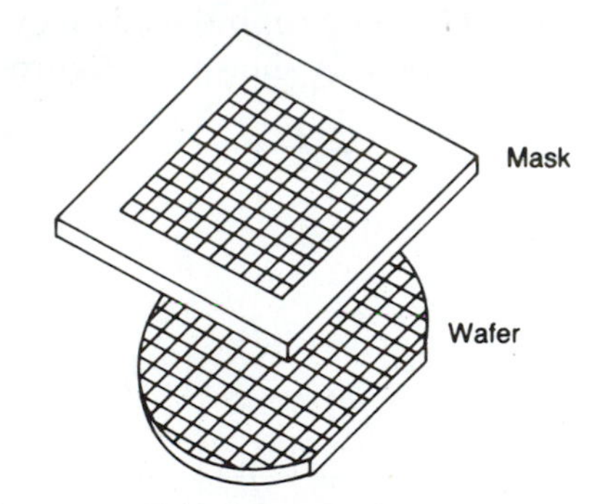


(F) Masking a Full Wafer

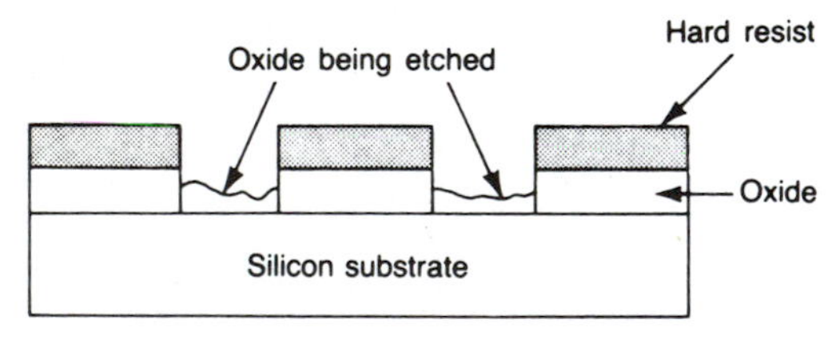


(G) Oxide Being Etched after Soft Resist Is Removed

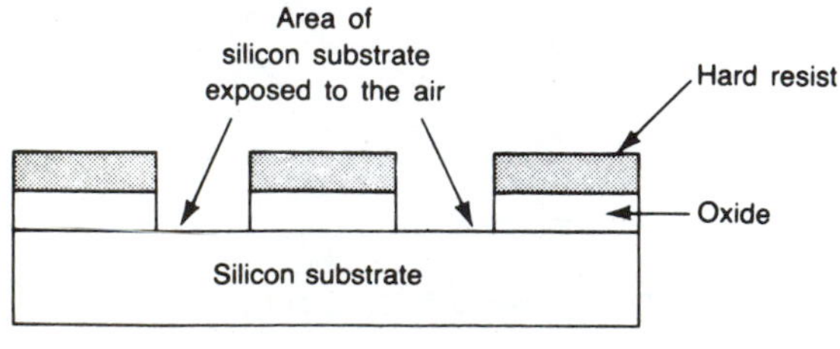


(H) Wafer after Etching

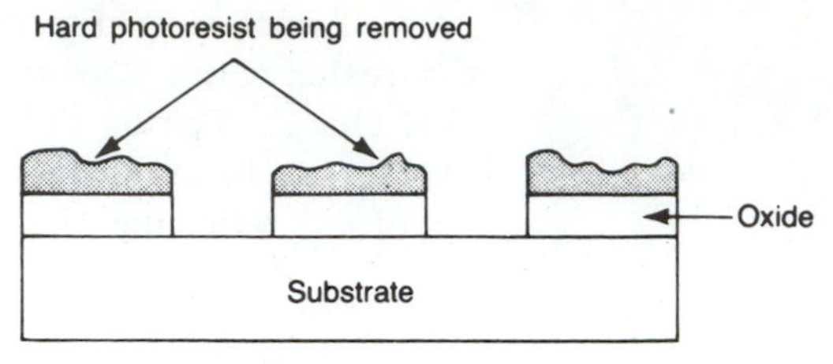


(I) Removal of the Hard Resist, No Longer Needed

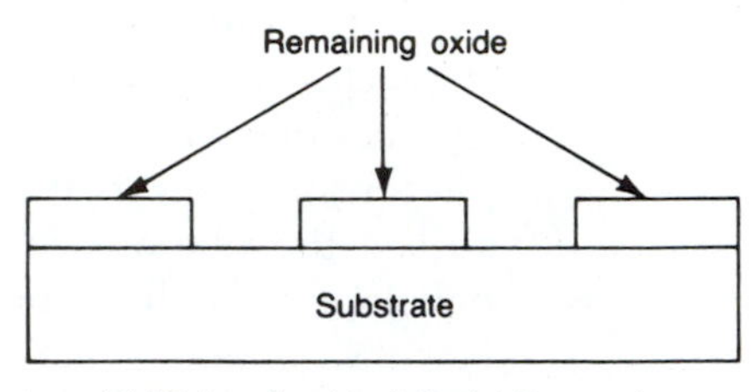


(J) Wafer after Hard Resist Removal, Ready for Dopant Application

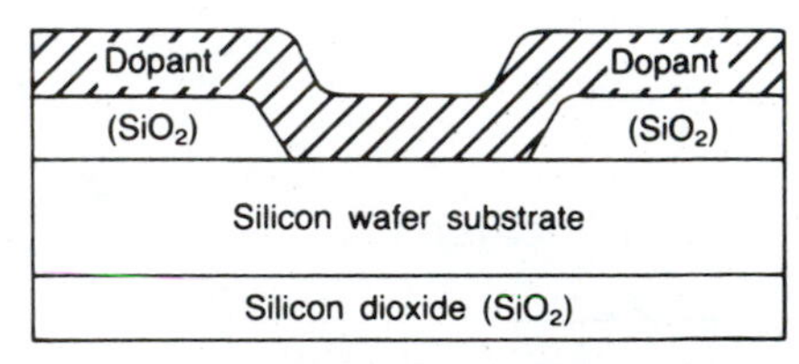


(K) Deposition of Dopants

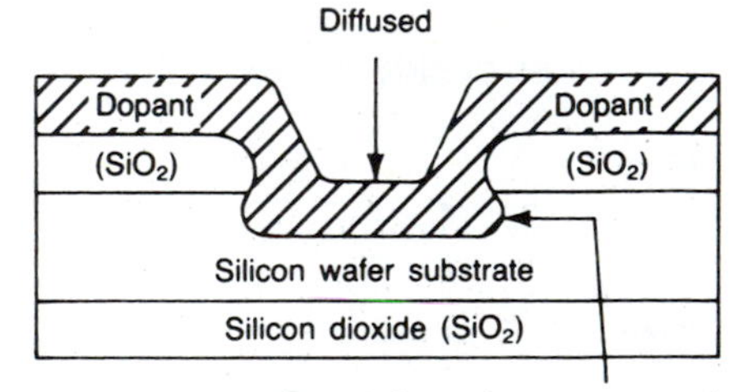


(L) Diffusion of the Dopant into Substrate

**Figure 1-3.** *Continued*

### Etching

After developing, the wafer is submitted to *oxide etching*. During this step, the masking procedure is completed. In the etching process, for example, oxide not protected by hard photoresist will be etched (worn away) (Figure 1–3G) and completely removed down to the bare silicon (Figure 1–3H). Then the wafers are cleaned of the photoresist on the oxide surface because it is no longer needed. The hard layer of photoresist is dissolved completely (Figure 1–3I) and the patterns from the mask remain (Figure 1–3J).

### Diffusion

There are two general operations performed in the *diffusion* process:

1. Deposition of dopants onto the wafer's surface.
2. Diffusion of the dopant to the desired depth in the wafer.

In the diffusion process, electrical characteristics are built into the silicon wafers step-by-step. Following each masking step (there may be five or more, depending on the product), the wafers are returned to diffusion, where N-type dopants carrying negative (−) charges or P-type dopants carrying positive (+) charges are deposited onto the wafer surfaces (Figure 1–3K). The wafers are then exposed to very high temperatures. The extreme heat supplies the needed energy to diffuse, or drive, the dopants to the desired depths in the wafers (Figure 1–3L). The electrical characteristics of the wafers are thereby changed in predetermined areas—areas that were formed by the mask pattern during the masking operation immediately preceding each diffusion.

The diffusion steps and the corresponding masking steps are used in combination to fabricate individual components of the semiconductors in the silicon wafers.

Once the individual components have been fabricated, they must be connected in order for them to operate as complete circuits. The circuits are completed when additional masking steps are done to perform two functions:

1. Link the individual components together.
2. Provide the necessary contacts that connect the circuit to the leads of the package that covers and protects the finished device.

Thus the fabrication of integrated circuits on a single wafer may require as many as ten or more masking steps plus the corresponding diffusion operations that are done after each of the steps.

### Final Processing

After fabrication, each circuit is probed by computer-controlled test instruments to distinguish good circuits from bad ones. If a circuit is defective, it is marked with an ink spot to indicate that it should be discarded. A single source of contamination during the fabrication process, such as a scratch, can cause a break in an electrical connection and ruin an entire circuit.

Following probing, the wafers are *scribed* (cut) and sectioned into individual dies, each of which is a complete electronic circuit. Individual dies with good circuits are then bonded into packages, with hairlike wires connected from the contacts on the die to the package leads. To make sure that the circuits have not been scratched or otherwise damaged, they are viewed under high- and low-power magnification. The packages are then marked or symboled with the company logo, an identification number, and a product type designation. When these processes are completed, the packages are sealed, put through final testing, and shipped for eventual customer use. The entire production flow, from wafer fabrication to shipment of the final product, may require three to four months.

## 1.7 Packaging of Linear Integrated Circuits

LICs come in a variety of packages. Package outlines and specifications, other than those shown in this section, are shown in Appendix A. Selection of the package type depends upon the application for which the IC is intended. Devices to be used in extreme environmental conditions obviously must have the packaging and high performance that provide the highest level of reliability and integrity, while other applications might use less rigorous specifications.

### Packaging Materials

There are three types of materials used for packaging ICs—plastic, metal, and ceramic. Each possesses certain characteristics that make it suitable for particular uses. It is important to know these characteristics in order to choose the appropriate package for a given application.

The least expensive of the packaging materials is a *plastic,* such as resin, epoxy, silicone, or a combination of these. The plastic is molded around the chip and leads using high temperature and pressure, thus offering good mechanical protection and some small protection from the surrounding environment. It is not hermetically sealed; therefore, it offers no protection against moisture or chemicals in the surrounding air. Because of this, the plastic package is unsuitable for use in damp or corrosive atmospheres. Plastic materials have the lowest thermal dissipation capability of all material types available. This limits the plastic package to use at lower temperatures and power ratings than any of the other materials. The higher thermal resistance of the plastic package results in higher chip temperatures than in other packages because the plastic does not allow as much heat to flow away from the chip.

The *metal-encased* chip provides a better hermetic seal because the joints can be soldered or welded, thus sealing the package. This isolation of the chip allows it to be used in environments that would melt or corrode a plastic-packaged chip. The metal package costs more to manufacture, but manufacturing in volume reduces that cost.

*Ceramic* packaging offers the highest reliability and integrity against harmful elements in the environment. The chip is hermetically sealed, preventing moisture and corrosive elements from reaching and damaging the chip. It also offers the highest thermal dissipation capability of the packaging materials, which allows its use at much higher temperatures than the other types.

## Metal Packages

*Metal cans,* similar to the familiar units used for discrete transistors, were used for packaging the first LICs and are still in use today. Figure 1–4 shows an 8-pin TO-5 package. Leads connected to the chip are brought out from the base of the package, just like transistors. These leads can be either pushed through holes in a *printed circuit board* (PCB) or inserted in special mounts.

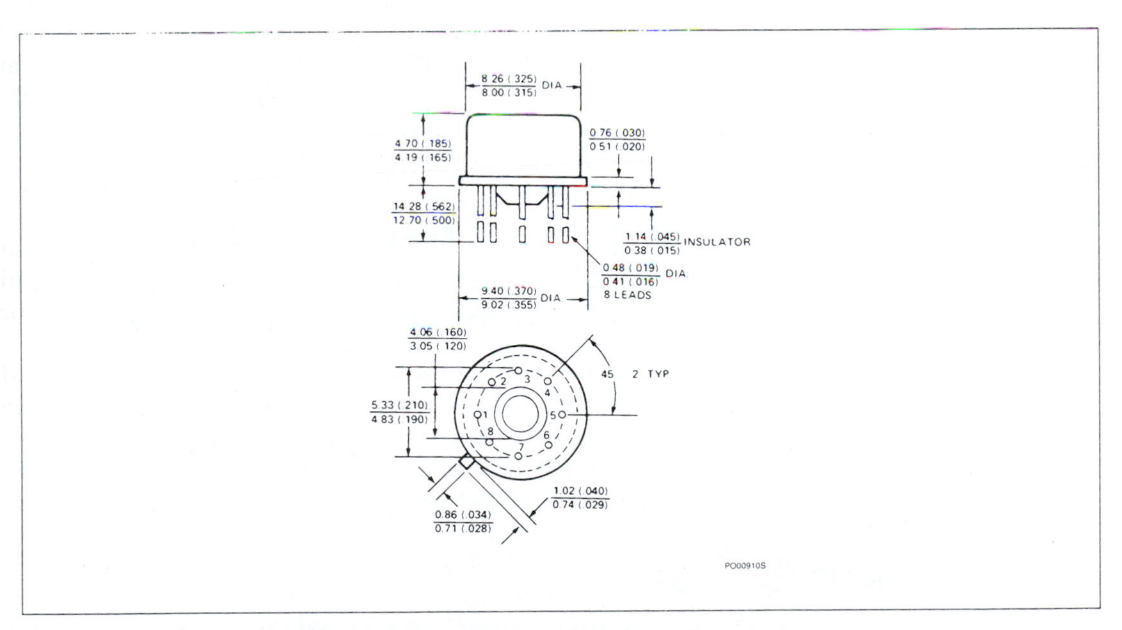


**Figure 1–4.** 8-Pin Hermetic TO-5 Header (H Package) (Courtesy of Signetics Company)

The cans are hermetically sealed by forcing inert gas into the cap to remove moisture or other corrosive materials and then welding the cap to the base. The seal prevents any harmful materials from entering and damaging the chip.

## Flat-Pack Packages

The *flat pack,* Figure 1–5, was the first package style that was designed specifically for the LIC. The leads extend from two or four sides of the flat pack, rather than from the base as in the can.

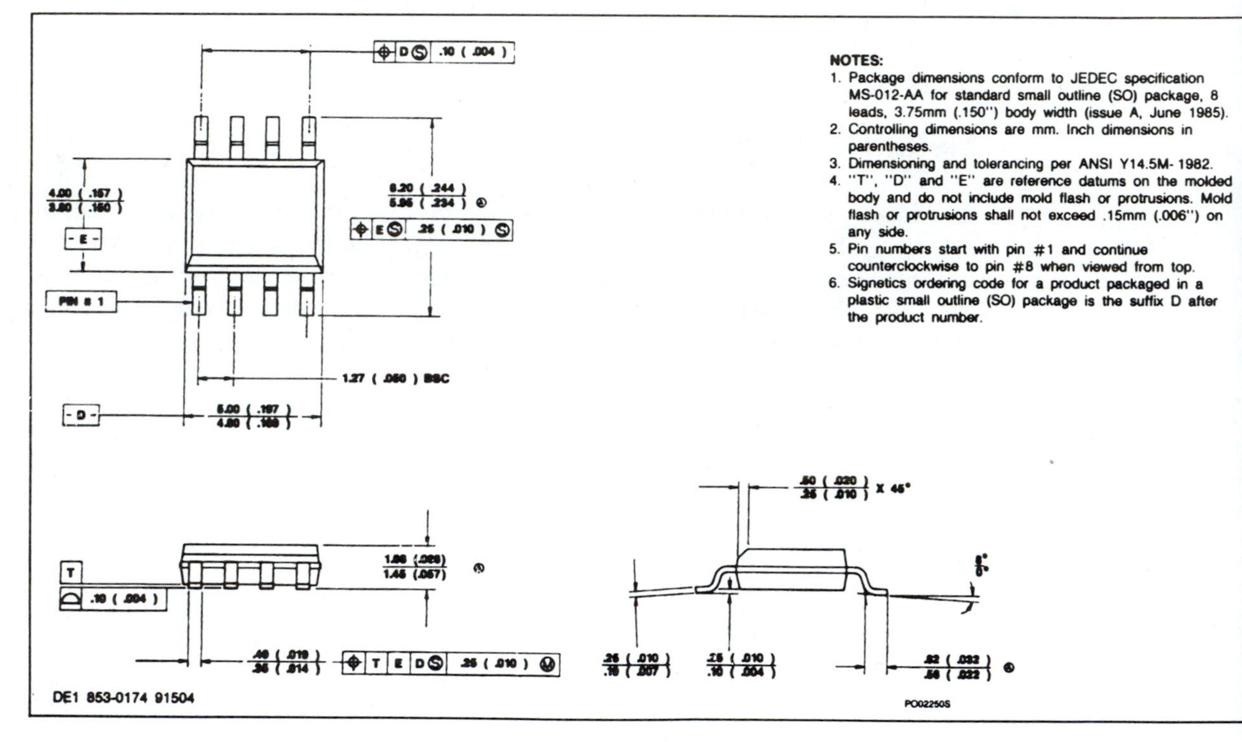


Figure 1–5. 8-Pin Plastic Small Outline (SO) (Courtesy of Signetics Company)

The flat pack is mounted flush on the surface of a PCB with the leads in contact with the copper on the surface rather than pushed through the board. Attaching the flat pack requires welding or resistance soldering. Because of the spread-out configuration, the flat pack requires more board space than the can; however, the reduced height of the flat pack allows more boards to be stacked.

## In-Line Packages

*In-line* packaging means that the pins extend out and down in lines along the sides of the package. In-line packaging was designed to improve the mounting and soldering capabilities of the LICs to PCBs. The package is easily inserted into the PCB by through-hole technology (pushing the leads through holes drilled in the board). The boards are then either dip- or wave-soldered. The more useful method of mounting such packages, particularly in consumer products, is on a mount that is soldered to the PCB.

The four types of in-line packages available are single, dual, single-bent-to-dual, and quad.

### Single In-Line Package

A 9-pin plastic *single in-line package* (SIP) is shown in Figure 1–6. The 9 pins extend from one side of the package in a single line. These pins can be pushed through a PCB and soldered, or inserted into a plug-in module. The metal strip along the top of the package is used for dissipating heat.

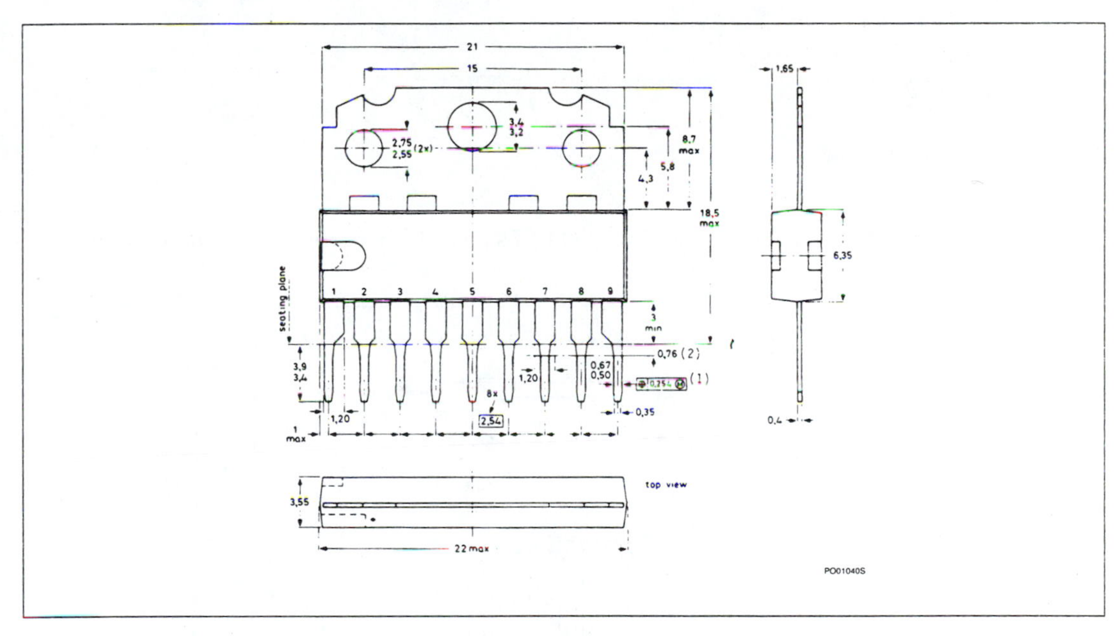


**Figure 1–6.** 9-Pin Plastic SIP (SOT-110B) (Courtesy of Signetics Company)

### Dual In-Line Package

A 14-pin plastic *dual in-line package* (DIP) is shown in Figure 1–7. Here, the 14 pins extend in two (dual) lines, 7 pins from each side of the package. Advantages of the DIP are they can be fabricated in many pin styles, with any number of pins, and at low cost due to the ease of fabrication.

### Single In-Line Bent to Dual In-Line

Figure 1–8 shows a 13-pin plastic power *single in-line package bent to dual in-line package* (SIP-Bent-to-DIP). Although the 13 pins extend from one side of the package, alternate pins are bent in such a manner that mounting to the PCB is in a dual line.

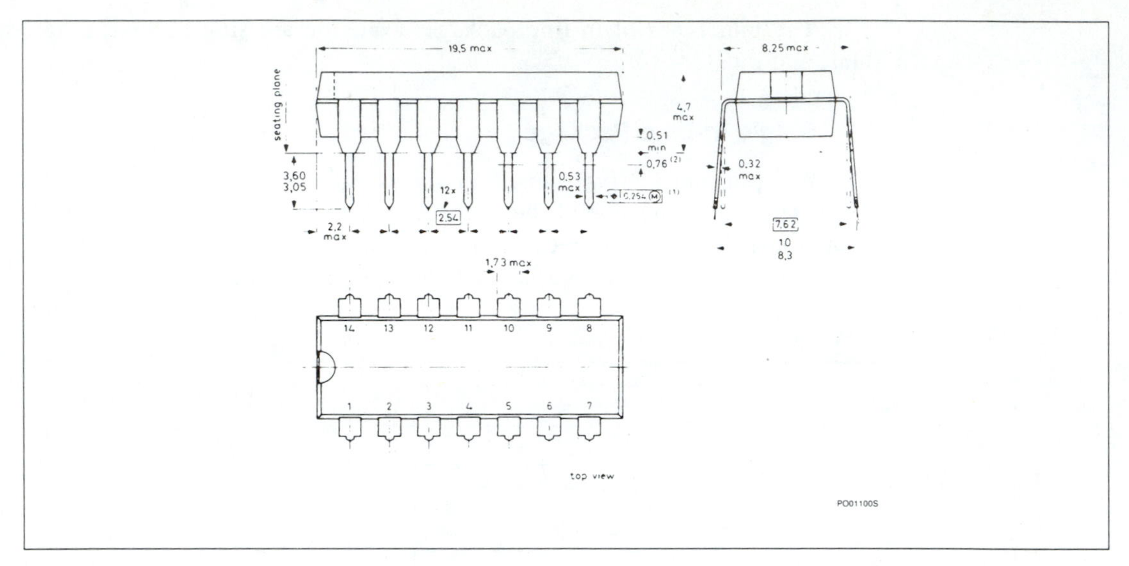


Figure 1–7. 14-Pin Plastic DIP (SOT-27K, M, T) (Courtesy of Signetics Company)

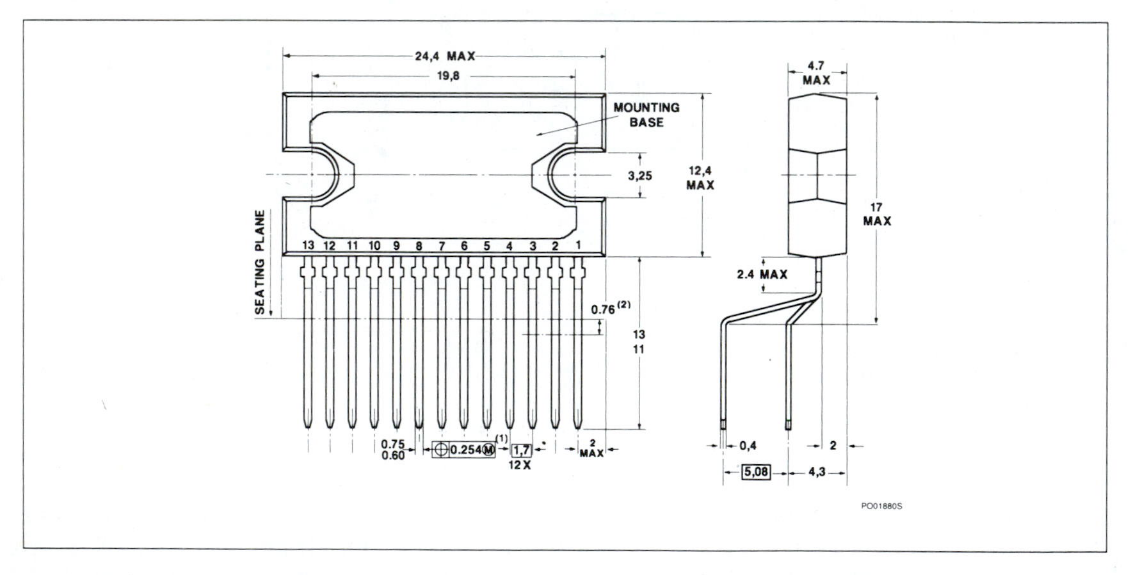


Figure 1–8. 13-Pin Plastic Power SIP-Bent-to-DIP (SOT-141BA) (Courtesy of Signetics Company)

### Quad In-Line Package

A 16-pin plastic *quad in-line package* (QIP), Figure 1–9, shows the pins extending from both sides of the package in such a way that four (quad) lines result.

As you have learned, selection of the right IC, the right package, and the right packaging material are equally important. The choices all depend upon the application for which the IC is intended.

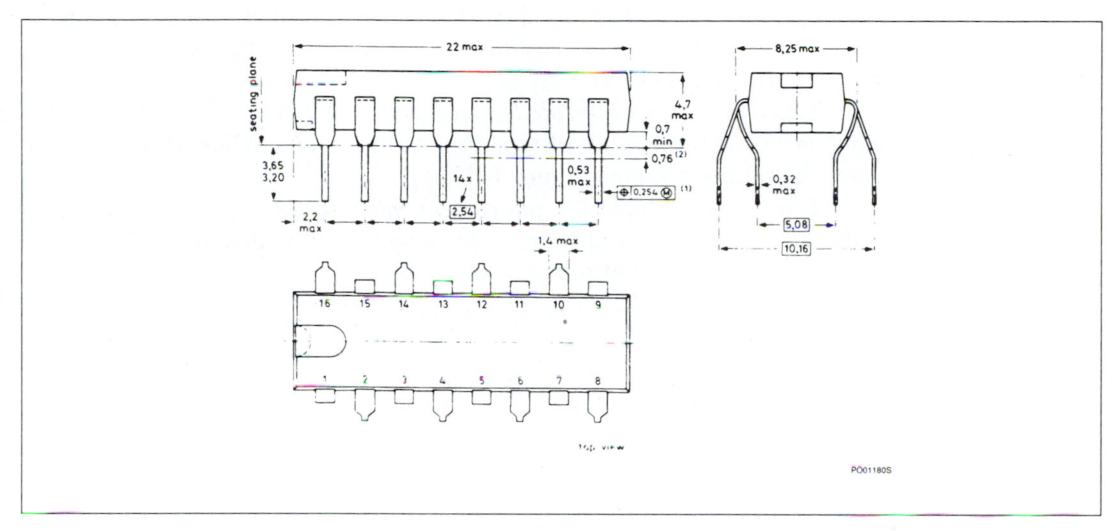


**Figure 1–9.** 16-Pin Plastic QIP (SOT-58) (Courtesy of Signetics Company)

## Surface-Mount Technology

*Surface-mount technology* is penetrating rapidly into all areas of modern electronic equipment manufacture—in professional, industrial, and consumer applications. Boards are made with conventional print-and-etch PCBs, multilayer boards with thick-film ceramic substrates, and with a host of new materials specially developed for surface-mount technology.

Surface-mount technology entails a totally new automated circuit assembly process using a new generation of electronic components; *surface-mounted devices* (SMDs). Smaller than conventional components, SMDs are placed on the surface of the PCB (substrate), not through it like through-hole (leaded) components. The fundamental difference between SMD assembly and conventional through-hole component assembly is SMD component positioning is relative, not absolute.

When a leaded component is inserted into a PCB, positioning is exactly determined by the holes provided. An SMD, however, is placed onto the substrate surface, its position only relative to the soldering surface, and placement accuracy is therefore influenced by variations in the substrate track patterns, component size, and placement machine accuracy.

### Substrate Configurations

Three SMD substrate configurations are shown in Figure 1-10, and are classified as follows:

Type 1: *Total-surface mount* (all-SMD); substrates with no through-hole components at all. SMDs of all types (surface-mounted ICs, discrete semiconductors, and passive devices) can be mounted either on one side or both sides of the substrate (Figure 1-10A).

Type 11A: *Double-sided mixed-print;* substrates with both through-hole components and SMDs of all types on the top, and smaller SMDs (transistors and passives) on the bottom (Figure 1-10B).

Type 11B: *Underside attachment mixed-print;* the top of the substrate is dedicated exclusively to through-hole components, with smaller SMDs (transistors and passives) on the bottom (Figure 1-10C).

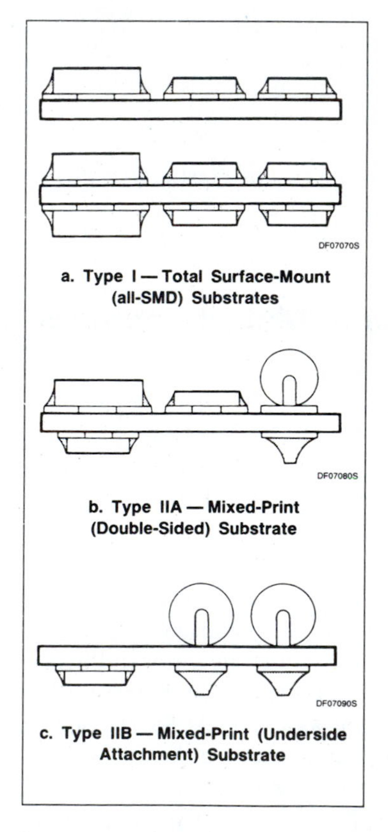


**Figure 1-10.** Surface-Mounted Device Configurations (Courtesy of Signetics Company)

The all-SMD substrate results in the cheapest and smallest variation since there are no through-hole components. However, many manufacturers are looking to the mixed-print substrate, for this technique enjoys most of the advantages of SMD assembly and overcomes the problem of nonavailability of some components in surface-mount form.

## 1.8 Classification and Numbering of Linear Integrated Circuits

Type classification and numbering of integrated circuits vary with manufacturers. All manufacturers assign code numbers to the devices they produce. In general, standard ICs such as the 741 and 301 are labeled as such by all manufacturers. However, other ICs may have different part numbers assigned, even if they are identical circuits manufactured by different vendors. You also should be aware that even if an IC from one vendor has the same basic circuit and the same part number as an IC from another vendor, the ICs may not be interchangeable.

Most vendors use an alphanumeric code to classify their ICs. The first two or three alphabetic characters identify the vendor. The letters are followed by a numeric designation to classify the type of IC. Some vendors assign a final alphabetic character to designate the package type and material.

Some representative designations from various manufacturers are listed in Table 1–3. Some examples of specific code numbers and products are also shown in the table. Many manufacturers provide a section in their data books and data sheets in which their code numbering system is explained. Two examples of these are shown in Figure 1–11.

**Table 1–3 Manufacturers' Type Identification Codes**

| Manufacturer | Identification Code | Example | |
|---|---|---|---|
| Advanced Micro Devices (AMD) | Am | Am 9614 | Dual differential line driver |
| Fairchild | μA | μA 741 | Frequency-compensated op amp |
| Harris | HA | HA 2000 | FET-input op amp |
| Motorola | MC, MFC | MC 4044P | Frequency phase detector |
| National Semiconductor | LH, LM | LM 311H | High-performance voltage comparator |
| RCA | CA, CD | CD 4529 | Dual 4-channel analog data selector |
| Signetics | SE/NE, N/S | NE 544N | Servo amplifier |
| Texas Instruments | SN | SN 52741 | High-performance op amp |

Signetics

# Ordering Information for Prefixes ADC, AM, AU, CA, DAC, ICM, LF, LM, MC, NE, SA, SE, SG, μA, UC

## Linear Products

Table 1 provides part number information concerning Signetics originated products.

Table 2 is a cross reference of both the old and new package suffixes for all presently existing types, while Tables 3 and 4 provide appropriate explanations on the various prefixes employed in the part number descriptions.

As noted in Table 3, Signetics defines device operating temperature range by the appropriate prefix. It should be noted, however, that an SE prefix (−55°C to +125°C) indicates only the operating temperature range of a device and *not* its military qualification status. The military qualification status of any Linear product can be determined by either looking in the Military Data Manual and/or contacting your local sales office.

**Table 1. Part Number Description**

| PART NUMBER | CROSS REF PART NO. | PRODUCT FAMILY | PRODUCT DESCRIPTION |
|---|---|---|---|
| N E 5 5 3 7 N | LF398 | LIN | Sample-and-Hold Amp |

Sample-and-Hold Amp → Description of Product Function

LIN → Linear Product Family

N (suffix) → Package Descriptions — See Table 2

5537 → Device Number

NE → Device Family and Temperature Range Prefix — See Tables 3 & 4

**Table 2. Package Descriptions**

| OLD | NEW | PACKAGE DESCRIPTION |
|---|---|---|
| A, AA | N | 14-lead plastic DIP |
| A | N-14 | 14-lead plastic DIP (selected analog products only) |
| B, BA | N | 16-lead plastic DIP |
| | D | Microminiature package (SO) |
| F | F | 14-, 16-, 18-, 22-, and 24-lead ceramic DIP (Cerdip) |
| I, IK | I | 14-, 16-, 18-, 22-, 28-, and 4-lead ceramic DIP |
| K | H | 10-lead TO-100 |
| L | H | 10-lead high-profile TO-100 can |
| NA, NX | N | 24-lead plastic DIP |
| Q, R | Q | 10-, 14-, 16-, and 24-lead ceramic flat |
| T, TA | H | 8-lead TO-99 |
| U | U | SIP plastic power |
| V | N | 8-lead plastic DIP |
| XA | N | 18-lead plastic DIP |
| XC | N | 20-lead plastic DIP |
| XC | N | 22-lead plastic DIP |
| XL, XF | N | 28-lead plastic DIP |
| | A | PLCC |
| | EC | TO-46 header |
| | FE | 8-lead ceramic DIP |

**Table 3. Signetics Prefix and Device Temperature**

| PREFIX | DEVICE TEMPERATURE RANGE |
|---|---|
| NE | 0 to +70°C |
| SE | −55°C to +125°C |
| SA | −40°C to +85°C |

**Table 4. Industry Standard Prefix**

| PREFIX | DEVICE FAMILY |
|---|---|
| ADC | Linear Industry Standard |
| AM | Linear Industry Standard |
| CA | Linear Industry Standard |
| DAC | Linear Industry Standard |
| ICM | Linear Industry Standard |
| LF | Linear Industry Standard |
| LM | Linear Industry Standard |
| MC | Linear Industry Standard |
| NE | Linear Industry Standard |
| SA | Linear Industry Standard |
| SE | Linear Industry Standard |
| SG | Linear Industry Standard |
| μA | Linear Industry Standard |
| UC | Linear Industry Standard |

(A) Numbering System for Signetics Company

**Figure 1–11.** Typical Manufacturer's Numbering System Explanation (Part A: Courtesy of Signetics Company; Part B: Courtesy of National Semiconductor Corporation)

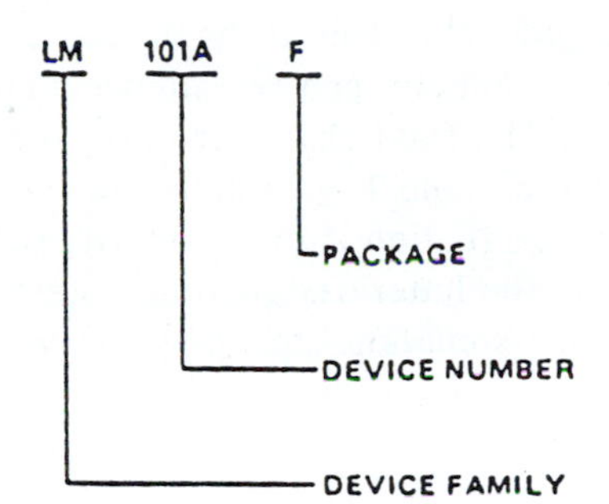


PACKAGE

D – Glass/Metal Dual-In-Line Package
F – Glass/Metal Flat Pack
H – TO-5 (TO-99, TO-100, TO-46)
J – Low Temperature Glass Dual-In-Line Package
K – TO-3 (Steel)
KC – TO-3 (Aluminum)
N – Plastic Dual-In-Line Package
P – TO-202 (D-40, Durawatt)
S – "SGS" Type Power Dual-In-Line Package
T – TO-220
W – Low Temperature Glass Flat-Pack
Z – TO-92

DEVICE NUMBER

3, 4, or 5 Digit Number Suffix Indicators:
A – Improved Electrical Specification
C – Commercial Temperature Range

DEVICE FAMILY

AD – Analog to Digital
AH – Analog Hybrid
AM – Analog Monolithic
CD – CMOS Digital
DA – Digital to Analog
DM – Digital Monolithic
LF – Linear FET
LH – Linear Hybrid
LM – Linear Monolithic
LX – Transducer
MM – MOS Monolithic
TBA – Linear Monolithic

(B) Numbering System for National Semiconductor Corporation

**Figure 1–11.** *Continued*

## Pro-Electron Numbering System

Because of the diversity of numbering systems for LICs, an organization called *Pro-Electron,* located in Belgium, has attempted to establish a standard identification system. All vendors would register their IC products by a code of six characters, three alphabetic and three numeric. Figure 1–12 shows an example of one such unit. Note that the first alphabetic character establishes the device type. The

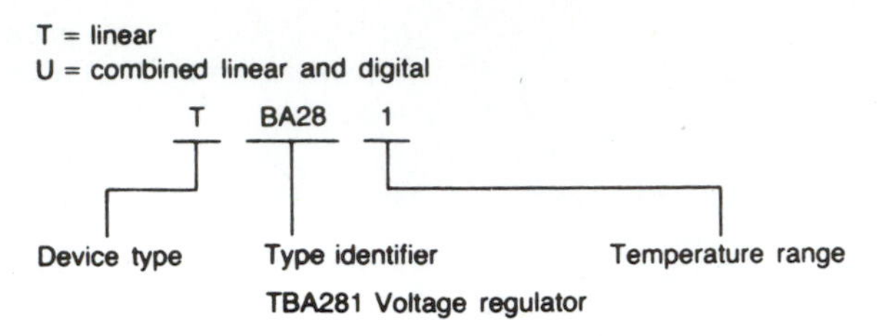


Figure 1–12. Pro-Electron Numbering System

letter *T* is used for LICs, and the letter *U* is used if the device contains both linear and digital circuits. The next two alphabetic and two numeric characters form an identification of the type of device. The final character designates the temperature range of operation, which is shown in Table 1–4. Finally, one additional character may be added to the six-character designation if there are different versions of the same device. (This is reminiscent of the letter designations following identifiers on the vacuum tube that indicated when some change was made in electrical characteristics or in the tube envelope.)

**Table 1–4 Temperature Range Code Used in the Pro-Electron System**

| Code Number | Temperature Range |
|---|---|
| 0 | No designation |
| 1 | 0° to +70°C |
| 2 | −55° to +125°C |
| 3 | −10° to +85°C |
| 4 | +15° to +55°C |
| 5 | −25° to +70°C |
| 6 | −40° to +85°C |

## Military Numbering System

There is one more important numbering system that you should be aware of—the military system. Devices that meet exacting military standards are assigned a Joint Army-Navy (JAN) part number. The JAN system of numbering involves several groups of characters, as shown in Figure 1–13. The system provides far more information about the device than the standard and Pro-Electron systems show. Most manufacturers who sell their devices to military sources provide a section in their data books that designates the JAN system and the steps taken to satisfy the military standards.

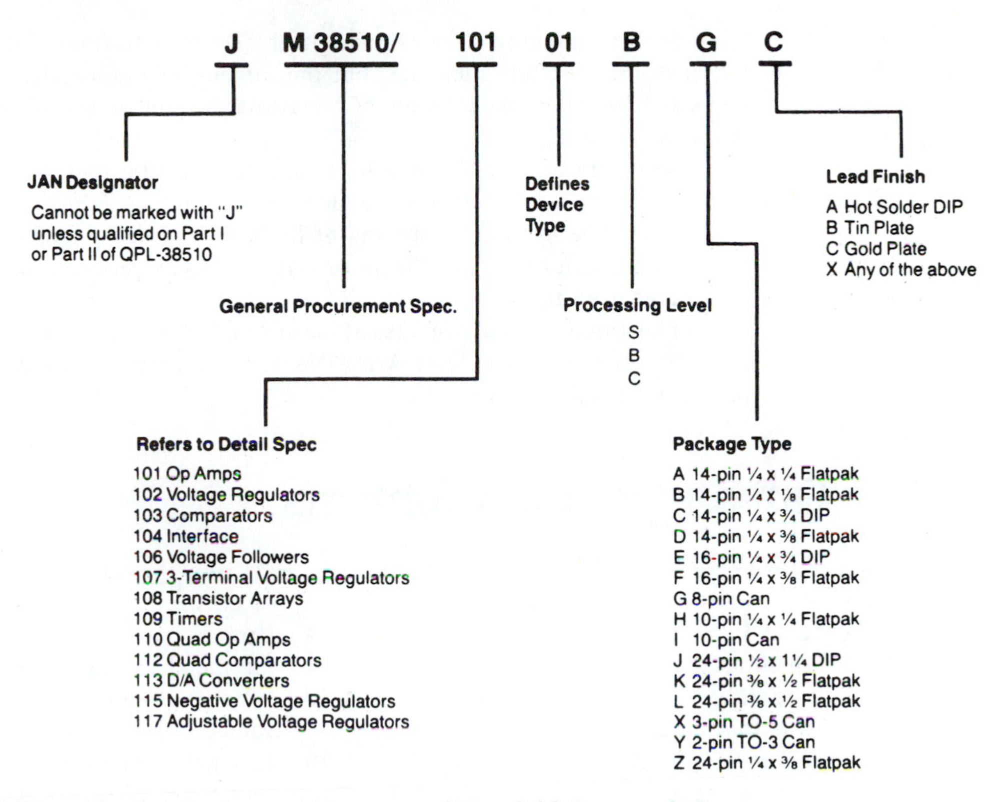


Figure 1–13. JAN Part Numbering System (Courtesy of Fairchild Camera & Instrument Corporation)

# 1.9 Summary

☐ Digital and analog signals are used to generate, process, and transfer information. Digital signals are characterized by a fast-switching, up-down action, and have two possible values. Analog signals can take any value of voltage or current and are characterized by a continuously changing signal. The most common type of analog signal is the sine wave.

☐ A linear system is defined as any unit or assembly of components that generates or processes analog signals. Applications such as power supplies, radio and television receivers, signal generators, and communication systems are examples of linear systems.

☐ Linear integrated circuits are fabricated in a variety of ways, primarily by diffusion. Using plastic, metal, and ceramic as packaging materials, the circuits are packaged in metal cans; flat-packs; single, dual, and quad in-lines; and surface-mounts. The packaging materials have different environmental tolerances, with plastic the least tolerant and ceramic the most tolerant.

☐ Surface-mounted devices come in three configurations. The cheapest to construct is the all-SMD substrate, but the mixed-print substrates offer more advantages and overcome the problem of nonavailability of some components in the form of surface-mounts.

☐ Selection of an LIC for a particular application must include the consideration of the environment, board space, and interchangeability.

☐ Classification and numbering of LICs vary with the manufacturer. Reference to the manufacturer's data sheets or data books is recommended when selecting replacement parts.

☐ The three methods of classification and numbering are the manufacturer's own, Pro-Electron, and Joint Army-Navy. Of the three methods, the JAN method provides the most information.

## 1.10 Questions and Problems

**1.1** Draw a waveform for a digital signal with a positive value of +5 V and a negative value of −2 V.

**1.2** Draw an analog signal waveform that varies smoothly between +10 V and −10 V. What is this type of waveform called?

**1.3** Explain the difference between a digital signal and an analog signal.

**1.4** Define *linear system.*

**1.5** Give three examples of linear systems.

**1.6** List five applications of LICs.

**1.7** What factors must you consider when selecting an IC?

**1.8** Explain monolithic IC construction.

**1.9** What is the primary purpose of the thin-film and thick-film construction techniques?

**1.10** In the thin-film technique, what determines the values for (a) resistors, and (b) capacitors?

**1.11** Explain the differences between the thin-film and thick-film techniques.

**1.12** Explain hybrid IC construction.

**1.13** List the three types of packaging materials and the advantages or disadvantages of each.

**1.14** How is damage to the chip prevented in the metal can package?

**1.15** What is the (a) advantage and (b) disadvantage of the flat-pack package?

**1.16** List and explain the configuration of the four types of in-line packages.

**1.17** Explain the fundamental difference between the assembly of SMDs and conventional leaded components.

**1.18** You are to use a linear system in a tropical coastal region. What type of packaging would you select for the ICs in that system? Why?

**1.19** You are to replace an IC in the audio section of a television receiver. You do not have an exact replacement, but you do have similar ICs in stock. What must you do in order to select the proper replacement?

**1.20** Which classification and numbering system offers the most information to the user?

Chapter 2

# The Operational Amplifier— A Building Block

## 2.1 Introduction

Operational amplifiers (*op amps*) were originally constructed with discrete components within a sealed package. They were used for analog computing circuits, instrumentation applications, and in control circuits. Basically, the term *op amp* was used to describe certain high-performance dc amplifiers.

Today's op amp is a conveniently packaged, very high-gain dc amplifier in the form of an integrated circuit. It is the building block for many circuits, because only a few externally connected components are required to provide feedback, which controls the op amp response.

Circuits based on the op amp as the active device come in many configurations. The most widely used in the electronics industry has a differential input and single output. This configuration will be used throughout this book.

In this chapter you will study the stages contained within the IC op amp, its dc and ac operating characteristics, power supplies, protective circuits, and manufacturers' data sheets.

# 2.2 Objectives

When you complete this chapter, you should be able to:

- ☐ List the operating characteristics of an op amp.
- ☐ Name the stages of an op amp and state the function of each.
- ☐ List the advantages and disadvantages of an op amp.
- ☐ Define input offset current, input offset voltage, slew rate, gain-bandwidth product, common-mode operation, and common-mode rejection ratio.
- ☐ Determine which op amp configurations allow use of a single power supply.
- ☐ Obtain valuable information from manufacturers' data sheets.

# 2.3 Op Amp Symbol

Before we examine practical op amp characteristics, let's look at the op amp schematic symbol and some of the identifiers usually found there. The standard symbol is shown in Figure 2–1. This symbol remains the same regardless of the type of op amp, package material, or package configuration being considered.

The triangular shape of the symbol in Figure 2–1 is representative of all amplifiers. The major difference in the op amp symbol is at the input. Note the two input terminals. The minus (−) sign indicates an *inverting* input. This means that any ac or dc signal applied to this input will be 180° out of phase at the output. The plus (+) sign indicates a *noninverting* input. Any ac or dc signal applied to this input will be in phase at the output.

Power supply terminals are shown above and below the triangle. These terminals may not always be shown on a schematic. The same applies to other terminals, such as those used for null or frequency compensation adjustments.

The manufacturer's part number or type of op amp is generally centered in the triangle. In some general-circuit schematics that do not indicate a specific type of op amp, the symbol may contain other identifiers, such as $A_1$, $A_2$, $A_3$ (representing amplifiers 1, 2, 3).

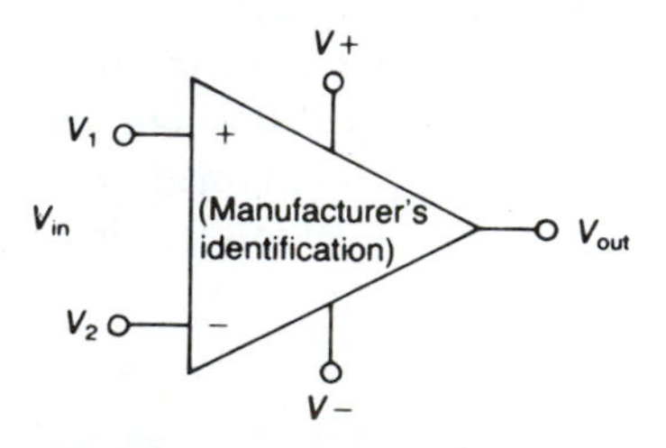


**Figure 2–1.** Standard Op Amp Schematic Symbol

# 2.4 Operating Characteristics of the Op Amp

LIC op amps provide amplifier characteristics that were unattainable in op amps constructed with discrete components. These unique characteristics are what make op amps so versatile and almost universally used in electronic circuits today.

The ideal op amp would have the following characteristics:

1. Infinite open-loop gain
2. Infinite input impedance
3. Zero output impedance
4. Infinite bandwidth
5. Zero offset, that is, zero output voltage when input voltage is zero
6. Zero drift with temperature change

In practice, there are no ideal op amps. However, the operating characteristics of modern op amps do make them an ideal base upon which to build many practical circuits.

## Reading the Data Sheets

Knowing the practical characteristics of LICs will provide you with the understanding necessary for you to service or design circuits in which they are used. There are many different LICs available, each with its own set of characteristics. Manufacturers provide data sheets that specify typical operating characteristics. It is important that you be able to read data sheets so that you can select the proper device for a specified application.

We will now study a typical data sheet—the μA741C data sheet shown in Figure 2–2. The *C* represents the commercial version of the μA741, a commonly used op amp. In the following discussion, specific sections in the data sheet are keyed by numbers.

The figure lists "absolute maximum ratings," ①. These are electrical limitations that must not be exceeded if the device is to operate properly and if possible damage to the device is to be minimized. These ratings are generally self-explanatory and need not be discussed at this point. Specifics will be examined when necessary as we proceed through this chapter.

Figure 2–2 also lists important parameters for specified electrical characteristics. Some special-purpose op amps may have an *input resistance* ($R_{IN}$), ②, (impedance) as high as 100 MΩ, but standard op amps are typically 1 MΩ or more. The 741C, for example, has an input resistance of 2 MΩ. The higher the input impedance, the better the op amp will perform. *Input capacitance* ($C_{IN}$), ③, may become an important factor when the op amp is to be operated at high frequencies. Typically, this capacitance is less than 2 pF. Our example has an input capacitance of 1.4 pF.

Signetics

# μA741/μA741C/SA741C

## General Purpose Operational Amplifier

***Product Specification***

**Linear Products**

### DESCRIPTION

The μA741 is a high performance operational amplifier with high open-loop gain, internal compensation, high common mode range and exceptional temperature stability. The μA741 is short-circuit-protected and allows for nulling of offset voltage.

### FEATURES

- **Internal frequency compensation**
- **Short circuit protection**
- **Excellent temperature stability**
- **High input voltage range**

### PIN CONFIGURATION

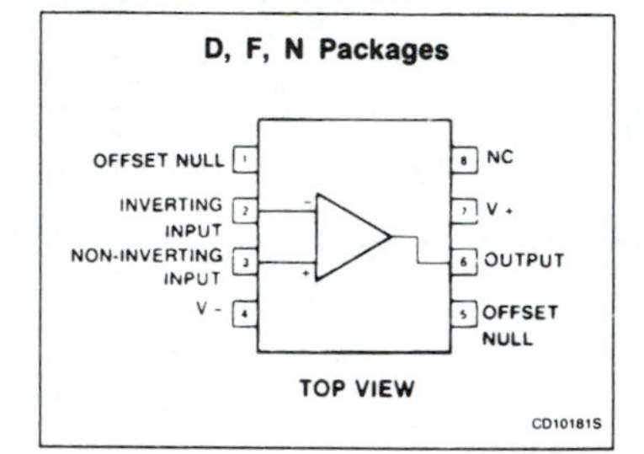


### ORDERING INFORMATION

| DESCRIPTION | TEMPERATURE RANGE | ORDER CODE |
|---|---|---|
| 8-Pin Plastic DIP | -55°C to +125°C | μA741N |
| 8-Pin Plastic DIP | 0 to +70°C | μA741CN |
| 8-Pin Plastic DIP | -40°C to +85°C | SA741CN |
| 8-Pin Cerdip | -55°C to +125°C | μA741F |
| 8-Pin Cerdip | 0 to +70°C | μA741CF |
| 8-Pin SO | 0 to +70°C | μA741CD |

### EQUIVALENT SCHEMATIC

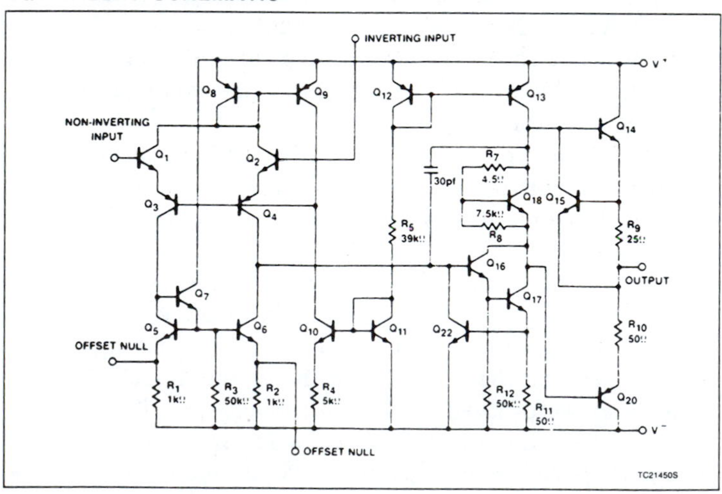


Figure 2–2. Typical Data Sheet (Courtesy of Signetics Company, Linear Division)

Signetics Linear Products — Product Specification

## General Purpose Operational Amplifier — μA741/μA741C/SA741C

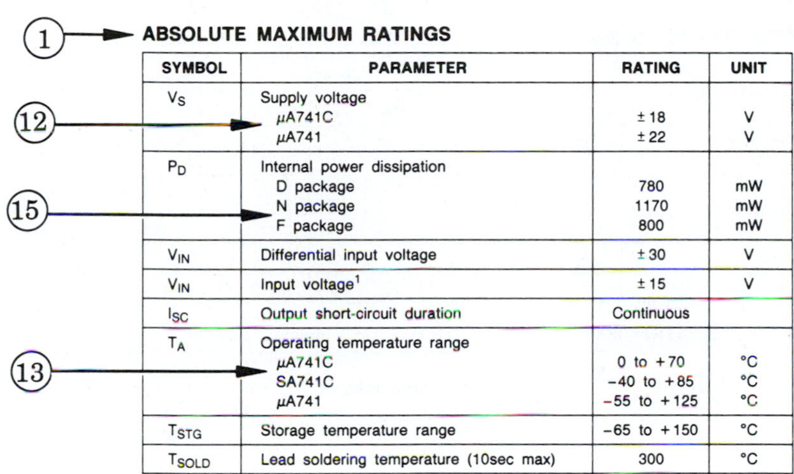


### ABSOLUTE MAXIMUM RATINGS

| SYMBOL | PARAMETER | RATING | UNIT |
|---|---|---|---|
| $V_S$ | Supply voltage | | |
| | μA741C | ±18 | V |
| | μA741 | ±22 | V |
| $P_D$ | Internal power dissipation | | |
| | D package | 780 | mW |
| | N package | 1170 | mW |
| | F package | 800 | mW |
| $V_{IN}$ | Differential input voltage | ±30 | V |
| $V_{IN}$ | Input voltage[1] | ±15 | V |
| $I_{SC}$ | Output short-circuit duration | Continuous | |
| $T_A$ | Operating temperature range | | |
| | μA741C | 0 to +70 | °C |
| | SA741C | −40 to +85 | °C |
| | μA741 | −55 to +125 | °C |
| $T_{STG}$ | Storage temperature range | −65 to +150 | °C |
| $T_{SOLD}$ | Lead soldering temperature (10sec max) | 300 | °C |

**NOTE:**
1. For supply voltages less than ±15V, the absolute maximum input voltage is equal to the supply voltage.

### DC ELECTRICAL CHARACTERISTICS (μA741, μA741C) $T_A = 25°C$, $V_S = \pm15V$, unless otherwise specified.

| SYMBOL | PARAMETER | TEST CONDITIONS | μA741 | | | μA741C | | | UNIT |
|---|---|---|---|---|---|---|---|---|---|
| | | | Min | Typ | Max | Min | Typ | Max | |
| $V_{OS}$ | Offset voltage | $R_S = 10kΩ$ | | 1.0 | 5.0 | | 2.0 | 6.0 | mV |
| | | $R_S = 10kΩ$, over temp. | | 1.0 | 6.0 | | | 7.5 | mV |
| $ΔV_{OS}/ΔT$ | | | | 10 | | | 10 | | μV/°C |
| $I_{OS}$ | Offset current | | | 20 | 200 | | 20 | 200 | nA |
| | | Over temp. | | | | | | 300 | nA |
| | | $T_A = +125°C$ | | 7.0 | 200 | | | | nA |
| | | $T_A = -55°C$ | | 20 | 500 | | | | nA |
| $ΔI_{OS}/ΔT$ | | | | 200 | | | 200 | | pA/°C |
| $I_{BIAS}$ | Input bias current | | | 80 | 500 | | 80 | 500 | nA |
| | | Over temp. | | | | | | 800 | nA |
| | | $T_A = +125°C$ | | 30 | 500 | | | | nA |
| | | $T_A = -55°C$ | | 300 | 1500 | | | | nA |
| $ΔI_B/ΔT$ | | | | 1 | | | 1 | | nA/°C |
| $V_{OUT}$ | Output voltage swing | $R_L = 10kΩ$ | ±12 | ±14 | | ±12 | ±14 | | V |
| | | $R_L = 2kΩ$, over temp. | ±10 | ±13 | | ±10 | ±13 | | V |
| $A_{VOL}$ | Large-signal voltage gain | $R_L = 2kΩ$, $V_O = \pm10V$ | 50 | 200 | | 20 | 200 | | V/mV |
| | | $R_L = 2kΩ$, $V_O = \pm10V$, over temp. | 25 | | | 15 | | | V/mV |
| | Offset voltage adjustment range | | | ±30 | | | ±30 | | mV |
| PSRR | Supply voltage rejection ratio | $R_S \leq 10kΩ$ | | | | | 10 | 150 | μV/V |
| | | $R_S \leq 10kΩ$, over temp. | | 10 | 150 | | | | μV/V |
| CMRR | Common-mode rejection ratio | | | | | 70 | 90 | | dB |
| | | Over temp. | 70 | 90 | | | | | dB |
| $I_{CC}$ | Supply current | | | 1.4 | 2.8 | | 1.4 | 2.8 | mA |
| | | $T_A = +125°C$ | | 1.5 | 2.5 | | | | mA |
| | | $T_A = -55°C$ | | 2.0 | 3.3 | | | | mA |

**Figure 2–2.** *Continued*

## General Purpose Operational Amplifier $\mu$A741/$\mu$A741C/SA741C

**DC ELECTRICAL CHARACTERISTICS** (Continued) ($\mu$A741, $\mu$A741C) $T_A = 25°C$, $V_S = \pm 15V$, unless otherwise specified.

| | SYMBOL | PARAMETER | TEST CONDITIONS | $\mu$A741 | | | $\mu$A741C | | | UNIT |
|---|---|---|---|---|---|---|---|---|---|---|
| | | | | Min | Typ | Max | Min | Typ | Max | |
| | $V_{IN}$ | Input voltage range | ($\mu$A741, over temp.) | ±12 | ±13 | | ±12 | ±13 | | V |
| (2) → | $R_{IN}$ | Input resistance | | 0.3 | 2.0 | | 0.3 | 2.0 | | MΩ |
| (14) → | $P_D$ | Power consumption | | | 50 | 85 | | 50 | 85 | mW |
| | | | $T_A = +125°C$ | | 45 | 75 | | | | mW |
| | | | $T_A = -55°C$ | | 45 | 100 | | | | mW |
| (4) → | $R_{OUT}$ | Output resistance | | | 75 | | | 75 | | Ω |
| | $I_{SC}$ | Output short-circuit current | | 10 | 25 | 60 | 10 | 25 | 60 | mA |

**DC ELECTRICAL CHARACTERISTICS** (SA741C) $T_A = 25°C$, $V_S = \pm 15V$, unless otherwise specified.

| SYMBOL | PARAMETER | TEST CONDITIONS | SA741C | | | UNIT |
|---|---|---|---|---|---|---|
| | | | Min | Typ | Max | |
| $V_{OS}$ | Offset voltage | $R_S = 10k\Omega$ | | 2.0 | 6.0 | mV |
| | | $R_S = 10k\Omega$, over temp. | | | 7.5 | mV |
| $\Delta V_{OS}/\Delta T$ | | | | 10 | | μV/°C |
| $I_{OS}$ | Offset current | | | 20 | 200 | nA |
| | | Over temp. | | | 500 | nA |
| $\Delta I_{OS}/\Delta T$ | | | | 200 | | pA/°C |
| $I_{BIAS}$ | Input bias current | | | 80 | 500 | nA |
| | | Over temp. | | | 1500 | nA |
| $\Delta I_B/\Delta T$ | | | | 1 | | nA/°C |
| $V_{OUT}$ | Output voltage swing | $R_L = 10k\Omega$ | ±12 | ±14 | | V |
| | | $R_L = 2k\Omega$, over temp. | ±10 | ±13 | | V |
| $A_{VOL}$ | Large-signal voltage gain | $R_L = 2k\Omega$, $V_O = \pm 10V$ | 20 | 200 | | V/mV |
| | | $R_L = 2k\Omega$, $V_O = \pm 10V$, over temp. | 15 | | | V/mV |
| | Offset voltage adjustment range | | | ±30 | | mV |
| PSRR | Supply voltage rejection ratio | $R_S \leq 10k\Omega$ | | 10 | 150 | μV/V |
| CMRR | Common mode rejection ration | | 70 | 90 | | dB |
| $V_{IN}$ | Input voltage range | Over temp. | ±12 | ±13 | | V |
| $R_{IN}$ | Input resistance | | 0.3 | 2.0 | | MΩ |
| $P_d$ | Power consumption | | | 50 | 85 | mW |
| $R_{OUT}$ | Output resistance | | | 75 | | Ω |
| $I_{SC}$ | Output short-circuit current | | | 25 | | mA |

**AC ELECTRICAL CHARACTERISTICS** $T_A = 25°C$, $V_S = \pm 15V$, unless otherwise specified.

| | SYMBOL | PARAMETER | TEST CONDITIONS | $\mu$A741, $\mu$A741C | | | UNIT |
|---|---|---|---|---|---|---|---|
| | | | | Min | Typ | Max | |
| | $R_{IN}$ | Parallel input resistance | Open-loop, f = 20Hz | 0.3 | | | MΩ |
| (3) → | $C_{IN}$ | Parallel input capacitance | Open-loop, f = 20Hz | | 1.4 | | pF |
| (10) → | | Unity gain crossover frequency | Open-loop | | 1.0 | | MHz |
| | | Transient response unity gain | $V_{IN} = 20mV$, $R_L = 2k\Omega$, $C_L \leq 100pF$ | | | | |
| | $t_R$ | Rise time | | | 0.3 | | μs |
| | | Overshoot | | | 5.0 | | % |
| (8) → | SR | Slew rate | $C \leq 100pF$, $R_L \geq 2k\Omega$, $V_{IN} = \pm 10V$ | | 0.5 | | V/μs |

**Figure 2–2.** *Continued*

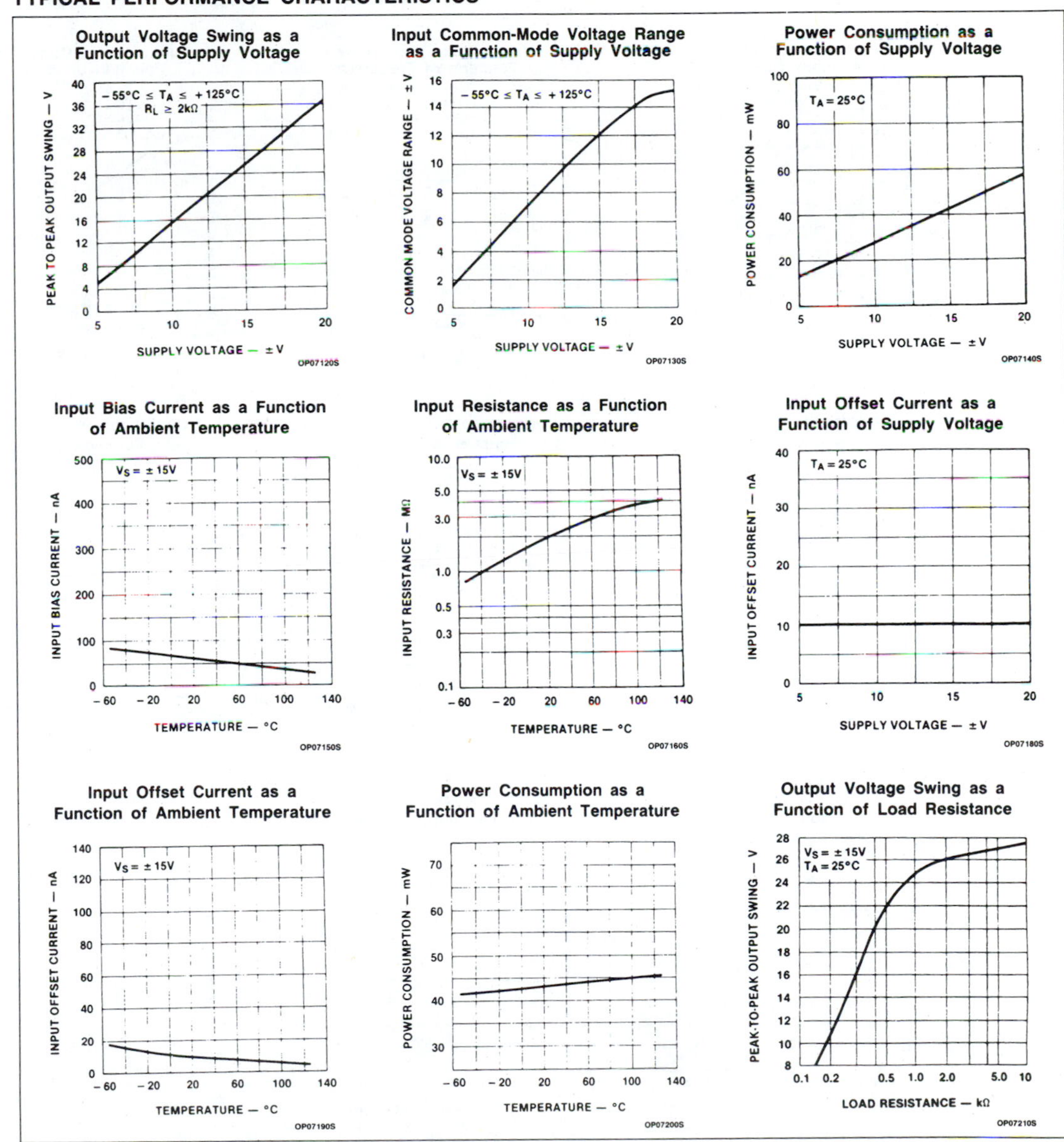


**Figure 2–2.** *Continued*

## General Purpose Operational Amplifier μA741/μA741C/SA741C

### TYPICAL PERFORMANCE CHARACTERISTICS (Continued)

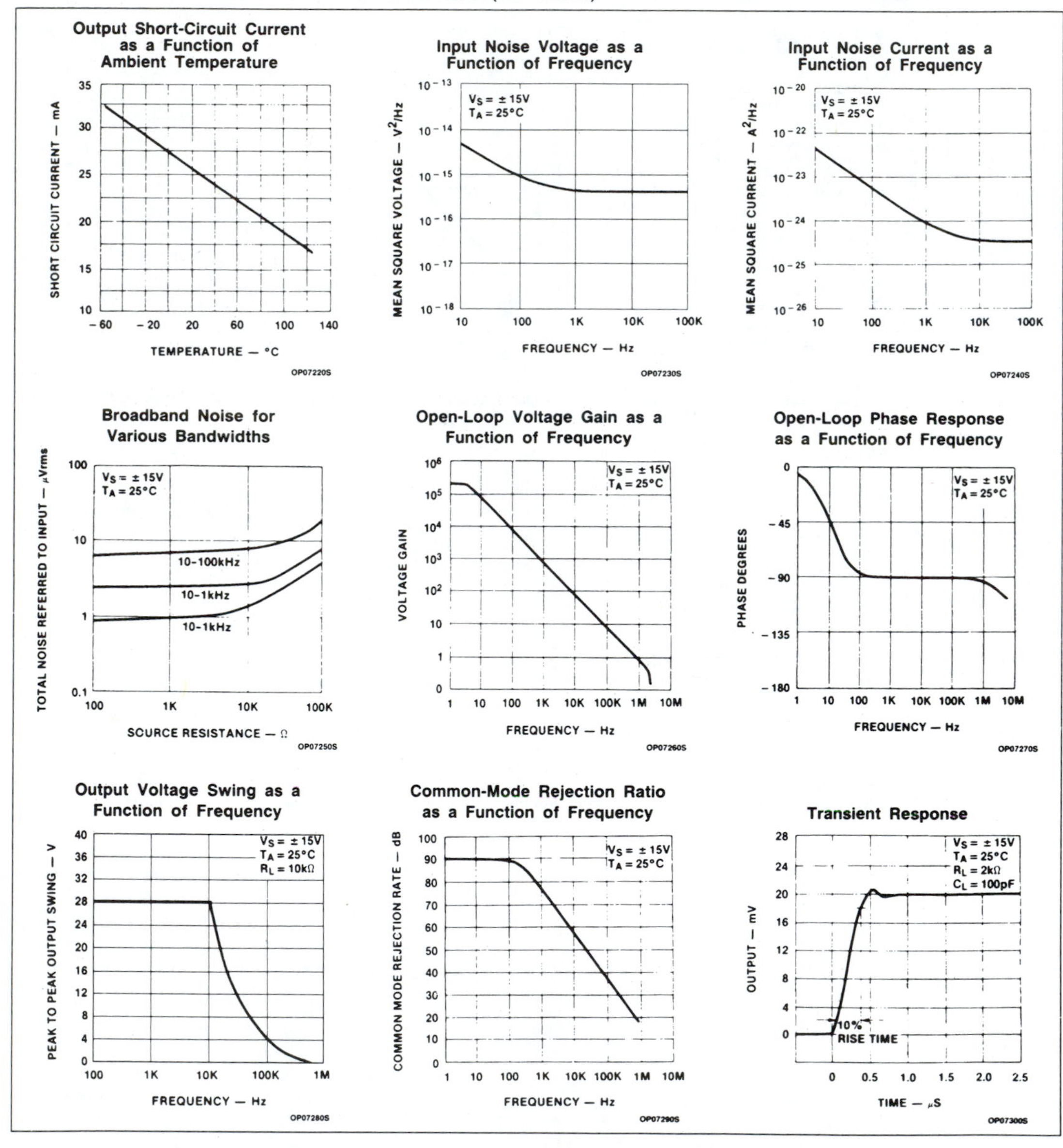


Figure 2-2. *Continued*

Signetics Linear Products Product Specification

General Purpose Operational Amplifier μA741/μA741C/SA741C

TYPICAL PERFORMANCE CHARACTERISTICS (Continued)

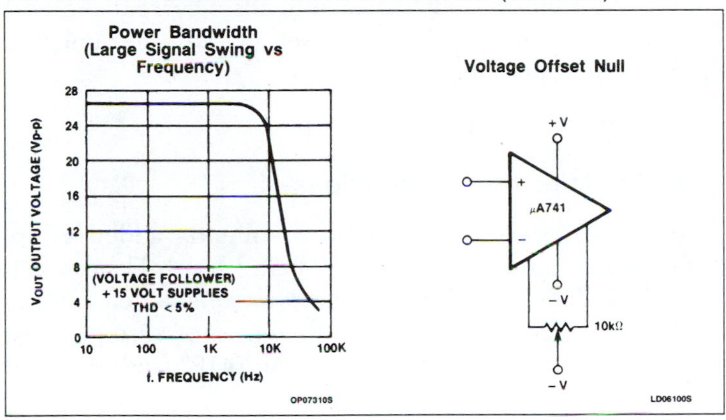


**Figure 2–2.** *Continued*

Since all op amps are different, *output resistances* ($R_{OUT}$), ④, (impedances) may vary from near zero to several thousand ohms. The 741C has a nominal output resistance of 75 Ω. With such low impedance, the output can function as a *voltage source* capable of providing current for a variety of load ranges. Also, with high input impedance and low output impedance, the op amp makes an excellent *impedance-matching* device.

### Characteristics Affecting dc Operations

*Input bias current* ($I_{BIAS}$), ⑤, the average of the two input currents (+ and −) with no signal applied, is held to microamperes or less by the high input impedance. This current can cause an imbalance in the op amp, and thus in the output. Using field effect transistors (FETs) at the inputs of the op amp will reduce the amount of input bias current and the imbalance in the op amp.

To obtain zero output voltage, both input currents should be equal. Since this is practically impossible, there must be an *input offset current* ($I_{OS}$), ⑥, the absolute value of the difference between the two input currents, to maintain a 0-volt output. In other words, one input may require more current than the other. This offset current may go as high as 20 mA in some ICs. The 741C has a typical input offset current of 20 nA.

The output voltage of an op amp should be zero when the voltage at both inputs is zero, but because of the inherent high gain in op amps, any circuit imbalance can cause some output voltage. The output voltage can be maintained at zero by applying a small *input offset voltage* ($V_{os}$), ⑦, the absolute value of the voltage between the input terminals, to one of the inputs. This is accomplished by applying a small reference voltage to one of the terminals. The 741C typically requires 2.0 mV.

Op amps, as all other solid-state devices, are susceptible to temperature changes. Alternating current circuits using op amps are less susceptible than are dc circuits. *Drift* is the term used for a change in offset current and offset voltage caused by temperature change. Drift will upset any adjusted imbalance in the op amp and produce output voltage errors.

### Characteristics Affecting ac Operations

*Frequency compensation* capacitors are frequently added to op amps to prevent high frequency output signals from being fed back to the input and causing undesired oscillations. These capacitors may be added externally or internally. The capacitor decreases the gain of the op amp as frequency increases, thus reducing the possibility of such feedback. The 741C uses internal frequency compensation.

*Slew rate* (SR), ⑧, is one parameter that does not carry minimum or maximum limits. Slew rate is one of the most important characteristics affecting the ac operation of an op amp, because it limits large-signal bandwidth (BW). Worst case conditions arise in the unity gain noninverting mode.

Figure 2–3A shows a typical bench setup for measuring the response of the output to a step (square wave) input. The input step frequency should be of a frequency low enough for the output of the op amp to have sufficient time to slew from limit to limit. Input voltage must be less than absolute maximum and the input wave should have good rise and fall times. The slew rate is then calculated from the slope of the output voltage versus time.

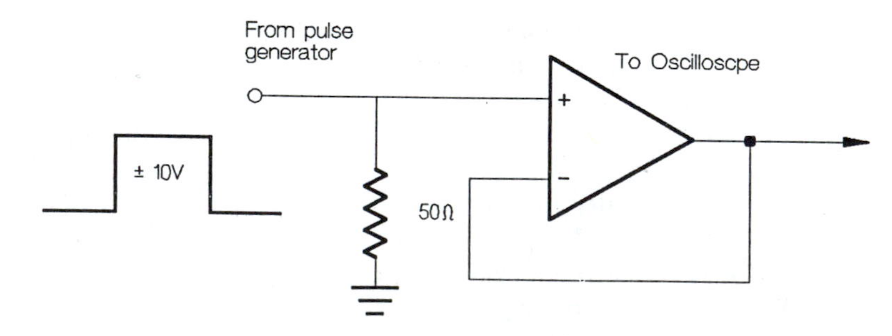


(A) Typical bench setup for measuring slew rate

**Figure 2–3.** Slew Rate Testing and Limiting Responses

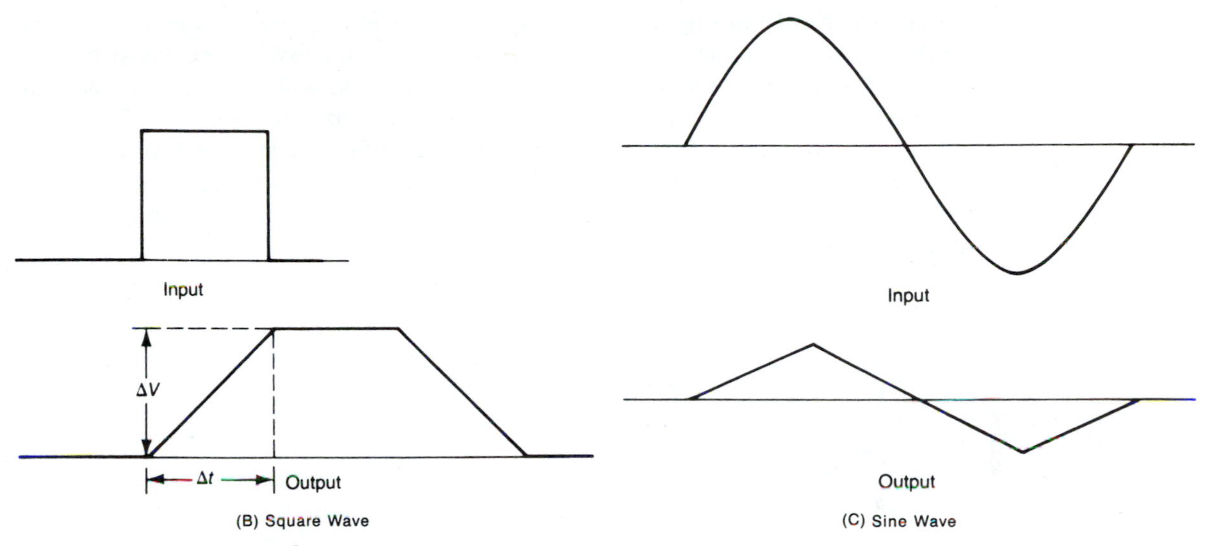


Figure 2-3. *Continued*

Typically, as the frequency of the input signal increases, the slew rate limiting becomes more pronounced. For an indication of how fast the output voltage of an op amp can change, slew rate is calculated as follows:

$$SR = \frac{\text{Maximum change in output voltage}}{\text{Change in time}} = \frac{\Delta V_{out(max)}}{\Delta t} \tag{2.1}$$

For example, the 741C has a slew rate of 0.5 volts per microsecond (V/μs), which means that output voltage can change no faster than 0.5 volts per microsecond, regardless of how fast the input voltage changes. The frequency compensation capacitor of the op amp, either internal or external, is the most common cause of slew rate limiting. Slew rate is normally specified at unity gain of the op amp.

The output signal in Figure 2-3B shows the limitation of the rapid rise and fall of the step input signal caused by slew rate limiting. Note the delayed output signal.

The sine wave input of Figure 2-3C results in the distorted output wave shown. This is, however, one way to produce a triangle waveform.

### Voltage Gain

The higher the voltage gain ($A_{VOL}$), ⑨, of an op amp, the better. Gains of 200,000 are common. This gain is *open-loop gain* (considered as infinite gain), generally at 0 hertz (Hz) or dc, and *without feedback.* Figure 2-4 shows the gain-frequency response curve for the 741C. Most of this gain is sacrificed in practical circuits, because the op amp is normally operated with heavy degenerative or regenerative feedback. Degenerative feedback drastically reduces the gain of the circuit, but it has an advantage, however, in that it increases the bandwidth (BW) of the circuit. Notice in Figure 2–4 that as the frequency is increased by one decade (tenfold) the gain is decreased by one decade, and the BW is increased by one decade. For example, the open-loop BW is about 10 Hz with an open-loop gain of 100,000. As gain is reduced to about 100, BW is increased to about 10 kHz. When gain is 10, BW is about 100 kHz. Unity gain, gain of one, occurs at 1 MHz, which is called the *unity-gain frequency,* ⑩ , for this example. Many manufacturers use the unity-gain frequency as a reference point for op amps. Different op amps may have different values than the 741.

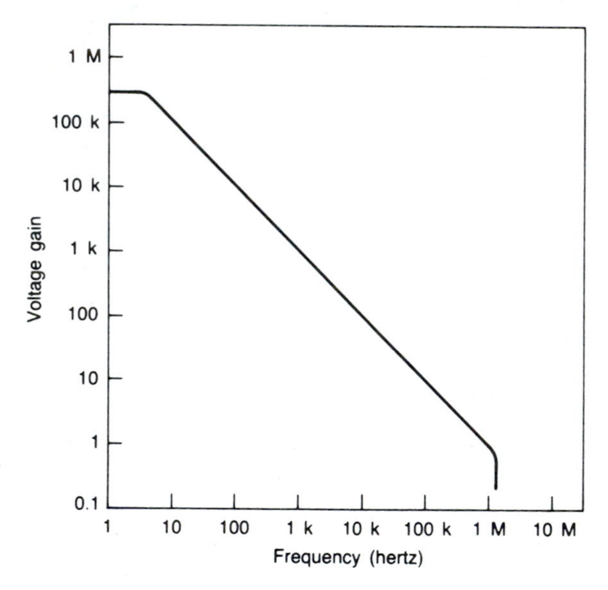


**Figure 2–4.** Gain Versus Frequency Response Curve

The *gain-bandwidth product* (*GBP*) is equal to the unity-gain frequency. The *GBP* not only indicates the upper useful frequency of a circuit, but it provides a means of determining the bandwidth for any given gain. Refer again to Figure 2–4. If you multiply the gain and bandwidth of any specific circuit, the product will equal the unity-gain frequency. This relationship can be expressed as follows:

$$GBP = A_v \times BW = \text{Unity-gain frequency} \quad \textbf{(2.2)}$$

where

$GBP$ = gain-bandwidth product
$A_v$ = amplifier voltage gain
$BW$ = bandwidth

## Example 2.1

An op amp circuit has a gain of 100 and a BW of 10 kHz. What is the GBP of the circuit?

*Solution*

Use Equation 2.2:

$$GBP = A_v \times BW = 100 \times 10{,}000\text{ Hz} = 1{,}000{,}000\text{ Hz} = 1\text{ MHz}$$

## Example 2.2

An op amp has a gain of 10 and a BW of 100 kHz. What is the GBP of the circuit?

*Solution*

Use Equation 2.2:

$$GBP = A_v \times BW = 10 \times 100\text{ kHz} = 1000\text{ kHz} = 1\text{ MHz}$$

By simple algebraic manipulation of Equation 2.2, you can determine the upper frequency limit or BW of any circuit with any given gain by dividing the unity-gain frequency of the op amp used by the given gain, as follows:

$$BW = \frac{\text{Unity-gain frequency}}{A_v} \tag{2.3}$$

## Example 2.3

You are using a general-purpose op amp that has a unity-gain frequency of 1 MHz. Your circuit is to have a gain of 50. What will be the BW of the circuit?

*Solution*

Use Equation 2.3:

$$BW = \frac{\text{Unity-gain frequency}}{A_v} = \frac{1{,}000{,}000\text{ Hz}}{50} = 20{,}000\text{ Hz} = 20\text{ kHz}$$

Likewise, the gain of the circuit can be determined by further manipulation of Equation 2.2:

$$A_v = \frac{\text{Unity-gain frequency}}{BW} \tag{2.4}$$

**Example 2.4**

You are using a 741 op amp, and your circuit has a BW of 3 kHz. What is the gain of the circuit?

*Solution*

Use Equation 2.4:

$$A_v = \frac{\text{Unity-gain frequency}}{BW} = \frac{1{,}000{,}000 \text{ Hz}}{3{,}000 \text{ Hz}} = 333$$

### Noise

Any electronic circuit is susceptible to noise, and op amps are no exception. External noise generated by electrical devices and inherent noise of electronic components can be minimized by proper construction techniques. Internal noise resulting from internal components, bias current, and drift can be minimized by keeping input and feedback resistor values as low in value as possible and still maintain circuit requirements. You should remember that undesired noise will be amplified along with the desired signal.

### Common-mode Rejection

The input circuit of the IC op amp is a differential amplifier stage. Associated with this input stage is a feature called *common-mode rejection.* This means that if the voltage applied to both inputs of the differential amplifier is in phase and has the same amplitude and frequency, the output will be zero. A voltage at the output can be produced only if a difference of potential is introduced at the inputs. The ability of an op amp to reject common-mode signals while amplifying differential signals is called the *common-mode rejection ratio* (CMRR), (11) . The CMRR is usually expressed in *decibels* (dB). The higher the rating, the better the common-mode rejection. Common-mode operation will be discussed in more detail later in this chapter.

## 2.5 The Three Stages in Op Amp Construction

While it is not necessary to know what is inside the op amp in order to use it, knowledge of the op amp and its operation is enhanced by understanding the three stages it consists of. Figure 2–5 is a block diagram of a typical op amp that shows the three stages—a differential amplifier, a high-gain voltage amplifier, and an output amplifier. It also shows the power supply and input and output connections.

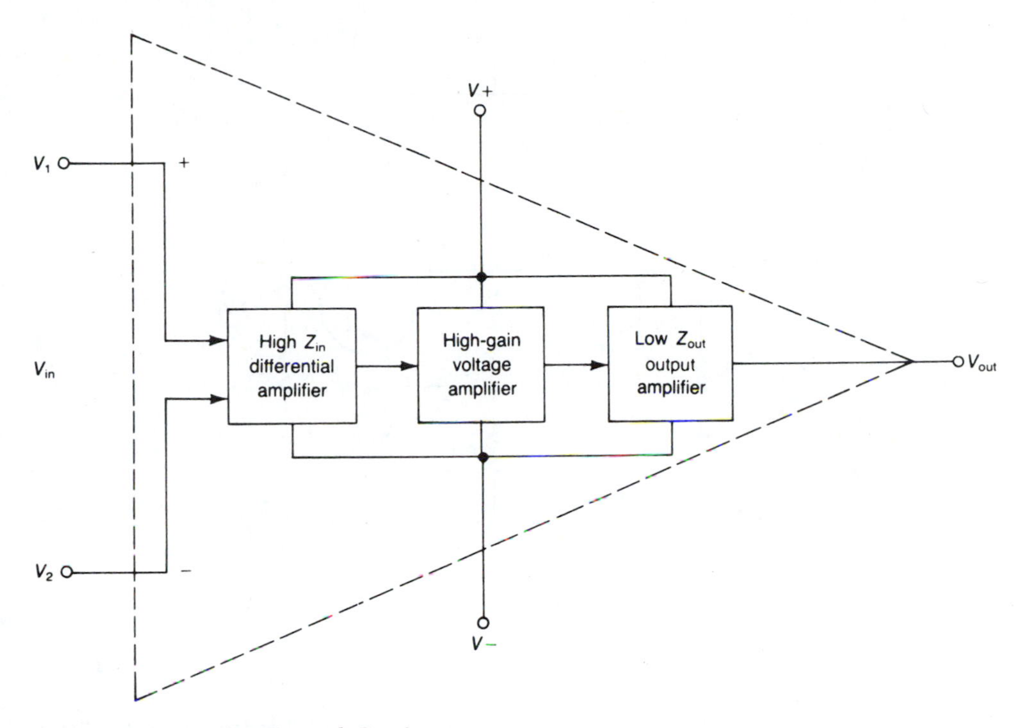


**Figure 2–5.** Block Diagram of a Typical Op Amp

## Review of Differential Amplifiers

The input stage of virtually all op amps is a *differential amplifier*. The differential amplifier is ideally suited to IC fabrication because it has no capacitors; it requires only transistors and resistors.

A simple differential amplifier is shown in Figure 2–6. The amplifier has two input ($V_1$ and $V_2$) and two output ($\pm V_{out}$) terminals. The differential output is taken between the two collectors. Resistor $R_E$ assures an equal voltage across the emitter-base junctions of both transistors. With discrete component construction, this circuit would require matched-pair transistors ($Q_1$ and $Q_2$) to provide the desired symmetry. However, transistors fabricated in ICs have almost identical characteristics, so the circuit is very nearly symmetrical. Transistor biasing is accomplished with base resistors ($R_B$) and collector resistors ($R_C$). The circuit amplifies the difference between the two inputs. Output voltage ($V_{out}$), taken between the two collectors, is equal to voltage input 1 ($V_1$) minus voltage input 2 ($V_2$) times the amplifier gain ($A_v$). Expressed in equation form,

$$V_{out} = A_v (V_1 - V_2) \tag{2.5}$$

In special cases, the two input signals will be identical except that they will be 180° out of phase. Because the signals are 180° out of phase, the differential

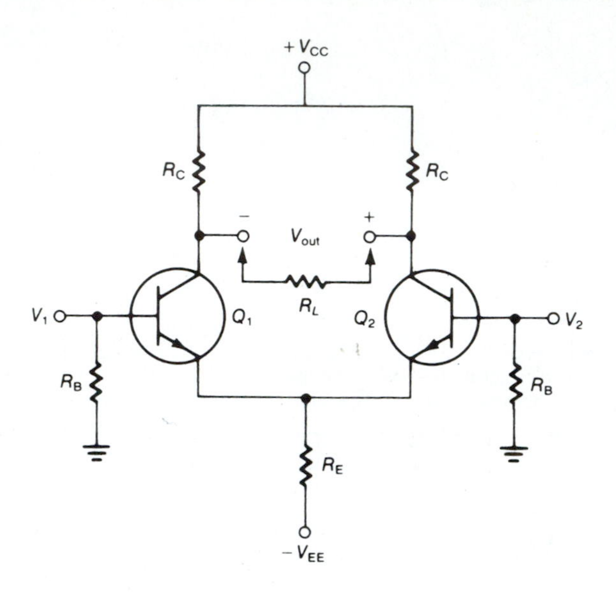


**Figure 2–6.** Basic Differential Amplifier Circuit

(difference) voltage ($V_D$) between the inputs is twice the amplitude of either input signal. This relationship can be stated as follows:

$$V_D = V_1 - V_2 \tag{2.6}$$

## Example 2.5

If input $V_1$ is equal to $+10$ mV and input $V_2$ is equal to $-10$ mV, what is the amplitude of the difference voltage?

*Solution*

Use Equation 2.6:

$$V_D = V_1 - V_2 = +10 \text{ mV} - (-10 \text{ mV}) = +10 \text{ mV} + 10 \text{ mV} = 20 \text{ mV}$$

Analyzing the effects of each input separately provides a better understanding of why the circuit responds as it does. Refer to Figure 2–6, and first consider the effects of $V_1$. A positive-going signal at the base of transistor $Q_1$ causes an amplified negative-going signal at the collector of $Q_1$. By emitter-follower action, $V_1$ also couples a positive-going signal to the emitter of transistor $Q_2$, which causes an amplified positive-going signal at the collector of $Q_2$.

$V_2$ swings negative at the same time $V_1$ swings positive. This places a negative-going signal on the base of $Q_2$, which drives the collector of $Q_2$ more positive. This means that $V_1$ and $V_2$ acting together tend to force the collector of $Q_2$ positive. Also, through emitter-follower action, $Q_2$ couples a sample of $V_2$ to the emitter of $Q_1$. The negative-going signal of $V_2$ on the emitter of $Q_1$ causes $Q_1$ to conduct more and drive its collector more negative. Thus, $V_1$ and $V_2$ acting together tend to force the collector of $Q_1$ more negative.

Acting together, $V_1$ and $V_2$ produce an amplified output signal. A load is normally connected between the two output terminals on the collectors. A large output voltage is developed across the load ($R_L$) because the signals at the two output terminals are 180° out of phase.

If desired, the load could be connected between either output terminal and ground. The voltage available at the remaining terminal would have the same amplitude but the opposite phase.

### Constant-current Source

Most differential amplifiers used in ICs have a *constant-current source* in the emitter circuit. Figure 2–7 shows a practical current source, using a zener diode to maintain a constant voltage ($V_Z$) across the emitter-base junction. Resistor $R_s$ is a current-limiting series resistor for zener diode protection.

Assume $Q_1$ is a silicon transistor with an emitter-to-base voltage drop ($V_{BE}$) of exactly 0.6 V. Then the voltage across the emitter resistor ($R_E$) is as follows:

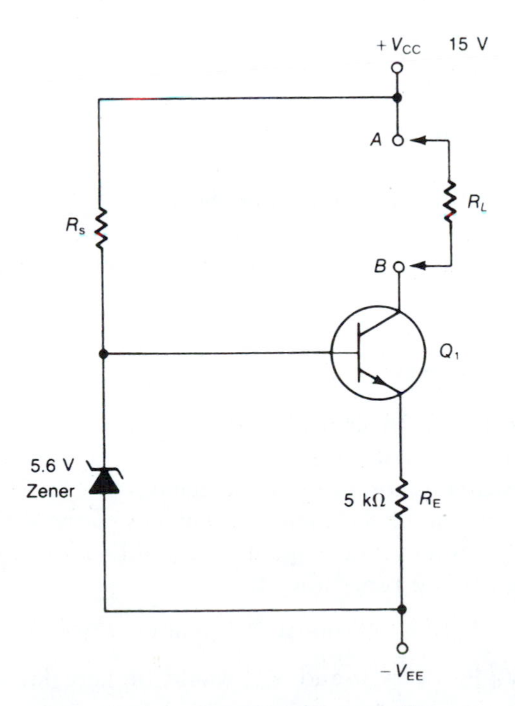


**Figure 2–7.** Constant-Current Source

$$V_{R_E} = V_Z - V_{BE}$$
$$= 5.6\text{ V} - 0.6\text{ V} = 5.0\text{ V} \quad \textbf{(2.7)}$$

The emitter current ($I_E$) is

$$I_E = \frac{V_{R_E}}{R_E}$$
$$= \frac{5\text{ V}}{5000\ \Omega} = 0.001\text{ A, or 1 mA} \quad \textbf{(2.8)}$$

Since $Q_1$ draws a very small amount of base current, for practical purposes, consider the collector current to be 1 mA. Shorting across points $A$ and $B$ will cause 1 mA of current to flow. Connecting a load resistor ($R_L$) across these same points will have little effect on current flow as long as $R_L$ is not made too large. For proper transistor action, the collector-base junction must be reverse biased, so the collector voltage must not drop lower than the base voltage. Figure 2–7 shows a base voltage of 5.6 V and a $V_{CC}$ of 15 V. Therefore, the collector voltage will be the same as base voltage when

$$V_L = V_{CC} - V_B$$
$$= 15\text{ V} - 5.6\text{ V} = 9.4\text{ V} \quad \textbf{(2.9)}$$

The collector current ($I_C$) is essentially 1 mA, so $R_L$ will drop 9.4 V when its value is

$$R_L = \frac{V_L}{I_C}$$
$$= \frac{9.4\text{ V}}{0.001\text{ A}} = 9400\ \Omega \quad \textbf{(2.10)}$$

Keeping $R_L$ well below this value will allow the circuit to act as a constant-current source. Figure 2–8 shows the current source circuit connected to the differential amplifier circuit. The current delivered to the differential amplifier is independent of changes in the input signals.

### Common-Mode Signal

A look at another characteristic of the differential amplifier will demonstrate the importance of the constant-current source.

As discussed earlier, a common-mode signal is one that is identical at both input terminals. Since one of the major advantages of the differential amplifier is its ability to reject common-mode signals, it should not respond to such signals. This becomes obvious if we remember that

$$V_{out} = A_v\,(V_1 - V_2) \qquad \text{(repeat of Equation 2.5)}$$

If $V_1 = V_2$, then $V_1 - V_2 = 0$ and $V_{out}$ would be zero for all common-mode signals.

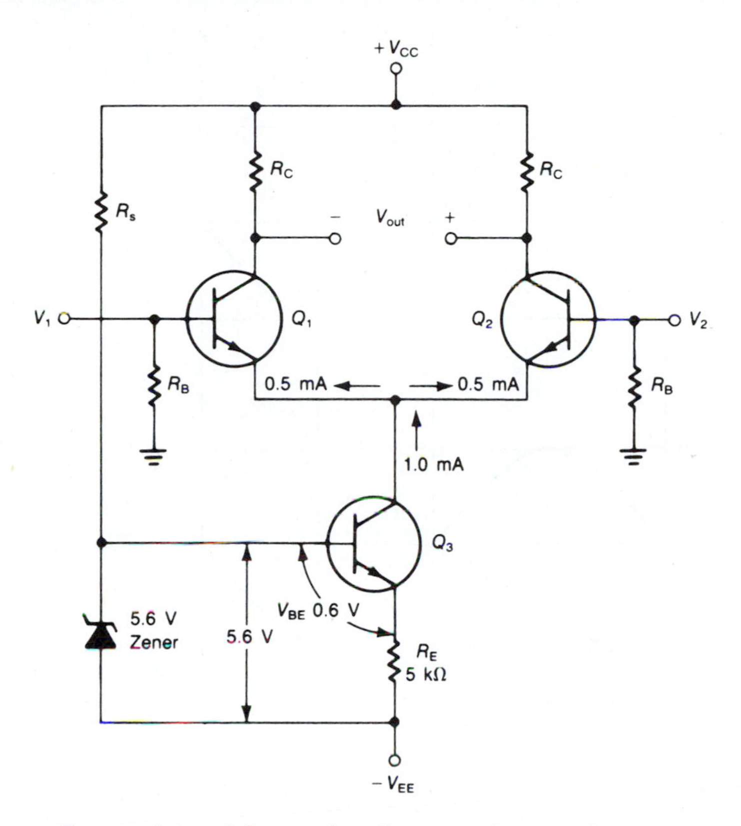


**Figure 2–8.** Differential Amplifier with a Constant-Current Source

Consider what happens when the two signals shown in Figure 2–9 are applied to the circuit shown in Figure 2–8. The two signals can be classified as common-mode signals because $V_1$ and $V_2$ are identical; that is, they are in phase and of equal amplitude.

Both $V_1$ and $V_2$ are at zero volts at time $T_0$. A constant 1 mA current is provided by the current source (see Figure 2–8). This current is split, 0.5 mA to each transistor. Both $V_1$ and $V_2$ swing positive to +100 mV at time $T_1$ (see Figure 2–9). Under ordinary circumstances you would expect both transistors to conduct more when the bases swing more positive. But remember that the current source is providing exactly 1 mA, and since both bases swing positive by the same amount, both transistors will still draw exactly the same 0.5 mA current as before. The current does not change and, therefore, the output voltage ($V_{out}$) does not change.

The resultant rejection of the common-mode signal and amplification of the differential signal is a ratio, called the CMRR. It is the ratio of difference gain ($A_D$) to common-mode gain ($A_{CM}$), usually expressed in dB.

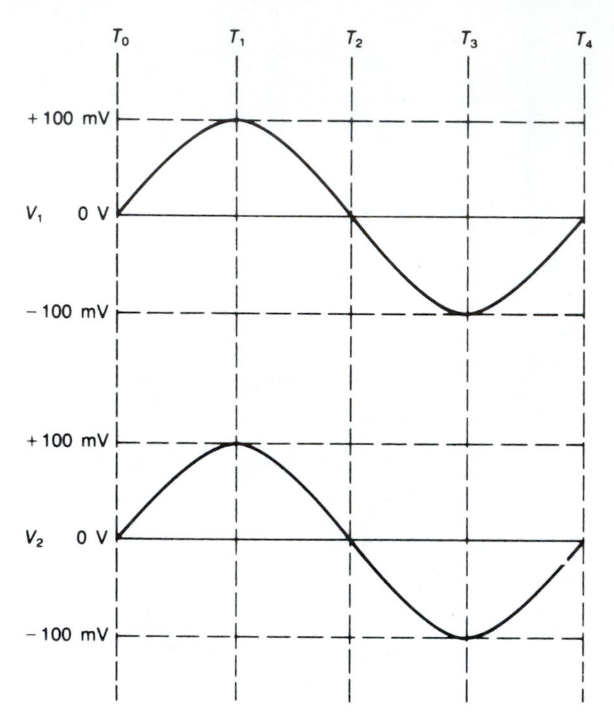


**Figure 2–9.** Common-Mode Signals

$$\text{CMRR} = 20 \log \frac{A_D}{A_{CM}} \tag{2.11a}$$

or

$$\text{CMRR} = 20 \log \frac{V_{out}(\text{CM})}{V_{in}(\text{CM})} \tag{2.11b}$$

Comparing a change in the output voltage ($\Delta V_{out}$) to a change in the differential input voltage ($\Delta V_D$) will provide the difference gain ($A_D$):

$$A_D = \frac{\Delta V_{out}}{\Delta V_D} \tag{2.12}$$

## Example 2.6

A change of 10 mV at the differential input causes a change of 10 V at the output. What is the difference gain? The common-mode gain? The CMRR?

*Solution*

Use Equation 2.12:

$$A_D = \frac{\Delta V_{out}}{\Delta V_D} = \frac{10 \text{ V}}{0.01 \text{ V}} = 1000$$

If a common-mode voltage change ($\Delta V_{CM}$) of 10 mV causes an output voltage change ($\Delta V_{out}$) of 1 mV, the common-mode gain ($A_{CM}$) is

$$A_{CM} = \frac{\Delta V_{out}}{\Delta V_{CM}} = \frac{1\ \text{mV}}{10\ \text{mV}} = 0.1 \quad \textbf{(2.13)}$$

Recall that CMRR is 20 log times the ratio of $A_D$ to $A_{CM}$. Therefore, using the previous calculations including Equation 2.11a,

$$\text{CMRR} = 20 \log \frac{A_D}{A_{CM}} = 20 \log \frac{1000}{0.1} = 80\ \text{dB}$$

In practice, the amplifier may not entirely reject the common-mode signal. But a good amplifier will amplify the desired signal by a large amount while attenuating the undesired frequency by a similar amount. Thus, the output may indeed result in the desired signal being 80 dB higher than the undesired signal, as the calculations in Example 2.6 show.

## Review of Darlington Pairs

The second stage of the op amp is a very high-gain voltage amplifier stage. This amplifier stage is usually constructed with *Darlington pairs* and provides most of the gain of the op amp.

Recall from earlier studies of transistor amplifiers that the dc beta ($\beta_{dc}$) of a transistor is known as the dc current gain and is designated on most data sheets as $h_{FE}$. The higher $\beta_{dc}$ is, the higher the dc current gain of the transistor.

Recall also that one way to increase beta is to connect two transistors, as illustrated in Figure 2–10. The overall beta of the two transistors is equal to the product of the individual betas, that is,

$$\beta = \beta_1\beta_2 \quad \textbf{(2.14)}$$

### Example 2.7

If $\beta_1 = \beta_2 = 100$, calculate the effective beta of the Darlington pair.

*Solution*

Use Equation 2.14:

$$\beta = \beta_1\beta_2 = 100 \times 100 = 10{,}000$$

Obviously, then, cascading additional Darlington pairs will result in the high gain for which op amps are noted. Remember that a typical op amp may have an open-loop voltage gain of 200,000 or more.

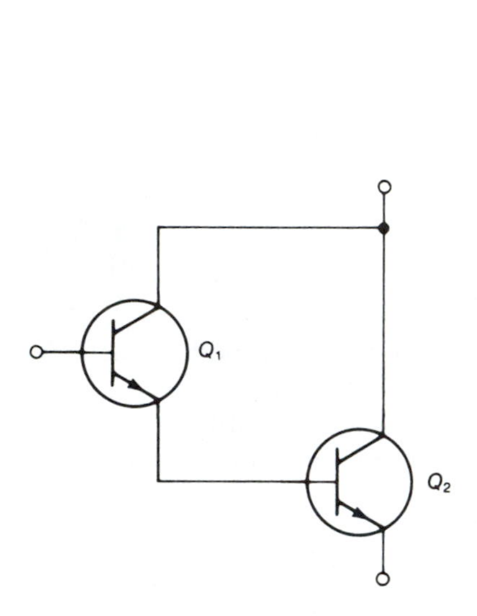

**Figure 2–10.** Darlington Pair

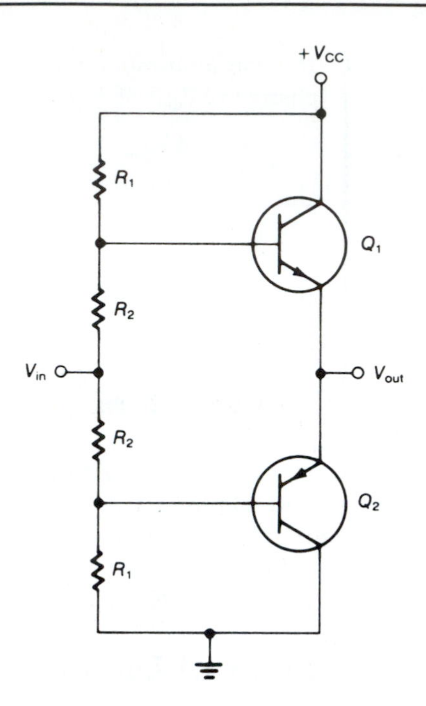


**Figure 2–11.** Class B Push-Pull Emitter Follower

### Review of Push-Pull Amplifiers

The third, or output, stage of the op amp is usually a class B push-pull emitter follower. Figure 2–11 shows a typical push-pull circuit. Assume biasing near cutoff, as you will recall from earlier studies.

The push-pull circuit action is obtained by combining the two emitter followers, with the output taken from the junction of the two emitters. The NPN transistor, $Q_1$, handles the positive half cycle of input voltage, while the PNP transistor, $Q_2$, handles the negative half cycle. The output voltage then combines the opposite halves and produces a complete sine wave.

In the typical op amp, equal positive and negative voltages provided by the split supply will ideally produce a *quiescent* (no input signal) output voltage of zero.

A typical schematic diagram for an op amp, identifying the three stages we have discussed, is shown in Figure 2–12. (This is an enlargement of the 741 shown in Figure 2–2.)

## 2.6 Power Supply Considerations

Most op amps require a dual power supply (equal positive and negative voltages) for proper operation. Such a power source allows the op amp output to swing

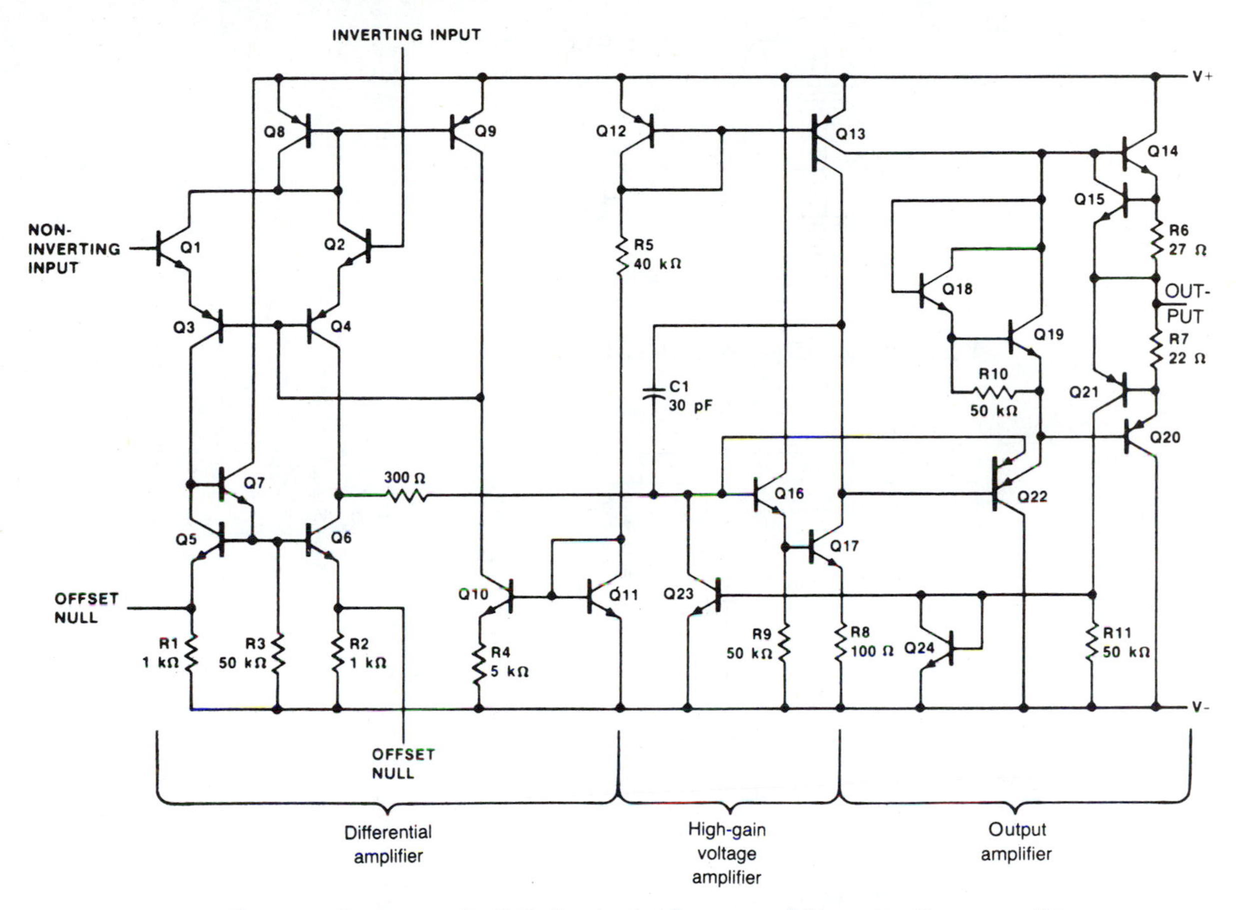


**Figure 2–12.** Schematic Diagram of a 741 Op Amp (Courtesy of Signetics Company, Linear Division)

positive and negative with respect to ground, a particularly useful feature in dc circuits and audio applications.

Dual power sources can be provided by connecting two batteries in series, as illustrated in Figure 2–13, with the common connection being reference ground. Such a supply is portable, but all batteries must be fresh if the circuit is to operate properly.

Battery-operated power supplies can provide the basic voltages required for op amp operation, but many op amp applications require very stable, noise-free, regulated voltages. For such applications, IC *dual tracking* voltage regulators may be used, as shown in Figure 2–14. Input capacitor $C_1$ provides steady positive and negative voltages to the regulator, and the two 0.1 μF capacitors are bypass capacitors that provide stability. Table 2–1 lists several such ICs.

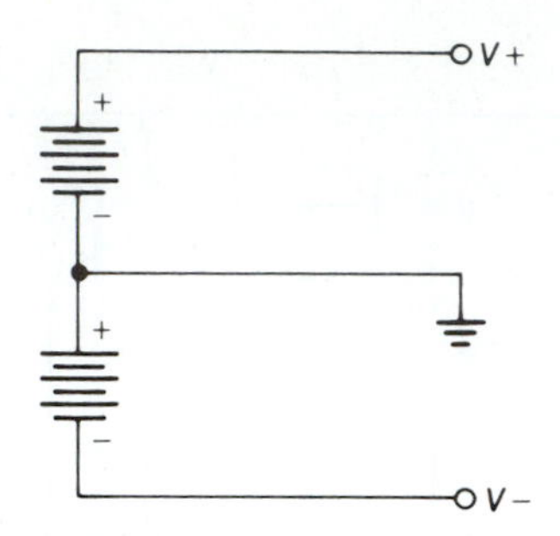


Figure 2–13. Dual ± Battery Power Source

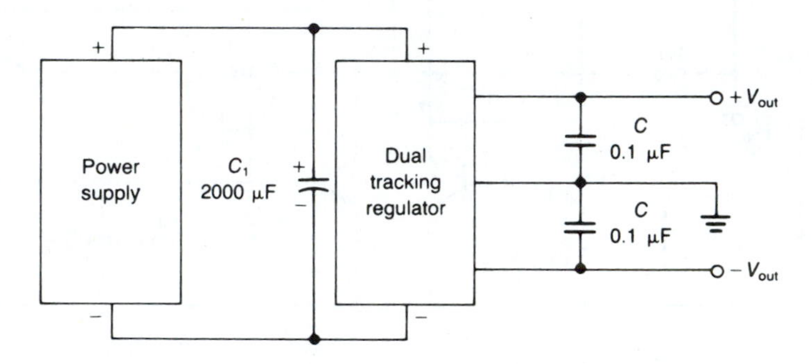


Figure 2–14. Dual Tracking IC Voltage Regulation

**Table 2–1 Dual Tracking Voltage Regulators**

| Type | Voltage Range |
|---|---|
| LM 125 | ±15 V |
| LM 126 | ±12 V |
| LM 127 | +5 V, −12 V |
| NE/SE 5551N | ±5 V |
| NE/SE 5552N | ±6 V |
| NE/SE 5553N | ±12 V |
| NE/SE 5554N | ±15 V |
| NE/SE 5555N | +5 V, −12 V |
| RM/RC 4195 | ±15 V |

Some special op amps are designed to operate from single-voltage power supplies. These supplies usually are positive voltage types. In the quiescent state the output voltage should measure approximately halfway between the maximum voltage supply and ground.

In some limited applications, standard op amps use a single power supply. We will study some of those applications later.

## Heat Effects

Because the IC chip is so small, special care must be taken to keep the heat generated by internal currents within limits specified by the manufacturer. Input and output conditions will vary these currents. With no input signal and no output load, the state of the transistors and resistors of the internal circuits determine the supply current. When input signals are applied and an output load is connected, the power supply current through the device increases. If the generated heat becomes excessive, the device may be destroyed.

Data sheets provide values of parameters related to power requirements. Refer back to Figure 2–2. Note, for example, *supply voltage* ($V_S$), (12), of the 741C op amp is limited to ±18 volts. Operating at this maximum power supply value for any length of time will degrade the characteristics of the device to a point where it cannot function properly, and exceeding this maximum value can destroy the device.

## Power Consumption with No Load

Remember that any device that carries current will generate heat and, therefore, will dissipate (consume) power. Refer again to the data sheet of Figure 2–2. Note that the *operating temperature range maximum* ($T_A$), (13), is +70 °C and the *maximum power consumption* ($P_D$), (14), is 85 mW. The total no-load power consumption ($P_{CC}$) of the device is the product of power supply voltage ($V_{CC}$) and current ($I_{CC}$). When using a dual power supply, the total no-load power consumption can be determined as follows:

$$P_{CC} = (+V_{CC})(+I_{CC}) + (-V_{CC})(-I_{CC}) \quad (2.15)$$

If a single power supply is used, the power consumption is

$$P_{CC} = (+V_{CC})(+I_{CC}) \quad (2.16)$$

### Example 2.8

A typical circuit has $V_{CC} = \pm 15$ V and $I_{CC} = \pm 2.8$ mA. Under no-load conditions, how much power is consumed in the device?

*Solution*

Use Equation 2.15:

$$\begin{aligned} P_{CC} &= (+V_{CC})(+I_{CC}) + (-V_{CC})(-I_{CC}) \\ &= (+15\text{ V})(+2.8\text{ mA}) + (-15\text{ V})(-2.8\text{ mA}) \\ &= 42\text{ mW} + 42\text{ mW} = 84\text{ mW} \end{aligned}$$

This is less than the maximum 85 mW indicated by the data sheet (Figure 2–2), and the device will not consume more power than this under the no-load condition.

## Power Consumption Under Load

More current from the power supply flows in the device under load conditions. The extra current includes current received from or delivered to the load. The total power dissipation ($P_D$) of the device under these conditions is the power delivered from the source ($P_s$) minus the power consumed in the load ($P_L$):

$$P_D = P_s - P_L \tag{2.17}$$

$P_L$ is the product of load voltage ($V_L$) and load current ($I_L$):

$$P_L = V_L I_L \tag{2.18}$$

### Example 2.9

A load drawing 17 mA at +8 V is connected to the circuit of Example 2.8. If the power supplied to the device is 250 mW, what is the internal power dissipation of the device?

*Solution*

Use Equation 2.18:

$$P_L = V_L I_L = 8\text{ V} \times 17\text{ mA} = 136\text{ mW}$$

Use Equation 2.17:

$$P_D = P_s - P_L = 250\text{ mW} - 136\text{ mW} = 114\text{ mW}$$

Note that *internal power dissipation* ($P_D$), ⑮, for the standard DIP (μA741CN, Figure 2–2) has a maximum rating of 1170 mW. Therefore, the device is well within specifications.

## Circuit Current Requirements

The *rated supply current* is an indication of the maximum current drawn from the power supply under no-load, no-signal conditions. The power supply must provide at least that amount, plus any increase in current requirements of the device caused by the addition of an input signal and a load. Therefore, the *maximum* current required by the circuit with an input signal and a load will determine the *minimum* power supply current requirements. A good practice is to select a power supply that will provide a current value of *at least 10% higher* than the maximum circuit requirements. This will ensure sufficient current for the circuit and will not load down the supply voltage.

## Example 2.10

A power supply is to be selected for the circuit of Example 2.9. The load draws 17 mA. What is the power supply current requirement? Use the typical $I_{CC}$ of the circuit in Example 2.8.

*Solution*

The maximum circuit current ($I_{CC(max)}$) requirement must include the load current ($I_L$) plus the maximum rated no-load device current ($I_{CC}$):

$$\begin{aligned} I_{CC(max)} &= I_L + I_{CC} \\ &= 17 \text{ mA} + 2.8 \text{ mA} = 19.8 \text{ mA} \end{aligned} \tag{2.19}$$

The minimum power supply current ($I_{PS(min)}$) should provide at least 10% greater current, therefore,

$$\begin{aligned} I_{PS(min)} &= I_{CC(max)} + 0.1(I_{CC(max)}) \\ &= 19.8 \text{ mA} + \approx 2 \text{ mA} = 21.8 \text{ mA} \end{aligned} \tag{2.20}$$

The power supply selected for this circuit, therefore, must be capable of providing a minimum of 22 mA while maintaining the required $\pm 15$ V.

## Other Factors to Consider

Factors other than voltage and current needs must be considered when selecting a power supply to ensure that the supply will not interfere with the proper operation of the device or damage it. Such considerations are *long-term stability*, *ac stability* of the device, and *protective circuits*.

Long-term stability is necessary to provide a constant output voltage over a long period of operation, assuming constant load conditions, ambient temperature, line voltage, and circuit output control adjustments. Such stability can be ensured by regulating the load with line voltage regulators, checking power supply filter circuits for minimum ripple voltage, or using a common ground to eliminate the possibility of ground loops.

The ac stability of the device may be improved by installing bypass capacitors between all power supply connections and ground. The capacitors should be greater than 0.01 μF and located as close to the device terminals as possible. Also, the leads should be kept short to reduce the possibility of resonance with lead inductance. This precautionary measure will prevent undesirable feedback through the internal impedance of the power supply.

The last factor mentioned that should be considered when choosing a power source was protective circuits. Protective circuits are so important that they will be discussed under a separate heading.

## Protective Circuits

Protective circuits should be provided for both the device and the power supply. The device, for example, can be destroyed if the power supply terminals are reversed, even momentarily. One method of protection from such a possibility is to place diodes ($D_1$ and $D_2$) in the connecting lines, as shown in Figure 2–15. The breakdown voltage of the diodes must be greater than the supply voltage used. Then, if the power supply is inadvertently connected with reversed polarity, the diodes will be reverse biased, act as open circuits, and prevent current from destroying the op amp.

### Overvoltage Protection

Most op amps operate with maximum values of about ±20 V. Therefore, to prevent performance degradation and the possibility of destroying the op amp, *overvoltage protection* should be considered. Figure 2–16 illustrates the use of zener diodes ($Z_1$ and $Z_2$) as overvoltage protective devices. The zener breakdown

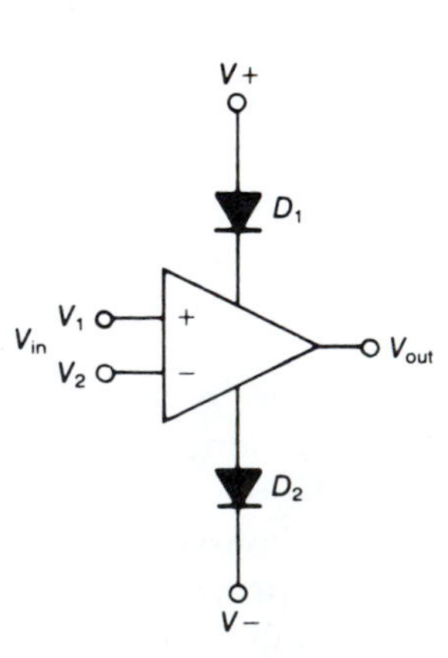


**Figure 2–15.** Diode Protection for Power Supply Connection Lines

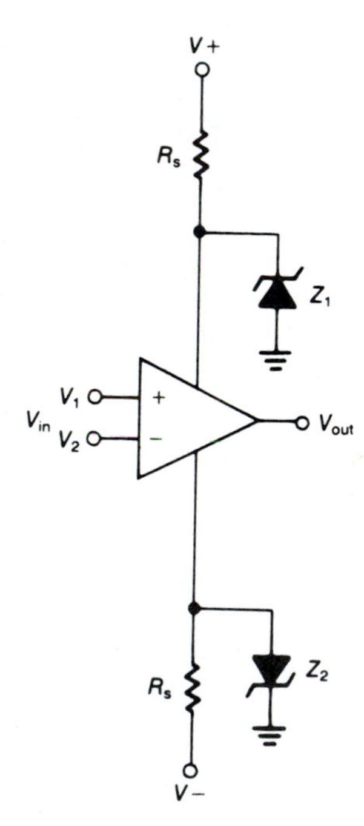


**Figure 2–16.** Zener Diode Overvoltage Protective Circuit

voltage should be slightly higher than that of the power supply operating value, but lower than the maximum allowable supply voltage.

The series resistors ($R_S$) are inserted to limit the zener current and prevent excessive current drain from the power supply. The proper resistance value can be determined as follows:

$$R_s = \frac{V - V_Z}{I_Z} \tag{2.21}$$

where

$V - V_Z$ = the voltage drop across the resistor
$I_Z$ = the desired zener current

## Example 2.11

Refer to Figure 2–16. What value of resistance ($R_s$) should be used to limit op amp voltage to 20 V? You also must limit power supply current drain to a safe level in case the voltage rises to 25 V.

*Solution*

First, select a zener diode that will satisfy the requirement. A look at a data book would indicate a possible choice of a 1N968, which has $V_Z$ = 20 V and $I_Z$ = 15 mA. Then, use Equation 2.21:

$$R_s = \frac{V - V_Z}{I_Z} = \frac{25\text{ V} - 20\text{ V}}{15\text{ mA}} = 333\ \Omega$$

You should use a standard 330-ohm resistor in the circuit.

### Load-Protection Circuits

Most commercial power supplies have built-in *load-protection* circuits that automatically reduce the voltage or limit the current of the supply if something goes wrong with the power supply (PS). These protection circuits are in addition to the standard fuses and circuit breakers.

One such circuit is the *overvoltage crowbar* circuit illustrated in Figure 2–17. If the output voltage exceeds a predetermined value, the comparator circuit generates a signal that causes the silicon-controlled rectifier (SCR) to conduct. The resistance of the SCR and the voltage drop across it are reduced to a low value. Therefore, the load is almost shorted and receives a very low voltage from the supply, thereby preventing damage to the load from overvoltage. The SCR series resistance *R* is a current-limiting resistor that prevents current from destroying the SCR.

Another common protective method protects both the load device and the power supply if the load current exceeds a predetermined value. Such excess

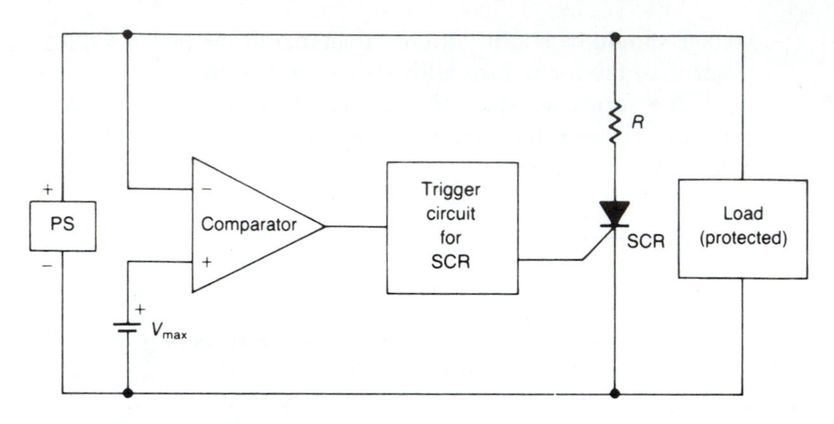


Figure 2–17. Block Diagram of an Overvoltage Crowbar Circuit for Load Protection

current may occur because of too many devices being powered by the source or because of a short of the outputs.

The block diagram in Figure 2–18 shows a simple *foldback current-limiting* circuit. When load impedance goes too low, the power supply voltage decreases, drawing excessive current from the supply. The excessive load current causes an increased voltage drop across $R_1$ that activates the sensing device. The sensing device causes $R_2$ to increase in value and thereby drop more of the input voltage, leaving less voltage for the load. The reduced output voltage causes the supply current to drop to a safe level. The power supply output voltage returns to its predetermined value when the conditions that caused the increased current load are removed.

The current-limiting resistance, $R_2$, of Figure 2–18 is actually a transistor in the op amp—$Q_{15}$ in Figure 2–12. The transistor acts as a variable resistor by varying the base voltage applied to it.

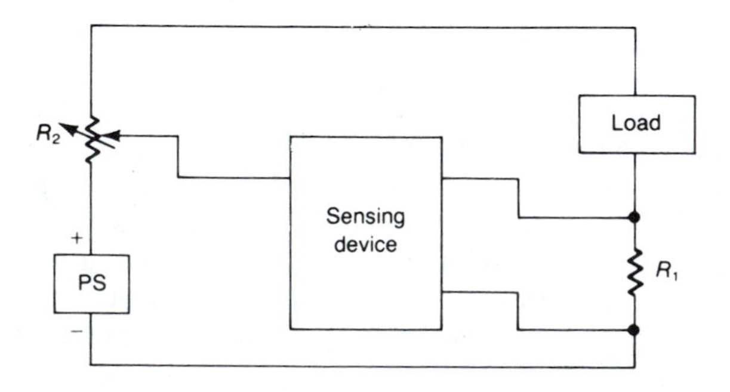


Figure 2–18. Block Diagram of a Foldback Current-Limiting Circuit

# 2.7 Single-Voltage Supply Systems

Op amps are designed specifically to operate using symmetrical dual-voltage supply systems. Most other LICs operate from a single positive- or single negative-power supply, although there are special circuits where op amps can also be operated from a single voltage supply. It should be obvious, however, that the single-voltage supply must provide the same amount of voltage as did the dual voltage supply it replaces. For example, a single supply to operate a μA741C must provide 36 V maximum to replace the ±18 V maximum provided by a dual supply normally used. In practice, operating voltages are limited to ±15 V, so the single supply must provide 30 V. If a single positive supply is used, it is connected to the positive-voltage terminal of the op amp, and the negative terminal is grounded. The inverse is true if a single-negative supply is used. Figure 2–19 shows the proper methods of connecting the single supply.

When using a single-voltage supply, the input signal should be isolated from the input terminals through a capacitor because the input terminals are above ground. Examples of circuits using single-supply systems will be studied in later chapters.

## Inverter and Converter Circuits

If the single power supply available to you cannot provide enough voltage for correct operation of the active device, inverter or converter circuits may be used. An *inverter* is a circuit that converts a dc input voltage to a higher or lower ac output voltage. The dc input voltage is used as the supply voltage for an oscillator, whose ac voltage output is then stepped up or down through transformer action. Similarly, a *converter* is a circuit that converts a dc input voltage to a higher or lower dc output voltage. The dc input voltage is used to supply voltage to an oscillator, whose ac voltage is stepped up or down through transformer action. However, the secondary of the transformer has a rectifier-filter circuit that converts the ac into the desired dc level.

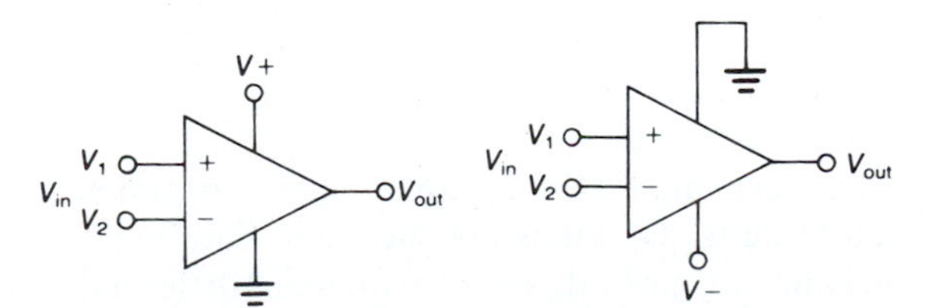


**Figure 2–19.** Block Diagram of Single-Voltage Power Supply Connections to Op Amp

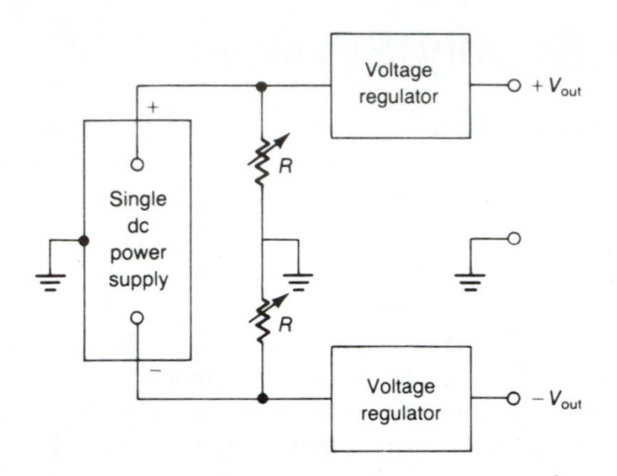


**Figure 2–20.** Block Diagram of a Single-Voltage Power Supply Converted to a Dual-Voltage Power Supply

## 2.8 Dual-Voltage Supply Systems

Power supplies for op amp circuits must provide exactly equal but opposite voltages. These exact values must be maintained during the operation of the device if it is to operate properly; that is, the device should provide true outputs for any given input. Each terminal of the supply must provide equal amounts of regulation if it is to maintain a balanced output.

Single power supplies can be converted to dual power supplies quite easily. One approach is to use a circuit like the one shown in Figure 2–20. Variable resistors in the voltage divider network ensure equal and opposite voltages for balance. While this circuit can be used with most single power supplies, be aware that the output of the power supply must be floating with respect to its case, which is grounded.

## 2.9 Summary

☐ Op amps are very high-gain dc amplifiers, conveniently packaged in the form of integrated circuits. Op amps are the basic building blocks for many modern electronic circuits. In general, op amps have a differential input and a single output.

☐ The distinguishing feature between the symbols for the op amp and the regular amplifier is the op amp shows two inputs and the regular amplifier shows a single input. A plus sign on the op amp symbol indicates a noninverting input, and a minus sign indicates an inverting input. Power supply terminals may not always be shown on the symbol.

☐ A manufacturer's data sheet provides much valuable information for the technician. Absolute maximum ratings shown therein must never be exceeded if damage to the op amp is to be avoided, and the operating characteristics must be considered in order to select the right device to accomplish the desired result in any given application.

☐ Slew rate, a parameter without minimum or maximum limits, is an important characteristic affecting the ac operation of an op amp. Slew rate is an indication of how fast the output voltage of an op amp can change, and is normally specified at the unity-gain frequency. This frequency is equal to the GBP and is used by manufacturers as a reference point for op amps.

☐ Noise is any unwanted signal on input lines or internal to the op amp. This unwanted signal is amplified along with the desired signal. The common-mode rejection feature rejects any unwanted signals that are in phase, of equal amplitude, and of the same frequency. The CMRR is the ratio of difference gain to common-mode gain, usually expressed in decibels.

☐ The three stages of an op amp are differential input, high-gain amplifier, and output. The high-gain stage provides most of the amplification and is generally constructed with Darlington pairs.

☐ Basic op amp circuits can be operated from battery power sources. Some special-application devices can operate from single-power supply sources, but most op amps require dual- (+ and –) power supply sources.

☐ Protective circuits should be provided for both the op amp and the power supply. Protective circuits can be internal or external to the device. Many modern op amps have built-in short-circuit protection.

## 2.10 Questions and Problems

**2.1** List the characteristics of an ideal op amp.

**2.2** Explain what the plus and minus signs on the op amp schematic symbol represent.

**2.3** Why is it necessary to be able to read a manufacturer's data sheet?

**(Refer to Figure 2–2 for Problems 2.4 through 2.8.)**

**2.4** Draw the logic symbol for the μA741 op amp. Show all connections with labels.

**2.5** For the SA741CN, list the (a) temperature range, (b) maximum offset voltage for RS = 10 kΩ over temperature, (c) minimum CMRR, and (d) typical offset voltage adjustment range.

**2.6** For the μA741C, list the (a) maximum output short-circuit current, (b) typical output voltage swing for RL = 10 kΩ, (c) maximum differential input voltage, and (d) output short-circuit duration.

**2.7** What is the absolute maximum input voltage for supply voltages less than ± 15 V?

**2.8** What is the parallel input resistance for

the condition of open-loop, f = 20 Hz?

**2.9** Explain how a differential amplifier works.

**2.10** What would you expect to happen if you momentarily applied reversed polarities to power an op amp?

**2.11** Calculate the bandwidth of a general-purpose op amp with a unity-gain frequency of 1 MHz and a circuit gain of 100.

**2.12** Calculate the voltage gain of a circuit with a BW of 5 kHz and a unity-gain frequency of 1.5 MHz.

**2.13** What is the amplitude of the difference voltage ($V_D$) of an op amp that has input $V_1$ equal to +2 mV and input $V_2$ equal to -2 mV?

**2.14** Calculate $V_{out}$ of a circuit with an amplifier gain of 100, $V_1$ = +2 mV, and $V_2$ = +1 mV.

**2.15** Given a difference gain ($A_D$) of 500 and common-mode gain ($A_{CM}$) of 0.1, calculate the CMRR.

**2.16** What is the effective gain (beta) of a Darlington pair if the beta of $Q_1$ = 80 and the beta of $Q_2$ = 100?

**2.17** A typical op amp circuit has $V_{CC}$ = ± 18 V and $I_{CC}$ = ±2.8 mA. Calculate the power consumption of the device under no-load conditions.

**2.18** A device with no-load power consumption of 100 mW is connected to a load drawing 20 mA at +9 V. If the power supplied to the device is 300 mW, what is the internal power dissipation of the device?

**2.19** What is the minimum current requirement of a power supply selected for the circuit described in Problem 18, where $I_{CC}$ = ±2.8 mA?

**2.20** What value of resistance would be required to limit op amp voltage to 18 V if it is desirable to limit power supply current drain to a safe level in case the voltage rises to 22 V? The zener diode used would have $V_Z$ = 18 V and $I_Z$ = 12 mA. Refer to Figure 2–16.

Chapter 3

# Basic Operational Amplifier Circuits

## 3.1 Introduction

In Chapter 2 we defined the op amp as the building block for linear circuit applications. In this chapter we will discuss basic op amp circuits, their performance, and their design. The basic linear applications discussed in this chapter are those in which the output signal is directly proportional to the input signal. Both dual-supply and single-supply circuits will be explored. The concept of virtual ground and degenerative feedback also will be examined.

## 3.2 Objectives

When you complete this chapter, you should be able to:

- ☐ Explain the difference between the inverting and noninverting inputs to the op amp.
- ☐ Identify the different types of amplifier circuits by observation.
- ☐ Design and predict the performance of the various types of amplifiers.
- ☐ Distinguish between dual-supply and single-supply op amp circuits.

- ☐ Discuss dc output offset voltage and know how to minimize its effect.
- ☐ Identify the input and feedback elements in the various types of amplifier circuits.
- ☐ Determine the effects of feedback upon the op amp circuits and how to alter circuit performance by altering the feedback elements.
- ☐ Discuss the concept of virtual ground.

## 3.3 Noninverting Amplifiers

The circuit in Figure 3–1 shows a common configuration of a *noninverting amplifier*. The name stems from the input signal being applied to the plus (+), or noninverting, input of the op amp. The output signal, then, is simply an amplified version of the input signal; that is, it is not inverted but always in phase. Resistor $R_1$ is the *input element*. Resistor $R_f$ is the *feedback element*, so called because part of the output voltage is diverted, or feeds back, to the minus (−), or inverting, input of the op amp. The feedback element is also the determining factor for gain and bandwidth, as you will discover later. For most amplifier circuits, a degenerative signal is fed back to the input. Regenerative feedback will be studied in a later chapter.

For reasons that will be explained later in this chapter, there is virtually zero voltage between the (+) and (−) input terminals of the op amp. Therefore, both terminals are at the same potential, that is, $V_{in}$, and $V_{in}$ dropped across $R_1$ causes current $I$ to flow. This current $I$ flows through $R_f$, developing a voltage drop $V_{R_f}$, determined as follows:

$$V_{R_f} = IR_f = \frac{R_f}{R_1}V_{in} \tag{3.1}$$

The output voltage $V_{out}$ of the noninverting amplifier is determined by adding

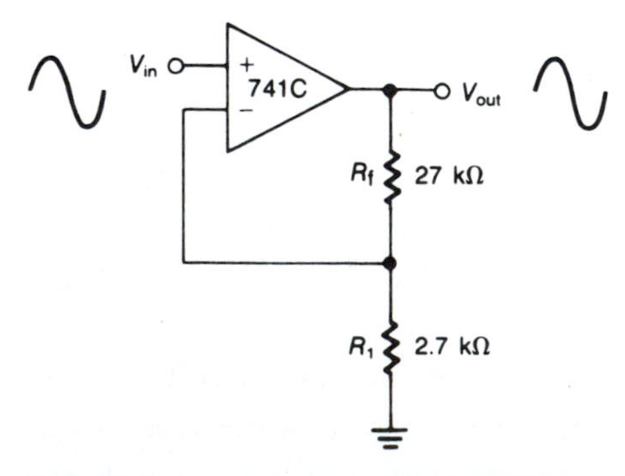


**Figure 3–1.** Noninverting Amplifier

$V_{in}$, which is the voltage drop across $R_1$, to $V_{R_f}$, which is the voltage drop across $R_f$, so that

$$V_{out} = V_{in} + \frac{R_f}{R_1}V_{in} \tag{3.2a}$$

Factoring out $V_{in}$, we have

$$V_{out} = \left(1 + \frac{R_f}{R_1}\right)V_{in} \tag{3.2b}$$

Recall that the voltage gain ($A_v$) of a circuit is the ratio of the output voltage ($V_{out}$) to the input voltage ($V_{in}$):

$$A_v = \frac{V_{out}}{V_{in}} \tag{3.3a}$$

Therefore, by algebraic manipulation of Equations 3.2b and 3.3a, the voltage gain of the noninverting amplifier is

$$A_v = 1 + \frac{R_f}{R_1} \tag{3.3b}$$

In practical terms, then, we can state that the voltage gain of a noninverting amplifier will always be greater than unity (1), regardless of the values assigned to resistors $R_1$ and $R_f$, except in the special case where $R_f = 0\ \Omega$ and $R_1 \approx \infty\ \Omega$. Under these conditions, $A_v$ is equal to 1. (The circuit shown in Figure 3–10 is such a special case.)

One of the intrinsic characteristics of an op amp without feedback is the *open-loop gain* ($A_{OL}$). With feedback present, we have the *closed-loop gain* ($A_{CL}$) that is equal to $A_v$. In terms of the noninverting amplifer,

$$A_{CL} = 1 + \frac{R_f}{R_1} \tag{3.3c}$$

The *loop gain* ($A_L$) is the ratio of the open-loop gain ($A_{OL}$) to the closed-loop gain ($A_{CL}$):

$$A_L = \frac{A_{OL}}{A_{CL}} \tag{3.4}$$

The open-loop input impedance $Z_{in}$ of the noninverting amplifier is, for all practical purposes, the intrinsic input resistance $Z_{iin}$ of the op amp itself. For closed-loop operation this impedance becomes

$$Z_{in} = Z_{iin} \times A_L \tag{3.5a}$$

In either case, this impedance is high enough to minimize input circuit loading.

The output impedance $Z_{out}$ is the ratio of the intrinsic output resistance $Z_{iout}$ of the op amp to the loop gain ($A_L$):

$$Z_{out} = \frac{Z_{iout}}{A_L} \tag{3.5b}$$

## Example 3.1

Refer to Figure 3–1 and to the 741C data sheet (Figure 2–2). Determine (a) closed-loop gain, (b) loop gain, (c) input impedance, and (d) output impedance.

*Solutions*

a. Using Equation 3.3c, the closed-loop gain is

$$A_{CL} = 1 + \frac{R_f}{R_1} = 1 + \frac{27\ \text{k}\Omega}{2.7\ \text{k}\Omega} = 11$$

b. The 741C data sheet (Figure 2–2) indicates the typical open-loop gain is 200,000. Using Equation 3.4, the loop gain of the circuit is

$$A_L = \frac{A_{OL}}{A_{CL}} = \frac{200{,}000}{11} = 18{,}200$$

c. The 741C data sheet indicates the typical input resistance is 2 MΩ, so from Equation 3.5a,

$$Z_{in} = Z_{iin} \times A_L$$
$$= 2\ \text{M}\Omega \times 18{,}200 = 3.64 \times 10^{10}\ \Omega$$

d. The 741C data sheet indicates the typical output resistance is 75 Ω. Using Equation 3.5b, the output impedance is

$$Z_{out} = \frac{Z_{iout}}{A_L} = \frac{75\ \Omega}{18{,}200} = 0.004\ \Omega$$

Example 3.1 demonstrates the results of adding feedback to the circuit. The loop gain is increased and the output impedance is decreased. Since the output impedance in such a circuit is so small, it can, in general, be ignored.

## Example 3.2

Refer to Figure 3–1. Assume $R_f = 100\ \text{k}\Omega$ and $R_1 = 1\ \text{k}\Omega$. Determine (a) $A_{CL}$ and (b) $A_L$.

*Solutions*

a. From Equation 3.3c,

$$A_{CL} = 1 + \frac{R_f}{R_1} = 1 + \frac{100\ \text{k}\Omega}{1\ \text{k}\Omega} = 101$$

b. From Equation 3.4,

$$A_L = \frac{A_{OL}}{A_{CL}} = \frac{200{,}000}{101} = 1980$$

# 3.4 Inverting Amplifiers

An *inverting amplifier* has the input signal applied to the minus (−) terminal of the op amp, as shown in Figure 3–2. In such an amplifier the output signal is 180° out of phase, or inverted, compared with the input signal. Resistor $R_1$ is the input element, and resistor $R_f$ is the feedback element.

The output voltage ($V_{out}$) of the inverting amplifier is

$$V_{out} = -\left(\frac{R_f}{R_1}\right)V_{in} \tag{3.6}$$

The − sign in Equation 3.6 indicates only that the circuit being evaluated is an inverting amplifier. It should not be construed as a negative output voltage, except in the case of a positive dc input voltage.

## Virtual Ground

To better understand the operation of an inverting amplifier, you must first understand the concept of *virtual ground*. Virtual ground is a point at which voltage is zero with respect to ground, yet is isolated from ground. In practical terms, this means that no current can flow into or out of this point.

The inverting input of Figure 3–2 acts as a virtual ground. In this circuit, a positive-going signal, $V_{in}$, is applied to the inverting input terminal of the op amp. An inverted output signal is then fed back to the inverting input terminal through feedback resistor $R_f$. The voltage between the inverting and noninverting input terminals is essentially zero, therefore the inverting terminal is at zero volts, or ground potential. However, it is not a true ground point because of the high input impedance of the op amp and should never be used as a true ground point. This is an important point to remember when analyzing op amp circuits.

In the previous section we learned that $A_v$ is equal to $V_{out}/V_{in}$ and that $A_{CL}$ is equal to $A_v$. Therefore, for the inverting amplifier,

$$A_{CL} = -\frac{R_f}{R_1} \tag{3.7}$$

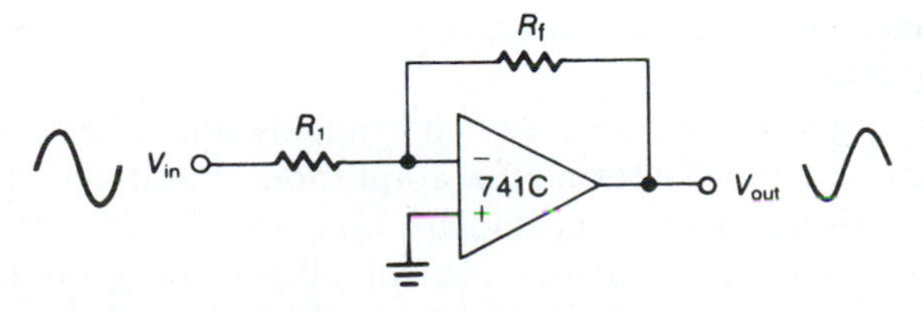


**Figure 3–2.** Inverting Amplifier

The ratio of $R_f$ to $R_1$, then, will determine whether the voltage gain is less than, equal to, or greater than unity (1).

Through algebraic manipulation of Equations 3.4 and 3.7, we can now state that the loop gain, $A_L$, for the inverting amplifier is

$$A_L = \frac{A_{OL}}{A_{CL}} = \frac{A_{OL}}{\left(-\frac{R_f}{R_1}\right)} = A_{OL}\left(-\frac{R_1}{R_f}\right) \quad (3.8)$$

The input impedance of the inverting amplifier is the value of the input element. In practice, this value will always be much less than that for a noninverting amplifier. The output impedance of the inverting amplifier is determined in the same manner as for the noninverting amplifier:

$$Z_{out} = \frac{Z_{iout}}{A_L} \quad \text{(repeat of Equation 3.5b)}$$

## Example 3.3

Refer to Figure 3–2. Assume $R_1 = 2\ \text{k}\Omega$, $R_f = 22\ \text{k}\Omega$, $V_{in} = 0.25$ V, and the op amp is a 741C. Determine (a) output voltage, (b) closed-loop gain, (c) loop gain, (d) input impedance, and (e) output impedance.

*Solutions*

a. Using Equation 3.6,

$$V_{out} = -\left(\frac{R_f}{R_1}\right)V_{in} = -\left(\frac{22\ \text{k}\Omega}{2\ \text{k}\Omega}\right)0.25\ \text{V} = -2.75\ \text{V}$$

b. Using Equation 3.7,

$$A_{CL} = -\frac{R_f}{R_1} = -\frac{22\ \text{k}\Omega}{2\ \text{k}\Omega} = -11$$

c. Using Equation 3.8,

$$A_L = A_{OL}\left(-\frac{R_1}{R_f}\right) = 200{,}000 \times -0.09 = -18{,}200$$

d. Input impedance is simply the value of $R_1$, or 2 k$\Omega$.

e. Using Equation 3.5b,

$$Z_{out} = \frac{Z_{iout}}{A_L} = \frac{75\ \Omega}{-18{,}200} = -0.004\ \Omega$$

NOTE: Minus signs used in solution e should be considered absolute values, not negatives.

The inverting amplifier can be easily and quickly adapted to invert the polarity of an input signal without altering the amplitude of that signal. Simply make $R_1$ and $R_f$ equal. This results in a *unity-gain inverter*.

Replacing $R_f$ with a variable resistor allows the closed-loop gain of the inverting amplifier to be controlled. Alternatively, switching in different values of feedback resistors, as shown in Figure 3–3, allows gain preselection. Any desired gain can be preselected by carefully selecting the values of the feedback resistors used.

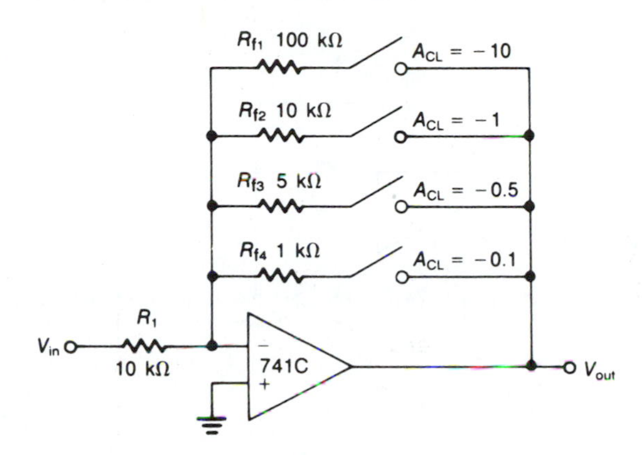


**Figure 3–3.** Switchable Feedback Resistances for Gain Preselection

## ac Operation with a Single Power Supply

The circuits discussed in Sections 3.3 and 3.4 have been based on the op amp being powered by a dual power supply. However, it is possible to operate the circuits from a single power supply. In order to do that, the *quiescent dc output voltage* of the op amp must be set to one-half the supply voltage. It is necessary that ac coupling capacitors for the input and output signals be used in such applications.

To properly operate from a single-supply voltage, the op amp circuit must be able to produce both negative- and positive-going signals. Properly biased, and with no input signal applied, the dc output voltage of the circuit will be one-half that of the supply voltage. This dc output voltage will be the quiescent, or resting, voltage. When an ac input signal is applied to the circuit, the output signal will vary about the quiescent voltage. The op amp output signal is a combination of the quiescent dc voltage and the amplified ac component superimposed upon it. The dc component can be considered to be an *output offset voltage* that should be removed before applying the desired output signal to another stage. (We will discuss dc output offset voltage more fully in the next section.)

Basic single-supply powered circuits for the inverting ac amplifier and the noninverting ac amplifier are shown in Figure 3–4A and B. Note that the supply connections are included in the diagram to indicate that single-supply voltage is used, and that the $V^-$ connection of the op amp goes directly to ground. The voltage divider network $R_2$–$R_3$ applies a dc voltage to the noninverting input terminal of the op amp to set the dc output voltage at one-half the supply voltage ($V^+/2$) with no input signal applied. Capacitor $C_1$ couples the input signal to the circuit. Capacitor $C_2$ removes the dc output offset voltage and couples the output signal from the circuit.

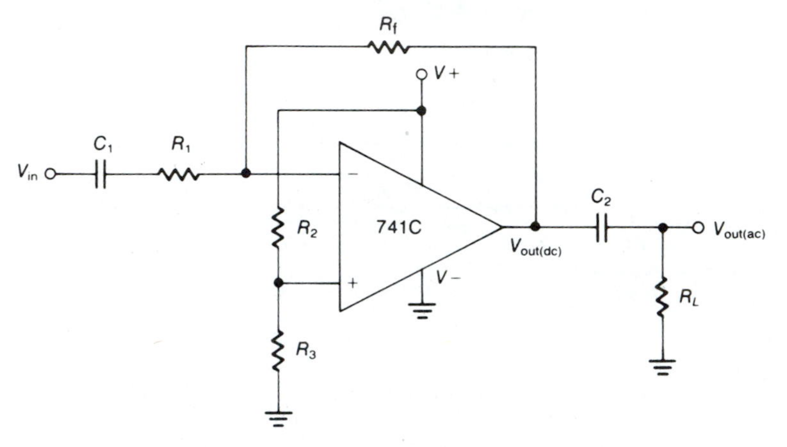


(A) Inverting ac Amplifier with Single-Supply Voltage

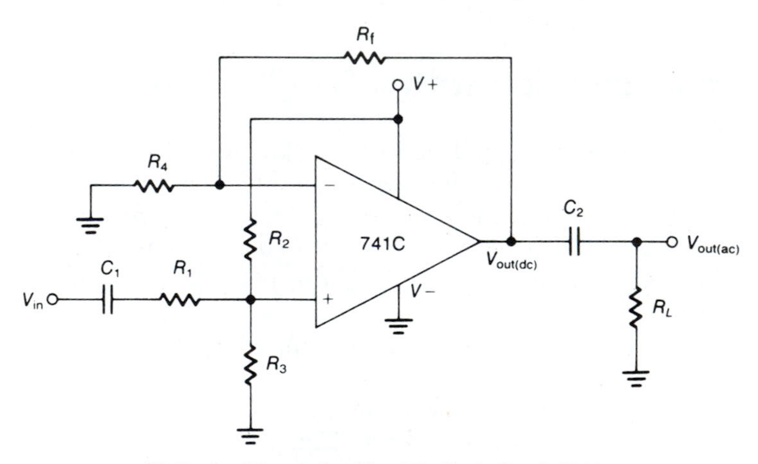


(B) Noninverting ac Amplifier with Single-Supply Voltage

**Figure 3–4.** Single-Supply Powered Amplifier Circuits

The quiescent dc output voltage can be determined as follows:

$$V_{out(dc)} = \frac{R_3}{R_2 + R_3}(V^+) \tag{3.9}$$

In general, however, resistors $R_2$ and $R_3$ are made equal to each other so that $V_{out(dc)} = V^+/2$. This being the case, the maximum peak-to-peak output voltage that can be expected without distortion cannot exceed that same value. Therefore, the maximum peak-to-peak input voltage ($V_{in(max,\ p\text{-}p)}$) that can be applied to the circuit is

$$V_{in(max,\ p\text{-}p)} = \frac{V^+}{A_{CL}} \tag{3.10}$$

The closed-loop gain, $A_{CL}$, of the single-supply inverting amplifier is determined by using Equation 3.7 (as for the dual-supply circuit), and for the noninverting amplifier by using Equation 3.3C.

The desired low-frequency response and the impedance of either the input or load determines the values for capacitors $C_1$ and $C_2$. These capacitors cannot be just any convenient value, and the ± tolerance of the capacitors must be considered when making the actual selection. The minimum value for input capacitor $C_1$ is determined by

$$C_1 = \frac{1}{2\pi f_c R_1} \tag{3.11}$$

where

$f_c$ = cutoff frequency
$R_1$ = input resistance

and the minimum value for output capacitor $C_2$ is determined by

$$C_2 = \frac{1}{2\pi f_c R_L} \tag{3.12}$$

where

$f_c$ = cutoff frequency
$R_L$ = load resistance

The cutoff frequency, $f_c$, is equal to approximately $0.1 f_{op}$, where $f_{op}$ is the lowest desired operating frequency.

## Example 3.4

Refer to Figure 3–4A. Design an inverting ac amplifier circuit with a closed-loop gain of 10, an input impedance of 10 kΩ, an operating frequency of 60 hertz (Hz), and a load of 20 kΩ. The single-supply voltage is to be $^+30$ V. Calculate the maximum peak-to-peak input voltage.

*Solution*

First, evaluate the information given.

Since the input impedance of the inverting circuit is equal to the value of the input resistor, $R_1$ must equal 10 kΩ. Given a gain of 10, $R_f$ must equal $10R_1$, or 100 kΩ. Resistors $R_2$ and $R_3$ must be equal. The general guide is to make them equal to $2R_f$. Therefore, $R_2$ and $R_3$ are both given values of 200 kΩ, which provides the $V^+/2$ value necessary for $V_{out(dc)}$.

Now calculate the values for the coupling and decoupling capacitors using Equations 3.11 and 3.12:

$$C_1 = \frac{1}{2\pi f_c R_1} = \frac{1}{6.28 \times 6 \text{ Hz} \times 10 \text{ k}\Omega}$$
$$= 2.65\ \mu\text{F} \qquad \text{(Select a standard 2.7 } \mu\text{F.)}$$
$$C_2 = \frac{1}{2\pi f_c R_L} = \frac{1}{6.28 \times 6 \text{ Hz} \times 20 \text{ k}\Omega}$$
$$= 1.33\ \mu\text{F} \qquad \text{(Select a standard 1.5 } \mu\text{F.)}$$

Finally, calculate the maximum peak-to-peak input voltage using Equation 3.10:

$$V_{in(max,\ p\text{-}p)} = \frac{V^+}{A_{CL}} = \frac{30\text{ V}}{-10} = 3\ V_{p\text{-}p}$$

Figure 3–5 shows the completed circuit design.

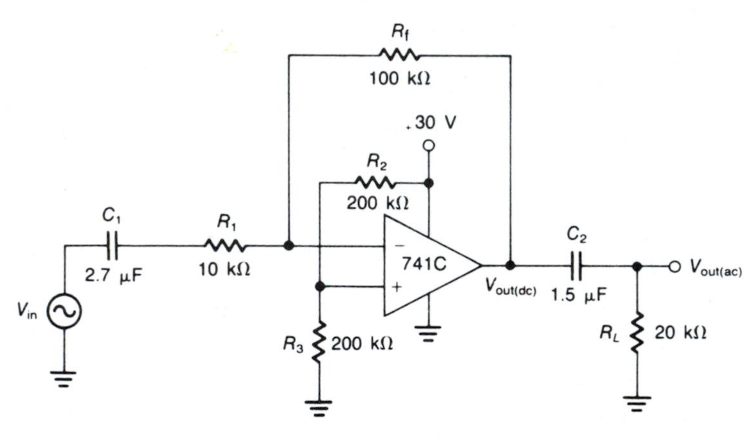


**Figure 3–5.** Circuit Design for Example 3.4

# 3.5 Offset Considerations

An ideal op amp will have zero output voltage when the input voltage is zero, but practical commercial op amps are not ideal. Therefore, in practical circuits there will be a small dc output voltage, even when there is zero input. This voltage is called *dc output offset voltage* ($V_{os}$) and results from the following op amp characteristics:

1. Input bias current
2. Input offset current
3. Input offset voltage
4. Drift

In op amp ac amplifier circuits, the coupling capacitors remove the dc output offset voltage. Therefore, the discussion in this section pertains to dc amplifier circuits.

## Input Bias Current

Since the input impedance of a practical op amp is not infinite, there will be a very small, but finite, amount of current flowing into the (+) and (−) input terminals. The average of these two currents, $I_{b+}$ and $I_{b-}$, is termed the *input bias current*, $I_b$:

$$I_b = \frac{I_{b+} + I_{b-}}{2} \tag{3.13}$$

With no signal applied to the circuit, $I_b$ flows through both the input and feedback resistors. The current flowing in the resistors develops a voltage that appears as a dc input voltage. This voltage is then amplified by the op amp and appears as the dc output offset voltage, $V_{os}$, which can be calculated by using the following equation:

$$V_{os} = I_b R_f \tag{3.14}$$

## Input Offset Current

In practical op amp circuits, the bias currents flowing into the two input terminals will not be equal. The difference between the two currents, $I_{b+}$ and $I_{b-}$, is termed the *input offset current*, $I_{oi}$, and is calculated as follows:

$$I_{oi} = I_{b+} - I_{b-} \tag{3.15}$$

The input offset current will cause a small output offset voltage:

$$V_{os} = I_{oi} R_f \tag{3.16}$$

A simple method of correction for $V_{os}$ resulting from input current imbalances is shown in Figure 3–6. Resistor $R_2$ is a *current-compensating resistor* placed in

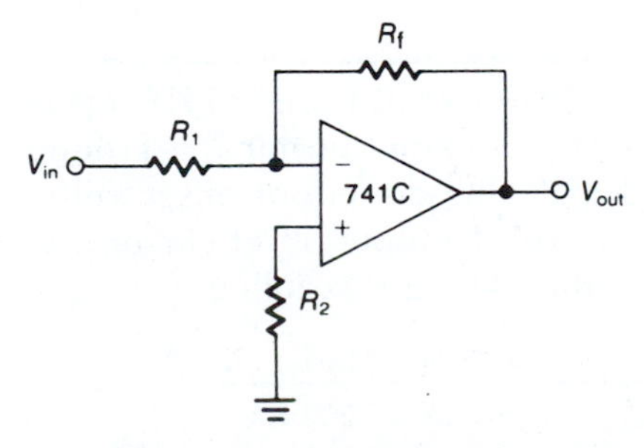


**Figure 3–6.** Input Current-Correction Circuit

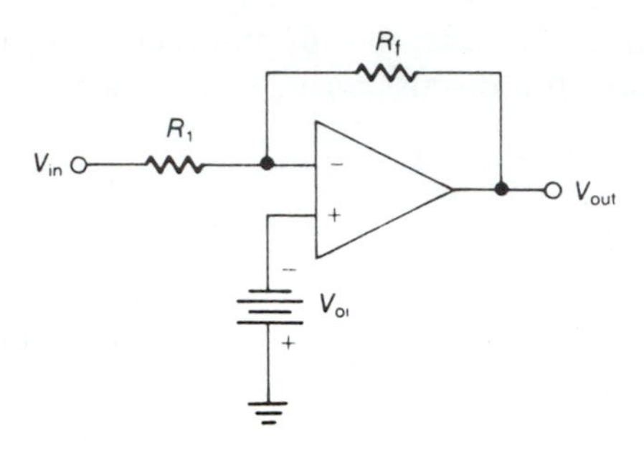


**Figure 3–7.** Input Offset Voltage Model

series with the (+) input terminal of the op amp. The value of $R_2$ should be equal to the parallel value of $R_1$ and $R_f$, or

$$R_2 = \frac{R_1 R_f}{R_1 + R_f} \tag{3.17}$$

The voltage developed across $R_2$ will be equal to that developed across $R_1$ and will therefore cancel, resulting in zero input voltage and zero output voltage.

## Input Offset Voltage

Imbalances in the input circuits of the practical op amp are caused by mismatches in the internal circuitry. Such imbalances present another error factor, called *input offset voltage,* $V_{oi}$, which can result in $V_{os}$. Figure 3–7 shows a model where $V_{oi}$ is represented by the battery. $V_{os}$ can be calculated as follows:

$$V_{os} = V_{oi}\left(1 + \frac{R_f}{R_1}\right) \tag{3.18}$$

A current-compensating resistor ($R_2$ in Figure 3–6) in series with the (+) input cannot correct this problem.

## Example 3.5

Assume an inverting amplifier circuit using a 741C op amp. Input impedance is 10 kΩ and gain is 10. The (+) input terminal is grounded. Using the typical parameters shown, calculate the dc output offset voltage ($V_{os}$) as a result of (a) the input bias current, (b) the input offset current, and (c) the input offset voltage. Parameters for the 741C are as follows:

| | **Typical** | **Maximum** |
|---|---|---|
| Input bias current ($I_b$) | 80 nA | 500 nA |
| Input offset current ($I_{oi}$) | 20 nA | 200 nA |
| Input offset voltage ($V_{oi}$) | 2.0 mV | 6.0 mV |

*Solutions*

First, evaluating the facts given should tell you that $R_f$ must be 100 kΩ. To solve (a), use Equation 3.14:

$$V_{os} = I_b R_f = 80 \text{ nA} \times 100 \text{ k}\Omega = (80 \times 10^{-9} \text{ A})(100 \times 10^3 \ \Omega) = 8 \text{ mV}$$

To solve (b), use Equation 3.16:

$$V_{os} = I_{oi} R_f = 20 \text{ nA} \times 100 \text{ k}\Omega = (20 \times 10^{-9} \text{ A})\ (100 \times 10^3 \ \Omega)$$
$$= 2 \times 10^{-3} \text{ V} = 2 \text{ mV}$$

To solve (c), use Equation 3.18:

$$V_{os} = V_{oi}\left(1 + \frac{R_f}{R_1}\right) = 2 \text{ mV} \times 11 = (2 \times 10^{-3} \text{ V})11$$
$$= 22 \times 10^{-3} \text{ V} = 22 \text{ mV}$$

## Example 3.6

In worst-case conditions, the maximum parameters of the 741C would have to be considered. For comparison purposes, repeat the calculations performed in Example 3.5, this time using the maximum parameters given.

*Solutions*

To solve (a), use Equation 3.14:

$$V_{os} = I_b R_f = 500 \text{ nA} \times 100 \text{ k}\Omega$$
$$= (500 \times 10^{-9} \text{ A})\ (100 \times 10^3 \Omega)$$
$$= 50 \text{ mV}$$

To solve (b), use Equation 3.16:

$$V_{os} = I_{oi} R_f = 200 \text{ nA} \times 100 \text{ k}\Omega = (200 \times 10^{-9} \text{ A})(100 \times 10^3 \ \Omega)$$
$$= 20 \text{ mV}$$

To solve (c), use Equation 3.18:

$$V_{os} = V_{oi}\left(1 + \frac{R_f}{R_1}\right) = 6 \text{ mV} \times 11 = (6 \times 10^{-3} \text{ V})11 = 66 \text{ mV}$$

In each case, $V_{os}$ shows a significant increase over the typical values.

## Example 3.7

Calculate the value required for a current-compensating resistor, $R_2$, to be placed in series with the (+) input terminal of the circuit described in Example 3.5.

*Solution*

Use Equation 3.17:

$$R_2 = \frac{R_1 R_f}{R_1 + R_f} = \frac{10\ \text{k}\Omega \times 100\ \text{k}\Omega}{10\ \text{k}\Omega + 100\ \text{k}\Omega} = 9.1\ \text{k}\Omega$$

Remember, however, that this resistor will correct only for input current imbalances, not for input offset voltage.

### Null and Balance Procedures

While the current-compensating resistor corrects for the dc output offset voltage that results from the input bias current, other methods must be used to correct errors that result from the input offset voltage. Many op amps, including the 741C, provide terminals used specifically for cancelling the dc output offset voltage. The terminals are generally identified on data sheets as *offset null,* or *balance,* terminals. These terminals are internally connected to affect the internal circuitry of the op amp. The offset voltage may be nulled by connecting a 10 kΩ potentiometer across null pins 1 and 5 as shown in Figure 3–8. The wiper arm of the potentiometer is connected to $V^-$. In general, a multi-turn, wire-wound potentiometer, capable of providing the small voltage changes necessary to affect the internal circuitry is used. The potentiometer is adjusted until the output voltage, with zero input, is also zero. This method of cancelling the dc output offset voltage is called *internal nulling.*

*External nulling* methods may be necessary for op amps that have no provisions for internal nulling. Data sheets usually indicate which method is used. Figure 3–9 shows typical external nulling circuits.

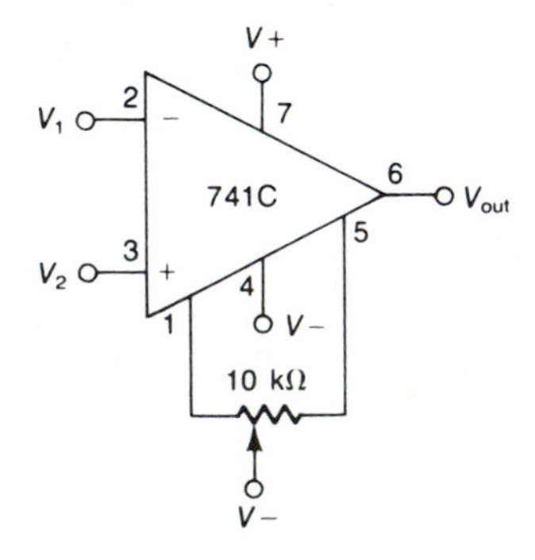


**Figure 3–8.** Offset Null Circuit for the μA741C Eight-Lead MINIDIP

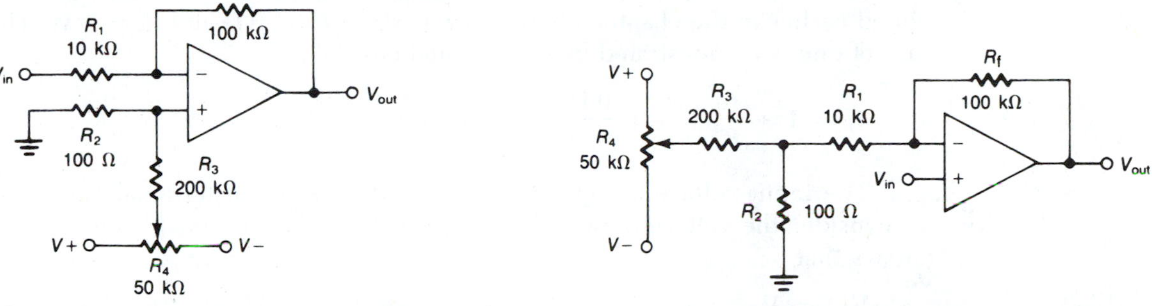


**Figure 3–9.** Typical External Nulling Circuits for Op Amps Having No Provisions for Internal Nulling (Courtesy of Signetics Company)

### Drift

Offset current and offset voltage changes can occur because of aging of the components, supply voltage changes, or temperature changes in the op amp. Changes resulting from temperature fluctuations are called *drift*. Drift cannot be eliminated, but it can be minimized by maintaining constant temperature around the circuit and by selecting op amps with offset characteristics that change minimally with variations in temperature.

## 3.6 Voltage Follower

The simple op amp circuit shown in Figure 3–10 is called a *voltage follower*, but it has many other names, each related to its application. The circuit is called voltage follower, or *source follower*, because the output voltage follows, or is an exact reproduction of, the input (source) voltage. The circuit is simply a unity-gain noninverting amplifier; therefore, it is sometimes referred to as a *unity-gain amplifier*. The circuit is also called a *buffer amplifier*, or an *isolation amplifier*, because its high input impedance and low output impedance buffers, or isolates, an input signal from its load.

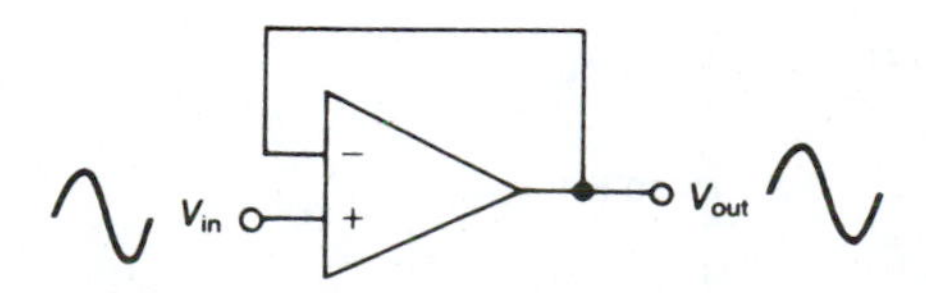


**Figure 3–10.** Voltage Follower Circuit

This circuit is that special case for noninverting amplifiers that was mentioned earlier in the chapter where $A_v = 1$, since $R_f = 0\ \Omega$ and $R_1 \approx \infty\ \Omega$. This gain of one is demonstrated by using Equation 3.3b:

$$A_v = 1 + \frac{R_f}{R_1} = 1 + \frac{0\ \Omega}{\infty\ \Omega} = 1 + 0 = 1$$

The input voltage is applied directly to the (+) input terminal, and since we consider the voltage between the (+) and (−) input terminals to be zero, it follows that

$$V_{out} = V_{in} \tag{3.19}$$

and

$$A_{CL} = \frac{V_{out}}{V_{in}} = 1 \tag{3.20}$$

With no input or feedback resistances in the circuit, the input impedance is extremely high and, for all practical purposes, is equal to the op amp's intrinsic input impedance. The output impedance is very low, because it is simply the intrinsic output of the op amp divided by the open-loop gain of the op amp.

## Example 3.8

Refer to Figure 3–11 and Figure 2–2. Determine (a) input impedance, $Z_{iin}$; (b) output impedance, $Z_{out}$; (c) output voltage, $V_{out}$; and (d) load current, $I_L$.

*Solutions*

**a.** Typical intrinsic input impedance of the 741C is 2 MΩ.

**b.** The output impedance of the 741C is 75 Ω, and the typical open-loop gain is 200,000, therefore,

$$Z_{out} = \frac{75\ \Omega}{200{,}000} = 0.000375\ \Omega$$

**c.** From Equation 3.19,

$$V_{out} = V_{in} = 5\ V_{p\text{-}p}$$

**d.** From Ohm's Law,

$$I_L = \frac{V_L}{R_L} = \frac{5\ V}{10\ k\Omega} = 0.5\ mA$$

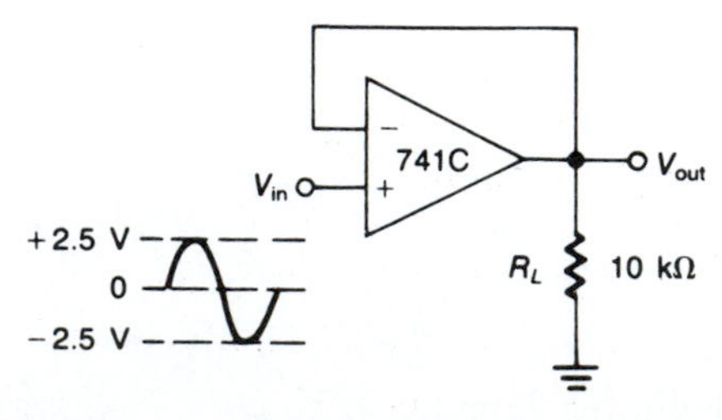


**Figure 3–11.** Voltage Follower Circuit for Example 3.8

# 3.7 Summing Amplifiers

A *summing amplifier*, shown in Figure 3–12, is constructed by connecting two or more resistors simultaneously to the (−) input terminal of an op amp. The output voltage of the circuit is the sum of the individual input voltages in terms of individual resistances:

$$V_{out} = -\left(\frac{R_f}{R_1}V_1 + \frac{R_f}{R_2}V_2 + \frac{R_f}{R_3}V_3\right) \quad (3.21a)$$

and simplified,

$$V_{out} = -R_f\left(\frac{V_1}{R_1} + \frac{V_2}{R_2} + \frac{V_3}{R_3}\right) \quad (3.21b)$$

If the feedback resistor is made equal to the value of the input resistors, the output voltage is

$$V_{out} = -(V_1 + V_2 + V_3) \quad (3.22)$$

By making the feedback resistor larger than any of the input resistors, each input voltage is increased by a set gain and the individual results are added. The input impedance for each input is simply the value for the corresponding input resistor, and the gain for each input can be adjusted individually by choosing the desired ratio between each corresponding input resistor and the feedback resistor.

## Audio Mixer

A summing amplifier is useful as an audio mixer. The input voltages developed across the input resistors do not interact, since they are common to the inverting input, which is at virtual ground. If the input voltages of Figure 3–12 were microphone inputs, the ac voltages developed by each microphone would be added (mixed) at the output, as demonstrated by Equation 3.22. If volume controls are placed between each of the input resistors and the summing point, the relative volume of the various inputs can be controlled for a more desirable output sound. Figure 3–13 shows this application, where each input affects the output according to the volume control setting, so the gain for each input is

$$A_v = -\frac{R_f}{R_1 + R_2} \quad (3.23)$$

## Averaging Amplifier

A variation of the summing amplifier is the *averaging amplifier* that gives an ouput voltage equal to the average of the input voltage (where $N$ = number of inputs).

$$V_{out(av)} = \frac{-(V_1 + V_2 + V_3 + \dots + V_n)}{N} \quad (3.24)$$

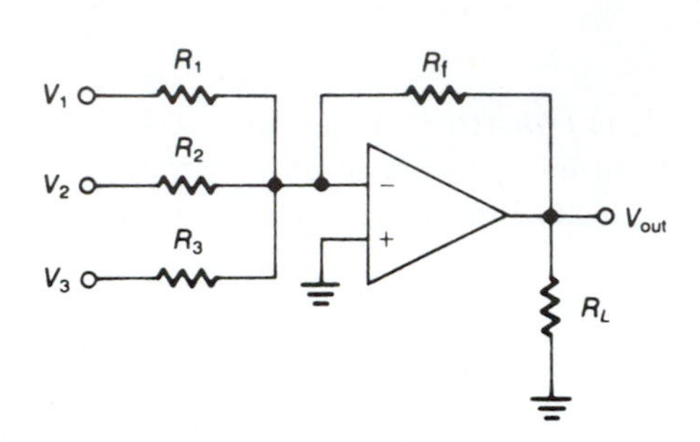


Figure 3–12. Three-Input Summing Amplifier

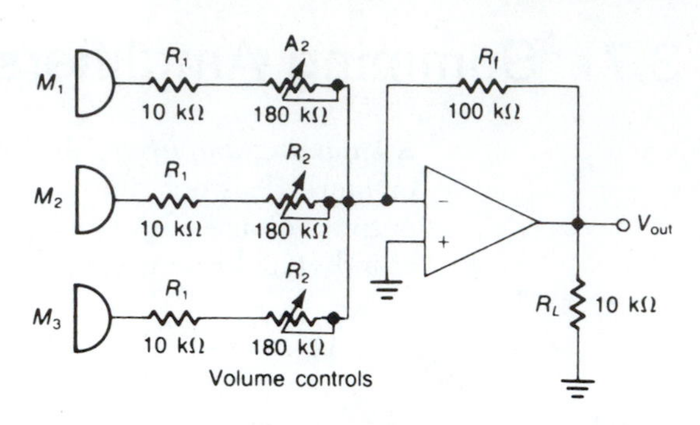


Figure 3–13. Three-Microphone Input Audio Mixer

For a three-input amplifier, as shown in Figure 3–12, averaging is accomplished by making each of the input resistors equal in value, and the feedback resistor equal to one-third the value of any input resistor.

## Example 3.9

Refer to Figure 3–12. Assume $R_1 = 5\text{ k}\Omega$, $R_2 = 10\text{ k}\Omega$, $R_3 = 20\text{ k}\Omega$, and $R_f = 50\text{ k}\Omega$. Find the output voltage if $V_1 = 0.5$ V, $V_2 = 0.75$ V, and $V_3 = 1$ V.

*Solution*

Using Equation 3.21b,

$$V_{out} = -R_f\left(\frac{V_1}{R_1} + \frac{V_2}{R_2} + \frac{V_3}{R_3}\right) = -50\text{ k}\Omega\left(\frac{0.5\text{ V}}{5\text{ k}\Omega} + \frac{0.75\text{ V}}{10\text{ k}\Omega} + \frac{1\text{ V}}{20\text{ k}\Omega}\right)$$

$$= -50\text{ k}\Omega(0.1\text{ mA} + 0.075\text{ mA} + 0.05\text{ mA}) = -11.25\text{ V}$$

## Example 3.10

Find the output voltage of the five-input averaging amplifier illustrated in Figure 3–14.

*Solution*

Using Equation 3.24,

$$V_{out(av)} = \frac{-(V_1 + V_2 + V_3 + V_4 + V_5)}{5}$$

$$= \frac{-(2.5\text{ V} + 1\text{ V} + 1.5\text{ V} + 2\text{ V} + 3\text{ V})}{5} = -2\text{ V}$$

Recall that the – sign in front of the equations in this chapter indicates only that the output signal is inverted from the input signal.

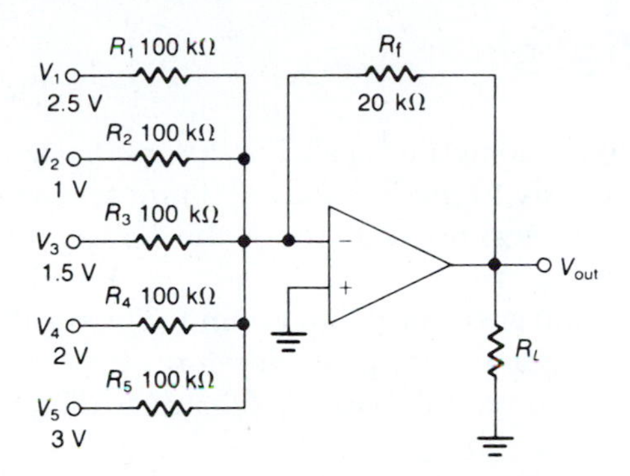


**Figure 3–14.** Five-Input Averaging Amplifier

## Single-Supply Summing Amplifier

A single-supply voltage op amp summing amplifier circuit is shown in Figure 3–15. Note the coupling capacitors, $C_1$ through $C_4$, and the grounded power terminal. Resistors $R_4$ and $R_5$ are made equal in value, forming a voltage divider network that biases the quiescent dc output voltage at one-half the $V^+$ supply voltage. The output voltage is determined in the same manner as for the dual-supply amplifier, that is, by using Equation 3.21b. The value for the input coupling capacitors is determined by using Equation 3.11, and for the output coupling capacitor by using Equation 3.12.

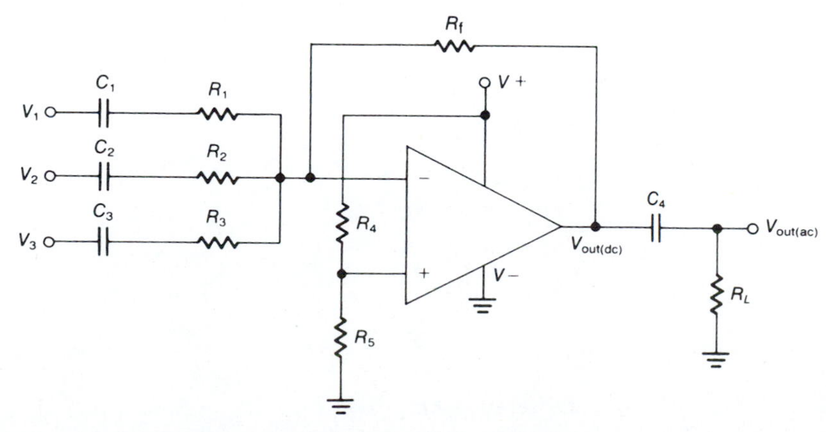


**Figure 3–15.** Summing Amplifier with Single-Supply Voltage

# 3.8 Difference Amplifiers

A basic *difference amplifier*, sometimes called a *differential amplifier*, has input signals applied simultaneously to the (+) and (−) input terminals of the op amp, as shown in Figure 3–16. A short analysis of the figure will simplify circuit understanding.

First, assume that point $A$ is shorted to ground. The resulting circuit would appear as a simple inverting amplifier and, therefore, the output voltage would be calculated by using Equation 3.6, with $V_1$ substituted for $V_{in}$:

$$V_{out} = -\left(\frac{R_f}{R_1}\right)V_1$$

Now assume the short at point $A$ is removed and input signal $V_1$ is shorted to ground. The result is essentially a noninverting amplifier that feels the voltage at point $A$ at the (+) terminal input. This voltage is produced by the voltage divider action of resistors $R_2$ and $R_3$ and the input voltage $V_2$:

$$V_{(A)} = \left(\frac{R_3}{R_2 + R_3}\right)V_2 \tag{3.25}$$

where

$V_{(A)}$ = voltage at point $A$

The noninverting output voltage for this circuit, then, is the product of the voltage at point $A$ and the voltage gain of a noninverting amplifier; that is,

$$V_{out} = A_v V_{(A)} = \left(1 + \frac{R_f}{R_1}\right)\left(\frac{R_3}{R_2 + R_3}\right)V_2 \tag{3.26}$$

The output voltage for the difference amplifier is the sum of the inverted output, Equation 3.6, and the noninverted output, Equation 3.26:

$$V_{out} = -\left(\frac{R_f}{R_1}\right)V_1 + \left(1 + \frac{R_f}{R_1}\right)\left(\frac{R_3}{R_2 + R_3}\right)V_2 \tag{3.27}$$

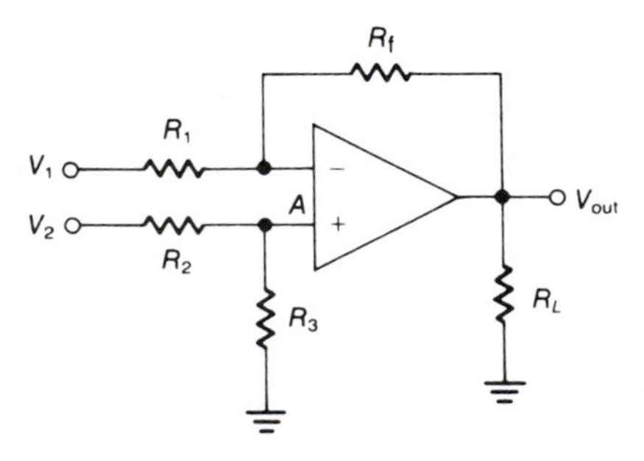


**Figure 3–16.** Difference Amplifier

The voltage gain of the difference amplifier ($A_{vD}$) is set by all four resistors, using $R_1$ as a base element with relationships where

$$R_2 = R_1 \tag{3.28a}$$

and

$$R_3 = R_f = R_1 A_{vD} \tag{3.28b}$$

A major advantage of the difference amplifier described here is its ability to sense a small differential voltage buried in a larger signal. It can thus measure as well as amplify that small signal.

## Subtractor

One variation of the difference amplifier is called a *unity-gain analog subtractor*, a *voltage subtractor*, or an *analog mathematical circuit*. This circuit is constructed simply by making all four resistors equal in value. When used in this application, the output voltage is the difference between $V_2$ and $V_1$:

$$V_{out} = V_2 - V_1 \tag{3.29}$$

## Designing a Difference Amplifier

Designing a difference amplifier or a unity-gain analog subtractor is very simple. For the difference amplifier, determine the desired gain of the circuit, then select any standard resistance value for $R_1$ and $R_2$, which are equal, and multiply that value times the desired gain, which will provide the resistance value for $R_3$ and $R_f$. For a unity-gain analog subtractor, simply set all resistance values equal.

### Example 3.11

Design a difference amplifier with a gain of 20.

*Solution*

First, select a value for $R_1$, say 5 kΩ. Next, determine the values for $R_2$, $R_3$, and $R_f$ using Equations 3.28a and 3.28b.

$$R_2 = R_1 = 5\ \text{k}\Omega$$
$$R_3 = R_f = R_1 A_{vD} = 5\ \text{k}\Omega \times 20 = 100\ \text{k}\Omega$$

The completed design circuit is shown in Figure 3–17. By replacing resistors $R_3$ and $R_f$ with 5 kΩ resistors, the circuit becomes a unity-gain analog subtractor.

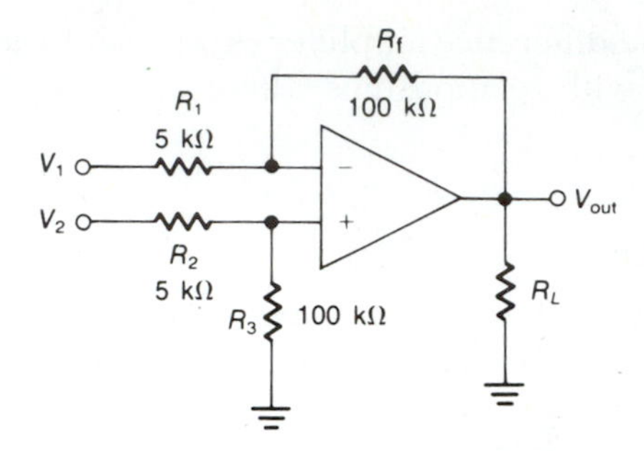


Figure 3–17. Difference Amplifier with Gain of 20

## Single-supply Difference Amplifier

A single-supply voltage op amp difference amplifier circuit is shown in Figure 3–18. The voltage gain is determined in the same manner as for the dual-supply circuit of Figure 3–16, that is, by using Equation 3.27. $R_4$ is made equal to $R_3$, forming the voltage divider for proper biasing, but with the added restriction that

$$\frac{R_f}{R_1} = \frac{R_3}{2R_2} \tag{3.30}$$

The value for input coupling capacitors is determined by using Equation 3.11, and for the output coupling capacitor by using Equation 3.12.

The circuit of Figure 3–18 can also be used for the unity-gain analog subtractor. For unity-gain, resistors $R_1$, $R_2$, and $R_f$ must all be equal. $R_4$ is equal to $R_3$, and the value for $R_3$ is determined by Equation 3.30.

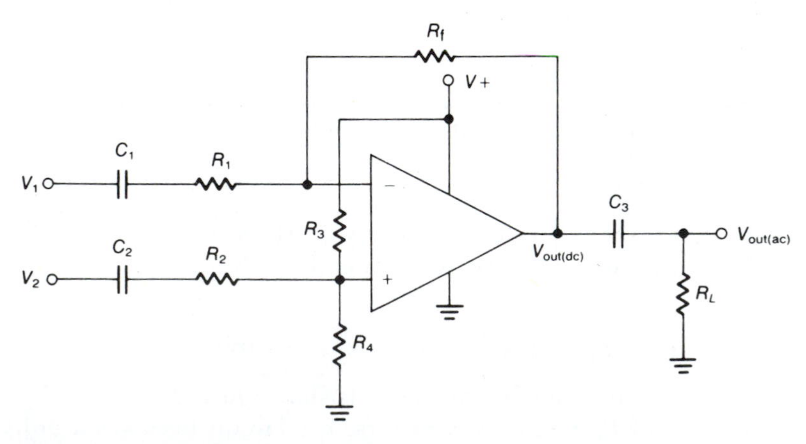


Figure 3–18. Difference Amplifier with Single-Supply Voltage

# 3.9 Norton Amplifiers

The Norton amplifier is designed for use in ac amplifier circuits that operate from a single power supply. It can, however, be used in many of the dc circuits previously discussed.

The schematic symbol for a single Norton amplifier is shown in Figure 3–19. The circled arrow pointing toward the noninverting (+) input identifies the Norton amplifier as a current-driven device. Power supply connections are not shown because there are usually two or four Norton amplifiers within a single IC package, sharing power supply and ground pins.

The operation of a Norton amplifier is different from that of a standard op amp. A standard op amp is a voltage-driven device that requires no significant input current, using instead negative feedback to keep the two inputs at the same potential (near zero volts). However, a Norton amplifier is a current-driven device whose input potentials are approximately 0.7 V (the bias voltage of the input transistors of the device). Negative feedback in this device is used to keep the two input currents equal, hence the Norton amplifier is sometimes called a *current-differencing* amplifier.

Single-supply operation of the Norton amplifier requires that the output be biased to a dc level equal to one-half that of the power supply. Standard current-mirror biasing, shown in Figure 3–20, sets the quiescent (resting) dc output voltage at one-half the $V^+$ supply level, with the power supply connected across pins 7 and 14. Values for $R_f$ and $R_b$ are typically in the megohm range

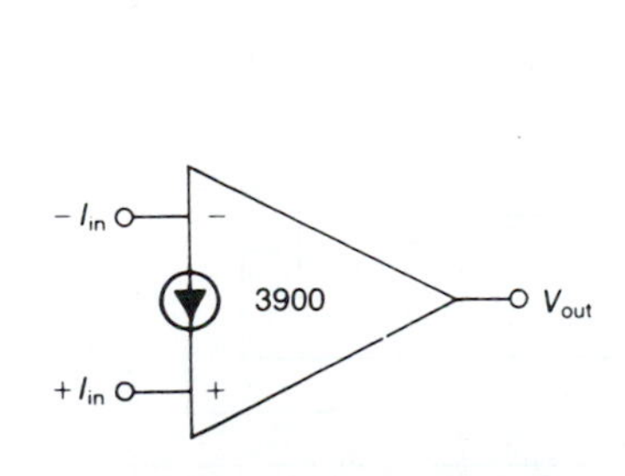


**Figure 3–19.** Symbol for the Norton Op Amp

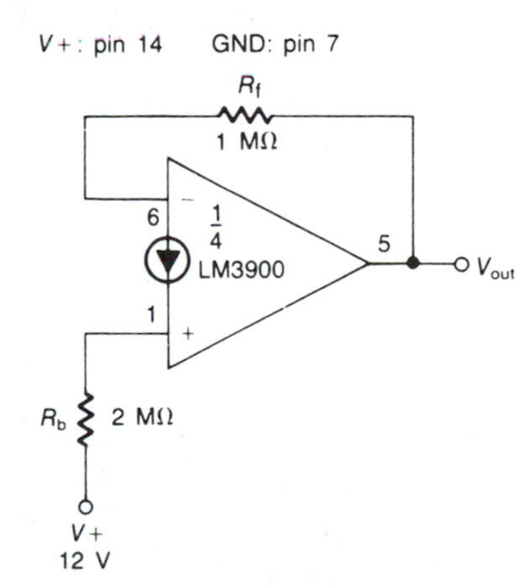


**Figure 3–20.** Current-Mirror Biasing for the Norton Op Amp (Courtesy of National Semiconductor Corporation)

because of the very small mirror current requirements of the Norton amplifier. In general practice, the biasing resistor $R_b$ is made twice the value of the feedback resistor $R_f$.

A commonly used Norton amplifier is the LM3900, a device that consists of four independent, dual input, internally frequency-compensated amplifiers, as illustrated in the package configuration of Figure 3–21. Such a device can be used in many of the applications of a standard op amp. Performance as a dc amplifier that uses only a single supply is not as precise as a standard op amp that operates with split supplies, but it is adequate in many less critical applications.

## Precautions

Certain precautions must be observed when using the current-differencing (Norton) amplifier. Since this is a current-driven device, series resistors must *always* be connected to each input. In biasing, resistors $R_f$ and $R_b$ provide the series resistance. Current into the noninverting (+) input is established by $V^+$, through $R_b$, so that

$$I^+ = \frac{V^+}{R_b} \tag{3.31}$$

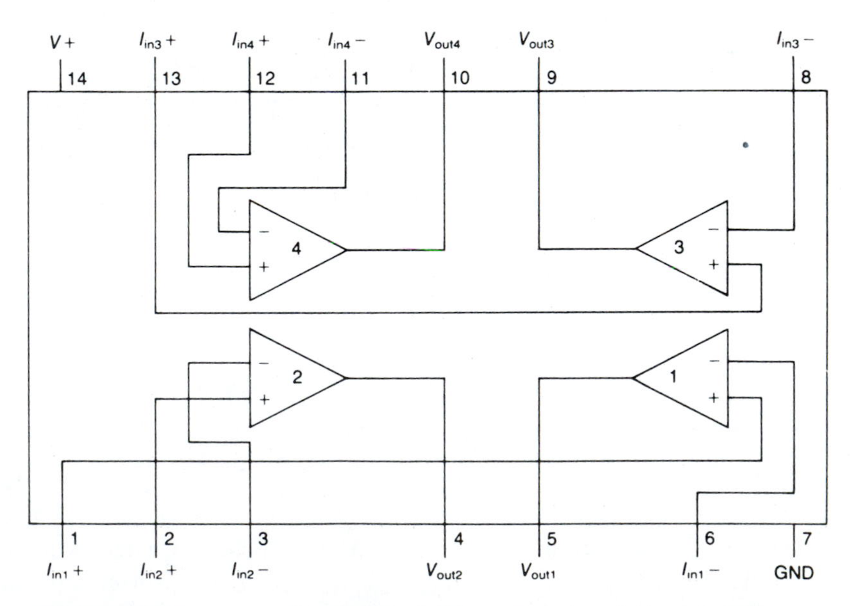


**Figure 3–21.** Package Configuration for the LM3900 Norton Amplifier IC (Courtesy of National Semiconductor Corporation)

Negative feedback through $R_f$ keeps the two input currents equal, so that

$$I^- = I^+ = \frac{V^+}{R_b} \tag{3.32}$$

Since the inverting (−) input is at ground (virtual ground) level, the dc output voltage $V_{out(dc)}$ is the voltage dropped across $R_f$ by $I^-$:

$$V_{out(dc)} = R_f I^- \tag{3.33}$$

Substitute Equation 3.32 into Equation 3.33:

$$V_{out(dc)} = \frac{V^+}{R_b}(R_f) \text{ or } V_{out(dc)} = \frac{R_f}{R_b}(V^+) \tag{3.34}$$

Thus, the output is biased to a dc level one-half that of the supply voltage.

## Example 3.12

Refer to Figure 3–20. Calculate the output dc voltage level.

*Solution*

Using Equation 3.34,

$$V_{out(dc)} = \frac{R_f}{R_b}(V^+) = \frac{1\text{ M}\Omega}{2\text{ M}\Omega}(12\text{ V}) = 6\text{ V}$$

The power supply for the Norton amplifier IC should never become reversed in polarity, nor should the device be inadvertently installed backwards in a socket. Such actions will result in an unlimited current surge within the device that will fuse the internal conductors and destroy the unit.

Output short circuits either to ground or to the positive power supply should be brief. The device can be destroyed by long-lasting short circuits because of excessive junction temperatures caused by the large increase in IC chip power dissipation.

Unintentional signal coupling from the output to the noninverting input can cause oscillations. High-value biasing resistors used in the noninverting input circuit make this input lead highly susceptible to unintentional ac signal pickup. A quick check for this condition is to bypass the noninverting input to ground with a capacitor. Careful lead dress or locating the noninverting input biasing resistor closer to the IC will help prevent this problem.

Operation of the Norton amplifier can best be understood by remembering that the input currents are differenced at the inverting input terminal and that this difference current flows through the external feedback resistor to produce a large voltage output swing. The maximum output voltage swing of the amplifier is approximately $(V^+ - 1)$ $V_{p\text{-}p}$.

## Inverting Norton Amplifier

An inverting Norton amplifier is shown in Figure 3–22. Capacitors $C_1$ and $C_2$ couple the ac signals into and out of the amplifier while blocking dc. The gain of the ac signal is set by resistors $R_f$ and $R_1$. Biasing is established by $V^+$ and resistor $R_b$ at the noninverting input. The input impedance $Z_{in}$ is equal to resistance $R_1$.

While this circuit looks very much like an inverting amplifier constructed with a standard op amp, there are several differences to be observed. The LM3900 has a typical output impedance of 8 kΩ and a typical output current of 5 mA or less. Therefore, it is necessary to take care to prevent loading. The output is loaded by both $R_f$ and load resistance $R_L$. Keeping both $R_f$ and $R_L$ as high in value as practicable will minimize this loading effect. This technique is opposite from that used for the standard op amp circuit where $R_f$ is to be kept reasonably low in value to minimize output offset voltages. Formulas for the inverting Norton amplifier are given by the following equations. Output ac voltage:

$$V_{out(ac)} = -\frac{R_f}{R_1}(V_{in(ac)}) \qquad \textbf{(3.35a)}$$

or

$$V_{out(ac)} = A_{ac} \times V_{in(ac)} \qquad \textbf{(3.35b)}$$

where

$A_{ac}$ = signal gain

ac gain:

$$A_{ac} = -\frac{V_{out(ac)}}{V_{in(ac)}} = -\frac{R_f}{R_1} \qquad \textbf{(3.36)}$$

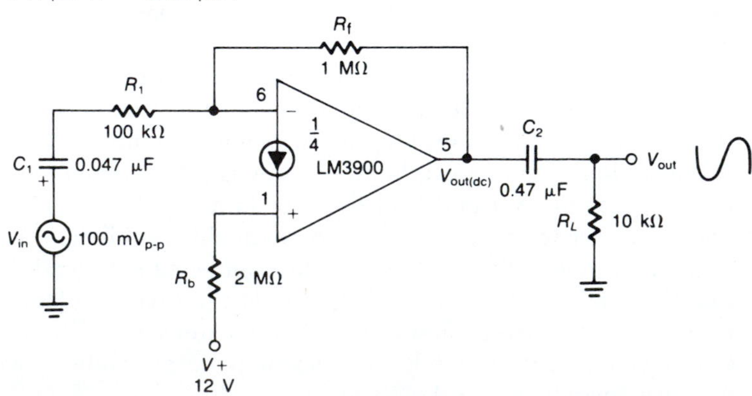


Figure 3–22. Inverting ac Norton Amplifier Circuit (Courtesy of National Semiconductor Corporation)

Output dc voltage level:

$$V_{out(dc)} = \frac{R_f}{R_b}(V^+) \qquad \text{(repeat of Equation 3.34)}$$

Input impedance:

$$Z_{in} = R_1 \tag{3.37}$$

Cutoff frequency:

$$f_c = \frac{1}{2\pi C_1 R_1} \tag{3.38}$$

Input capacitor $C_1$:

$$C_1 = \frac{1}{2\pi f_c R_1} \qquad \text{(repeat of Equation 3.11)}$$

Output capacitor $C_2$:

$$C_2 = \frac{1}{2\pi f_c R_L} \qquad \text{(repeat of Equation 3.12)}$$

## Example 3.13

Refer to Figure 3–22. Calculate (a) $V_{out(dc)}$, (b) $V_{out(ac)}$, (c) $f_c$, (d) $Z_{in}$, and (e) $A_{ac}$.

*Solutions*

**a.** Using Equation 3.34,

$$V_{out(dc)} = \frac{R_f}{R_b}(V^+) = \frac{1\ \text{M}\Omega}{2\ \text{M}\Omega}(12\ \text{V}) = 6\ \text{V}$$

**b.** Using Equation 3.35a,

$$V_{out(ac)} = -\frac{R_f}{R_1}(V_{in(ac)}) = -\frac{1\ \text{M}\Omega}{100\ \text{k}\Omega}(100\ \text{mV}_{p\text{-}p}) = -\frac{1000\ \text{k}\Omega}{100\ \text{k}\Omega}(0.1\ \text{V}_{p\text{-}p})$$
$$= -1\ \text{V}_{p\text{-}p}$$

**c.** Using Equation 3.38,

$$f_c = \frac{1}{2\pi C_1 R_1} = \frac{1}{6.28 \times 0.047\ \mu\text{F} \times 100\ \text{k}\Omega}$$
$$= \frac{1}{(6.28)(47 \times 10^{-9}\ \text{F})(100 \times 10^3\ \Omega)} = 34\ \text{Hz}$$

(Note that this is cutoff frequency, equal to approximately 0.1 $f_{op}$. Therefore, the circuit should be operated at 340 Hz or higher to prevent signal attenuation.)

**d.** Using Equation 3.37,

$$Z_{in} = R_1 = 100\ \text{k}\Omega$$

**e.** Using Equation 3.36,

$$A_{ac} = -\frac{R_f}{R_1} = -\frac{1\ \text{M}\Omega}{100\ \text{k}\Omega} = -\frac{1000\ \text{k}\Omega}{100\ \text{k}\Omega} = -10$$

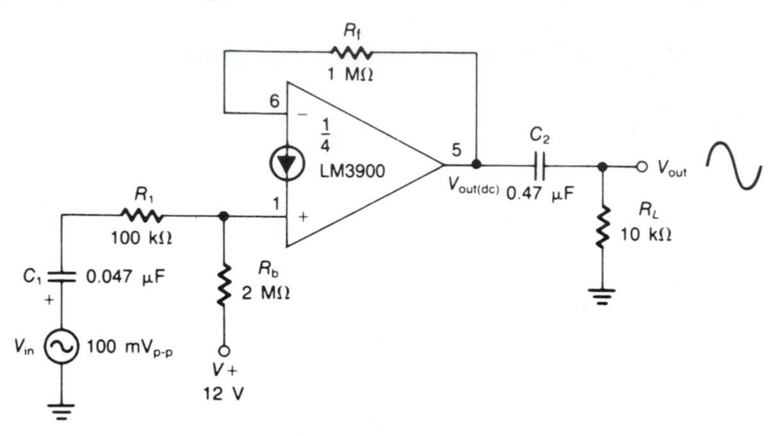


Figure 3–23. Noninverting ac Norton Amplifier Circuit (Courtesy of National Semiconductor Corporation)

### Noninverting Norton Amplifier

A noninverting Norton amplifier is produced by simply moving the input source, coupling capacitor $C_1$, and series resistor $R_1$ to the noninverting input terminal, as shown in Figure 3–23. Note that biasing does not change and that negative feedback current still flows through $R_f$. However, in this circuit the input signal current *adds* to the bias current at the noninverting input terminal.

There are some differences to be observed between the noninverting amplifier constructed with the Norton amplifier and that constructed with the standard op amp. The noninverting Norton circuit is more closely related to the inverting Norton amplifier than to the noninverting op amp circuit. For example, in the noninverting Norton amplifier, the source sees an input impedance $Z_{in}$ equal to $R_1$. The signal gain $A_{ac}$ is equal to $R_f/R_1$, which produces the same magnitude as does the inverting amplifier. Therefore, for the noninverting Norton amplifier,

$$Z_{in} = R_1 \tag{3.39}$$

and

$$A_{ac} = \frac{R_f}{R_1} \tag{3.40}$$

All precautions observed for the inverting Norton amplifier must also be observed for the noninverting Norton amplifier.

Other applications for the Norton amplifier will be presented in a later chapter. The data sheets in Appendix B provide many additional applications.

## 3.10 Wideband Amplifiers

*Wideband* op amps are capable of amplifying a wide range of frequencies, extending from dc up to several hundred megahertz (MHz). In practical terms, this means that the device can amplify radio frequency (RF), very high frequency (VHF), and ultrahigh frequency (UHF) signals.

The μA733, with a bandwidth (*BW*) of dc to 120 MHz, is used in such typical wideband applications as communications systems, as a video or pulse amplifier, and as a read-head amplifier for magnetic tape, drum, or disc memories. Another wideband device is the HA2620 with a *BW* of dc to 100 MHz. Figure 3–24 shows the HA2620 used as a simple video amplifier.

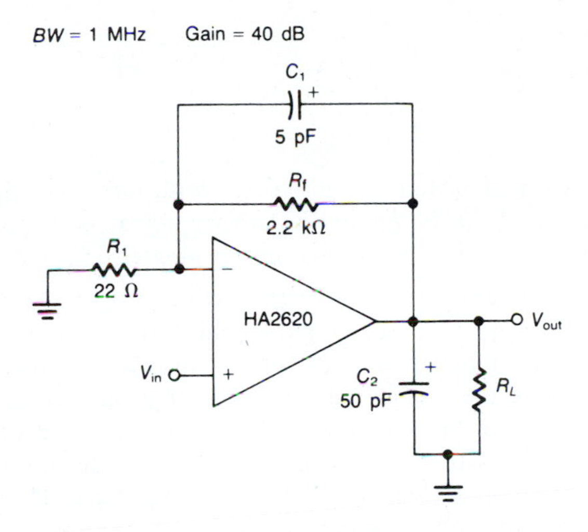


**Figure 3–24.** Video Amplifier Circuit Using the HA2620 Wideband Op Amp (Courtesy of Harris Semiconductor Products Division)

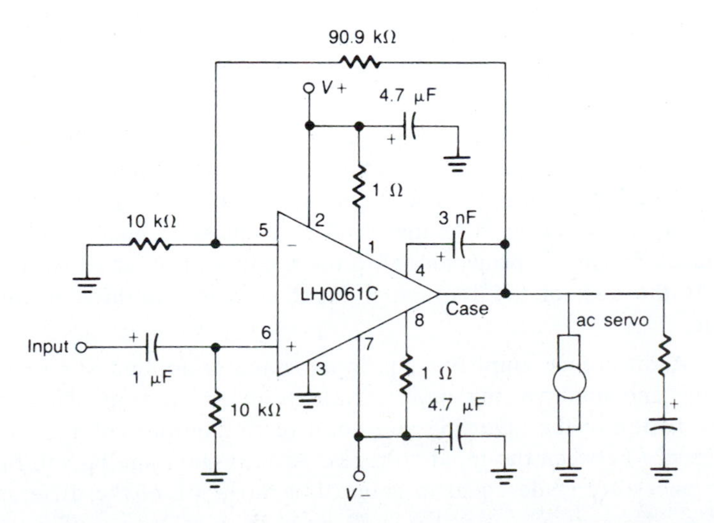


**Figure 3–25.** ac Servo Amplifier Circuit (Courtesy of National Semiconductor Corporation)

The LH0061C is a wideband, high-speed op amp capable of supplying currents in excess of 0.5 A at voltage levels of ±12 V. The wide BW (1 MHz), high slew rate (70 V/μs), and high output power capability makes the LH0061C ideal for such applications as ac servo amplifiers, deflection yoke drivers, capstan drivers, and audio amplifiers. Figure 3–25, page 87, illustrates the LH0061C used as an ac servo amplifier.

There are many other devices available with lower *BW*s for applications where *BW* requirements are not so wide. Reference to manufacturers' data sheets will provide the necessary information about such devices.

# 3.11 Summary

☐ A noninverting amplifier is constructed with the input signal applied in the (+) input terminal of the op amp. The output signal is in phase with the input signal. The closed-loop gain is always one or greater, since $A_{CL} = 1 + (R_f/R_1)$.

☐ An inverting amplifier is constructed with the input signal applied to the (–) input terminal of the op amp. The output signal is 180° out of phase with the input signal. The closed-loop gain can be less than, equal to, or greater than unity, since $A_{CL} = (R_f/R_1)$, where the – sign indicates only that the output signal is inverted. The input impedance of the inverting amplifier is the value of the input element and will always be less than the input impedance of the noninverting amplifier, which is the intrinsic input impedance of the op amp itself. The inverting amplifier can be used as a unity-gain inverter when the input and feedback elements are equal in value.

☐ Error voltage, called dc output offset voltage, is the result of certain op amp characteristics: input bias current, input offset current, input offset voltage, and drift. Null circuits can eliminate all but drift.

☐ A voltage follower circuit has no input or feedback resistances. The output signal follows exactly the input, or source, signal. The circuit is also called source follower, unity-gain amplifier, buffer amplifier, and isolation amplifier.

☐ A summing amplifier has two or more resistors connected simultaneously to the inverting input terminal of an op amp. The output voltage is the sum of the individual input voltages. An averaging amplifier is a variation of the summing amplifier in which each of the input resistances is made equal in value, and the feedback resistor is proportional to the number of input resistors. The output voltage is the sum of the individual input voltages, divided by the number of the inputs.

☐ A difference amplifier has input voltages applied simultaneously to the inverting and noninverting input terminals of an op amp. The output voltage is a combination of the inverting and noninverting output voltages, or, effectively, the difference between the input voltages. A unity-gain analog subtractor, in which all resistances are made equal in value, is a variation of the difference amplifier.

☐ Many standard op amp circuits can be constructed with a single-supply power source if certain external components are added. The Norton op amp is designed specifically to operate from a single power supply. It is used in a multitude of applications. It is sometimes called a current-differencing amplifier, because it operates on the difference of currents flowing into the (+) and (-) input terminals.

☐ The gain of an amplifier can be controlled by the feedback resistance. Virtual ground is a point at which voltage with respect to true ground is zero. It cannot be used as a ground point because it is isolated from true ground.

☐ Wideband amplifiers are those capable of amplifying frequencies over the range from dc to several hundred megahertz (MHz).

# 3.12 Questions and Problems

**3.1** Several op amp amplifiers are shown in Figure 3-26A–L. Identify each amplifier in the figure from the following list of labels (some figure parts may have more than one label):

1. Unity-gain amplifier
2. Averaging amplifier
3. Voltage follower
4. Inverting amplifier
5. Summing amplifier
6. Noninverting amplifier
7. Unity-gain inverter
8. Difference amplifier
9. Unity-gain analog subtractor
10. Comparator

**3.2** Calculate the closed-loop gain for the amplifier in Figure 3–26A.

**3.3** Calculate the output voltage for Figure 3–26L.

**3.4** Calculate the values for each of the capacitors in Figure 3–26K. Assume a frequency response of 30 Hz.

**3.5** Assume an inverting amplifier circuit where the input resistance is 10 kΩ and the gain is 20. Using the μA741C typical parameters from Figure 2–2, calculate the dc output offset voltage as a result of (a) the change in input bias current over a 20° change in temperature, (b) the change in input offset current over a 10° change in temperature, and (c) the change in input offset voltage over a 30° change in temperature.

**3.6** Calculate the value required for a current-compensating resistor to be placed in series with the positive (+) input terminal of the circuit described in Problem 3.5.

**3.7** How is output offset voltage that results from drift minimized?

**3.8** Explain *offset null* and how it works.

**3.9** List four precautions that must be taken when using Norton amplifiers.

**3.10** What type of biasing is standard for the Norton amplifier?

**3.11** What is the advantage of a difference amplifier?

**3.12** Design a difference amplifier with a gain of 10. Draw the completed design circuit.

**3.13** Using the circuit designed in Problem 3.12, design and draw the completed circuit for a unity-gain subtractor.

**3.14** Design and draw the completed circuit for a noninverting amplifier whose gain is 21. The input element is 5 kΩ.

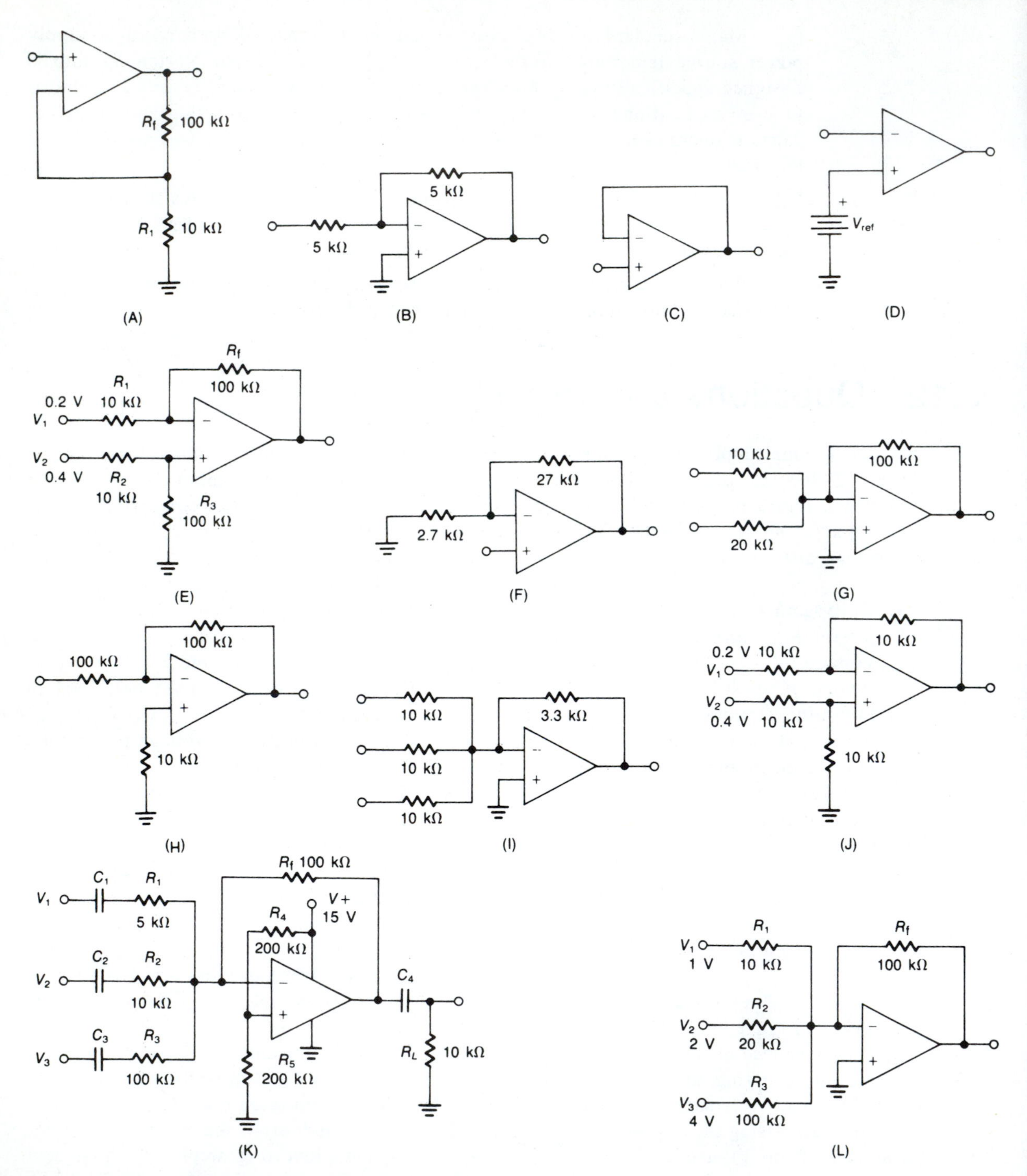


**Figure 3-26.** Figures for Questions and Problems

Chapter 4

# The Comparator

## 4.1 Introduction

A comparator is a circuit that evaluates two or more signals, and indicates whether or not the signals are matched in some particular way. Comparators may test for amplitude, frequency, phase, voltage, current, waveform type, or numerical value. A number comparator has three outputs: greater than (>), equal to (=), and less than (<). A phase or frequency comparator may have an output voltage that varies depending on which quantity is leading or lagging, larger or smaller.

The characteristic definitions for the comparator indicate that many similarities exist between op amps and the amplifier section of voltage comparators. A comparator is a linear integrated circuit (LIC), but it is not defined as an op amp. Although the comparator is not an op amp in the true sense, op amps can be used to implement the comparator function at low frequencies.

This chapter presents the comparator as a sensing device, with several applications.

## 4.2 Objectives

When you complete this chapter, you should be able to:

- ☐ Define the comparator and explain its operation.
- ☐ Describe the differences between a comparator and a standard op amp.
- ☐ Discuss the effects of saturation voltage and slew rate on a comparator circuit.
- ☐ Discuss the comparator as a sensing device.
- ☐ Understand definitions relating specifically to comparators.
- ☐ Design a simple level detector and explain its operation.
- ☐ Design a level detector with hysteresis and explain its operation.
- ☐ Design a zero-crossing detector and explain its operation.
- ☐ Explain the operation of the Schmitt trigger and distinguish the difference between it and a comparator.
- ☐ Design a peak detector and explain its operation.
- ☐ Design a window detector and explain its operation.

## 4.3 Correlation

*Correlation* is important in determining the causes of various kinds of circuit malfunctions. Often, two quantities that might intuitively seem correlated are actually not significantly correlated. Sometimes, two quantities that do not seem correlated actually are.

Correlation, a mathematical expression for the relationship between two quantities, may be positive, zero, or negative. *Positive correlation* indicates that an increase in one parameter is accompanied by an increase in the other. *Negative correlation* indicates that an increase in one parameter occurs with a decrease in the other. *Zero correlation* means that variations in the two parameters are unrelated. The *coefficient of correlation* between two variables is usually expressed as a number in the range between −1 (the most negative) and +1 (the most positive). For example, we might say that the failure rate of a certain component (such as an IC) is correlated with the temperature; the higher the temperature the more frequently the component fails. A statistical sampling can determine this correlation, and assign it a correlation coefficient.

Figure 4–1 shows examples of positive correlation, zero correlation, and negative correlation. The closer the correlation coefficient is to +1, the more nearly the points lie along a straight line with the slope of a positive value (Figure 4–1A). When the two parameters are not correlated, or have zero correlation, the points are randomly scattered (Figure 4–1B). When the parameters are negatively

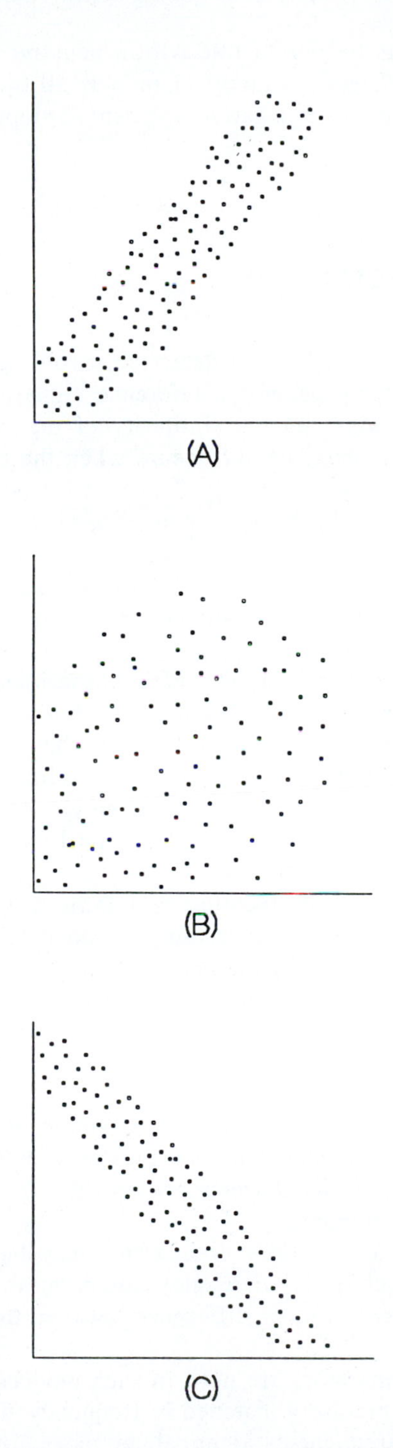


**Figure 4–1.** Examples of Correlation: (A) Positive, (B) Zero, (C) Negative

correlated, the points lie near a line with a negative slope (Figure 4–1C). When the correlation coefficient is exactly -1 or +1, all the points lie along a perfectly straight line having either a negative or positive slope, respectively.

## Correlation Detection

*Correlation detection* is a form of detection involving the comparison of an input signal with an internally generated reference signal. The output of a correlation detector varies depending on the similarity of the input signal to the internally generated signal; maximum output occurs when the two signals are identical.

### Phase Comparator

A *phase comparator* is an example of a correlation detector. In-phase signals produce maximum output; if the signals are not perfectly in phase, the output is reduced.

A phase comparator is a part of a phase-locked-loop (PLL) circuit. The phase comparator does exactly what its name implies; it compares two signals in terms of their phase. The output voltage of the phase comparator depends on whether the signals are in phase or not. In an oscillator circuit, for example, if the signals are not in phase, the voltage from the comparator causes the phase of the oscillator signal to be shifted back and forth until its signal is exactly in phase with a signal from a reference oscillator.

In some types of radiolocation and radionavigation systems, the relative phase of two signals is used to determine the position of the ship or airplane. The relative phase is determined by the phase comparator, the output of which is fed to a computer, which in turn determines the position.

### Frequency Comparator

A *frequency comparator* is another example of a correlation detector. Two signals of the same frequency produce maximum output. The greater the difference between the input signal frequency and the internally generated signal frequency, the less the output.

The output of a frequency comparator may be in video (visual) or audio (audible) form. Basically, all frequency comparators operate on the principle of *beating*. The *beat frequency,* or difference between the two input frequencies, determines the output.

Frequency comparators are used in such devices as PLL circuits, where the two signals must be precisely matched in frequency. The output of the comparator is used to keep the frequencies identical, by means of error-sensing circuits.

## Error-Sensing Circuits

An *error-sensing* circuit is a means of regulating the output current, power, or voltage of a device. In a power supply, for example, an error-sensing circuit might be used to keep the output voltage at a constant value; such a circuit requires a standard reference with which to compare the output of the supply. If the power supply output voltage increases, the error-sensing circuit produces a signal to reduce it. If the power supply voltage decreases, the error-sensing circuit produces a signal to raise it. An error-sensing circuit is, therefore, a form of amplitude comparator.

An error-sensing circuit does not necessarily have to provide regulation for the circuit to which it is connected; some error-sensing circuits merely alert the equipment operator that something is wrong by causing a bell to ring or a warning light to flash.

# 4.4 Voltage Comparators

*Voltage comparators* are high-gain, differential input-logic output devices, specifically designed for open-loop operation with a minimum of delay time. A basic comparator circuit is shown in Figure 4–2. Although variations of the comparator are used in a host of applications, all uses depend upon the basic transfer function. Device operation is simply a change of output voltage dependent upon whether the signal input is above or below the threshold voltage.

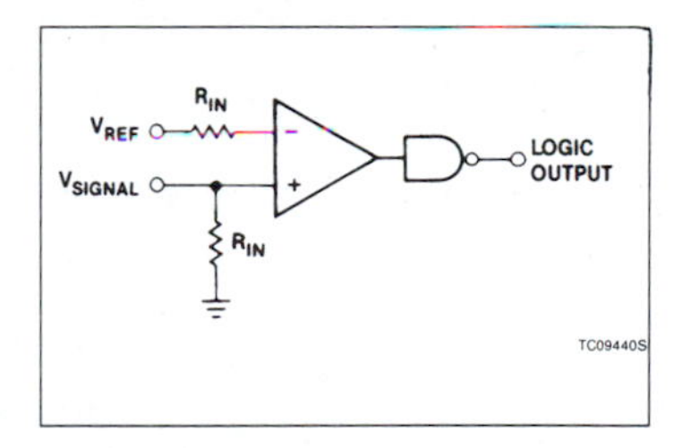


**Figure 4–2.** Basic Comparator Circuit (Courtesy of Signetics Company)

## The Comparator as a Sensing Device

When an op amp is used as a comparator, it does not amplify signals. Instead, it is used as a *sensing device* to determine if a voltage is greater than or less than a given reference voltage level. Since a comparator operates in the open-loop mode, even the smallest difference in input voltages will cause the output of the amplifier to saturate.

### Simple Level Detector

A simple op amp application as a *comparator level detector* is illustrated in Figure 4–3A. A reference voltage $V_{ref}$ is connected between the (+) input terminal and ground. The input voltage $V_{in}$ connected to the (–) input terminal is compared with the reference voltage. If $V_{in}$ is less than $V_{ref}$, then $V_{out}$ goes positive. If $V_{in}$ becomes greater than $V_{ref}$, then $V_{out}$ switches to negative. The results are shown in Figure 4–3B, an idealized characteristic curve.

The reference voltage need not be connected to the (+) terminal. If it is connected to the (–) terminal, then the polarity of the output is reversed.

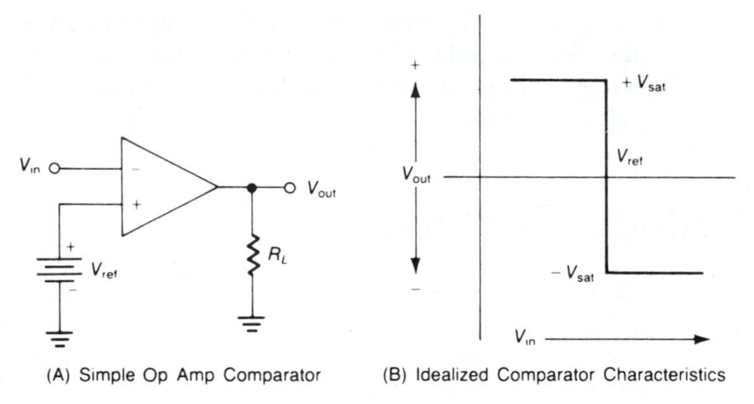


**Figure 4–3.** Simple Sensing Circuit Using a Comparator as a Level Detector

## Zero-Crossing Detector

By grounding the (+) input terminal (0 V) of the comparator, we have a *zero-crossing detector.* Figure 4–4 shows the relationship between the input and output voltages. With a small sine wave applied to $V_{in}$, note that $V_{out}$ is negative when the sine wave is positive, and vice versa. The switch between positive and negative occurs when the input signal crosses the reference voltage, or zero.

## Saturation Voltage

Saturation voltage, $\pm V_{sat}$, is normally about 1 V less than the applied power supply voltages of the op amp. For example, the op amp in Figure 4–5A has supply voltages of +15 V and –15 V applied to it. The saturation voltage would then be approximately $\pm 14$ V, as shown in Figure 4–5B.

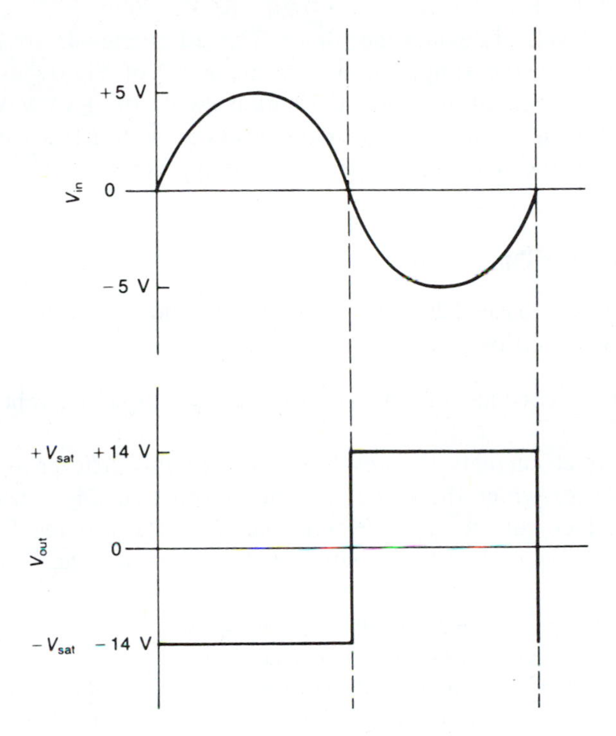


**Figure 4–4.** Zero-Crossing Detector Input/Output Relationships

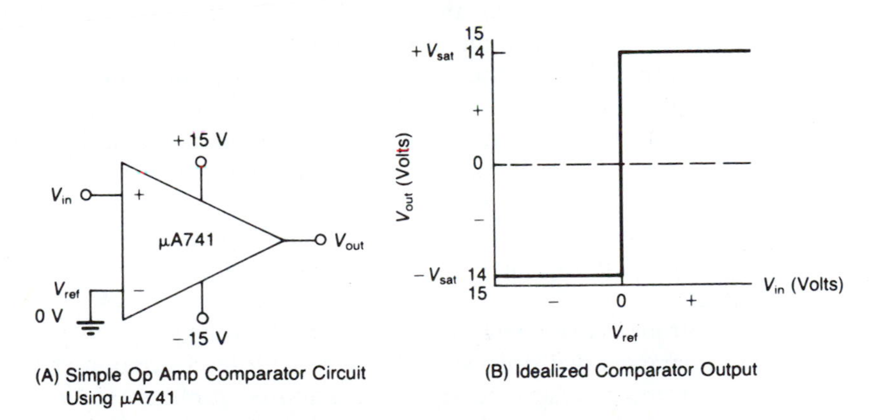


**Figure 4–5.** Simplified Comparator Circuit

## Constructing Comparator Circuits

A comparator can be constructed with any op amp operated in the *open-loop* mode, that is, with no feedback. However, many applications of comparators require interfacing with digital devices, so their output swing must be compatible with the logic levels of devices they drive. Those logic levels are normally between 0 V and +5 V, so the simple op amp comparator of Figure 4–5A could not be used in such an application, since the output swing of ± 14 V would destroy the driven digital device. Therefore, manufacturers provide LICs with special characteristics, designed especially to operate as comparators.

### Comparator Definitions

Comparators have their own set of terms and definitions. Several of those definitions are as follows:

Input bias current—the average of the two input currents with no signal applied.

Input offset current—the absolute value of the difference between the two input currents for which the output will be driven to change states.

Input offset voltage—the absolute value of the voltage between the input terminals required to make the output voltage greater than or less than some specified value.

Input voltage range—the range of voltage on the input terminals (common-mode) over which the offset specifications apply.

Logic threshold voltage—the voltage at the output of the comparator at which the driven logic circuitry changes its digital state. This definition relates to the operation of digital devices.

Response time—the interval between the application of an input step function and the time when the output crosses the logic threshold voltage. (This parameter is similar to the propagation delay of standard op amps.)

Strobe current—the current out of the strobe terminal when it is at the zero logic level.

Strobed output level—the dc output voltage, independent of input conditions, with the voltage on the strobe terminal equal to or less than the specified low state.

Strobe "ON" voltage—the maximum voltage on either strobe terminal required to force the output to the specified high state independent of the input voltage.

Strobe "OFF" voltage—the minimum voltage on the strobe terminal that will guarantee that it does not interfere with the operation of the comparator.

Strobe release time—the time required for the output to rise to the logic threshold voltage after the strobe terminal has been driven from zero to the high logic level.

Strobing—a method used to enable or to disable the comparator. When strobed OFF, the comparator is disabled and the output will not respond to an input signal. Conversely, when strobed ON, the comparator is enabled and the output responds to an input signal. Figure 4–6 shows a strobed LM311 comparator. Additional definitions may be obtained from manufacturer's data books. (The data sheet for the LM311 is in Appendix I.)

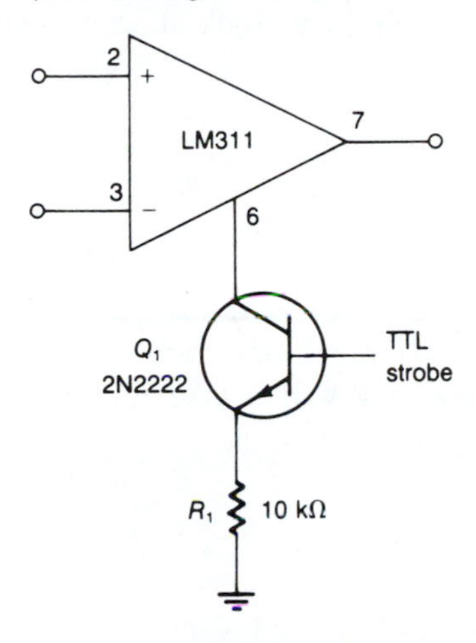


**Figure 4–6.** Strobed LM311 Comparator (Courtesy of National Semiconductor Corporation)

### Effects of Slew Rate

The open-loop operation of an op amp increases the BW and unity gain of the device. No compensation circuit is needed; therefore, the device has an increased slew rate that results in a faster output change. The faster change in output means the comparator has the ability to more closely follow rapid changes of signals at the input. The time it takes for a comparator output to change from one supply voltage to the other is determined by dividing the change in output voltage, $\Delta V$, by the slew rate (SR) of the device. As a formula, this relationship is expressed by the following equation:

$$t = \frac{\Delta V}{SR} \tag{4.1}$$

## Example 4.1

Refer to Figure 4–5A. The slew rate of the μA741C is 0.5 $V/\mu s$. With the supply voltages in the figure, how long will it take for the output to swing from $+V_{sat}$ to $-V_{sat}$?

*Solution*

Use Equation 4.1:

$$t = \frac{\Delta V}{SR} = \frac{+14\text{ V} - (-14\text{ V})}{0.5\text{ V}/\mu\text{s}} = \frac{28\text{ V}}{0.5\text{ V}}\,\mu\text{s} = 56\ \mu\text{s}$$

The rate of change in Example 4.1 is very slow, so the circuit using the μA741C would be limited. Now let us look at an op amp with a higher slew rate.

## Example 4.2

An LM318 op amp has a typical slew rate of 70 $V/\mu s$. If supply voltages of ±15 V are applied, how long will it take for the output to swing from $+V_{sat}$ to $-V_{sat}$?

*Solution*

Use Equation 4.1:

$$t = \frac{\Delta V}{SR} = \frac{+14\ \text{V} - (-14\ \text{V})}{70\ \text{V}/\mu\text{s}} = \frac{28\ \text{V}}{70\ \text{V}}\ \mu\text{s} = 0.40\ \mu\text{s}$$

Obviously, the LM318 is preferable over the μA741C for use as a comparator, because the time required for the output swing is about 140 times faster in the LM318. It should be just as obvious then that devices designed specifically to operate as comparators, with their faster response times, would be even more desirable.

### Evaluating a Comparator Circuit

As previously discussed, a comparator is used to detect a changing voltage on one input with a reference voltage on the other input. Refer to Figure 4–7A. Assume that a sine wave is applied to the noninverting input terminal and the inverting input terminal is grounded, resulting in a 0 V reference voltage. When the input signal is positive, the output is at $+V_{sat}$. When the input signal swings through the zero point going negative, the output swings to $-V_{sat}$. The output is in phase with the input. Figure 4–7B shows the input/output relationship. If the input signals were reversed, the output would be 180° out of phase with the input.

## Zero Detector

The circuit in Figure 4–7A is sometimes referred to as a *zero detector,* because each time the input signal crosses the zero point, the output swings to the opposite polarity. The circuit also is sometimes called a *squaring circuit,* because the sine wave at the input is converted to a square wave output.

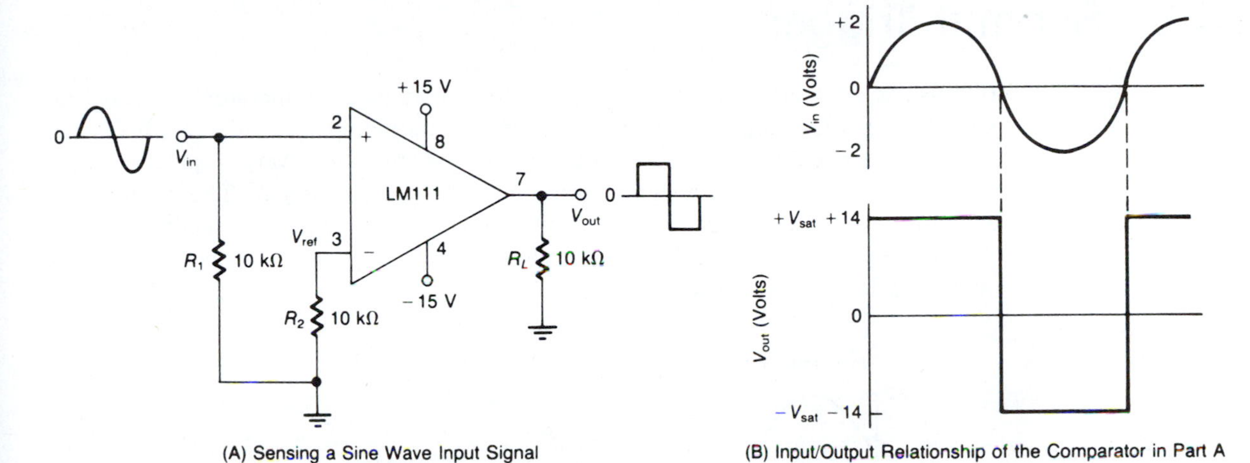


(A) Sensing a Sine Wave Input Signal

(B) Input/Output Relationship of the Comparator in Part A

**Figure 4–7.** Noninverting Comparator Circuit (Part A: Courtesy of National Semiconductor Corporation)

## Phase Difference Detection

The common-mode rejection characteristic of the comparator provides the ability to detect phase differences. If two input signals are of the same frequency, as shown in Figure 4–8A, the output results detect the phase differences caused by a differential voltage at the inputs whenever the two signals are out of phase, as shown in Figure 4–8B. When $V_2$ is more positive than $V_1$, output will be $+V_{sat}$, and when $V_1$ is more positive than $V_2$, the output will be $-V_{sat}$. When the two input signals are in phase, the common-mode rejection ratio (CMRR) characteristic causes the output to be zero.

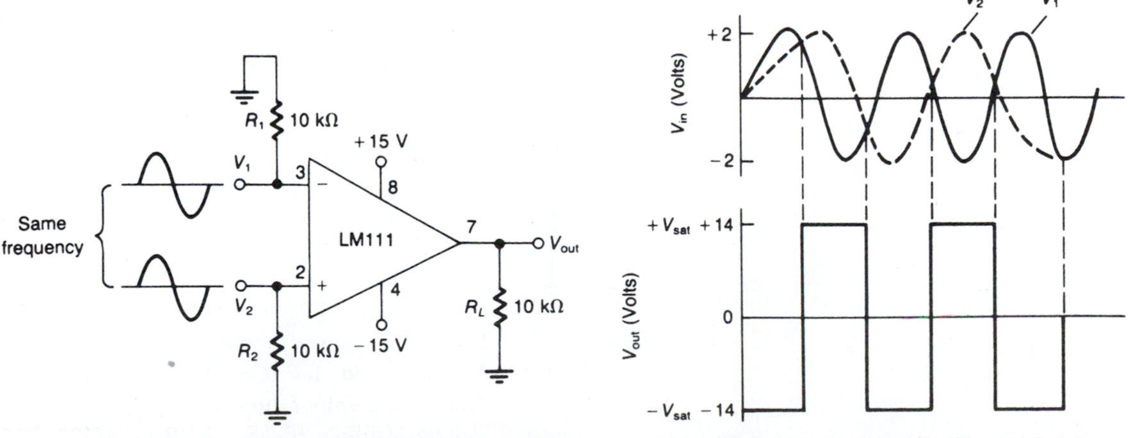


(A) LM111 Phase-Difference Detector with Common-mode Input Signal

(B) Input/Output Relationship of the Circuit in Part A

**Figure 4–8.** Phase-Difference Detector Circuit (Part A: Courtesy of National Semiconductor Corporation)

## 4.5 Schmitt Trigger

If noise is present at either of the input terminals of a comparator, false output signals are generated. Figure 4-9 shows the output results of a noninverting zero-crossing comparator with noise riding on the sine wave input. Positive feedback, shown in Figure 4-10, overcomes the noise problem. The feedback is accomplished by taking a small fraction of the output voltage and feeding it back to the (+) terminal, creating a reference voltage dependent upon $V_{out}$. Such a configuration is called a *Schmitt trigger.*

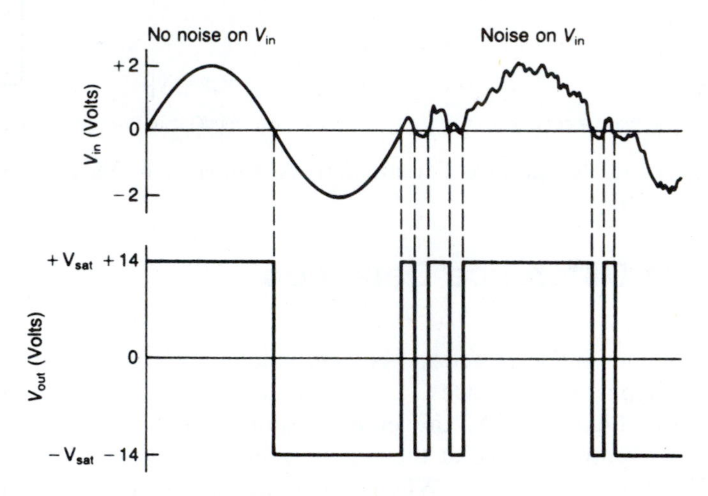


**Figure 4-9.** Effect of Noise on a Zero-Crossing Detector

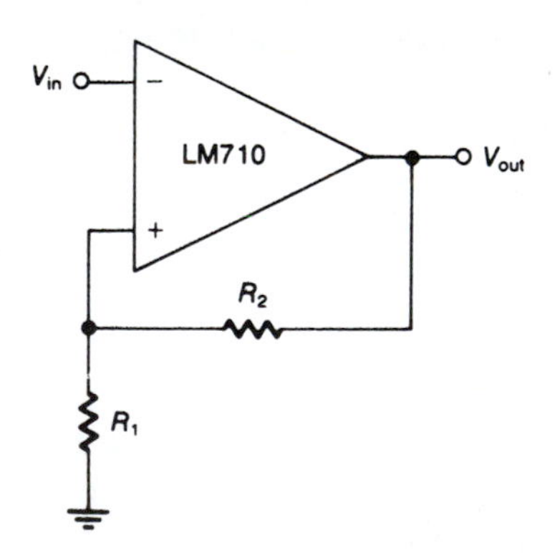


**Figure 4-10.** LM710 Comparator with Positive Feedback (Courtesy of National Semiconductor Corporation)

When $V_{out} = +V_{sat}$, the positive feedback voltage is called the *upper threshold voltage*, $+V_{uth}$, and is expressed as follows:

$$V_{uth} = V_{out}\left(\frac{R_1}{R_1 + R_2}\right) + V_{ref} \quad \textbf{(4.2)}$$

Conversely, when $V_{out} = -V_{sat}$, the positive feedback voltage is called the *lower threshold voltage*, $V_{lth}$, and is expressed as follows:

$$V_{lth} = V_{ref} - V_{out}\left(\frac{R_1}{R_1 + R_2}\right) \quad \textbf{(4.3)}$$

## Example 4.3

Refer to Figure 4–10. Assume $R_1 = 1\ \text{k}\Omega$, $R_2 = 100\ \text{k}\Omega$, $V_{ref} = 0$ V, and $\pm V_{sat} = \pm 5$ V. Find $V_{uth}$ and $V_{lth}$.

*Solution*

Use Equation 4.2:

$$V_{uth} = V_{out}\left(\frac{R_1}{R_1 + R_2}\right) + V_{ref} = 5\ \text{V}\left(\frac{1\ \text{k}\Omega}{1\ \text{k}\Omega + 100\ \text{k}\Omega}\right) + 0\ \text{V}$$
$$= 49.5\ \text{mV}$$

Use Equation 4.3:

$$V_{lth} = V_{ref} - V_{out}\left(\frac{R_1}{R_1 + R_2}\right) = 0\ \text{V} - 5\ \text{V}\left(\frac{1\ \text{k}\Omega}{1\ \text{k}\Omega + 100\ \text{k}\Omega}\right)$$
$$= -49.5\ \text{mV}$$

## Hysteresis

The difference in voltage between the upper threshold voltage and the lower threshold voltage is called the *hysteresis voltage*, $V_H$. It is expressed in equation form as follows:

$$V_H = V_{uth} - V_{lth} \quad \textbf{(4.4)}$$

Hysteresis produces two switching points instead of one. The resultant switching action is sometimes called *snap action*, because of the rapid change of output states when the threshold voltages have been reached; in effect, the output voltage snaps to the opposite state.

Hysteresis can be illustrated on a single graph, as in Figure 4–11. In this graph we show $V_{in}$ on the horizontal axis and $V_{out}$ on the vertical axis. When $V_{in}$

is less than $V_{uth}$, $V_{out} = +V_{sat}$. When $V_{in}$ increases to $V_{uth}$, the output snaps to $-V_{sat}$. When $V_{in}$ drops back to $V_{lth}$, $V_{out}$ snaps back to $+V_{sat}$. With 0 V reference and symmetrical $\pm V_{sat}$, the output voltages will be symmetrical.

By designing the circuit so that hysteresis voltage is greater than the peak-to-peak noise voltage of the input signal, false output signal swings are eliminated. Therefore, we can say that $V_H$ provides the *limits* of the noise levels that the circuit can withstand.

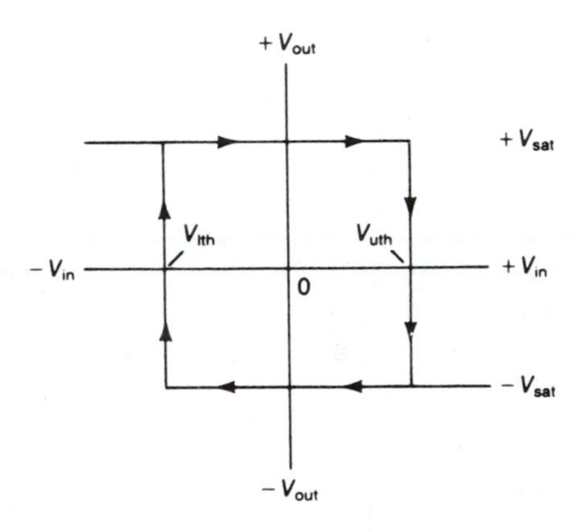


**Figure 4–11.** Input/Output Characteristics of Hysteresis Voltage

## Example 4.4

Refer to Example 4.3. Calculate $V_H$.

*Solution*

Use Equation 4.4:

$$V_H = V_{uth} - V_{lth} = 49.5 \text{ mV} - (-49.5 \text{ mV}) = 99 \text{ mV}$$

Therefore, the circuit of Figure 4–10 can withstand a maximum peak-to-peak noise voltage of 99 mV superimposed on the input signals without false output signal switching.

### Level Detector with Hysteresis

Normally saturated high or low, the amplifiers used in voltage comparators are seldom held in their threshold region. If the compared voltages remain at or near the threshold for long periods of time, the comparator may oscillate or respond to noise pulses. To avoid this oscillation in the linear range, hysteresis can be employed from output to input.

Figure 4–12A shows the NE521 voltage comparator used as a *level detector with hysteresis*. Hysteresis occurs because a small portion of the high-level output voltage is fed back in phase and added to the input signal. This feedback aids the signal in crossing the threshold. When the signal returns to the threshold, the positive feedback must be overcome by the signal before switching can occur. This assures the switching process and oscillations cannot occur. The threshold dead zone created by this method (± 10 mV/cm section in Figure 4–12B), prevents output chatter with signals having slow or erratic zero-crossings.

As shown, the voltage feedback in this circuit is calculated as follows:

$$V_{\text{hyst}} = \frac{E_{\text{out}} \times R_{\text{in}}}{R_{\text{in}} + R_{\text{f}}} \tag{4.5}$$

where $E_{\text{out}}$ is the logic gate high output voltage.

If symmetrical hysteresis is desired, the NE521 requires an additional inverting gate in the output. Care must be taken in the selection of the inverter that propagation delay is minimum.

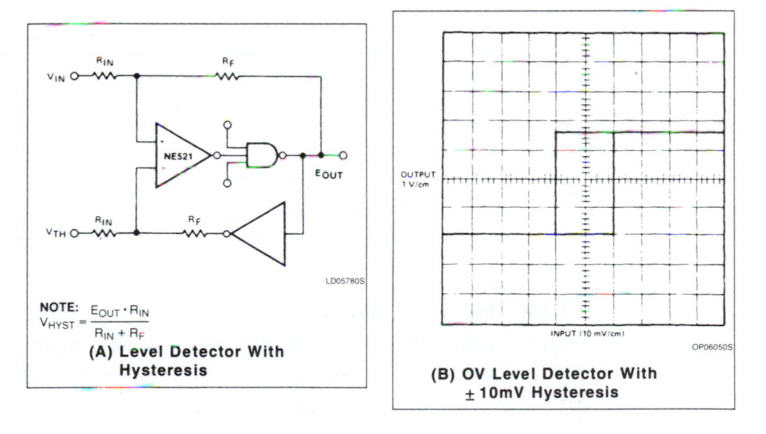


**Figure 4–12.** Application for NE521 Voltage Comparator (Courtesy of Signetics Company)

# 4.6 Peak Detector

The *peak detector* shown in Figure 4–13A is a circuit that *detects* and *remembers* the peak value of an input signal. Momentarily shorting capacitor $C_1$ to ground with RESET switch $S_1$ sets the output to zero. When a positive voltage is applied to the noninverting input terminal, the output voltage of the op amp forward biases diode $D_1$, and $C_1$ is charged until the inverting input voltage equals the noninverting input voltage.

The time that the peak value is held is determined by the time constant of capacitor $C_1$ and load impedance. Placing a buffer amplifier with its high input impedance between the peak detector and the load, as shown in Figure 4–13B, will extend the holding time. (Peak detector action is similar to that of the sample-and-hold (S & H) circuit to be discussed in a later chapter.)

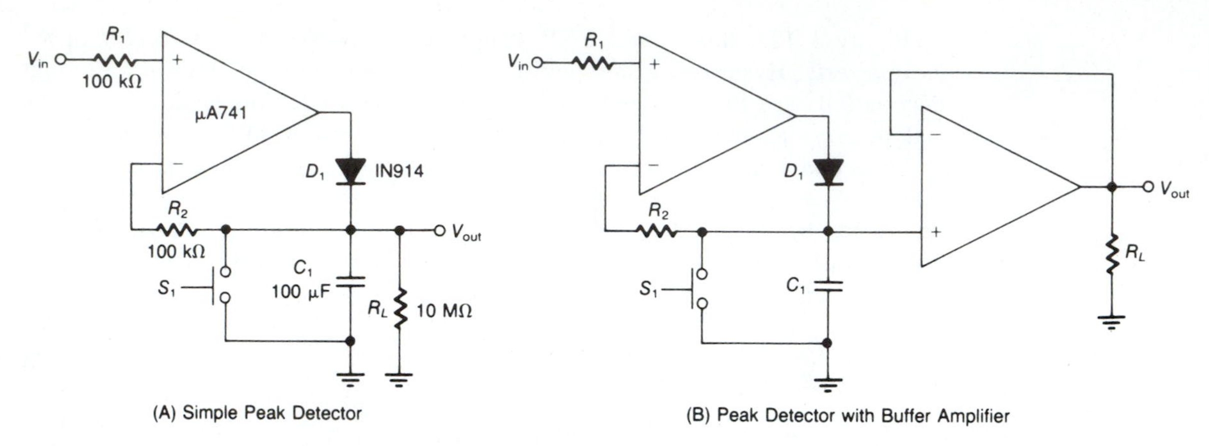


**Figure 4–13.** Peak Detectors

## Example 4.5

Refer to Figure 4–13A. Calculate the peak value holding time.

*Solution*

Using the standard time constant formula $t = RC$,

$$t = 10\ \text{M}\Omega \times 100\ \mu\text{F} = (10 \times 10^{6}\ \Omega)(100 \times 10^{-6}\ \text{F}) = 1000\ \text{s}$$

Peak detectors can provide either positive or negative peak detection, as shown in Figure 4–14. Here we use LM311 voltage comparators with LM310 voltage followers to construct the circuits. Note that the diode used in the basic circuits previously discussed is not required for these circuits.

# 4.7 Window Detector

A circuit called a *window detector,* designed to monitor an input voltage and indicate predetermined upper and lower limits, is shown in Figure 4–15. Such a circuit is sometimes referred to as a *double-ended limit detector.*

A typical application for the window detector in Figure 4–15A is for limiting IC logic power supplies to the standard +5.0 V and 0 V. For this circuit, $V_{out} = +5$ V when $V_{in}$ is greater than $V_{lth}$ and less than $V_{uth}$, and $V_{out} = 0$ V when $V_{in}$ is less than $V_{lth}$ and greater than $V_{uth}$. The output of this single monolithic chip is compatible with resistor-transistor logic (RTL), diode-transistor logic (DTL), and transistor-transistor logic (TTL). The device is also capable of driving lamps and relays at currents up to 25 mA.

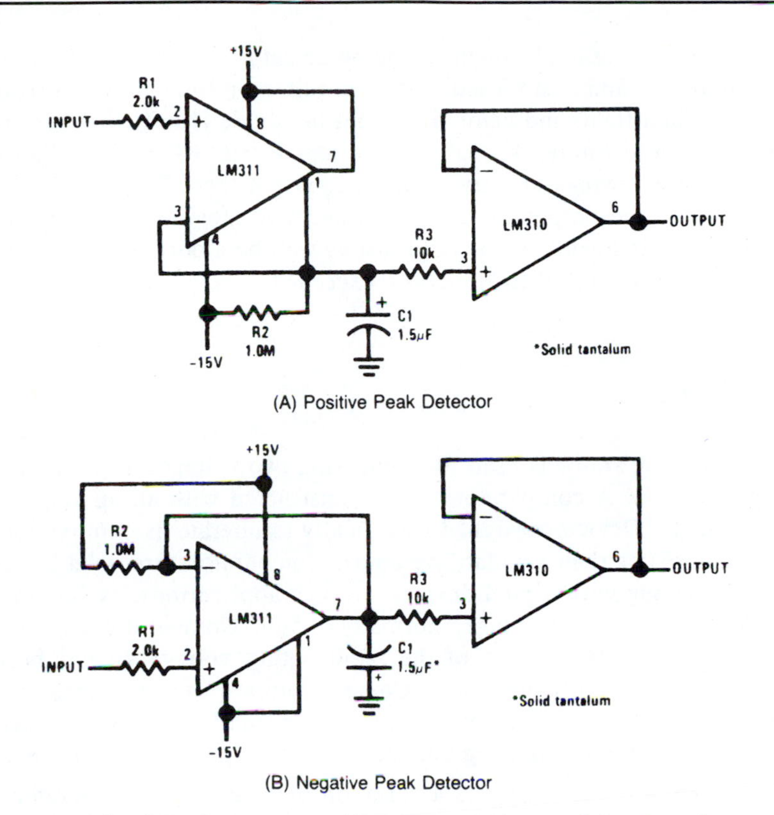

(A) Positive Peak Detector

(B) Negative Peak Detector

**Figure 4–14.** Standard Peak Detectors (Courtesy of National Semiconductor Corporation)

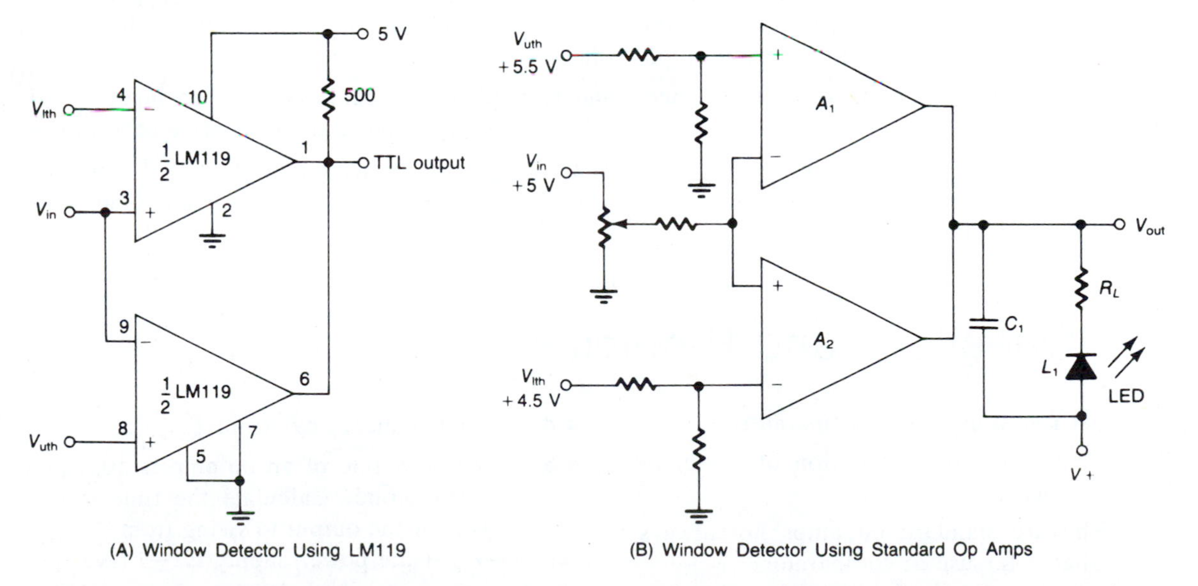

(A) Window Detector Using LM119

(B) Window Detector Using Standard Op Amps

**Figure 4–15.** Window Detectors (Part A: Courtesy of National Semiconductor Corporation)

Another example of the window detector is shown in Figure 4–15B, where standard op amps are used. This circuit monitors an input voltage of +5 V, ±10%, and lights indicator light-emitting diode (LED) $L_1$ when the input voltage exceeds these limits. $V_{uth}$ is +5.5 V and $V_{lth}$ is +4.5 V. If $V_{in}$ exceeds $V_{uth}$, the output of $A_1$ swings negative, forward biasing the LED, and $L_1$ lights. If $V_{in}$ falls below $V_{lth}$, the output of $A_2$ swings negative, forward biasing the LED, and $L_1$ lights. The indicator lamp could just as well be a correction voltage or a switch to turn OFF a system when limits are exceeded.

## 4.8 Summary

☐ A comparator is used for comparing two voltages at the input, not for amplifying signals. A comparator can be constructed with an op amp used in the open-loop mode. Devices designed specifically to operate as comparators have different characteristics than standard op amps. Comparators are used in applications such as interfacing with digital devices. In such applications, the comparator must provide a stable drive voltage, normally +5 V. Comparators are sometimes called squaring circuits because of the rapid output voltage swings between $+V_{sat}$ and $-V_{sat}$ that square off the peaks. Comparators are used to sense voltage variations, that is, to determine if a voltage is larger or smaller than a reference voltage. Level detectors and zero-crossing detectors are two applications of the comparator.

☐ The Schmitt trigger is a variation of a standard comparator. It has positive feedback that prevents swings in output voltages caused by noise at the input. Hysteresis voltage in a Schmitt trigger is the difference between the upper threshold voltage and the lower threshold voltage. It is this hysteresis voltage that causes the snap action of the output voltage.

☐ A peak detector detects and remembers the peak value of an input signal. Peak detectors can be constructed to detect either positive or negative peak values.

☐ A window detector is a circuit designed to indicate predetermined upper and lower limits of an input signal. Another name for such a circuit is a double-ended limit detector.

## 4.9 Questions and Problems

**4.1** List five uses of the comparator.

**4.2** What is the basic function of the comparator?

**4.3** Why are standard op amps not always suitable for use as comparators?

**4.4** What is meant by strobing?

**4.5** The slew rate of an op amp is 50 V per microsecond. Calculate the time it will take for the output to swing from $+V_{sat}$ to $-V_{sat}$ if the power supply is ±15 V.

**4.6** A comparator with a sine wave input signal produces what output waveform?

**4.7** What possible problem is created in the output of a comparator by noise on either of the input terminals?

**4.8** What method is used to eliminate the problem referred to in Problem 7?

**4.9** Define upper and lower threshold voltage.

**4.10** Define hysteresis voltage and explain its purpose.

**4.11** Refer to Figure 4–10. Assume that $R_1$ = 10 kΩ, $R_2$ = 100 kΩ, $V_{ref}$ = 2 V, and $\pm V_{sat} = \pm 5$ V. Calculate $V_{uth}$ and $V_{lth}$.

**4.12** Are the upper and lower threshold voltages calculated in Problem 11 symmetrical? Explain.

**4.13** Calculate $V_H$ for Problem 11.

**4.14** A peak detector has a 10 μF capacitor and a 20 MΩ load impedance. What is the peak value-holding time?

**4.15** Assume a need to have a peak detector hold a peak value for five minutes. Assume a load impedance of 12 MΩ. Calculate the capacitor value needed.

**4.16** A circuit constructed with standard op amps must light a green indicator lamp when the input voltage exceeds +5 V and light a red indicator lamp when the input voltage drops to +4.5 V. What is such a circuit called? (HINT: It has two names.)

# Chapter 5

# Regulator Circuits

## 5.1 Introduction

Several problems can arise in an operating LIC when variations occur in the supply voltage of a circuit. For example, a change in supply voltage may shift the operating point of an audio amplifier into the nonlinear region, resulting in a distorted output, or it could cause a change in the offset voltage, which would produce an undesirable signal at the output. Such variations in supply voltage are a result of poor regulation.

Current regulation circuits provide a constant current to a load. The current is independent of voltage changes across the load, so such voltage variations will not affect the operation of the circuit.

In this chapter we will discuss methods, LIC devices, and circuits that can be used for voltage and current regulators.

## 5.2 Objectives

When you complete this chapter, you should be able to:

- ☐ Define regulator terms.
- ☐ Explain basic voltage and current regulator concepts.

- ☐ Compare series, shunt, and switching regulators.
- ☐ Explain how op amps are used as comparators in regulator circuits.
- ☐ Discuss types of available monolithic regulators and their advantages.
- ☐ Define and discuss hybrid regulators.
- ☐ Define and discuss regulator protection circuits.

# 5.3 Fundamentals of Regulators

For proper linear device operation to occur, supply voltage variations, whether caused by temperature, changing load conditions, or power line conditions, must be controlled. Voltage regulator circuits can reduce the amplitude of those output voltage variations to a level that will not interfere with the operation of the device. Voltage regulators can be built into the power supply design, or they can be added externally. If external regulators are added, they may be placed either at the power supply output or at the operating device power terminals (local regulation).

Voltage regulators come in many forms. Some use diodes, resistors, transistors, or op amps as active devices. A monolithic voltage regulator is in the form of an LIC, that is, one single active device that performs the regulating function normally performed by several discrete components.

## Regulation

*Line regulation,* sometimes called *static regulation,* refers to the changes in the output (as a percent of nominal or actual value) as the input ac is varied slowly from its rated minimum value to its rated maximum value (for example, from 105 $V_{RMS}$ to 125 $V_{RMS}$).

*Load regulation,* sometimes called *dynamic regulation,* refers to changes in output (as a percent of nominal or actual value) when the load conditions are suddenly changed (for example, from minimum load to full load).

*Thermal regulation* refers to changes caused by ambient variations or thermal drift.

NOTE: The combination of line (static) and load (dynamic) regulation is cumulative. Therefore, care should be taken when referring to the regulation characteristics of a power supply.

## Types of Regulator Circuits

There are three basic approaches for obtaining regulated dc voltage from an ac power source: series, shunt, and switching. Although all three types perform the same function and each uses a different approach to regulation, they all have two things in common; all require ac-power line rectification and all use feedback to regulate output voltage.

### Series Regulator

The *series regulator* gets its name from the fact that the variable control element is in series with the load. When a change in load or power supply occurs, the variable element automatically adjusts to offset those changes, effectively maintaining a constant load voltage.

A practical series voltage regulator using discrete components is shown in Figure 5-1. The unregulated output of the power supply is the input to the regulator circuit. Transistor $Q_1$ provides rapid response to load or power supply changes. $Q_1$ is operated in its active region and carries the load current. The load current results in a power loss, a major disadvantage of the series regulator circuit. The power dissipated in the transistor can be calculated as follows:

$$P_D = V_{CE} I_C \tag{5.1}$$

where

$P_D$ = power dissipated in the transistor
$V_{CE}$ = collector-emitter voltage
$I_C$ = collector current (which is also the load current in this case)

Because of this power loss, the circuit in Figure 5-1 is called a *dissipative voltage regulator* circuit, and is inefficient.

Transistor $Q_2$ senses any change in load voltage and adjusts the voltage at the base of $Q_1$, which causes $Q_1$ to change its collector-emitter voltage $V_{CE}$, returning the load voltage to its original level. $Q_2$ is biased by the voltage divider

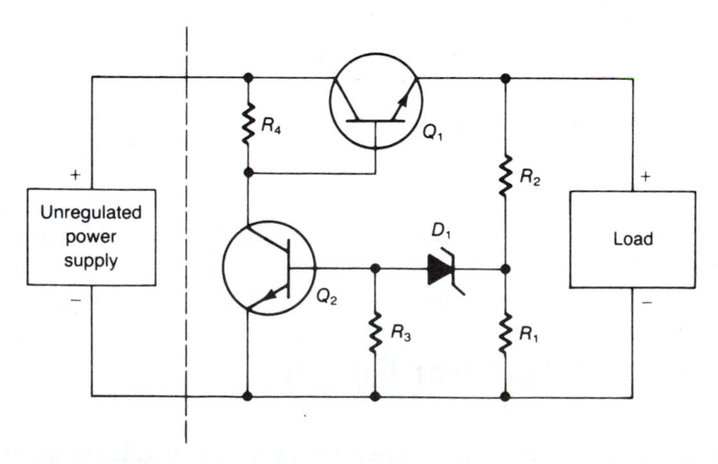


**Figure 5-1.** Series Voltage Regulator Using Discrete Components (Courtesy of RCA Corporation)

network of $R_1$ and $R_2$, the zener diode $D_1$, operated in its breakdown mode, and $R_3$. There is a constant voltage drop across $D_1$, so any change across $R_1$ also is across $R_3$. This means that the low-input impedance of $Q_2$ will not load $R_1$; therefore, $R_1$ can be a large value, which limits the bleeder current drawn from the load current. Then any change across $R_3$ changes the bias on $Q_2$.

### Shunt Regulator

In the *shunt regulator* circuit, the variable control element is in parallel with the load. Figure 5-2 shows a shunt regulator circuit using discrete components. As in the series regulator, the unregulated output of the power supply is the input to the regulator circuit. Transistor $Q_1$ is the shunt element. The amount of current through the circuit is controlled by $Q_2$, which acts as a variable resistor and controls the amount of base current at $Q_1$. The voltage divider network of $R_1$ and zener diode $D_1$ provides the biasing for $Q_2$.

Regulator output voltage is determined by the bias circuit of $Q_2$, one part of which is the zener diode, $D_1$. To help maintain a constant voltage drop across

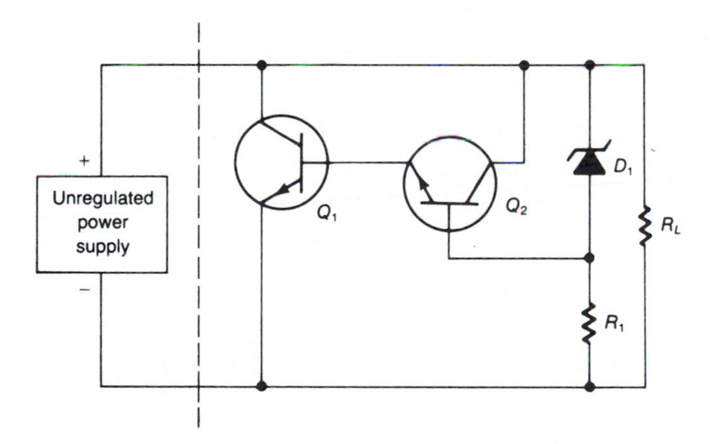


**Figure 5-2.** Shunt Voltage Regulator Using Discrete Components (Courtesy of RCA Corporation)

the diode with changes in temperature, select a zener diode with a low temperature coefficient and limit diode current flow.

The shunt regulator circuit, just as the series circuit, is inefficient because of power dissipation in the control transistor. In Figure 5-2 transistor $Q_1$ is operated in the active region and therefore dissipates power. Such power dissipation in the control transistor prevents maximum power transfer to the load, resulting in low efficiency.

### Switching Regulator

The *switching regulator* overcomes the power dissipation disadvantage by switching the control transistor between the cutoff mode and the saturation mode.

Recall that at cutoff, the transistor has only a small leakage current flow but high voltage. In the saturation mode, current is high but voltage is very low. In either mode, then, the resultant product of voltage and current is relatively low. This results in low power dissipation in the control transistor and provides high circuit efficiency.

A typical switching regulator circuit using discrete components is shown in Figure 5-3. Again, the unregulated output voltage of the power supply is the input to the regulator circuit. Applying a control signal to the base of the transistor, as shown, will cause it to alternately switch between cutoff and saturation. This alternation allows the input voltage to be periodically applied to the filter circuit, where it is measurable across the diode. The control signal should be high enough in frequency to allow low values to be used for the inductor and capacitor in the filter circuit. Low values of inductance and capacitance result in reduced component size and weight, always a consideration in electronic circuit design, but is especially important in power supplies.

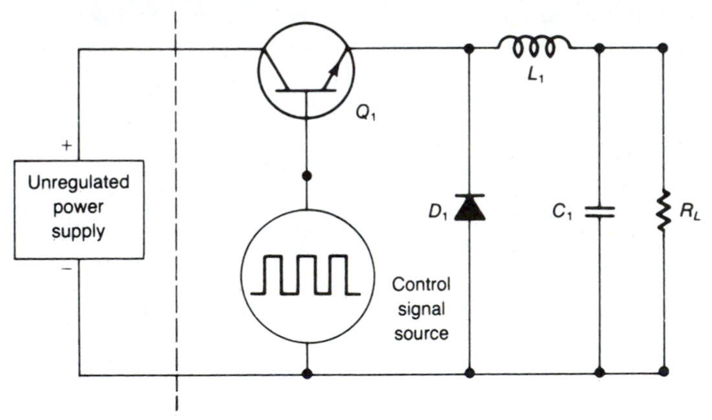


**Figure 5-3.** Switching Voltage Regulator Using Discrete Components

## Duty Cycle

The dc load voltage is determined by the ratio of the amount of time the transistor is in the saturated mode ($t_{on}$) to the period of the control signal ($T$). This ratio is called the *duty cycle* and is expressed as a percentage. As an equation,

$$\text{Percent duty cycle} = \frac{t_{on}}{T} \times 100\% \tag{5.2}$$

The average voltage across the diode is the value of the dc component of the output of the circuit ($V_{dc}$) and is directly related to the duty cycle. This relationship may be shown as follows:

$$V_{dc} = \frac{V_D(\%\text{ duty cycle})}{100\%} \tag{5.3}$$

where

$V_D$ = maximum voltage across the diode

## Example 5.1

Assume the control signal frequency in Figure 5–3 is 30 kHz and the transistor ON time is 8 μs. Calculate the duty cycle.

*Solution*

First, recall that the period $T = 1/f$ and solve for $T$:

$$T = \frac{1}{f}$$

where $T$ is in seconds and $f$ is in hertz (Hz)

$$T = \frac{1}{30 \times 10^3 \text{ Hz}} = 33.3 \times 10^{-6} \text{ s} = 33.3 \ \mu\text{s}$$

Then solve as follows using Equation 5.2:

$$\% \text{ duty cycle} = \frac{t_{on}}{T} \times 100\% = \frac{8 \ \mu\text{s}}{33.3 \ \mu\text{s}} \times 100\%$$

$$= \frac{8 \times 10^{-6} \text{ s}}{33.3 \times 10^{-6} \text{ s}} \times 100\% = 24\%$$

## Example 5.2

Assume the maximum voltage across the diode in Figure 5–3 is 50 V. Using the duty cycle calculated in Example 5.1, calculate the dc component of the output voltage of the circuit.

*Solution*

$$V_{dc} = \frac{V_D(\% \text{ duty cycle})}{100\%} = \frac{50 \text{ V} \times 24\%}{100\%} = 12 \text{ V}$$

It should now be obvious that if we wish to change the output voltage, the duty cycle must change. The switching regulator accomplishes this by using a sensing device to sample the output voltage. The value of the output voltage is then used to vary the duty cycle of the control signal. In Section 5.5, we will consider how and why an op amp is used for the sensing device.

# 5.4 Heat Sinking

To fully utilize the various available regulator packages, sufficient attention must be paid to proper heat removal. For efficient thermal management, you must rely on important parameters supplied by the manufacturer. The device temperature depends on the power dissipation level, the means for removing the heat generated by this power dissipation, and the temperature of the body (heat sink) to which this heat is removed.

Remember that any semiconductor device that carries current will dissipate power, generally in the form of heat. This heat is generated at the junction of the device and must flow through the package to ambient (surrounding) air. All packaging material presents some opposition to this flow of heat, designated on data sheets as *thermal resistance* ($\theta_{JA}$). This opposition causes the temperature of the junction ($T_J$) to rise. If the junction temperature becomes too high, the device will be damaged, or thermal shutdown will occur.

RESISTANCES LISTED AS FOLLOWS $\theta_{JC(TYP)}$ $\theta_{JC(MAX)}$ / $\theta_{JA(TYP)}$ $\theta_{JA(MAX)}$ in C/W

| REG TYPE | DEVICE NO./SERIES | $I_{OUT}$ (A) | TO-3 K | 4-LEAD TO-3 K | TO-220 U | POWER WATT U1* | TO-39 H | TO-92 W | TO-99 8-LEAD TO-5 H | TO-100 10-LEAD TO-5 H | TO-116 14-PIN PLASTIC D | TO-116 14-PIN CERAMIC D | 8 PIN MINIDIP T |
|---|---|---|---|---|---|---|---|---|---|---|---|---|---|
| POS. 3-TERM. | μA78LXX | 0 1 | | | | | 20 40<br>140 190 | 160 180 | | | | | |
| | μA78MXX | 0 5 | | | 3 0 5 0<br>62 70 | | 18 25<br>120 185 | | | | | | |
| | μA78CXX | 0 5 | | | | 6 8<br>75 80 | | | | | | | |
| | μA109, μA209, μA309, 5 V | 1 | 3 5 5 5<br>40 45 | | | | | | | | | | |
| | μA78XX | 1 | 3 5 5 5<br>40 45 | | 3 0 5 0<br>60 65 | | | | | | | | |
| | μA78CB | 2 | 3 5 5 5<br>40 45 | | 3 0 5 0<br>60 65 | | | | | | | | |
| | 78H05, 5V | 5 | 1 5 2 0<br>37 40 | | | | | | | | | | |
| | μA78HXX | 5 | 2 0 2 5<br>32 . 38 | | | | | | | | | | |
| NEG. 3-TERM. | μA79MXX | 0 5 | | | 3 0 5 0<br>62 70 | | 18 25<br>120 185 | | | | | | |
| | μA79XX | 1 | 3 5 5 5<br>40 45 | | 3 0 5 0<br>60 65 | | | | | | | | |
| POS. ADJ. | μA105/305/376 | 0 012 to 0 045 | | | | | | | 25 40<br>150 190 | | | | 160 190 |
| | μA723 | 0 125 | | | | | | | | 25 50<br>150 190 | 150 190 | 125 160 | |
| | μA78MG 4-TERM. | 0 5 | | | | 6 8<br>75 80 | | | | | | | |
| | μA78G 4-TERM. | 1 | | 4 0 6 0<br>44 47 | | 6 8<br>75 80 | | | | | | | |
| | μA78HG | 5 | 2 0 2 5<br>32 38 | | | | | | | | | | |
| NEG. ADJ. | μA104/304 | 0 020 | | | | | | | | 25 50<br>150 190 | | | |
| | μA79MG 4-TERM. | 0 5 | | | | 6 8<br>75 80 | | | | | | | |
| | μA79G 4-TERM. | 1 | | 4 0 6 0<br>44 47 | | 6 8<br>75 80 | | | | | | | |

**Figure 5-4.** Thermal Resistance ($\theta_{JC}$, $\theta_{JA}$) by Device and Package (Courtesy of Fairchild Camera & Instrument Corporation, Linear Division)

A tabulation by package of various regulators is shown in Figure 5–4. The figure also lists the average and maximum values of thermal resistance for the regulator chip-package combinations and can be used as a guide for selecting a suitable package when designing a regulator circuit.

## Determining Heat Sinking Requirements

Thermal characteristics of voltage regulator chips and packages determine that some form of heat sinking is mandatory whenever the power dissipation exceeds the following values at 25°C ambient temperature:

0.67 W for the TO-39 package
0.69 W for the TO-92 package
1.56 W for the Power Watt (similar to TO-202 packages)
1.8 W for the TO-220 package
2.8 W for the TO-3 package

At ambient temperatures above 25°C, heat sinking becomes necessary at lower power levels.

To choose a heat sink, you must determine the following regulator parameters:

$P_{D(\max)}$—Maximum power dissipation:

$$P_{D(\max)} = (V_{\text{in}} - V_{\text{out}})\, I_{\text{out}} + V_{\text{in}} I_Q$$

where

$V_{\text{in}}$ = input voltage
$V_{\text{out}}$ = output voltage
$I_{\text{out}}$ = output current
$I_Q$ = quiescent current

$T_{A(\max)}$—Maximum ambient temperature the regulator will encounter during operation

$T_{J(\max)}$—Maximum operating junction temperature, specified by the manufacturer

$\theta_{JC}$, $\theta_{JA}$—Junction-to-case and junction-to-ambient thermal resistance values, specified by the manufacturer

$\theta_{CS}$—Case-to-heat-sink thermal resistance

$\theta_{SA}$—Heat-sink-to-ambient thermal resistance, specified by the heat sink manufacturer

The maximum permissible dissipation without a heat sink is:

$$P_{D(\max)} = \frac{T_{J(\max)} - T_{A(\max)}}{\theta_{JA}} \tag{5.4}$$

If the device dissipation $P_D$ exceeds this amount, a heat sink is required. The total required thermal resistance is then calculated as follows:

$$\theta_{JA(tot)} = \theta_{JC} + \theta_{CS} + \theta_{SA} = \frac{T_{J(max)} - T_{A(max)}}{P_D} - \theta_{JC} \tag{5.5}$$

Case-to-sink and sink-to-ambient thermal resistance information on commercially available heat sinks is normally provided by the heat sink manufacturer. A summary of some commercially available heat sinks is given in Table 5–1.

## Example 5.3

Determine the heat sink required for a μA7806C regulator to be used in a circuit that has the following system requirements:

Operating ambient temperature range: 0°C–60°C ($T_A$)
Maximum junction temperature: 125°C ($T_J$)
Maximum output current: 800 mA ($I_{out}$)
Maximum input to output differential: 10 V

*Solution*

A review of the data sheet for the μA7800 series (Appendix C) shows that the device is available in either TO-39 or TO-220 packages. The TO-220 package has a lower cost and better thermal resistance, therefore, it should be selected for this regulator requirement.

$\theta_{JC}$ = 5 °C/W maximum (from data sheet or Figure 5–4, μA78xx)

$$\theta_{JA(tot)} = \theta_{JC} + \theta_{CS} + \theta_{SA} = \frac{T_J - T_A}{P_D} - \theta_{JC}$$

$$\theta_{CS} + \theta_{SA} = \frac{125°C - 60°C}{0.8\ A \times 10\ V} - 5°C = 3.13°C/W$$

Assuming $\theta_{CS}$ = 0.13 °C/W, then $\theta_{SA}$ = 3 °C/W. This thermal resistance value can be achieved with any of the TO-220 package heat sinks selected from Table 5–1.

## Tips for Heat Sinking

Here are some tips for better regulator heat sinking:

**1.** Avoid placing heat-dissipating components such as power resistors next to regulators.

**2.** When using low dissipation packages, such as TO-5, TO-39, and TO-92, keep lead lengths to a minimum and use the largest possible area of the printed

**Table 5-1**

## Heat Sink Selection Guide

This list is only representative. No attempt has been made to provide a complete list of all heat sink manufacturers. All values are typical as given by manufacturer or as determined from characteristic curves supplied by manufacturer.

| $\theta_{SA}$ Approx. (°C/W) | Manufacturer and Type |
|---|---|
| **TO-3 Packages** | |
| 0.4 (9" length) | Thermalloy (Extruded) 6590 Series |
| 0.4–0.5 (6" length) | Thermalloy (Extruded) 6660, 6560 Series |
| 0.56–3.0 | Wakefield 400 Series |
| 0.6 (7.5" length) | Thermalloy (Extruded) 6470 Series |
| 0.7–1.2 (5–5.5" length) | Thermalloy (Extruded) 6423, 6443, 6441, 6450 Series |
| 1.0–5.4 (3" length) | Thermalloy (Extruded) 6427, 6500, 6123, 6401, 6403, 6421, 6463, 6176, 6129, 6141, 6169, 6135, 6442 Series |
| 1.9 | IERC E2 Series (Extruded) |
| 2.1 | IERC E1, E3 Series (Extruded) |
| 2.3–4.7 | Wakefield 600 Series |
| 4.2 | IERC HP3 Series |
| 4.5 | Staver V3-5-2 |
| 4.8–7.5 | Thermalloy 6001 Series |
| 5–6 | IERC HP3 Series |
| 5–10 | Thermalloy 6013 Series |
| 5.6 | Staver V3-3-2 |
| 5.9–10 | Wakefield 680 Series |
| 6 | Wakefield 390 Series |
| 6.4 | Staver V3-7-224 |
| 6.5–7.5 | IERC UP Series |
| 8 | Staver V1-5 |
| 8.1 | Staver V3-5 |
| 8.8 | Staver V3-7-96 |
| 9.5 | Staver V3-3 |
| 9.5–10.5 | IERC LA Series |
| 9.8–13.9 | Wakefield 630 Series |
| 10 | Staver V1-3 |
| 11 | Thermalloy 6103, 6117 Series |
| **TO-220 Packages** (See Note 1) | |
| 4.2 | IERC HP3 Series |
| 5–6 | IERC HP1 Series |
| 6.4 | Staver V3-7-225 |
| 6.5–7.5 | IERC VP Series |
| 7.1 | Thermalloy 6070 Series |
| 8.1 | Staver V3-5 |
| 8.8 | Staver V3-7-96 |
| 9.5 | Staver V3-3 |
| 10 | Thermalloy 6032, 6034 Series |
| 12.5–14.2 | Staver V4-3-192 |
| 13 | Staver V5-1 |
| 15 | Thermalloy 6030 Series |
| 15.1–17.2 | Staver V4-3-128 |
| 16 | Thermalloy 6072, 6106 Series |
| 18 | Thermalloy 6038, 6107 Series |
| 19 | IERC PB Series |
| 20 | Staver V6-2 |
| 20 | Thermalloy 6025 Series |
| 25 | IERC PA Series |

1. Most TO-3 heat sinks can also be used with TO-220 packages with appropriate hole patterns.
2. Most TO-220 heat sinks can be used with the Power Watt package.

| $\theta_{SA}$ Approx. (°C/W) | Manufacturer and Type |
|---|---|
| **TO-92 Packages** | |
| 30 | Staver F2-7 |
| 46 | Staver F5-7A, F5-8-1 |
| 50 | IERC RUR Series |
| 57 | Staver F5-7D |
| 65 | IERC RU Series |
| 72 | Staver F1-7 |
| 85 | Thermalloy 2224 Series |
| **Mini Batwing** | |
| 10 | Thermalloy 6069 Series |
| 10.6 | Thermalloy 6068 Series |
| 11.7 | Thermalloy 6067 Series |
| 13 | Thermalloy 6066 Series |
| 20 | Thermalloy 6062 Series |
| 26 | Thermalloy 6064 Series |
| **TO-5 and TO-39 Packages** | |
| 12 | Thermalloy 1101, 1103 Series |
| 12–16 | Wakefield 260-5 Series |
| 15 | Staver V3A-5 |
| 22 | Thermalloy 1116, 1121, 1123 Series |
| 22 | Thermalloy 1130, 1131, 1132 Series |
| 24 | Staver F5-5C |
| 25 | Thermalloy 2227 Series |
| 26–30 | IERC Thermal Links |
| 27–83 | Wakefield 200 Series |
| 28 | Staver F5-5B |
| 34 | Thermalloy 2228 Series |
| 35 | IERC Clip Mount Thermal Link |
| 39 | Thermalloy 2215 Series |
| 41 | Thermalloy 2205 Series |
| 42 | Staver F5-5A |
| 42–65 | Wakefield 296 Series |
| 46 | Staver F6-5, F6-5L |
| 50 | Thermalloy 2225 Series |
| 50–55 | IERC Fan Tops |
| 53 | Thermalloy 2211 Series |
| 55 | Thermalloy 2210 Series |
| 56 | Thermalloy 1129 Series |
| 58 | Thermalloy 2230, 2235 Series |
| 60 | Thermalloy 2226 Series |
| 68 | Staver F1-5 |
| 72 | Thermalloy 1115 Series |
| **Power Watt (similar to TO-202) Packages (See Note 2)** | |
| 12.5–14.2 | Staver V4-3-192 |
| 13 | Thermalloy 6063 Series |
| 13 | Staver V5-1 |
| 15.1–17.2 | Staver V4-3-128 |
| 19 | Thermalloy 6106 Series |
| 20 | Staver V6-2 |
| 24 | Thermalloy 6047 Series |
| 25 | Thermalloy 6107 Series |
| 37 | IERC PA1-7CB with PVC-1B Clip |
| 40–42 | Staver F7-3 |
| 40–43 | Staver F7-2 |
| 42 | IERC PA2-7CB with PVC-1B Clip |
| 42–44 | Staver F7-1 |

IERC: 135 W. Magnolia Blvd., Burbank, CA 91502
Staver Co. Inc.: 41-51 N. Saxon Ave., Bay Shore, N.Y. 11706
Thermalloy Inc.: 2021 W. Valley View Lane, Dallas, TX 75234
Wakefield Engineering, Inc.: Audubon Rd., Wakefield, MA 01880

Source: Courtesy of Fairchild Camera & Instrument Corporation, Linear Division.

circuit board (PCB) traces or mounting hardware to provide a heat dissipation path for the regulator.

**3.** When using larger packages, be sure the heat sink surface is flat and free from ridges and high spots. Check the regulator package for burrs or peened-over corners. Regardless of the smoothness and flatness of the contact between the package and heat sink, air pockets between them are unavoidable unless a lubricant is used. Therefore, for good thermal conduction, use a thin layer of thermal lubricant.

**4.** In some applications, especially with negative regulators, it is desirable to insulate electrically the regulator case from the heat sink. Hardware kits for this purpose are commercially available for such packages as the TO-3 and TO-220. The kits generally consist of a thin piece of mica or bonded fiberglass that electrically isolates the two surfaces, yet provides a thermal path between them. The thermal resistance will increase, but some improvement can be realized by using thermal lubricant on each side of the insulator material.

**5.** If the regulator is mounted on a heat sink with fins, the most efficient heat transfer takes place when the fins are in a vertical plane, because this type of mounting forces the heat transfer from fin to air in a combination of radiation and convection.

**6.** If it is necessary to bend any of the regulator leads, handle them carefully to avoid straining the package. Furthermore, lead bending should be restricted since bending will fatigue and eventually break the leads.

## 5.5 Voltage Regulators Using Op Amps

In Section 5.3 we learned that a voltage regulator circuit uses a transistor as a sensing device to detect any change in the output voltage and to drive a second transistor that, in turn, controls the voltage or current of the load. The types of regulator circuits discussed in Section 5.3 offer some good regulation features, but they also have some problems. One problem is the low sensitivity caused by the low gain of the sensing transistor. This low sensitivity prevents slight changes in the load voltage from providing large enough changes in the output of the sensing transistor to affect the operation of the control transistor. Another problem is the slow response time of the sensing transistor. Another is the loading effect that may occur in the sensing divider circuit because of the low input impedance of the sensing transistor. These are problems that cannot be tolerated in circuits that require a high degree of regulation.

Replacing the sensing transistor with an op amp solves these problems for us. The op amp has the high gain necessary to provide a high level of sensitivity and a rapid response to changes in the load voltage. The high input impedance of the op amp prevents loading down devices that are connected to it. The op amp quickly senses slight changes in load voltage, then drives the control transistor that quickly returns the load voltage to its desired value.

## Op Amp Series Regulator

Figure 5-5 shows a series voltage regulator with an op amp as the sensing element. The output of the op amp drives the base of the control transistor, $Q_1$. The op amp is connected as a *voltage level detector.* The amplitude and polarity of the output voltage of the op amp are determined by the more positive of the two input voltages. Recall from Chapter 2 that

$$V_{out} = A_v(V_1 - V_2) \qquad \text{(repeat of Equation 2.5)}$$

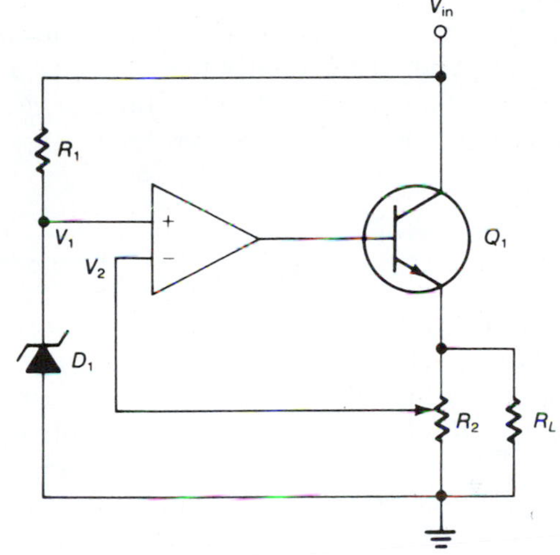


**Figure 5-5.** Series Voltage Regulator with an Op Amp as the Sensing Element

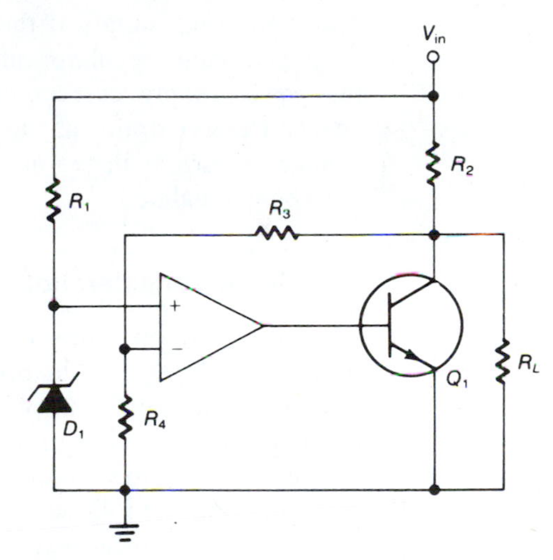


**Figure 5-6.** Shunt Voltage Regulator with an Op Amp as the Sensing Element

Input signal $V_1$ is the reference voltage across the zener diode. Input signal $V_2$ is that portion of the load voltage developed across the potentiometer, $R_2$. Any change in load voltage will cause a change in the differential voltage, which will cause a change in the output of the detector. Emitter-follower action of $Q_1$ then causes the load voltage to readjust to its desired value.

The resistance of potentiometer $R_2$ should be only of high enough value compared with $R_L$, to limit the amount of current and not load down the output. Resistor $R_1$ limits the current through the zener diode, reducing the temperature rise of the diode and thereby providing increased stability in the circuit. Reducing the gain of the amplifier will also increase the stability of the circuit.

## Op Amp Shunt Regulator

A shunt regulator using an op amp as the sensing element is shown in Figure 5-6. The differential voltage between the reference voltage of the zener diode and that portion of output voltage across $R_4$ determines the output of the comparator, which drives the base of the control transistor, $Q_1$. The current amplification process of the transistor causes the collector current to change, causing the transistor to act as a variable resistance. The current in $Q_1$ is thus automatically adjusted inversely to compensate for any load current change. The effect is to keep the supply current constant, which in turn keeps the load voltage constant.

The divider network formed by resistors $R_3$ and $R_4$ senses load voltage changes. Current-limiting resistor $R_1$ helps maintain a steady temperature in the zener diode. Resistor $R_2$ offers circuit protection by limiting the maximum current drawn from the supply if the load is shorted.

The shunt regulator offers an advantage over the series regulator because the power supply current output is kept constant. This means that there is a constant power drain on the supply and the ac lines feeding it. With little or no change in current in the ac source lines feeding the supply, the lines themselves are better regulated.

### Some Considerations

Although op amps do eliminate certain problems, as discussed earlier, there are other possible problems that can arise. To prevent or reduce these new problems, some simple factors must be considered. An unstable circuit can result from the high gain of the op amp and the feedback through the control transistor. Choosing the proper transistor characteristics overcomes this problem. Also, the emitter-follower circuit of the series regulator is susceptible to oscillations. Reducing the lead length reduces the inductance and removes possible unwanted feedback paths.

Another consideration is the stability of the zener reference voltage. Changes in both load and temperature affect zener voltage. To overcome the problem of temperature changes, place a forward-biased zener in series with the reverse-biased reference zener. This works because reverse-biased zeners operating in the breakdown mode have positive temperature coefficients, while forward-biased zeners have negative coefficients. Therefore, temperature compensation occurs, providing a stable reference.

Another important consideration is the response time of the circuit, which determines how rapidly the regulator can adjust the output back to its original value. The slew rate of the op amp affects the response time, so selecting an op amp with a high slew rate will offer the best response time, resulting in best regulation.

## Op Amp Switching Regulator

Figure 5–7 shows an op amp comparator that is used in a switching regulator. The comparator replaces the control signal source shown in Figure 5–3. The value of load voltage compared to the zener reference voltage determines the

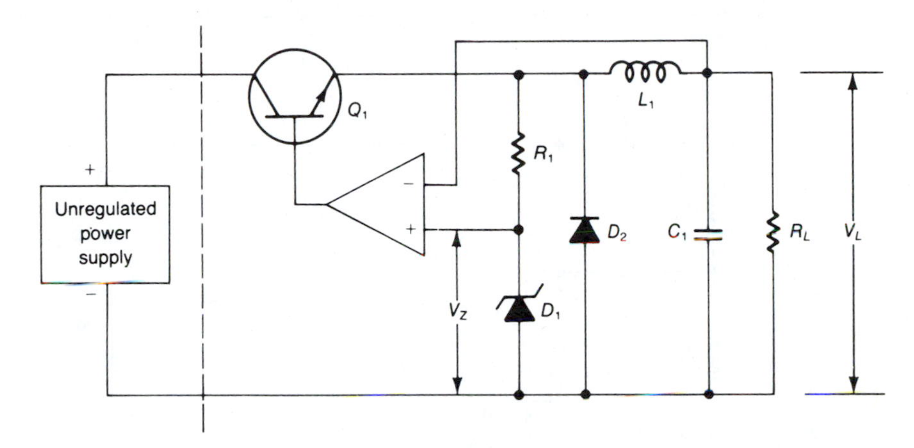


**Figure 5-7.** Switching Voltage Regulator with Op Amp Comparator (Courtesy of John Wiley & Sons, Inc.)

output polarity of the comparator and, therefore, the duty cycle of the transistor. Zener diode $D_1$ provides a reference voltage ($V_Z$) to the noninverting input, and the load voltage ($V_L$) supplies an input to the inverting terminal. When $V_L$ increases to a value greater than $V_Z$, the output of the comparator swings negative and transistor $Q_1$ is cut off. This results in a decrease in current through coil $L_1$ and a decrease in voltage across capacitor $C_1$ discharging through forward-biased $D_2$. With $C_1$ in parallel with the load, $V_L$ is also reduced, thereby holding the load current constant.

If $V_L$ decreases to a value less than $V_Z$, the output polarity of the comparator swings positive, driving $Q_1$ into saturation. This increases current through $L_1$, charging $C_1$ to a higher voltage, and thus increasing $V_L$, again stabilizing the load current. Diode $D_2$ is reverse biased at this time to allow charging of $C_1$.

## Dual-Voltage Regulator

Since op amps require two supply voltages of equal value but opposite polarity, it is necessary to regulate both negative and positive outputs of the supply. A dual-voltage regulator using op amps is illustrated in Figure 5-8.

The two op amps, $A_1$ and $A_2$, drive complementary transistors $Q_1$ and $Q_2$. A reference voltage is supplied by zener diode $D_1$ and fed into the (+) inputs of both op amps from the junction of the divider network of $R_3$ and $R_4$. Any change in the regulated output is fed back through feedback resistor $R_f$ to the (−) input of the respective op amp.

The regulated negative voltage, $-V_2$, is developed across capacitor $C_1$, and the regulated positive voltage, $+V_2$, is developed across capacitor $C_2$.

The zener diode also applies equal and opposite voltages to the (−) inputs of the op amps, and the op amp outputs in turn place equal and opposite voltages

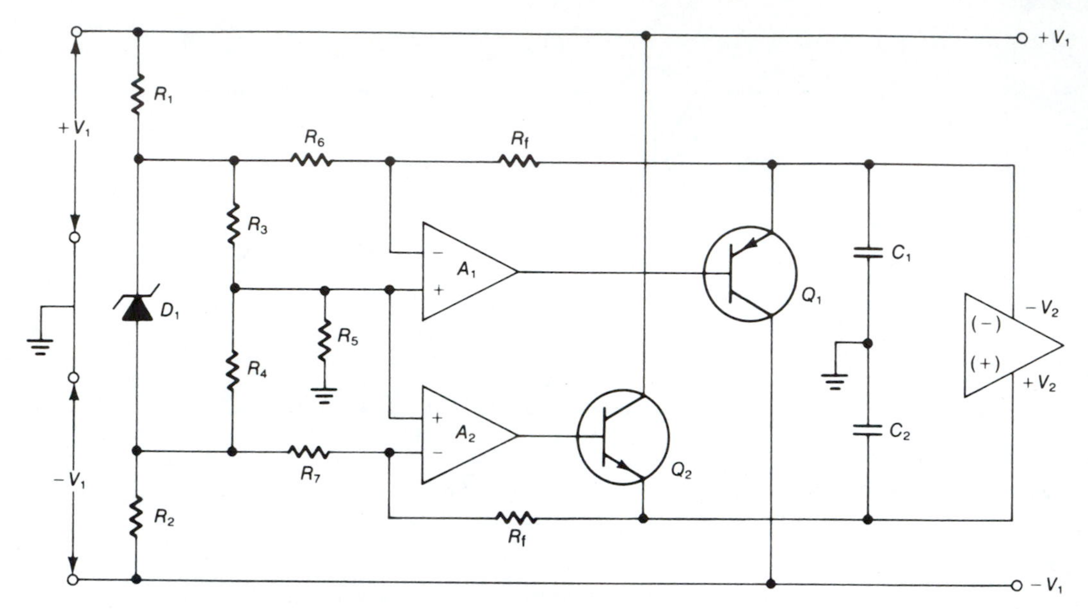


**Figure 5-8.** Dual-Voltage Regulator Constructed with Op Amps (Courtesy of John Wiley & Sons, Inc.)

at the bases of the transistors. The feedback resistors determine the value of these voltages, and these voltages determine the emitter voltages and, therefore, the load voltage.

## Monolithic Regulators

By incorporating all the necessary components on a single chip by the monolithic construction method, voltage regulators can be greatly reduced in size. This smaller regulator, in the form of a single IC, can then be used for *local regulation;* that is, it can be located physically at any point in the system where it is needed. This local on-card regulation capability eliminates the requirement to regulate the entire power supply, thereby reducing greatly the cost of regulation.

Most monolithic regulators require that some external components be added to provide overload protection, to channel desired currents and voltages to selected loads, and possibly to convert to a switching regulator.

There are three main categories of monolithic regulators:

1. Three-terminal devices with fixed positive or negative outputs
2. Adjustable regulators with positive or negative outputs
3. Dual-polarity tracking regulators

Within each category there are a great many regulators available with a variety of parameters.

### Fixed-Voltage Regulators

*Fixed-voltage regulators* are excellent devices for use at points in a system that requires a single, fixed, well-regulated voltage. The devices are small, require no external components, and have only three terminals. These characteristics permit these devices to be located directly on the PCB that requires the specified regulated voltage. Fixed voltage regulators are ideal for use in limited space.

The LM340 series are typical three-terminal fixed positive-voltage regulators. The 340 series offer output voltages of 5, 6, 8, 10, 12, 15, 18, and 24 volts. A built-in reference voltage drives the (+) terminal of an amplifier. An internal voltage divider, preset to one of the output voltages listed earlier, provides feedback voltage to the (–) terminal of the amplifier. The chip also has a control, *pass,* transistor capable of handling more than 1.5 A of load current if properly heat sinked, and includes *current limiting* and *thermal shutdown.* Figure 5–9 shows the block diagram of the LM340 series.

### Thermal Shutdown

When the internal temperature of the device reaches 175°C, the regulator automatically turns off and prevents any further increase in internal temperature. This thermal shutdown is a precaution against excessive power dissipation in the device.

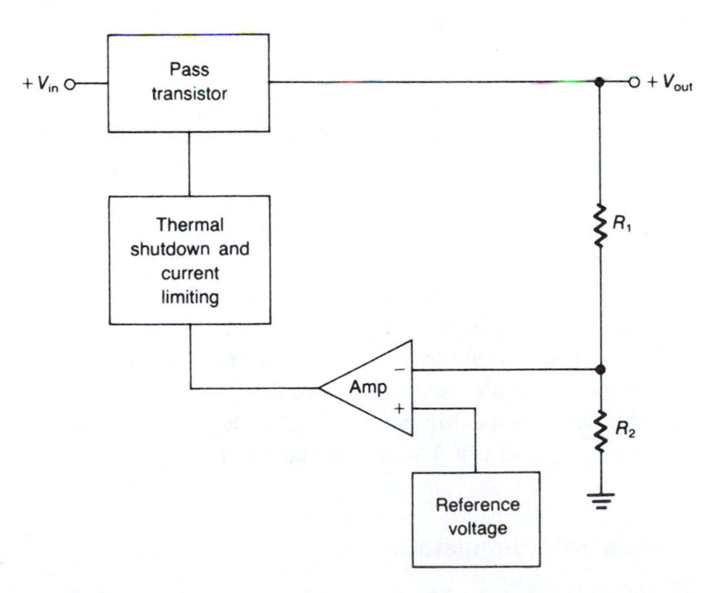


**Figure 5–9.** Block Diagram of a Typical Three-Terminal Fixed Positive Voltage Regulator (Courtesy of National Semiconductor Corporation)

A fixed-voltage regulator in its simplest configuration is illustrated in Figure 5–10. The unregulated voltage of the power supply is fed to pin 1, the regulated output is taken from pin 2, and pin 3 is ground. In addition to regulating voltage, this type of device also attenuates ripple. Typical ripple rejection ranges from 55 decibels (dB) for the 340-5 to 44 dB for the 340-24.

### Bypass Capacitors

Power supply feedback may produce oscillations within the IC through lead inductance if the device is more than a few inches from the power supply filter circuit. Figure 5–11 shows a typical bypassed regulator circuit. Bypass capacitor $C_1$ provides circuit stability. Although no output bypass capacitor ($C_2$) is needed for stability, one would improve the transient response. If used, $C_2$ would typically be a 0.1 μF ceramic disc.

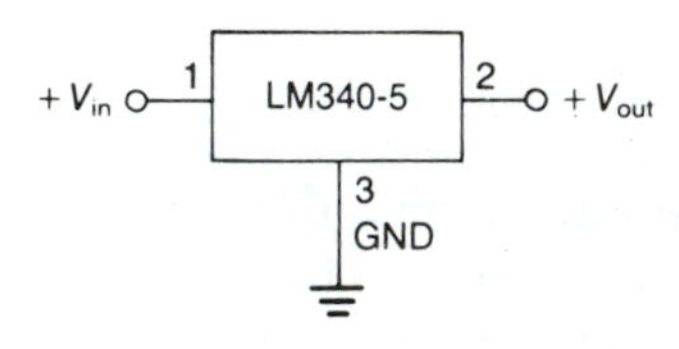


**Figure 5–10.** Simple Configuration of a Fixed Positive Voltage Regulator (Courtesy of National Semiconductor Corporation)

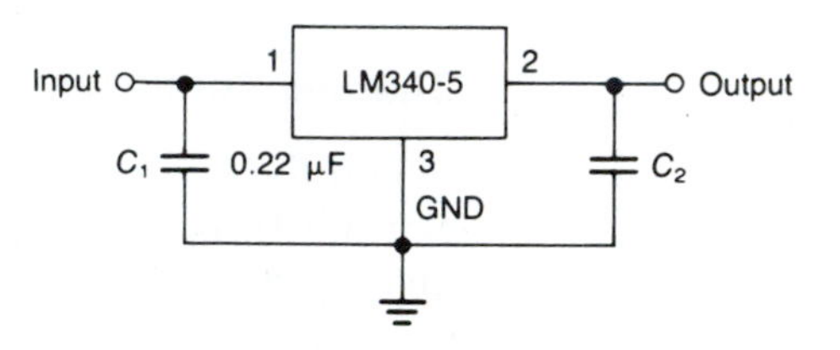


**Figure 5–11.** Typical Bypassed Regulator Circuit (Courtesy of National Semiconductor Corporation)

If the junction temperature is maintained at 25 °C, then with an input range of 7–20 V, the 340-5 will provide a regulated output of 5 V. An input range of 27–38 V will result in a regulated output of 24 V from the 340-24. Any device in the 340 series requires an input voltage of 2–3 V greater than the regulated output.

The LM320 series are three-terminal fixed negative-voltage regulators, offering output voltages ranging from –5 to –24 V. The 320 series are similar to the 340 series, including the same features of current-limiting, thermal shutdown, and good ripple rejection. Figure 5–12 illustrates the major difference between the series—the pin numbering. Note that pin 1 is ground, pin 2 is the regulated output, and pin 3 is the unregulated input.

### Adjustable Regulators

Adjustable output voltages can be obtained by adding a few external components to the fixed regulator. Figure 5–13A shows a 340-5 arranged as an

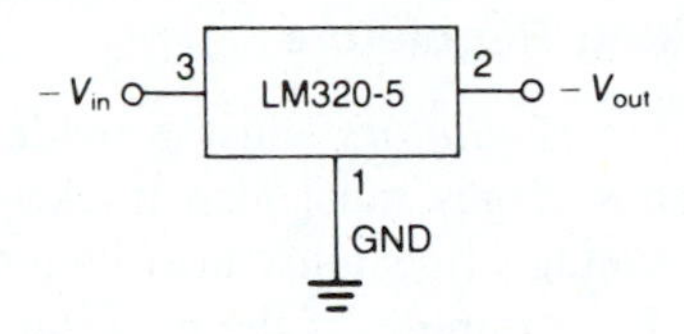


**Figure 5–12.** Simple Configuration of a Fixed Negative Voltage Regulator (Courtesy of National Semiconductor Corporation)

*adjustable positive regulator.* Note that pin 3 of the 340 is not grounded, but instead is tied to the junction of $R_1$ and $R_2$. This places the regulated output $V_{reg}$ across $R_1$. Quiescent current $I_Q$ flows through $R_2$. The output voltage is from pin 2 to ground, and

$$V_{out} = V_{reg} + \left(\frac{V_{reg}}{R_1} + I_Q\right) R_2 \tag{5.6}$$

Typical $I_Q$ of the 340-5 is 7 mA, with a maximum change $\Delta I_Q$ of 1.5 mA over line and load changes.

Figure 5–13B shows a typical *adjustable negative regulator* constructed with the LM320-5 and external components. Capacitor $C_2$ is optional, but it does improve transient response and ripple rejection. Output is taken between pin 2 and ground, and

$$V_{out} = V_{reg}\left(\frac{R_1 + R_2}{R_2}\right) \tag{5.7}$$

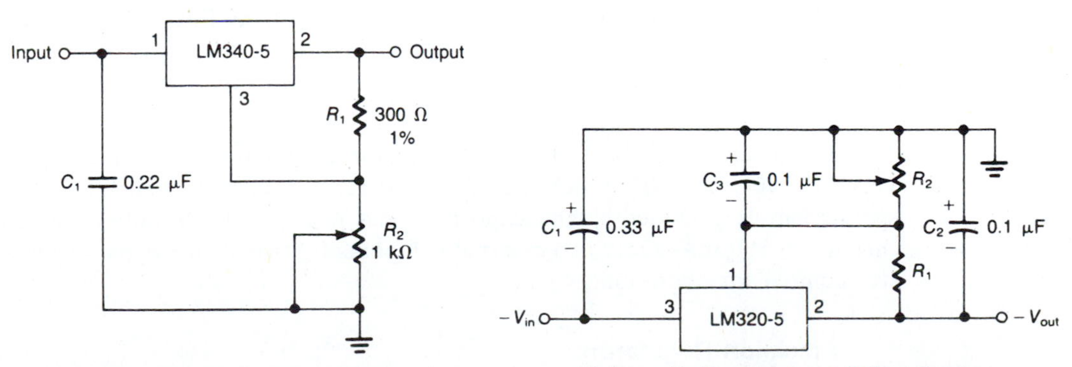


**Figure 5–13.** Three-Terminal Voltage Regulators (Courtesy of National Semiconductor Corporation)

### Dual-Polarity Tracking Regulators

*Dual-polarity tracking regulators* must provide both plus and minus regulated voltages, and these voltages must also track. That is, if the plus voltage line changes, the minus voltage line must also change in the opposite direction and by the same amount. For example, if the plus line of a dual tracking regulator suddenly decreases from +24 V to +23 V, the minus line will simultaneously increase from -24 V to -23 V. This tracking ensures that the magnitude of both of the line voltages to ground is constant.

Several methods may be used to provide tracking for dual voltages. One method is to use op amps and external components, as discussed earlier (see Figure 5-8). Another method is to use two single-voltage regulators, as demonstrated in Figure 5-14. This circuit uses a 7815 (+15 V output) and a 7915 (-15 V output), combined with the proper external circuitry, to provide dual regulated voltages of ± 15 V at 1.0 A. The use of selected single-voltage regulators in this manner can supply a wide range of voltages at high currents.

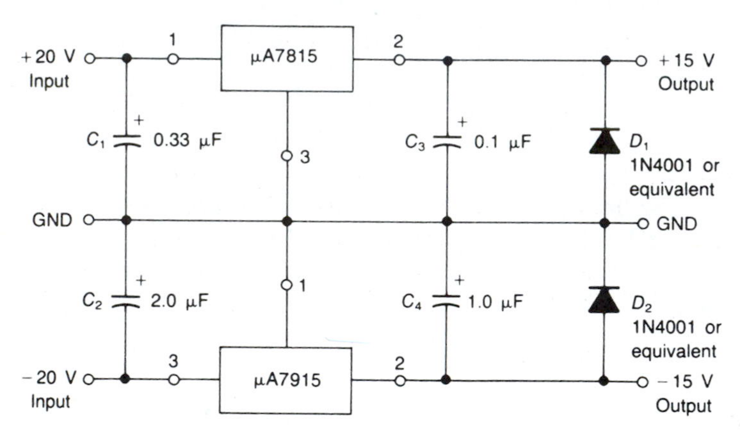


**Figure 5-14.** Dual-Supply Voltage Regulator Using Two Single-Voltage Regulators (Courtesy of Fairchild Camera & Instrument Corporation)

A typical dual-polarity tracking regulator circuit is shown in Figure 5-15. This 4195 regulator IC is designed for local regulation and intended for ease of application. It provides balanced positive and negative 15 V output voltages at currents to 100 mA. Only two external components, two 10 μF bypass capacitors, are required for operation.

### Precision Regulators

Figure 5-16 shows the schematic diagram of the μA723, a monolithic precision voltage regulator capable of operation in positive and negative power supplies as a series, shunt, or switching regulator.

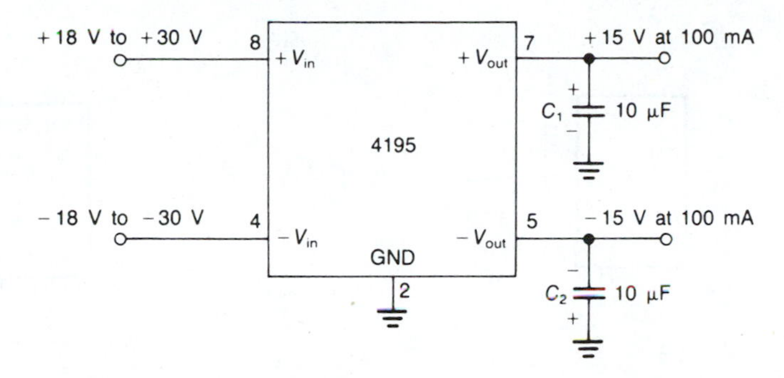


**Figure 5-15.** Dual-Polarity Tracking Regulator (Courtesy of Raytheon Company, Semiconductor Division)

The 723 provides 0.01% line and load regulation. Its output voltage is adjustable from 2 V to 37 V, and output current can range to 150 mA without the need for an external pass transistor. Five typical applications for the 723 are shown in Figure 5-17.

## Precision References

In many signal processing and conditioning applications, standard discrete reference zener diodes are not adequate to provide the precision reference needed. Therefore, monolithic ICs have been developed to overcome this problem. Figure 5-18 shows the schematic and the functional block diagram of the LM3999 *precision reference* IC. Constructed on a single monolithic chip, the circuit

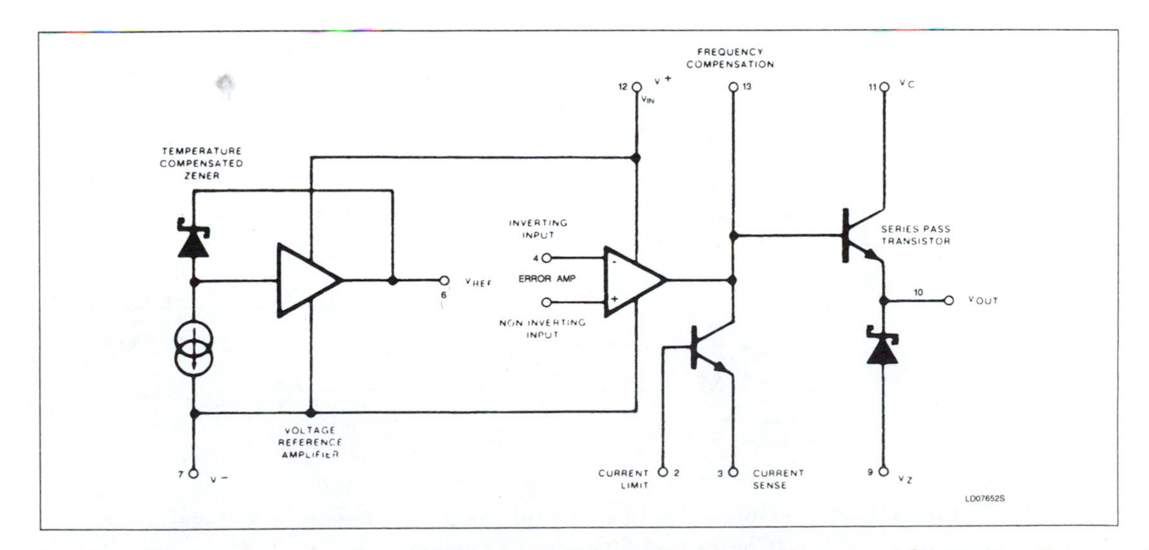


**Figure 5-16.** Schematic of μA723 Precision Voltage Regulator (Courtesy of Signetics Company)

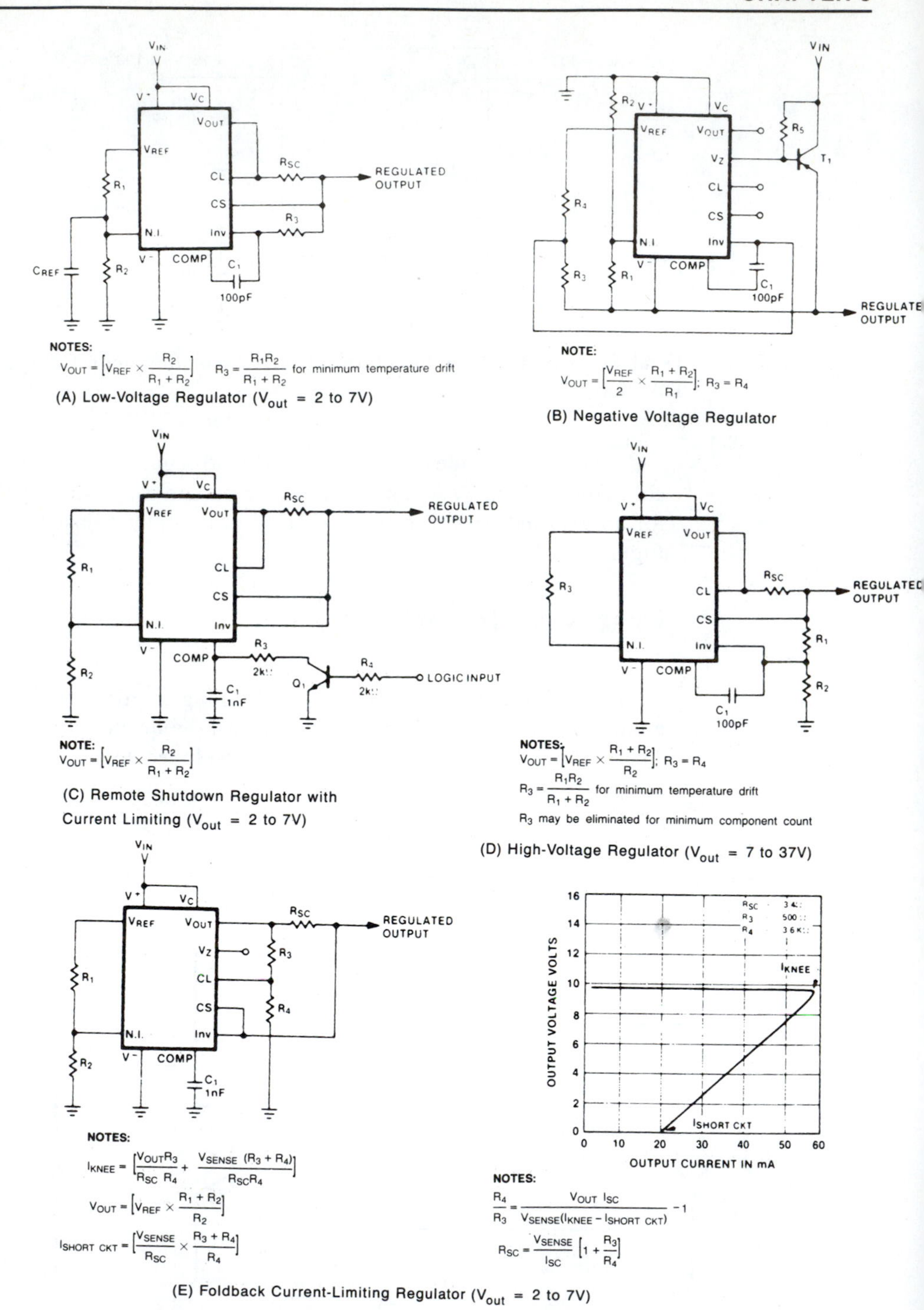


Figure 5-17. Typical Applications for the μA723 Precision Voltage Regulator (Courtesy of Signetics Company)

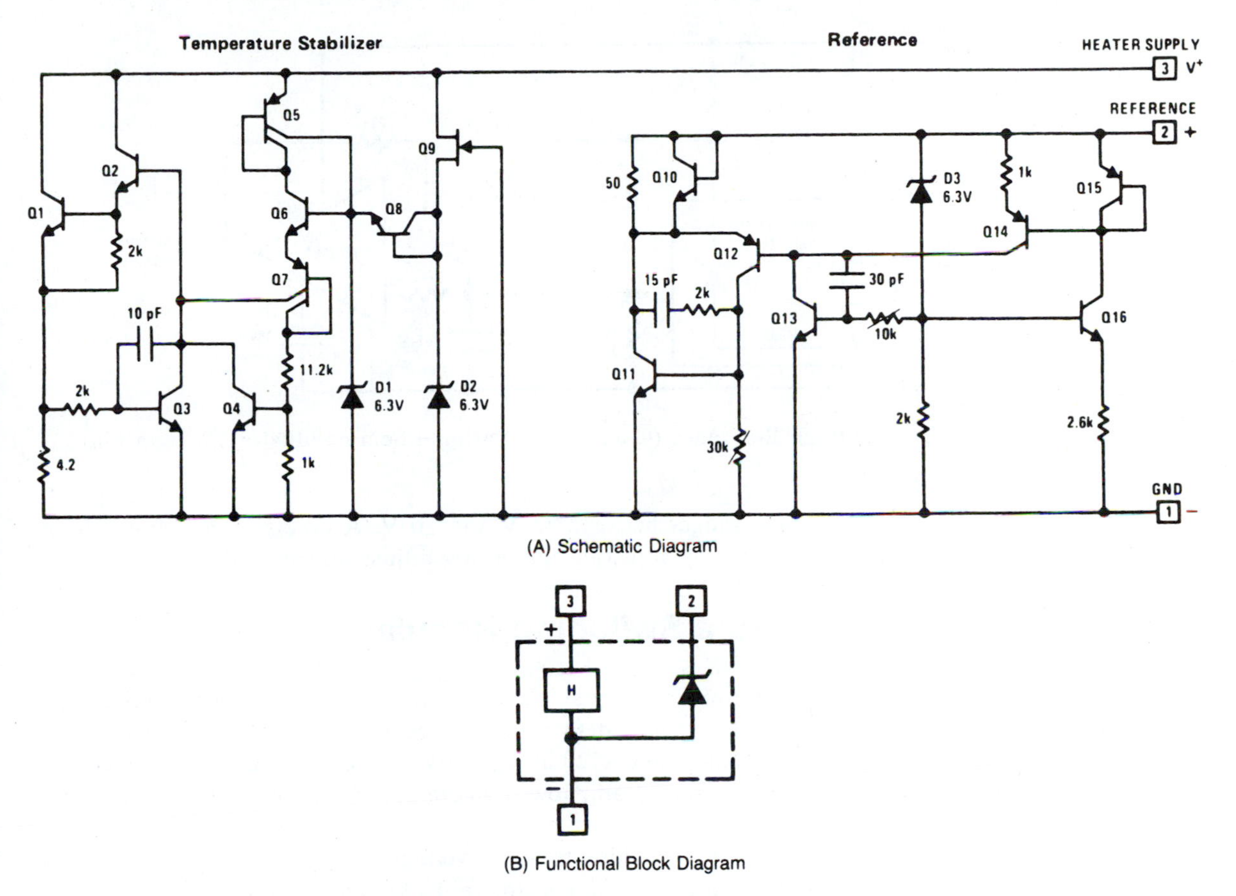


(A) Schematic Diagram

(B) Functional Block Diagram

**Figure 5-18.** LM3999 Precision Reference (Courtesy of National Semiconductor Corporation)

consists of a temperature stabilizer circuit and an active reference zener circuit. The active circuitry reduces the dynamic impedance of the zener to about 0.5 Ω and allows the zener to operate over the current range of 0.5 mA to 10mA with essentially no change in voltage or temperature coefficient.

The IC configuration offers many advantages over the discrete zener. It is easy to use, it is free of voltage shifts due to lead stress, and warm-up time is short because the unit is temperature stabilized. Further, in many cases the LM3999 can replace references in existing equipment with a minimum of wiring changes.

The LM3999 precision reference can be used in almost any application in place of ordinary zeners and improve performance. Some ideal applications are analog-to-digital (ADC) converters, precision voltage or current sources, and precision power supplies. Figure 5-19 shows one such typical application, that of a precision voltage reference. In this case the precision output voltage is 1.01 V with

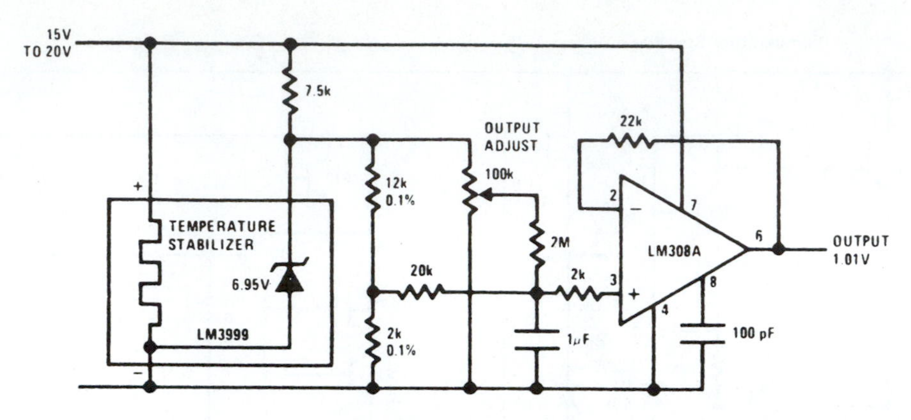


**Figure 5-19.** Precision Voltage Reference (Courtesy of National Semiconductor Corporation)

an input voltage that ranges between 15 V and 20 V. A precision 5 V logic drive voltage can be obtained by adjusting the output adjust potentiometer.

## Conversion to a Switching Regulator

Through the use of proper external circuitry, many of the standard regulators available today may be converted to switching regulators. Figure 5-20 presents one such conversion using the μA723 as a positive switching regulator. The device features low standby current drain, low temperature drift, and high ripple rejection.

In Figure 5-20, $V_{in}$ provides the $+V$ voltage for the device (pin 12) and the $V_{CC}$ voltage for the series-pass transistor (pin 11). The $V_{in}$ also provides biasing through resistors $R_5$ and $R_7$, for external transistors $Q_1$ and $Q_2$, which are PNP devices configured as a Darlington pair. Capacitor $C_1$ is a bypass capacitor to provide stability. A reference voltage from pin 6 is provided through a voltage divider, resistors $R_1$ and $R_2$, where the collector output voltage from the Darlington pair is compared with the reference voltage at pin 5, the noninverting input to the error amplifier. Frequency compensation pin 13 is not used in this application. Output voltage pin 10, current limit pin 2, current sense pin 3, and the inverting input to the error amplifier pin 4, are connected together through current-limiting resistor $R_6$. The regulated output positive voltage is taken across the filter network of inductor $L_1$ and capacitor $C_2$.

## Hybrid Regulators

Hybrid regulators use both the monolithic chip and discrete components in chip form (thin-film or thick-film construction methods), all mounted on a ceramic material substrate. Gold wires connect the chips together and the ceramic material insulates the chips from each other. Packaged as a single device, this form

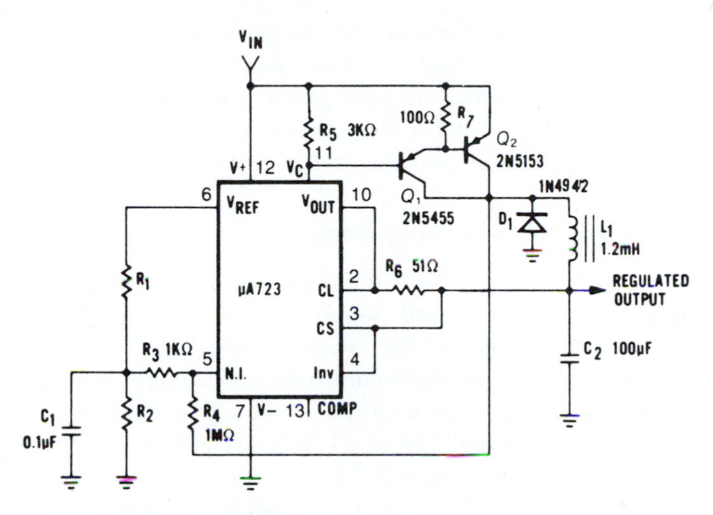


Figure 5-20. Switching Regulator Circuit Constructed with a Standard Regulator and External Components (Courtesy of Fairchild Camera & Instrument Corporation)

of regulator offers matched components that provide a desired output. The package usually consists of a monolithic regulator chip and several transistor chips, with resistors and capacitors formed on the substrate. Because of this multilayered construction, hybrid devices are not as small as monolithic regulators, but they can handle larger currents and dissipate more power.

One example of the hybrid technique is demonstrated by the Fairchild μA78P05, shown in block diagram form in Figure 5-21. This is a three-terminal positive 5 V hybrid voltage regulator capable of delivering 10 A. It is virtually

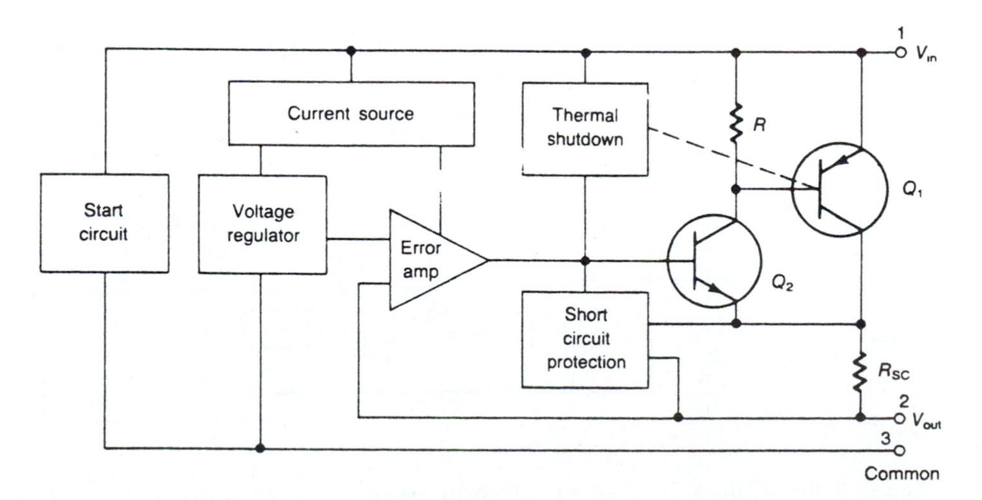


Figure 5-21. Three-Terminal Hybrid Voltage Regulator (Courtesy of Fairchild Camera & Instrument Corporation)

blowout proof and contains all the protection features inherent in monolithic regulators, such as internal short circuit current-limiting through resistor $R_{SC}$, thermal overload, and safe area protection. This hybrid consists of a monolithic control chip that drives a rugged Mesa transistor and it is contained in a hermetically sealed steel package that provides 70-watt power dissipation.

### Step-Down Switching Regulator

The SH1605 is a high-efficiency step-down switching regulator capable of supplying 5 A of regulated output current over an adjustable range from 3.0 V to 30.0 V of output voltage, with up to 150 W of output power. The device incorporates a temperature-compensated voltage reference, a duty-cycle-controllable oscillator, error amplifier, high-current high-voltage output switch, and a power diode. A simplified block diagram is shown in Figure 5–22, and a design example is shown in Figure 5–23.

In this circuit (Figures 5–22 and 5–23) when power is first applied, the output voltage $V_{out}$ is low, thus forcing the comparator output into a HIGH state. As a result, the oscillator freely toggles the output switch (set-reset (SR) flip-flop) on and off at a rate determined by the charge and discharge rate of the timing capacitor $C_T$. This is a temporary condition that continues until $V_{out}$ has

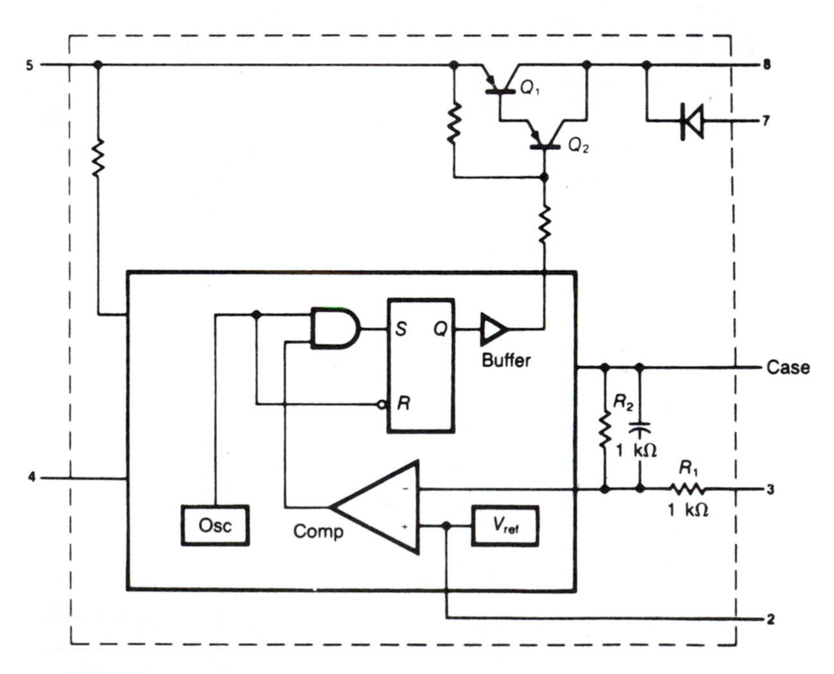


Figure 5–22. Block Diagram of a Hybrid Switching Regulator (Courtesy of Fairchild Camera & Instrument Corporation)

exceeded the reference voltage ($V_{ref}$) level times the factor set by $R_1$, $R_2$, and $R_3$. The output voltage is expressed as

$$V_{out} = V_{ref}\frac{(R_1 + R_2 + R_3)}{R_1} \tag{5.8}$$

Since the values of $R_1$ and $R_2$ (1 kΩ each) inside the SH1605 are established, $R_3$ can be determined as follows:

$$R_3 = \frac{(R_1 + R_2)(V_{out} - V_{ref})}{V_{ref}} \tag{5.9}$$

where

$V_{ref} = 2.5$ V

Equilibrium is reached at the completion of the ON cycle when the comparator input has exceeded the reference level. When the comparator output goes LOW, the oscillator output is disabled and $Q_1$ switches OFF. $V_{out}$ then begins to fall at a rate determined by the ratio of the output voltage to the inductor value:

$$\frac{\Delta I_{L_1}}{t_{off}} = \frac{V_{out} + V_{in}}{L_1} \tag{5.10}$$

where

$I_{L_1}$ = inductor current
$L_1$ = inductor
$t_{off}$ = time $Q_1$ is in the OFF state

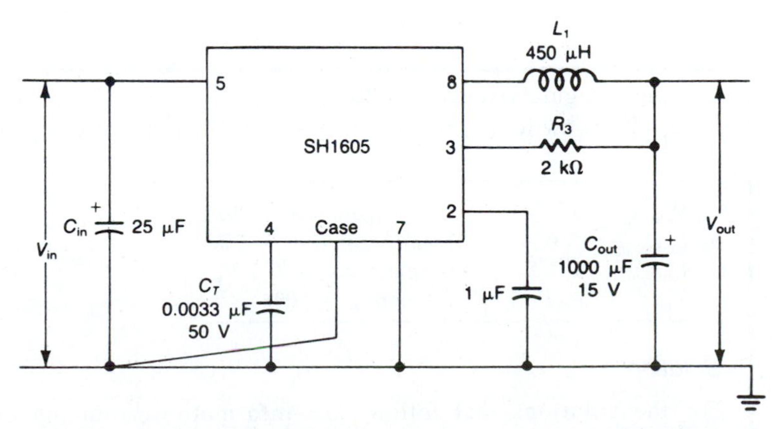


**Figure 5-23.** Typical Application of a Hybrid Switching Regulator (Courtesy of Fairchild Camera & Instrument Corporation)

Whenever $V_{out}$ falls to the level specified by Equation 5.8, the comparator changes state and the output switches ON. It remains in this state until the voltage across the timing capacitor, $C_T$, reaches a positive threshold level. The rate of $C_T$ charge is determined by the size of the timing capacitor and the magnitude of the constant-current source inside the oscillator. Charging current is typically 25 μA and discharging current is 225 μA. From the following formulas describing ON and OFF duration, the frequency of oscillation can be determined as follows:

$$t_{on} = \frac{C_T \Delta V}{I_C} \tag{5.11}$$

$$t_{off} = \frac{\Delta I_{L_1} \times L_1}{V_{out} + V_{in}} \tag{5.12}$$

where

$C_T$ = timing capacitor
$I_C$ = oscillator charging current
$V$ = 0.5 V

Then nominal frequency is calculated by the following equation:

$$\frac{1}{\dfrac{C_T \Delta V}{I_C} + \dfrac{\Delta I_{L_{1(nom)}} \times L_1}{V_{out} + V_{in}}} = \frac{1}{t_{on} + t_{off}} = \frac{1}{T} \tag{5.13}$$

For improved system efficiency, the operating period should always be many times longer than the device transition times. A nominal value is eight times longer. A trade-off must be sought between inductor size and efficiency when selecting the frequency of operation.

## Example 5.4

A typical regulator design follows. Refer to Figure 5–23, a design circuit for a step-down switching regulator. Assume the nominal design objectives to be:

| | |
|---|---|
| $V_{out}$ = +5 V | Line regulation = 2% |
| $I_{out(max)}$ = 5.0 A | Load regulation = 2% |
| $I_{out(min)}$ = 1.0 A | Ripple (max) = 0.1 $V_{p\text{-}p}$ |
| $V_{in}$ = 12 to 18 V | Efficiency = 70% |

*Solution*

For the solutions that follow, use information from the discussion in this section. First calculate $R_3$ from Equation 5.9:

$$R_3 = \frac{(R_1 + R_2)(V_{out} - V_{ref})}{V_{ref}} = \frac{(2 \times 10^3\ \Omega)(5\ V - 2.5\ V)}{2.5\ V} = 2\ k\Omega$$

Since the required $I_{out(min)}$ is 1 A to maintain continuous operation, the peak-to-peak current excursion (twice the value of $I_{out(min)}$) must be equal to 2 A or less, so that

$$\Delta I_{L_1} = 2(I_{out(min)})$$

To calculate the value of $L_1$, assume the nominal ON time of the system as 60 μs. This value is chosen keeping the efficiency/component size trade-off in mind. Therefore,

$$L_1 = \frac{V_{in(nom)} - V_{out}}{\Delta I_{L_1}}(t_{on})$$

$$= \frac{10\ V}{2\ A}(60 \times 10^{-6}\ s) = 300\ \mu H \qquad \textbf{(5.14)}$$

where

$V_{in(nom)} = 15$ V
$t_{on} = 60$ μs
$\Delta I_{L_1} = 2$ A

An important element in achieving the optimum performance in a switching regulator is to keep the inductor below the specified saturation limits.

Since the timing capacitor controls the 60 μs ON time, $C_T$ can be determined by manipulation of Equation 5.11:

$$C_T = \frac{t_{on} I_C}{\Delta V} = \frac{(60 \times 10^{-6}\ s)(25 \times 10^{-6}\ A)}{0.5\ V} = 3000\ pF$$

where

$I_C = 25$ μA(nominal) from the data sheet

The final step is to determine the requirements for the output capacitor, $C_{out}$, to obtain the desired value of ripple voltage ($V_{ripple}$). Consideration must be given to the absolute value of $C_{out}$ as well as to the internal effective series resistance (ESR). Since capacitor size is inversely proportional to the operating frequency, the lowest frequency of operation must be calculated. Minimum operating frequency can be determined by substituting $\Delta I_{L1(max)}$ for $\Delta I_{L1(nom)}$ in Equation 5.13, so that minimum frequency is calculated as follows:

$$f_{min} = \frac{1}{\dfrac{C_T \Delta V}{I_C} + \dfrac{\Delta I_{L1(max)} L_1}{V_{out} + V_{in}}}$$

where

$$\Delta I_{L1(\max)} = I_{\text{out}(\max)} - I_{\text{out}(\min)}$$

Inserting values from our circuit, we find that the minimum frequency is

$$\frac{1}{\dfrac{3 \times 10^{-9}\ \text{F} \times 0.5\ \text{V}}{25 \times 10^{-6}\ \text{A}} + \dfrac{4\ \text{A} \times 3 \times 10^{-4}\ \text{H}}{5\ \text{V} + 12\ \text{V}}} \approx 7.7\ \text{kHz}$$

The output capacitor minimum value can now be determined using the following equation:

$$C_{\text{out}(\min)} = \frac{\Delta I_{L_1}}{8 f_{(\min)} V_{\text{ripple}(\max)}}$$

$$= \frac{2\ \text{A}}{8 \times 7.7 \times 10^3\ \text{Hz} \times 0.1\ \text{V}} = 325\ \mu\text{F} \qquad \textbf{(5.15)}$$

The maximum acceptable ESR is, therefore,

$$\text{ESR}_{(\max)} = \frac{V_{\text{ripple}(\max)}}{\Delta I_{L1(\max)}} = \frac{0.1\ \text{V}}{4\ \text{A}} = 0.025\ \Omega$$

Normally the minimum capacitance value for $C_{\text{out}}$ should be increased considerably if a low ESR capacitor is not used. As a final step for minimizing switching transients at the input of the device, a low ESR capacitor $C_{\text{in}}$ must be used for decoupling purposes between the input terminal and ground.

### Universal Switching Regulator

The μA78S40 Universal Switching Regulator Subsystem shown in Appendix D, while not a hybrid circuit, is an excellent example of the available switching regulators. It offers step-up, step-down, and inverting options as well as use as a series pass regulator. The data sheets show the design formulas for the various options and typical designs and operational performance results.

## 5.6 Current Regulation

*Current regulator* circuits maintain a constant current in a load independent of changes in that load. This means that variations in load impedances or voltage changes across the load impedances do not affect the circuit operation. An op amp current regulator circuit is shown in Figure 5–24.

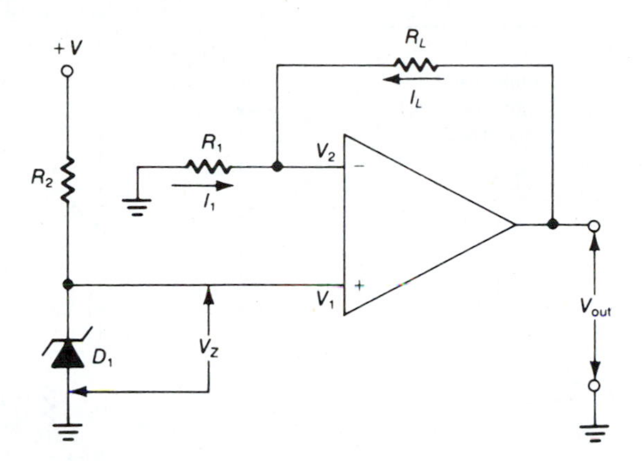


**Figure 5-24.** Op Amp Current Regulator (Courtesy of John Wiley & Sons, Inc.)

Assuming an ideal device, $V_1$ and $V_2$ are equal to $V_Z$, and the current in $R_1$ is

$$I_1 = \frac{V_Z}{R_1} \quad \textbf{(5.16)}$$

Since an ideal device has zero current entering or leaving the (–) terminal, $I_1$ must equal $I_L$. From Equation 5.16 we note that $I_1$ is determined by two constant terms, $V_z$ and $R_1$. Therefore, $I_1$ and $I_L$ must also be constant. Also note that $R_L$ is not considered in the equation. This means that $I_L$ is independent of $R_L$. Any change in $R_L$ will change the gain of the circuit, causing $V_{out}$ to vary, which maintains a constant value of $I_L$. $R_2$ is a current-limiting resistor for the zener diode.

## Constant-Current Regulator

Any three-terminal voltage regulator can be used as a *constant-current regulator,* as illustrated in Figure 5-25. The output current ($I_{out}$) dictates the regulator type to be used and is determined by the following equation:

$$I_{out} = \frac{V_{out}}{R_1} + I_Q \quad \textbf{(5.17)}$$

where

$V_{out}$ = the regulator output voltage
$I_Q$ = the quiescent current

The input voltage ($V_{in}$) must be high enough to accommodate the dropout voltage at the low end, but must not exceed the maximum input voltage rating at the high end.

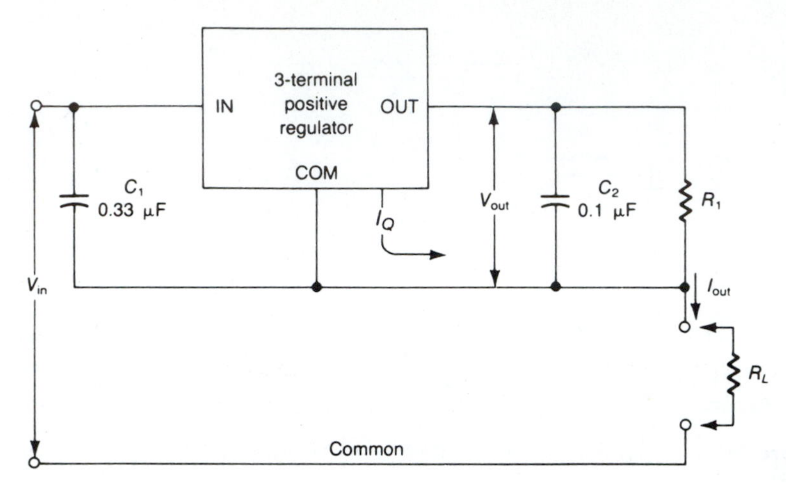


**Figure 5–25.** Three-Terminal Voltage Regulator Used as a Constant-Current Regulator (Courtesy of Fairchild Camera & Instrument Corporation)

# 5.7 Constant-Current Source

An op amp can be used as a *constant-current source* when arranged as in Figure 5–26. The input voltage can be a battery or some other stable reference voltage that delivers a constant current through the input resistor, $R_1$. This current also flows through the feedback resistor, which functions as the load, $R_L$. The current through $R_1$ and $R_L$ is

$$I_1 = \frac{V_{ref}}{R_1} \tag{5.18}$$

Since $V_{ref}$ and $R_1$ are constant, $I_1$ must remain constant. Changes in $R_L$ cannot alter this fact.

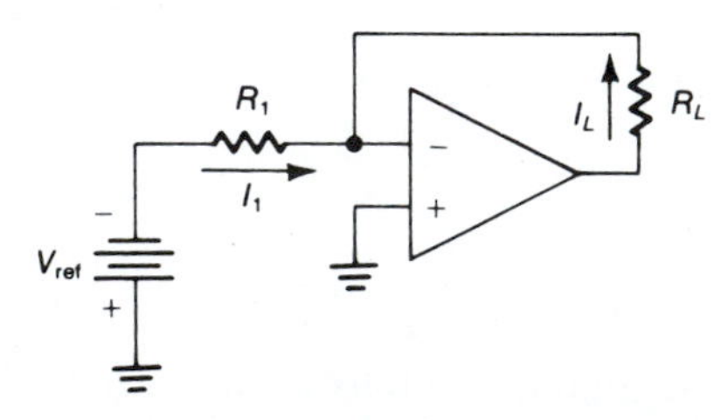


**Figure 5–26.** Op Amp Constant-Current Source

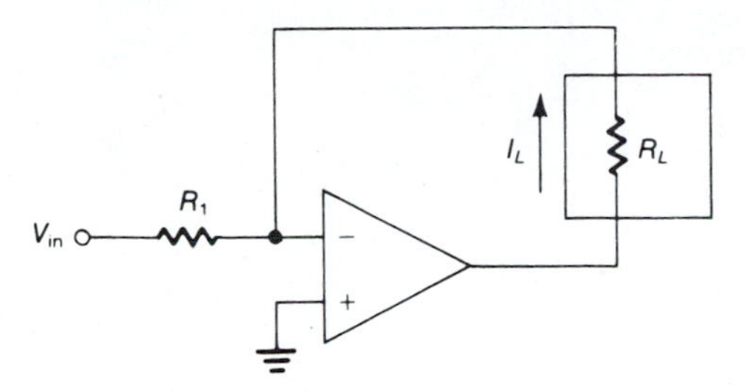


(A) Inverting Voltage-to-Current Converter for Floating Loads

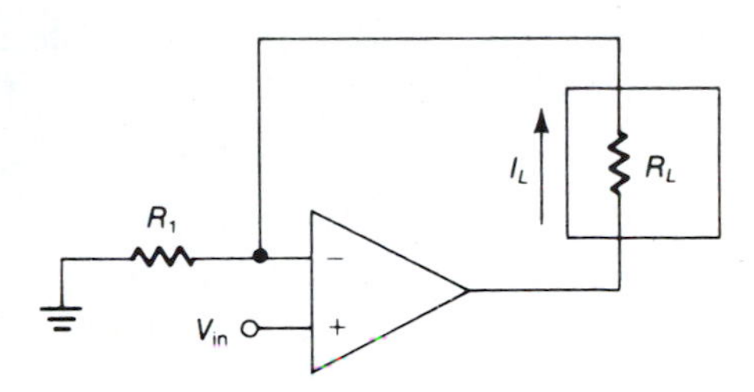


(B) Noninverting Voltage-to-Current Converter for Floating Loads

**Figure 5–27.** Voltage-to-Current Converters

# 5.8 Voltage-to-Current Converters

*Voltage-to-current converters,* also called *transmittance amplifiers,* are frequently used for applications such as driving relays and analog meters. Such converters can drive either floating or grounded loads, depending upon the specific application.

## Floating Loads

For floating loads, either the inverting converter (Figure 5–27A) or the noninverting converter (Figure 5–27B) may be used. These circuits are similar in form to the inverting and noninverting amplifiers. The major difference is that in the converter, the load element, $R_L$, is a relay coil or an analog meter with an internal resistance representing $R_L$.

Also, note the similarity between the inverting converter of Figure 5–27A and the constant-current source of Figure 5–26. The current flowing through the floating load is

$$I_L = \frac{V_{in}}{R_1} \tag{5.19}$$

and is independent of the value of $R_L$. The load current for the noninverting converter of Figure 5–27B is determined in the same manner as that for the inverting converter of Figure 5–27A.

## Grounded Loads

A different configuration must be used for loads that are grounded on one side. Figure 5–28 shows such a circuit. The load current is controlled by the input voltage ($V_{IN}$) and is determined as follows:

$$I_L = \frac{V_{in}}{R_3} \qquad (5.20)$$

when

$$\frac{R_4}{R_3} = \frac{R_2}{R_1} \qquad (5.21)$$

# 5.9 Current-to-Voltage Converters

Figure 5–29 shows a basic *current-to-voltage converter.* Essentially, the input is fed directly to the inverting input of the op amp. This input current also flows through the feedback resistor, therefore, the output voltage ($V_{out}$) is

$$V_{out} = I_{in}R_f \qquad (5.22)$$

However, care should be taken with such a circuit, because the op amp's bias current ($I_b$) is added to the input current ($I_{in}$) and $V_{out}$ becomes

$$V_{out} = (I_{in} + I_b)R_f \qquad (5.23)$$

Therefore, the input current should be kept high compared to the input bias current.

# 5.10 Regulator Protection Circuits

*Protective circuits* are added on chips to improve reliability and to make regulators immune to certain types of overloads. They protect the regulators against short circuit conditions (current limit), against excessive input-output differential conditions (safe operating area limit), and against excessive junction temperatures (thermal limit). Figure 5–30 shows a basic regulator protection circuit.

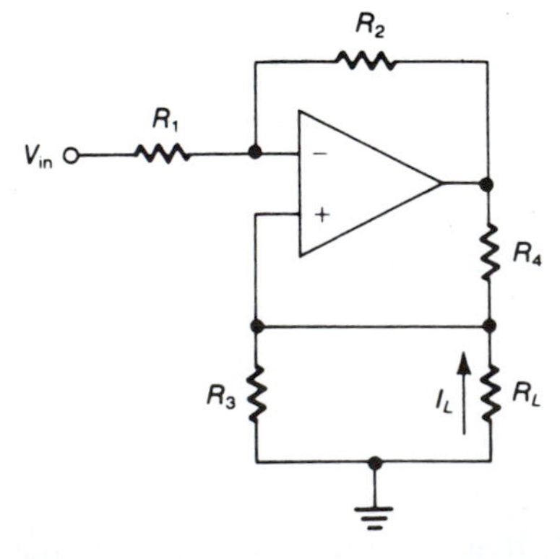


**Figure 5–28.** Voltage-to-Current Converter for Grounded Loads

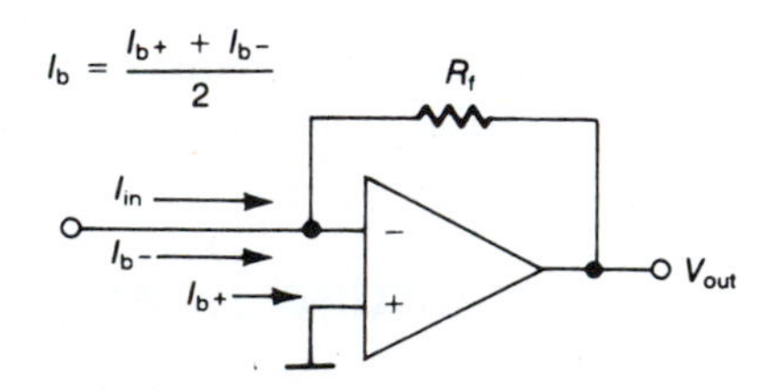


**Figure 5–29.** Current-to-Voltage Converter

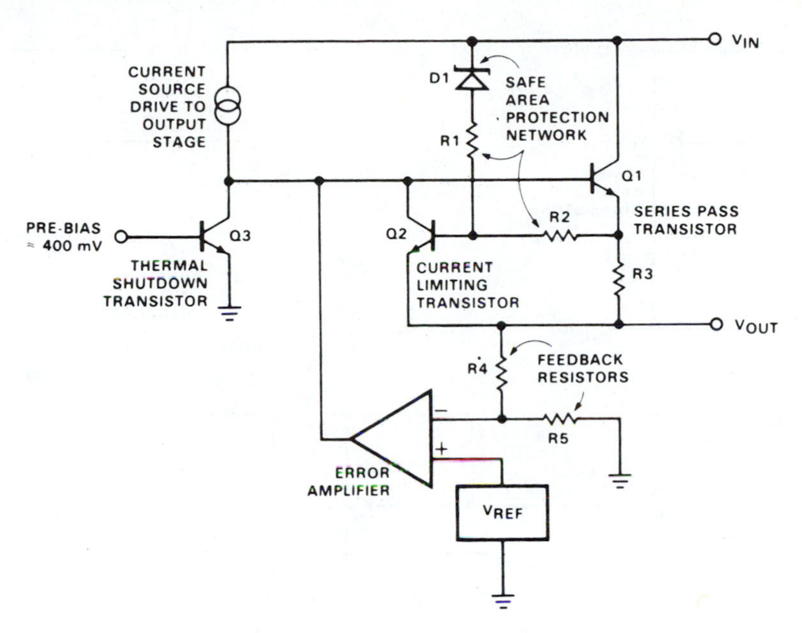


Figure 5–30. Basic Regulator Protection Circuit (Courtesy of Fairchild Camera & Instrument Corporation)

## Current-Limiting Protection

The most commonly used protection scheme is *current limiting* for guarding the output series-pass transistor against excessive output currents or short-circuit conditions. With low input-output conditions, zener diode $D_1$ is not conducting, and there is no current flowing through $R_1$. The current-limiting transistor, $Q_2$, senses the voltage drop across the current-limiting resistor, $R_3$. As the output current increases, the drop across $R_3$ increases. As a result, the $Q_2$ base-emitter voltage increases until $Q_2$ begins to conduct, thereby removing the base drive of the series-pass transistor, $Q_1$. No additional output current can be pulled out since any increase in the output current will cause $Q_2$ to conduct harder. However, this current-limiting circuit has a slight disadvantage—the voltage developed across $R_3$ adds to the regulator dropout voltage and degrades the load regulation and output impedance.

A simple way of getting around this problem is to prebias the base of $Q_2$ with a fraction of the base-emitter voltage of $Q_1$ through $R_6$ and $R_7$, as illustrated in Figures 5–31A and B. In these current-limiting schemes, the output current decreases with increasing temperature since the base-emitter voltage of $Q_2$ is the threshold for preventing the flow of additional regulator output current.

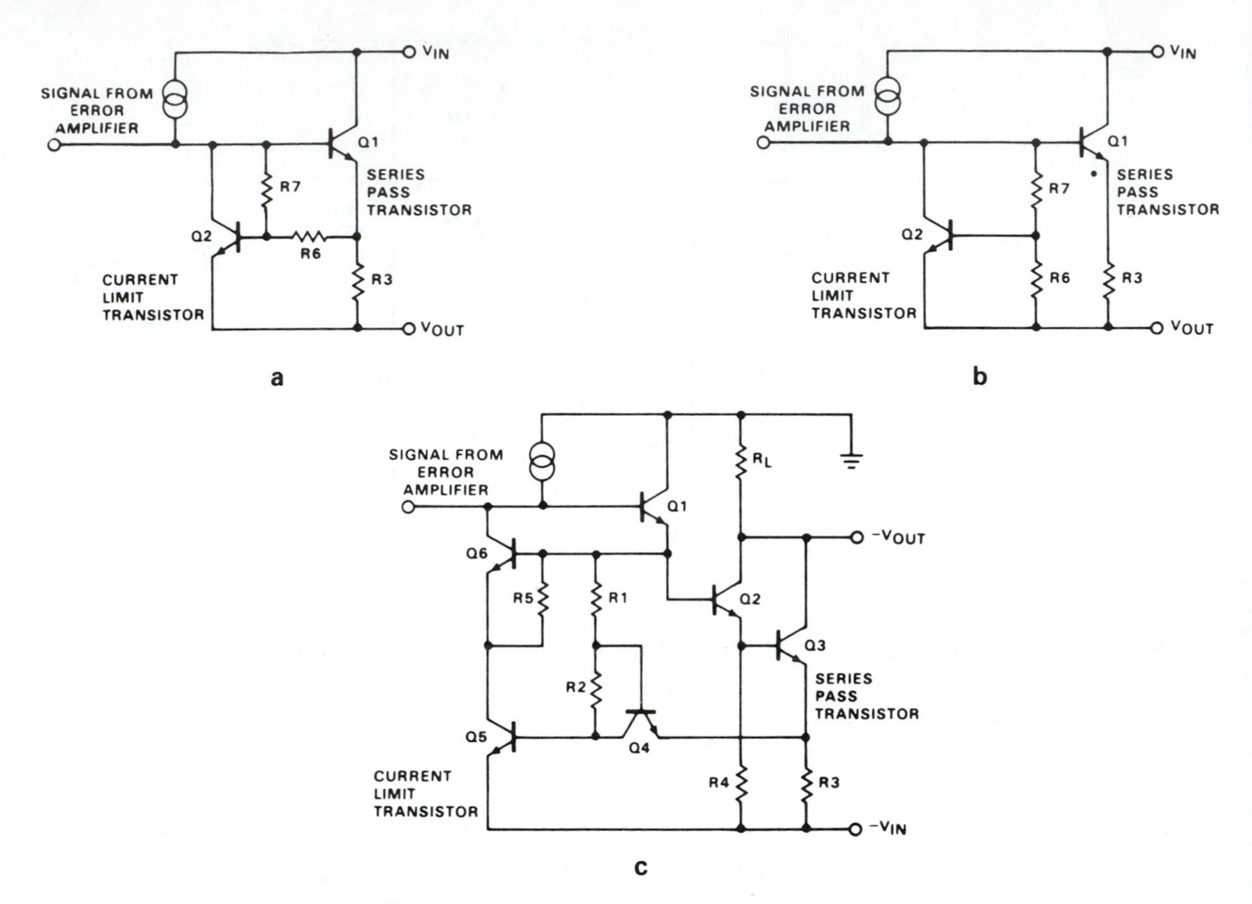


Figure 5-31. Alternate Current-Limiting Circuits (Courtesy of Fairchild Camera & Instrument Corporation)

## Temperature-Independent Current Protection

*Temperature-independent* short circuit current protection can be achieved with a slight increase in circuit complexity. Figure 5-31C shows the circuit used in the μa79M00 series of negative regulators to obtain temperature-independent peak output current. At low to medium output current levels, $Q_4$ is on and $Q_5$ and $Q_6$ are off. When the voltage drop across $R_3$ reaches a predetermined level, $Q_4$ begins to turn off and current is diverted to $Q_5$, which in turn causes $Q_6$ to turn on. $Q_1$ and, consequently, $Q_2$ base currents are thus diverted and the output current is prevented from increasing further.

## Safe Operating Area Protection

*Safe operating area* (SOA) protection is included in IC regulators to protect the series-pass transistor against excessive power dissipation by reducing the collector current as the collector-emitter voltage is increased (Figure 5-30). When the input-output voltage differential exceeds the breakdown voltage differential of $D_1$, current flows through $D_1$, $R_1$, $R_2$, and $R_3$. The voltage drop across $R_3$ and, therefore, the $Q_2$ base voltage both become a function of not only the output current but also the input-output voltage differential. Hence, maximum output current is available when the input-output voltage differential is less than the $D_1$ breakdown voltage. The safe area protection network reduces the available output current ($I_{out}$) as the input-output differential $\Delta(V_{in} - V_{out})$ increases at a rate determined by the following equation:

$$\frac{\Delta I_{out}}{\Delta(V_{in} - V_{out})} = -\frac{R_2}{R_1 R_3} \tag{5.24}$$

The safe area protection network thus reduces the available output current as the input-output differential increases, and limits the regulator operation to within the SOA of the series-pass transistor. The SOA network is lumped in with the short circuit protection network, and consequently both have the same temperature characteristics.

When selecting a regulator to operate with high input voltage or with high input-output voltage differential, it must be remembered that output current decreases with increased input-output voltage differential. Under heavy load and high input-output conditions, the SOA protection circuit may cause a high output voltage device to *latch up* (lock on to the high voltage-high current condition) after a momentary short, since the input voltage becomes the input-output differential during the short. The regulator may not be able to supply as much current after the fault condition as before. *Latching* will not damage the regulator. Interrupting the power, reducing the load current, or reducing the input voltage momentarily will restore normal operation.

A discrete regulator usually relies on current limiting for overload protection since there is no practical way to sense junction temperature in a separate series-pass transistor. The dominant failure mechanism of this type of regulator, then, is excessive heating of the series-pass transistor. In a monolithic regulator, the series-pass transistor is contained within the thermal overload protection circuit where its maximum junction temperature is limited, independent of input voltage, type of overload, or degree of package heat sinking. It is, therefore, considerably more effective than current limiting by itself. An added bonus to combined thermal and current overload protection is that a higher output current level under normal conditions can be considered, since there is no excessive regulator heating when a load fault occurs.

### Thermal Limit Protection

The base-emitter junction of a transistor placed as close as practical to the series-pass transistor is used to sense the chip temperature. The thermal shutdown transistor, $Q_3$ in Figure 5–30, is normally biased below its activation threshold so that it does not affect normal operation of the circuit. However, if the chip temperature rises above its maximum limit due to an overload, inadequate heat sinking, or other condition, the thermal shutdown transistor turns on, removes the base drive to the output transistor $Q_1$, and shuts down the regulator to prevent any further chip heating.

## 5.11 Summary

☐ Voltage regulators control supply voltage variations so that proper linear device operation can occur. There are three major types of voltage regulators: series, shunt, and switching. In a series regulator, the variable control element is in series with the load. In the shunt regulator, the variable control element is in parallel with the load. In the switching regulator, the control transistor is switched between the cutoff and saturation modes. The series and shunt regulators have the disadvantage of high power dissipation in the control transistor. The switching regulator has low power dissipation in the control transistor, thereby providing higher circuit efficiency.

☐ Heat sinking is required for many applications to prevent damage to devices and circuits due to excessive temperatures.

☐ Using op amps as sensing elements in regulator circuits helps to overcome the problems of low sensitivity, slow response time, and loading effects.

☐ Monolithic regulators are constructed with all components on a single chip. This allows the regulator to be used for local "on-card" regulation. Most monolithic regulators require that some external components be added in practical applications. The three main categories of monolithic regulators are fixed, adjustable, and dual tracking. Adjustable output voltages may be obtained by adding external components to fixed regulators. Fixed and adjustable regulators can be obtained with either positive or negative outputs. Dual tracking regulators provide both positive and negative voltages that track (follow) each other. Many standard regulators may be converted to switching regulators through the use of proper external circuitry.

☐ Hybrid regulators are a combination of monolithic and thin-film or thick-film construction methods that are packaged as a single device. Hybrid devices are larger than monolithic devices, but they are capable of handling larger currents and they dissipate more power.

☐ Any three-terminal voltage regulator can be used as a constant-current regulator. An op amp can be used as a constant-current source when properly arranged.

☐ Voltage-to-current converters can be used to drive relays or analog meters, with either floating or grounded loads. Current-to-voltage converters operate in a manner similar to constant-current sources.

☐ Improved reliability and immunity from certain types of overloads are provided by adding protective circuits to regulators. Current limiting is protection against short circuit conditions and is the most commonly used protection scheme. Safe operating area (SOA) limit is protection against excessive input-output differential conditions. Thermal limit is protection against excessive junction temperatures.

☐ Precision reference ICs are used in place of ordinary zener diodes to provide improved performance and accurate references that are free from voltage shifts due to lead stress and temperature variations.

# 5.12 Questions and Problems

**5.1** What might be the result in an audio amplifier if there were a sudden shift in the supply voltage?

**5.2** What causes variations in supply voltages?

**5.3** Define a monolithic voltage regulator.

**5.4** List the three major types of regulator circuits.

**5.5** What is the feature that all three types of regulators have in common?

**5.6** Explain why the circuit in Figure 5–1 is called a dissipative voltage regulator circuit.

**5.7** What distinguishes the shunt regulator circuit from the series regulator circuit?

**5.8** What disadvantage do the series and shunt regulator circuits have in common?

**5.9** How does the switching regulator overcome the disadvantage described in Problem 5.8?

**5.10** What are the advantages of the switching regulator over the series and shunt regulators?

**5.11** Refer to Figure 5–3. Assume the control signal frequency to be 50 kHz and the transistor ON time to be 5 μs. Calculate the duty cycle.

**5.12** Assume that the maximum voltage across the diode in Figure 5–3 is 60 V. Using the duty cycle calculated in Problem 5.11, calculate the dc component of the output voltage ($V_{dc}$) of the circuit.

**5.13** List three problems that arise in regulators that use transistors as sensing devices.

**5.14** What is one simple solution to the problems described in Problem 5.13?

**5.15** Discuss some possible problems that may arise from the solution stated in Problem 5.14 and describe some solutions to those problems.

**5.16** What is meant by local regulation?

**5.17** Describe the three main categories of monolithic regulators.

**5.18** What is the major advantage of the fixed voltage regulator?

**5.19** Refer to Figure 5–13A. Assume $V_{reg}$ =

5 V, $I_Q$ = 7 mA, $R_1$ = 100 Ω, and $R_2$ = 175.4 Ω. Calculate $V_{out}$.

**5.20** Refer to Figure 5–13B. Assume $V_{reg}$ = 5V, $R_1$ = 3 kΩ, and $R_2$ = 1kΩ. Calculate $V_{out}$.

**5.21** Specify two methods of providing tracking for dual voltages.

**5.22** State the advantages of the 4195 regulator.

**5.23** Explain how hybrid regulators are constructed.

**5.24** Explain how current regulators work.

**5.25** List the three major protective circuits used in regulators and state what kind of protection each provides.

**5.26** List three advantages of precision reference ICs over standard reference zener diodes.

Chapter 6

# Basic Oscillators and Wave-Shaping Circuits

## 6.1 Introduction

Oscillators are as important in electronics as are amplifiers. Oscillators can be found in almost every field of electronics. They are used in computers, communications systems, industrial control and process handling, and radio and television sets. For example, the high-frequency oscillator in the television tuner selects the channel to be viewed. This is perhaps the most common use of the high-frequency oscillator.

Basically, an oscillator generates a continuously repetitive output signal that can be used for timing or synchronizing operations. The output signals can be either sinusoidal (sine) waves or nonsinusoidal (square, triangle, sawtooth) waves. The oscillator is an electronic generator that operates from a dc power supply, has no moving parts, and can produce ac signal frequencies ranging into millions of hertz (Hz). A good oscillator will have a uniform output, varying in neither frequency nor amplitude.

In this chapter we will discuss oscillator fundamentals, various types of oscillators, and the design of simple sine wave oscillators. Waveforms other than the sine wave will also be discussed. The descriptions of the waveforms are based on the shape of the waves, and the signal generators used to generate the waveforms are classified by the wave shapes they produce. We will look at single-form generators and multiform generators, and we will examine the design of generators using op amps.

## 6.2 Objectives

When you complete this chapter, you should be able to:

- ☐ Recognize the names of oscillators by observation.
- ☐ Determine the effects on frequency when reactive elements are varied.
- ☐ Determine if a crystal is operating in the series-resonant or parallel-resonant mode.
- ☐ Identify the three basic types of oscillators.
- ☐ Identify the frequency-determining components of various oscillators.
- ☐ Explain how each of the oscillators operates.
- ☐ Identify and discuss the two types of multivibrators used in analog systems.
- ☐ Identify and discuss the differences between differentiators and integrators.
- ☐ Identify the various types of generators by observation.
- ☐ Design several types of wave-shaping generators using op amps and external components.

## 6.3 Fundamentals of Oscillators

You will recall from your study of resonant circuits that excitation of a tank circuit by a dc source tends to cause oscillations in the circuit. These oscillations, or back-and-forth oscillatory motions, are the result of circulating current flow inside the tank circuit. If there were no resistance within the circuit, the oscillations would continue indefinitely. However, there is resistance, and this resistance dissipates energy and damps the oscillations.

For continued oscillation, the energy lost due to resistance must be replaced. In a simple amplifier circuit, this is accomplished through regenerative (positive) feedback obtained by simply feeding back to the input a portion of the amplifier output. However, the feedback signal must be in phase with the input signal in order for the circuit to oscillate. A *phase-shifting network,* as illustrated in Figure 6–1, may be necessary, depending upon the type of amplifier circuit being considered.

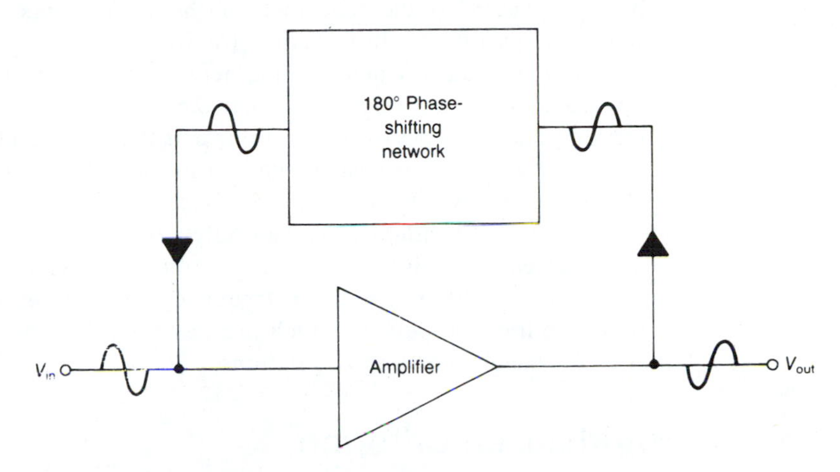


**Figure 6–1.** Feedback Loop with 180° Phase-Shifting Network

Once the amplifier begins to operate, the input signal is no longer required. The circuit will continue to oscillate even if the input signal is removed, because the gain of the amplifier replaces the lost energy in the circuit through positive feedback, thus sustaining oscillations. There is a matter for concern, however, because the circuit may be triggered by external noise pulses that cause undesired changes in the oscillating frequency.

For an oscillator to be useful, it must have a constant output. This stability is accomplished through proper selection of components in the *frequency-determining network*. One such network is shown in Figure 6–2. The parallel inductor-capacitor *(LC)* network in the positive feedback loop resonates at a frequency determined by the values of the components. The desired 180° phase

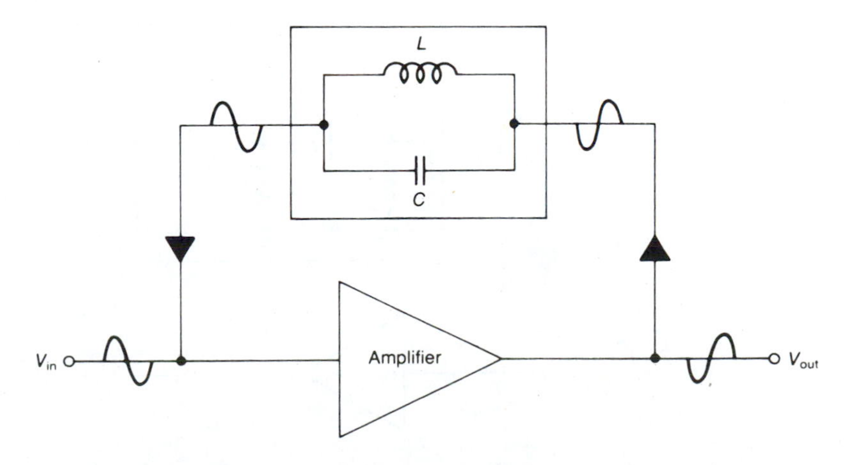


**Figure 6–2.** *LC* Frequency-Determining Network

shift is produced by the reactance of the components, and the amplifier gain replaces the energy lost in the tank circuit.

Another frequency-determining network is shown in Figure 6-3. The values of the components are selected so that the resistor-capacitor *(RC)* time constants determine the frequency of oscillation, and the desired phase shift is determined by the number of *RC* sections contained in the feedback loop. The amplifier gain replaces the energy lost across the *RC* sections.

In practice, oscillators are naturally *self-starting.* When an amplifier circuit is first turned on, noise pulses are generated. These pulses are fed back to the input of the amplifier through a frequency-determining network. The amplifier action amplifies the pulses, which are again fed back to the input. This process builds up, thereby creating oscillations.

## Barkhausen Criterion

An evaluation of the basic oscillator circuit illustrated in Figure 6-4 will show that certain conditions must be met for the oscillator to be self-starting and self-sustaining. For the oscillator to produce its own input signal continuously, the product of the amplifier gain $(A_v)$ and the fractional feedback factor $(B_v)$, a small fraction of the output signal that is fed back to the input, must equal one. This condition is called the *Barkhausen Criterion.* Mathematically, oscillator stage gain $(A_{osc})$ is expressed as follows:

$$A_{osc} = \frac{A_v}{1 - A_v B_v} \tag{6.1}$$

When the condition of $A_v B_v = 1$ is met, the oscillator stage gain is infinite, which satisfies one of the requirements for an oscillator—an output signal must be present without an input signal.

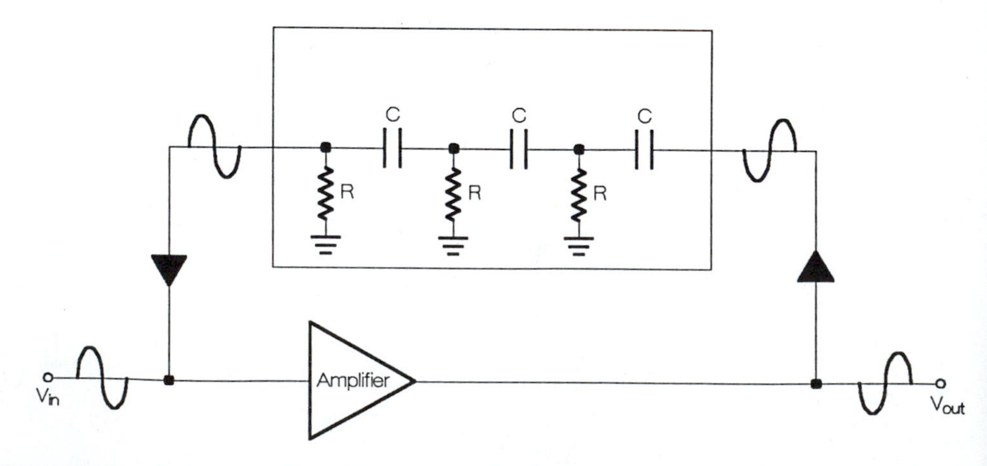


**Figure 6-3.** *RC* Frequency-Determining Network

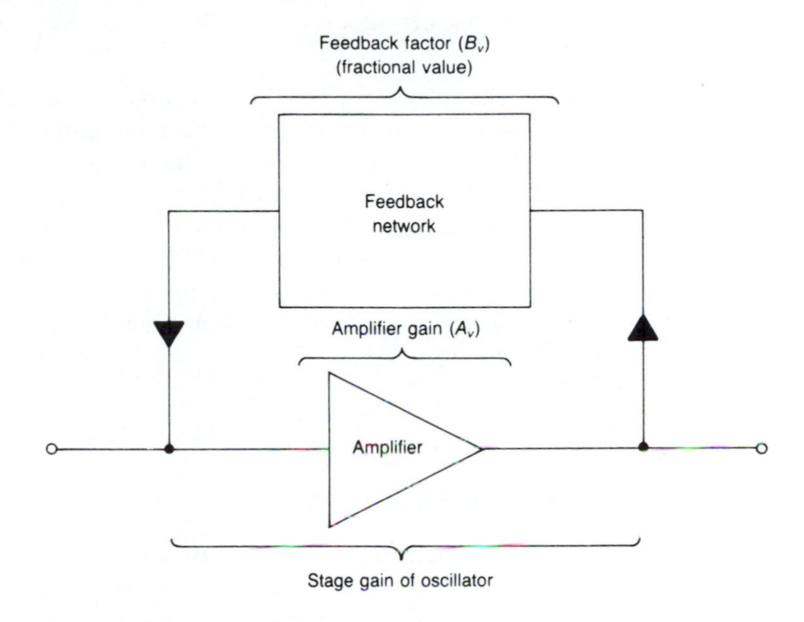


**Figure 6–4.** Concept of Basic Oscillator Stage Gain

When the oscillator first starts, the output increases to a point where $A_vB_v$ is slightly higher than unity. After a few oscillations, however, $A_vB_v$ is reduced to unity. This is accomplished in one of two ways: by amplifier saturation, which damps the signal to its required level, or by placing a nonlinear resistor in the feedback loop, which reduces the feedback factor.

## 6.4 Classifications of Oscillators

Oscillators are generally classified according to the components used in the frequency-determining networks. The three basic classifications are *LC oscillators, RC oscillators,* and *crystal oscillators*. These classifications will be discussed individually.

Briefly summarized, the basic fundamentals common to all oscillators are as follows:

1. Amplification is required to replace circuit losses.
2. A frequency-determining network is required to set the desired frequency of oscillation.
3. A regenerative (positive) feedback signal is required to sustain oscillation.
4. The oscillator is required to be self-starting, with no input signal.

## *LC* Oscillators

The frequency-determining network in the *LC* oscillator is a tuned circuit that consists of capacitors and inductors, connected either in series or in parallel. Figure 6–5A shows a block diagram of a basic *LC* oscillator using an op amp. The feedback network is formed by $Z_1$, $Z_2$, and $Z_3$. Their values determine the oscillating frequency. One or more of the impedances can be made variable if a range of frequencies is desired. The two basic *LC* oscillators are the *Hartley* and the *Colpitts*. If $Z_1$ and $Z_2$ are inductors, the circuit is a Hartley oscillator. If $Z_1$ and $Z_2$ are capacitors, the circuit is a Colpitts oscillator.

### Hartley Oscillator

A Hartley oscillator is illustrated in Figure 6–5B. The inductor is tapped to form two coils, $L_{1A}$ and $L_{1B}$, corresponding to $Z_1$ and $Z_2$ in Figure 6–5A. A capacitor, $C_1$, is placed across coil $L_1$, making the entire coil part of a tuned circuit. Current flow through $L_{1A}$ replaces energy lost in the tank, thus providing the positive feedback necessary for oscillation. The amount of feedback can be controlled by adjusting the position of the coil tap. The *tapped coil* is the identifying feature of the Hartley oscillator.

Oscillating frequency, $f_o$, is approximated by the following equation:

$$f_o = \frac{1}{2\pi\sqrt{L_{(eq)}C}} \tag{6.2a}$$

where

$L_{(eq)} = L_{1A} + L_{1B} + 2L_M$
$L_M$ = mutual inductance

and

$$L_M = K\sqrt{L_{1A} \times L_{1B}} \tag{6.2b}$$

where

$K$ = coefficient of coupling

For $L_{1A}$ and $L_{1B}$ on paper or plastic form, $K = 0.1$; for $L_{1A}$ wound over $L_{1B}$, $K = 0.3$; and for $L_{1A}$ and $L_{1B}$ on the same iron core, $K = 1$.

Two disadvantages of the Hartley configuration are that (1) the coils tend to be mutually coupled, which makes the frequency of oscillation differ slightly from the calculated frequency; and (2) the oscillating frequency cannot easily be varied over a wide range. The second disadvantage stems from the difficulty of changing the value of an inductor. Some limited frequency variation can be made in the Hartley oscillator by making capacitor $C_1$ variable.

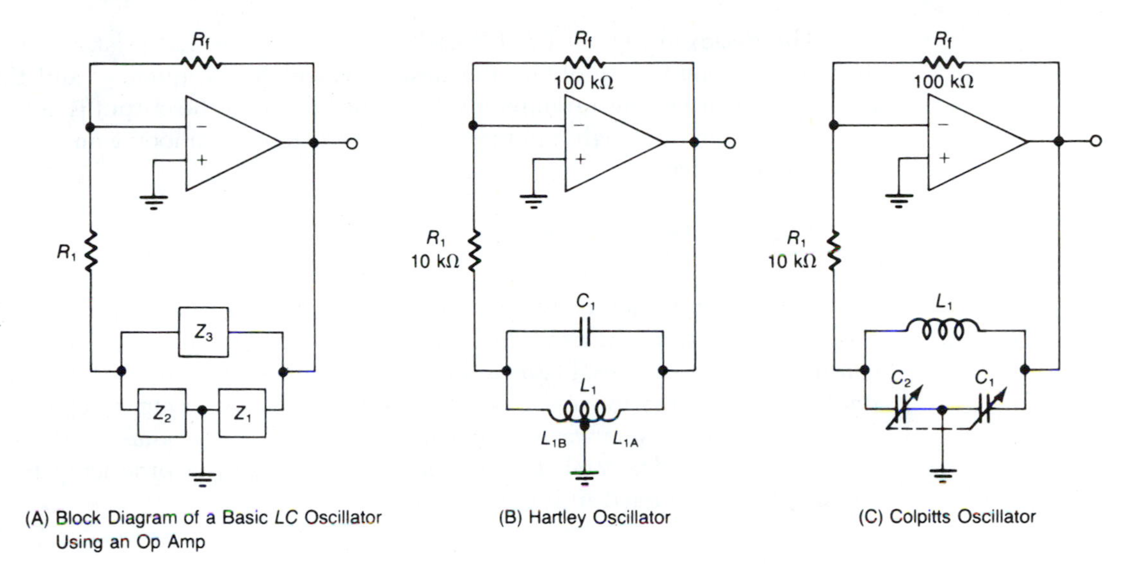


(A) Block Diagram of a Basic *LC* Oscillator Using an Op Amp

(B) Hartley Oscillator

(C) Colpitts Oscillator

**Figure 6-5.** *LC* Oscillators

### Colpitts Oscillator

A Colpitts oscillator is shown in Figure 6-5C. Note its similarity to the Hartley oscillator. The identifying feature of the Colpitts oscillator is the *tapped capacitor* arrangement. Frequency is determined by inductor $L_1$ and the series combination of capacitors $C_1$ and $C_2$, so that

$$f_o = \frac{1}{2\pi\sqrt{LC_{(eq)}}} \tag{6.3}$$

where

$$C_{(eq)} = \frac{C_1C_2}{C_1 + C_2}$$

The disadvantages of the Hartley oscillator are overcome in the Colpitts oscillator. Variable capacitors are readily available. By providing a wide range of capacitance, a wide range of frequencies can be obtained. Note that both capacitors in the Colpitts oscillator in Figure 6-5C are variable, allowing the user to vary the frequency over a wide range.

The Colpitts oscillator is used extensively in amplitude modulation (AM) and frequency modulation (FM) radio receivers. The variable capacitors are ganged so that the tuning dial selects both the mixer/oscillator frequency and the resonant frequencies of the RF amplifier stages. For TV reception in older model TV sets, the two capacitors are fixed, and different values of inductance are switched into the circuit. This is accomplished by mounting a fixed-value inductor for each channel on a shaft that is rotated. Each inductor is put into the oscillator tank circuit one at a time as the shaft is rotated, thereby changing the frequency of the circuit to the desired channel frequency.

The disadvantages of the *LC* oscillator circuits are that relatively pure inductance is required to obtain the desired calculated frequency, and that the input impedance of the amplifier must be infinite so that the output is not loaded. The op amp, with its high input impedance and large open loop gain, overcomes these disadvantages.

## *RC* Oscillators

The *RC* oscillator uses resistance-capacitance networks to determine oscillator frequency. Resistors and capacitors are in great supply and relatively inexpensive, consequently the *RC* oscillator is an inexpensive, easily constructed, relatively stable circuit design for use in low- and audio-frequency ranges. There are basically two types of sine wave-producing *RC* oscillators: the *phase-shift* oscillator and the *Wien-bridge* oscillator. The nonsinusoidal output-producing *RC* oscillators will be discussed in the next chapter.

### Phase-Shift Oscillator

As the name implies, the phase-shift oscillator shown in Figure 6–6 employs a phase-shifting *RC* feedback network that provides the 180° shift necessary to produce the required regenerative feedback. This type of oscillator is typically used in fixed-frequency applications.

The phase difference in an *RC* circuit is a function of the capacitive reactance, $X_C$, and the resistance, $R$, of the network. Resistance does not change with frequency, so the capacitor is the frequency-sensitive component. By careful selection of the components, the amount of phase shift across an *RC* section can be controlled. Stability can be improved by increasing the number of *RC* sections, thereby reducing the phase shift across each network. Typically, three *RC* sections are used.

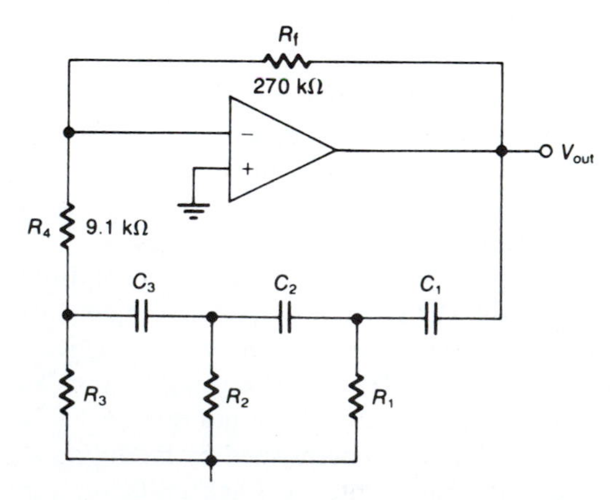


**Figure 6–6.** Phase-Shift Oscillator

The frequency of oscillation, $f_o$, can be obtained as follows:

$$f_o = \frac{1}{2\pi RC\sqrt{6}} \text{ (Hz)} \tag{6.4}$$

where

$R$'s are identical in each of the three sections
$C$'s are identical in each of the three sections
$B_v = 1/29$
Total phase shift = 180°

For oscillator action to start, $A_v$ must be greater than 29 (Barkhausen Criterion). This gain is set by the ratio of resistors $R_f$ and $R_4$.

## Example 6.1

Refer to Figure 6–6. Assume each $RC$ section has values of $R = 1 \text{ k}\Omega$ and $C = 0.1\ \mu\text{F}$. Calculate the oscillating frequency, $f_o$.

*Solution*

Use Equation 6.4:

$$f_o = \frac{1}{2\pi RC\sqrt{6}} \text{ (Hz)} = \frac{1}{2(3.14) \times (1 \text{ k}\Omega) \times (0.1\ \mu\text{F}) \times (2.45)}$$

$$= \frac{1}{6.28 \times 1000\ \Omega \times (0.1 \times 10^{-6}\text{F}) \times 2.45} = 650 \text{ Hz}$$

Designing any oscillator is quite simple. Capacitors are available in many fixed values, so the key to the design is to determine what frequency you wish to operate with, then select a standard fixed-value capacitor. By algebraic manipulation of Equation 6.4, we can solve for $R$:

$$R = \frac{1}{2\pi f_o C\sqrt{6}} \ (\Omega) \tag{6.5}$$

The value of $R$ can be found by substituting into Equation 6.5 the values of the frequency and the capacitor you have chosen. The reasoning here is that resistance values are more easily varied than are capacitance values; that is, a wider range of resistor values are available than are capacitor values.

**Example 6.2**

Design a phase-shift oscillator with three *RC* sections and an oscillating frequency of 1 kHz.

*Solution*

Select standard 0.02 μF capacitors for the *RC* sections, then calculate *R*. From Equation 6.5,

$$R = \frac{1}{2\pi f_o C\sqrt{6}}\,(\Omega) = \frac{1}{2(3.14) \times (1\text{ kHz}) \times (0.02\ \mu\text{F}) \times (2.45)}$$

$$= \frac{1}{6.28 \times 1000\text{ Hz} \times (0.02 \times 10^{-6}\text{F}) \times 2.45} = 3250\ \Omega$$

To approximate the designed frequency, you could use standard 3300 ohm resistors. For accuracy, you should use precision 1% 3250 ohm resistors.

### Wien-Bridge Oscillator

Like the phase-shift oscillator, the Wien-bridge oscillator uses *RC* networks. However, in the Wien-bridge oscillator, the *RC* networks are part of a *bridge circuit* that provides both regenerative and degenerative feedback. The networks select the frequency at which the feedback occurs, but they do not shift the phase of the feedback signal.

Feedback is applied to both inputs of the op amp, as shown in Figure 6–7. The frequency-selective network, sometimes called the *lead-lag* network, consists of $C_1$—$R_1$ and $C_2$—$R_2$ and provides regenerative feedback to the noninverting input terminal. Degenerative feedback is developed across $R_3$ and $R_4$ and is applied to the inverting input terminal. $R_4$ is made variable so that negative feedback can be reduced, because the positive feedback must be greater than the negative feedback in order for the circuit to sustain oscillations. The setting of $R_4$ is such that the circuit will start oscillating. The ratio of $R_4$ to $R_3$ must be 2:1 for proper operation.

Since resistance values in the degenerative feedback path do not change with frequency, degenerative feedback remains constant. However, regenerative feedback depends on the frequency response of the frequency-selective network, which is frequency sensitive. If oscillator frequency begins to increase, the reactance of $C_2$ will decrease and shunt some positive feedback to ground, decreasing regenerative feedback. Likewise, if the frequency begins to decrease, the reactance of $C_1$ becomes greater, causing less voltage to be developed across the $C_2$–$R_2$ network, and thus reducing regenerative feedback. Therefore, this network forces the oscillator to stay on its operating frequency.

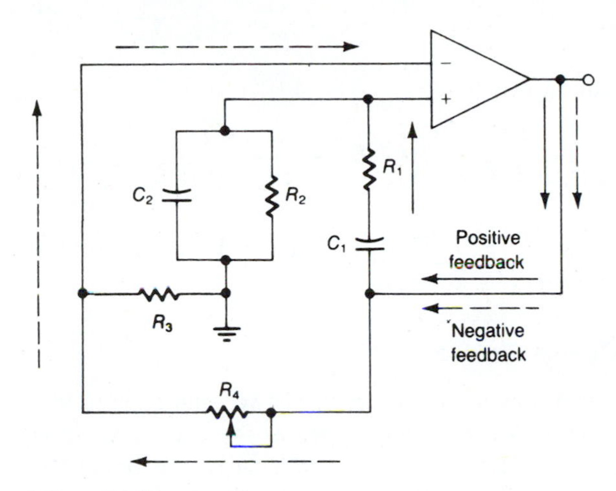


**Figure 6–7.** Wien-Bridge Oscillator

The circuit output frequency is determined by the values of $C_1$, $C_2$, $R_1$, and $R_2$ and can be calculated as follows:

$$f_o = \frac{1}{2\pi\sqrt{R_1R_2C_1C_2}} \text{ (Hz)} \tag{6.6}$$

However, if $R_1 = R_2$ and $C_1 = C_2$, then

$$f_o = \frac{1}{2\pi R_1C_1} \text{ (Hz)} \tag{6.7}$$

## Example 6.3

Refer to Figure 6–7. Assume $R_1 = R_2 = 20\ \text{k}\Omega$ and $C_1 = C_2 = 1$ nF. Calculate $f_o$.

*Solution*

From Equation 6.7,

$$f_o = \frac{1}{2\pi R_1C_1} \text{ (Hz)} = \frac{1}{2(3.14) \times (20\ \text{k}\Omega) \times (1\ \text{nF})}$$

$$= \frac{1}{6.28 \times (20 \times 10^3\ \Omega) \times (1 \times 10^{-9}\ \text{F})} = 8\ \text{kHz}$$

## Example 6.4

Refer to Figure 6–7. Assume $R_1 = 10\ \text{k}\Omega$, $R_2 = 20\ \text{k}\Omega$, $C_1 = 0.5$ nF, $C_2 = 1$ nF. Calculate $f_o$.

*Solution*

From Equation 6.6,

$$f_o = \frac{1}{2\pi\sqrt{R_1R_2C_1C_2}}\ (\text{Hz}) = \frac{1}{2(3.14)\times\sqrt{(10\ \text{k}\Omega)\times(20\ \text{k}\Omega)\times(0.5\ \text{nF})\times(1\ \text{nF})}}$$

$$= \frac{1}{6.28\times\sqrt{(10\times10^3\ \Omega)\times(20\times10^3\ \Omega)\times(0.5\times10^{-9}\ \text{F})\times(1\times10^{-9}\ \text{F})}}$$

$$= 1600\ \text{Hz} = 1.6\ \text{kHz}$$

### Sine-Cosine Oscillator

It is sometimes desirable to have two sine waves, 90° out of phase, in an electronic system. A circuit that provides the two waves is known as a *sine-cosine* oscillator, or a *quadrature* oscillator. Basically, the circuit consists of two op amps connected as integrators with positive feedback. In Figure 6–8 a dual op amp is used to construct the oscillator. The output of the first op amp is the sine wave, while the output of the second op amp is the cosine wave, 90° out of phase with the sine wave.

The value of resistor $R_1$ is made slightly less than that of $R_2$ to ensure that the oscillator circuit starts oscillating. If $R_1$ has too low a value, the outputs may be clipped. Using a potentiometer for $R_1$ would allow adjustment for minimum output distortion.

Another possible problem is saturation of the op amps. If this occurs, two zener diodes connected anode-to-anode across $C_2$ will limit the cosine output at the zener voltage. This limiting circuit is illustrated in Figure 6–8 by $Z_1$ and $Z_2$.

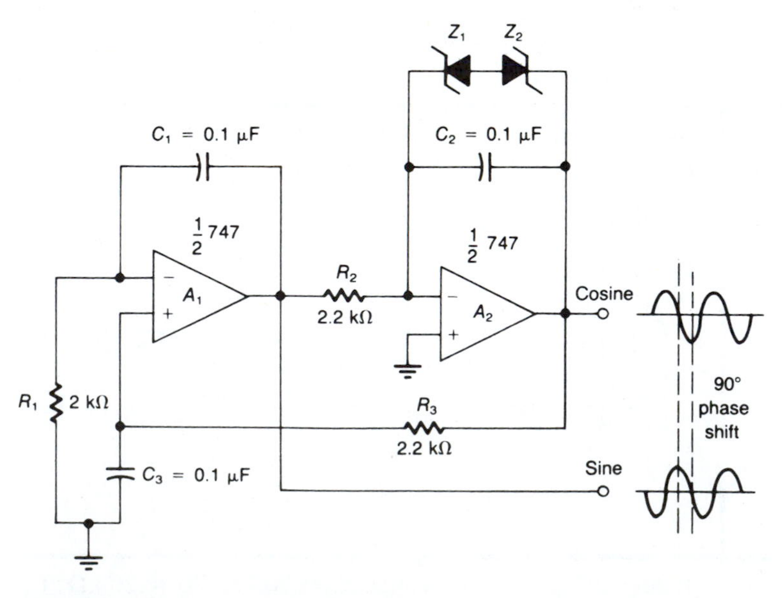


**Figure 6–8.** Sine-Cosine (Quadrature) Oscillator

The output frequency of the sine-cosine oscillator is

$$f_o = \frac{1}{2\pi R_2 C_2} \text{ (Hz)} \quad (6.8)$$

where

$R_1 < R_2$
$R_2 = R_3$
$C_1 = C_2 = C_3$

## Example 6.5

Calculate the output frequency of the sine-cosine oscillator in Figure 6–8.

*Solution*

From Equation 6.8,

$$f_o = \frac{1}{2\pi R_2 C_2} \text{ (Hz)} = \frac{1}{2(3.14) \times (2.2 \text{ k}\Omega) \times (0.1 \ \mu\text{F})}$$

$$= \frac{1}{6.28 \times (2.2 \times 10^3 \ \Omega) \times (1 \times 10^{-7} \text{ F})} = 720 \text{ Hz}$$

## Example 6.6

Design a sine-cosine oscillator with an output frequency of 1.5 kHz.

*Solution*

First, select a standard capacitor, say 0.047 μF. Then, adapt Equation 6.8 to the requirements of the example:

$$R_2 = \frac{1}{2\pi f_o C_2} (\Omega) = \frac{1}{2(3.14) \times (1.5 \text{ kHz}) \times (0.047 \ \mu\text{F})}$$

$$= \frac{1}{6.28 \times (1.5 \times 10^3 \text{ Hz}) \times (47 \times 10^{-9} \text{ F})} = 2260 \ \Omega$$

To approximate the designed frequency, you could use a standard 2200 ohm resistor. For accuracy, you should use a precision 1% 2260 ohm resistor.

## Crystal Oscillators

Oscillator instability is a problem common to all the oscillators we have discussed. This problem stems from several sources: temperature changes, aging of components, quality ($Q$) of the circuits, and circuit design. The use of crystals in the oscillator circuits provides the desired stability.

A crystal used in oscillator circuits must have the property of *piezoelectricity;* that is, the qualities of (1) generating a difference of potential across its faces when subjected to mechanical pressure, and (2) compressing when a difference of potential is applied across its faces. A crystal has a *natural frequency of vibration,* $f_n$, that provides an electrical signal from the crystal. This natural frequency of vibration is extremely constant, which makes the crystal ideal for oscillator circuits.

The natural frequency of a crystal is normally determined by its thickness. The thinner the crystal, the higher its natural frequency. Conversely, the thicker the crystal, the lower its natural frequency. There are practical limits, however, on how thin a crystal can be cut without becoming so fragile that it is easily fractured. The highest fundamental frequency of a common quartz crystal is in the range of 15 to 20 MHz. Above this frequency range, *harmonics* must be used. A harmonic frequency is any multiple of the fundamental frequency. For example, if the fundamental frequency is 20 MHz, the second harmonic is 40 MHz; the third harmonic is 60 MHz, and so on.

## Overtone Crystals

Crystal oscillators may operate either at the fundamental frequency of the crystal, or at one of the harmonic frequencies. Crystals designed especially for operation at a harmonic frequency are called *overtone* crystals. Overtone crystals are almost always used at frequencies above 20 MHz, because a fundamental-frequency crystal would be too thin at these frequencies, and might easily crack.

### Series-Resonant Crystal Circuit

An unmounted crystal, as in Figure 6–9A, appears electrically to be a series-resonant circuit with minimum impedance. In the equivalent circuit, illustrated in Figure 6–9B, the crystal mass that causes vibration is represented by inductor $L$, crystal stiffness is represented by capacitor $C$, and the electrical equivalent of internal resistance caused by friction is represented by resistor $R$.

### Parallel-Resonant Circuit

Normally a crystal is mounted between two metal plates that secure the crystal and provide electrical contact. The normal schematic symbol for the crystal is shown in Figure 6–10A. The symbol is derived from the manner in which the

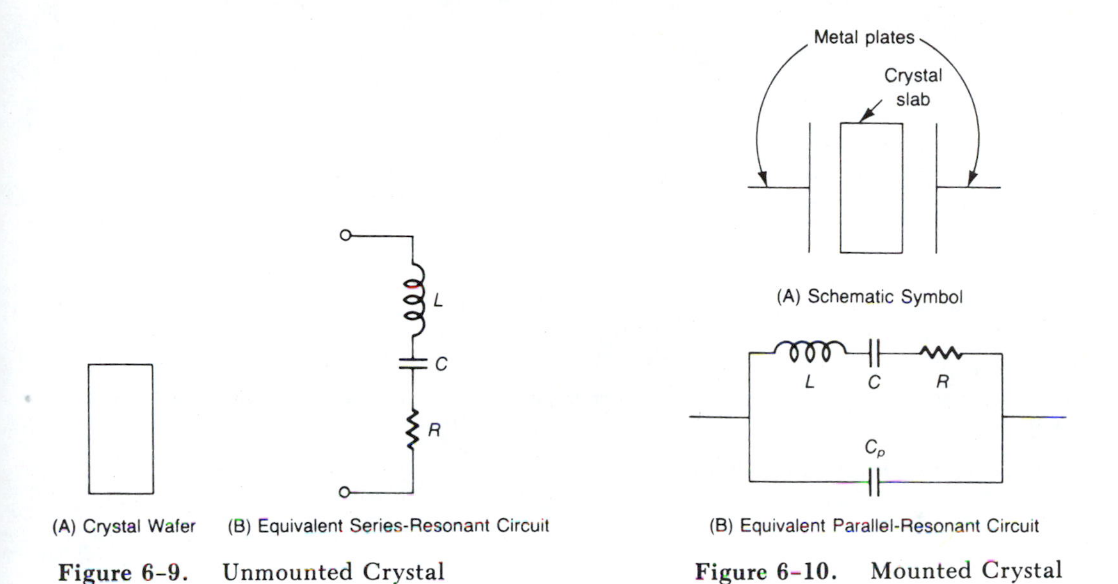


(A) Crystal Wafer (B) Equivalent Series-Resonant Circuit

**Figure 6–9.** Unmounted Crystal

(A) Schematic Symbol

(B) Equivalent Parallel-Resonant Circuit

**Figure 6–10.** Mounted Crystal

crystal is mounted and represents the crystal wafer held between two plates. When mounted in this manner, the crystal appears electrically to be a parallel-resonant circuit with maximum impedance. In the equivalent circuit, illustrated in Figure 6–10B, capacitor $C_p$ represents the metal plates in parallel with the series-resonant circuit of the crystal wafer. The value of this equivalent capacitance is relatively high and at low frequencies has little effect on the series-resonant circuit of the crystal. However, at frequencies above the series-resonant frequency of the crystal, the inductive reactance of the crystal increases, the capacitive reactance decreases, and the crystal appears to be inductive. At a point where the frequency is slightly higher than the series-resonant frequency of the crystal, inductive reactance of the crystal equals the capacitive reactance of the plates, and the circuit is parallel-resonant.

### Hartley Circuit

A Hartley crystal-controlled oscillator (CCO) is shown in Figure 6–11A. The crystal is connected in series with the feedback path. The *LC* network is tuned to the series-resonant frequency of the crystal; therefore, the crystal operates at its series-resonant frequency.

When the oscillator is operating at the crystal frequency, the equivalent circuit offers minimum impedance. Therefore, there is minimum opposition to current flow, and the feedback is maximum. If the oscillator drifts away from the crystal frequency, the impedance increases, reducing feedback, and thereby forcing the oscillator to return to the crystal frequency. The conclusion can be drawn that when the crystal is series-connected, it controls the amount of feedback.

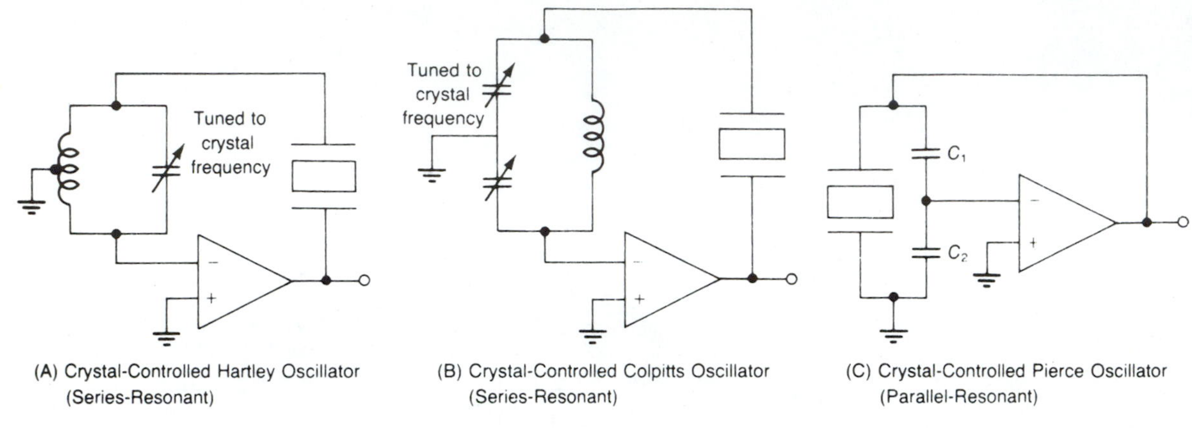


(A) Crystal-Controlled Hartley Oscillator (Series-Resonant)

(B) Crystal-Controlled Colpitts Oscillator (Series-Resonant)

(C) Crystal-Controlled Pierce Oscillator (Parallel-Resonant)

**Figure 6–11.** Crystal-Controlled Oscillators

### Colpitts Circuit

A Colpitts CCO is shown in Figure 6–11B. In this circuit the crystal is again connected in series with the feedback path. The *LC* network is tuned to the crystal frequency, and the series-connected crystal controls the amount of feedback, just as in the Hartley CCO.

### Pierce Oscillator

A variation of the Colpitts CCO, sometimes called a *Pierce oscillator,* is illustrated in Figure 6–11C. In this circuit the crystal replaces the inductor of a standard Colpitts oscillator circuit. The crystal operates at its parallel-resonant frequency, which is slightly higher than the series-resonant frequency of the crystal, and appears as an inductance.

Positive feedback results through the 180° phase shift provided by the voltage divider arrangement of capacitors $C_1$ and $C_2$. The ratio of these two capacitors also determines the feedback ratio and the crystal excitation voltage. The crystal will provide a very stable output, since its response is extremely sharp, vibrating over a very narrow range of frequencies.

Since the crystal operates in its parallel-resonant mode, it controls the impedance of the tuned circuit. At resonance, circuit impedance is maximum and a large feedback voltage is developed across capacitor $C_1$. If frequency drifts away from resonance, crystal impedance decreases, and feedback decreases. Thus, by controlling tuned circuit impedance, the crystal effectively determines the amount of feedback and provides a highly stable oscillator.

A final word on crystals: in some applications, such as transmitters, crystals may be mounted inside ovens in which temperature is tightly controlled to insure the high stability and tight frequency control required.

## Voltage-Controlled Oscillators

In the oscillators discussed up to this point, changing the frequency of oscillation requires changing the value of one or more of the frequency-determining components. Such an operation requires some manual manipulation by the operator, with a resultant loss of operating time. A device that makes changes in the frequency-determining components automatically and quickly is the *voltage-controlled oscillator* (VCO).

The frequency of the VCO is controlled by a signal voltage; the amount of change in frequency is directly proportional to the input voltage level. The basic circuit in the operation of the VCO is a sine wave oscillator in which the feedback is varied electronically, thus changing the output frequency. Another technique is the generation of a square wave rather than a sine wave, in which case a low-pass (LP) filter is used to remove all unwanted harmonic frequencies, leaving the desired sine wave.

One version of a VCO that is shown in Figure 6–12 uses part of a μA3403, a quad op amp device. Amplifier $A_1$ is an integrator. It has a triangle wave output whose slope is determined by the input terminal with the more positive input voltage. Amplifier $A_2$ is a comparator whose output is a constant positive or negative voltage, determined by its input terminal with the more positive input voltage.

If the output of the comparator is initially HIGH, the transistor will be driven to saturation. If $V_{CE(sat)}$ is ignored, then the voltage at the inverting input terminal of $A_1$ will be equal to $1/3V_c$ (control voltage) due to the voltage divider arrangement of $R_1$–$R_2$. The 51 kΩ resistor divider sets the noninverting input voltage

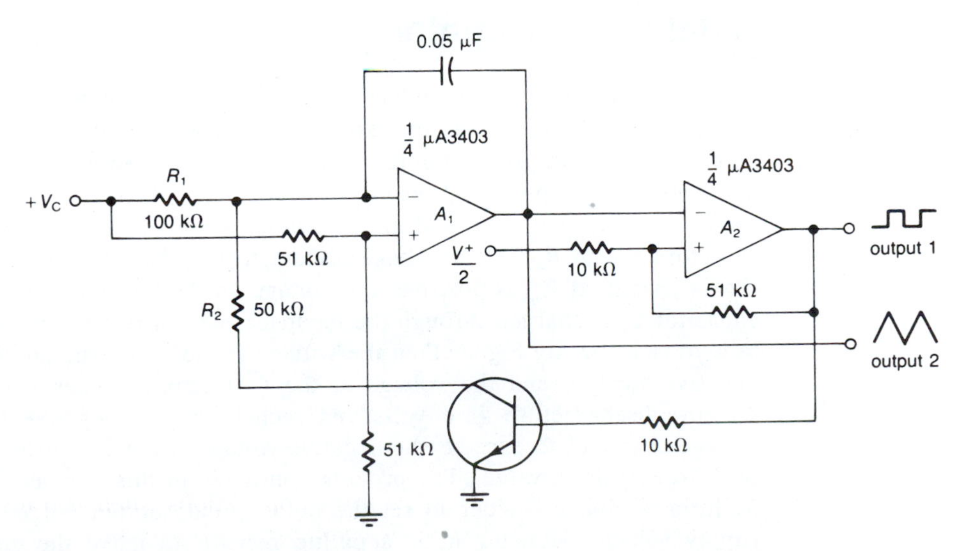


**Figure 6–12.** Voltage-Controlled Oscillator

to $1/2V_c$. This generates a negative-going ramp at output 2. When the integrator output voltage goes below the voltage at the (+) terminal of $A_2$, the output of $A_2$ goes HIGH and the cycle repeats.

The (+) terminal of the comparator has an input voltage determined by $V^+/2$ and a feedback voltage from its output. As the output changes between two values, so does the (+) terminal voltage of $A_2$. It can therefore be deduced that the period of the triangle wave and, therefore, its frequency, and the frequency of the comparator output, are determined by the value of $V_c$. If $V_c$ varies, then the output frequency will change. $V_c$ can vary from 0 V to $2(V^+ - 1.5\text{ V})$ and will provide a wide range of frequencies.

The resulting square wave at output 1 can be filtered through an LP filter that removes the unwanted frequency components, and thus provides a desired sine wave. (Filters are discussed at length in Chapter 7.)

There are many forms and uses for the VCO. VCOs are used in FM systems to control the carrier frequency, in phase-locked loops (PLLs), and in many wave-shaping circuits. These applications will be discussed in the following chapters.

# 6.5 Multivibrators

There are two types of multivibrators used in analog systems. They are the free-running, or *astable*, multivibrator, and the one-shot, or *monostable*, multivibrator. Another multivibrator, the bistable, is used primarily in digital systems and will not be discussed here.

## Astable Multivibrator

A free-running multivibrator has a square wave output whose frequency is determined by the values of the externally connected components. Figure 6–13 shows a 100 kHz free-running multivibrator with an LM311 op amp voltage comparator (see Appendix I) as the active device. This is a self-starting circuit and requires no external input voltage. The input voltage is replaced with capacitor $C_1$. Resistors $R_4$ and $R_2$ form a voltage divider to feed back a portion of the output to the (+) input. If $V_{out}$ is positive, the voltage on the (+) input is also positive, and capacitor $C_1$ is charged through the feedback path of resistor $R_3$. When $C_1$ charges to a voltage slightly higher than the voltage at the (+) terminal, $V_{out}$ switches to a negative voltage, and the voltage at the (+) terminal goes negative. Capacitor $C_1$ now discharges to zero volts and recharges to a negative value. When the voltage charge of $C_1$ exceeds the negative voltage at the (+) terminal, $V_{out}$ switches back to a positive value. The process continues in this manner. Resistors $R_1$ and $R_2$ form a voltage divider to set the quiescent dc output voltage at one-half the supply voltage. Resistor $R_5$ is a pullup resistor to make the circuit compatible with transistor-transistor logic (TTL) and diode-transistor logic (DTL) circuits.

### Simplified Circuit

A simplified astable multivibrator is shown in Figure 6–14. Evaluation of this circuit is the same as for the more complicated circuit of Figure 6–13. The advantage of the simplified version is the ease of design.

The op amp is used as a comparator in this circuit. To determine oscillating frequency, capacitor charge and discharge times are needed. These times are determined by the values of $R_f$ and $C_1$. If we let the charge time equal $t_1$ and the discharge time equal $t_2$, then $t_1 + t_2 = T$, the period of one complete cycle. If we set the value of $R_f = R_2$ and $R_1 = 0.86R_2$, the period, $T$, is found as follows:

$$T = 2R_fC_1 \tag{6.9a}$$

Since frequency, $f$, is the reciprocal of the period, $T$, then

$$f = \frac{1}{T} = \frac{1}{2R_fC_1} \tag{6.9b}$$

where

$T$ is expressed in seconds
$f$ is expressed in hertz
$R_f$ is expressed in ohms
$C_1$ is expressed in farads

NOTE: A nonstandard resistor may be required for $R_1$.

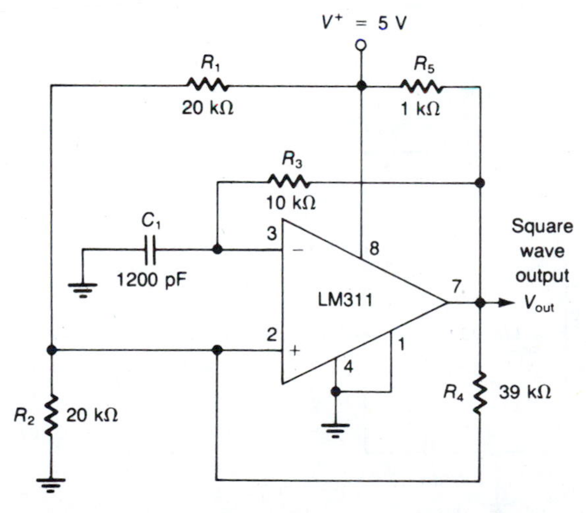


Figure 6–13. 100 kHz Free-Running Multivibrator (Courtesy of National Semiconductor Corporation)

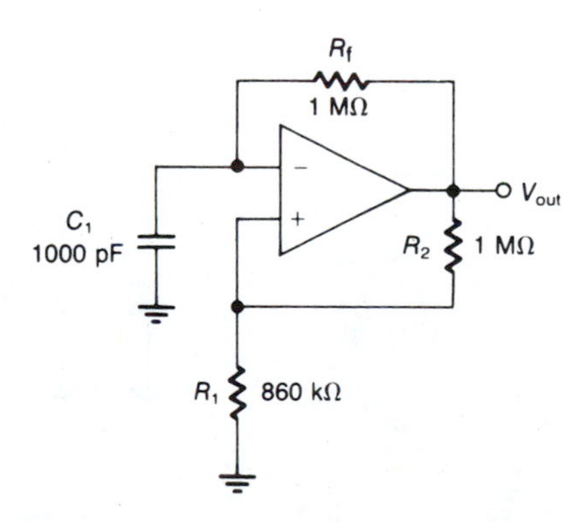


Figure 6–14. Simplified Astable Multivibrator

## Example 6.7

Refer to Figure 6–14. Calculate (a) the period and (b) the frequency of oscillation.

*Solution*

a. Use Equation (6.9a):

$$T = 2R_fC_1 = (2)(1\ \text{M}\Omega)(1000\ \text{pF}) = 2(1 \times 10^6\ \Omega) \times (1 \times 10^{-9}\ \text{F})$$
$$= 2 \times 10^{-3}\ \text{s} = 2\ \text{ms}$$

b. Use Equation (6.9b):

$$f = \frac{1}{T} = \frac{1}{2 \times 10^{-3}\ \text{s}} = 500\ \text{Hz}$$

## Monostable Multivibrator

A one-shot multivibrator generates a single output pulse whose pulse width *(PW)* is determined by the values of external components. A trigger signal at one of the input terminals of the op amp is required to generate the output pulse. Figure 6–15 shows a monostable multivibrator with a reference voltage ($V^+$) using one-half of an LM193 op amp comparator.

The output voltage is compared with the input trigger voltage at the inputs. The output state of the multivibrator is determined by the larger of the input

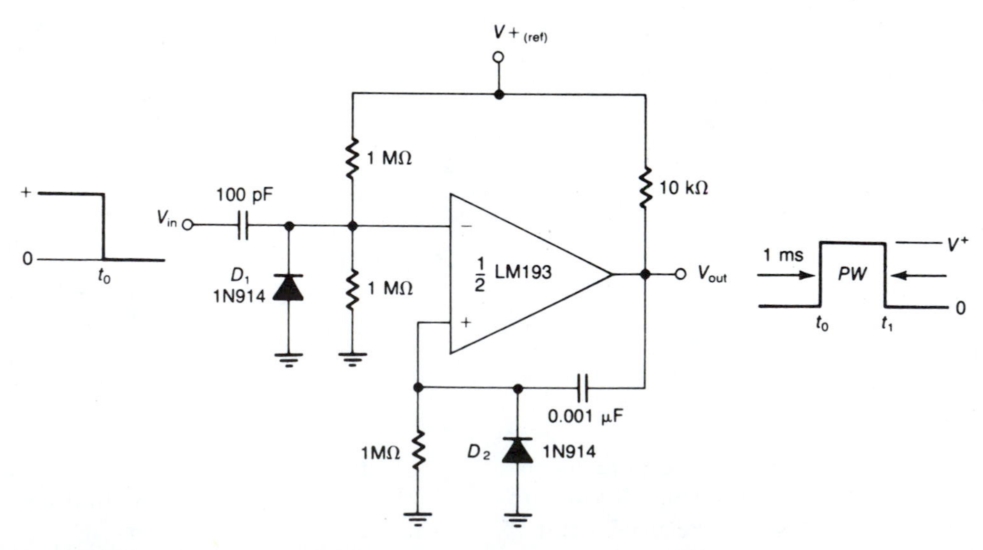


Figure 6–15. One-Shot Multivibrator with $V+$ Reference Voltage (Courtesy of National Semiconductor Corporation)

voltages. The trigger pulse is applied to the (-) input terminal through an *RC* high-pass (HP) filter circuit. Initially, the output is at zero volts, the normal state for this multivibrator. When a negative-going triggering pulse is applied to the (-) terminal, the output pulse rises to the $V^+$ reference voltage. This condition is known as the *unstable,* or *timing,* state; that is, it is not the normal state for the circuit. The output pulse will stay in this unstable state for 1 millisecond (ms), the period of time determined by one time constant of the 0.001 μF capacitor and the 1 MΩ resistor:

$$PW = RC \text{ time constant} \qquad \textbf{(6.10)}$$

The output will then return to its stable state until another input pulse is applied.

The 1N914 diode, $D_1$, prevents the device from triggering on the positive-going edge of the input pulse. Diode $D_2$ prevents the output signal from swinging negative.

## Example 6.8

Refer to Figure 6–15. Assume that $R$ = 100 kΩ and $C$ = 100 μF in the feedback line to the (+) terminal. What will be the PW of the output pulse?

*Solution*

Use Equation 6.10:

$$PW = RC = 100 \text{ k}\Omega \times 100\ \mu\text{F} = (1 \times 10^5\ \Omega) \times (1 \times 10^{-4}\ \text{F}) = 10 \text{ s}$$

Example 6.8 should clarify the relationship between component values and pulse duration.

### Circuit without Reference Voltage

A monostable multivibrator that requires no $V^+$ reference voltage is shown in Figure 6–16. In this circuit the negative input pulse is applied to the (+) terminal, resulting in a negative output pulse. The op amp comparator of this circuit has normal output approximately equal to the value of the power supply $V^+$ voltage. When a negative trigger pulse is applied, diode $D_1$ conducts, making the (+) terminal less positive than the (-) terminal, and the output switches to the $V^-$ supply voltage level. Diode $D_2$ is cut off and $C_1$ is charged toward $V^-$ through feedback resistor $R_f$. Resistors $R_1$ and $R_2$ hold the (+) terminal at the negative voltage level until the voltage at the (-) terminal becomes less negative than the voltage at the (+) terminal. At that point when the (-) terminal voltage drops below the (+) terminal voltage, the output switches back to the $V^+$ voltage level and completes the pulse duration cycle.

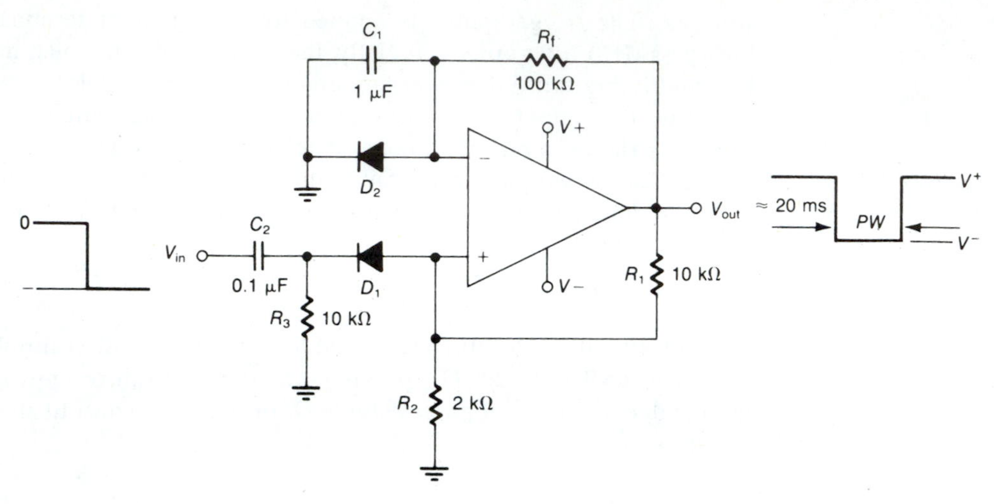


**Figure 6–16.** Monostable Multivibrator with No Reference Voltage

The pulse width of this circuit is determined as follows:

$$PW = R_f C_1 \ln\left(\frac{R_1 + R_2}{R_1}\right) \qquad \textbf{(6.11)}$$

where

ln = natural logarithm

A close approximation of the pulse duration can be obtained by the following equation:

$$PW = \frac{R_f C_1}{x} \qquad \textbf{(6.12a)}$$

where

$$x = \frac{R_1}{R_2}$$

so that

$$PW = \frac{R_f C_1 R_2}{R_1} \qquad \textbf{(6.12b)}$$

## Example 6.9

Refer to Figure 6–16. Calculate the pulse duration using (a) Equation 6.11 and (b) Equation 6.12b.

*Solution*

a.

$$PW = R_f C_1 \ln\left(\frac{R_1 + R_2}{R_1}\right) = 100\ \text{k}\Omega \times 1\ \mu\text{F} \times \ln\left(\frac{10\ \text{k}\Omega + 2\ \text{k}\Omega}{10\ \text{k}\Omega}\right)$$

$$= (1 \times 10^5\ \Omega) \times (1 \times 10^{-6}\ \text{F}) \times \ln\left(\frac{(1 \times 10^4\ \Omega) + (2 \times 10^3\ \Omega)}{1 \times 10^4\ \Omega}\right)$$

$$= 1.82 \times 10^{-2}\ \text{s} = 18.2\ \text{ms}$$

b.

$$PW = \frac{R_f C_1 R_2}{R_1} = \frac{100\ \text{k}\Omega \times 1\ \mu\text{F} \times 2\ \text{k}\Omega}{10\ \text{k}\Omega}$$

$$= \frac{(1 \times 10^5\ \Omega) \times (1 \times 10^{-6}\ \text{F}) \times (2 \times 10^3\ \Omega)}{1 \times 10^4\ \Omega} = 2 \times 10^{-2}\ \text{s}$$

$$= 20\ \text{ms}$$

Example 6.9 shows that the approximation formula (Equation 6.12b) provides accuracy to within 10%.

Reversing the diodes in the circuit of Figure 6–16 will produce a positive output pulse when a positive input pulse is applied to the (+) input terminal.

# 6.6 Differentiators

A *differentiator* is a short time-constant circuit whose output signal is the derivative of its input waveform. The circuit is used in analog computations and wave shaping, but is not limited to those applications.

A simple *RC* differentiator circuit is illustrated in Figure 6–17A. If $V_{out}$ is much less than $V_{in}$, ($V_{out} < V_{in}$), then the voltage across the resistance will be much less than the voltage across the capacitance, and

$$V_{out} = RC\frac{dV_{in}}{dt} \tag{6.13a}$$

where

$\frac{dV_{in}}{dt}$ = the change in input voltage over a specified time interval

To insure that $V_{out} < V_{in}$, either or both $R$ and $C$ must be made very small to cause very small values for $V_{out}$. Using an op amp in the circuit, as shown in Figure 6–17B, overcomes this problem. The desired result that $V_{out} < V_{in}$ is obtained in this circuit because the input impedance of the op amp with feedback resistor $R_f$ is a resistance equal to $R/(1 + A_r)$. This resistance occurs because the inverting input is at virtual ground, so the capacitor is isolated from $R_f$. The result is a very small effective resistance comparable to $R$ in Figure 6–17A, but the output voltage is no longer small because of the gain of the op amp.

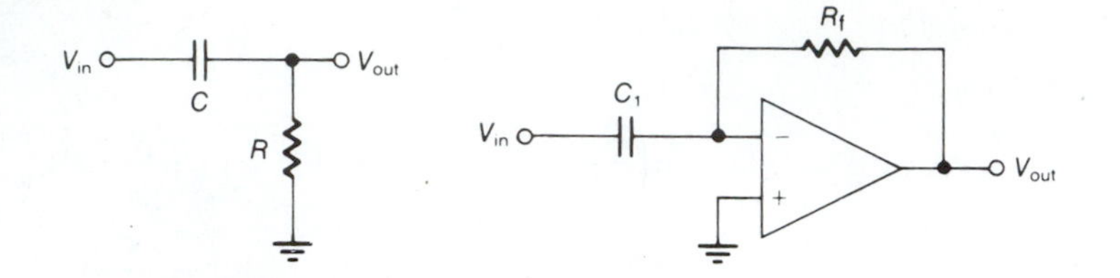


**Figure 6–17.** Differentiator Circuits

The op amp differentiator circuit is similar to the inverting amplifier, but the input element is a capacitor. Since the input is applied to the (−) terminal, the output voltage is

$$V_{out} = -R_f C_1 \frac{dV_{in}}{dt} \qquad \textbf{(6.13b)}$$

## Reducing High-Frequency Gain Disadvantages

There are two major disadvantages associated with the basic circuit illustrated in Figure 6–17B. These occur because the reactance of the capacitor varies inversely with frequency. Therefore, as frequency increases, capacitive reactance decreases and stage gain increases, creating a tendency for the circuit to go into undesirable oscillations at the high-frequency end of operation. The increase in gain of high frequencies also allows amplification of high-frequency noise signals that may override the desired output signal. Adding external components will reduce the effect of these disadvantages.

Placing a resistor in series with the capacitor, as shown in Figure 6–18A, limits the high-frequency gain to the ratio of $R_f/R_1$. The output voltage is still calculated by Equation 6.13b. This circuit only acts as a differentiator, however, for input frequencies less than

$$f_{op} = \frac{1}{2\pi R_1 C_1} \qquad \textbf{(6.14)}$$

For input frequencies *greater* than those determined by Equation 6.14, the circuit begins to take on the characteristics of an inverting amplifier with the following output voltage gain:

$$A_v = -\frac{V_{out}}{V_{in}} = -\frac{R_f}{R_1} \qquad \textbf{(6.15)}$$

In practice, $R_1$ is usually around 100 Ω. With reference to Equation 6.13b, the time constant of $R_f C_1$ should be approximately equal to the period of the input signal to be differentiated.

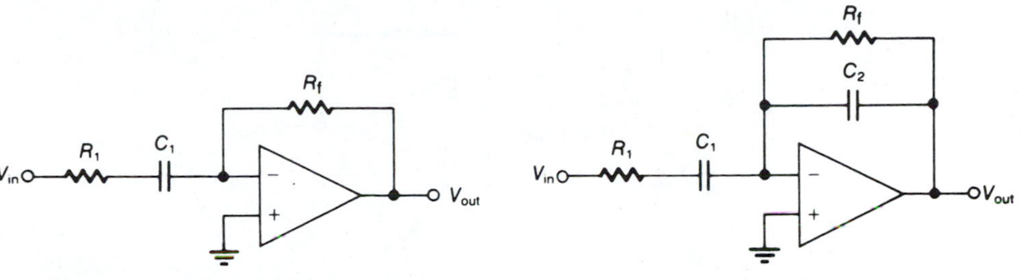


**Figure 6–18.** Differentiator Circuits for Limiting Gain Disadvantages

Figure 6-18B shows the addition of a capacitor across the feedback resistor. This capacitor causes the gain of the differentiator to drop off rapidly above a predetermined frequency, thus further reducing the possibility of high-frequency noise signals and undesirable oscillations. The sharp drop in gain also reduces the opportunity for the circuit to take on the characteristics of an inverting amplifier.

For this circuit, there are two calculations to be made: the input frequency, $f_{op}$, and the upper cutoff frequency, $f_c$. They may be calculated as follows:

$$f_{op} = \frac{1}{2\pi R_1 C_1} \qquad \text{(repeat of Equation 6.14)}$$

and

$$f_c = \frac{1}{2\pi R_f C_2} \tag{6.16}$$

## Example 6.10

Design an op amp circuit that will differentiate a 60 Hz input signal. Limit the high-frequency gain to 20.

*Solution*

With $f_{op}$ = 60 Hz, $T$ = 1/60 = 16.7 ms; therefore, the time constant $R_f C_1$ from Equation 6.13b must = 16.7 ms. Selecting $R_1$ = 100 Ω, then $R_f$ must = 2 kΩ, and

$$C_1 = \frac{TC}{R_f}$$

$$= \frac{16.7 \text{ ms}}{2 \text{ k}\Omega} = \frac{16.7 \times 10^{-3} \text{ s}}{2 \times 10^3 \ \Omega} = 8.35 \times 10^{-6} \text{ F} = 8.35 \ \mu\text{F} \tag{6.17}$$

The final practical circuit is shown in Figure 6–19.

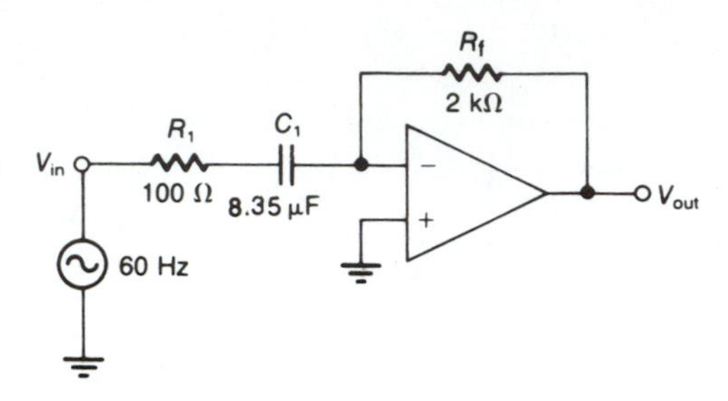


Figure 6–19. 60 Hz Differentiator Circuit with High-Frequency Gain Limit of 20

## Example 6.11

Using the practical circuit of Figure 6–19, determine what value capacitor would be required across $R_f$ to cut off frequencies at 400 Hz.

*Solution*

Use Equation 6.16 and substitute known values from Figure 6–19:

$$C_2 = \frac{1}{2\pi R_f f_c}$$

$$= \frac{1}{6.28 \times 2\ \text{k}\Omega \times 400\ \text{Hz}} = \frac{1}{6.28 \times (2 \times 10^3\ \Omega) \times (4 \times 10^2\ \text{Hz})}$$

$$= 1.99 \times 10^{-7}\ \text{F} \approx 0.2\ \mu\text{F}$$

## Sine Wave Differentiator

For a sine wave input signal, the differentiator output voltage is a function of time and can be calculated as follows:

$$V_{out} = -\omega R_f C V_m \cos(\omega t) \tag{6.18}$$

where

$\omega$ = input frequency in radians/second (rad/s) = $2\pi f$
$V_m$ = peak voltage of the input wave

The output signal is a cosine wave, shifted 90° from the input sine wave, but also inverted because the input signal is applied to the (−) input terminal. Therefore, the total phase shift is 270° from the sine wave input signal. The peak output voltage is

$$V_{out(peak)} = \omega R_f C V_m \tag{6.19}$$

The input and output waveforms for the sine wave differentiator are shown in Figure 6–20.

## Example 6.12

Refer to Figure 6–19. If the 60 Hz input signal is a 1 V peak sine wave, what will be the peak output voltage? Assume $C_1 = 1\ \mu F$.

*Solution*

Use Equation 6.19:

$$\begin{aligned} V_{out(peak)} &= \omega R_f C V_m = 6.28 \times 60\text{ Hz} \times 2\text{ k}\Omega \times 1\ \mu\text{F} \times 1\text{ V} \\ &= 6.28 \times 60\text{ Hz} \times (2 \times 10^3\ \Omega) \times (1 \times 10^{-6}\text{ F}) \times 1\text{ V} \\ &= 0.75\text{ V} \end{aligned}$$

## Triangle Wave Differentiator

The input and output waveforms for a triangle wave input differentiator are shown in Figure 6–21. The output is a square wave with the following peak value:

$$V_{out(peak)} = \frac{\pm R_f C(2V_m)}{t_1 \text{ or } t_2} \tag{6.20}$$

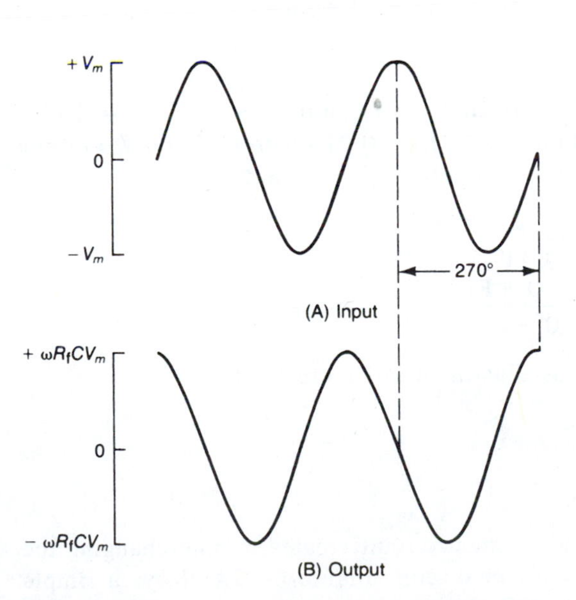


**Figure 6–20.** Input/Output Waveforms for Sine Wave Input Differentiator

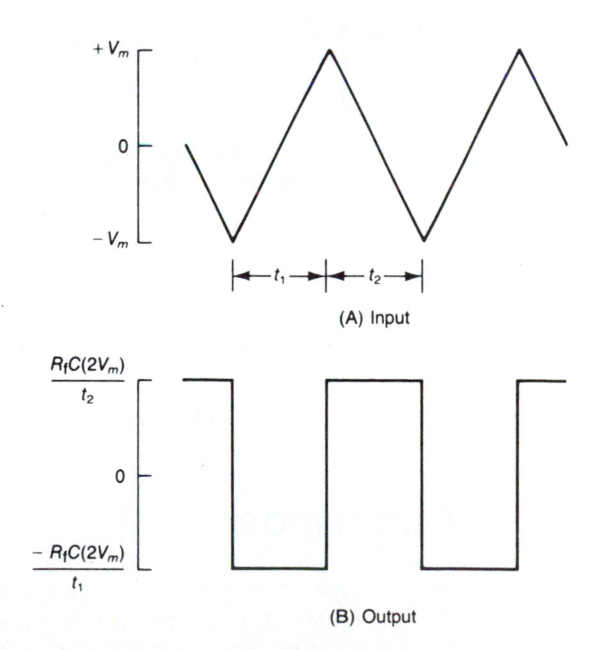


**Figure 6–21.** Input/Output Waveforms for Triangle Wave Input Differentiator

For a triangle wave input signal, the input frequency is

$$f = \frac{1}{t_1 + t_2} \tag{6.21}$$

The voltages for the input signals during time periods $t_1$ and $t_2$ are given by the following equations:

$$V_{\text{in}(t_1)} = -V_m + 2\frac{V_m}{t_1}T \tag{6.22}$$

$$V_{\text{in}(t_2)} = V_m - 2\frac{V_m}{t_2}T \tag{6.23}$$

The output voltages for the two time periods are as follows:

$$V_{\text{out}(t_1)} = \frac{-R_fC(2V_m)}{t_1} \tag{6.24}$$

$$V_{\text{out}(t_2)} = \frac{R_fC(2V_m)}{t_2} \tag{6.25}$$

**Example 6.13**

Refer to Figure 6–19. Assume a symmetrical 2 V peak-to-peak 50 Hz triangle wave input. What will be the peak output voltage? Assume $C_1 = 1\ \mu\text{F}$.

*Solution*

Since the triangle wave is symmetrical, $t_1 = t_2$, and $t_1 + t_2 = T = 1/f = 1/50\ \text{Hz} = 0.02$ seconds. Therefore, $t_1 = t_2 = 0.01$ seconds. Solve for either the (+) or (−) output voltage. We will solve for (+), that is, $V_{\text{out}(t_2)}$:

$$V_{\text{out}(t_2)} = \frac{R_fC(2V_m)}{t_2} = \frac{2\ \text{k}\Omega \times 1\ \mu\text{F} \times 2 \times 1\ \text{V}}{0.01\ \text{s}}$$

$$= \frac{(2 \times 10^3\ \Omega) \times (1 \times 10^{-6}\ \text{F}) \times 2 \times 1\ \text{V}}{1 \times 10^{-2}\ \text{s}} = +0.4\ \text{V}$$

$V_{\text{out}(t1)}$ would then be −0.4 volts, as illustrated in Figure 6–21.

# 6.7 Integrators

An op amp *integrator* is a long time-constant circuit created by interchanging the resistor and capacitor of the differentiator circuit. Figure 6–22A shows a simple *RC* integrator circuit, and Figure 6–22B shows the basic op amp circuit. In this circuit, the resistor is the input element and the capacitor is the feedback element. The integrator circuit is, effectively, the inverse of the differentiator circuit.

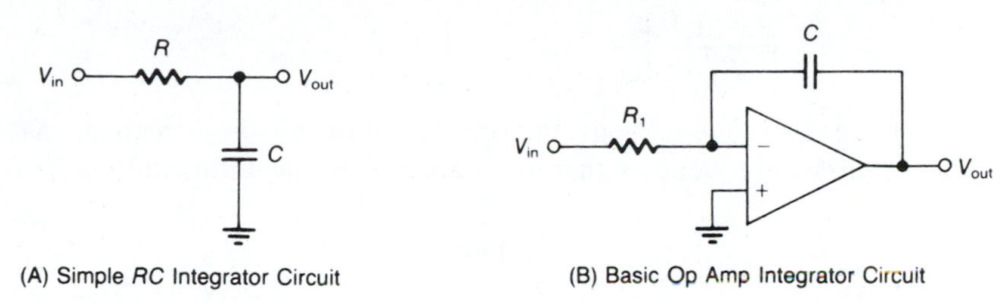


**Figure 6–22.** Integrators

The integrator output voltage as a function of time is expressed by the following equation:

$$V_{out} = -\frac{1}{R_1 C} t V_{in} \tag{6.26}$$

where

$V_{in}$ = peak input voltage
$t = t_1$ or $t_2$

The time constant $R_1C$ is made approximately equal to the period of the input signal to be integrated. A more practical integrator circuit can be obtained by placing a shunt resistor, $R_{sh}$, across feedback capacitor $C$, as shown in Figure 6–23A. In general, $R_{sh}$ is made approximately ten times the value of $R_1$. The shunt resistor limits the low-frequency gain of the circuit, which prevents integration of the dc offset voltage over the integration period. Resistor $R_2$ minimizes the dc offset voltage caused by the input bias current. The value for $R_2$ is

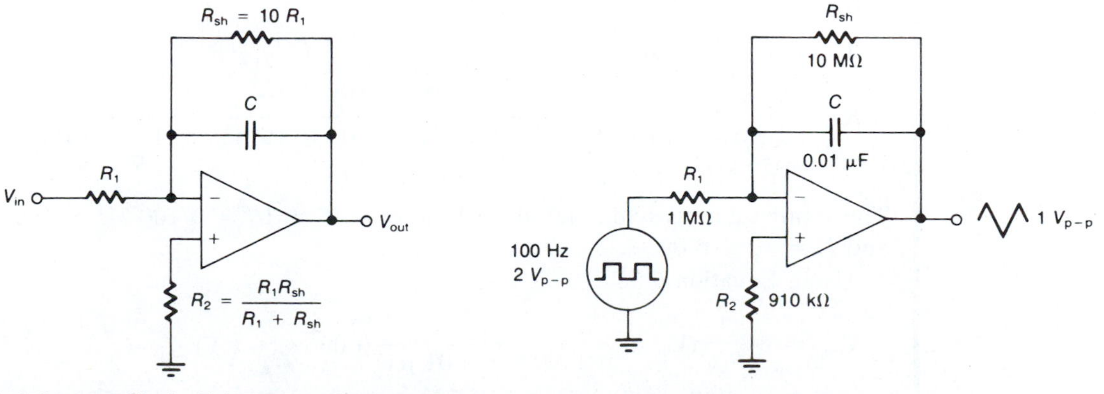


**Figure 6–23.** Integrator Circuits with Limited Low-Frequency Gain

$$R_2 = \frac{R_1 R_{sh}}{R_1 + R_{sh}} \quad \textbf{(6.27)}$$

The low-frequency gain-limiting action of the shunt resistor makes Equation 6.26 valid for frequencies that are *greater* than the following cutoff frequency:

$$f_c = \frac{1}{2\pi R_{sh} C} \quad \textbf{(6.28)}$$

For input frequencies *lower* than the $f_c$ of Equation 6.28, the circuit begins to take on the characteristics of the inverting amplifier with a voltage gain of

$$-\frac{V_{out}}{V_{in}} = -\frac{R_{sh}}{R_1} \quad \textbf{(6.29)}$$

## Example 6.14

Refer to Figure 6–23A. Assume a symmetrical 2 V peak-to-peak 100 Hz square wave input signal. Determine values for $R_1$, $R_2$, $R_{sh}$, $C$, and the peak output voltage.

### *Solution*

The time constant $R_1C$ should be approximately equal to the period of the input signal, so select a standard value for $C$, say 0.01 μF, and calculate $R_1$.

$$R_1 = \frac{1}{f_{op}C} = \frac{1}{100\text{ Hz} \times 0.01\ \mu\text{F}} = \frac{1}{(1 \times 10^2\text{ Hz}) \times (1 \times 10^{-8}\text{ F})}$$
$$= 1 \times 10^6\ \Omega = 1\text{ M}\Omega$$

then

$$R_{sh} = 10R_1 = 10(1\text{ M}\Omega) = 10\text{ M}\Omega$$

Using Equation 6.27,

$$R_2 = \frac{1\text{ M}\Omega \times 10\text{ M}\Omega}{1\text{ M}\Omega + 10\text{ M}\Omega} = \frac{(1 \times 10^6\ \Omega) \times (10 \times 10^6\ \Omega)}{(1 \times 10^6\ \Omega) + (10 \times 10^6\ \Omega)} = \frac{10 \times 10^{12}\ \Omega}{11 \times 10^6\ \Omega}$$
$$= 910\text{ k}\Omega$$

The input signal is symmetrical, so $t_1 + t_2 = T = 1/f = 1/100\text{ Hz} = 0.01$ s, and $t_1 = t_2 = 0.005$ s.

Using Equation 6.26,

$$V_{out} = -\frac{1}{R_1C} tV_{in} = -\frac{1}{1\text{ M}\Omega \times 0.01\ \mu\text{F}}(0.005\text{ s})(-1\text{ V})$$
$$= -(100)(-1)(0.005) = +0.5\text{ V}$$

The completed circuit is shown in Figure 6–23B.

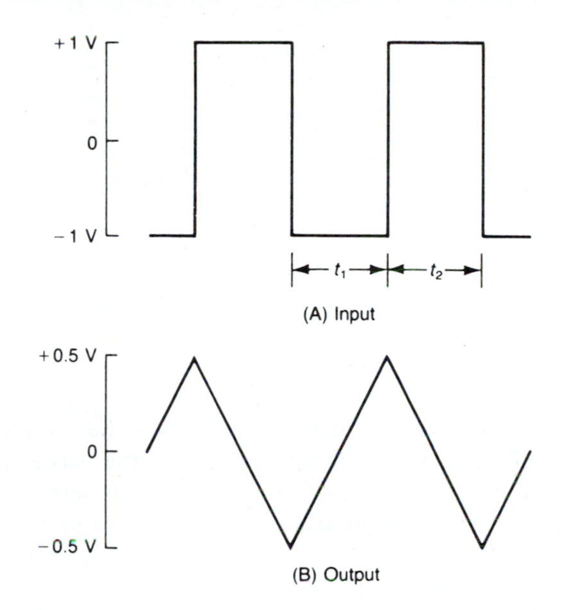


**Figure 6–24.** Input/Output Waveforms for Square Wave Input Integrator

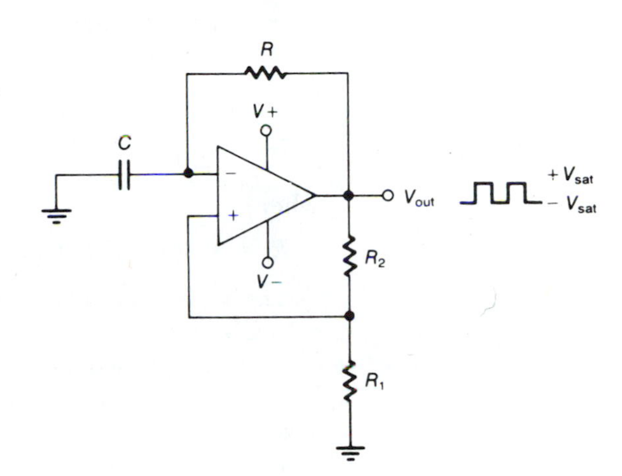


**Figure 6–25.** Basic Square Wave Generator

The input and output waveforms for Example 6.14 are shown in Figure 6–24. Note that the negative alternation of the output triangle wave is of equal amplitude but opposite polarity to the positive alternation; that is, $V_{out}$ for $t_2$ equals -0.5 V.

# 6.8 Square Wave Generator

The *square wave generator* provides an output that is constantly changing states between high and low with no input signal applied. It is, therefore, an oscillator circuit, or a multivibrator. The square wave generator is sometimes called a *relaxation oscillator* because the output frequency is set by the charging and discharging of a capacitor. Figure 6–25 shows a basic square wave generator. Note that there are two feedback paths for this circuit. The *RC* combination at the inverting input terminal determines the fundamental operating frequency of the circuit, and resistors $R_1$ and $R_2$ form a voltage divider that produces a reference voltage to the noninverting input terminal. There are two methods for determining the output frequency, depending upon the values given the resistors:

$$f_{op} = \frac{1}{2RC \ln\left(\frac{2R_1}{R_2} + 1\right)} \tag{6.30}$$

where

$R_1 = R/3$
$R_2 = 2$ to 10 times $R_1$
ln = natural logarithm

and

$$f_{op} = \frac{1}{2RC} \tag{6.31}$$

where

$R_1 = 0.86R_2$ (Figure 6–25)

The reference voltage at the noninverting input terminal causes the circuit to behave as a voltage-level detector. The output of the op amp will swing between $+V_{sat}$ and $-V_{sat}$, based on the charge-discharge of the capacitor, which sets the threshold voltage $\pm V_{th}$ at the (–) input terminal of the op amp. The saturation voltages $\pm V_{sat}$ are always about 2 V less than the supply voltages $V^+$ and $V^-$. The action of the capacitor voltage versus the op amp output voltage is illustrated in Figure 6–26.

$t_{rise}$ = rise time = $RC$ $\quad$ $t_{fall}$ = fall time = $RC$
$T = t_{rise} + t_{fall} = 2\ RC$

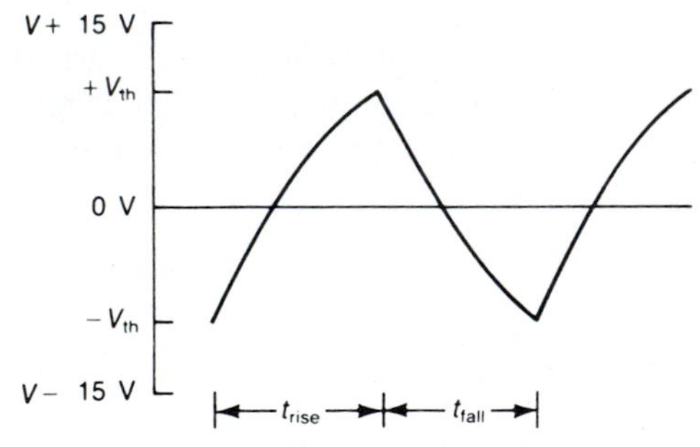


(A) Capacitor Charge-Discharge Voltage

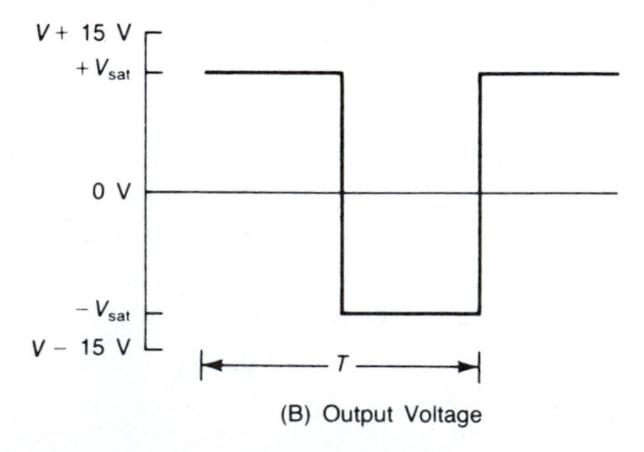


(B) Output Voltage

**Figure 6–26.** Capacitor Charge/Discharge Voltage Versus Op Amp Output Voltage

**Example 6.15**

Design a square wave generator whose output frequency is 1 kHz.

*Solution 1*

First, select a standard value capacitor, say 0.05 μF. Then by manipulation of Equation 6.31, you can solve for R:

$$R = \frac{1}{2f_{op}C} = \frac{1}{2 \times 1\text{ kHz} \times 0.05\ \mu\text{F}} = \frac{1}{2(1 \times 10^3\text{ Hz}) \times (5 \times 10^{-8}\text{ F})}$$
$$= 10 \times 10^3\ \Omega = 10\text{ k}\Omega$$

Next select $R_2 = 10R$, or 100 kΩ, and $R_1 = 0.86R_2$, or 86 kΩ. This completes the circuit design using Equation 6.31.

*Solution 2*

Select any standard value for $C$, $R_1$, and $R_2$. We will select $C = 0.01$ μF, $R_1 = 27$ kΩ, and $R_2 = 47$ kΩ. Manipulation of Equation 6.30 gives us

$$R = \frac{1}{2f_{op}C \ln\left(\dfrac{2R_1}{R_2} + 1\right)}$$
$$= \frac{1}{2 \times 1\text{ kHz} \times 0.01\ \mu\text{F} \times \ln\left(\dfrac{2 \times 27\text{ k}\Omega}{47\text{ k}\Omega} + 1\right)}$$
$$= \frac{1}{2 \times (1 \times 10^3\text{ Hz}) \times (1 \times 10^{-8}\text{ F}) \times \ln\left(\dfrac{2 \times (2.7 \times 10^3\ \Omega)}{4.7 \times 10^3\ \Omega} + 1\right)}$$
$$\approx 6.5 \times 10^4\ \Omega = 65\text{ k}\Omega$$

Either of the designed circuits will produce a 1 kHz square wave output.

# 6.9 Triangle Wave Generator

The output of the *triangle wave generator* consists of positive and negative ramp voltages and sometimes is called a *ramp generator.* The output of an integrator circuit is a triangle wave when the input is a square wave. Therefore, the basic triangle wave generator circuit, shown in Figure 6–27, is a square wave generator, $A_1$, wired as a comparator, connected to an integrator, $A_2$.

The output waveforms shown in Figure 6–28 demonstrate the action of the combination circuit. When the output of the square wave generator goes positive, the output of the triangle wave generator goes negative. Conversely, when the output of the square wave generator goes negative, the output of the

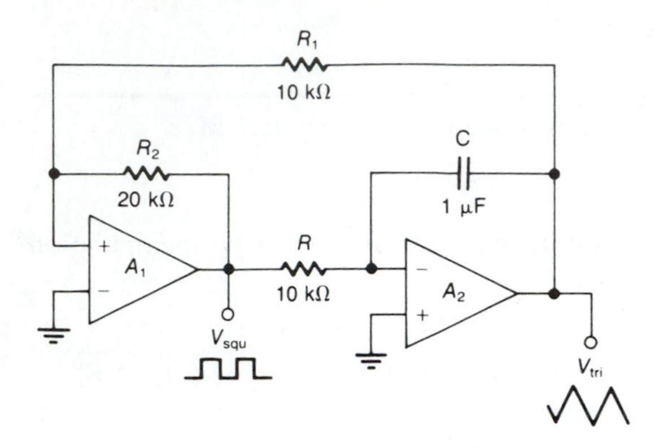


**Figure 6–27.** Basic Triangle Wave Generator

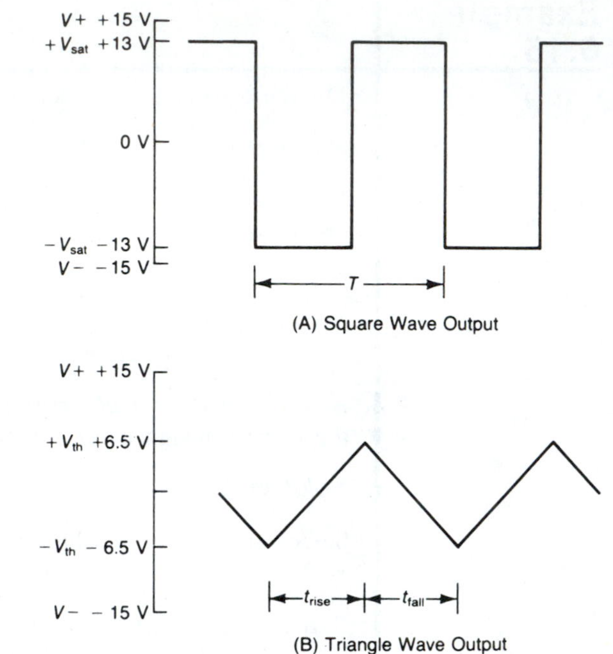


**Figure 6–28.** Output Waveforms of Square/Triangle Wave Generator

triangle wave generator goes positive. A generator that produces two or more different output waveforms, as this one does, is called a *function generator.* We will discuss this concept in more detail later in this chapter.

## Determining Output Amplitudes

The output amplitude of the square wave generator is determined by the output swing of the generator, that is, $+V_{sat}$ to $-V_{sat}$. The output amplitude of the triangle wave generator is determined by the ratio $R_1/R_2$ times the output amplitude of the square wave generator. The frequency of oscillation for both waveforms is

$$f_{op} = \frac{1}{4RC}\left(\frac{R_2}{R_1}\right) \tag{6.32}$$

## Example 6.16

Refer to Figure 6–27. Assume that $A_1$ and $A_2$ are (1/2)747 with a $\pm$15V supply. Determine the output amplitude of each section of the generator and the oscillation frequency.

*Solution*

$$V_{out(squ)} = \pm V_{sat} = \pm 13 \text{ V}$$

$$V_{out(tri)} = \frac{R_1}{R_2}(V_{out(squ)}) = \frac{10 \text{ k}\Omega}{20 \text{ k}\Omega}(\pm 13 \text{ V}) = \pm 6.5 \text{ V}$$

$$f_{op} = \frac{1}{4RC}\left(\frac{R_2}{R_1}\right) = \frac{1}{4 \times 10 \text{ k}\Omega \times 1 \text{ }\mu\text{F}}\left(\frac{20 \text{ k}\Omega}{10 \text{ k}\Omega}\right)$$

$$= \frac{1}{4\ (1 \times 10^3\ \Omega) \times (1 \times 10^{-6}\ \text{F})}\left(\frac{20 \times 10^3\ \Omega}{10 \times 10^3\ \Omega}\right) = 50 \text{ Hz}$$

# 6.10 Sawtooth Wave Generator

A *sawtooth wave generator* is basically a *nonsymmetrical* triangular wave generator. The time length of the negative ramp of the wave, $t_2$ in Figure 6–29A, is extremely short compared to the positive ramp, $t_1$. The shorter $t_2$ is compared to $t_1$, the more ideal the sawtooth wave signal.

One method of constructing a sawtooth wave generator is illustrated in Figure 6–29B. A single op amp connected as an integrator is used with a junction field effect transistor (JFET). An external negative trigger pulse is required to provide the input signal.

A negative voltage at the inverting input causes a positive ramp at the output. When the input pulse returns to zero volts, the JFET turns on, effectively shorting the capacitor, which causes the output voltage to drop rapidly toward zero. The period of the output signal is determined by the timing of the input pulses.

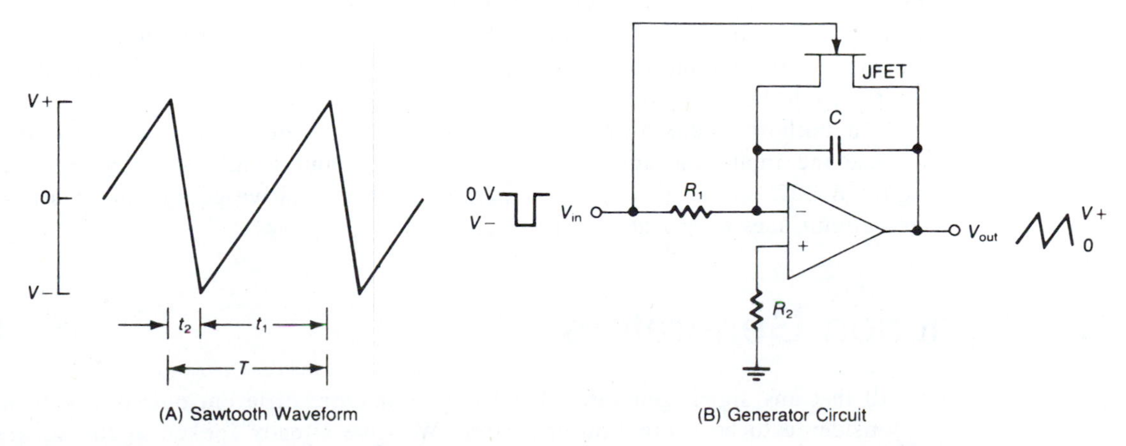


**Figure 6–29.** Sawtooth Wave Generator

### Application of the Sawtooth Waveform

The sawtooth waveform finds application in any device that uses a *cathode ray tube* (CRT). Principal among such devices are oscilloscopes and TV sets. The positive ramp ($t_1$) sweeps the electron beam across the face of the CRT and causes a display to appear. At the end of the display sweep, the negative ramp ($t_2$) returns the beam to its starting point so that the next display sweep may begin. Time interval $t_2$ is called the *flyback*, or *retrace*, portion of the wave. This portion of the wave is "blanked out" since it contains no information and is not observable on the screen.

## 6.11 Staircase Generator

A simple circuit for a *staircase generator* is shown in Figure 6–30A. A square wave input signal charges capacitor $C_1$ to a charge *(Q)* equal to

$$Q = C_1(V_{in} - 0.7 \text{ V}) \tag{6.33}$$

When the reset switch is open, capacitor $C_2$ is charged by each input pulse, through $C_1$ and $D_1$, in equal steps as voltage changes ($\Delta V_{out}$) so that

$$\Delta V_{out} = (V_{in} - 1.4 \text{ V}) \frac{C_1}{C_2} \tag{6.34}$$

In this manner the voltage across $C_2$ is built up one step at a time. At the end of a set number of pulses, or when the voltage across $C_2$ reaches a predetermined value, another pulse closes the switch, shorting $C_2$ and causing its voltage and the output voltage to go to zero. When the pulse is removed, the switch opens and the cycle begins again.

Figure 6–30B shows a similar staircase generator that uses a unijunction transistor (UJT) as a reset switch. When the charge on capacitor $C_2$ reaches the firing potential of the UJT emitter, the UJT conducts, discharging capacitor $C_2$ and driving the output voltage to zero. The UJT now acts like an open switch and the cycle begins again.

In both the generators just discussed, the output signal went negative because the input was applied to the inverting input terminal of the op amp. Figure 6–30C shows a staircase generator with a positive-going output signal. This circuit uses a separate reset pulse.

## 6.12 Function Generators

Recall that any signal generator that has two or more different output waveforms is considered to be a function generator. We have already looked at the square/triangle wave generator. Figure 6–31 shows a *multifunction* waveform generator

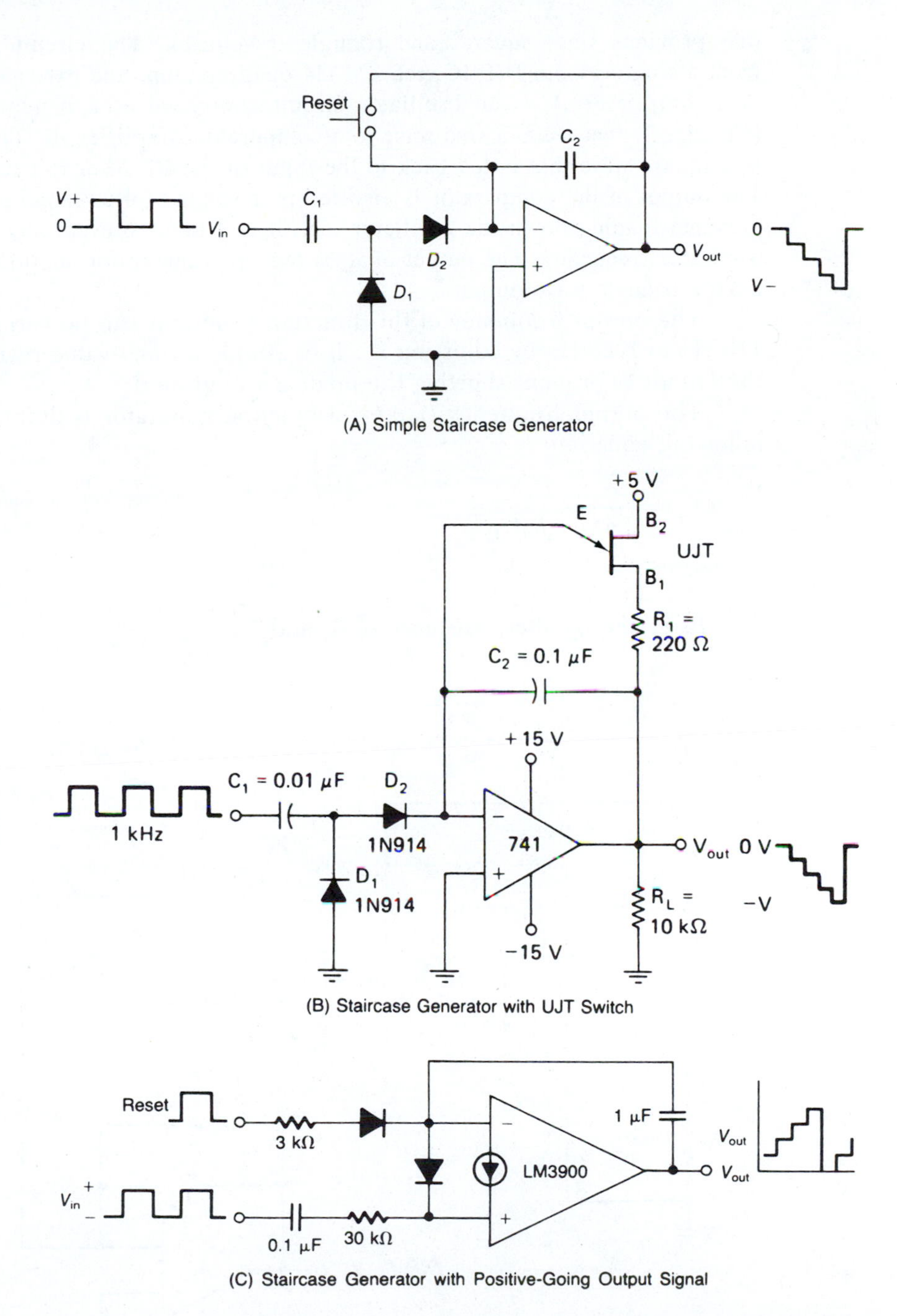


(A) Simple Staircase Generator

(B) Staircase Generator with UJT Switch

(C) Staircase Generator with Positive-Going Output Signal

**Figure 6-30.** Staircase Generators (Parts B and C: Courtesy of National Semiconductor Corporation)

that provides sine, square, and triangle waveforms. The circuit is constructed from a single 14-pin DIP IC, the LM324 quad op amp, and external components.

Amplifier $A_1$ is an oscillator circuit connected as a bandpass (BP) filter (Chapter 7) that feeds a sine wave to a comparator, amplifier $A_2$. The output of $A_2$ is a square wave that is fed back to the input of the BP filter to cause oscillation. The output of the comparator is also fed to a voltage follower, amplifier $A_3$. This prevents loading down the oscillator circuit and helps prevent any change in the oscillator frequency. The output of $A_3$ is fed to an integrator, amplifier $A_4$, to produce a triangle wave output.

The output frequency of this function generator can be varied from about 110 Hz to 1125 Hz by adjusting $R_1$. $R_2$ is simply a small-value resistor placed in the circuit to prevent shorting the feedback to ground.

The output frequency ($f_{op}$) for a function generator is determined by the following equation:

$$f_{op} = \frac{1}{2\pi\sqrt{R_p R_4 C_1 C_2}} \tag{6.35}$$

where

$$R_p = \text{the parallel resistance of } R_1 \text{ and } R_3 = \frac{R_1 R_3}{R_1 + R_3}$$

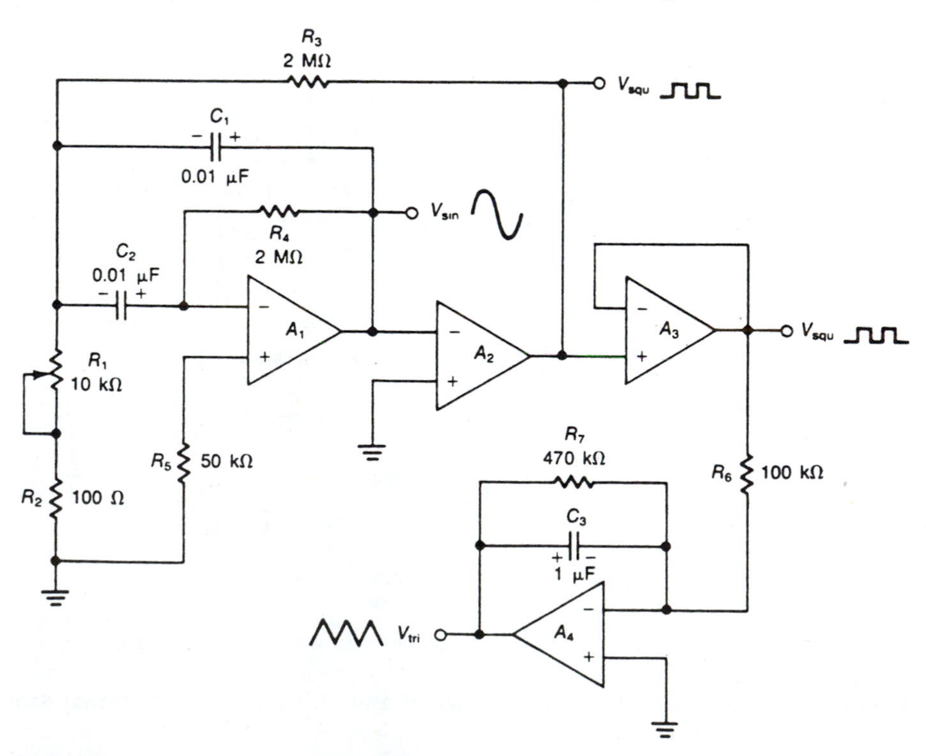


Figure 6–31. Low-Frequency Multifunction Generator

# 6.13 Summary

☐ Oscillators are as important to electronics as are amplifiers. An oscillator can provide either sinusoidal or nonsinusoidal output waveforms. Oscillators have no moving parts and produce ac signals. Good oscillators produce uniform output signals with constant amplitude and frequency. Frequency of operation of an oscillator is determined by the components selected for the feedback path. All oscillator circuits must (a) have amplification, (b) have a frequency-determining network, (c) have positive feedback, and (d) be self-starting. An oscillator must meet the requirements of the Barkhausen Criterion in order to be self-starting and self-sustaining; that is, $A_{\upsilon}B_{\upsilon} = 1$.

☐ *LC* oscillators use capacitors and inductors, connected either in series or parallel, for the frequency-determining network. A Hartley oscillator is identified by its tapped inductor arrangement. The Hartley oscillator has two disadvantages: inductors are mutually coupled, and changing frequencies is difficult. A Colpitts oscillator is identified by its tapped capacitor arrangement. The operating frequency of the Colpitts oscillator can be easily changed by the use of variable capacitors in the frequency-determining network. The high input impedance and large open-loop gain of the op amp overcome the disadvantages of the standard *LC* oscillator circuits.

☐ Resistance-capacitance networks are used for frequency-determining in *RC* oscillators. *RC* oscillators are used primarily in low- and audio-frequency ranges. The two basic types of sine wave-producing *RC* oscillators are the phase-shift oscillator and the Wien-bridge oscillator. The phase-shift oscillator is used principally in fixed frequency applications. Stability in the *RC* phase-shift oscillator is improved by increasing the number of *RC* sections used to provide the necessary 180° phase shift in the feedback network. Three sections are typically used. The Wien-bridge oscillator uses both positive and negative feedback. A lead-lag network determines frequency and applies the positive feedback to the noninverting input, while the negative feedback is applied to the inverting input.

☐ A sine-cosine oscillator, sometimes called a quadrature oscillator, consists of two op amps connected as integrators with positive feedback. The output signals from the two op amps are 90° out of phase.

☐ Voltage-controlled oscillators (VCOs) are used in a variety of applications. The frequency of the VCO is directly proportional to the input voltage level.

☐ Using crystals in oscillator circuits provides a high degree of frequency stability. This stability occurs because of the crystal's natural frequency of vibration. The natural frequency limit is about 20 MHz, above which harmonics must be used. Crystals designed especially to operate on harmonics are called overtone crystals.

☐ Two types of multivibrators used in analog systems are the astable and monostable multivibrators. An astable multivibrator has a square wave output and does not require an input trigger pulse. A monostable multivibrator produces a single output pulse for a predetermined time interval and requires an input trigger pulse.

☐ A differentiator circuit produces an output signal that is a derivative of its input signal. If the input signal is a sine wave, the output signal is a cosine wave that is 270° out of phase with the input signal. If the input signal is a triangle wave, the output signal is a square wave. An integrator circuit is the inverse of the differentiator circuit. It is constructed simply by interchanging the resistors and capacitors in a differentiator circuit. A square wave input signal produces a triangle wave output signal.

☐ A square wave generator is basically a multivibrator, sometimes called a relaxation oscillator because the output frequency is set by the charging and discharging of a capacitor.

☐ A triangle wave generator, sometimes called a ramp generator, produces an output wave with positive and negative ramp voltages. The basic circuit is an integrator whose input is a square wave from a comparator.

☐ A sawtooth generator output signal is a nonsymmetrical triangular wave whose negative-going ramp time interval is very short compared with its positive-going ramp time interval. The sawtooth wave is applied principally as a sweep circuit in oscilloscopes and TV sets.

☐ Function generators are signal generators that have two or more different output waveforms.

## 6.14 Questions and Problems

**6.1** What types of waveforms can oscillators produce?

**6.2** What is the Barkhausen Criterion?

**6.3** What determines the stability of the output of an oscillator?

**6.4** State the basic fundamental requirements of all oscillators.

**6.5** Match each oscillator in the following list with its proper schematic diagram in Figure 6–32.

1. *LC* oscillator
2. *RC* oscillator
3. Crystal oscillator
4. Hartley oscillator
5. Colpitts oscillator
6. Sine-cosine oscillator
7. Phase-shift oscillator
8. Wien-bridge oscillator
9. Colpitts crystal oscillator

**6.6** State the identifying features of (a) the Hartley oscillator and (b) the Colpitts oscillator.

**6.7** What are the disadvantages of the Hartley oscillator?

**6.8** What determines the frequency of operation of an oscillator?

**6.9** What advantages do op amp circuits offer over standard *LC* oscillator circuits?

**6.10** Explain the feedback networks of the Wien-bridge oscillator.

**6.11** How may the stability of the phase-shift oscillator be improved?

**6.12** What oscillator uses both positive and negative feedback?

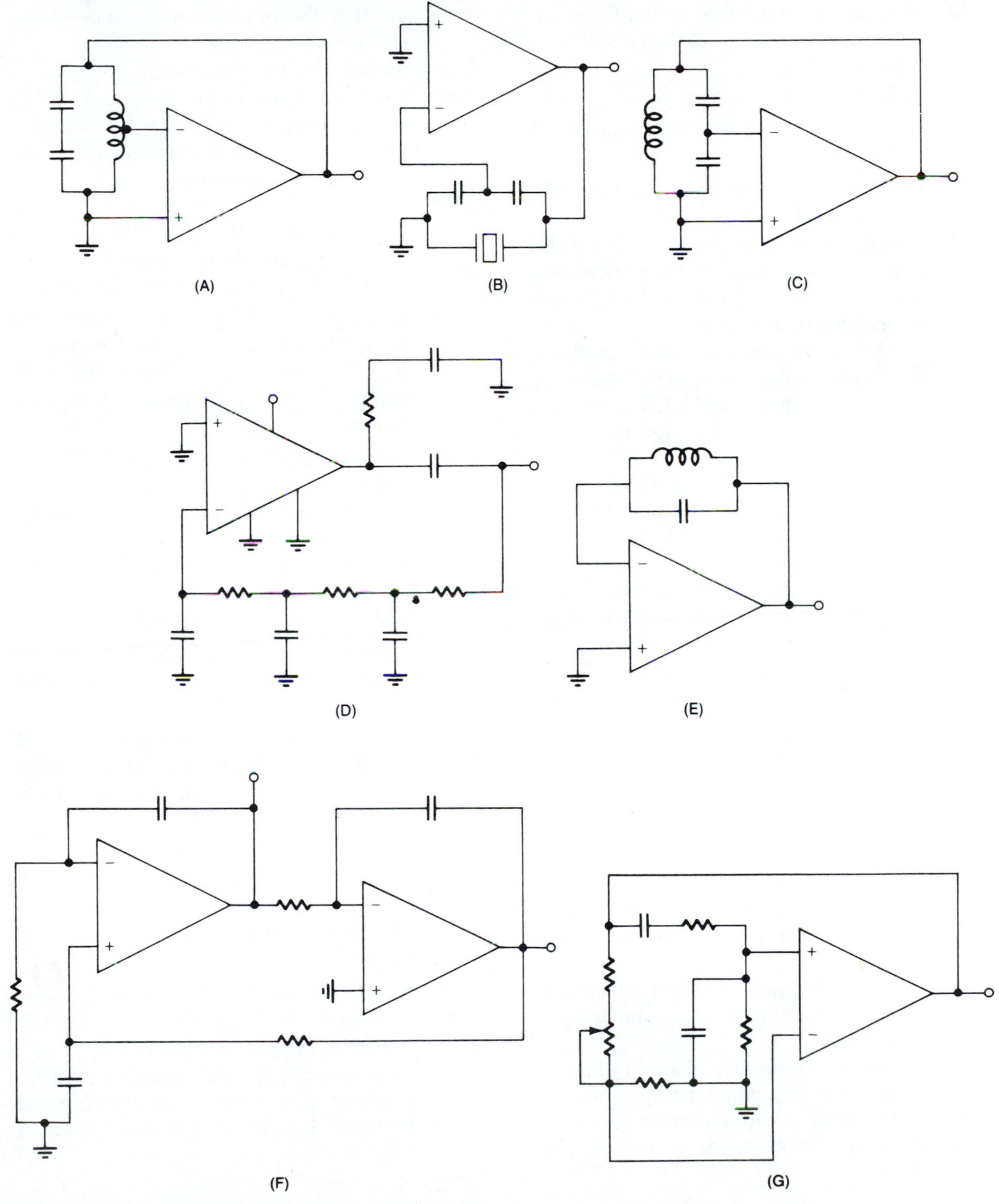


**Figure 6–32.** Circuit Diagrams for Problem 6.5

6.13 Why are crystals used in oscillator circuits? What property of the crystal gives it an advantage?

6.14 Define harmonic frequency.

6.15 What is the fifth harmonic frequency of 12 MHz?

6.16 Define overtone crystals and explain why they are used.

6.17 Design an op amp phase-shift oscillator that will oscillate at 600 Hz. Draw the schematic diagram and label all components with their values.

6.18 Design an op amp quadrature oscillator that will oscillate at 3 kHz. Draw the schematic diagram and label all components with their values. Show the output waveforms.

6.19 Design an op amp Wien-bridge oscillator that will oscillate at 10 kHz. Draw the schematic diagram and label all components with their values.

6.20 An electronic system application requires an oscillator operating at 60 MHz. What would you use to construct the oscillator to satisfy the system requirements?

6.21 List the two types of multivibrators used in analog systems.

6.22 Which type of multivibrator requires an input trigger pulse?

6.23 The astable multivibrator produces what kind of output waveform?

6.24 Describe the operation of the monostable multivibrator.

6.25 Describe the output signal of a differentiator circuit whose input signal is a sine wave.

6.26 Describe the basic circuitry differences between the differentiator and integrator circuits.

6.27 The output frequency of a relaxation oscillator is set by which components?

6.28 Describe the ramp generator operation.

6.29 Describe the sawtooth generator operation.

6.30 What is the principal application of the sawtooth generator?

6.31 Define function generators.

6.32 Refer to Figure 6–14. Assume $R_1$ = 86 kΩ, $R_f$ = $R_2$ = 100 kΩ, and $C_1$ = 1000 pF. Determine (a) the period and (b) the frequency of oscillation.

6.33 Assume a simplified astable multivibrator where $R_2$ = $R_f$ = 3.3 MΩ and $C_1$ = 100 pF. Determine the values for (a) $R_1$, (b) T, and (c) frequency of oscillation.

6.34 A simplified astable multivibrator has $R_1$ = 330 kΩ and $C_1$ = 1200 pF. Assume $R_f$ = $R_2$. Determine the values for (a) $R_2$ and $R_f$, (b) T, and (c) frequency of oscillation.

6.35 Refer to Figure 6–15. Assume $R$ = 1 MΩ and $C$ = 10 μF in the feedback line to the (+) terminal. Determine the output PW.

6.36 Refer to Figure 6–15. Assume PW = 5 s and the positive feedback line R = 3.3 MΩ. Calculate the value for C.

6.37 Assume PW = 200 ms and C = 0.02 μF in the positive feedback line of the circuit in Figure 6–15. Calculate R.

6.38 Refer to Figure 6–16. Assume $R_f$ = 100 kΩ, $R_1$ = 20 kΩ, $R_2$ = 4 kΩ, and $C_1$ = 2 μF. Use the approximation method to solve for the output PW.

6.39 Refer to Figure 6–16. Assume $R_f$ = 330 kΩ, $R_1$ = 33 kΩ, $R_2$ = 6.8 kΩ, and $C_1$ = 0.5 μF. Calculate PW using (a) the accurate method and (b) the approximation method.

6.40 Refer to Figure 6–18A. Assume $R_1$ = 100 Ω, $R_f$ = 3 kΩ, and $C_1$ = 5.0 μF. Calculate (a) gain and (b) $f_{op}$.

6.41 Determine what value capacitor will be required across $R_f$ in the circuit from Problem 6.40 to cut off frequencies at 200 Hz.

**6.42** Design an op amp circuit that will differentiate a 1 kHz input signal while limiting the high frequency gain to 10.

**6.43** If the 1 kHz input signal of Problem 6.42 is a 0.5-V peak sine wave, what will be the peak output voltage?

**6.44** Refer to Figure 6–23B. Assume a symmetrical 3-V peak-to-peak 400 Hz square wave input signal. Determine values for $R_1$, $R_2$, $R_{sh}$, $C$, and the peak output voltage.

**6.45** Use two methods of calculation to design a square wave generator whose output frequency is 400 Hz.

**6.46** Refer to Figure 6–27. Assume $R$ = 20 kΩ, $R_1$ = 20 kΩ, $R_2$ = 40 kΩ, and $C$ = 2 μF. $A_1$ and $A_2$ are (1/2)747 with a ±15 V supply. Determine the output amplitude of each section of the generator and the frequency of oscillation.

# Chapter 7

# Active Filters

## 7.1 Introduction

Electrical filters are designed to either attenuate unwanted frequencies or to amplify desired frequencies. Attenuation is accomplished by using passive filters that use only resistors, capacitors, and inductors. Amplification is accomplished by using active filters that use resistors and capacitors around an active device, usually an op amp.

In this chapter we will review the fundamentals of filters, types of filters, filter characteristics, and filter configurations. We will distinguish between passive and active filters. We will examine first-, second-, and high-order filters. Multiple-feedback and state-variable filters will be introduced. Finally, design criteria and techniques will be thoroughly discussed.

## 7.2 Objectives

When you complete this chapter, you should be able to:

- ☐ Define and explain active filters.
- ☐ Explain the many uses of active filters.
- ☐ Define and explain a normalized response curve in terms of (a) cutoff fre-

quency, (b) rolloff, (c) center frequency, (d) bandwidth, (e) passband, (f) stopband, (g) order, (h) decibel, (i) decade, and (j) octave.

- ☐ Define Butterworth response.
- ☐ Distinguish between Butterworth and other types of filters in terms of damping.
- ☐ Design and explain (a) the various orders of low-pass and high-pass filters, (b) multiple-feedback low-pass, high-pass, and bandpass filters, (c) notch filters, (d) wideband filters, and (e) state-variable filters.

# 7.3 Review of Filter Fundamentals

A *filter* is a device that screens out certain frequencies or passes electric current of only certain frequencies. Basically, there are two categories of filters—passive filters and active filters.

*Passive filters* are constructed with resistors, capacitors, and inductors. They contain no active devices such as transistors, diodes, vacuum tubes, or op amps. *Active filters* are constructed with a network of resistors and capacitors around an active device, usually an op amp. Inductors are seldom used in active filters.

Active filters offer the following advantages over passive filters:

1. *Low cost*. Inductors, which are usually expensive components, are seldom used in active filters, because of their susceptability to stray noise pickup.
2. *Ease of adjustment*. In general, active filters can be adjusted over a wide range of frequencies without appreciable change in the desired output response.
3. *No insertion loss*. The op amp in the active filter circuit overcomes any loss or attenuation of the circuit by providing amplification.
4. *Isolation*. Because of the very high input impedance and low output impedance of the op amp, there is practically no interaction between the filter and the source or load.

There are also some limitations and disadvantages in active filter circuits. They are as follows:

1. *Frequency response*. All op amps have a frequency response limit, so the type of op amp selected for the filter circuit will be the deciding factor in filter frequency response.
2. *Power supply*. Active filters require voltage power supplies to power the op amp. Many op amps require two separate supplies, or dual power supplies, although some circuits may be operated with a single power supply.

Filters are classified according to the range of frequencies they reject or allow to pass. A *low-pass* (LP) filter allows frequencies below a given value to pass and rejects frequencies above that value. A *high-pass* (HP) filter rejects lower frequencies and allows only frequencies above a given value to pass. A *bandpass* (BP) filter allows a band of frequencies between a designated high value and a designated low value to pass and rejects frequencies above and below those values. A filter that prevents a band of frequencies between two designated values from passing is variously called a *notch, band-reject,* or *band-stop* filter. Graphic representations of these various filters show the relationship between filter output in *decibels* (dB) and frequency in hertz (Hz). Figure 7–1 illustrates generalized characteristics of the filters we have mentioned, assuming a constant amplitude for all input signal frequencies. Note that unwanted frequencies are not completely rejected, but roll off in a sharp curve. The $f_c$ designation is the frequency cutoff and is the limiting frequency for the filter.

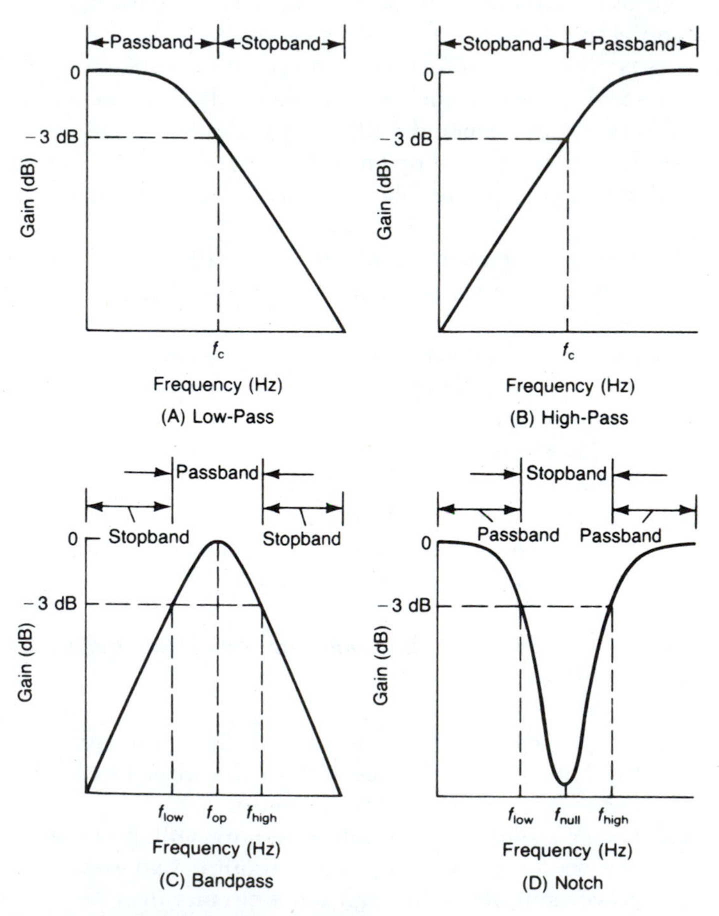


**Figure 7–1.** Generalized Filter Characteristics

The range of frequencies passing through a filter with maximum gain or minimum attenuation is called the *passband*. The value of the cutoff frequency determines the width of the passband.

The *transfer function* of a filter is considered to be its *gain* or *amplitude response*. The transfer function is represented by the relationship between the filter's output voltage and its input voltage at various frequencies, expressed as $V_{out}/V_{in}$, a function of frequency. This ratio is more conveniently expressed in terms of dB to eliminate the use of decimal voltage values. As a formula

$$A_{dB} = 20 \log_{10} A_v \tag{7.1}$$

where

$$A_v = \frac{V_{out}}{V_{in}}$$

$A_v$ may be equal to, greater than, or less than unity. A resistive network will attenuate a signal, which results in a loss, or less than unity gain. This is shown by a negative dB value. An amplifier in the system may provide a positive dB value, or greater than unity gain. Unity gain is represented by 0 dB.

## Example 7.1

An LP filter circuit has $V_{in} = 1\ V_{p\text{-}p}$ and $V_{out} = 3\ V_{p\text{-}p}$. What is the dB gain of the circuit?

*Solution*

Use Equation 7.1:

$$A_{dB} = 20 \log_{10} A_v = 20 \log_{10} 3 = 20(0.47712) = 9.5 \text{ dB}$$

## Example 7.2

An HP filter circuit has $V_{in} = 1\ V_{p\text{-}p}$ and $V_{out} = 0.5\ V_{p\text{-}p}$. What is the dB gain of the circuit?

*Solution*

$$A_{dB} = 20 \log_{10} \frac{0.5 \text{ V}}{1 \text{ V}} = 20 \log_{10} 0.5 = 20(-0.301) \approx -6 \text{ dB}$$

### Example 7.3

A BP filter circuit has unity gain. What is the dB gain of the circuit?

*Solution*

$$A_{dB} = 20 \log_{10}1 = 20(0) = 0 \text{ dB}$$

### Example 7.4

A notch filter circuit has $V_{in} = 1 \text{ V}_{p\text{-}p}$ and $V_{out} = 1.12 \text{ V}_{p\text{-}p}$. What is the dB gain of the circuit?

*Solution*

$$A_{dB} = 20 \log_{10}1.12 = 20(0.04921) = 1 \text{ dB}$$

The frequency at which the voltage gain of the filter circuit drops to 0.707 of its maximum value is the *cutoff frequency*, $f_c$. The frequency or frequencies at which this occurs identifies the filter's passband limits. The cutoff frequency point is variously called the *0.707 point*, the *3 dB point*, the *break point*, and the *half-power point*. Expressed as an equation,

$$f_c \text{ point} = \text{dB gain}_{(max)} - 3 \text{ dB} \tag{7.2}$$

or

$$f_c \text{ point} = 0.707 \frac{V_{out}}{V_{in}} \text{ (in volts)} \tag{7.3}$$

where

$$\frac{V_{out}}{V_{in}} = A_{v(max)}$$

Figure 7–2 and some examples will demonstrate the origin of these designations.

### Example 7.5

The frequency response curve for Example 7.1 is shown in Figure 7–2A. Determine the $f_c$ point for the curve.

*Solution*

Using Equation 7.2,

$$f_c \text{ point} = \text{dB gain}_{(max)} - 3 \text{ dB} = 9.5 \text{ dB} - 3 \text{ dB} = 6.5 \text{ dB}$$

Using Equation 7.3,

$$f_c \text{ point} = 0.707A_{v(max)} = 0.707(3) = 2.12 \text{ V}$$

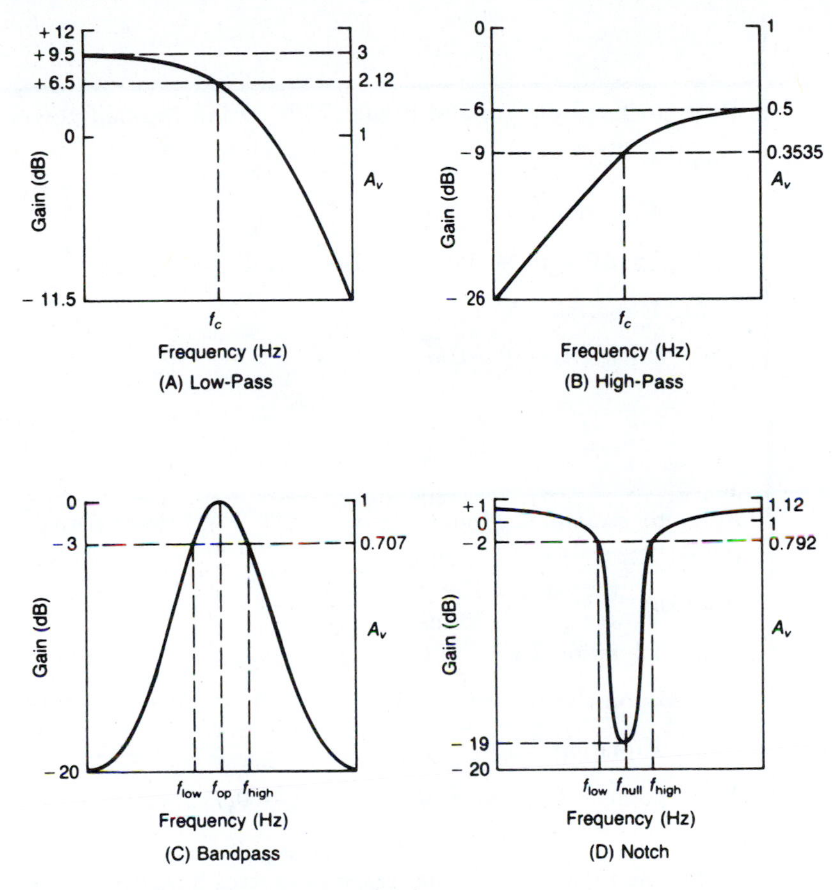


**Figure 7–2.** Frequency Response Curves of Filters

## Example 7.6

Refer to Example 7.2 and Figure 7–2B. Determine the break point.

*Solution*

Using Equation 7.2,

$$\text{Break point} = \text{dB gain} - 3\text{ dB} = -6\text{ dB} - 3\text{ dB} = -9\text{ dB}$$

Using Equation 7.3,

$$\text{Break point} = 0.707A_v = 0.707(0.5) = 0.3535\text{ V}$$

**Example 7.7**

Refer to Example 7.3 and Figure 7–2C. Find the half-power point ($P_{(1/2)}$).

*Solution*

Using Equation 7.2,

$$P_{(1/2)} = \text{dB gain} - 3\text{ dB} = 0\text{ dB} - 3\text{ dB} = -3\text{ dB}$$

Using Equation 7.3,

$$P_{(1/2)} = 0.707A_v = 0.707(1) = 0.707\text{ V}$$

**Example 7.8**

Refer to Example 7.4 and Figure 7–2D. Find the 3 dB point and the 0.707 point.

*Solution*

Using Equation 7.2,

$$3\text{ dB point} = \text{dB gain} - 3\text{ dB} = 1\text{ dB} - 3\text{ dB} = -2\text{ dB}$$

Using Equation 7.3,

$$0.707\text{ point} = 0.707A_v = 0.707(1.12) = 0.792\text{ V}$$

Note that the BP and notch filter response curves have two frequency cutoff points. Also note that *the 3 dB point is always –3 dB from the maximum dB gain,* regardless of the filter type.

# 7.4 Order of Filters

As discussed in the previous section, the passband is that range of wanted frequencies where the amplitude response is relatively constant. The *stopband* is the range of frequencies outside the passband. In the LP filter, for example, the stopband is the range of frequencies above the filter's cutoff frequency.

## Normalized Frequency Response Curve

A *normalized* frequency response curve is one in which the maximum amplitude response is set to unity, or 0 dB. Using a normalized frequency graph allows us to compare the relative selectivity of different orders of filters. The simplistic graph in Figure 7–3 shows the rolloff of several different orders of LP filters with the gain (or loss) in dB as a function of logarithmic change in frequency. Note that the amplitude response in the stopband (above 1 kHz) decreases linearly as frequency increases logarithmically. The *order* of the filter defines the rate of this decrease, called the *rolloff* or *falloff*. Note also that the rolloff of a *first-order* filter is –20 dB between the maximum dB gain and 10 kHz. This can be stated as a rolloff of –20 dB per decade, or –6 dB per octave.

### Decade and Octave Defined

A *decade* is defined as a tenfold increase or decrease in frequency. An *octave* is defined as a doubling or halving of frequency. Table 7–1 demonstrates these relationships, showing several decades and octaves above and below a cutoff frequency of 1 kHz.

### Graph Evaluation

Returning to Figure 7–3, we can now more easily evaluate the graph. Note that increasing the order of the filter increases the rolloff. Also note that the designated order of the filter indicates a multiple of the first-order rolloff value.

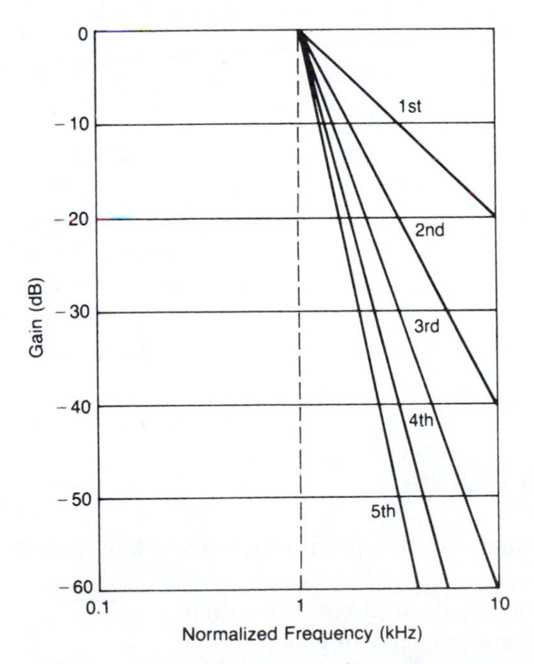


**Figure 7–3.** Rolloff for Various Orders of Filters

Table 7–1 **Relationships between Frequency, Decade, and Octave**

| | | | | ($f_c$) | | | |
|---|---|---|---|---|---|---|---|
| Frequency (Hz) | 1 | 10 | 100 | 1 k | 10 k | 100 k | 1000 k |
| Decade | −3 | −2 | −1 | 0 | +1 | +2 | +3 |
| | | | | ($f_c$) | | | |
| Frequency (Hz) | 125 | 250 | 500 | 1 k | 2 k | 4 k | 8 k |
| Octave | −3 | −2 | −1 | 0 | +1 | +2 | +3 |

For example, a *second-order* filter has a −40 dB per decade (or −12 dB per octave) value, a *third-order* filter has a −60 dB per decade (or −18 dB per octave) value, and so on. The sharper the rolloff, the more efficient the filter.

## Butterworth Response

A more realistic frequency response graph is shown in Figure 7–4A. Here we see that the cutoff frequency is 3 dB down (−3 dB) from the assumed 0 dB unity gain. Note that the passband amplitude response, for all orders of filters, is relatively constant, or flat, with no oscillations. Such a passband response is called a *Butterworth* response.

A *Chebyshev* response is shown in Figure 7–4B. Note the oscillation, or ripple, in the passband.

### Damping

The determining factor in the shaping of the filter's passband is *damping,* represented by the Greek letter alpha, $\alpha$. The higher the damping, the flatter the passband. Therefore, the Butterworth response represents a highly damped filter, and a Chebyshev response represents a filter with less damping. For the LP and HP filters, we will discuss the design and evaluation of only the Butterworth response.

## Filter Design Rules

To simplify the design of active filters, here are a few basic rules to follow:

1. Determine the desired cutoff frequency, $f_c$.
2. Select a standard capacitor value.
3. Determine the value of the resistance to be used for impedance level by the following equation:

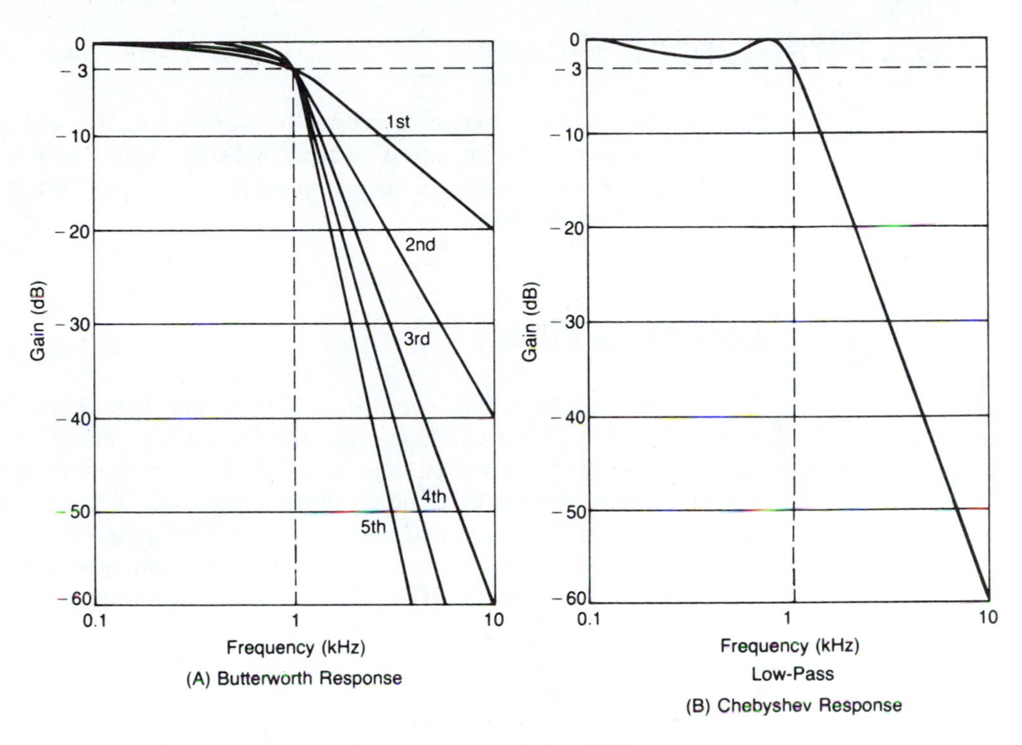


**Figure 7–4.** Amplitude Response Curves for Various Orders of Filters

$$R = \frac{1}{2\pi f_c C} \quad \text{(For first-order filters)} \tag{7.4}$$

NOTE: The calculated value may not produce a standard resistor value. For accuracy, the calculated value should be used. For most purposes, however, the nearest standard value will suffice.

## Example 7.9

Design an LP filter in which $f_c = 600$ Hz.

*Solution*

First, select a standard capacitor, say 0.1 μF. Then determine the value for $R$.

$$R = \frac{1}{2\pi f_c C} = \frac{1}{(6.28)(600\ \text{Hz})(0.1 \times 10^{-6}\ \text{F})}$$

$$= 2.65 \times 10^3\ \Omega = 2.65\ \text{k}\Omega \quad \text{(Use a standard 2.7 k}\Omega\text{ resistor.)}$$

# 7.5 First-Order Filters

The simplest forms of active filters are first-order LP, HP, and unity-gain filters. Such filters are built with a passive filter network, which consists of one resistor and one capacitor, connected to an op amp. First-order filters are easily constructed and inexpensive.

## Low-Pass Filters

A first-order LP filter has a rolloff of -20 dB per decade, or -6 dB per octave. Figure 7–5 shows the basic first-order LP filter design circuit. It is simply an LP passive filter ($R_1C_1$) connected to an active device, a noninverting amplifier, having a cutoff frequency of 1 kHz and an input impedance of 5 kΩ. The resistive feedback network of $R_2$ and $R_3$ determines the passband gain.

Starting with the basic design circuit, we can now make the necessary calculations to provide the desired filter characteristic response.

### Example 7.10

Convert the standard 1 kHz filter circuit of Figure 7–5 to one with a cutoff frequency of 2 kHz.

*Solution*

Refer to the basic rules of filter design that we established. The first item has been predetermined; that is, the desired frequency is 2 kHz. Next, select a standard capacitor value, say 0.02 μF for $C_1$. Now use Equation 7.4 to determine the value for resistor $R_1$.

$$R_1 = \frac{1}{2\pi f_c C} = \frac{1}{(6.28)(2 \times 10^3\ \text{Hz})(0.02 \times 10^{-6}\ \text{F})}$$

$$= 3.98 \times 10^3\ \Omega = 3.98\ \text{k}\Omega \qquad \text{(Use a standard 3.9 k}\Omega\text{ resistor.)}$$

Figure 7–6 shows the completed circuit design and amplitude response curve.

## High-Pass Filters

An HP filter circuit is identical to the LP circuit, except that the frequency-determining resistors and capacitors ($R_1C_1$) are interchanged. The frequency response curve is also identical, except it is reversed. The design of a first-order HP filter is accomplished in the same manner as the LP design.

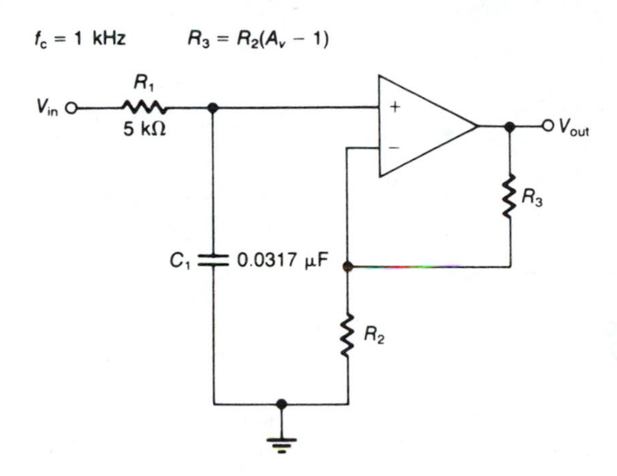


**Figure 7–5.** Basic Design Circuit for a First-Order Low-Pass Filter

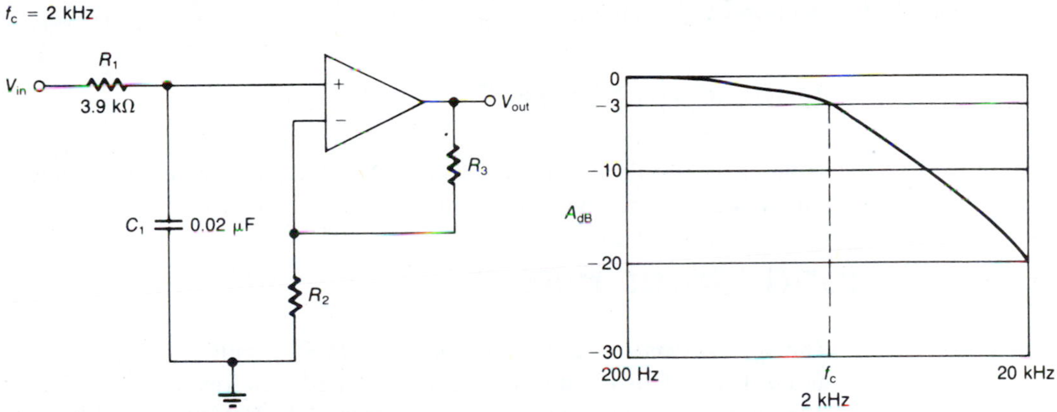

(A) Completed Design Circuit

(B) Amplitude Response Curve

**Figure 7–6.** Solution to Example 7.10

## Example 7.11

Design a first-order HP filter with a cutoff frequency of 750 Hz.

*Solution*

Applying the basic rules, select a standard capacitor value and determine the resistance value. We will select a 0.1 μF capacitor for $C_1$, then

$$R_1 = \frac{1}{2\pi f_c C} = \frac{1}{(6.28)(750\ \text{Hz})(0.1 \times 10^{-6}\ \text{F})} = 2.12 \times 10^3\ \Omega$$

$$= 2.12\ \text{k}\Omega \qquad \text{(Use a standard 2 k}\Omega\text{ or 2.2 k}\Omega\text{ resistor.)}$$

Figure 7–7 shows the completed circuit design and amplitude response curve.

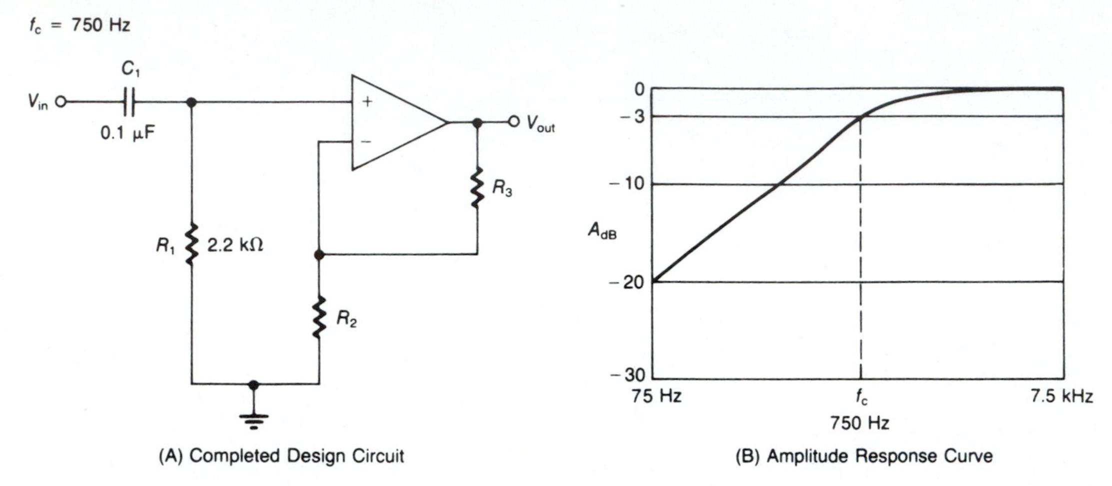

(A) Completed Design Circuit
(B) Amplitude Response Curve

**Figure 7–7.** Solution to Example 7.11

### Unity-Gain Filters

If unity gain is desired in either the LP or HP filter, a voltage follower is used for the active device. Figure 7–8 illustrates such circuits.

## 7.6 Second-Order Filters

The *voltage-controlled voltage source* (VCVS) circuit is the simplest second-order active filter. This circuit is sometimes called a *Sallen and Key* (S & K) filter. In this section we will examine second-order LP, HP, and equal-component VCVS filters. You will observe that the passive network consists of two resistors and two capacitors.

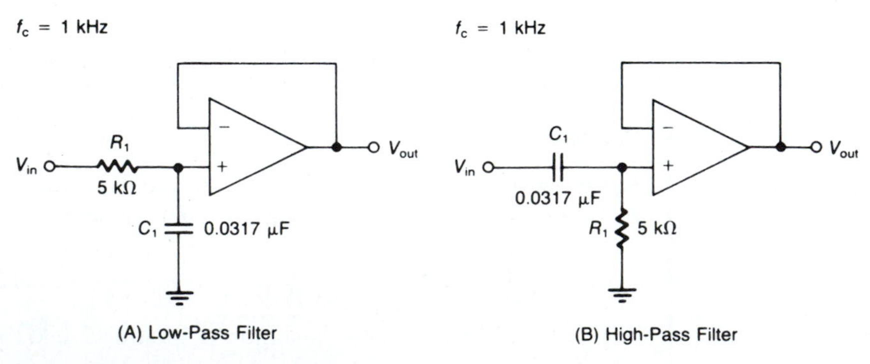

(A) Low-Pass Filter
(B) High-Pass Filter

**Figure 7–8.** Unity-Gain First-Order Filters

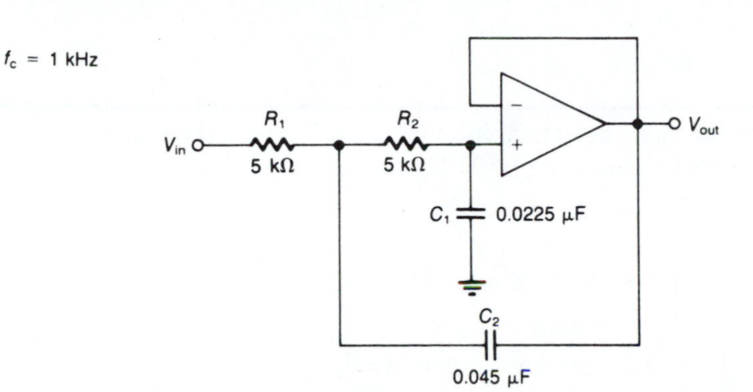


**Figure 7–9.** Basic Design Circuit for a Second-Order VCVS Low-Pass Filter

**Table 7–2** **Scaling Procedure**

| | Low-Pass | High-Pass |
|---|---|---|
| 1. | $\frac{\text{Basic Frequency}}{\text{Desired Frequency}} = X$ | $\frac{\text{Basic Frequency}}{\text{Desired Frequency}} = X$ |
| 2. | $\frac{\text{Selected } C_1}{\text{Basic } C_1} = Y$ | $\frac{\text{Selected } C_1}{\text{Basic } C_1} = Y$ <br> *where:* $C_1 = C_2$ |
| 3. | $XR_1 = \text{New } R$ <br> *where:* $R_1 = R_2$ | $X\,(\text{Basic } R_2) = \text{New } R_2$ |
| 4. | $\frac{\text{New } R}{Y} = \text{Required } R \text{ for } R_1 = R_2$ | $\frac{\text{New } R_2}{Y} = \text{Required } R_2$ |
| 5. | Make $C_2 = 2\,(\text{Selected } C_1)$ | Make $R_1 = 2R_2$ |

## Low-Pass VCVS Circuit

The cutoff frequency for the VCVS second-order LP filter is

$$f_c = \frac{1}{2\pi\sqrt{R_1R_2C_1C_2}} \tag{7.5}$$

Figure 7–9 illustrates a basic VCVS second-order LP filter design circuit with a $f_c$ of 1 kHz, an impedance level of 5 kΩ, and unity gain. Note that $C_2 = 2C_1$ and $R_1 = R_2$.

### Scaling Techniques

A convenient method of design for the VCVS second-order filter is *scaling*. This is accomplished by first multiplying all the frequency-determining resistors of the basic circuit design by the ratio of the basic design frequency to the new desired frequency. Then select a standard value of capacitor and divide the new resistance values by the ratio of the selected capacitor value to the basic design value. Table 7–2 will help to clarify the procedure.

## Example 7.12

Convert the basic design circuit of Figure 7–9 to a second-order VCVS LP filter with a cutoff frequency of 500 Hz and unity gain.

*Solution*

Follow the steps in Table 7–2. Determine the frequency ratio $X$:

$$X = \frac{\text{basic frequency}}{\text{desired frequency}} = \frac{1000 \text{ Hz}}{500 \text{ Hz}} = 2$$

Select a standard capacitor value and determine the capacitor ratio $Y$:

$$Y = \frac{\text{selected } C_1}{\text{basic } C_1} = \frac{0.05 \ \mu\text{F}}{0.0225 \ \mu\text{F}} = 2.22$$

Scale the resistances to determine the new value for $R$:

$$\text{new } R = XR_1 = 2(5 \text{ k}\Omega) = 10 \text{ k}\Omega$$

Determine the required value of $R$ for $R_1$ and $R_2$:

$$\text{required } R = \frac{\text{new } R}{Y} = \frac{10 \text{ k}\Omega}{2.22} = 4.5 \text{ k}\Omega$$

Finally,

$$C_2 = 2(\text{selected } C_1) = 2(0.05 \ \mu\text{F}) = 0.1 \ \mu\text{F}$$

Standard 4.3 kΩ resistors for $R_1$ and $R_2$ will produce a cutoff frequency within 10% of the desired 500 Hz frequency for this circuit. Verification can be obtained by using Equation 7.5 and the new values of components, and is left as an exercise.

Figure 7–10 shows the completed circuit design and amplitude response curve.

## Equal-Component Circuits

An *equal-component* second-order VCVS LP filter, whose basic circuit is shown in Figure 7–11, requires that the passband gain be fixed at $A_v = 1.59$.

This value is derived from $A_v = 3 - \alpha$, where $\alpha = 1.41$ for a second-order Butterworth response. The circuit will not produce a Butterworth response at any other passband gain. For the configuration of Figure 7–11, using the non-inverting amplifier, the value of feedback resistor $R_4$ must equal 0.59 times the value of input resistor $R_3$.

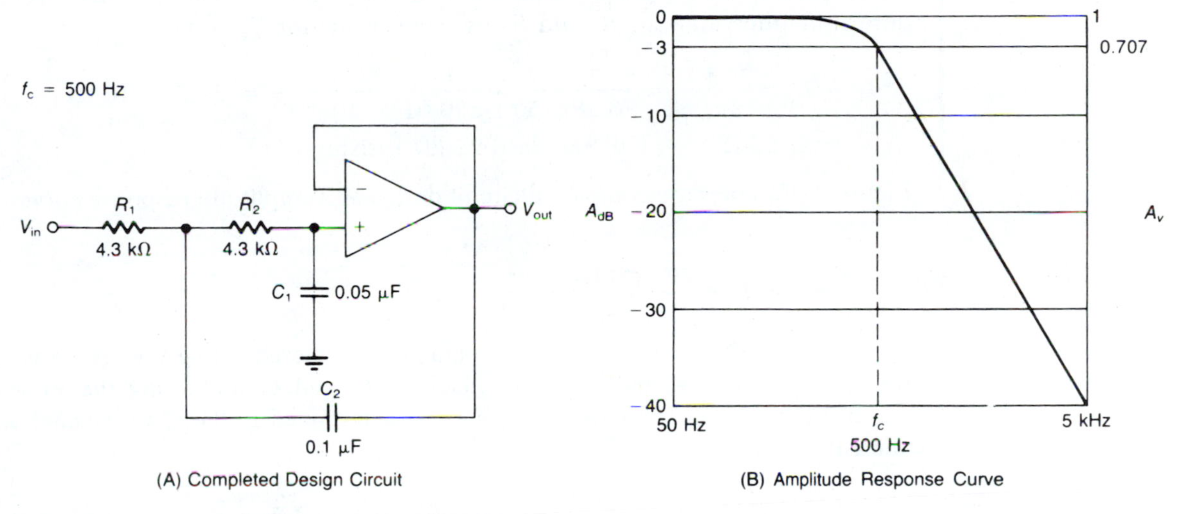


**Figure 7–10.** Solution to Example 7.12

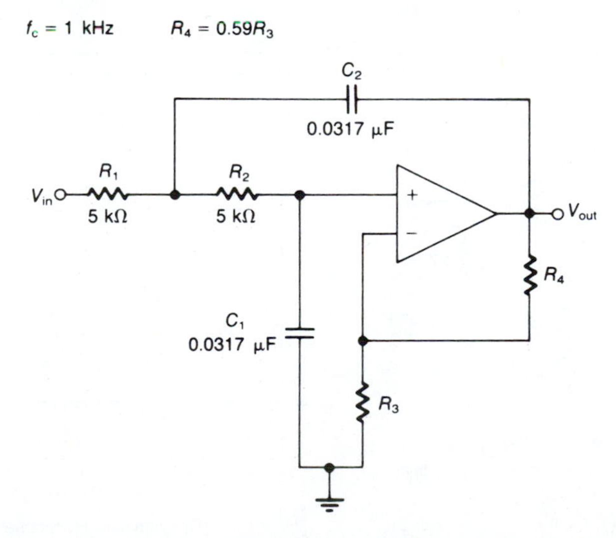


**Figure 7–11.** Basic Design Circuit for an Equal-Component VCVS Low-Pass Filter

**Example 7.13**

Design an equal-component second-order VCVS LP filter with a cutoff frequency of 750 Hz.

*Solution*

Select a standard capacitor value for $C_1$ and $C_2$. We will use 0.01 μF. Then determine the value for $R_1$ and $R_2$ by using Equation 7.4:

$$R_1 = R_2 = \frac{1}{2\pi f_c C} = \frac{1}{(6.28)(750\ \text{Hz})(0.01 \times 10^{-6}\ \text{F})} = 2.12 \times 10^4\ \Omega$$
$$= 21.2\ \text{k}\Omega \qquad \text{(Use standard 22 k}\Omega\text{ resistors.)}$$

Figure 7–12 shows the completed circuit design and amplitude response curve.

## High-Pass VCVS Circuit

A second-order VCVS HP filter can be obtained by interchanging the frequency-determining resistors $R_1$ and $R_2$ with capacitors $C_1$ and $C_2$ and using the values shown in Figure 7–13. The cutoff frequency is determined in the same manner as for the LP:

$$f_c = \frac{1}{2\pi\sqrt{R_1 R_2 C_1 C_2}} \qquad \text{(repeat of Equation 7.5)}$$

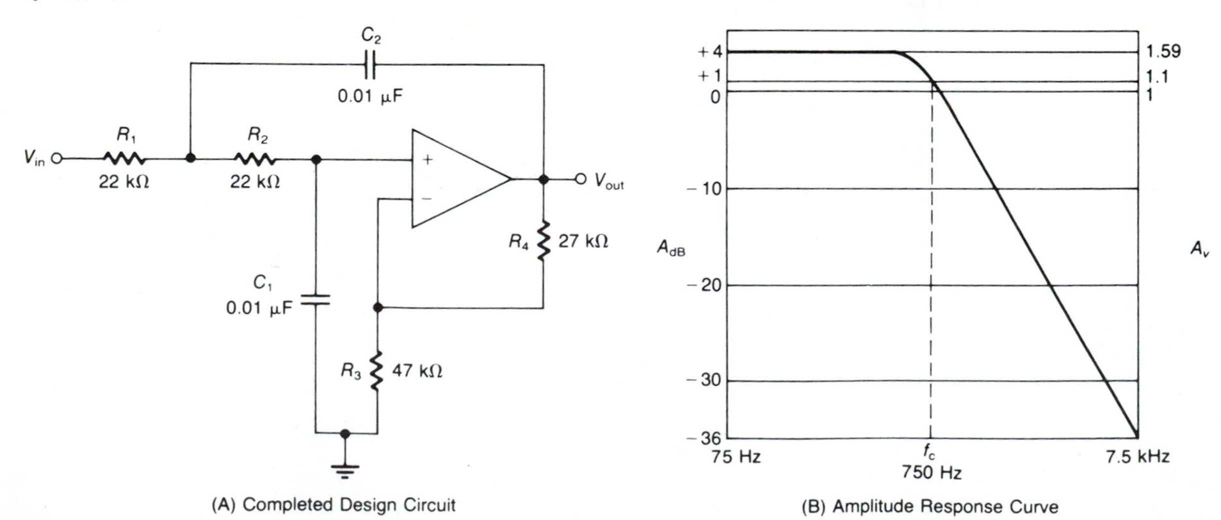


**Figure 7–12.** Solution to Example 7.13

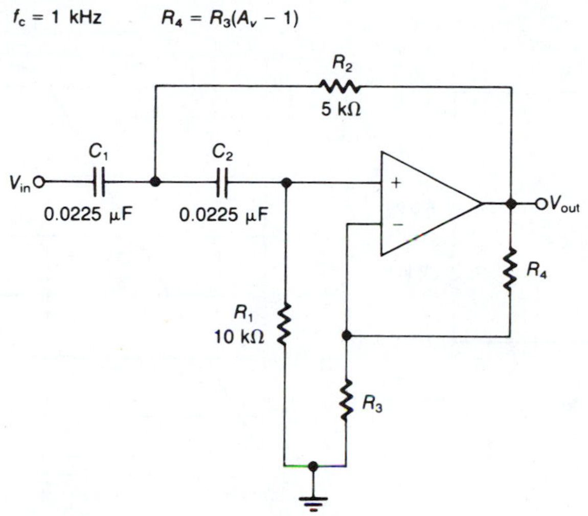

$f_c = 1$ kHz $\quad R_4 = R_3(A_v - 1)$

Figure 7–13. Basic Design Circuit for a Second-Order VCVS High-Pass Filter

Note, however, that now $C_1 = C_2$ and $R_1 = 2R_2$.

Design of the second-order HP VCVS filter is accomplished in the same manner as for the LP, by scaling.

## Example 7.14

Convert the filter circuit of Figure 7–13 to have a $f_c = 3$ kHz and $A_v = 2$. Refer to Table 7–2 as necessary.

*Solution*

We will select $C_1 = C_2 = 0.01\ \mu\text{F}$, so that the capacitor ratio ($Y$) is $0.01/0.0225 = 0.44$. The frequency ratio ($X$) is 1 kHz/3 kHz = 0.33. Therefore,

$$\text{New } R_2 = 5\text{ k}\Omega(0.33) = 1.65\text{ k}\Omega$$

$$\text{Required } R_2 = \frac{1.65\text{ k}\Omega}{0.44} = 3.75\text{ k}\Omega$$

and

$$R_1 = 2R_2 = 2(3.75\text{ k}\Omega) = 7.5\text{ k}\Omega$$

Substituting these values into Equation 7.5 gives us

$$f_c = \frac{1}{2\pi\sqrt{R_1R_2C_1C_2}} = \frac{1}{(6.28)\sqrt{(7.5\text{ k}\Omega)(3.75\text{ k}\Omega)(0.01\ \mu\text{F})(0.01\ \mu\text{F})}}$$
$$= 3\text{ kHz}$$

This verifies the calculations since 3 kHz is the desired frequency. Figure 7–14 illustrates the completed design and amplitude response curve.

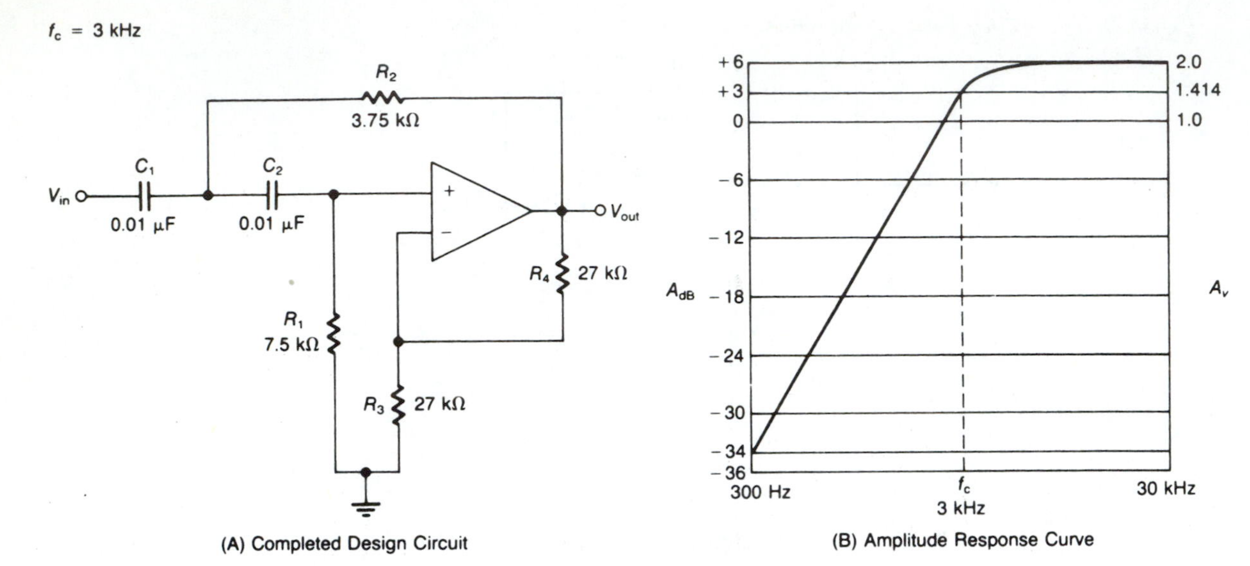


**Figure 7–14.** Solution to Example 7.14

### Equal-Component High-Pass VCVS Circuit

A basic design circuit for the equal-component second-order HP VCVS filter is made by interchanging the components in Figure 7–11. As with the equal-component VCVS LP filter, the passband gain is fixed at 1.59 in order to obtain a Butterworth response.

An obvious advantage of the equal-component filter over the unity-gain filter is the simplified component selection process. Also, $f_c = 1/(2\pi RC)$ as for the first-order filters, since $R_1 = R_2$ and $C_1 = C_2$. This simplified calculation further simplifies circuit design.

## Example 7.15

Design an equal-component HP VCVS filter with a cutoff frequency of 2 kHz.

*Solution*

We will select $C_1 = C_2 = 0.01\ \mu\text{F}$ and determine the value for $R_1 = R_2$.

$$R_1 = R_2 = \frac{1}{2\pi f_c C} = \frac{1}{(6.28)(2000\ \text{Hz})(0.01 \times 10^{-6}\ \text{F})} = 8 \times 10^3\ \Omega$$
$$= 8000\ \Omega$$

The completed circuit design and amplitude response curve are shown in Figure 7–15.

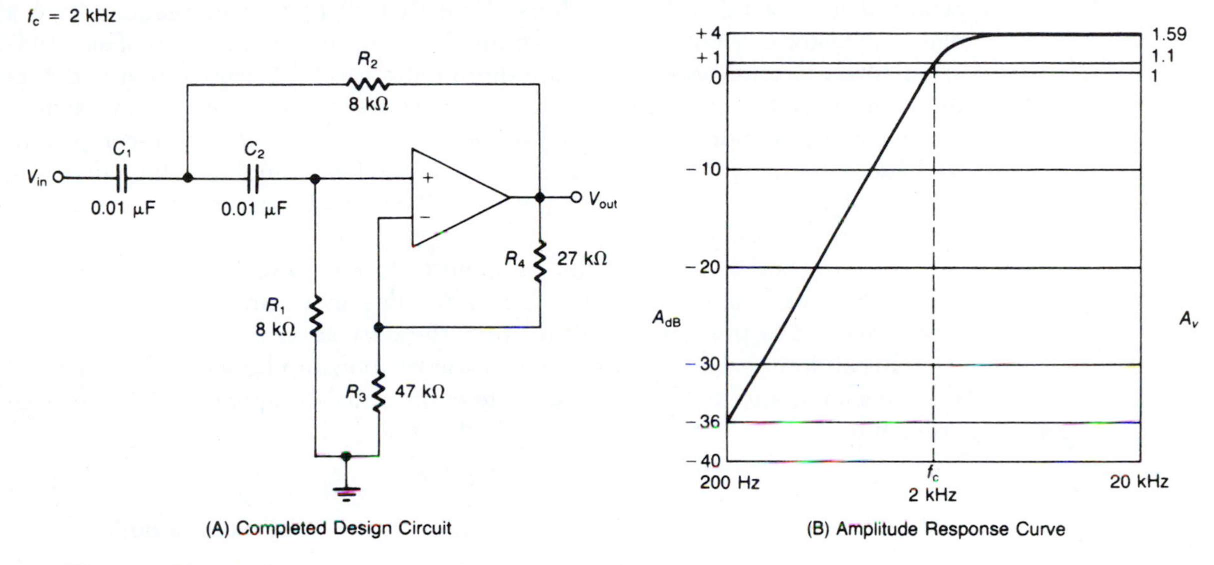


(A) Completed Design Circuit

(B) Amplitude Response Curve

**Figure 7–15.** Solution to Example 7.15

# 7.7 Developing High-Order Filters

It is not practical to construct efficient filters higher than second-order from a single op amp. Therefore, high-order filters must be formed by *cascading* first- and second-order filter sections. For example, a third-order filter would consist of one first-order filter and one second-order filter; a fourth-order filter would require two second-order filters; a fifth-order filter would have one first-order filter and two second-order filters, and so on.

When two or more second-order filter sections are cascaded, the sections are not identical. Each section must have a different damping factor if the filter is to produce a Butterworth response. Table 7–3 shows the damping factors

**Table 7–3** **High-Order Filter Damping Factors (α)**

| Order | First Section | Second Section | Third Section |
|---|---|---|---|
| 3 | 1.00 | 1.00 | — |
| 4 | 1.85 | 0.77 | — |
| 5 | 1.00 | 1.62 | 0.62 |
| 6 | 1.93 | 1.41 | 0.52 |

required for several orders of filters. Note that all first-order sections have a damping factor of 1.00, so there is no problem with the first section of any odd-order filter. Second-order sections require more careful consideration to determine the component values necessary to produce the correct filtering action. Necessary calculations can be greatly reduced by using only equal-component VCVS second-order filter sections. However, because of the damping factors involved in higher-order filters using equal-component VCVS second-order sections, stage gains will not always equal 1.59.

The method of deriving the damping factors is beyond the scope of this text. Advanced mathematics texts and filter design engineering manuals are recommended to those who wish to study the derivations.

Recall from earlier discussions that the relationship between the passband voltage gain $A_v$ and the damping factor ($\alpha$) of the equal-component VCVS second-order filter is

$$A_v = 3 - \alpha \tag{7.6}$$

and that the voltage gain for this type of filter is the same as for a noninverting amplifier:

$$A_v = 1 + \frac{R_f}{R_{in}} \tag{7.7}$$

Combining Equations 7.6 and 7.7 produces a simple formula for determining the values of $R_f$ and $R_{in}$, as expressed by the following equation:

$$\frac{R_f}{R_{in}} = 2 - \alpha \tag{7.8}$$

where

$\alpha$ = the damping factor (specified in Table 7–3)

## Example 7.16

Determine the values for $R_f$ and $R_{in}$ for the second section of a third-order LP filter.

*Solution*

Table 7–3 shows that the damping factor of the second section is 1.00. Using Equation 7.8,

$$\frac{R_f}{R_{in}} = 2 - \alpha = 2 - 1.00 = 1$$

Therefore, $R_f$ must be equal to $R_{in}$ for this filter section. Any standard resistor values can be used. Figure 7–16 shows a typical third-order LP filter, in which $R_5 = R_{in}$ and $R_6 = R_f$. Note that the second-order section is an equal-component VCVS circuit, selected for design simplification.

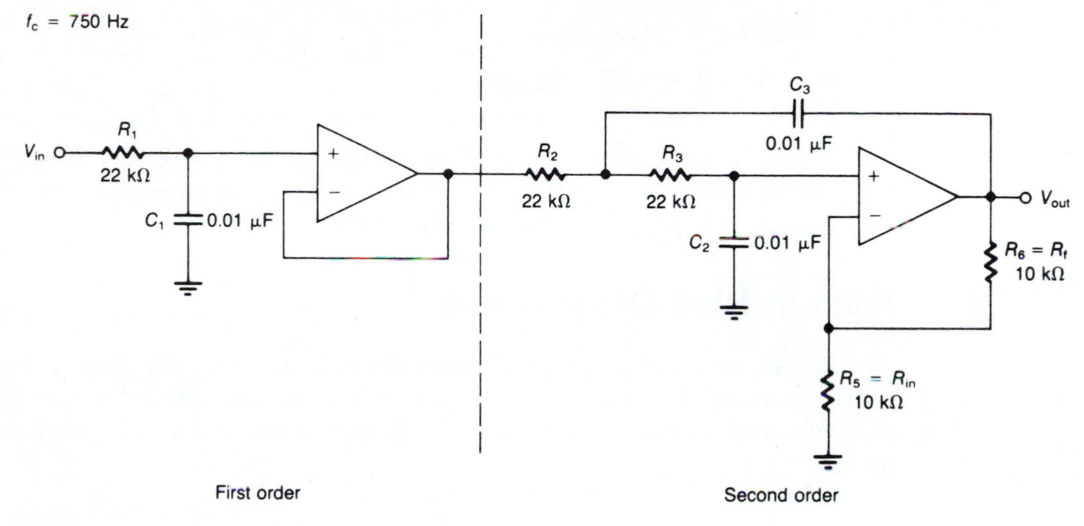


**Figure 7–16.** Third-Order Low-Pass Filter

## Example 7.17

Determine the values for $R_f$ and $R_{in}$ for the first, second, and third sections of a sixth-order filter.

*Solution*

Remember that a sixth-order filter is made up of three second-order filter sections, and that we are using only equal-component VCVS sections.

First section:

$$\frac{R_f}{R_{in}} = 2 - \alpha = 2 - 1.93 = 0.07$$

If we let $R_{in} = 100\ \text{k}\Omega$, then

$$R_f = 100\ \text{k}\Omega \times 0.07 \approx 7.0\ \text{k}\Omega \qquad \text{(Use a standard 6.8 k}\Omega\text{ resistor.)}$$

Second section:

$$\frac{R_f}{R_{in}} = 2 - \alpha = 2 - 1.41 = 0.59$$

If we let $R_{in} = 47\ \text{k}\Omega$, then

$$R_f = 47\ \text{k}\Omega \times 0.59 \approx 27\ \text{k}\Omega$$

Third section:

$$\frac{R_f}{R_{in}} = 2 - \alpha = 2 - 0.52 = 1.48$$

If we let $R_{in} = 15\ k\Omega$, then

$$R_f = 15\ k\Omega \times 1.48 \approx 22\ k\Omega$$

Figure 7–17 illustrates a typical sixth-order LP filter using the values calculated in Example 7.17. Interchanging the frequency-determining components would make this circuit a sixth-order HP filter.

## Gain in High-Order Filters

The overall passband gain of high-order filters is based on the gain of the individual sections. The overall voltage gain is the PRODUCT of the voltage gain of the individual sections, and the overall dB gain is the SUM of the individual dB gains.

### Example 7.18

Determine the overall voltage and dB gains of the circuit in Figure 7–17.

*Solutions*

The voltage gain can be determined from either Equation 7.6 or 7.7 and the calculations of Example 7.17. In this case the simplest method is to use Equation 7.7. Since we have already calculated $R_f/R_{in}$, we can simply add *one* to the results found in Example 7.17:

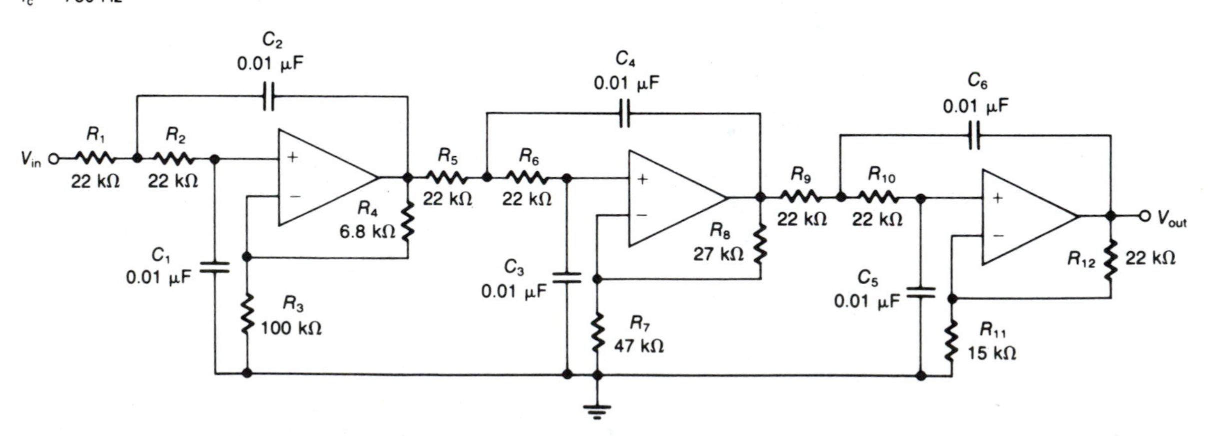


**Figure 7–17.** Sixth-Order Low-Pass Filter

First section $A_v = 1 + 0.07 = 1.07$
Second section $A_v = 1 + 0.59 = 1.59$
Third section $A_v = 1 + 1.48 = 2.48$

Multiply to find the overall voltage gain:

$$A_v = 1.07 \times 1.59 \times 2.48 = 4.22$$

Now use Equation 7.1 to determine the overall dB gain:

$A_{dB} = 20 \log_{10} A_v$
First section $A = 20 \log_{10}(1.07) = 0.587$ dB
Second section $A = 20 \log_{10}(1.59) = 4.028$ dB
Third section $A = 20 \log_{10}(2.48) = 7.889$ dB

Summing and rounding, we have

$$A = 0.587 + 4.028 + 7.889 = 12.504 = 12.5 \text{ dB}$$

Using the methods demonstrated in Examples 7.17 and 7.18 will enable you to evaluate and design any order of filter.

### Order of Filter Determination

Determining which order of filter to use is an important consideration when designing and constructing circuits. Practically, you should use only the order of filter necessary to meet circuit requirements. A higher order filter than necessary may be more expensive and require more space.

## 7.8 Bandpass Filters

A BP filter passes a specified range of frequencies and rejects frequencies above and below that range. Figure 7–1C shows a typical amplitude response curve for a BP filter. For evaluation, we are interested in the *center operating frequency* ($f_{op}$) and *bandwidth (BW)*.

In general, the $f_{op}$ is the highest point of the amplitude response curve, or the point at which the maximum voltage gain is reached. The BW is the difference between the upper frequency ($f_{high}$) and the lower frequency ($f_{low}$) where the voltage gain is down 3 dB (–3 dB), or 0.707 times its maximum voltage gain value.

$$BW = f_{high} - f_{low} \tag{7.9}$$

In general terms, if the BW is less than 10% of the $f_{op}$, the filter is considered to be a *narrow-bandpass* filter. Conversely, if the BW is greater than 10% of $f_{op}$, the filter is considered to be a *wide-bandpass* filter.

## Selectivity and Circuit $Q$

A narrow BW means high selectivity, and a wide BW means low selectivity. The amount of selectivity is expressed by the letter $Q$, the quality factor, of the circuit. The relationship between the BP filter's BW and $f_{op}$ is expressed as follows:

$$Q = \frac{f_{op}}{BW} \tag{7.10}$$

By manipulation of the equation, we have

$$BW = \frac{f_{op}}{Q} \tag{7.11a}$$

and

$$f_{op} = Q(BW) \tag{7.11b}$$

In general, the higher the $Q$ of the circuit, the larger the output voltage. A comparison between high-$Q$ and low-$Q$ BP filter frequency response curves is illustrated in Figure 7–18.

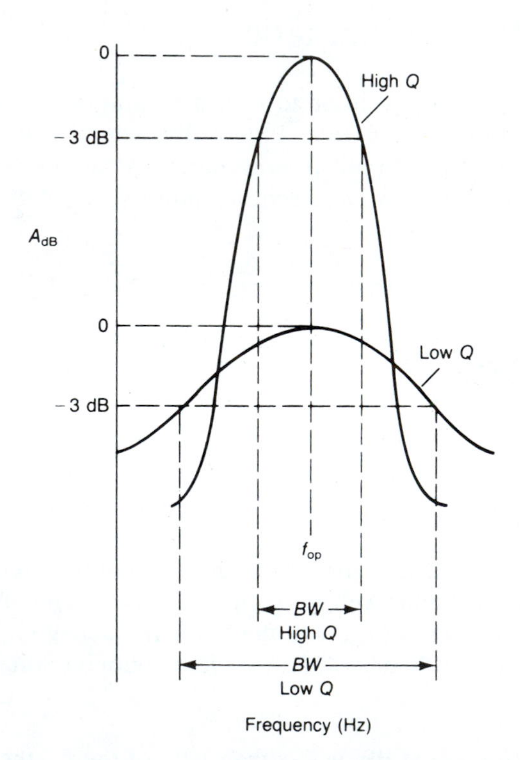


**Figure 7–18.** Comparison between High-$Q$ and Low-$Q$ Bandpass Filters

## Bandpass Filter Construction

By combining LP and HP filter networks with selected frequencies, an active BP filter can be constructed, as shown in Figure 7–19. LP filtering is accomplished by $R_1C_1$, and HP filtering is accomplished by $R_2C_2$. Feedback resistor $R_3$ establishes the gain of the circuit, but it also affects the $Q$ of the circuit and the $f_{op}$. A large value for $R_3$ will result in a high $f_{op}$ and a low $Q$. The $f_{op}$ may be found as follows:

$$f_{op} = \frac{1}{2\pi\sqrt{R_p R_3 C_1 C_2}} \tag{7.12}$$

where

$$R_p = \frac{R_1 R_2}{R_1 + R_2}$$

If $C_1 = C_2$, then the $Q$ of the circuit is

$$Q = 0.5\sqrt{\frac{R_3}{R_p}} \tag{7.13}$$

The $f_{op}$ voltage gain is

$$A_v = \frac{R_3}{R_1\left(1 + \dfrac{C_1}{C_2}\right)} \tag{7.14}$$

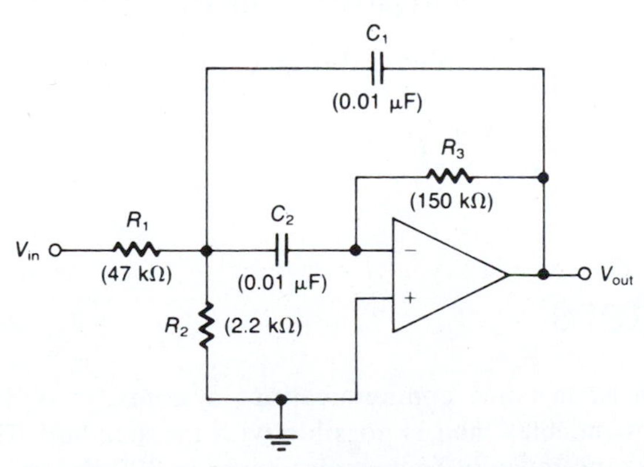


**Figure 7–19.** Active Bandpass Filter (The values in parentheses are the solutions to Example 7.19)

Designing an active BP filter is simplified by making $C_1 = C_2$, selecting standard capacitor values, and preselecting $f_{op}$, $Q$, and $A_v$. Then resistor values may be calculated as follows:

$$R_1 = \frac{Q}{2\pi f_{op} C A_v} \tag{7.15}$$

$$R_2 = \frac{Q}{2\pi f_{op} C(2Q^2 - A_v)} \tag{7.16}$$

$$R_3 = \frac{2Q}{2\pi f_{op} C} \tag{7.17}$$

**Example 7.19**

Design a BP filter with $f_{op} = 900$ Hz, $Q = 4$, and $A_v = 1.5$. Refer to Figure 7–19.

*Solution*

First, select a standard value for $C_1$ and $C_2$, say 0.01 μF. Then determine the resistor values using Equations 7.15, 7.16, and 7.17.

$$R_1 = \frac{Q}{2\pi f_{op} C A_v} = \frac{4}{(6.28)(900 \text{ Hz})(0.01 \times 10^{-6} \text{ F})(1.5)} \approx 47 \text{ k}\Omega$$

$$R_2 = \frac{Q}{2\pi f_{op} C(2Q^2 - A_v)} = \frac{4}{(6.28)(900 \text{ Hz})(0.01 \times 10^{-6} \text{ F})2(16) - 1.5}$$
$$\approx 2.3 \text{ k}\Omega$$

$$R_3 = \frac{2Q}{2\pi f_{op} C} = \frac{2(4)}{(6.28)(900 \text{ Hz})(0.01 \times 10^{-6} \text{ F})} \approx 141 \text{ k}\Omega$$

Figure 7–19 shows the completed design circuit with standard resistor values given in parentheses.

## 7.9 Wideband Filters

Sometimes, such as in voice communications systems, it is desirable to pass a wider range of frequencies than is possible with the standard BP filter. Typically, such systems pass the range of frequencies between 300 Hz and 3 kHz. Designing such a filter is simple; cascade an LP filter with a $f_c$ of 3 kHz and an HP filter with a $f_c$ of 300 Hz. The design is simplified further by using equal-component VCVS filters. It makes no difference which filter comes first in the cascading process.

## Example 7.20

Design a wideband filter that will pass the range of frequencies between 500 Hz and 2.5 kHz.

*Solution*

Using equal-component second-order VCVS filters, we will first select a standard capacitor value, such as 0.022 μF. We can use the same value for both the LP and HP sections. Then, solving for $R$,

$$R_{\text{HP}} = \frac{1}{2\pi f_c C} = \frac{1}{(6.28)(500 \text{ Hz})(0.022 \times 10^{-6} \text{ F})} \approx 14.5 \text{ k}\Omega$$

$$R_{\text{LP}} = \frac{1}{2\pi f_c C} = \frac{1}{(6.28)(2500 \text{ Hz})(0.022 \times 10^{-6} \text{ F})} \approx 3 \text{ k}\Omega$$

Recall that the voltage gain of equal-component VCVS filters is set at 1.59. Now we can select standard values for the input and feedback resistors, such as 47 kΩ and 27 kΩ, respectively. We will use standard 15 kΩ resistors for $R_{\text{HP}}$. Figure 7–20 shows the completed design.

# 7.10 Notch Filters

The notch filter, sometimes referred to as a band-reject filter, operates in reverse of the BP filter; that is, it rejects a certain frequency or band of frequencies while

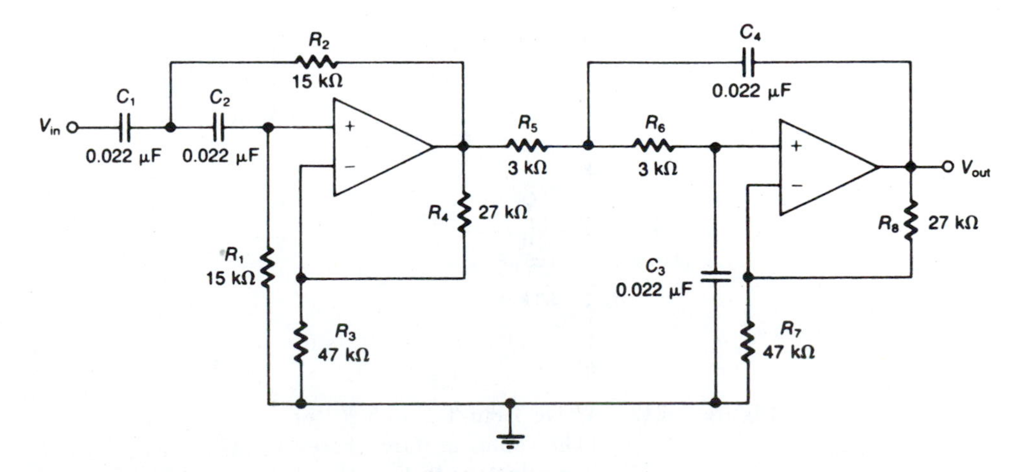


**Figure 7–20.** Completed Design Circuit for Example 7.20

passing all others. Typical applications are in audio and instrumentation systems where a 60 Hz line frequency or a 400 Hz motor-generated frequency can cause interference noise and on telephone lines to null out control frequencies.

## Twin-T Notch Filter

A popular notch filter is the *twin-T* notch filter shown in Figure 7–21. The circuit consists of a passive twin-T filter connected to an op amp voltage follower. In general, this circuit is only good for frequencies below 1 kHz, because of stray capacitances at higher frequencies. The circuit has unity gain at frequencies above and below the *null,* or center, frequency.

### Null Frequency

The null frequency is that frequency at which maximum rejection occurs, and is determined as follows:

$$f_{\text{null}} = \frac{1}{2\pi R_1 C_1} \tag{7.18}$$

where

$$R_1 = R_2 = 2R_3$$
$$C_1 = C_2$$
$$C_3 = 2C_1$$

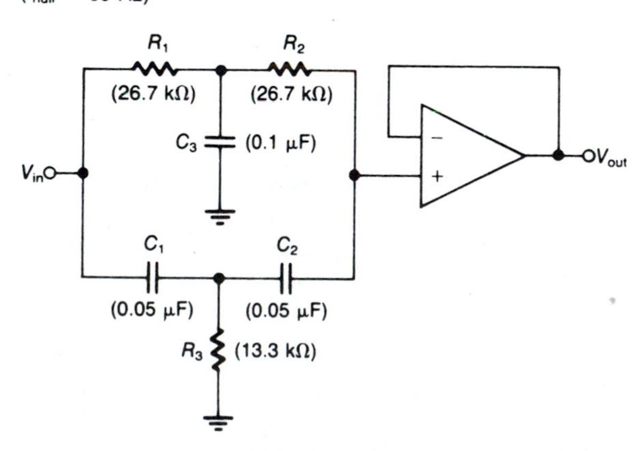


Figure 7–21. Active Twin-T Notch Filter (The values in parentheses are the solutions to Example 7.21)

## Example 7.21

Design a twin-T active notch filter with $f_{null}$ = 120 Hz.

*Solution*

For ease of design, select some standard capacitor value for $C_3$ that will allow another standard value to be used for $C_1$ and $C_2$. Such values could be $C_1 = C_2 = 0.05\ \mu F$ and $C_3 = 2C_1 = 0.1\ \mu F$. Next, solve for $R_1$:

$$R_1 = \frac{1}{2\pi f_{null} C_1} = \frac{1}{(6.28)(120\ \text{Hz})(0.05 \times 10^{-6}\ \text{F})} = 26.5\ \text{k}\Omega$$

(Use a 1% 26.7 kΩ resistor.) Since $R_1 = R_2 = 2R_3$, $R_3 \approx 13.3$ kΩ. Use a 1% 13.3 kΩ resistor. Figure 7-21 shows the completed design circuit with the values calculated in this example given in parentheses.

## Other Notch Filter Circuits

If gain other than unity is desired, the circuit of Figure 7–22 may be used. In this circuit, gain would be determined as for any noninverting amplifier; that is, $A_v = 1 + (R_f/R_{in})$.

Another notch filter design is shown in Figure 7–23 in which amplifier circuit $A_1$ is a basic BP filter, and amplifier circuit $A_2$ operates as a summing amplifier. At point $X$ the input signal to $A_2$ is subtracted from the output signal of $A_1$. A

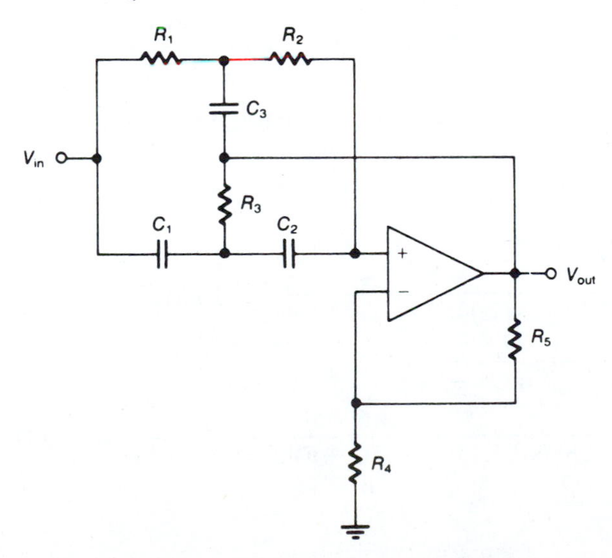


Figure 7–22. Active Twin-T Notch Filter with Gain

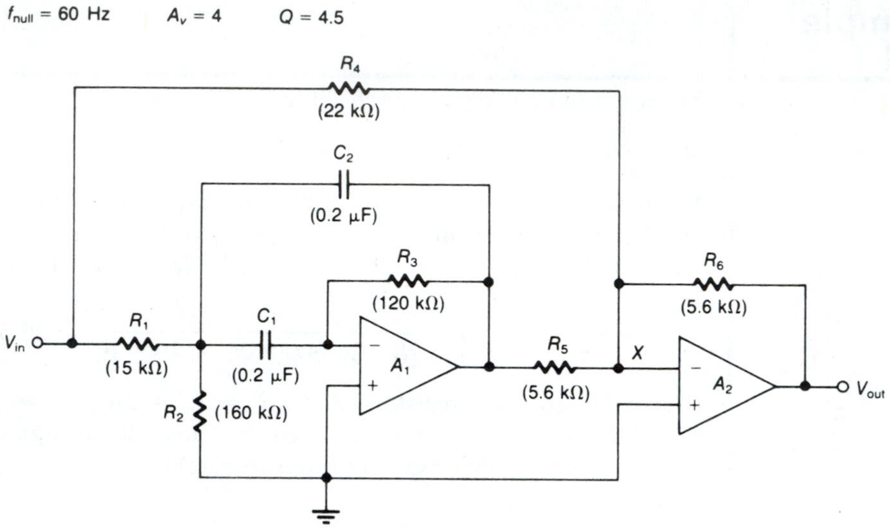

$f_{null} = 60$ Hz $A_v = 4$ $Q = 4.5$

Figure 7–23. Two-Op Amp Notch Filter (The values in parentheses are the solutions to Example 7.22)

deep null can be obtained from this circuit by setting $C_1 = C_2$ and setting the passband gain of $A_1 = R_4/R_5$.

## Example 7.22

Use Figure 7–23 to design a notch filter in which $f_{null} = 60$ Hz, $A_v = 4$, and $Q = 4.5$.

*Solution*

First we will select a standard capacitor value, such as 0.2 μF. Then using Equations 7.15, 7.16, and 7.17 where $f_{op} = f_{null}$, we solve for resistor values:

$$R_1 = \frac{Q}{2\pi f_{null} C A_v}$$

$$= \frac{4.5}{(6.28)(60\text{ Hz})(0.2 \times 10^{-6}\text{ F})(4)} \approx 15\text{ k}\Omega$$

$$R_2 = \frac{Q}{2\pi f_{null} C(2Q^2 - A_v)}$$

$$= \frac{4.5}{(6.28)(60\text{ Hz})(0.2 \times 10^{-6}\text{ F})[2(20.25) - 4]} \approx 160\ \Omega$$

$$R_3 = \frac{2Q}{2\pi f_{null} C}$$

$$= \frac{2(4.5)}{(6.28)(60\text{ Hz})(0.2 \times 10^{-6}\text{ F})} \approx 120\text{ k}\Omega$$

Now, since the passband gain of $A_1$ is $A_v = 4$, the ratio $R_4/R_5$ must also be 4. Therefore, we will select standard resistor values such that $R_4$ can equal $4R_5$. We will select $R_5 = 5.6\ \text{k}\Omega$ and $R_4 = 22\ \text{k}\Omega$. Also we will want unity gain in $A_2$, so we will make $R_6 = R_5 = 5.6\ \text{k}\Omega$. The completed circuit design shown in Figure 7–23 gives the values for this example in parentheses.

# 7.11 Multiple-Feedback Filters

*Multiple-feedback* filters derive their name from the fact that they have an additional feedback path. Figure 7–24 shows a basic second-order LP multiple-feedback filter design circuit, in which $R_3$ is the additional feedback path. Also note that the input is connected to the inverting terminal of the op amp. The cutoff frequency for this circuit is 1 kHz and is determined by the following equation:

$$f_c = \frac{1}{2\pi\sqrt{R_2R_3C_1C_2}} \tag{7.19}$$

The passband gain is

$$A_v = \frac{R_3}{R_1} \tag{7.20}$$

## Designing the Multiple-Feedback Filter

The additional feedback component does not necessarily complicate the design of the multiple-feedback filter. For this circuit to have a Butterworth response, $A_v$ must equal unity. Therefore, all three resistors will be equal in value. Also note that $C_1$ is equal to approximately $4.5C_2$ in Figure 7–24. Scaling will produce the desired component values.

### Example 7.23

Refer to Figure 7–24. Design a second-order multiple-feedback filter with $f_c = 250$ Hz and $A_v = 1$. (Refer to Table 7–2 as necessary.)

*Solution*

Since 1 kHz/250 Hz = 4, we multiply all frequency-determining resistors by 4, so that $R_1 = R_2 = 20\ \text{k}\Omega$, and with unity gain, $R_3$ also will equal 20 kΩ. Then we will select capacitor values such that $C_1 = 4.5C_2$, or $C_2 = 0.022\ \mu\text{F}$ and $C_1 = 0.1\ \mu\text{F}$. Then 0.022 μF/0.015 μF = 1.47, so we divide the new resistor values by that ratio, or 20 kΩ/1.47 = 13.6 kΩ. Using 1% 13.7 kΩ resistors would satisfy the circuit requirements. Figure 7–25 shows the completed circuit design.

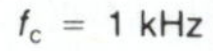


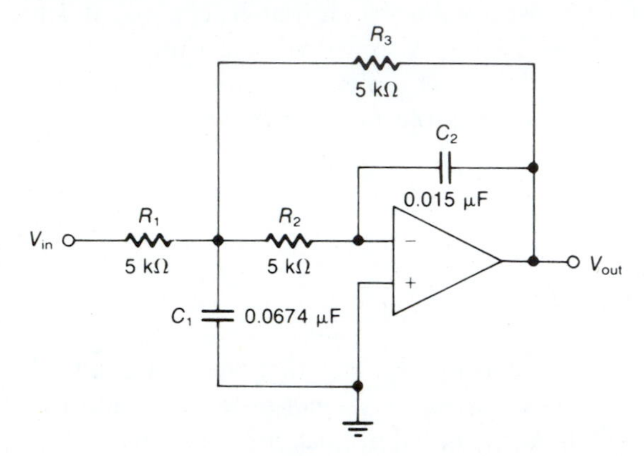


**Figure 7–24.** Basic Design Circuit for a Second-Order Low-Pass Multiple-Feedback Filter

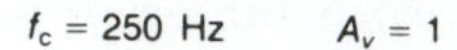


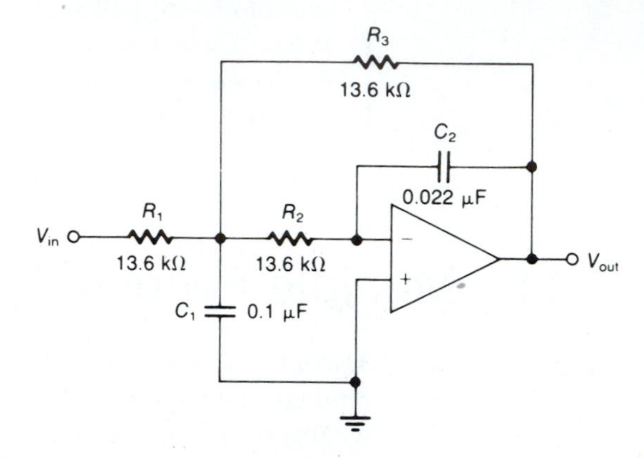


**Figure 7–25.** Completed Design Circuit for Example 7.23

The multiple-feedback HP filter is formed by simply interchanging the components of the LP filter, as shown in Figure 7–26. Here, the additional feedback path is through $C_3$, and $R_2 = 4.5R_1$. The cutoff frequency is determined by the following equation:

$$f_c = \frac{1}{2\pi\sqrt{R_1R_2C_2C_3}} \tag{7.21}$$

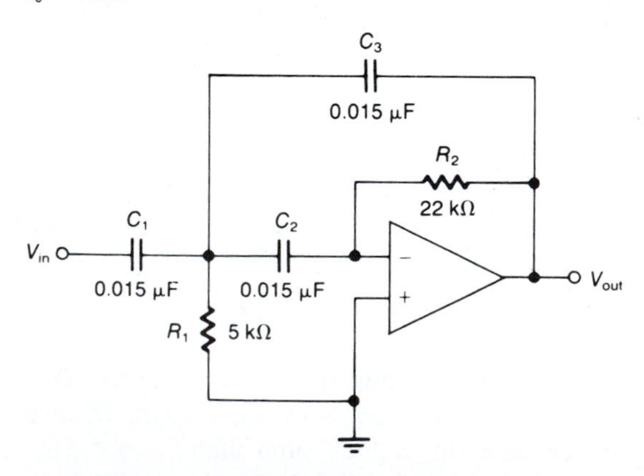


**Figure 7–26.** Basic Design Circuit for a Second-Order High-Pass Multiple-Feedback Filter

The passband gain of this circuit is determined by the ratio of two capacitors rather than by two resistors, which makes this circuit unique in op amp circuits. The gain is

$$A_v = \frac{C_3}{C_1} \tag{7.22}$$

Circuit design is accomplished in the same manner as was the design of the LP multiple-feedback filter, that is, by scaling the basic filter design circuit shown in Figure 7–26.

# 7.12 State-Variable Filters

The *state-variable* filter is a different type of multiple-feedback filter, because it uses three or four op amps, has one input, and has three outputs that provide LP, HP, and BP filtering simultaneously. Figure 7–27 shows the block diagram of the basic state-variable filter that consists of a summing amplifier, two integrators, and a damping network.

## Designing the State-Variable Filter

For a Butterworth response from this circuit, damping is set by the resistor network, $R_3$–$R_4$. The basic design circuit for a state-variable filter is shown in Figure 7–28. Designing such a circuit may look difficult, but by making a few assumptions, we can greatly simplify the process. The best procedure is to choose standard values so that $C = C_1 = C_2$. And if we let $R_5 = R_6$ and $R_1 = R_2 = R_3 = R_7 = R_8 = R_9$, we have the following formulas,

$$R_5 = \frac{1}{2\pi f_{op} C} \tag{7.23}$$

$$R_4 = R_3(3Q - 1) \tag{7.24}$$

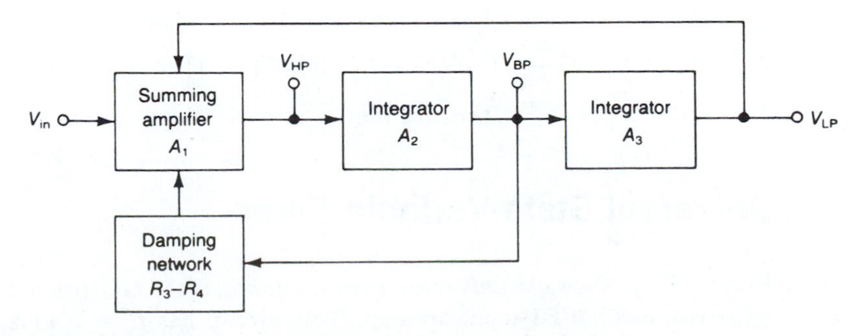


**Figure 7–27.** Block Diagram of a Basic State-Variable Filter

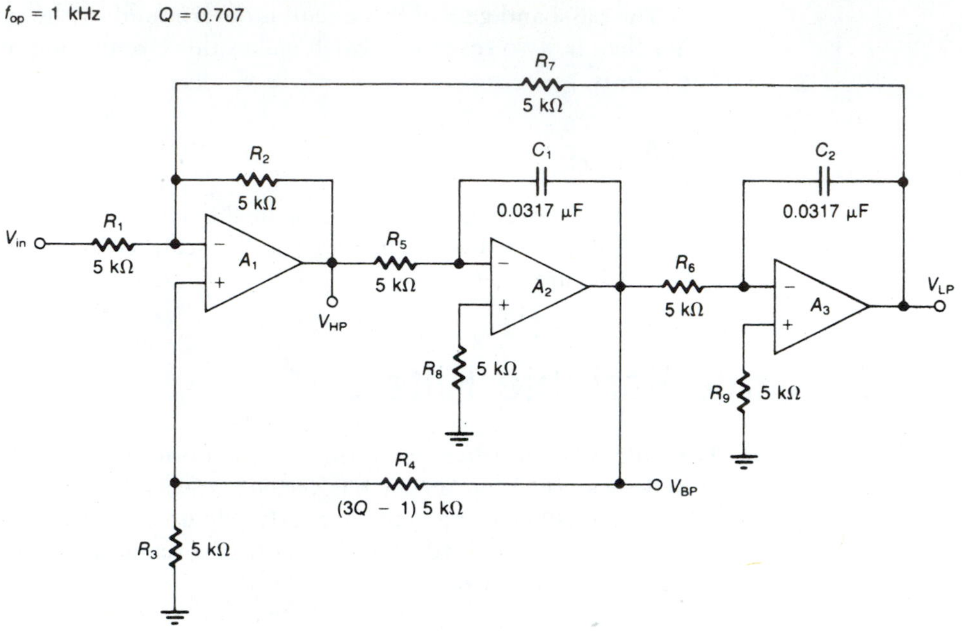

Figure 7–28. Basic Design Circuit for a State-Variable Filter

## Example 7.24

Design a state-variable filter with a Butterworth response in which $f_{op} = 500$ Hz, $Q = 0.707$, $C = 0.22\ \mu\text{F}$, and $R_7 = 15\ \text{k}\Omega$.

*Solution*

From Equation 7.23,

$$R_5 = R_6 = \frac{1}{2\pi f_{op}C} = \frac{1}{(6.28)(500\ \text{Hz})(0.22 \times 10^{-6}\ \text{F})} \approx 1.5\ \text{k}\Omega$$

From Equation 7.24 and knowing $Q$ must equal 0.707,

$$R_4 = R_3(3Q - 1) = (15\ \text{k}\Omega)[3(0.707) - 1] = 16.8\ \text{k}\Omega$$

Figure 7–29 shows the completed circuit design.

## Universal State-Variable Filter

Figure 7–30 shows a *universal state-variable* filter constructed with the LF347 wide BW quad JFET-input op amp. This circuit has $f_{op} = 3$ kHz, $f_{null} = 9.5$ kHz, $Q = 3.4$, and passband gains are: $A_{\upsilon(HP)} = 0.1$, $A_{\upsilon(LP)} = 1$, $A_{\upsilon(BP)} = 1$, and $A_{\upsilon(null)} = 10$.

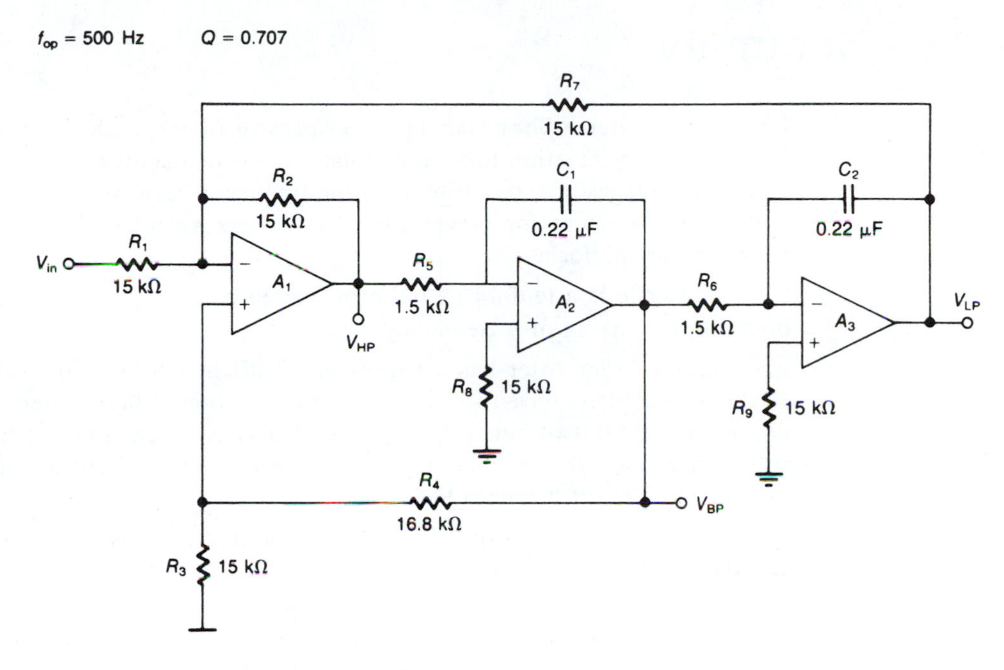


**Figure 7–29.** Completed Design Circuit for Example 7.24

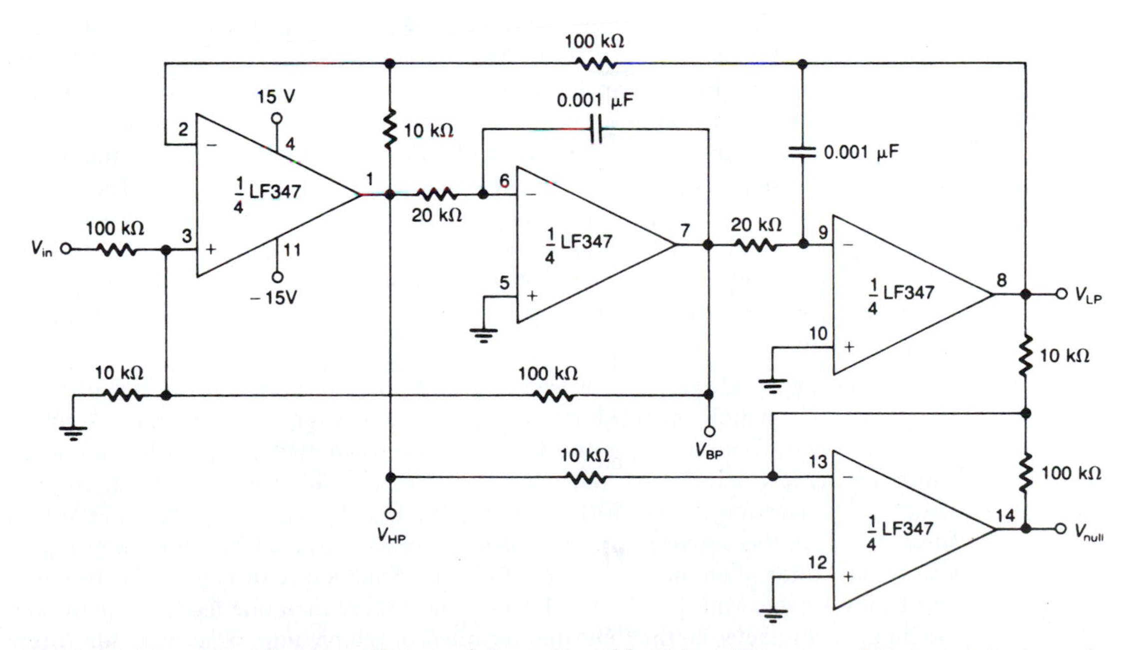


**Figure 7–30.** Four-Op Amp Universal State-Variable Filter (Courtesy of National Semiconductor Corporation)

# 7.13 Summary

☐ Active filters offer advantages over passive filters, such as lower cost, ease of adjustment, no insertion loss, and isolation between source and load. Limitations and disadvantages of active filters are the frequency response limits of the op amps and the requirement for power supplies. Filters are classified according to the function they perform.

☐ A decade is a tenfold increase or decrease in frequency, and an octave is a doubling or halving of a frequency.

☐ A first-order filter has a rolloff of 20 dB per decade, or 6 dB per octave. Higher-order filters have rolloffs in direct proportion to their order; for example, a second-order has two times the rolloff of a first-order, and a fifth-order has a rolloff five times that of the first-order. These rolloffs would result in 40 dB per decade and 100 dB per decade, respectively.

☐ The Butterworth response, a relatively flat response at frequencies above cutoff, occurs in highly damped filters.

☐ Most filter circuit designs can be simplified by first selecting standard capacitor values and then calculating resistor values. Scaling is a method of circuit design that starts with a basic circuit normalized with an input impedance of 5 kΩ and a $f_c$ of 1 kHz. This design method is convenient in multiple-feedback circuit design.

☐ A restriction on equal-component VCVS second-order filters is that the passband gain be fixed at 1.59 if a Butterworth response is to be obtained. The simplest second-order filter circuit design is the equal-component second-order VCVS filter, LP or HP, since component values are equal. It is not practical to use a single op amp in filters of a higher order than second-order. Such high-order filters are constructed by cascading first- and second-order filter sections. The overall passband gain of high-order filters depends on the gain of the individual filter sections. The voltage gain is the product of the voltage gains of the individual sections, and the dB gain is the sum of the dB gains of the individual sections. The order of filter to be used in any application should be determined by circuit requirements only.

☐ A narrow-bandpass filter is one in which the BW is less than 10% of $f_{op}$. It has a high $Q$, which means high selectivity, and high voltage gain. A wide-bandpass filter is one in which the BW is greater than 10% of $f_{op}$. It has a low $Q$, which means low selectivity, and low voltage gain. Wideband filters can be constructed by cascading an LP filter and an HP filter. It makes no difference which filter is first in the cascade. Notch filters can be constructed by connecting a passive twin-T filter to an op amp voltage follower. Such a circuit is good for frequencies below 1 kHz. Multiple-feedback filters have more than one feedback path, and the input is connected to the inverting terminal of the op amp. State-variable filters use three op amps and provide three outputs that consist of LP, HP, and BP filtering. The universal state-variable filter uses four op amps and provides four outputs that consist of LP, HP, BP, and notch filtering.

# 7.14 Questions and Problems

**7.1** Define *active filter.*

**7.2** State the distinguishing features of (a) the LP filter and (b) the HP filter.

**7.3** Design a first-order LP filter with $f_c$ = 100 Hz and $A_v$ = 3. Determine the dB gain of the circuit and the break point.

**7.4** Design a first-order HP filter with $f_c$ = 400 Hz and $A_v$ = 2. Determine the dB gain of the circuit and the break point.

**7.5** Design a second-order VCVS LP filter with $f_c$ = 100 Hz. What is the dB gain and the break point?

**7.6** Design a second-order HP VCVS filter with $f_c$ = 400 Hz. What is the dB gain and the break point?

**7.7** A second-order VCVS HP filter section with $f_c$ = 300 Hz is cascaded with a second-order LP filter section with $f_c$ = 3 kHz. (a) What kind of filter circuit is this? (b) What is the overall voltage gain? (c) What is the overall dB gain? (d) What is the BW? (e) What is the overall order of the circuit?

**7.8** What would you use to construct a sixth-order HP filter?

**7.9** What would be the overall voltage gain and dB gain of the circuit of Problem 7.8? (HINT: Refer to Table 7-3.)

**7.10** A BP filter has $Q$ = 5 and $BW$ = 200 Hz. What is $f_{op}$?

**7.11** A BP filter has $f_{op}$ = 2 kHz and $f_{low}$ = 1.8 kHz. What is (a) $f_{high}$ and (b) $Q$?

**7.12** Describe the state-variable filter and identify its sections by function.

**7.13** A 60 Hz notch filter has unity gain and a BW of 14 Hz. What is the lower cutoff frequency and the upper cutoff frequency?

**7.14** What is the dB break point for the notch filter of Problem 7.13?

**7.15** Design a state-variable filter with a Butterworth response where $f_{op}$ = 750 Hz. Select your own values for components.

**7.16** Explain how a multiple feedback filter differs from a standard active filter.

**7.17** What parameter is necessary to assure that a multiple feedback filter has a Butterworth response?

**7.18** Design a multiple feedback filter with a Butterworth response and $f_c$ = 2 kHz. Show all calculations and draw the completed design circuit.

Chapter 8

# Signal Processing Circuits

## 8.1 Introduction

Signal processing and conditioning are important functions of electronic circuits. The op amp, with its associated external components, is an excellent device to use as a basic signal processing and conditioning circuit. We have studied many such circuits in previous chapters, but in this chapter we will look at the types of circuits used in process control systems. Such systems require that certain signals be conditioned, or processed, in some form so that the resulting signal can be used as a control signal.

*Instrumentation amplifiers* (IAs), used in precision measurement and control, are constructed with differential amplifier circuits. Used in conjunction with a transducer, the instrumentation amplifier is used in a wide variety of applications, several of which are examined in this chapter.

# 8.2 Objectives

When you complete this chapter, you should be able to:

- ☐ Define transducer.
- ☐ Discuss transducers, their uses, and list several types of transducers.
- ☐ Define instrumentation amplifier.
- ☐ Discuss instrumentation amplifiers and their uses.
- ☐ Explain how CMRR and gain can be boosted in an instrumentation amplifier.

# 8.3 Transducers

A *transducer* is a device that converts one form of energy or disturbance into another. In electronics, transducers convert alternating current (ac) or direct current (dc) into sound, light, heat, radio waves, or other forms. Transducers can also convert these forms into alternating or direct current—in other words, transducers can be either input devices or output devices.

Common examples of electrical or electronic transducers include buzzers, speakers, microphones, piezoelectric crystals, light-emitting and infrared-emitting diodes, photocells, and radio antennas. Other transducer devices include strain gauges, thermistors, photoconductive cells, and thermocouples.

## Mechanical Transducers

One example of a *mechanical transducer* is an ordinary light switch, which can be manually actuated. Another is one that is actuated by some piece of machinery, such as a limit switch controlling the motion of a lathe or drill.

Figure 8–1 shows a mechanical transducer for tripping a switch in a pipe used for either liquid flow or pressure. The manufacturer of such transducers provides instructions for calibrating the ON-OFF points.

## Strain Gauges

*Stress* is any static force exerted against a body. Normally measured in pounds or kilograms per unit of cross-sectional area, stress can occur in three different ways:

1. Compression (pushing in)
2. Tension (pulling apart)
3. Laterally (sideways)

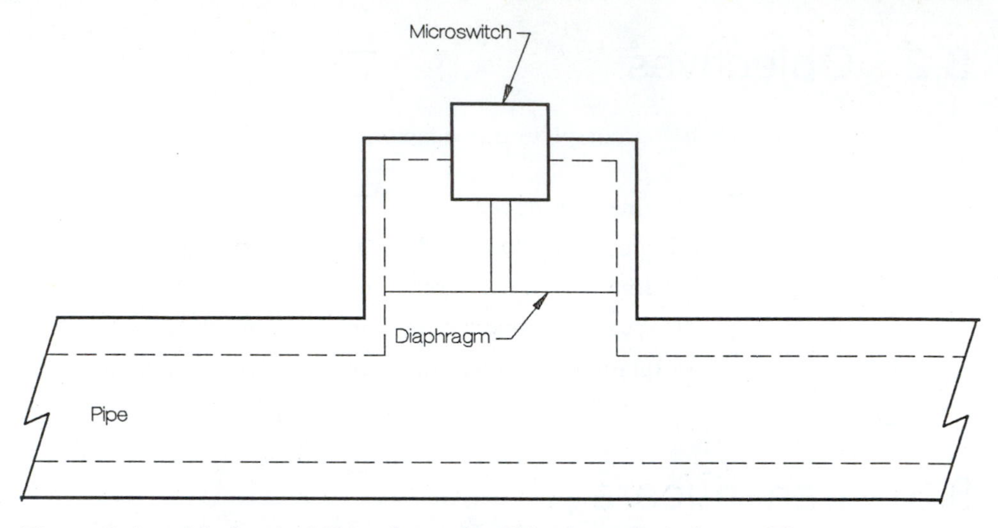


**Figure 8-1.** Mechanical Transducer for Tripping a Switch in a Flow Pipe

When an object is subjected to such physical forces, that object may change in shape. *Strain* is an expression of the extent to which an object changes shape under stress. A *strain gauge* is a device capable of quantitatively measuring those changes. The basis of many transducers is the *resistance strain gauge.* Load cells for use in electric weighing, fluid pressure transducers, and pressure-difference transducers are typical examples of resistance strain gauge transducers.

One common method for determining strain is to attach one or more resistive wires to the object under test as shown in Figure 8-2. Bonded to the body of the object under test, the wires stretch (Figure 8-2A) or contract (Figure 8-2B) along with the body as it changes shape. The change in the resistance of the wires produces the same amount of percent change as the object, causing a change in the known dc resistance. The amount of strain can be determined by measuring this dc resistance.

The strain of some objects can be measured by means of light interference patterns. A change in the physical shape of an object affects the way it reflects visible light. The amount of stress can be directly measured in some cases, and the strain determined indirectly from the force on the object under test.

## Piezoelectric Transducers

Certain crystalline or ceramic substances can act as transducers at audio and radio frequencies. When subjected to mechanical stress, these materials produce electric

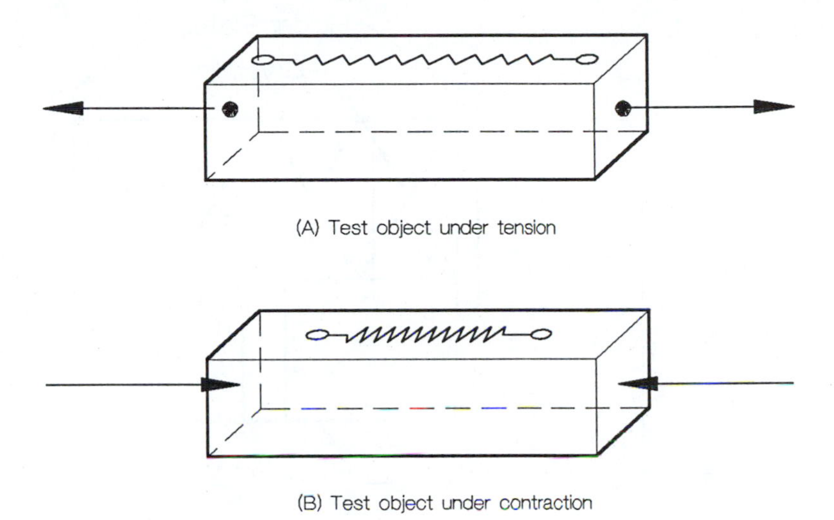

(A) Test object under tension

(B) Test object under contraction

**Figure 8–2.** Simple Resistance Strain Gauge

currents; when subjected to an electric voltage, the substances will vibrate. This effect is known as the *piezoelectric effect.*

Piezoelectric crystals are often used to convert mechanical vibrations into electrical impulses, and vice versa. Crystals, either natural or synthetic, make excellent transducers for this reason. Such piezoelectric substances are used in microphones, phonograph cartridges, earphones, and buzzer devices. The acquisition tones now used on many devices with keyboards are generated by tiny piezoelectric crystals.

Types of piezoelectric transducers are crystal transducers, ceramic microphones, ceramic pickups, and crystal microphones.

### Crystal Transducer

Figure 8–3 shows the operation of a simple *crystal transducer.* If alternating currents are applied to the holder plates, the crystal will vibrate, producing sound or ultrasound. Conversely, if sound or ultrasound is impinged on the diaphragm, an alternating current will appear between the holder plates. Crystal transducers can operate at frequencies well above the range of human hearing.

### Ceramic Microphone

When subjected to the stresses of mechanical vibration, certain ceramic materials generate electrical impulses. A *ceramic microphone* uses a ceramic cartridge to transform sound energy into electrical impulses.

Ceramic microphones display a high output impedance and excellent audio-frequency response. They must be handled with care, because they are easily damaged by impact.

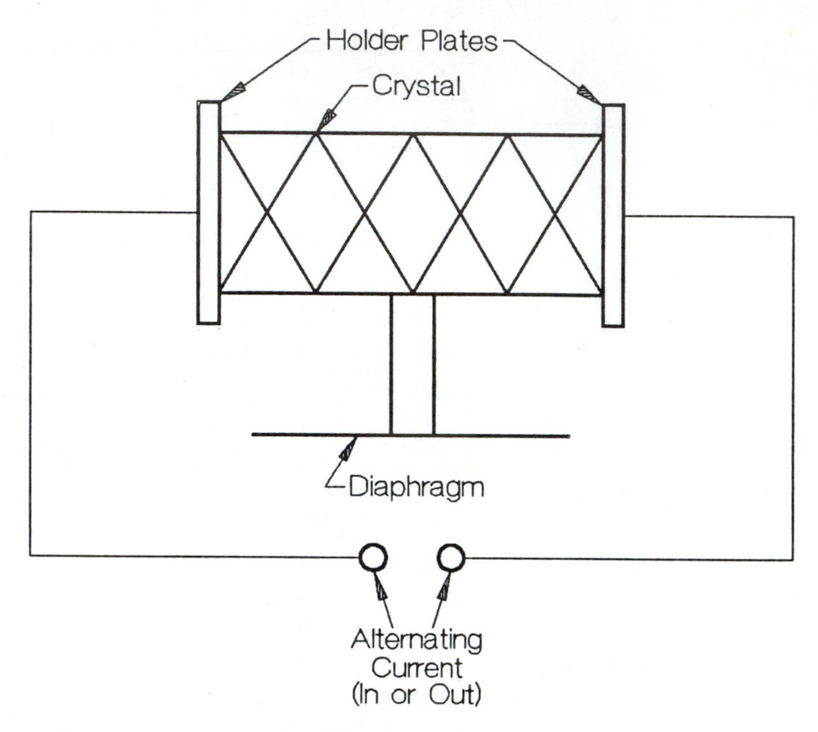


**Figure 8-3.** Construction of a Typical Crystal Transducer

### Ceramic Pickup

A *ceramic pickup* is a cartridge for phonograph use, in which a ceramic crystal is used to translate the vibrations of the stylus into electrical impulses. It operates according to the piezoelectric effect, in the same way that a ceramic microphone operates. An artificially polarized ceramic crystal generates electrical impulses over a wide range of frequencies.

A ceramic pickup is sometimes called a crystal pickup, in spite of the technical differences between the two. A ceramic pickup is somewhat more durable than a conventional crystal type, and has excellent sound-reproduction quality and excellent dynamic range.

### Crystal Microphone

A *crystal microphone* is a device that uses a piezoelectric crystal to convert sound vibrations into electrical impulses. These impulses may then be amplified for use in public address or communications circuits.

In the crystal microphone, which operates in a manner similar to the ceramic microphone, vibrating air molecules set a metal diaphragm in motion. The diaphragm, connected physically to the crystal, puts mechanical stress on the piezoelectric substance. This, in turn, results in small currents at the same frequency or frequencies as the sound.

Crystal microphones have excellent fidelity characteristics, but are rather fragile. If the microphone is dropped, the crystal may break, and the microphone will be ruined.

## Thermal Transducers

All substances conduct heat to some extent, but some conduct heat better than others. The extent to which a substance can transfer heat efficiently from one place to another can be expressed quantitatively. This expression is known as the *thermal conductivity* of the material.

*Thermodynamics* is a branch of physics concerned with the interaction between thermal energy and mechanical energy. In particular, thermodynamics involves the behavior of matter under conditions of variable temperature.

There are a variety of temperature-sensitive transducers, most of which are constructed with bimetal elements. In such transducers, two dissimilar metals with different coefficients of expansion, are bonded together. When the temperature changes, the difference in coefficient of expansion causes the unit to bend. Transducers made in this way can be used to open or close a circuit for a change in temperature.

We will examine three temperature-sensitive transducers in this section: thermostat, thermistor, and thermocouple.

### Thermostat

A *thermostat* is a temperature-sensitive transducer used to actuate some apparatus for regulation or protection. The most common applications of a thermostat is for temperature control in buildings. A thermostat can be constructed in various ways, but they all act as switches, opening and closing a circuit according to the temperature.

Figure 8–4 shows a simplified bimetal-strip thermostat. There are two modes: heating and cooling. In the heating mode, the furnace is switched ON when the temperature falls below a preset point. In the cooling mode, an air conditioner is switched ON when the temperature rises above a preset point. There is a certain amount of hysteresis, or sluggishness, in the thermostat, so that it will not cycle ON and OFF rapidly or continuously.

A thermostat can be used for thermal-protection purposes. For example, if a device gets too hot, the thermostat actuates a switch that removes power.

Thermistors and thermocouples can be used to sense temperature in electronic thermostats.

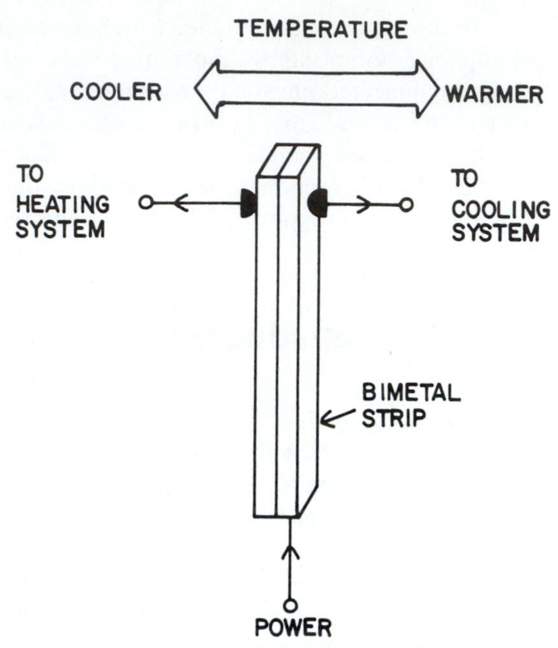


**Figure 8-4.** Bimetal-Strip Thermostat (From book #2000 *Encyclopedia of Electronics* © **TAB Books**, a division of McGraw-Hill)

### Thermistor

A resistor, designed especially to change value with temperature, is called a *thermistor.* The term thermistor is a contraction of **THERM**ally sensitive res**ISTOR.**

Thermistors are made from semiconductor materials, the most common of which are oxides of metals. The resistance may increase as the temperature rises, or it may decrease as the temperature rises. The resistance is a precise function of the temperature. Usually, the *resistive temperature coefficient* (the tendency of a component to change in value with temperature) is large and negative (where the value of the component decreases as the temperature rises). The construction of a typical thermistor is shown in Figure 8-5.

Thermistors are operated at low current levels, so that the resistance is affected only by the ambient temperature, and not by heating caused by the applied current itself. The characteristics of the thermistor make it ideal for use in electronic thermostats and thermal protection circuits.

### Thermocouple

A *thermocouple* is a device that generates a voltage by the heating of a junction between two metal electrodes, consisting of dissimilar metals, placed in physical contact. Figure 8-6 shows the construction of a typical thermocouple.

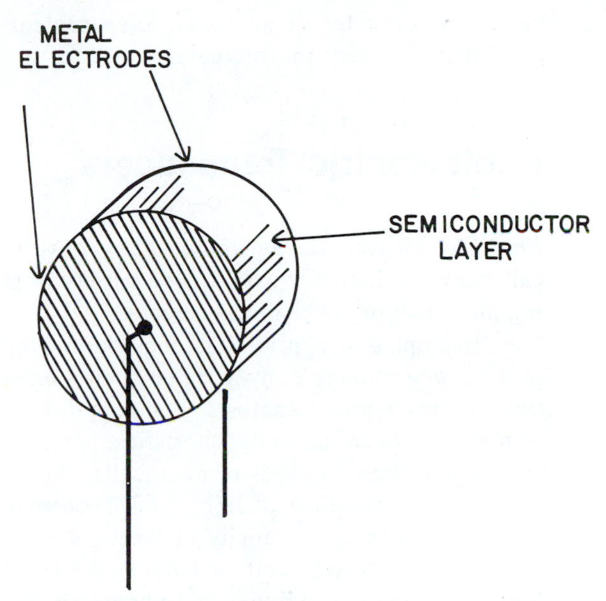


**Figure 8–5.** Construction of a Typical Thermistor (From book #2000 *Encyclopedia of Electronics* © **TAB Books**, a division of McGraw-Hill)

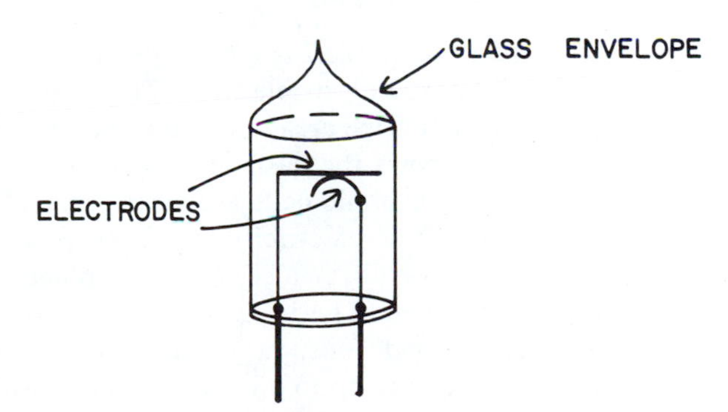


**Figure 8–6.** Construction of a Typical Thermocouple (From book #2000 *Encyclopedia of Electronics* © **TAB Books**, a division of McGraw-Hill)

Although a thermocouple normally produces only a few millivolts of potential at 100 °C, some can be used at temperatures of more than 1000 °C, and generate almost 0.5 V. In general, the voltage produced by a given thermocouple is proportional to the temperature.

Two or more thermocouples can be connected in series to obtain higher voltages than a single thermocouple delivers. Such a series connection is called a *thermopile*.

Thermopiles are commonly used in thermocouple type thermometers, which consist of an even number of thermocouples and a remote indicating device, usually a voltmeter or ammeter. Such devices can survive much higher temperatures than ordinary thermometers.

## Photoelectric Transducers

*Photoelectric transducers (photocells)* convert luminous (light) energy into electrical impulses. Industry uses a wide variety of photoelectric transducers for regulating electrical or mechanical devices.

Examples of applications include turning equipment ON and OFF, counting objects on a moving conveyor belt, color indexing and matching, equipment positioning, or chemical analysis. Each application and each type of photocell has its own requirements and characteristics.

There are two kinds of photocells: those that generate a current all by themselves in the presence of light, and those that simply cause a change in effective resistance when the intensity of the light is varied. The first type of photocell is called a photovoltaic cell or solar cell. The latter type is found in a variety of forms, such as photodiodes and photoresistors.

### Photovoltaic Cell

A *photovoltaic cell* is a semiconductor device that generates a direct current when it is exposed to visible light. These cells generally consist of a P-N junction having a large surface area, and a transparent housing to allow light to enter easily. Figure 8–7A shows the basic construction of such a device.

Most photovoltaic cells are made from silicon, and require no external bias; they generate their electricity by themselves. When visible light (and, in some cases, infrared or ultraviolet energy) impinges on the P-N junction, electron-hole pairs are produced. The intensity of the current from the photovoltaic cell, under constant load conditions, varies in linear proportion to the brightness of the light striking the device, up to a certain point. Beyond that point, the increase is more gradual, and it finally levels off at a maximum current called the saturation current, as shown in Figure 8–7B.

Photovoltaic cells are used in many different electronic devices, including modulated-light systems and solar-power generators. One important application is in power generation. The cells, in conjunction with storage batteries, can provide continuous electric power from sunlight.

### Photodiode

A *photodiode* is a diode that exhibits a variable resistance depending on the intensity of visible light that lands on its P-N junction.

A photodiode is specially designed so that its P-N junction is readily exposed to visible light. Its conductivity generally increases in the reverse direction (nega-

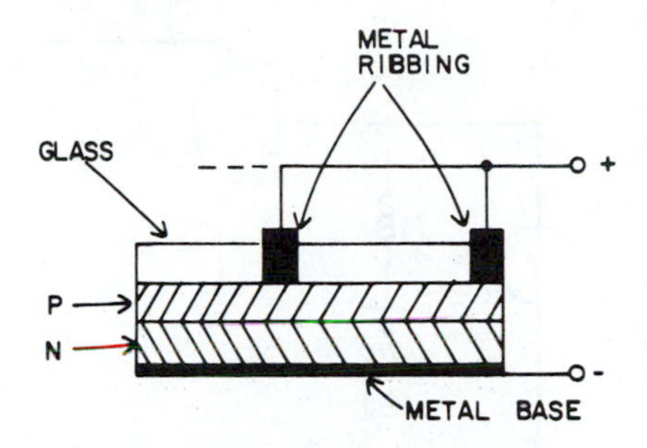


(A) Construction of a Typical Silicon Cell

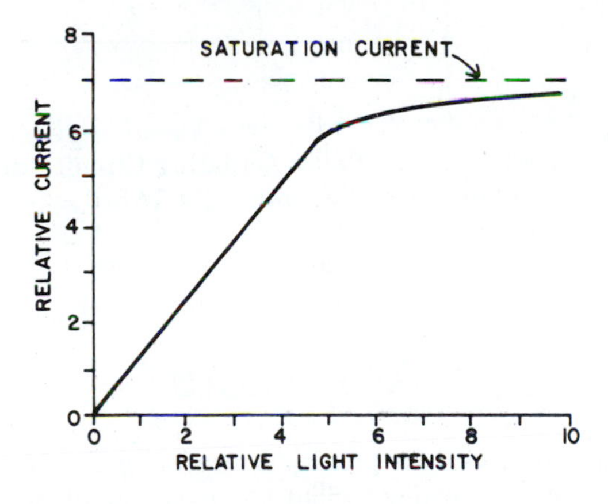


(B) Current Versus Brightness Characteristic

Figure 8–7. Photovoltaic Cell (From book #2000 *Encyclopedia of Electronics* © **TAB Books**, a division of McGraw-Hill)

tive coefficient) as the impinging light gets brighter, but a limiting factor to the drop in forward resistance is its saturation value.

A photodiode is connected in series with the circuit to be controlled, and is reverse-biased. Care must be taken to insure that the photodiode does not carry too much current. Figure 8–8 shows a typical circuit incorporating a photodiode for the purpose of opening and closing a relay.

### Photoresistor

A *photoresistor* is a device that exhibits a variable resistance, depending on the amount of light that strikes it. Normally, the resistance is a certain finite value in total darkness, but as the light increases, the resistance decreases up to a certain point. If the intensity of the light increases further than the maximum limit of the device, the resistance does not drop any further.

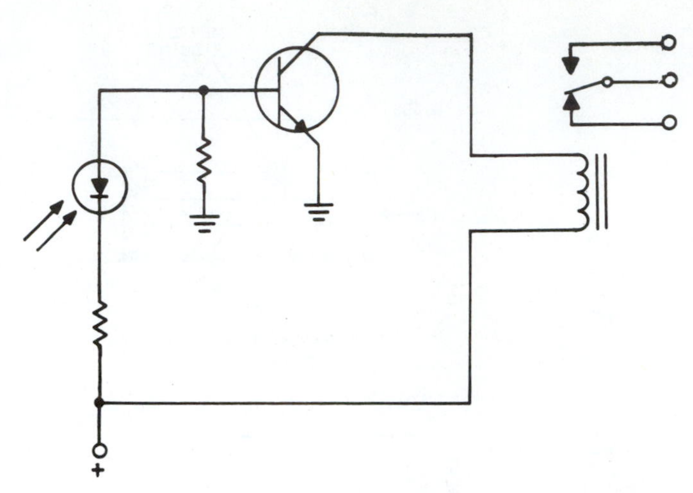


**Figure 8–8.** Schematic of a Simple Relay-Actuating Circuit Using a Photodiode (From book #2000 *Encyclopedia of Electronics* © **TAB Books**, a division of McGraw-Hill)

# 8.4 Instrumentation Amplifiers

A most useful circuit for precision measurement and control is the instrumentation amplifier (IA). A buffered-input IA consisting of two input voltage followers, precision resistors, and a differential amplifier is shown in Figure 8–9A, and the standard schematic symbol is shown in Figure 8–9B. The voltage followers offer extremely high input impedance with low error to the differential amplifier, which provides gain and high common-mode rejection ratio (CMRR). If $R_5$ is made variable, any common-mode voltage can be balanced out. A single resistor, $R_1$, is used to set the voltage gain of the circuit. The input impedance does not change as gain is varied, and $V_{out}$ depends only on the difference between the input voltages of the differential amplifier.

## LH0036 Instrumentation Amplifier

The equivalent circuit and connection diagram for the LH0036 IA is shown in Figure 8–10. Designed for precision differential signal processing, this device offers extremely high accuracy because it has an input impedance of 300 MΩ and 100 dB CMRR. Gain is programmable with one external resistor from 1 to 1000. Operating over a temperature range from -55 °C to +125 °C, its wide power supply range (±1 V to ±18 V) and low power consumption (90 μW), adjustable input bias current and output bandwidth (BW), make the LH0036 extremely versatile.

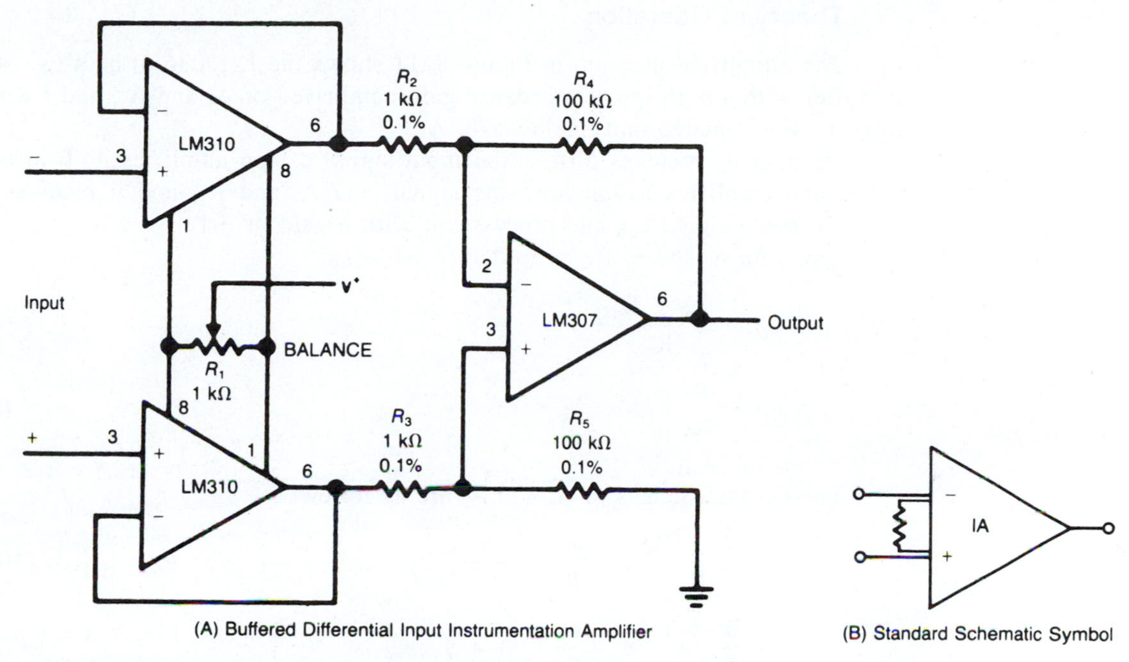


(A) Buffered Differential Input Instrumentation Amplifier

(B) Standard Schematic Symbol

**Figure 8–9.** Instrumentation Amplifier (Part A: Courtesy of National Semiconductor Corporation)

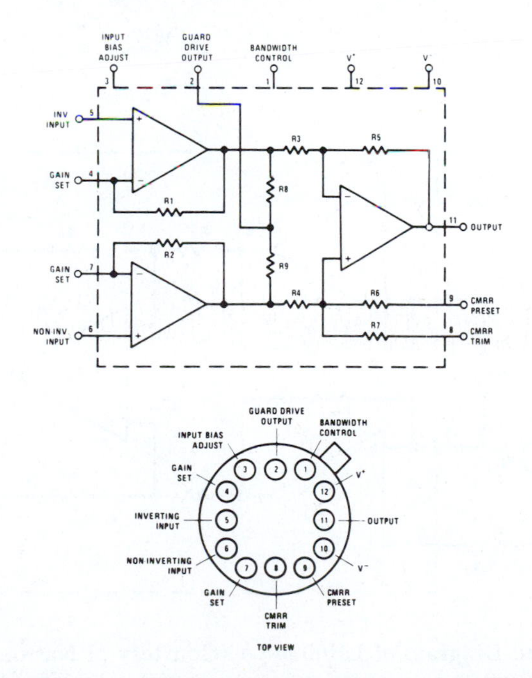


**Figure 8–10.** Equivalent Circuit and Connection Diagrams (Reprinted with permission of National Semiconductor Corporation)

### Theory of Operation

The simplified diagram in Figure 8–11 shows the LH0036 to be a two-stage amplifier with a high input impedance gain comprised of $A_1$ and $A_2$, and a differential to single-ended unity gain stage, $A_3$.

Op amp $A_1$ receives differential input signal $e_1$ and amplifies it. It also receives and amplifies $e_2$, an inverting signal, via $A_2$ and $R_2$, and it receives the common mode signal $e_{CM}$ and processes it with a gain of +1.

Gains for $e_1$ and $e_2$ are as follow:

$$A(e_1) = \frac{R_1 + R_G}{R_G} \tag{8.1}$$

$$A(e_2) = \frac{R_1}{R_G} \tag{8.2}$$

The output voltages of $A_1$ and $A_2$ are as follow:

$$V_1 = \frac{R_1 + R_G}{R_G} e_1 - \frac{R_1}{R_G} e_2 + e_{CM} \tag{8.3}$$

$$V_2 = \frac{R_2 + R_G}{R_G} e_2 - \frac{R_2}{R_G} e_1 + e_{CM} \tag{8.4}$$

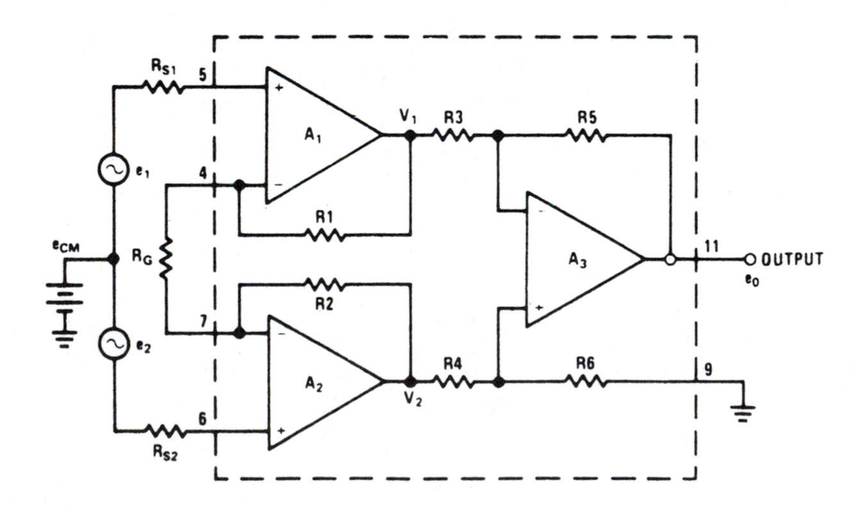


Figure 8–11. Simplified Diagram of LH0036 IA (Courtesy of National Semiconductor Corporation)

For the condition $R_1 = R_2$:

$$V_2 - V_1 = \left[\left(\frac{2R_1}{R_G}\right) + 1\right](e_2 - e_1) \tag{8.5}$$

For the condition $R_3 = R_4 = R_5 = R_6$, the gain of $A_3 = 1$ and:

$$e_o = (1)(V_2 - V_1) = (e_2 - e_1)\left[1 + \left(\frac{2R_1}{R_G}\right)\right] \tag{8.6}$$

As can be seen, for the condition of identically matched resistors, $e_{CM}$ is canceled out, and the differential gain is dictated by Equation 8.6, which reduces to:

$$A_{VCL} = \frac{e_o}{e_2 - e_1} = 1 + \frac{50\ k\Omega}{R_G} \tag{8.7}$$

The loop gain ($A_{VCL}$) may be set to any value from 1 ($R_G = \infty$) to 1000 ($R_G = 50\ \Omega$). Equation 8.7 rearranged in more convenient form can be used to select $R_G$ for a desired gain:

$$R_G = \frac{50\ k\Omega}{A_{VCL} - 1} \tag{8.8}$$

**For examples 8.1 through 8.5, refer to Figure 8–11. Assume $e_1$ = 8 mV, $e_2$ = 15 mV, and $e_{CM}$ = 5 V.**

## Example 8.1

Assume $R_1 = 22\ k\Omega$, $R_G = 10\ k\Omega$. Calculate (a) $A(e_1)$ and (b) $A(e_2)$.

*Solution*

a. Use Equation 8.1:

$$A(e_1) = \frac{R_1 + R_G}{R_G}$$

$$= \frac{22\ k\Omega + 10\ k\Omega}{10\ k\Omega}$$

$$= 3.2$$

b. Use Equation 8.2:

$$A(e_2) = \frac{R_1}{R_G} = \frac{22\ \text{k}\Omega}{10\ \text{k}\Omega} = 2.2$$

## Example 8.2

Let $R_1 = 25\ \text{k}\Omega$, $R_2 = 50\ \text{k}\Omega$, $R_G = 10\ \text{k}\Omega$, $e_1 = 8$ mV, $e_2 = 15$ mV, $e_{CM} = 5$ V. Calculate (a) $V_1$ and (b) $V_2$.

*Solution*

a. Use Equation 8.3:

$$V_1 = \frac{R_1 + R_G}{R_G} e_1 - \frac{R_1}{R_G} e_2 + e_{CM}$$

$$= \frac{25\ \text{k}\Omega + 10\ \text{k}\Omega}{10\ \text{k}\Omega} 8\ \text{mV} - \frac{25\ \text{k}\Omega}{10\ \text{k}\Omega} 15\ \text{mV} + 5\ \text{V}$$

$$= 0.028 - 12.54 = -12.5\ \text{V}$$

b. Use Equation 8.4:

$$V_2 = \frac{R_2 + R_G}{R_G} e_2 - \frac{R_2}{R_G} e_1 + e_{CM}$$

$$= \frac{50\ \text{k}\Omega + 10\ \text{k}\Omega}{10\ \text{k}\Omega} 15\ \text{mV} - \frac{50\ \text{k}\Omega}{10\ \text{k}\Omega} 8\ \text{mV} + 5\ \text{V}$$

$$= 0.09 - 25.04 = -24.95\ \text{V}$$

## Example 8.3

Assume $R_1 = R_2 = 25\ \text{k}\Omega$. Calculate (a) $V_1$, (b) $V_2$, and (c) $V_2 - V_1$.

*Solution*

a. Use Equation 8.3:

$$V_1 = \frac{R_1 + R_G}{R_G} e_1 - \frac{R_1}{R_G} e_2 + e_{CM}$$

$$= \frac{25\ \text{k}\Omega + 10\ \text{k}\Omega}{10\ \text{k}\Omega} 8\ \text{mV} - \frac{25\ \text{k}\Omega}{10\ \text{k}\Omega} 15\ \text{mV} + 5\ \text{V}$$

$$= 0.028 - 12.54 = -12.5 \text{ V}$$

b. Use Equation 8.4:

$$V_2 = \frac{R_2 + R_G}{R_G} e_2 - \frac{R_2}{R_G} e_1 + e_{CM}$$

$$= \frac{25 \text{ k}\Omega + 10 \text{ k}\Omega}{10 \text{ k}\Omega} 15 \text{ mV} - \frac{25 \text{ k}\Omega}{10 \text{ k}\Omega} 8 \text{ mV} + 5 \text{ V}$$

$$= 0.0525 - 12.52 = -12.468 \text{ V}$$

c. Use Equation 8.5:

$$V_2 - V_1 = \left[\left(\frac{2R_1}{R_G}\right) + 1\right](e_2 - e_1)$$

$$= \left[\left(\frac{(2)\ 25 \text{ k}\Omega}{10 \text{ k}\Omega}\right) + 1\right](15 \text{ mV} - 8 \text{ mV})$$

$$= (6)(7 \text{ mV}) = 42 \text{ mV}$$

## Example 8.4

Assume $R_3 = R_4 = R_5 = R_6 = 100\text{ k}\Omega$. Calculate (a) $e_O$ and (b) $A_{VCL}$.

*Solution*

a. Use Equation 8.6:

$$e_O = (1)(V_2 - V_1) = (e_2 - e_1)\left[1 + \left(\frac{2R_1}{R_G}\right)\right]$$

$$= (15 \text{ mV} - 8 \text{ mV})\left[1 + \left(\frac{(2)\ 25 \text{ k}\Omega}{10 \text{ k}\Omega}\right)\right]$$

$$= 42 \text{ mV}$$

b. Use Equation 8.7:

$$A_{VCL} = \frac{e_O}{e_2 - e_1} = 1 + \frac{50 \text{ k}\Omega}{R_G}$$

$$= 1 + \frac{50 \text{ k}\Omega}{10 \text{ k}\Omega}$$

$$= 6$$

## Example 8.5

Calculate the value for $R_G$ required to achieve gains of (a) 10, (b) 100, (c) 500, and (d) 1000.

*Solution*

Use Equation 8.8:

a. $$R_G = \frac{50\ k\Omega}{A_{VCL} - 1} = \frac{50\ k\Omega}{10 - 1} = \frac{50\ k\Omega}{9} = 5.56\ k\Omega$$

b. $$R_G = \frac{50\ k\Omega}{100 - 1} = \frac{50\ k\Omega}{99} = 505\ \Omega$$

c. $$R_G = \frac{50\ k\Omega}{500 - 1} = \frac{50\ k\Omega}{499} = 100\ \Omega$$

d. $$R_G = \frac{50\ k\Omega}{1000 - 1} = \frac{50\ k\Omega}{999} = 50\ \Omega$$

### Use of Bandwidth Control (Pin 1)

In the standard configuration, pin 1 is simply grounded. The amplifier's slew rate in this configuration is typically 0.3 V/μs and small signal BW 350 kHz for $A_{VCL} = 1$. In some applications, especially at low frequency, it may be desirable to limit BW in order to minimize the overall noise BW of the device. A resistor ($R_{BW}$) may be placed between pin 1 and ground to accomplish this task. Figure 8–12A shows typical small BW versus $R_{BW}$.

Large-signal BW and slew rate can also be adjusted down by use of $R_{BW}$. Figure 8–12B shows a plot of slew rate versus $R_{BW}$.

## Offset Voltage and CMRR Considerations

An important consideration in using IAs is the adjustment of offset voltage ($V_{OS}$) and CMRR. Manufacturers' data sheets provide information for adjusting these parameters for optimum operation of the IA. This section deals with adjustments for the LH0036.

### Use of CMRR Preset, Pin 9

For nominal operations, pin 9 (CMRR Preset) should be grounded. An internal factory trimmed resistor ($R_6$) will yield a CMRR in excess of 80 dB (for $A_{VCL}$ = 100). If a higher CMRR is desired, leave pin 9 open and use the procedure that follows.

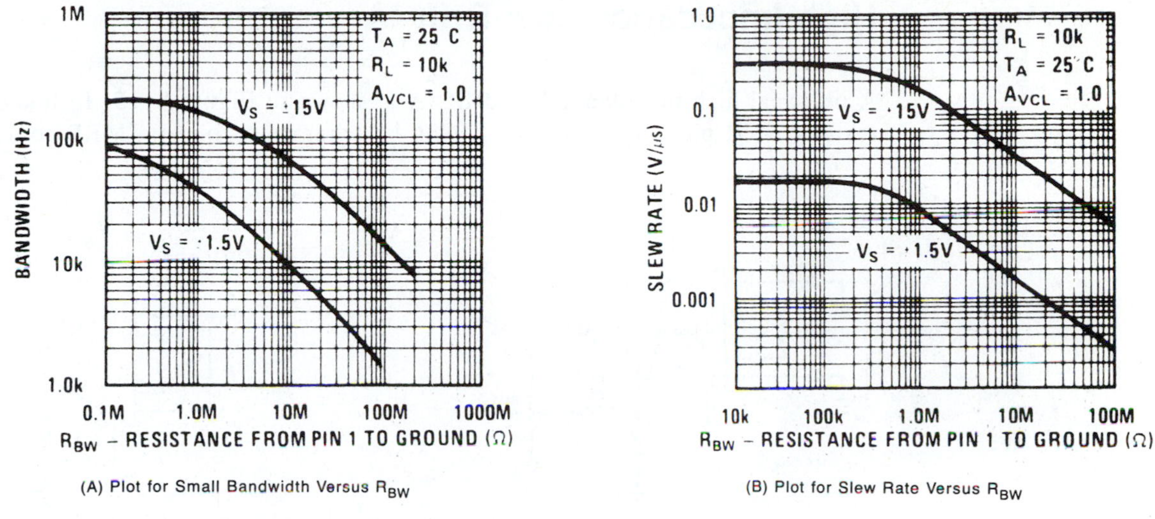


**Figure 8–12.** Plots for Bandwidth and Slew Rate Resistor Adjustments (Courtesy of National Semiconductor Corporation)

### dc Offset Voltage and Common Mode Rejection Adjustments

Offset voltage may be nulled at pin 8 by using the circuit in Figure 8–13A. Pin 8 is also used to improve the CMRR as shown in Figure 8–13B. Null is achieved by alternately applying $\pm 10$ V (for the condition V+ and V– = 15 V) to the inputs and adjusting $R_1$ for minimum change at the output.

To accomplish both $V_{OS}$ and CMRR null, combine the circuits in Figures 8–13A and 8–13B, as shown in Figure 8–13C. However, the $V_{OS}$ and CMRR adjustments are interactive, and several repeated adjustments are required. When using this circuit, the procedure should start with the inputs grounded.

For $V_{OS}$ null, adjust $R_2$. An input of +10 V is then applied and $R_1$ is adjusted for CMRR null. The procedure is repeated until the optimum adjustment is achieved.

The circuit in Figure 8–13D overcomes the adjustment interaction. In this case, $R_2$ is adjusted first for output null of the LH0036. Then $R_1$ is adjusted for output null with +10 V input. It is always a good idea to check CMRR null with a –10 V input. The optimum null achievable will yield the highest CMRR over the amplifier's common mode range.

### ac CMRR Considerations

The circuit shown in Figure 8–14 can be used to improve the ac CMRR. After adjusting $R_1$ for the best dc CMRR as before, $R_2$ should be adjusted for minimum peak-to-peak voltage at the output while applying an ac common mode signal at the maximum amplitude and frequency of interest.

## High-Impedance Low-Drift IA

A practical high-impedance low-drift IA is shown in Figure 8–15. In this circuit, resistor $R_3$ at pin 3 of amplifier $A_3$ can be trimmed to boost CMRR to 120 dB.

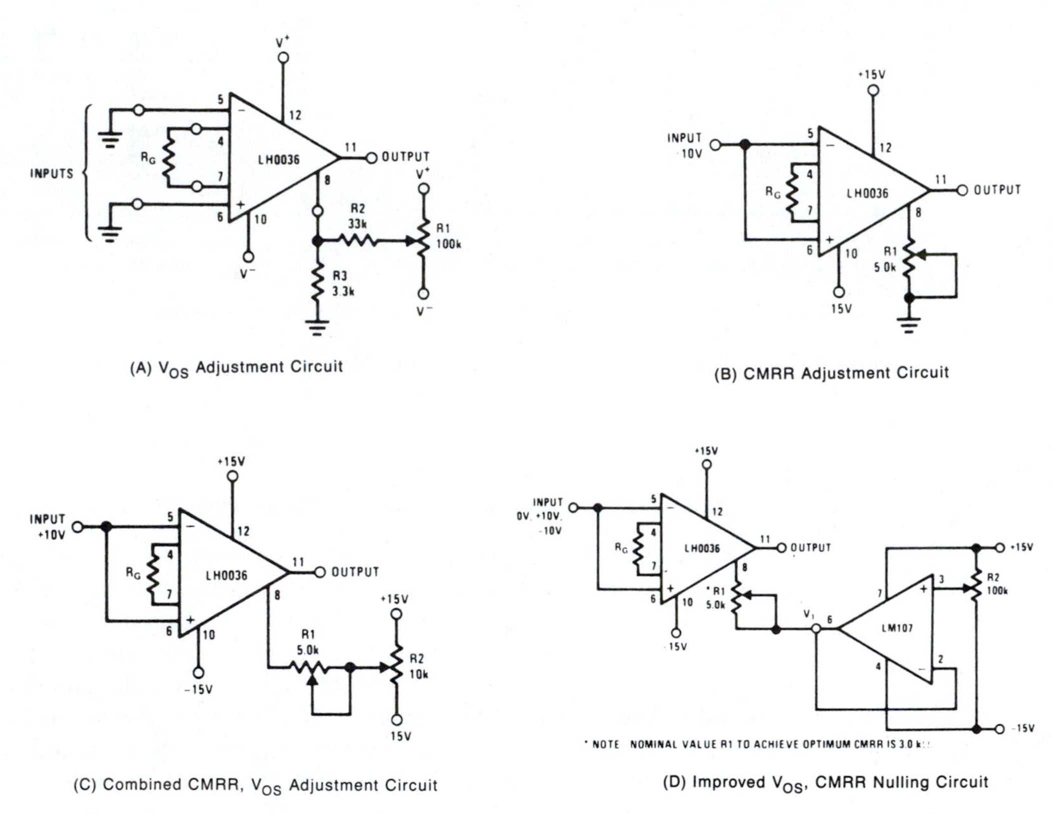


(A) $V_{OS}$ Adjustment Circuit

(B) CMRR Adjustment Circuit

(C) Combined CMRR, $V_{OS}$ Adjustment Circuit

(D) Improved $V_{OS}$, CMRR Nulling Circuit

**Figure 8–13.** Circuits for Adjusting Offset Voltage and CMRR (Courtesy of National Semiconductor Corporation)

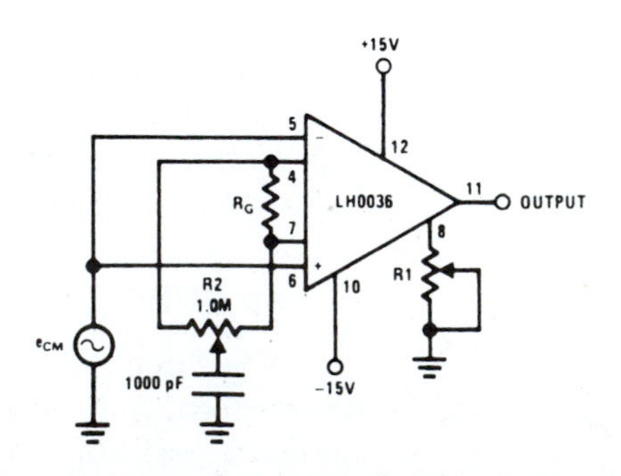


**Figure 8–14.** Circuit for Improving ac CMRR (Courtesy of National Semiconductor Corporation)

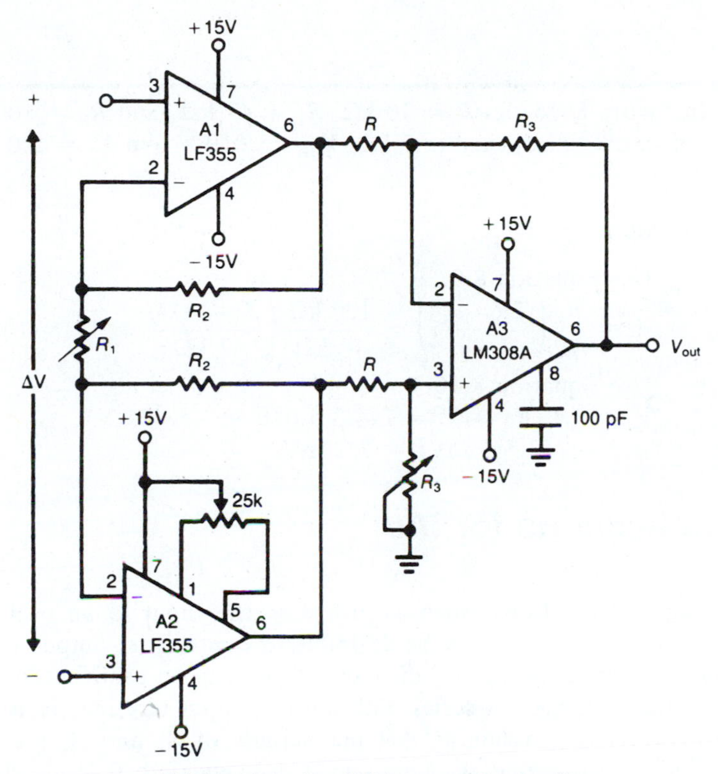


**Figure 8-15.** Instrumentation Amplifier (Courtesy of National Semiconductor Corporation)

Circuit offset voltage $V_{OS}$ is adjusted by the 25 kΩ $V_{OS}$ adjust resistor between pins 1 and 5 of amplifier $A_2$. Circuit voltage gain $A_v$ is adjusted by resistor $R_1$, and is determined as follows:

$$A_v = \frac{R_3}{R}\left(\frac{2R_2}{R_1} + 1\right) \tag{8.9}$$

Output voltage of the circuit is determined as follows:

$$V_{out} = A_v(V_1 - V_2) \tag{8.10}$$

## Example 8.6

In Figure 8–15, let $R$ = 10 kΩ, $R_2$ = 47 kΩ, and $R_3$ = 100 kΩ. Assume that resistor $R_1$ is adjusted to 7 kΩ. $V_1$ = 5.010 V and $V_2$ = 5.005 V. Calculate (a) $A_\upsilon$ and (b) $V_{out}$.

*Solution*

a. Use Equation 8.9:

$$A_\upsilon = \frac{R_3}{R}\left(\frac{2R_2}{R_1} + 1\right) = \frac{100\text{ k}\Omega}{10\text{ k}\Omega}\left(\frac{2(47\text{ k}\Omega)}{7\text{ k}\Omega} + 1\right) = 10(14.43) = 144.3$$

b. Use Equation 8.10:

$$V_{out} = A_\upsilon(V_1 - V_2) = 144.3(5.010\text{ V} - 5.005\text{ V}) = 144.3(5\text{ mV}) \approx 722\text{ mV}$$

## Applications for IAs

In some applications, such as providing the input to an oscillograph (chart recorder) trace pen, it may be desirable to control the output voltage of the IA to some reference level other than zero. This can be easily accomplished by adding a reference voltage in series with the (+) input terminal resistor of amplifier $A_3$ in Figure 8–15, assuming that the outputs of $A_1$ and $A_2$ are equal to 0 V. We can then assume that the inputs to $A_3$ are equal to 0 V, as shown in Figure 8–16. We must also set $R$ = $R_3$ in this application.

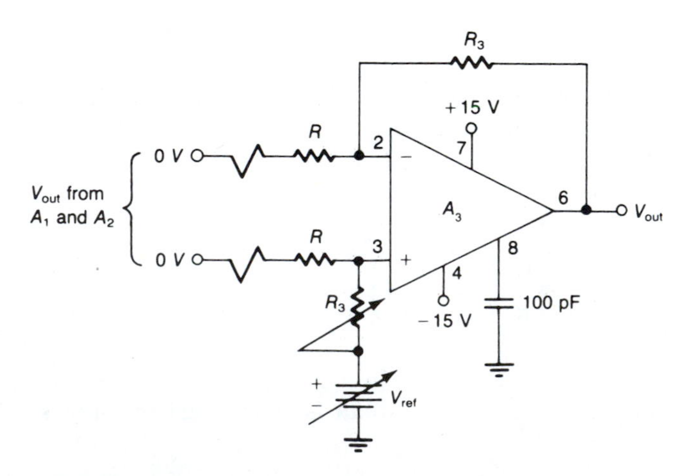


Figure 8–16. Referenced Output Voltage

In this circuit, $V_{ref}$ is inserted in series with $R_3$, which is equal to $R$, so that $V_{ref}$ is divided by two and is applied to the (+) input terminal of $A_3$. The noninverting amplifier provides a gain of 2, and $V_{out} = V_{ref}$. Adjusting $V_{ref}$ will then provide any desired $V_{out}$ reference level, allowing placement of the trace pen of the oscillograph to any desired starting point. In practice, for stability and accuracy in such a circuit, the output of a voltage follower circuit is used as $V_{ref}$.

Another typical IA application is shown in Figure 8–17. This circuit is used to monitor a medical patient and provides protection from electrical shock. The input circuit, that is, the electronics contacting the patient, is floating (not grounded). The output of the floating circuit is coupled to the IA by an optoisolator. This assures that the power supply voltage is safely isolated from the patient.

Other typical applications of IAs are the thermocouple amplifier illustrated in Figure 8–18 and the bridge amplifiers shown in Figure 8–19 and 8–20. The cold junction adjustment in Figure 8–18 allows setting the thermocouple amplifier system to some predetermined temperature level prior to making measurements of the physical changes under evaluation. Measuring the nozzle temperatures of a rocket engine is an example of this application.

The bridge circuit in Figure 8–19 provides a method of measuring electrical signals due to changes in a physical property, such as temperature, pressure, stress, or strain, that cause changes in a transducer. Variation of the physical property causes a change in $R_x$, resulting in an unbalanced bridge. The output of the IA feeds a monitoring or recording device, such as an oscillograph, data recorder, or an analog or digital meter.

Figure 8–20 shows a temperature control circuit. In this circuit, increases in temperature cause an increase in the temperature-dependent resistor of the

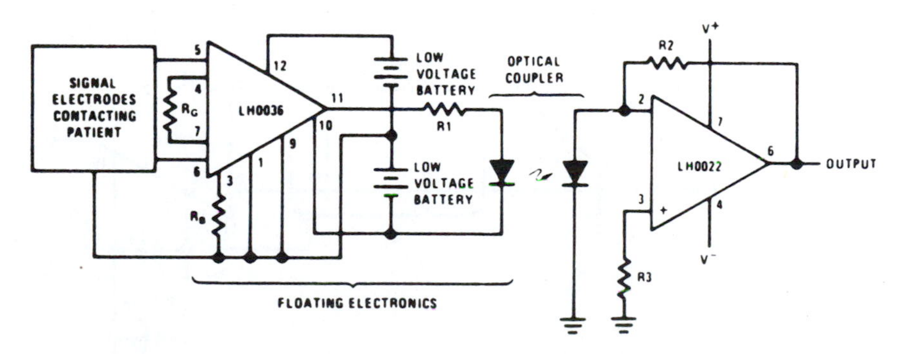


**Figure 8–17.** Isolation Amplifier for Medical Telemetry (Courtesy of National Semiconductor Corporation)

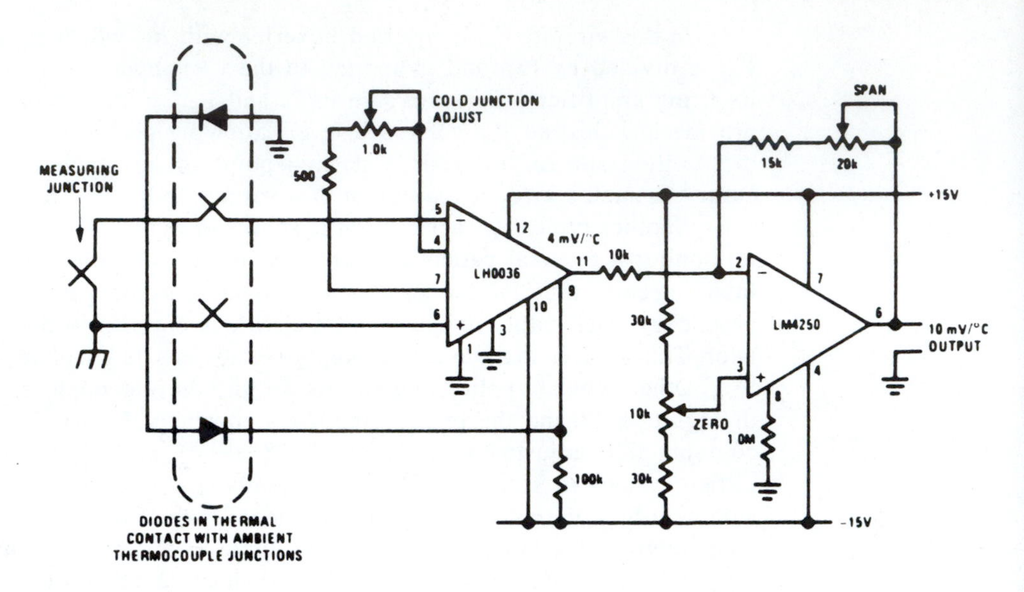


Figure 8–18. Thermocouple Amplifier with Cold-Junction Compensation (Courtesy of National Semiconductor Corporation)

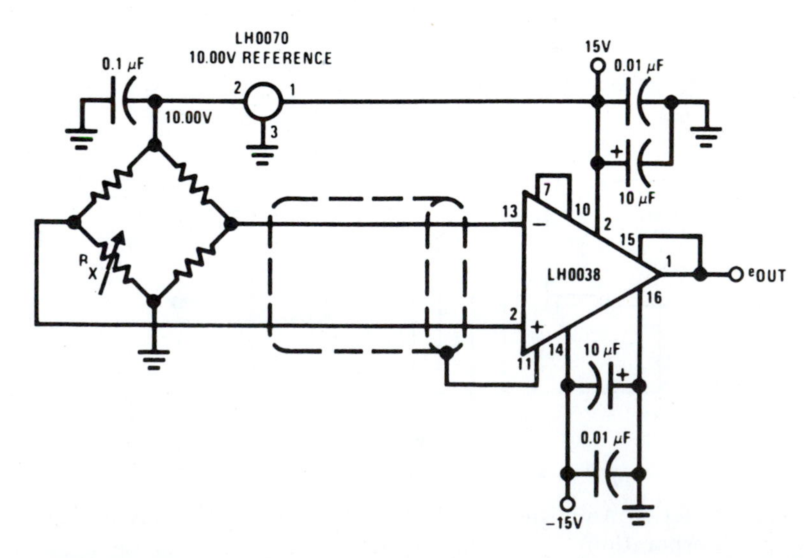


Figure 8–19. X-1000 Bridge Amplifier (Courtesy of National Semiconductor Corporation)

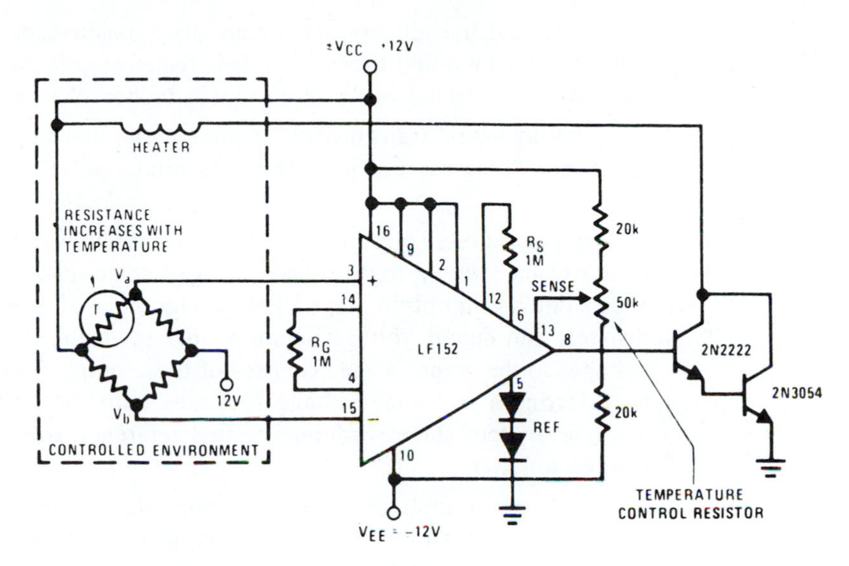


**Figure 8-20.** Temperature Control Circuit (Courtesy of National Semiconductor Corporation)

bridge circuit. Under balanced conditions, $V_{SENSE} - V_{ref}$ appears across $R_S$ (series resistor), $V_a - V_b$ appears across $R_G$ (gain resistor) and $I_{R_G} = I_{R_S}$.

$V_{SENSE}$, voltage applied to pin 13, is fixed by the temperature control resistor and $R_G/R_S$ is constant. The LF152 is used as a comparator with a feedback loop closed through the heater and the temperature dependent resistor. If $V_a - V_b$ is greater than $V_{SENSE}$ $(R_G/R_S)$, the output goes high, turning "ON" the heater. If $V_a - V_b$ is less than $V_{SENSE}$ $(R_G/R_S)$, the output goes low, turning "OFF" the heater.

# 8.5 Summary

☐ Transducers convert one form of energy or disturbance into another. Transducers may be mechanical, electrical, piezoelectric, photoelectric, or thermal, among others. The basis of many transducers is the resistance strain gauge.

☐ Mechanical transducers are generally simple switches that can be actuated either manually or by some piece of equipment, such as a pressure switch in a flow pipe.

☐ Piezoelectric transducers are constructed with crystals or ceramics. Piezoelectric transducers include ceramic microphones, ceramic pickups, and crystal microphones.

☐ Thermal transducers are temperature-sensitive devices and include thermostats, thermistors, and thermocouples. A thermopile consists of two or more thermocouples connected in series to obtain higher voltages.

☐ Photoelectric transducers, or photocells, convert light energy into electrical impulses. These devices include photovoltaic cells, photodiodes, and photoresistors.

☐ IAs are useful circuits for precision measurement and control. A typical IA is constructed with a high impedance, two-stage input and a differential to single-ended unity gain output stage. The voltage gain of a typical IA is set by a single resistor, and output voltage depends only upon the difference between the input voltages to the output stage. Control of these output voltages can be accomplished by inserting a reference voltage in series with the (+) input of the differential amplifier output stage. Practically, this reference voltage will be the output of a voltage follower.

☐ Typical IA applications include providing the input to an oscillograph, monitoring a medical patient while providing protection from electrical shock, thermocouple and bridge amplifiers, and temperature control systems.

# 8.6 Questions and Problems

**8.1** Define transducer.
**8.2** List twelve examples of transducers.
**8.3** Define strain gauge and explain its use.
**8.4** Explain the different ways in which a strain gauge can measure stress.
**8.5** Define the piezoelectric effect.
**8.6** Describe the operation of a crystal transducer.
**8.7** Explain how a ceramic microphone works.
**8.8** What is a ceramic pickup and how does it operate?
**8.9** Describe the crystal microphone and its operation.
**8.10** What precautions must be observed when handling ceramic and crystal microphones? Why?
**8.11** Define (a) thermal conductivity and (b) thermodynamics.
**8.12** Describe the use and operation of a thermostat.
**8.13** Explain the operation of a thermistor.
**8.14** Define (a) thermocouple and (b) thermopile.
**8.15** Describe how a thermopile is constructed and explain its use.
**8.16** Explain the two types of photocells.
**8.17** Describe the construction of a photovoltaic cell and explain its use.
**8.18** Define and explain the operation of (a) photodiode and (b) photoresistor.
**8.19** Define and explain the use of IAs.
**8.20** Describe the construction of the LH0036 IA.

**For Problems 8.21 through 8.25, refer to Figure 8–11.**

Assume $e_1 = 8$ mV, $e_2 = 15$ mV, $e_{CM} = 5$ V
**8.21** Assume $R_1 = 47$ kΩ and $R_G = 22$ kΩ. Calculate (a) $A(e_1)$ and (b) $A(e_2)$.
**8.22** Assume $R_1 = 18$ kΩ, $R_2 = 47$ kΩ, $R_G = 5$ kΩ, $e_1 = 4$ mV, $e_2 = 6$ mV, $e_{CM} = 10$ V. Calculate (a) $V_1$ and (b) $V_2$.
**8.23** Assume $R_1 = R_2 = 68$ kΩ, $R_G = 22$ kΩ. Calculate (a) $V_1$, (b) $V_2$, and (c) $V_2 - V_1$.
**8.24** Assume $R_3$ through $R_6 = 56$ kΩ. Calculate (a) $e_O$ and (b) $A_{VCL}$.

**8.25** Calculate the $R_G$ values necessary to achieve gains of (a) 25, (b) 50, (c) 200, (d) 350, and (e) 750.

**8.26** Refer to Figure 8–12A. For $V_S = \pm 15$ V, determine the $R_{BW}$ values for BWs of (a) 110 kHz, (b) 60 kHz, and (c) 8 kHz.

**8.27** From Figure 8–12A, for $V_S = \pm 1.5$ V, determine the BWs for $R_{BW}$ values of (a) 10 MΩ, (b) 3 MΩ, and (c) 100 kΩ.

**8.28** Refer to Figure 8–12B. For $V_S = \pm 15$ V, determine the $R_{BW}$ values for slew rates of (a) 0.2 V/μs, (b) 0.1 V/μs, and (c) 0.03 V/μs.

**8.29** From Figure 8–12B, for $V_S = \pm 1.5$ V, determine the slew rates for $R_{BW}$ values of (a) 10 MΩ, (b) 2 MΩ, and (c) 700 kΩ.

**8.30** List four typical applications of IAs.

**8.31** Why is it necessary to use an optoisolator between the electronics connected to a patient and the IA in medical applications?

**8.32** Refer to Figure 8–15. Assume $R_1$ = 22 kΩ, $R_2$ = 68 kΩ, $R$ = 27 kΩ, $R_3$ = 180 kΩ, $V_1$ = 5.005 V, and $V_2$ = 5.0025 V. Calculate (a) $A_v$ and (b) $V_{out}$.

**8.33** What is the purpose of inserting a reference voltage in series with the noninverting input of the differential amplifier stage of the IA?

# Chapter 9

# Conversion between Technologies

## 9.1 Introduction

In today's electronic world few systems are purely analog or purely digital. The proliferation of microcomputer process control systems demands a marriage of the two technologies, in which each may be used to its fullest efficiency. As you will recall from earlier discussions, the two technologies are unique in their characteristics.

Large systems are comprised of many different subsystems, all of which must interface to complete the system. All types of circuits, including linear, digital, and discrete are often used in the subsystems.

Interface circuits provide the necessary function of tying the parts of a system together. These circuits are usually not purely linear or digital but contain both types of circuit functions.

Interfacing is accomplished with *analog-to-digital converters* (ADCs) and *digital-to-analog converters* (DACs).

In this chapter we will examine the DAC first, because many ADCs use DACs as part of the conversion process. We will also examine a circuit that is important to the conversion process, the *sample-and-hold* (S & H) circuit. Finally, we will look at several applications of the converters.

# 9.2 Objectives

When you complete this chapter, you should be able to:

- ☐ Discuss the need to convert analog signals to digital form and digital signals to analog form.
- ☐ Describe the principles of operation of the DAC and the ADC.
- ☐ State the various types of converters available.
- ☐ Describe the principles of operation of sample-and-hold circuits.
- ☐ List the advantages and disadvantages of the various types of converters.
- ☐ Define the major terminology used in the description of converters and S & H circuits.

# 9.3 Digital-to-Analog Converters (DACs)

To accommodate today's industrial process control systems, it is necessary to combine analog and digital techniques. *Process control* involves the measurement of such physical quantities as temperature, liquid flow rate, pressure, light intensity, speed, strain, and vibration. The state of those quantities is then converted to electrical signals by transducers. The transduced signals are in analog form and must be converted to binary form so that a digital computer can process the information. The output signals of the computer are in binary form and may have to be converted back into analog form to be useful in making adjustments in a system that requires corrective action.

## Data Acquisition

A general data acquisition system block diagram is shown in Figure 9–1. The transduced analog signal is amplified and filtered. If more than one transduced signal is being converted, a multiplexer is used and the output is again amplified. To provide a steady input to the ADC, an S & H circuit is used. The ADC output is then digitally processed and the processor output goes to the final stage, the DAC.

### Numbering System

Figure 9–2 shows a typical parts numbering system for DACs and ADCs. This is the National Semiconductor system, but most manufacturers use a similar system.

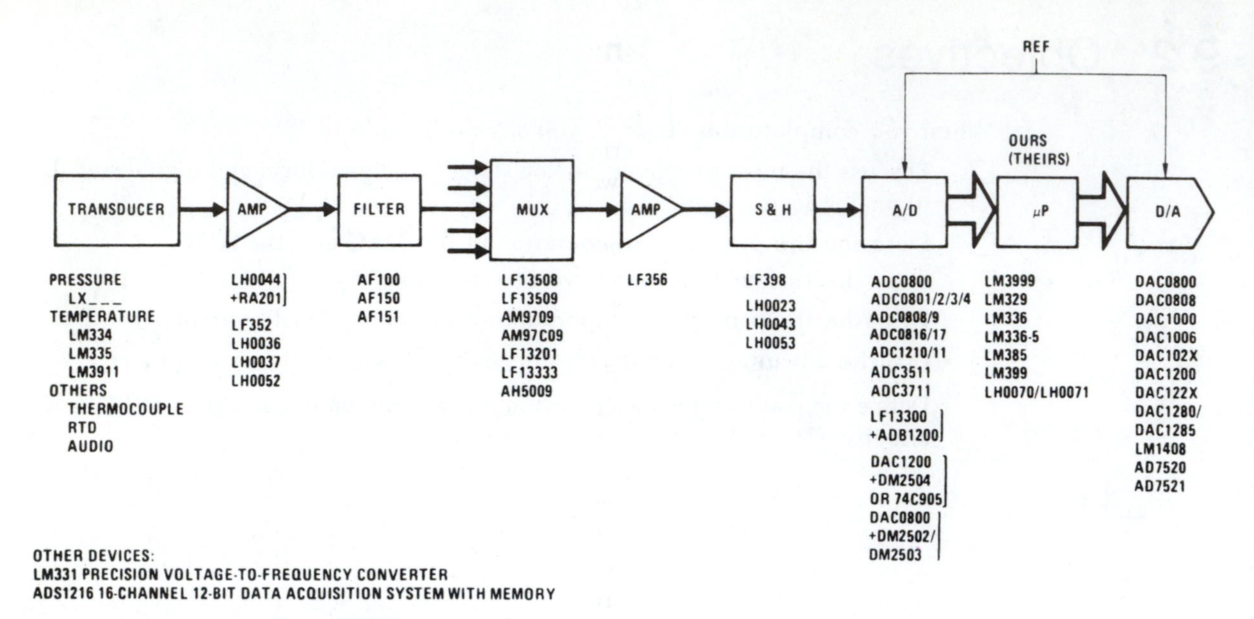


**Figure 9–1.** General Data Acquisition System Block Diagram (Courtesy of National Semiconductor Corporation)

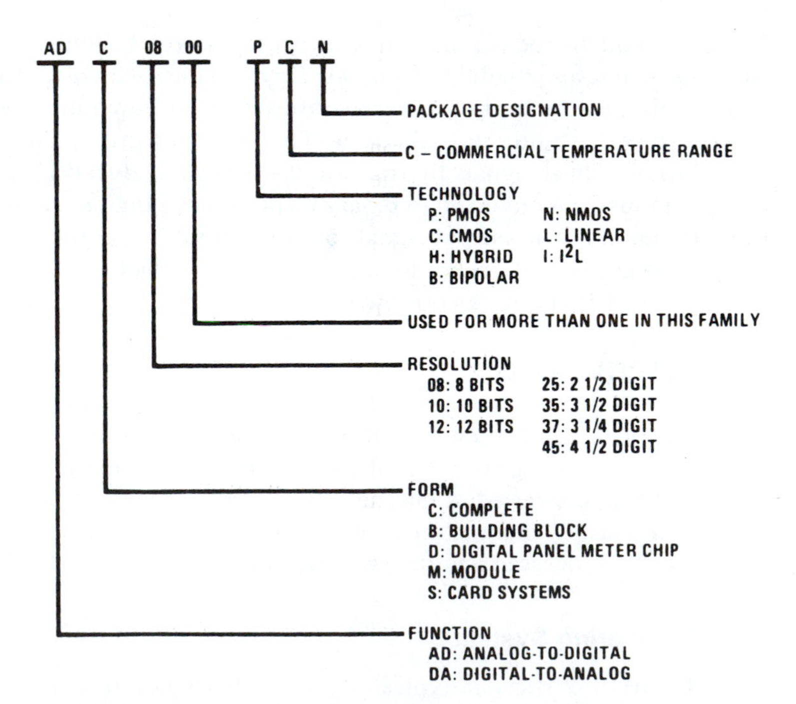


**Figure 9–2.** Converter Products Part Numbering System (Courtesy of National Semiconductor Corporation)

## Key Specifications and Terms for DACs

Key specifications for DACs are speed, settling time, error correction, and stability. Before proceeding to the various methods and circuits involved in the digital-to-analog conversion process, we will define key specifications and terms for DACs.

*Speed*—the conversion process should represent the input signal with the highest fidelity and minimal lag in time (real-time applications).

*Settling time*—a measure of a converter's speed and is defined as the elapsed time after a code transition for DAC output to reach final value within specified limits, usually ±1/2 least significant bit (LSB) (see Figure 9–3A).

*Offset error*—the output voltage of DAC with zero code input. Offset can be and usually is trimmed to zero with an offset zero adjust potentiometer (see Figure 9–3B).

*Gain error*—deviation in output voltage from correct level when the input calls for a full-scale output. This error may be trimmed to zero (see Figure 9–3C).

*Relative accuracy*—the maximum deviation of the DAC output relative to an ideal straight line drawn from zero to full scale (FS) minus one LSB (FS - 1 LSB) (see Figure 9–3D).

*Differential nonlinearity*—incremental error from any ideal LSB analog output change when the digital input is changed one LSB (see Figure 9–3E).

*Monotonicity*—as the input code is incremented from one code to the next in sequence, the analog output will either increase or remain constant (see Figure 9–3F).

*Stability*—a measure of the independence of converter parameters with respect to variations in external conditions such as temperature and supply voltage.

*Temperature coefficient*—the effects of temperature changes of the output, specified as %FS change.

*Supply rejection*—ability to resist changes in the output with supply changes, specified as %FS change.

*Long-term stability*—measure of how stable the output is over a long period of time.

*Resolution*—as an indication of the number of possible analog output levels a DAC will produce, resolution is often expressed as the number of input bits. For example, a 12-bit binary DAC will have $2^{12} = 4096$ possible output levels (including zero) and it has a resolution of 12 bits. As an indication of *step size,* resolution is always equal to the weight of the LSB, since it is the amount $V_{out}$ will change as the digital input goes from one step to the next. Thus, resolution can also be expressed as the amount of voltage or current per step. Although many manufacturers specify DAC resolution as the number of bits, a more useful method of expression is as a percent of full-scale range (FSR) or of the total number of steps:

$$\% \text{ resolution} = \frac{\text{Step size}}{\text{FSR}} \times 100\% \qquad \textbf{(9.1a)}$$

$$\% \text{ resolution} = \frac{1}{2^N - 1 \text{ steps}} \times 100 \quad \textbf{(9.1b)}$$

*Compliance*—compliance voltage is the maximum output voltage range that can be tolerated and still maintain the specified accuracy.

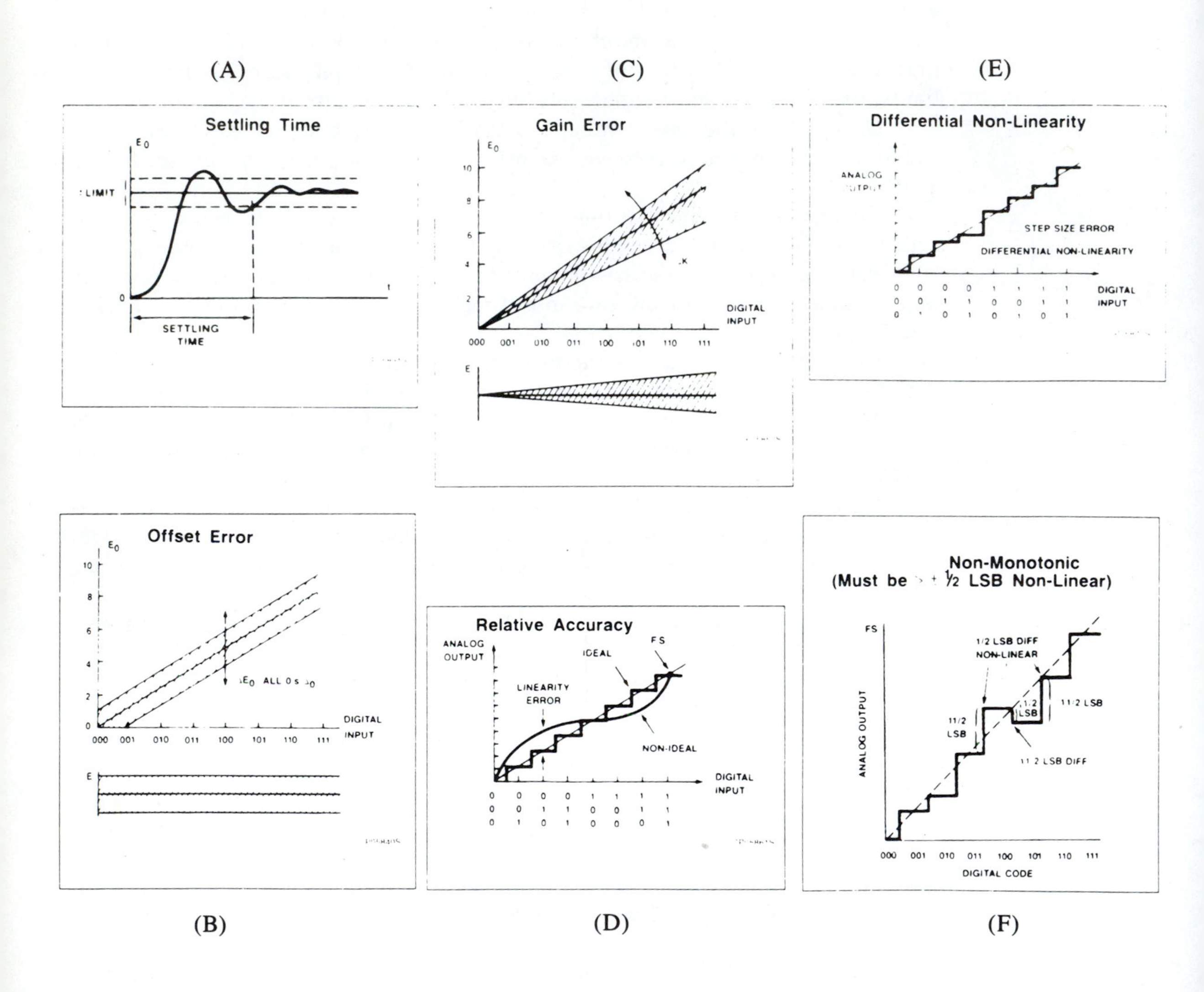


Figure 9–3. Waveforms for DAC Specifications and Terms: (A) Settling Time, (B) Offset Error, (C) Gain Error, (D) Relative Accuracy, (E) Differential Non-Linearity, (F) Non-monotonic (must be > ±1/2 LSB Non-linear) (Courtesy of Signetics Company)

## Four-Bit DAC

A simplified block diagram of a four-bit DAC is shown in Figure 9–4. The analog output can be either voltage or current. Assuming $V_{out}$ = 1 V for the LSB, the results of the four-bit DAC are shown in Table 9–1. The table shows that each digital input contributes a different amount to the analog output. Each input is weighted according to its position in the binary number, with *A* being the LSB and *D* being the most significant bit (MSB). It should be obvious, then, that the analog output is the weighted sum of the digital inputs.

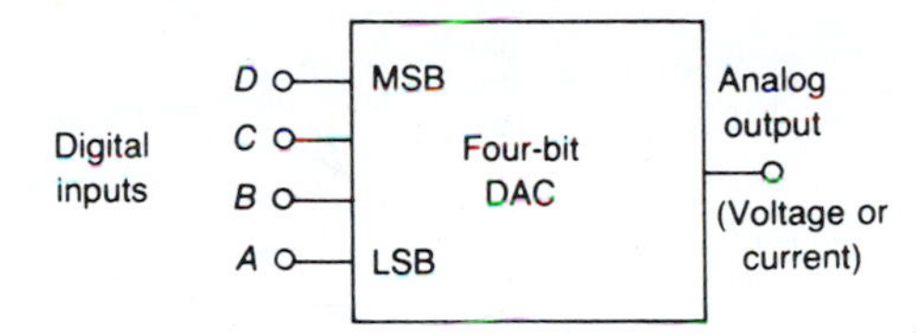


**Figure 9–4.** Simplified Four-Bit DAC

**Table 9–1** **Voltage Output Table for a Four-Bit DAC**

| *D* | *C* | *B* | *A* | $V_{out}$ (Volts) |
|---|---|---|---|---|
| 0 | 0 | 0 | 0 | 0 |
| 0 | 0 | 0 | 1 | 1 |
| 0 | 0 | 1 | 0 | 2 |
| 0 | 0 | 1 | 1 | 3 |
| 0 | 1 | 0 | 0 | 4 |
| 0 | 1 | 0 | 1 | 5 |
| 0 | 1 | 1 | 0 | 6 |
| 0 | 1 | 1 | 1 | 7 |
| 1 | 0 | 0 | 0 | 8 |
| 1 | 0 | 0 | 1 | 9 |
| 1 | 0 | 1 | 0 | 10 |
| 1 | 0 | 1 | 1 | 11 |
| 1 | 1 | 0 | 0 | 12 |
| 1 | 1 | 0 | 1 | 13 |
| 1 | 1 | 1 | 0 | 14 |
| 1 | 1 | 1 | 1 | 15 |

## Example 9.1

Assume a four-bit DAC produces $V_{out} = 0.5$ V for a digital input of 0001. Find the value of $V_{out}$ for the digital inputs of (a) 0011, (b) 1010, and (c) 1111.

*Solutions*

Since 0.5 V is the weight of the LSB, bit $A$, then the weights of the other bits are bit $B = 1$ V, bit $C = 2$ V, and bit $D = 4$ V. Therefore,

**a.** 0011 = 0.5 V + 1 V = 1.5 V
**b.** 1010 = 1 V + 4 V = 5 V
**c.** 1111 = 0.5 V + 1 V + 2 V + 4 V = 7.5 V

## Example 9.2

Determine the percentage of resolution for the DAC of Example 9.1.

*Solution*

Use Equation 9.1a:

$$\% \text{ resolution} = \frac{\text{Step size}}{\text{FSR}} \times 100\% = \frac{0.5 \text{ V}}{7.5 \text{ V}} \times 100\% = 6.67\%$$

Use Equation 9.1b:

$$\% \text{ resolution} = \frac{1}{2^N - 1 \text{ steps}} \times 100\%$$

$$\% \text{ resolution} = \frac{1}{15} \times 100\% = 6.67\%$$

The following points should be obvious now:

1. Increasing the number of bits increases the number of steps to reach full scale.
2. With each increase in the number of steps, each step is a smaller part of the FSR.
3. The larger the number of bits, the smaller the percent resolution (this means *better* resolution).

## Summing Amplifier DAC

One of the simplest DACs is a circuit with which you are already familiar, an op amp summing amplifier. Figure 9–5A shows the circuitry needed to implement the DAC of Figure 9–4. Recall that a summing amplifier multiplies each input voltage by the ratio of the feedback resistor $R_f$ to the corresponding input resistor $R_{in}$. Therefore, assuming a 5 V input voltage for Figure 9–5A we have the following general equation:

$$V_{out} = -V_{in}\frac{R_f}{R_{in}} \quad (9.2)$$

And

$$\text{Input } A = 5\text{ V}(2\text{ k}\Omega/10\text{ k}\Omega) = -1\text{ V}$$
$$\text{Input } B = 5\text{ V}(2\text{ k}\Omega/5\text{ k}\Omega) = -2\text{ V}$$
$$\text{Input } C = 5\text{ V}(2\text{ k}\Omega/2.5\text{ k}\Omega) = -4\text{ V}$$
$$\text{Input } D = 5\text{ V}(2\text{ k}\Omega/1.25\text{ k}\Omega) = -8\text{ V}$$

The total output voltage is equal to the sum of the individual output voltages.

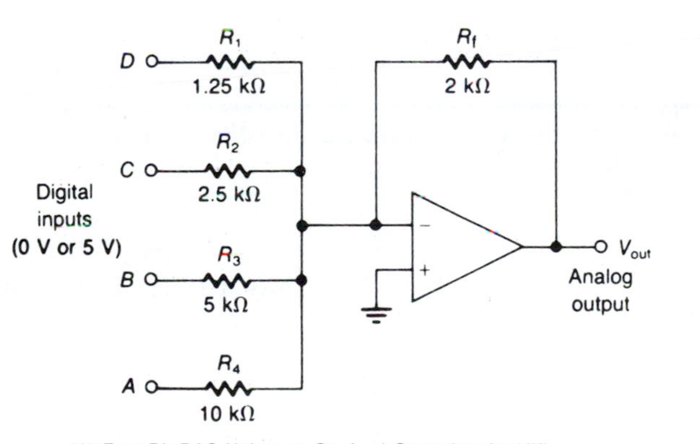


(A) Four-Bit DAC Using an Op Amp Summing Amplifier

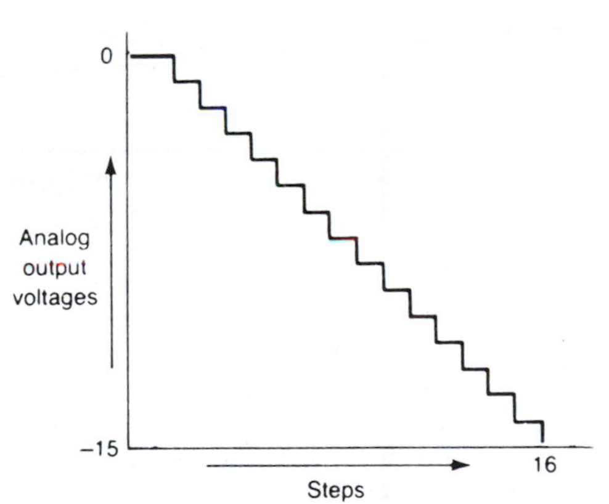


(B) Ideal Analog Output Voltage Steps for Part A

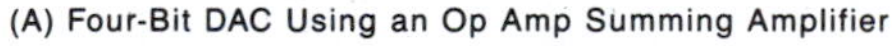

**Figure 9–5.** Four-Bit DAC

Recall that the minus sign only indicates that the input to the op amp is applied to the inverting terminal. Table 9–2 and Figure 9–5B represent the ideal output values for the circuit shown in Figure 9–5A.

Table 9–2

**Ideal Input/Output Conditions for Figure 9–5A**

| Digital Inputs | | | | Individual $V_{out}$ (Volts) | | | | |
|---|---|---|---|---|---|---|---|---|
| *D* | *C* | *B* | *A* | *D* | *C* | *B* | *A* | Sum of $V_{out}$ |
| 0 | 0 | 0 | 0 | 0 | 0 | 0 | 0 | 0 |
| 0 | 0 | 0 | 1 | 0 | 0 | 0 | 1 | −1 |
| 0 | 0 | 1 | 0 | 0 | 0 | 2 | 0 | −2 |
| 0 | 0 | 1 | 1 | 0 | 0 | 2 | 1 | −3 |
| 0 | 1 | 0 | 0 | 0 | 4 | 0 | 0 | −4 |
| 0 | 1 | 0 | 1 | 0 | 4 | 0 | 1 | −5 |
| 0 | 1 | 1 | 0 | 0 | 4 | 2 | 0 | −6 |
| 0 | 1 | 1 | 1 | 0 | 4 | 2 | 1 | −7 |
| 1 | 0 | 0 | 0 | 8 | 0 | 0 | 0 | −8 |
| 1 | 0 | 0 | 1 | 8 | 0 | 0 | 1 | −9 |
| 1 | 0 | 1 | 0 | 8 | 0 | 2 | 0 | −10 |
| 1 | 0 | 1 | 1 | 8 | 0 | 2 | 1 | −11 |
| 1 | 1 | 0 | 0 | 8 | 4 | 0 | 0 | −12 |
| 1 | 1 | 0 | 1 | 8 | 4 | 0 | 1 | −13 |
| 1 | 1 | 1 | 0 | 8 | 4 | 2 | 0 | −14 |
| 1 | 1 | 1 | 1 | 8 | 4 | 2 | 1 | −15 |

## Example 9.3

Design a four-bit DAC using an op amp summing amplifier. Resolution = LSB = 0.250 V and $V_{in}$ = 5 V.

*Solution*

Select a standard value for the feedback resistor $R_f$, say 10 kΩ. Determine the ratio of the LSB and the input voltage:

$$\frac{0.250\text{ V}}{5\text{ V}} = 0.05$$

From this ratio, the value for input *A* resistor $R_{in(A)}$ can be determined.

$$\frac{R_f}{0.05} = \frac{10\text{ k}\Omega}{0.05} = 200\text{ k}\Omega$$

Since we know that each succeeding input must double the voltage of the LSB (because of resolution), then each succeeding resistor must be one-half the value of the preceding resistor. Therefore, to determine the values for the remaining input resistors, divide successive resistors by two:

$$R_{in(B)} = \frac{R_{in(A)}}{2} = \frac{200\text{ k}\Omega}{2} = 100\text{ k}\Omega$$

$$R_{in(C)} = \frac{R_{in(B)}}{2} = \frac{100\ k\Omega}{2} = 50\ k\Omega$$

$$R_{in(D)} = \frac{R_{in(C)}}{2} = \frac{50\ k\Omega}{2} = 25\ k\Omega$$

The completed circuit is shown in Figure 9–6, and Table 9–3 shows the input/output results.

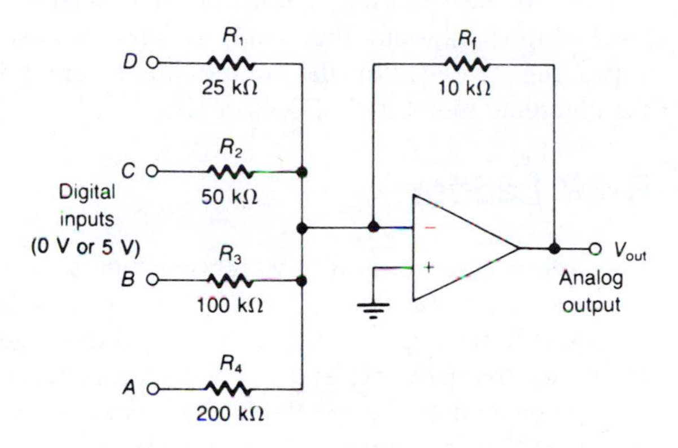


**Figure 9–6.** Completed Circuit for Example 9.3

**Table 9–3** Input/Output Results for Example 9.3

| *D* | *C* | *B* | *A* | $V_{out}$ (Volts) |
|---|---|---|---|---|
| 0 | 0 | 0 | 0 | 0 |
| 0 | 0 | 0 | 1 | -0.250 |
| 0 | 0 | 1 | 0 | -0.500 |
| 0 | 0 | 1 | 1 | -0.750 |
| 0 | 1 | 0 | 0 | -1.000 |
| 0 | 1 | 0 | 1 | -1.250 |
| 0 | 1 | 1 | 0 | -1.500 |
| 0 | 1 | 1 | 1 | -1.750 |
| 1 | 0 | 0 | 0 | -2.000 |
| 1 | 0 | 0 | 1 | -2.250 |
| 1 | 0 | 1 | 0 | -2.500 |
| 1 | 0 | 1 | 1 | -2.750 |
| 1 | 1 | 0 | 0 | -3.000 |
| 1 | 1 | 0 | 1 | -3.250 |
| 1 | 1 | 1 | 0 | -3.500 |
| 1 | 1 | 1 | 1 | -3.750 |

## Precision-Level Amplifier

We have been discussing DACs with ideal input/output results, assuming that the input and feedback resistors are precision values and that the input voltages are precisely 0 V or 5 V. While precision resistors can be obtained, input voltages taken directly from the outputs of flip-flops (FF) or logic gates are not precise values but vary over a given range. Therefore, it is necessary to place a *precision-level amplifier* (PLA) between each digital input and its corresponding input resistor to the summing amplifier. A precision 5 V reference supply feeds the level amplifiers, and the result is very precise output levels of 0 V or 5 V, depending on whether the digital inputs are LOW or HIGH. Figure 9–7 shows the complete four-bit DAC with PLAs.

## *R-2R* Ladder

Two approaches to binary-weighting can be employed in DACs. Figure 9–8A shows a binary-weighted resistor ladder employing *voltage switching.* The disadvantages of this approach are: (1) a wide range of resistor values are used in weighting the network; and (2) nodal capacitances which are charged/discharged during conversion. Figure 9–8B shows an *R-2R ladder* network employing *current switching.* The advantages of this type of network are: (1) no need for a wide range of resistor values; and (2) current switching eliminates transients in nodal parasitic capacitances.

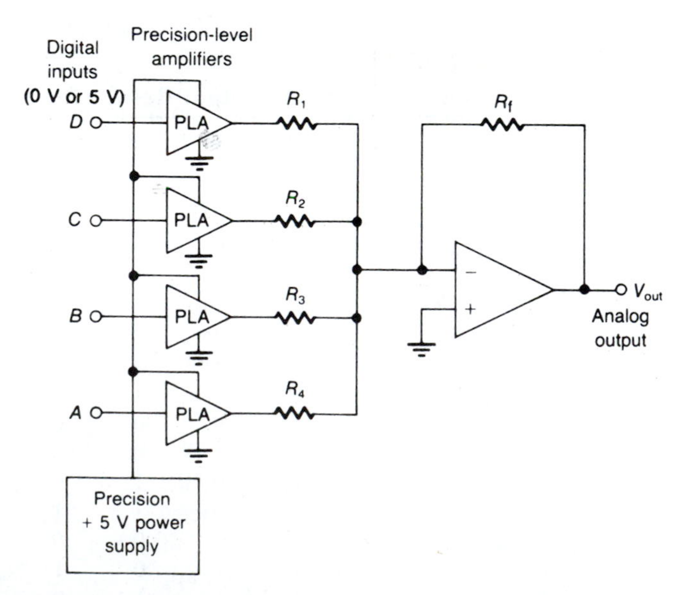


**Figure 9–7.** Four-Bit DAC with Precision Amplifiers

BINARY-WEIGHTED LADDER
EMPLOYING VOLTAGE SWITCHING

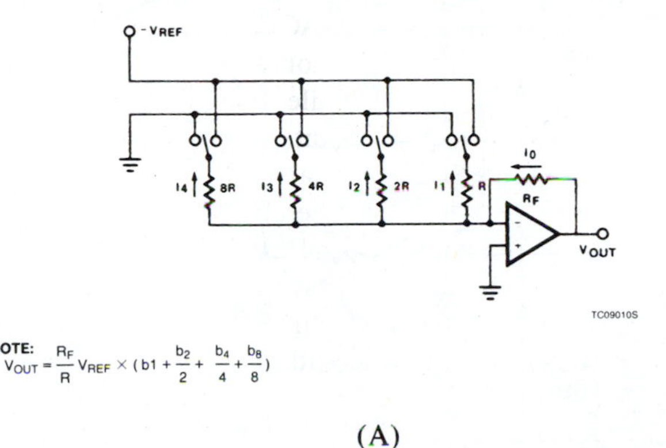


(A)

R-2R LADDER NETWORK
EMPLOYING CURRENT SWITCHING

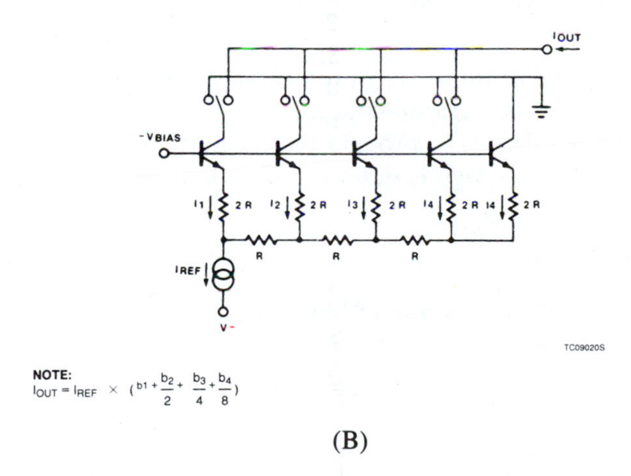


(B)

**Figure 9–8.** Binary-Weighting Approaches for DACs:
(A) Binary-Weighted Resistor Ladder
(B) R-2R Ladder Network
(Courtesy of Signetics Company)

Figure 9–9 illustrates a popular approach to the summing amplifier DAC, in which the resistive network has been transformed into the *R-2R* ladder form. This type circuit is particularly useful where a large number of digital inputs is involved. This circuit finds popular application in constant-input/constant-output audio and RF level controls.

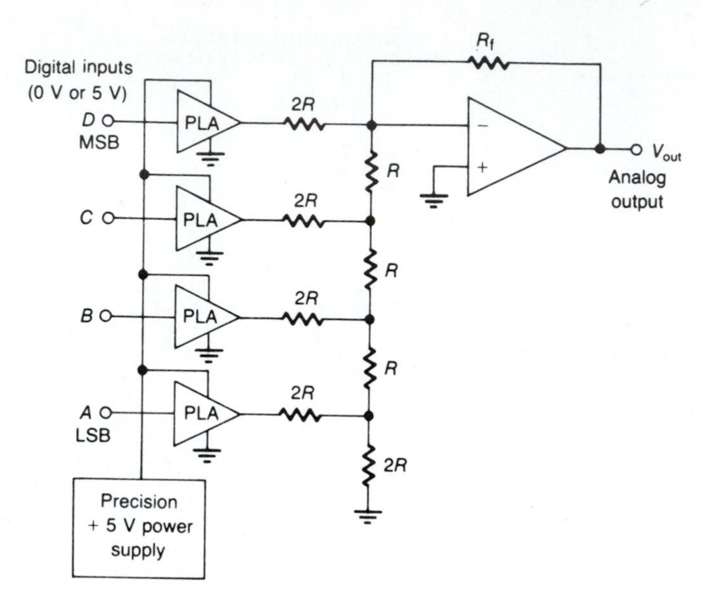


**Figure 9–9.** Ladder-Type Summing Amplifier

Note that there are only two values for the resistors in the ladder, $R$ and $2R$. This is true regardless of the number of stages. Feedback resistor $R_f$ can be any value selected to provide the desired output scale level.

Output voltage is determined as follows:

$$V_{out} = K(V_{ref})(F) \qquad \textbf{(9.3a)}$$

where

$K$ = decimal value of binary input
$F$ = scale factor

$$F = \frac{R_f}{R} \qquad \textbf{(9.3b)}$$

A short explanation will acquaint you with the method for determining the decimal value of a binary number. We will look only at the fractional portion of the number, that is, the portion of the number to the right of the binary point. Each bit has a specified weight as follows:

| | |
|---|---|
| LSB | 1 = 1 |
| binary point | . = . |
| bit 1 | 1 = 0.5 |
| bit 2 | 1 = 0.25 |
| bit 3 | 1 = 0.125 |
| bit 4 | 1 = 0.0625 |

To determine the decimal value of a binary number, add together all weights for bits that contain a binary 1. For example, the decimal value of the binary number 0.0101 is calculated as follows:

$$0.0 + 0.25 + 0 + 0.0625 = 0.3125$$

### Example 9.4

Refer to Figure 9–9. Assume $R_f = R = 10\ \text{k}\Omega$, then $2R = 20\ \text{k}\Omega$. Further assume a digital input of 0.0110. What is the output voltage?

*Solution*

Converting the binary input to decimal value, we have 0.375. Then using Equation 9.3a,

$$V_{out} = K(V_{ref})(F) = 0.375(5\text{ V})(1) = 1.875\text{ V}$$

The answer in Example 9.4 holds true, since $F = 1$ in this case. Now assume $R_f$ is changed to 5 kΩ and all other factors remain the same. Now,

$$F = \frac{R_f}{R} = \frac{5\ \text{k}\Omega}{10\ \text{k}\Omega} = 0.5$$

Therefore,

$$V_{out} = (0.375)(5\text{ V})(0.5) = (1.875\text{ V})(0.5) \approx 0.938\text{ V}$$

## 9.4 Sample-and-Hold Circuits

Converting a voltage from analog to digital form requires that the voltage be sampled periodically and then held constant as an input to an ADC. A circuit that performs this function is called an S & H circuit.

Figure 9–10A shows a basic S & H circuit, with the op amp connected as a voltage follower. When the switch is closed, the capacitor charges to $V_{in(max)}$. After the switch is opened, the capacitor remains charged and $V_{out}$ will be at the same potential as the capacitor. The sampled voltage will be held temporarily, the time being determined by leakage in the circuit.

The basic circuit cannot sample rapidly changing voltages, so circuits using field effect transistor (FET) switches, such as that shown in Figure 9–10B, are used. When a pulse is present at the sample input, the FETs are turned on and act as low resistances to the input signal. When the sample pulse is absent, the FETs are turned off and act as high impedances. The desired voltage is then held by capacitor $C_1$, which is isolated from the output by the high input impedance op amp. The capacitor cannot discharge significantly between pulses.

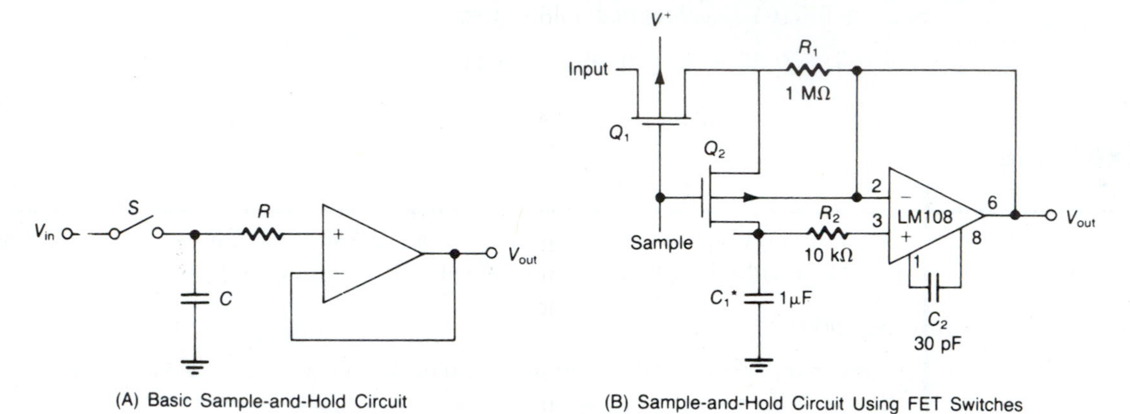


**Figure 9–10.** Sample-and-Hold Circuits (Part B: Courtesy of National Semiconductor Corporation)

A typical monolithic S & H circuit is the LF198 shown in the functional diagram in Figure 9–11. This device utilizes bipolar field effect transistor (BIFET) technology to obtain ultra-high dc accuracy with fast acquisition of signal and low droop rate. The overall design guarantees no feed-through from input to output in the hold mode even for input signals equal to the supply voltages. The device will operate from ±5 V to ±18 V supplies.

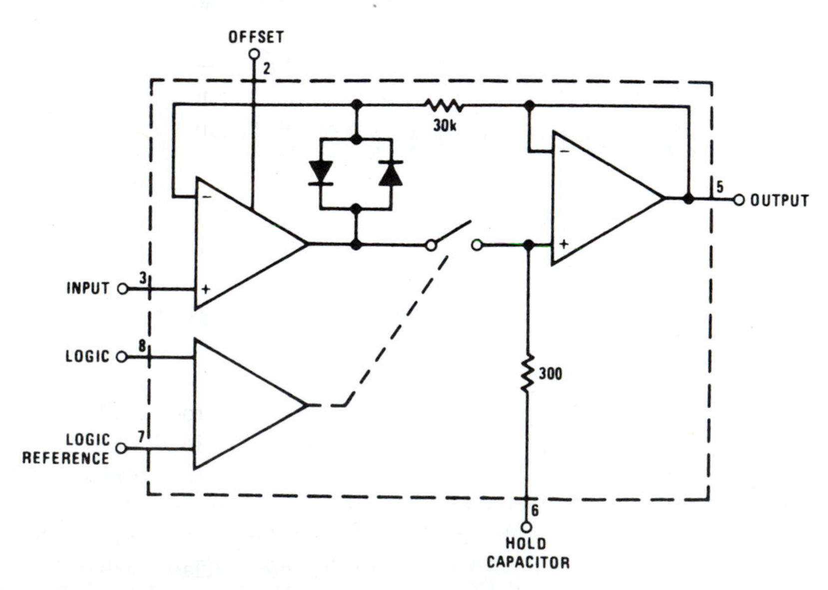


**Figure 9–11.** Functional Diagram of the LF198 Monolithic Sample-and-Hold Circuit (Courtesy of National Semiconductor Corporation)

### Sample-and-Hold Definitions

A brief description of terminology will help you to more fully understand the S & H circuits.

*Mode control signal*—the signal pulse that allows the FETs to turn on, thus allowing an input signal to be sampled. This signal is usually a clock pulse set at some predetermined frequency.

*Acquisition time*—the time required to acquire a new analog voltage with an output step of 10 V. The acquisition time is not just the time required for the output to settle, but also includes the time required for all internal nodes to settle so that the output assumes the proper value when switched to the hold mode.

*Aperture time*—the delay required between "hold" command and an input analog transition, so that the transition does not affect the held output.

*Hold settling time*—the time required for the output to settle within 1 mV of final value after the "hold" logic command.

*Dynamic sampling error*—the error introduced into the held output due to a changing analog input at the time the "hold" command is given. Error is expressed in millivolts with a given hold capacitor value and input slew rate. Note that this error term occurs even for long sample times.

*Gain error*—the ratio of output voltage swing to input voltage swing in the sample mode expressed as a percent difference.

*Hold step*—the voltage step at the output of the S & H when switching from sample mode to hold mode with a steady (dc) analog input voltage. Logic swing is 5 V.

*Droop (hold voltage drift)*—the variation or drift due to the charge leakage out of the holding capacitor through the amplifier input terminals and the switch. Droop rate is usually expressed in millivolts per second (mV/s).

## 9.5 Analog-to-Digital Converters (ADCs)

The ADC is used in the process of converting an analog signal to an equivalent digital signal. The conversion process is much more complex than the converse DAC operation discussed in Section 9.3.

Analog-to-digital conversion schemes generally fall into one of three categories:

1. *Multiple comparator*—(flash)
2. *Feedback*—counting, tracking (up/down), successive approximation
3. *Integrating*—single slope, dual slope, triple slope

A number of different types of ADCs have been developed. The type of converter chosen for a given application depends upon many things, some of which are: (1) the conversion speed necessary; (2) the necessary immunity to noise; and (3) cost of the device. This section examines four of those types: comparator, digital-ramp, successive-approximation, and dual-slope.

## Key Considerations and Terms for ADCs

Key considerations for ADCs include: (1) analog input signal range and resolution required; (2) linearity requirement and stability; (3) conversion speed required; (4) monotonicity requirement (can missing codes be tolerated?); (5) character of the input signal (is it noisy, sampled, filtered, slowly varying?); and (6) transfer characteristics (type of coding).

Following are key terms pertaining to ADCs. Note that some of these terms are identical to those used for DACs, but the meaning is slightly different.

*Resolution*—the input change required to increment the output between the two adjacent codes. This term also refers to the number of bits in the output word and, hence, the number of discrete output codes the input analog signal can be broken into (expressed in "bits" resolution).

*Transfer characteristic*—the relationship of the output digital word (code) to the input analog signal, binary coded decimal (BCD), and so on.

*Conversion speed*—the speed at which an ADC can make repetitive data conversions.

*Quantizing error*—an inherent error in the conversion process due to finite resolution (discrete output) (see Figure 9–12A).

*Quantization uncertainty*—a direct consequence of the resolution of the converter. All analog voltages within a given range are represented by a single digital output code. There is, therefore, an inherent conversion error even in a perfect ADC. Quantization uncertainty can only be reduced by increasing resolution.

*Offset error*—in a practical ADC the offset from the ideal ADC output (generally −1 LSB) (see Figure 9–12B).

*Gain error*—deviation in the output voltage from the ideal ADC output (see Figure 9–12C).

*Relative accuracy*—the deviation of an actual bit transition from the ideal transition value at any level over the range of the ADC (see Figure 9–12D).

*Hysteresis error*—the code transition voltage dependence relative to the direction (polarity) from which the transition is approached.

*Monotonicity*—when the output code either increases or remains the same for increasing or decreasing analog input signals.

*Missing codes*—a code combination that is skipped (see Figure 9–12E).

## Comparator (Flash) ADC

As the name implies, the *comparator* or *flash* ADC uses comparators to convert analog signals to digital form. A comparator ADC constructed with op amp

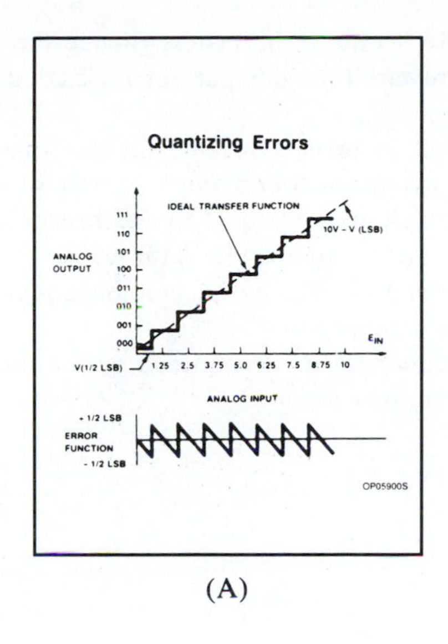


(A)

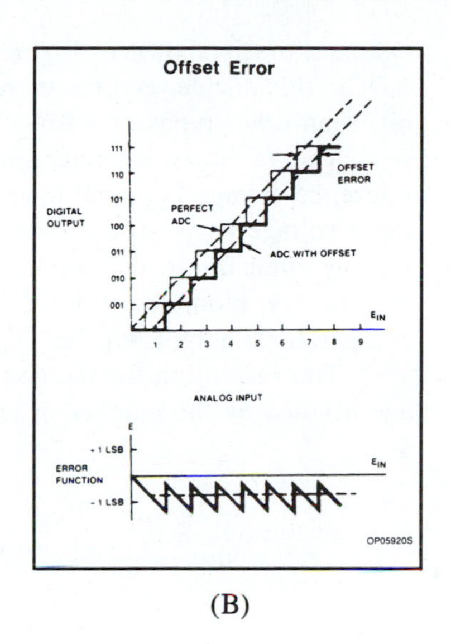


(B)

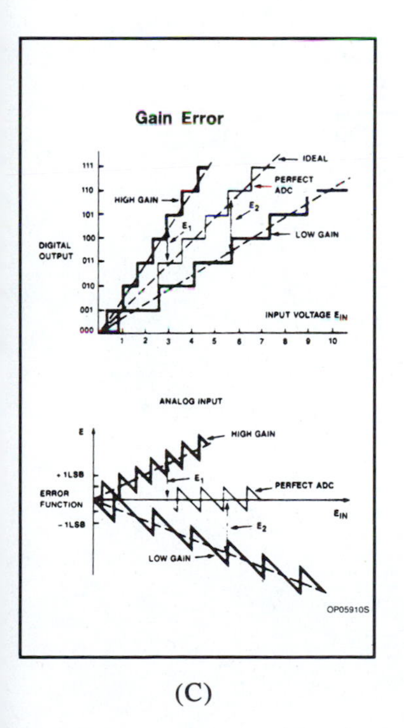


(C)

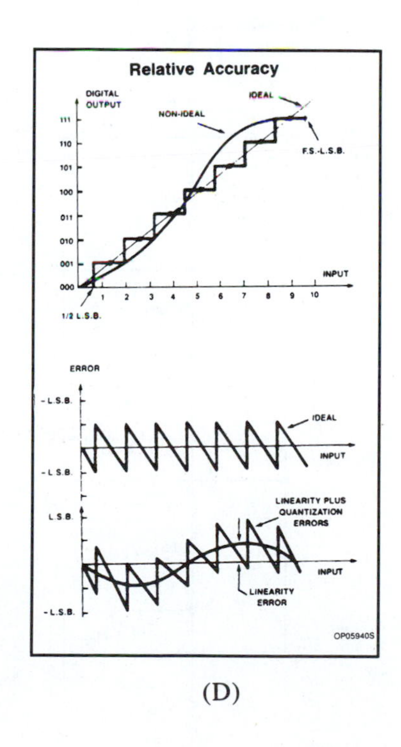


(D)

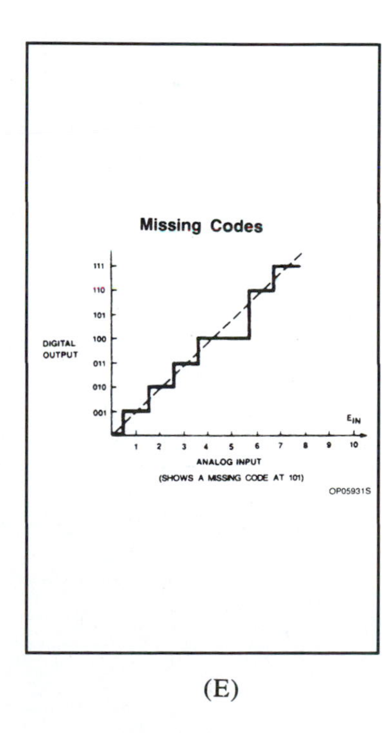


(E)

Figure 9-12. Waveforms for ADC Terms: (A) Quantizing Errors, (B) Offset Error, (C) Gain Error, (D) Relative Accuracy, (E) Missing Codes (Courtesy of Signetics Company)

comparators is shown in Figure 9–13. Among the fastest (hence the name *flash*) ADCs, this device requires more hardware (one comparator for each digital output bit) than other types of ADCs.

The resistors are interconnected to form a voltage divider that establishes reference voltage $V_{ref}$ input to each comparator. In op amp comparator operation, if input voltage $V_{in}$ is greater than $V_{ref}$ by an amount equal to the threshold voltage $V_{th}$ for any comparator, the output of that comparator will saturate to $+V_{sat}$, resulting in a binary 1 output. If $V_{in}$ is smaller than $V_{ref}$ by an amount equal to $V_{th}$, that comparator's output will be $-V_{sat}$, or a binary 0.

The resolution for the comparator ADC is determined by the reference voltage divided by the number of comparators used:

$$\text{Resolution} = \frac{V_{ref}}{\#\ \text{comparators}} \tag{9.4}$$

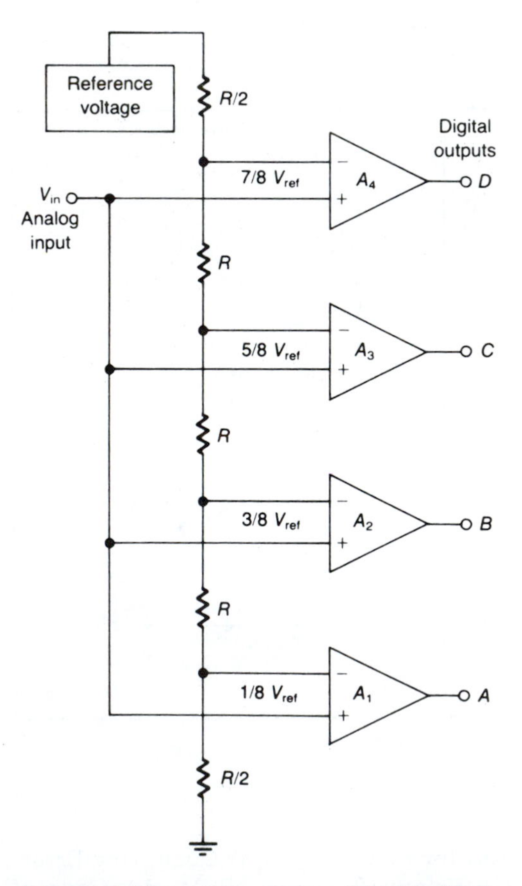


**Figure 9–13.** Comparator (Flash) ADC

For example, the ADC in Figure 9–13 has a resolution of $V_{ref}/4$; if eight comparators were used the resolution would be $V_{ref}/8$. This points out that increasing the number of comparators improves resolution.

### The ADC0820 Half-Flash Device

The block diagram of the ADC0820 is shown in Figure 9–14. By using a *half-flash* conversion technique, this eight-bit CMOS ADC offers 1.5 μs conversion time while dissipating a maximum 75 mW of power. The half-flash technique consists of 31 comparators, a most-significant four-bit ADC and a least-significant four-bit ADC.

The input of the ADC0820 is tracked and held by the input sampling circuitry, eliminating the need for an external S & H for signals slewing at less than 100 mV/μs.

For ease of interface to microprocessors, the ADC0820 has been designed to appear as a memory location or input/output (I/O) port without the need for external interfacing logic.

## Digital-Ramp ADC

One of the simplest methods used for analog-to-digital conversion is the *digital-ramp* ADC. Also called a *counting* ADC or *staircase-ramp* ADC, it is the slowest of all ADCs, but it requires the least hardware. A block diagram of the digital-ramp ADC is shown in Figure 9–15.

In this circuit, a positive START pulse resets the counter to zero and inhibits the AND gate. In this way no clock pulses (clp) can pass through to the counter while the START pulse is HIGH. When the START pulse goes LOW, the AND gate is enabled and pulses are allowed to pass to the counter.

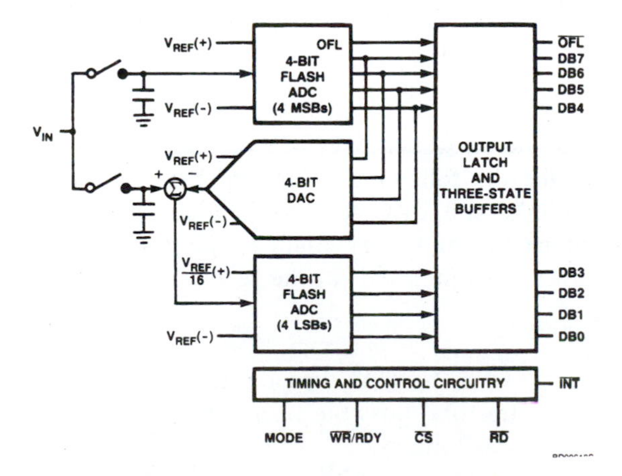


**Figure 9–14.** Block Diagram for the ADC0820 Half-Flash ADC (Courtesy of Signetics Company)

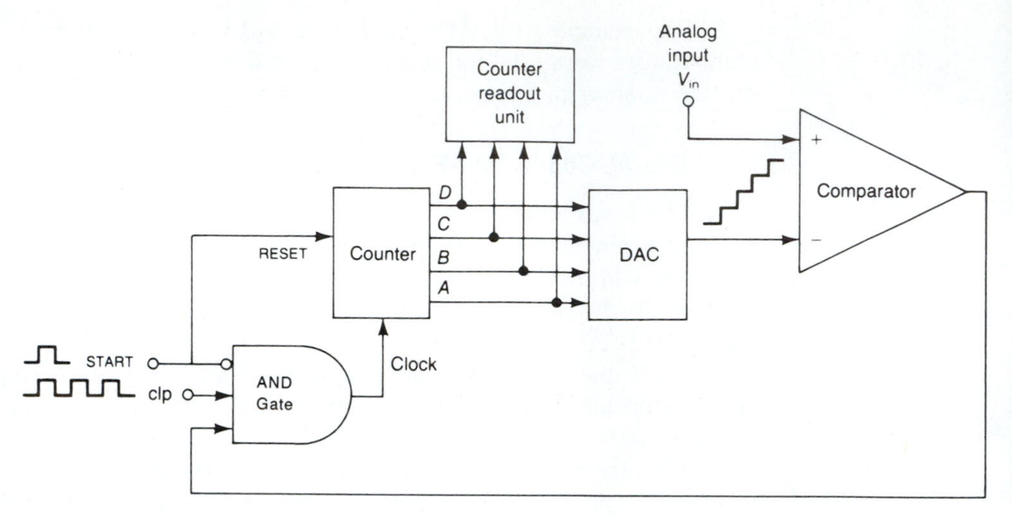


**Figure 9–15.** Block Diagram of the Digital-Ramp ADC

Now assume some positive value for $V_{in}$ that causes the output of the comparator to go HIGH and place a binary 1 at one of the inputs to the AND gate. This means the clock pulse determines the output of the AND gate, and, therefore, the input to the counter. Each clock pulse increases the counter output by an increment of 1, which causes the DAC output to increase by the amount of voltage associated with the LSB, in equal steps. When the DAC output voltage reaches a value equal to or greater than $V_{in}$, the comparator output goes LOW, placing a binary 0 on the AND gate input and turning it OFF. The binary counter output terminals then reflect the value of the signal at $V_{in}$, and the conversion is completed.

The time necessary for the conversion to take place is determined by the frequency of the clock pulse ($f_{clp}$). The frequency is limited by the speed with which the counter and the DAC can respond to a pulse.

## Example 9.5

Assume the following values for the ADC of Figure 9–15: $f_{clp}$ = 1 MHz, $V_{th}$ = 1 mV, and DAC full-scale output = 2.55 V with an eight-bit input. Determine (a) the binary count of the counter output when $V_{in}$ = 1.28 V, (b) the conversion time, and (c) the resolution of the ADC.

*Solutions*

**a.** Since the DAC has an eight-bit input and a full-scale output of 2.55 V, the total possible number of steps is $2^8 - 1 = 255$, and the step size is as follows:

$$\text{Step size} = \frac{\text{FSR}}{2^N - 1 \text{ steps}} \quad \frac{2.55 \text{ V}}{255} = 10 \text{ mV} \qquad \textbf{(9.5)}$$

Therefore, the DAC output increases by 10 mV per step as the counter counts up from 0. With $V_{in}$ = 1.28 V, the DAC output must reach 1.281 V or more before the comparator will switch LOW, therefore,

$$\frac{1.281\ \text{V}}{10\ \text{mV}} = 128.1 = 129\ \text{steps}$$

You must go to the next full step when there is a decimal fraction in the calculation. The binary count will be 10000001, which is the binary equivalent of $129_{10}$, representing the desired digital equivalent of $V_{in}$ = 1.28 V.

**b.** With $f_{clp}$ = 1 MHz, each clock pulse requires 1 μs. Since 129 steps were required at 1 μs per step, the total conversion time is 129 μs.

**c.** Resolution = step size of the DAC = 10 mV. As a percent, it is expressed as follows:

$$\frac{10\ \text{mV}}{2.55\ \text{V}} \times 100\% = 0.39\%$$

## Successive-Approximation ADC

The *successive-approximation* ADC is perhaps the most widely used ADC, mainly because of its excellent trade-offs in resolution, speed, accuracy, and cost. It goes through a sequence of approximations to obtain the digital representation of the analog input voltage. The conversion time is much shorter than for the counter types and is a fixed value independent of the value of $V_{in}$. However, this type of ADC does require more complex control circuitry. A block diagram of a four-bit successive-approximation ADC is shown in Figure 9–16.

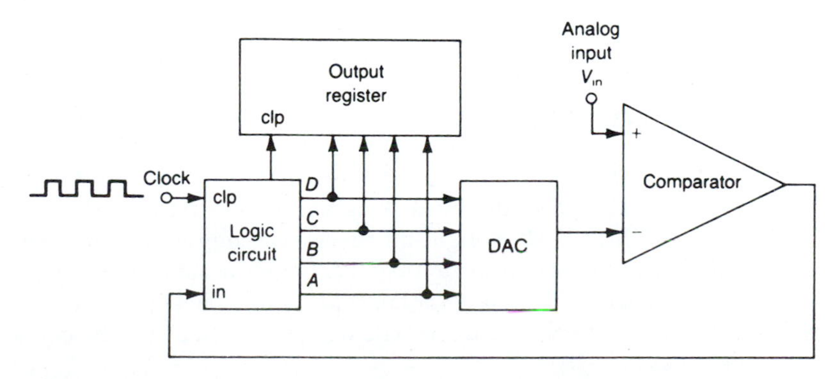


**Figure 9–16.** Block Diagram of a Successive-Approximation ADC

This type of ADC uses a register instead of a counter to provide the input for the DAC. The binary values of the register are operated on in the following manner:

1. A binary number is entered into the register with the MSB set to 1 and all other bits set to 0. This produces a value at the DAC output equal to the weight of the MSB. If this value is greater than $V_{in}$, the comparator goes LOW, resetting the MSB to 0. If the value is less than $V_{in}$, the MSB stays at 1.
2. The next significant bit is set to 1 and the comparison is again made as in the first step. If the new value is greater than $V_{in}$, this bit is reset to 0; otherwise, it remains at 1.
3. This sequence continues for all bits in the binary number of the register. The process requires one clock pulse period per bit.
4. After all bits have been tested, the output of the logic circuit provides a binary 1 to the clock pulse terminal of the register that records the digital representation of the analog input voltage.

A comparison of the maximum conversion times of an eight-bit digital-ramp ADC to an eight-bit successive-approximation ADC, both using a clock frequency of 1 MHz, will demonstrate the difference in speeds. The digital-ramp ADC has a maximum conversion time of

$$2^N - 1 \times 1\ \mu s = 2^8 - 1 \times 1\ \mu s = 255\ \mu s$$

where

$N$ = number of bits

The successive-approximation ADC has a maximum conversion speed of the number of bits times the clock period, or

$$8 \times 1\ \mu s = 8\ \mu s$$

It is obvious that the successive-approximation ADC is the choice for high-speed applications. Also, the higher the number of bits to be operated on, the greater the advantage in speed.

## The ADC0803/4/5-1

Figure 9–17 shows the block diagram for the ADC0803 family. This family is a series of three CMOS eight-bit successive-approximation ADCs, using a resistive ladder and capacitive array together with an auto-zero comparator. These ADCs are designed to operate with microprocessor-controlled buses using a minimum of external circuitry. The three-state output data lines can be connected directly to the data bus.

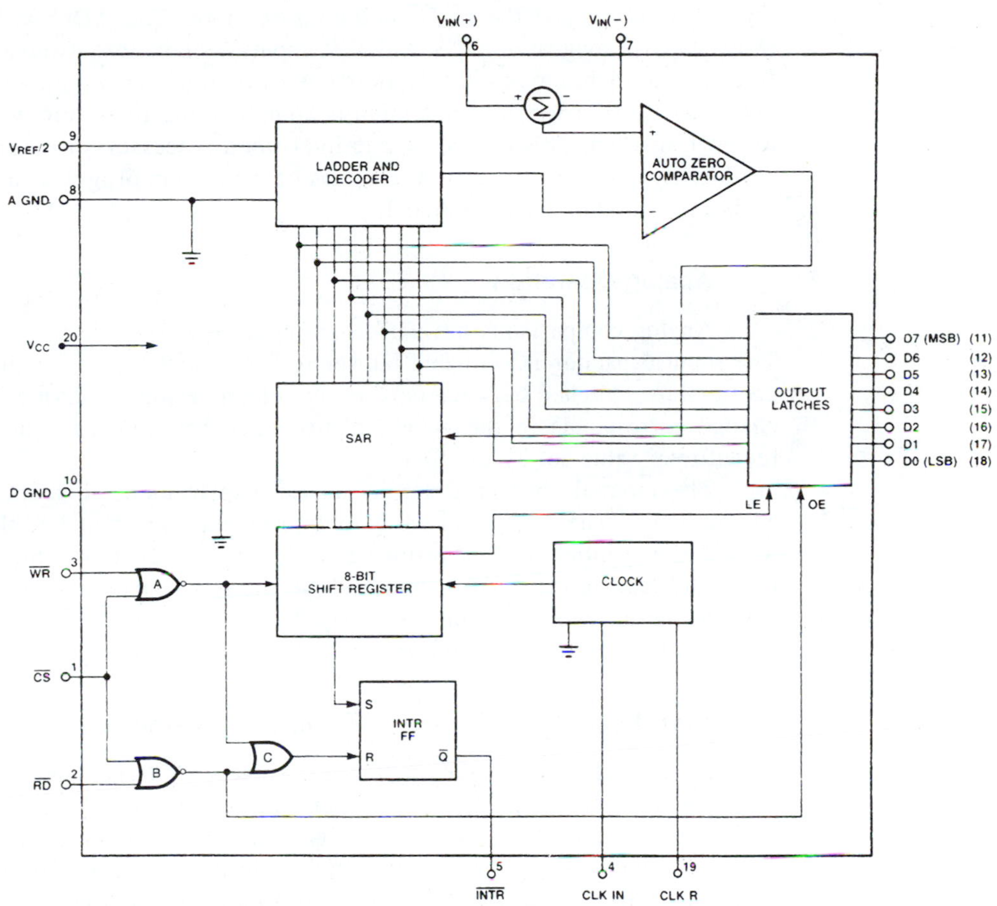


Figure 9–17. Block Diagram for the ADC0803 Family of ADCs (Courtesy of Signetics Company)

The differential analog voltage input allows for increased common mode rejection and provides a means to adjust the zero-scale offset. Additionally, the voltage reference input provides a means of encoding small analog voltages to the full eight bits of resolution.

### Functional Description

In the ADC0803 family, analog switches are closed sequentially by successive approximation logic until the input to the auto-zero comparator $[V_{IN}(+) - V_{IN}(-)]$ matches the voltage from the decoder. After all bits are tested and determined, the eight-bit binary code corresponding to the input voltage is transferred to the output latch. Conversion begins with the arrival of a pulse at the $\overline{WR}$ input if the $\overline{CS}$ input is low. On the high-to-low transition of the signal at the $\overline{WR}$ or the $\overline{CS}$ input, the successive-approximation register (SAR) is initialized, the shift register is reset, and the $\overline{INTR}$ output is set high. The ADC will remain in the reset state as long as the $\overline{CS}$ and $\overline{WR}$ inputs remain low. Conversion will start from one to eight clock periods after one or both of these inputs makes a low-to-high transition. After the conversion is complete, the $\overline{INTR}$ pin will make a high-to-low transition. This can be used to interrupt a processor, or otherwise signal the availability of a new conversion result. A conversion in progress can be interrupted by issuing another start command.

### Analog Operation

Analog comparisons are performed by a capacitive charge summing circuit. The input capacitor is switched between $V_{IN}$ (+) and $V_{IN}$ (-), while reference capacitors are switched between taps on the reference voltage divider string. The net change corresponds to the weighted difference between the input and the most recent total value set by the SAR.

The internal switching action causes displacement currents to flow in the analog inputs. The voltage on the on-chip capacitance is switched through the analog difference input voltage, resulting in proportional currents entering the $V_{IN}$ (+) input and leaving the $V_{IN}$ (-) input. These transient currents occur at the leading edge of the internal clock pulses. They decay rapidly so do not inherently cause errors as the on-chip comparator is strobed at the end of the clock period.

### Input Bypass Capacitors and Source Resistance

Bypass capacitors at the input will average the charges mentioned in the preceding section, causing a dc and an ac current to flow through the output resistance of the analog signal sources. This charge pumping action is worse for continuous conversion with the $V_{IN}$ (+) input at full scale. This current can be a few microamps, so bypass capacitors should not be used at the analog inputs of the $V_{REF}/2$ input for high resistance sources ($>1$ kΩ). If input bypass capacitors are desired for noise filtering and a high input resistance is desired to minimize capacitor size, detrimental effects of the voltage drop across the input resistance can be eliminated by adjusting the full scale with both the input resistance and the input bypass capacitor in place. This is possible because the magnitude of the input current is a precise linear function of the differential voltage.

Large values of source resistance where an input bypass capacitor is not used will not cause errors as the input currents settle out prior to the comparison time. If a low-pass (LP) filter is required in the system, use a low-valued series resistor ($<1$ kΩ) for a passive RC section, or add an op amp active LP filter. For applications with source resistances at or below 1 kΩ, a 0.1 μF bypass capacitor at the inputs will prevent noise pickup due to series lead inductance or a long wire.

If an op amp is used, a 100 Ω series resistor can be used to isolate the capacitor (both the resistor and the capacitor should be placed out of the feedback loop) from the output of the op amp.

### Analog Differential Voltage Inputs and Common Mode Rejection

The differential voltage input of these ADCs produce additional flexibility. The $V_{IN}$ (–) input (pin 7) can be used to subtract a fixed voltage from the input reading (this is called *tare correction*). Common mode noise can also be reduced by the use of the differential input.

The time interval between sampling $V_{IN}$ (+) and $V_{IN}$ (–) is 4.5 clock periods. The maximum error due to this time difference is given as:

$$V_{max} = (V_p)(2\pi f_{CM})\left(\frac{4.5}{f_{CLK}}\right) \tag{9.6}$$

where

$V$ = error voltage due to sampling delay
$V_p$ = peak value of common mode voltage
$f_{CM}$ = common mode frequency

By algebraic manipulation, we have:

$$V_p = \frac{(V_{max})(f_{CLK})}{(2\pi f_{CM})(4.5)} \tag{9.7}$$

## Example 9.6

Assume $f_{CM}$ = 60 Hz, $f_{CLK}$ = 1 MHz, and $V_{max}$ = 5 mV. Calculate the common mode voltage, $V_p$.

*Solution*

Use Equation 9.7:

$$V_p = \frac{(V_{max})(f_{CLK})}{(2\pi f_{CM})(4.5)}$$

$$= \frac{(5\text{ mV})(1\text{ MHz})}{(6.28)(60\text{ Hz})(4.5)} = 2.95\text{ V}$$

### Noise and Stray Pickup

Keeping the leads of the analog inputs as short as possible will minimize input noise coupling and stray signal pickup. The source resistance for these inputs should generally be kept below 5 kΩ to help avoid undesired noise pickup. Input bypass capacitors at the analog inputs can create errors as described previously. Full-scale adjustment with any input bypass capacitors in place will eliminate these errors.

### Reference Voltage

For application flexibility, these ADCs have been designed to accommodate fixed-reference voltages of 5 V to pin 20 and 2.5 V to pin 9, or an adjusted reference voltage at pin 9. The reference can be determined by the supply voltage (pin 20), or can be set by forcing it at $V_{REF}/2$ input (pin 9), as shown in Figure 9–18A.

Note that the $V_{REF}/2$ voltage is either 1/2 the voltage applied to the $V_{CC}$ supply pin, or is equal to the voltage that is externally forced at pin 9. In addition to allowing for flexible references and full span (range) voltages, this also allows for a ratiometric voltage reference. The internal gain of the $V_{REF}/2$ input is 2, making the full-scale differential input voltage twice the voltage at pin 9.

For example, a dynamic voltage range of the analog input voltage that extends from 0 to 4 V gives a span of 4 V (4 – 0), so the $V_{REF}/2$ voltage can be made equal to 2 V (half the 4 V span) and full-scale output will correspond to 4 V at the input.

On the other hand, if a dynamic input voltage has a range of 0.5 V to 3.5 V, the span or dynamic input range is 3 V (3.5 – 0.5). To encode this 3 V span with the 0.5 V yielding a code of zero, the minimum expected input (0.5 V in this case) is applied to the $V_{IN}$ (–) pin to account for the offset, and the $V_{REF}/2$ pin is set to 1/2 the 3 V span, or 1.5 V. The ADC will then encode the $V_{IN}$ (+) signal between 0.5 V and 3.5 V with 0.5 V at the input corresponding to a code of zero and 3.5 V at the input producing a full-scale output code. The full eight bits of resolution are thus applied over the reduced input voltage range. The required connections for this configuration are shown in Figure 9–18B.

### Operating Modes

The ADC0803 series can be operated in two modes:

1. Absolute mode
2. Ratiometric mode

In *absolute mode* applications, Figure 9–19, both the initial accuracy and the temperature stability of the reference voltage are important factors in the accuracy of the conversion. For $V_{REF}/2$ voltage of 2.5 V, initial errors of ±10 mV will cause conversion errors of ±1 LSB due to the gain of 2 at the $V_{REF}/2$ input. In reduced

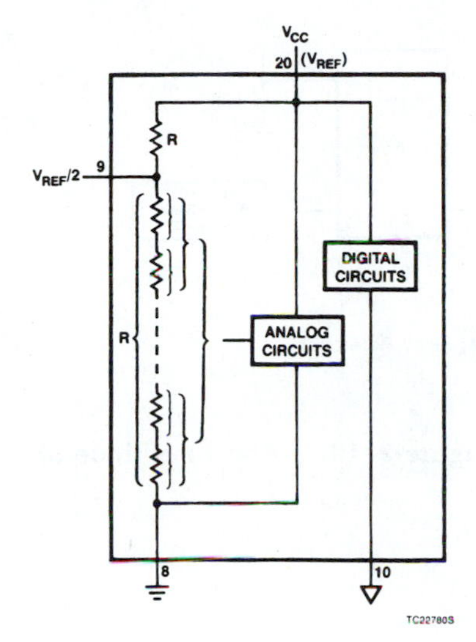


NOTE:
The $V_{REF}/2$ voltage is either ½ the $V_{CC}$ voltage or is that which is forced at Pin 9.

(A) Internal Reference Design

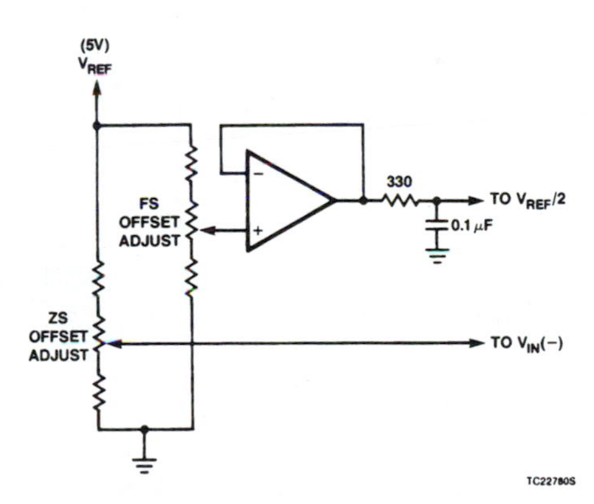


(B) Offsetting the Zero-Scale and Adjusting the Input Range (Span)

**Figure 9–18.** Reference Voltage Inputs and Adjustments (Courtesy of Signetics Company)

span applications, the initial value and stability of the $V_{REF}/2$ input voltage become even more important as the same error is a larger percentage of the $V_{REF}/2$ nominal voltage.

In *ratiometric* converter applications, Figure 9–20, the magnitude of the reference voltage is a factor in both the output of the source transducer and the output of the ADC. Therefore, $V_{REF}/2$ cancels out in the final digital code.

Generally, the reference voltage will require an initial adjustment. Errors due to an improper reference voltage value appear as full scale errors in the analog-to-digital transfer function.

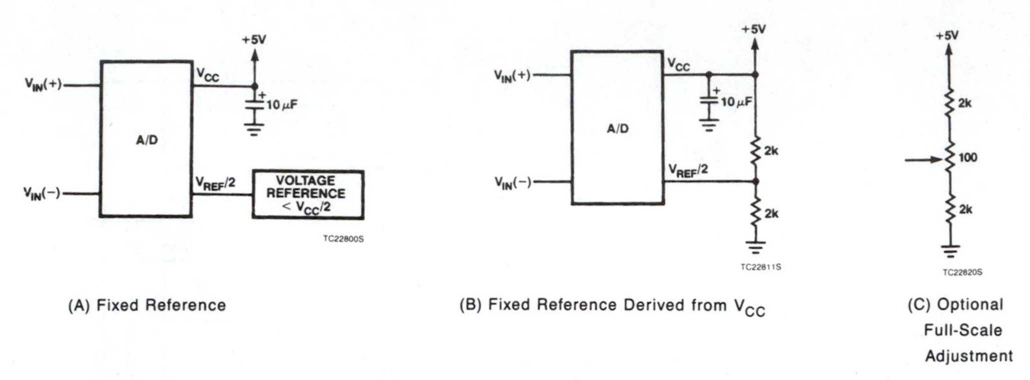


(A) Fixed Reference (B) Fixed Reference Derived from $V_{CC}$ (C) Optional Full-Scale Adjustment

**Figure 9–19.** Absolute Mode of Operation (Courtesy of Signetics Company)

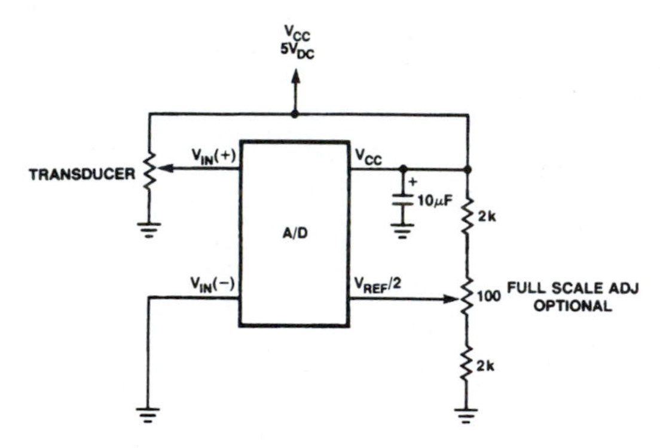


**Figure 9–20.** Ratiometric Mode of Operation with Optional Full-Scale Adjustment (Courtesy of Signetics Company)

## Errors and Input Span Adjustments

There are many sources of error in any data converter, some of which can be adjusted out. Although inherent errors, such as relative accuracy, cannot be eliminated, such errors as full-scale and zero-scale offset errors can be eliminated quite easily, as shown in Figure 9–18B.

Zero-scale error of an ADC is the difference of potential between the ideal 1/2 LSB value (9.8 mV for $V_{REF}/2 = 2.5$ V) and that input voltage that just causes an output transition from code 0000 0000 to 0000 0001.

If the minimum input value is not ground potential, a zero offset can be made. The converter can be made to output a digital code of 0000 0000 for the minimum expected input voltage by biasing the $V_{IN}$ (–) input to that minimum value expected at the $V_{IN}$ (+) input. This uses the differential mode of the converter. Any offset adjustment should be done prior to full-scale adjustment.

Full-scale gain is adjusted by applying any desired offset voltage to $V_{IN}$ (−), then applying to the $V_{IN}$ (+) input a voltage that is 1-1/2 LSB less than the desired analog full-scale voltage range. Then adjust the magnitude of $V_{REF}/2$ input voltage (or the $V_{CC}$ supply if there is no $V_{REF}/2$ input connection) for a digital output code that just changes from 1111 1110 to 1111 1111. The ideal $V_{IN}$ (+) voltage for full-scale adjustment is given by:

$$V_{IN}(+) = V_{IN}(-) - 1.5 \times \frac{V_{max} - V_{min}}{255} \quad \textbf{(9.8)}$$

where

$V_{max}$ = high end of analog input range (ground referenced)
$V_{min}$ = low end (zero offset) of analog input (ground referenced)

## Example 9.7

Assume an ADC0803 circuit where (a) $V_{max}$ = 15.0 V, $V_{min}$ = 0.5 V, and $V_{IN}$ (−) = 2.5 V and (b) $V_{max}$ = 10 V, $V_{min}$ = 0.5 V, and $V_{IN}$ (−) = 1.5 V. Determine the ideal $V_{IN}$ (+) voltage required for full-scale adjustment.

*Solution*

Use Equation 9.8:

$$V_{IN}(+) = V_{IN}(-) - 1.5 \times \left(\frac{V_{max} - V_{min}}{255}\right)$$

(a) $$V_{IN}(+) = 2.5\text{ V} - 1.5 \times \left(\frac{15\text{ V} - 0.5\text{ V}}{255}\right) = 2.415\text{ V}$$

(b) $$V_{IN}(+) = 1.5\text{ V} - 1.5 \times \left(\frac{10\text{ V} - 0.5\text{ V}}{255}\right) = 1.44\text{ V}$$

### Clocking Option

The clock signal for these ADCs can be derived from external sources, such as a system clock; or self-clocking can be accomplished by adding an external resistor and capacitor, as shown in Figure 9–21.

Heavy capacitive or dc loading of the CLK R pin should be avoided as this will disturb normal converter operation. Loads less than 50 pF are allowed. This permits driving up to seven ADC CLK IN pins of this family from a single CLK R pin of one converter. For larger loading of the clock line, a CMOS or low-power transistor-transistor logic (TTL) buffer or PNP input logic should be used to minimize the loading on the CLK R pin.

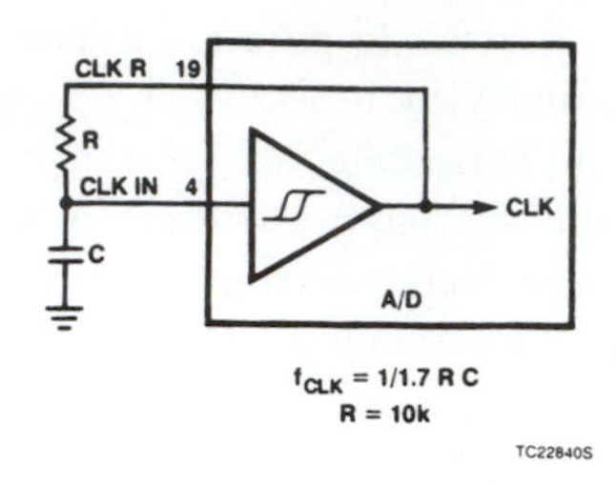


**Figure 9–21.** Self-Clocking the Converter (Courtesy of Signetics Company)

### Restart During a Conversion

A conversion in process can be halted and a new conversion begun by bringing the $\overline{CS}$ and $\overline{WR}$ inputs low and allowing at least one of them to go high again. The output data latch is not updated if the conversion in progress is not completed; the data from the previously completed conversion will remain in the output data latches until a subsequent conversion is completed.

### Continuous Conversion

To provide continuous conversion of input data, the $\overline{CS}$ and $\overline{RD}$ inputs are grounded and $\overline{INTR}$ output is tied to the $\overline{WR}$ input. The $\overline{INTR}/\overline{WR}$ connection should be momentarily forced to a logic low upon power up to insure circuit operation. One way to accomplish this is shown in Figure 9–22.

### Power Supplies

Noise spikes on the $V_{CC}$ line can cause conversion errors as the internal comparator will respond to them. A low inductance filter capacitor should be used close to the converter $V_{CC}$ pin and values of 1 μF or greater are recommended. A separate 5 V regulator for the converter (and other 5 V linear circuitry) will greatly reduce digital noise on the $V_{CC}$ supply and the attendant problems.

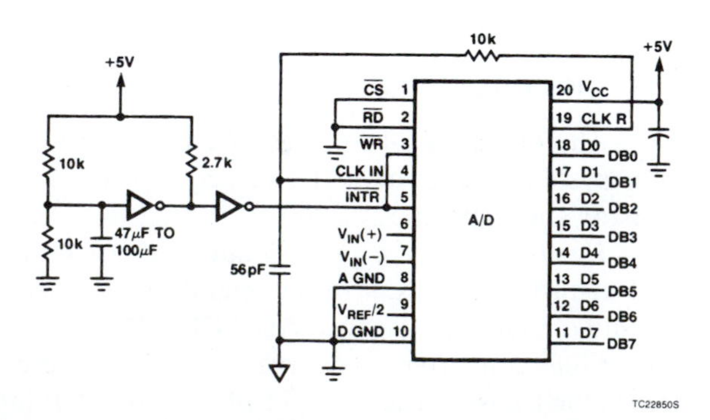


**Figure 9–22.** Connection for Continuous Conversion (Courtesy of Signetics Company)

### Wiring and Layout Precautions

Digital wire wrap sockets and connections are not satisfactory for this (or any) ADC. Sockets on printed circuit boards (PCBs) can be used. All logic signal wires and leads should be grouped or kept as far as possible from the analog signal leads. Single-wire analog input leads may pick up undesired hum and noise, requiring the use of shielded leads to the analog inputs in many applications.

A single-point analog ground separate from the logic or digital ground points should be used. The power supply bypass capacitor and the self-clocking capacitor, if used, should be returned to digital ground. Any $V_{REF}/2$ bypass capacitor, analog input filter capacitors, and any input shielding should be returned to the analog ground point. Proper grounding will minimize zero-scale errors that are present in every code. Zero-scale errors can usually be traced to improper PCB layout and wiring.

## Dual-Slope ADC

The *dual-slope ADC* is shown in the block diagram in Figure 9–23. This type of ADC requires no DAC, but it does require a close tolerance reference voltage. A constant $V_{in}$ is obtained by using S & H circuits, as discussed earlier in this chapter. This circuit makes use of both an integrator and a comparator.

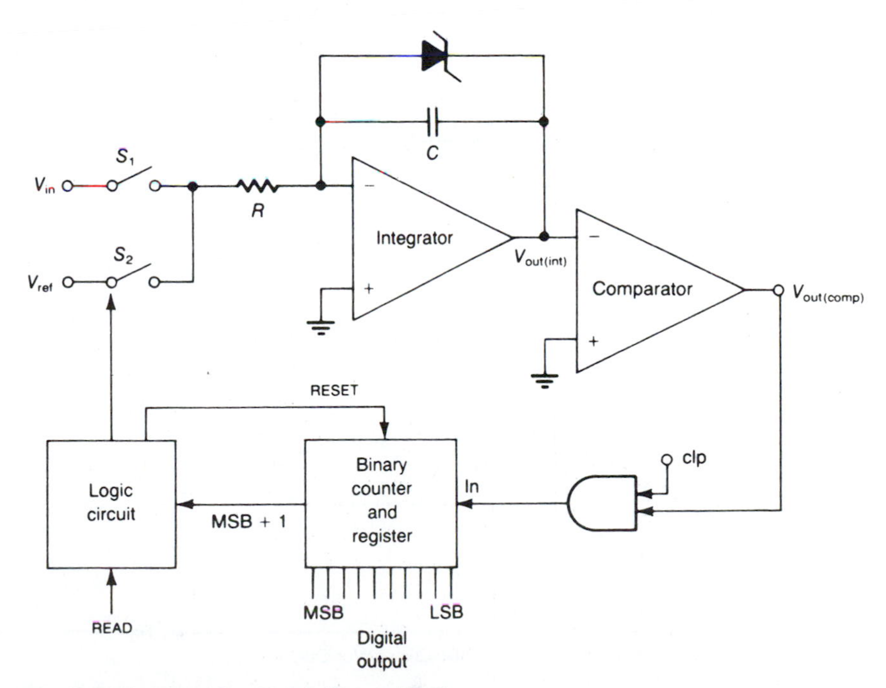


**Figure 9–23.** Block Diagram of a Dual-Slope ADC

The logic circuit controls electronic switches $S_1$ and $S_2$, and resets the counter. A READ signal is applied to the logic circuit, causing the counter to reset, $S_1$ to close at the $V_{in}$ position, and $S_2$ to open. The integrator output is

$$V_{out(int)} = -\frac{V_{in}t}{RC} \tag{9.9}$$

Since $V_{in}$ is held constant by an S & H circuit, $V_{out(int)}$ will have the waveform shown in Figure 9–24A. This voltage is then applied to the comparator input. Since the reference voltage is 0, the comparator output becomes $+V_{sat}$, which is adjusted to represent a binary 1. This voltage is then one input of the AND gate, and the output of the AND gate will be the same as the clock pulse. The counter counts the clock pulses and the process continues until all bits, with the exception of the MSB + 1, are set to 0. The binary 1 at the MSB + 1 causes the logic circuit to throw $S_1$ to the $V_{ref}$ position, $S_2$ to close, and at this instant,

$$V_{out(int)} = -\frac{V_{in}t_1}{RC} \tag{9.10}$$

as shown in Figure 9–24A. The integrator now integrates the $V_{ref-}$ voltage and causes the voltage to rise linearly toward 0 (Figure 9–24A). When the comparator input becomes slightly greater than 0, the comparator output switches to $-V_{sat}$, or binary 0, as shown in Figure 9–24B, closing the AND gate and stopping the clock pulse count. We can now read the output bit count that is stored in the register.

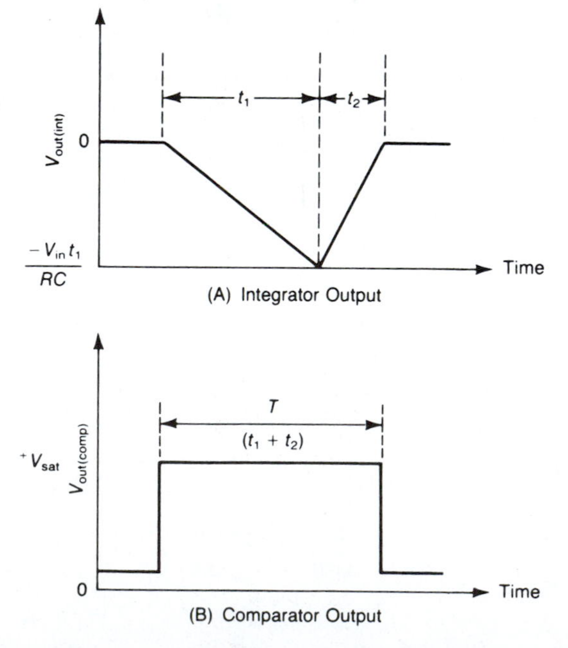


**Figure 9–24.** Integrator and Comparator Output Waveforms

The bit count is proportional to $t_2 - t_1$. Time $t_1$ is a fixed interval, the time it takes to go from 000 . . . 0 to 100 . . . 0 (with the 1 being $N + 1$). If the number of output bits ($2^N$) and the clock frequency ($f_{cl}$) are known, then

$$t_1 = \frac{(2^{N-1} - 1)}{f_{cl}} \quad \textbf{(9.11)}$$

and

$$t_2 = \frac{V_{in}}{V_{ref-}}(t_1) \quad \textbf{(9.12)}$$

Also, the number of clock pulses (#clp), or the binary output, will be

$$\#clp = t_2 f_{cl} \quad \textbf{(9.13)}$$

**Example 9.8**

Refer to Figure 9–23. The counter of this ADC has a 10-bit output and a clock frequency of 1 MHz. $V_{ref-} = 6$ V and $V_{in} = 4$ V. Find (a) $t_1$, (b) $t_2$, and (c) the binary output.

*Solutions*

**a.** Use Equation 9.11:

$$t_1 = \frac{(2^{N-1} - 1)}{f_{cl}} = \frac{(2^{10-1} - 1)}{1\text{ MHz}} = \frac{511}{1 \times 10^6} = 511\ \mu\text{s}$$

**b.** Use Equation 9.12:

$$t_2 = \frac{V_{in}}{V_{ref-}}(t_1) = \frac{4\text{ V}}{6\text{ V}}(511\ \mu\text{s}) = 341\ \mu\text{s}$$

**c.** Use Equation 9.13:

$$\#clp = t_2 f_{cl} = 341\ \mu\text{s} \times 1\text{ MHz} = 341 = 0101010101$$

# 9.6 Applications

There are a great number of DAC and ADC applications. In this section we will examine just five common ones—digital voltmeter, process control, amplifier/attenuator, blending system, and transistor curve tracer.

## Digital Voltmeter

The dual-slope ADC is widely used in digital voltmeters (DVMs) where high conversion speeds are not important to the operation. DVMs can be constructed

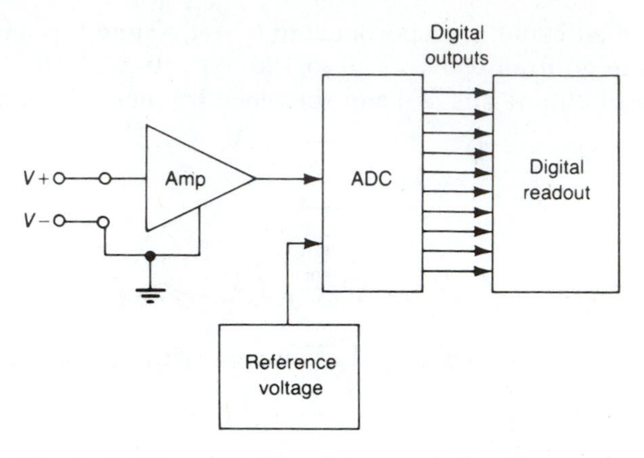


**Figure 9–25.** Block Diagram of a Digital Voltmeter (DVM)

with any of the ADCs we have discussed, but the dual-slope is most frequently used. This ADC is especially useful for making low-level voltage measurements because the input signal is averaged. This means that any high-frequency noise pulses will be averaged out, resulting in high accuracy. Also, this circuit does not require precision resistors to attain its high accuracy. A typical DVM block diagram is shown in Figure 9–25.

## Process Control

A typical *closed-loop process control system* is shown in Figure 9–26. Notice the transduced analog signal. This signal is converted by the ADC into binary form, which is processed by the computer. The computer provides two output signals: one output produces a visual display, and the other output feeds binary signals to the DAC. The DAC output is an analog control voltage that can cause the heater in the oven to increase or decrease in temperature as required to provide the exact desired oven temperature.

## Amplifier/Attenuator

An unusual application of the DAC is the *digitally controlled amplifier/attenuator* illustrated in Figure 9–27. The input voltage is applied through an on-chip feedback resistor. Op amp $A_1$ automatically adjusts the $V(\text{ref})_{\text{in}}$ voltage such that $I_{\text{out 1}}$ is equal to the input current ($V_{\text{in}}/R_{\text{f}}$). The magnitude of this $V(\text{ref})_{\text{in}}$ voltage depends on the digital word that is in the DAC register. $I_{\text{out 2}}$ then depends upon both the magnitude of $V_{\text{in}}$ and the digital word. Op amp $A_2$ converts $I_{\text{out 2}}$ to a voltage, $V_{\text{out}}$, that is given by the following equation:

$$V_{\text{out}} = V_{\text{in}}\left(\frac{1023 - N}{N}\right) \qquad \textbf{(9.14)}$$

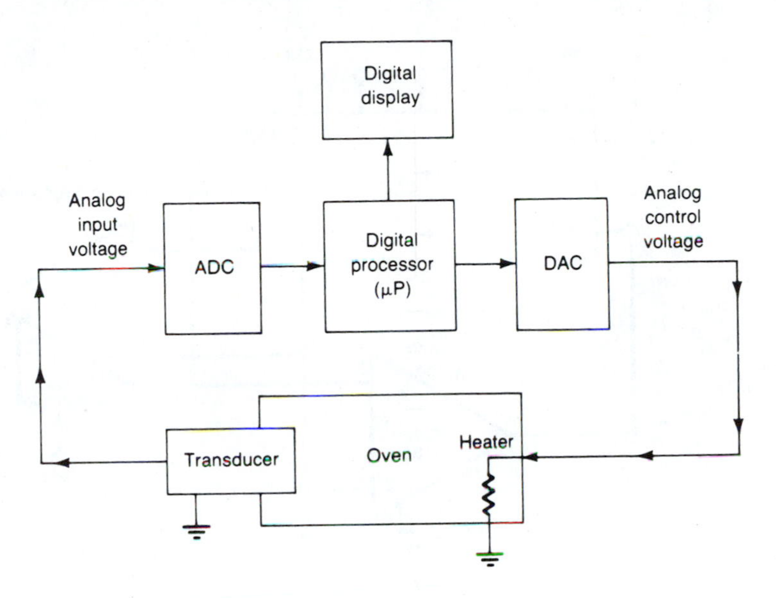


**Figure 9–26.** Closed-Loop Digital Controlled System

where

$0 < N < 1023$

Note that $N = 0$ (or a digital code for all 0s) is not allowed because this will cause the output amplifier to saturate at either $\pm V_{sat}$ depending on the sign of $V_{in}$.

To provide a digitally controlled divider, op amp $A_2$ can be eliminated. Ground the $I_{out\,2}$ pin of the DAC and $V_{out}$ will be taken from the op amp $A_1$ (which also drives the $V_{ref}$ input of the DAC). The expression of $V_{out}$ will now be

$$V_{out} = -\frac{V_{in}}{M} \tag{9.15}$$

where

$M$ = digital input (expressed as a fractional binary number)
$0 < M < 1$

## Blending System

Another process control system is the blending system shown in Figure 9–28. The control system on the left includes the computer, the DAC, and the ADC. On the right is the controlled object, the blender. The two analog outputs of the

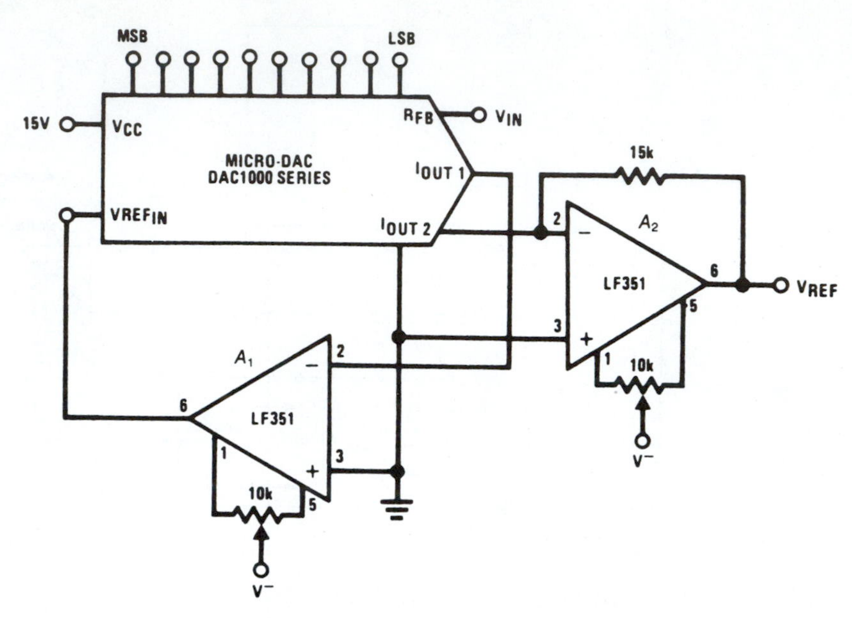


Figure 9–27. Digital Controlled Amplifier/Attenuator (Courtesy of National Semiconductor Corporation)

DAC actuate the HOT and COLD pumps. The blended fluid flows through a discharge tube that contains a flowmeter and a temperature probe. The controlled variables, flow and temperature, provide the analog inputs to the ADC. An external clock regulates the sampling times.

## Transistor Curve Tracer

A block diagram of a transistor curve tracer using a DAC is illustrated in Figure 9–29. The output of the Schmitt trigger cycles the four-bit counter continuously from 0 through 15, providing the capability of 16 curves of increasing levels of base current $I_B$.

The DAC can be adjusted for a desired level per bit by varying the voltage level, $V^+$. For example, adjusting $V^+$ for 5 μA per bit will produce $I_B$ characteristic curves ranging from $I_B = 0$ to $I_B = 75$ μA in 16 equal 5 μA steps. After the count of 15, $I_B$ returns to 0, and the curves retrace themselves, providing a continuous visual display on the oscilloscope.

## Microprocessor Interfacing

The ADC0803 family of ADCs was designed for easy microprocessor interfacing. These converters can be memory-mapped with appropriate memory address decoding for $\overline{CS}$ (read) input. The active-low write pulse from the processor is then

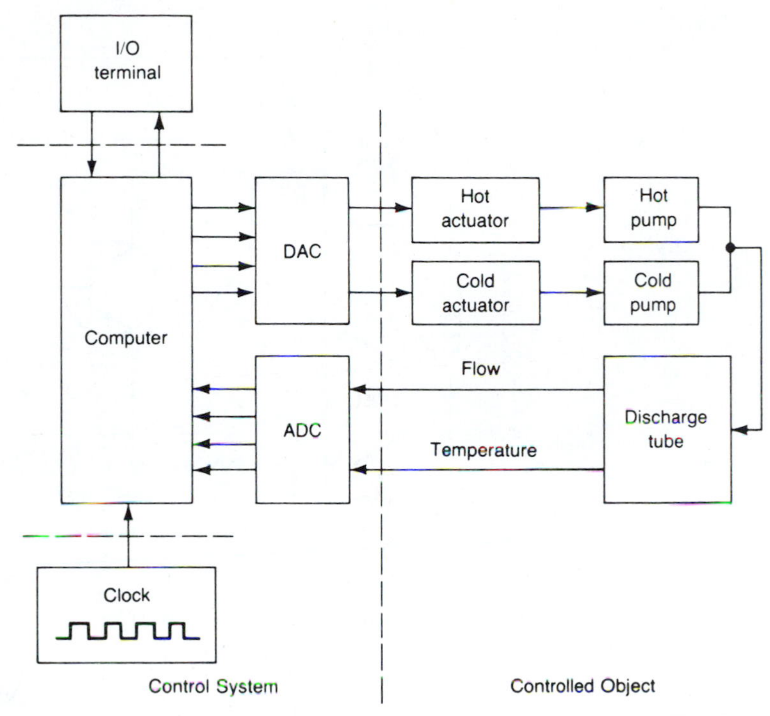


**Figure 9–28.** Blending Control System

connected to the $\overline{WR}$ input of the ADC, while the processor active-low read pulse is fed to the converter $\overline{RD}$ input to read the converted data. If the clock signal is derived from the microprocessor system clock, the user should be sure that there is no attempt to read the converter until 74 converter clock pulses after the start pulse goes high. Alternatively, the $\overline{INTR}$ pin may be used to interrupt the processor to cause reading of the converted data.

Figure 9–30 shows how the converter can be connected and addressed as a peripheral. A bus driver should be used as a buffer to the ADC output, as shown in Figure 9–31, in large microprocessor systems where the data leaves the PCB and/or must drive capacitive loads in excess of 100 pF.

Figure 9–32 shows the simplicity of interfacing the ADC to the SCN8051 microprocessor. Since the SCN8051 has 24 I/O lines, one of them (shown here as bit 0 or port 1) can be used as the $\overline{CS}$ signal to the converter, eliminating the need for an address decoder. The $\overline{RD}$ and $\overline{WR}$ signals are generated by reading from and writing to a dummy address.

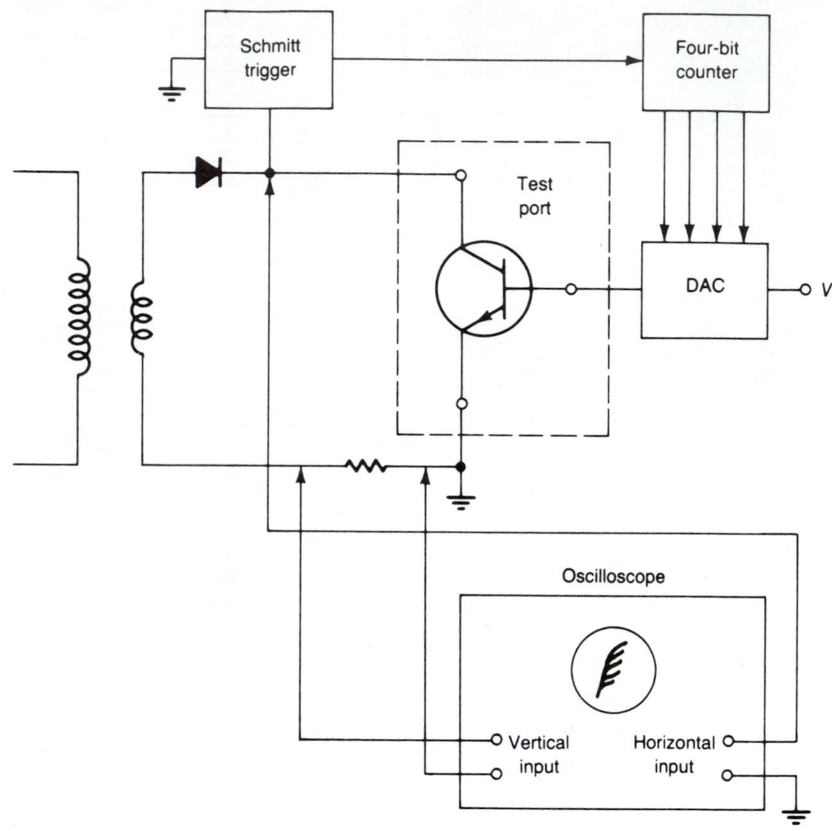


**Figure 9–29.** Block Diagram of a Typical Transistor Curve Tracer

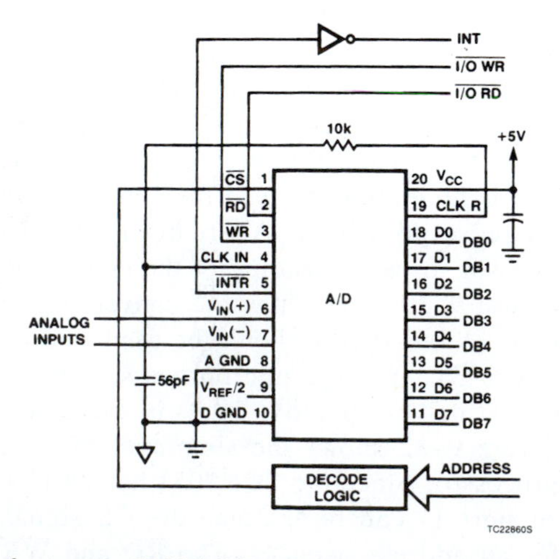


**Figure 9–30.** Interfacing to 8080A Microprocessor (Courtesy of Signetics Company)

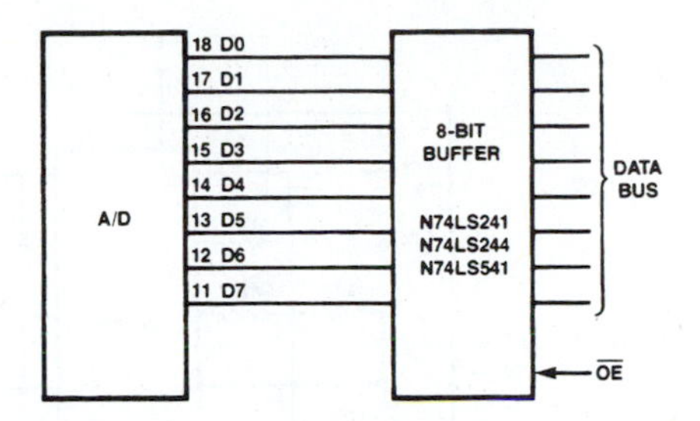


**Figure 9-31.** Buffering the A/D Output to Drive High Capacitance Loads and for Driving Off-Board Loads (Courtesy of Signetics Company)

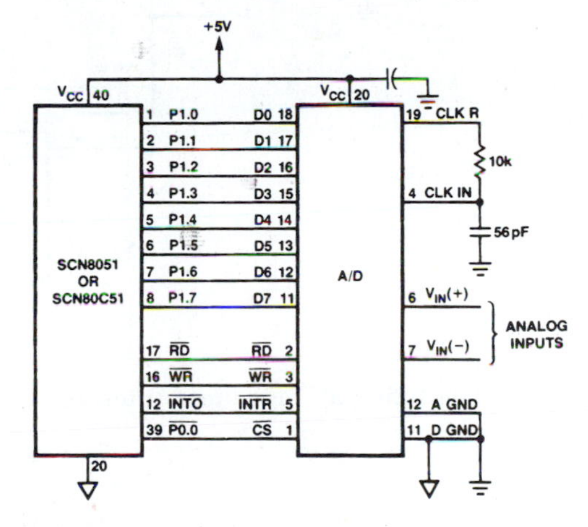


**Figure 9-32.** SCN8051 Interfacing (Courtesy of Signetics Company)

# Digitizing a Transducer Interface Output

Figure 9-33 shows an example of digitizing transducer interface output voltage. In this case, the transducer interface is the NE5521, a Linear Variable Differential Transformer (LVDT) Signal Conditioner. The diode at the ADC input is used to insure that the input to the ADC does not go excessively beyond the supply voltage of the ADC.

To adjust the full-scale and zero-scale of the ADC, determine the range of the voltages that the transducer interface output will take on. Set the LVDT core for null and set the Zero-Scale Adjust ($V_{IN}$ (−)) potentiometer for a digital output from the ADC of 1000 0000. Set the LVDT core for maximum voltage from the interface and set the Full-Scale Adjust potentiometer so that ADC output is just barely 1111 1111.

## A Digital Thermostat

The schematic for a digital thermostat is shown in Figure 9-34. The ADC digitizes the output of the LM35, a temperature transducer IC with an output of

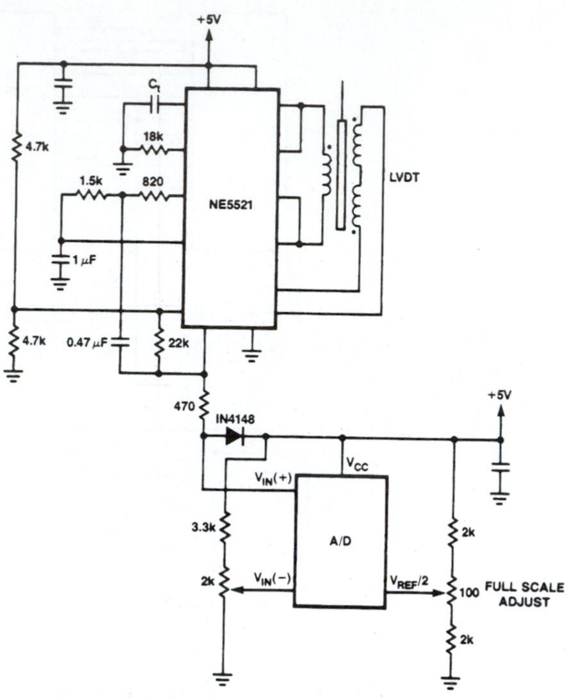


**Figure 9–33.** Digitizing a Transducer Interface Output (Courtesy of Signetics Company)

10 mV per °C. With $V_{REF}/2$ set for 2.56 V, this 10 mV corresponds to 1/2 LSB and the circuit resolution is 2 °C. Reducing $V_{REF}/2$ to 1.28 V yields a resolution of 1 °C. Of course, the lower $V_{REF}/2$ is, the more sensitive the ADC will be to noise.

The desired temperature is set by holding either of the set buttons closed. The SCC80C451 programming could cause the desired (set) temperature to be displayed while either button is depressed and for a short time after it is released. At other times the ambient temperature could be displayed.

The set temperature is stored in an SCN8051 internal register. The ADC conversion is started by writing anything at all to the ADC with port pin 10 set high. The desired temperature is compared with the digitized actual temperature, and the heater is turned on or off by clearing setting port pin P12. If desired, another port pin could be used to turn on or off an air conditioner.

The display drivers are NE587s if common-anode light-emitting diode (LED) displays are used, and NE589s if common-cathode LED displays are used. It is also possible to interface to LCD displays.

There are many more uses for DAC and ADC circuits than we have the space to list. Manufacturers' data books will provide ample information on applications for the selection of devices for specific purposes.

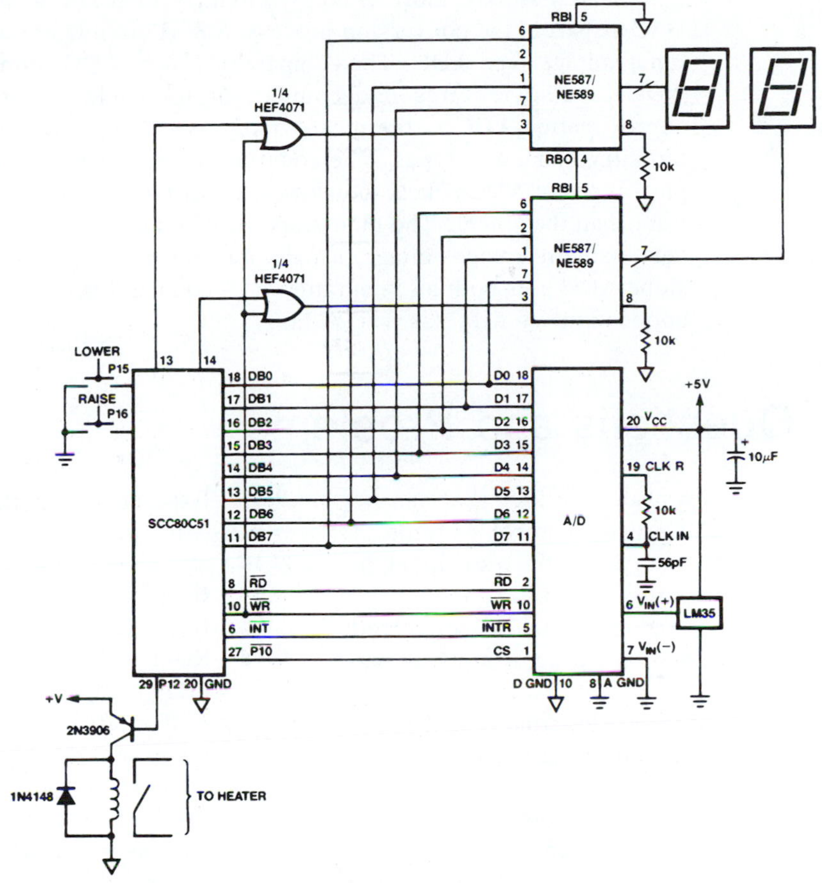


**Figure 9-34.** Digital Thermostat (Courtesy of Signetics Company)

# 9.7 Summary

☐ Combining analog and digital signals is necessary to accommodate industrial process control systems.

☐ The DAC is a relatively simple circuit that provides an analog output of either voltage or current. The analog output of a DAC is the weighted sum of the digital inputs. One of the simplest DACs is an op amp summing amplifier. A variation of the summing amplifier is the *R*-2*R* ladder circuit. The *R*-2*R* circuit is popular in audio and RF level control circuits.

☐ ADCs require more complex circuitry than DACs, and generally require a DAC as part of the conversion process. S & H circuits are used to provide a steady input voltage to an ADC. The comparator (flash) ADC is one of the fastest types of ADCs, but it requires one comparator for each output bit. The successive-approximation ADC is the most widely used type, but it requires more complex circuitry than other types. The counting, or digital-ramp, ADC is one of the simplest types of ADCs. It is the slowest of all types, but it requires much less hardware than the others. The dual-slope ADC requires no DAC, but it does require a very stable reference voltage, usually attained by using an S & H circuit. The dual-slope ADC uses both an integrator and a comparator. One of the most widely used applications of ADCs is in DVMs.

## 9.8 Questions and Problems

**9.1** List five physical quantities that can be measured in process control systems.

**9.2** What is the purpose of a transducer in a process control system?

**9.3** Why must analog signals be converted to binary form in a process control system?

**9.4** Draw a simplified block diagram of a process control system using both DACs and ADCs.

**9.5** In a binary DAC, what output level shift is created by the MSB?

**9.6** What is the resolution of a 10-bit binary DAC?

**9.7** What is the possible number of output levels with a 10-bit binary DAC?

**9.8** How can quantization undertainty be reduced in an ADC?

**9.9** An eight-bit DAC produces 0.25 V for a digital input of 00000001. Find the value of $V_{out}$ for the digital inputs of (a) 00001010, (b) 00000111, (c) 00001111, (d) 00010010, and (e) 11111111.

**9.10** What is the percentage of resolution for the DAC in Problem 9?

**9.11** What is the step size of the DAC in Problem 9?

**9.12** Design a four-bit DAC using an op amp summing amplifier, where $V_{in}$ = 5 V and the resolution is 0.5 V.

**9.13** Explain the operation of an S & H circuit.

**9.14** Which of the several ADCs described in this chapter offers the fastest conversion time? the slowest?

**9.15** Refer to Figure 9–15. Assume $f_{clp}$ = 100 kHz, comparator $V_{th}$ = 1 mV, and DAC FSR = 10.23 V with a 10-bit input. Determine (a) the binary count of the counter output when $V_{in}$ = 3.41 V, (b) the conversion time, and (c) the resolution of the ADC.

**9.16** What is the maximum conversion time for a 10-bit digital-ramp ADC with a clock frequency of 100 kHz?

**9.17** What is the maximum conversion time for a 10-bit successive-approximation ADC with a clock frequency of 100 kHz?

**9.18** What conclusion can you make when comparing the conversion times of the ADCs of Problems 16 and 17?

**9.19** What type of ADC requires no DAC?

**9.20** An ADC of the type referred to in Problem 19 has a 10-bit output and a clock frequency of 100 kHz. $V_{ref}$ = 4 V and $V_{in}$ = 2 V. Find (a) $t_1$, (b) $t_2$, and (c) the binary output.

**9.21** List three items to consider when selecting an ADC for a given application.

**9.22** Why is the successive-approximation

ADC perhaps the most widely used ADC?

**For Problems 9.23 through 9.35, refer to the ADC0803 family of ADCs.**

**9.23** When does conversion begin?

**9.24** What occurs at the $\overline{\text{INTR}}$ pin upon completion of the conversion?

**9.25** How can a conversion in progress be interrupted?

**9.26** Explain how analog comparisons are made.

**9.27** What detrimental effects occur when using bypass capacitors at the input?

**9.28** The use of input bypass capacitors for noise filtering and a high input resistance is desired. How can the effects described in Problem 9.27 be overcome?

**9.29** Define tare correction.

**9.30** Assume $f_{CM}$ = 120 Hz, $f_{CLK}$ = 2 MHz, and $V_{max}$ = 4.5 mV. Calculate $V_p$.

**9.31** What are important factors in the absolute mode applications?

**9.32** Explain what happens if improper reference voltages are applied in ratiometric conversion applications.

**9.33** Assume a desired maximum digital output of 30 V and an LSB of 2 V. Determine the ideal input voltage for full-scale adjustment.

**9.34** What will be the result if a conversion in progress is halted before it is completed?

**9.35** What precautions can be taken to minimize noise spikes on $V_{CC}$ lines?

Chapter 10

# Modulation and Demodulation Circuits

## 10.1 Introduction

The concepts of modulation and demodulation are important for the electronics technician. Circuits that involve these concepts are used in many of the systems and devices with which you will be required to work.

This chapter will introduce several methods of modulation and demodulation, with the emphasis on demodulation. Modulation will be covered in greater detail in communications courses. We will examine various devices used in the processes of modulation and demodulation. The phase-locked loop (PLL) will be explored, and compandors will be introduced. Finally, various applications of modulation and demodulation circuits will be discussed.

## 10.2 Objectives

When you complete this chapter, you should be able to:

- ☐ Define the differences between amplitude modulation (AM), frequency modulation (FM), and phase modulation (PM).
- ☐ Draw a block diagram of a typical AM transmitter and explain the operation of its modulator circuit.

- ☐ Draw a block diagram of a typical AM receiver and explain the operation of its detector circuit.
- ☐ Draw a block diagram of a typical FM transmitter and explain the operation of the modulator.
- ☐ Draw a block diagram of a typical FM receiver and explain the operation of the discriminator.
- ☐ List several applications for compandors.
- ☐ Draw a block diagram of a PLL and describe the operation of its four functional blocks.
- ☐ List the three modes of a PLL and describe the operating ranges of the PLL.
- ☐ Describe the operation of the VCO.
- ☐ Discuss the differences between digital and analog PLLs.
- ☐ Design a frequency synthesizer from a standard PLL circuit.
- ☐ List several applications for PLL circuits.

# 10.3 Modulation

The term *modulation* implies variation or shaping, and the term *demodulation* implies removal of variation or shaping. For example, audio signals can be transmitted by changing some characteristics of a higher-frequency carrier wave. If this change is in the amplitude, with frequency held constant, the process is called *amplitude modulation* (AM). If the change is in frequency or phase angle, with amplitude held constant, it is called *frequency modulation* (FM), or *phase modulation* (PM), respectively.

A communications system transmits intelligence (information) from one point to another. Circumstances often dictate the frequency of transmission. For example, the basic telephone system uses a direct current in a circuit as a carrier current, and if the carrier is varied in amplitude at a rate of 500 hertz (Hz) per second, it is said to be amplitude-modulated.

If the transmitted signal is a radio frequency (RF) ac carrier wave of some frequency made to vary in amplitude at a certain rate, we again have AM. If the RF ac carrier wave is made to vary in frequency, then we have FM. These are the basic principles of AM and FM broadcast transmissions.

## Amplitude Modulation

Let us first explore AM. The block diagram of Figure 10–1 shows an AM transmitter with audio modulation used for radiotelephone communication. Radiotelephone indicates voice communications, but it also can include music. The

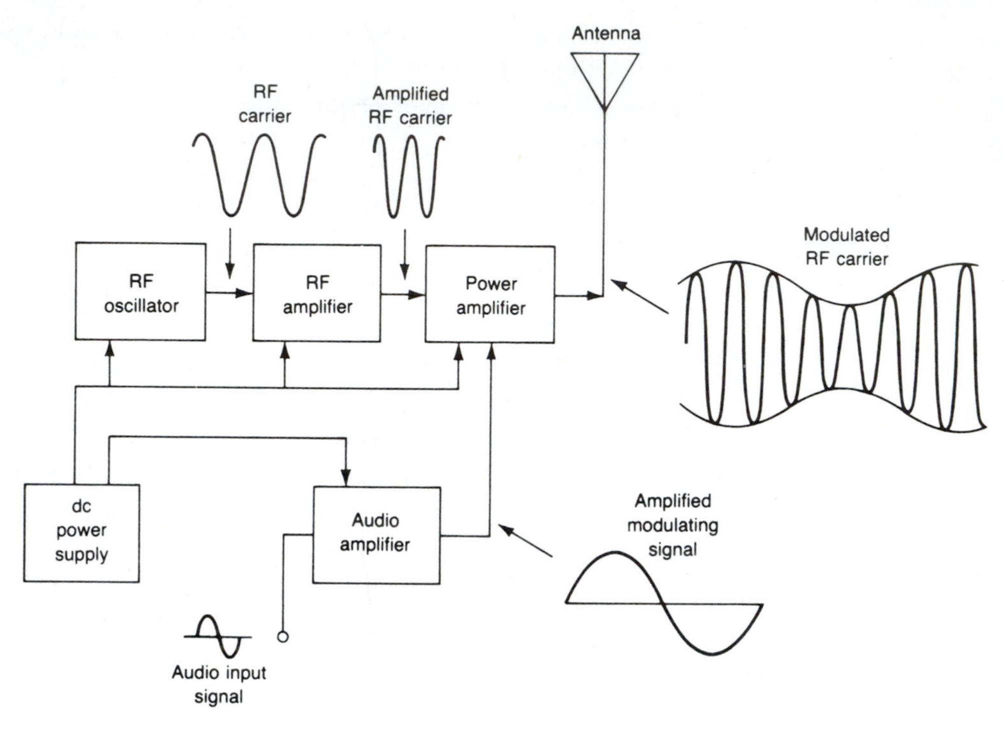


**Figure 10–1.** Block Diagram of a Radiotelephone AM Transmitter

oscillator stage generates the RF carrier wave. Crystal-controlled oscillators (CCOs) are typically used for this stage. The RF amplifier increases the frequency of the RF carrier signal.

The output of the oscillator is amplified by the RF amplifier stage. This stage performs two functions: first, it provides a large enough signal to drive the power amplifier (PA); and second, it acts as a buffer that separates the oscillator from the PA to reduce loading on the oscillator circuit and to increase frequency stability.

The PA drives the antenna with the amount of current needed for the desired power output. The more current that is provided, the stronger the radiated signal, and, therefore, the greater the distance the signal can be transmitted. Also, the PA is modulated by the audio signal from the audio amplifier. Therefore, the amplitude variations of the modulated RF signal are present in the output of the PA and, consequently, in the antenna circuit.

A radio frequency must be used for the transmitted carrier frequency because audio frequencies cannot be transmitted over any great distance, unless huge antennas are used. But by means of modulation, the carrier provides the transmission of the RF wave, while the audio modulating signal has the desired intelligence. Figure 10–2 shows the relationships between the unmodulated RF carrier and the modulated signal being transmitted. Simply stated, the basic sine waves of the RF carrier have been converted into a complex waveform. The waveform is still

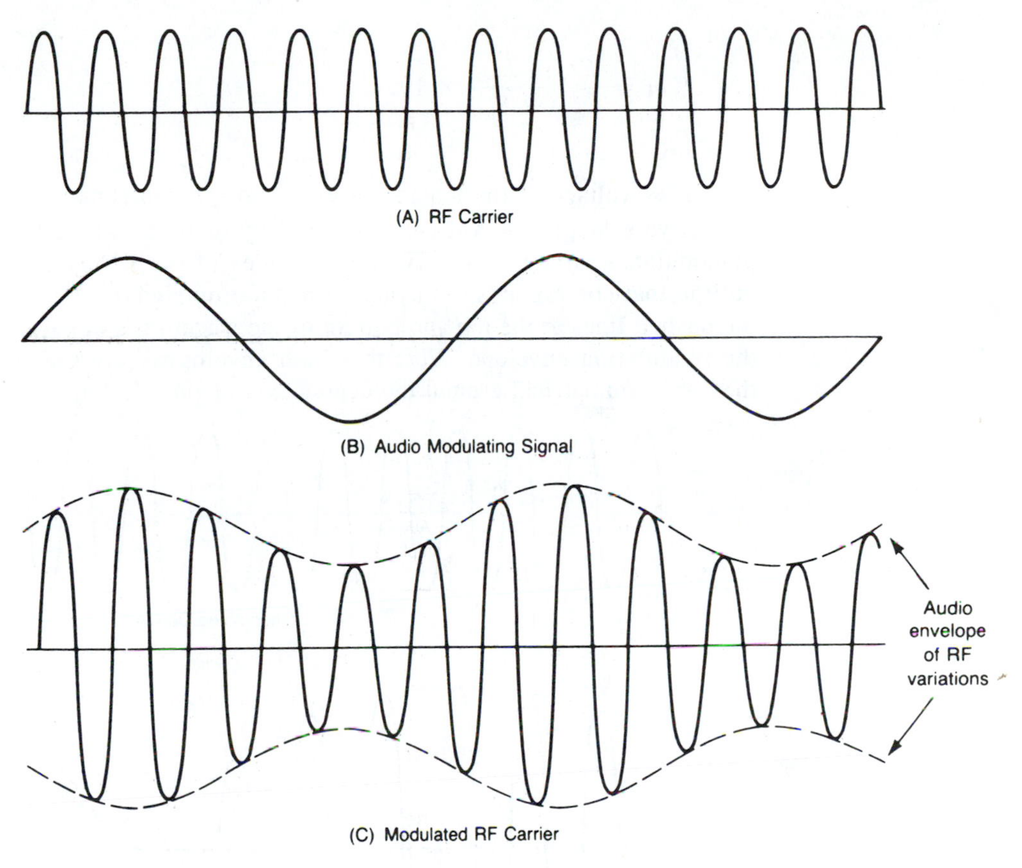


**Figure 10–2.** Relationships between Unmodulated and Modulated Signals

an RF signal, but it contains modulating signal variations. The RF signal allows long-distance transmission, and the modulation has the desired intelligence.

### Modulation Factor

The variation in the AM signal compared with the unmodulated carrier is called the *modulation factor*, or *index of modulation*, designated by the letter $m$. Multiplying $m$ by 100 results in the percent modulation, expressed as follows:

$$\%m = \frac{I_{max} - I_{min}}{I_{max} + I_{min}} \times 100 \qquad \textbf{(10.1a)}$$

where

$I$ = current of the modulated carrier on the antenna

or

$$\%m = \frac{E_{max} - E_{min}}{E_{max} + E_{min}} \times 100 \qquad \textbf{(10.1b)}$$

where

$E$ = voltage of the modulated carrier on the antenna

Two examples of AM are shown in Figure 10–3, where Part A illustrates the unmodulated carrier. The varying amplitudes of the RF carrier wave result in an outline that corresponds to the audio modulating signal. The outline, as shown by the dashed lines at the top and bottom of the waveforms in Parts B and C, is called the modulation envelope. Note that both envelopes have the same variations, so they are symmetrical around the center axis of the carrier.

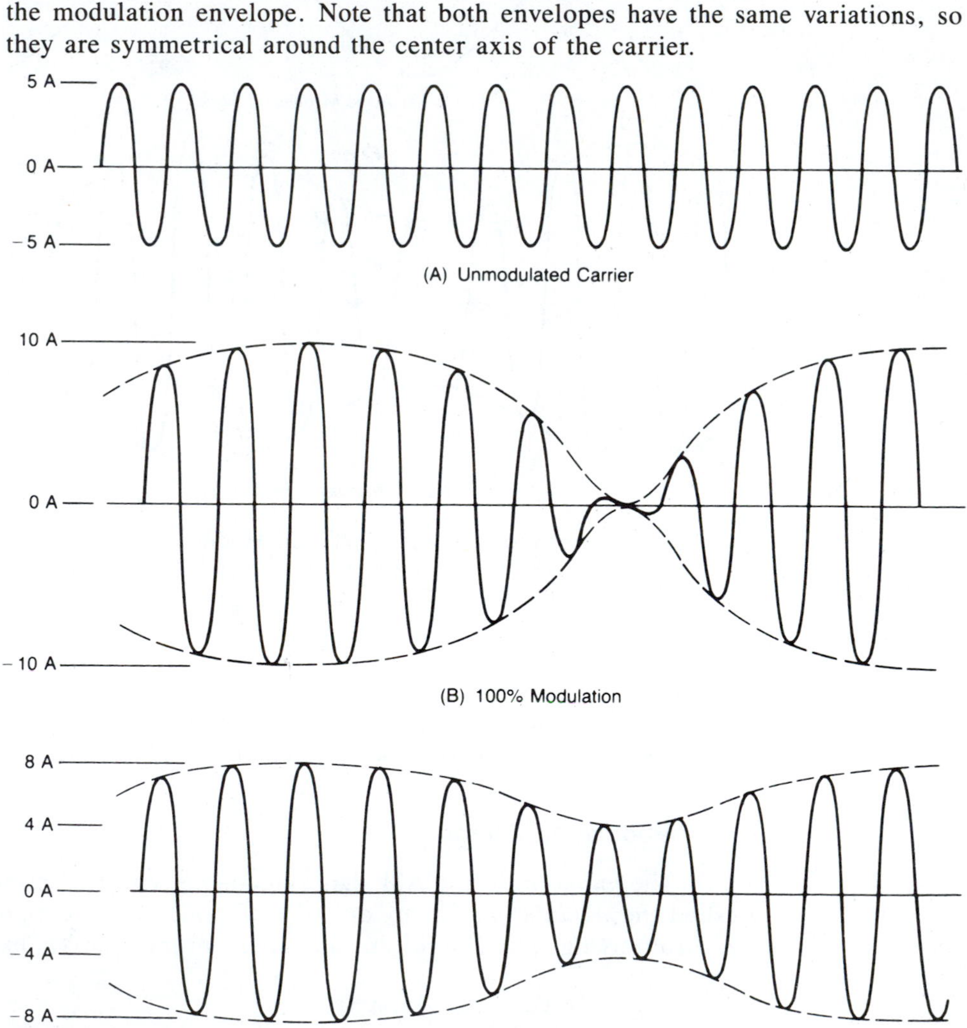


**Figure 10–3.** Relationships between Various Percent Modulation Signals

## Example 10.1

Assume a carrier current of 5 A. If the maximum current is 10 A and the minimum current is 0 A, what is the percent modulation?

*Solution*

Use Equation 10.1:

$$\%m = \frac{I_{max} - I_{min}}{I_{max} + I_{min}} \times 100 = \frac{10\text{ A} - 0\text{ A}}{10\text{ A} + 0\text{ A}} \times 100 = \frac{10\text{ A}}{10\text{ A}} \times 100$$
$$= 1 \times 100 = 100\%$$

Note that the RF signal for 100% modulation shown in Figure 10–3B varies from zero level to twice the level of the unmodulated carrier.

## Example 10.2

Now assume $I_{max} = 8$ A and $I_{min} = 4$ A. What is the $\%m$?

*Solution*

Use Equation 10.1:

$$\%m = \frac{I_{max} - I_{min}}{I_{max} + I_{min}} \times 100 = \frac{8\text{ A} - 4\text{ A}}{8\text{ A} + 4\text{ A}} \times 100 = \frac{4\text{ A}}{12\text{ A}} \times 100$$
$$= 0.33 \times 100 = 33\%$$

Figure 10–3C shows the results of Example 10.2.

### Overmodulation

In order for the most audio signal to be recovered by a receiver circuit, called the demodulator, or detector, it is important to have a high percentage modulation at the transmitter. For that reason modulation is generally maintained close to but not at 100%. However, the complexity of audio signals can cause the modulation to exceed 100%, a situation called *overmodulation*. This situation is demonstrated by Figure 10–4. Here the carrier signal is intermittently turned on and off, the effect of which is the generation of new frequencies that interfere with nearby channels and cause distortion of the transmitted intelligence.

## Frequency Modulation

FM differs from AM in the following way. The FM signal has a constant amplitude but varying frequencies above and below the center frequency of the carrier, and

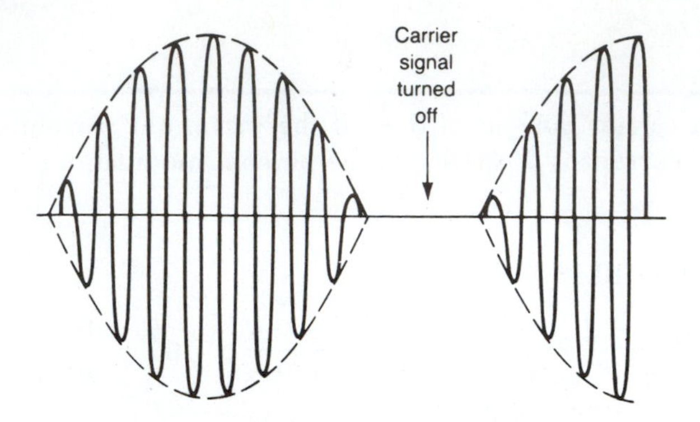


**Figure 10–4.** Overmodulated Signal

the AM signal has a varying amplitude but a constant carrier frequency. Figure 10–5 shows an FM signal. The changes in frequency of the RF carrier are produced by an audio-modulating voltage. The amount of change, called frequency deviation, increases with an increase in the audio-modulating voltage. The modulating voltage amplitude, not its frequency, determines the frequency deviation.

### Frequency Deviation

*Frequency deviation* is the amount of change from the center frequency. The center frequency, also called the rest frequency, is the frequency of the transmitted RF carrier without modulation, that is, when the modulating voltage is at its zero value. The peak audio-modulating voltage produces the peak frequency deviation. The total frequency deviation above and below center frequency is called the *frequency swing*.

A major advantage of FM transmission is that the signal is practically noise free, since most types of noise produce AM variations in a signal. In FM the desired signal is in the variations of frequency, therefore the noise does not affect the signal appreciably. However, to recover the desired signal without noise at

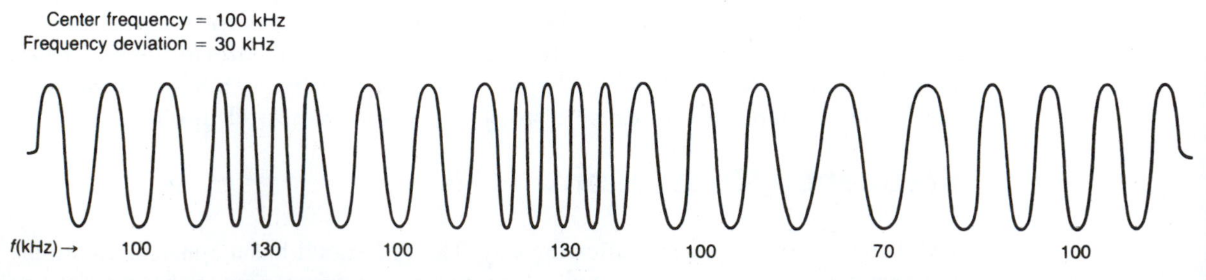


**Figure 10–5.** FM Signal

the receiver, you must have an FM detector circuit that responds to the changes in carrier frequency to recover the original audio signal, and a limiting circuit to remove amplitude variations in the FM signal. These circuits will be discussed later in this chapter.

### Phase Modulation

PM is, in effect, a modified version of FM, because any change in phase is equivalent to a change in frequency. The PM method generally uses CCOs to shift the phase angle of the RF output in step with the audio-modulating voltage. The RF carrier from the crystal oscillator is coupled into a reactive network where the audio input voltage shifts the bias above and below its average dc value, causing the phase angle of the RF carrier to be shifted from its average value. This action results in PM.

## 10.4 Demodulation

The modulated signal is transmitted, then is captured by a receiver in which it is demodulated, or detected. This process involves stripping the carrier wave from the intelligence information (audio signal) and amplifying the original signal to some usable level.

At the receiver, the information contained in the modulated signal must be separated from the carrier and passed on to a speaker or display device such as a television screen. This separation process is called demodulation, or detection. The circuits used for this process depend on the form of modulation used in the transmission of the signal. For AM signals, the circuit is called an *envelope detector*, and for FM signals, the circuit is referred to as a *discriminator*. Television signals are composites in which the audio information is FM, and the video information is AM. At the receiver, the audio and video signals are separated and channeled to different sections of the receiver to their proper audio and video demodulators.

A block diagram of an AM superheterodyne receiver is illustrated in Figure 10–6. The AM signal that operates in the 540–1600 kHz range is received by the antenna, is coupled into the RF amplifier, then coupled to the mixer where it is combined with a signal that is generated by a local oscillator. The oscillator and mixer stages combine to form the frequency-converter stage that produces an *intermediate frequency* (IF) of 455 kHz. Converting the received RF signal to an IF simplifies the design of the amplifier portion of the receiver. The IF signal is amplified and fed to the detector circuit, where the desired information is removed from the IF carrier.

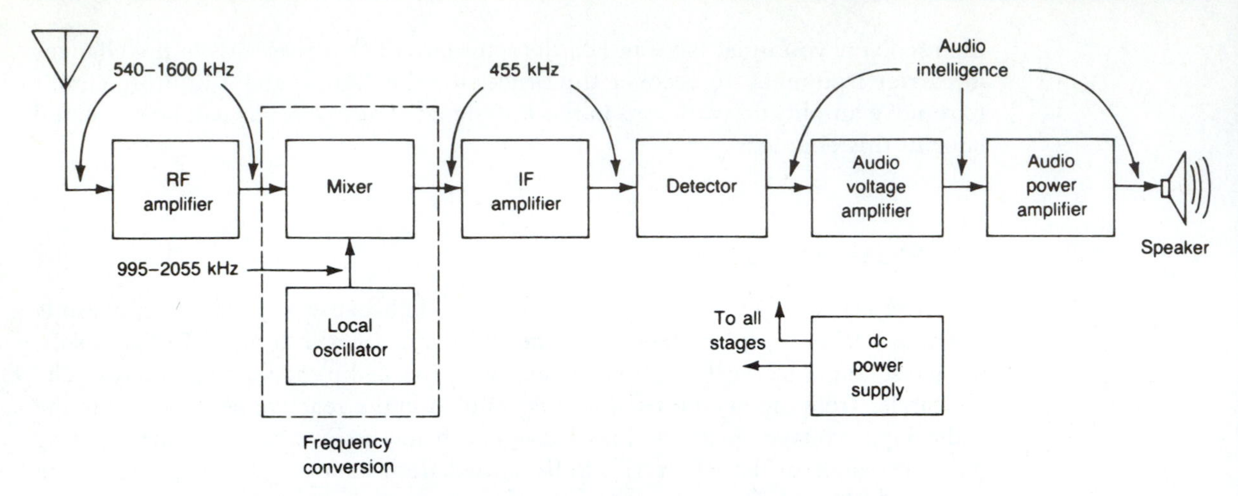


**Figure 10–6.** Block Diagram of an AM Superheterodyne Radio Receiver

## Diode Detector

A commonly used AM detector circuit is the *diode detector,* shown in Figure 10–7. The diode is used as a half-wave rectifier of the AM IF signal. The low-pass (LP) filter, composed of the RF parallel network, filters out the 455 kHz signal, but allows the audio signal to pass to the audio amplifier. As a result, the detector output has the desired audio intelligence.

## Automatic Gain Control

In addition to the intelligence signal, there is an average value dc level at the detector output that is used as a control signal for an *automatic gain control* (AGC) system. A feedback signal is sent to the RF amplifier to increase the amplifier gain when the dc level is low, or to decrease the gain when the dc level is high. In this manner, the output signal at the speaker is held relatively constant.

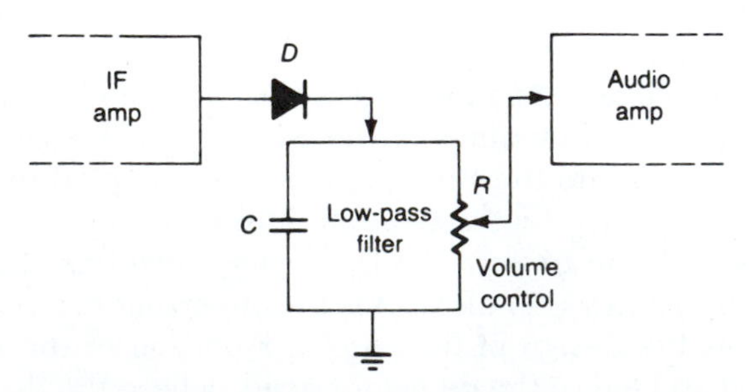


**Figure 10–7.** Diode Detector Circuit

## Single-Chip Radio Device

Many of the functions that previously required many discrete components and much board space are available now in a single LIC. One such single package device is the LM3820, illustrated in Figure 10–8. This is a three-stage AM radio IC that consists of an RF amplifier, an oscillator, a mixer, an IF amplifier, an AGC detector, and a zener regulator. The device was originally designed for use in slug-tuned automobile radio applications, but is also suitable for capacitor-tuned portable radios.

The circuit shown in Figure 10–9 that uses the LM3820 can be used as a starting point for portable radio designs. Loopstick antenna L1 is used in place of L0, and the RF amplifier is used with a resistor load to drive the mixer. A double-tuned circuit at the output of the mixer provides selectivity, while the remainder of the gain is provided by the IF section, which is matched to the diode through a unity-turn ratio transformer. $R_{AGC}$ may be used in place of $C_{AGC}$ to bypass the internal AGC detector and to provide more recovered audio. Coil specifications for Figure 10–9 are given in Figure 10–10.

An AM automobile radio design using the LM3820 is shown in Figure 10–11 (page 312). Tuning of both the input and the output of the RF amplifier and the mixer is accomplished with variable inductors (dashed lines). Selectivity is improved through the use of double-tuned interstage transformers. Input circuits are inductively tuned to prevent *microphonics* (electrical noises caused by vibrations in electronic components) and to provide a linear tuning motion to facilitate push-button operation.

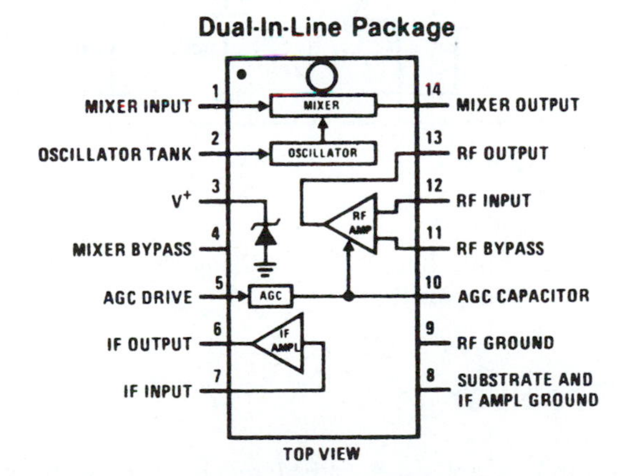


**Figure 10–8.** Connection Diagram of an LM3820 AM Radio Receiver System (Courtesy of National Semiconductor Corporation)

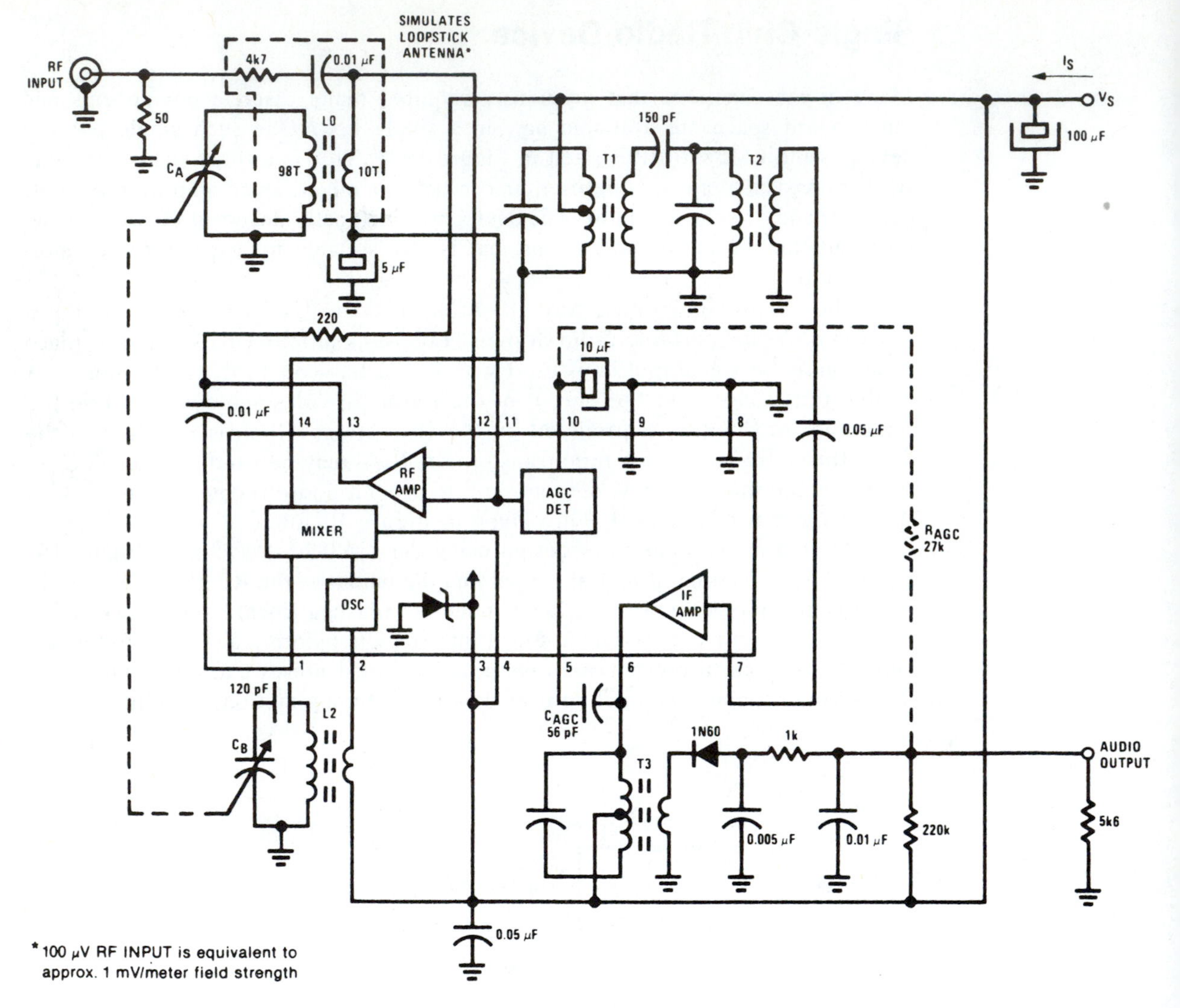


Figure 10–9. Capacitor-Tuned AM Portable Radio Design Using the LM3820 (Courtesy of National Semiconductor Corporation)

## FM Radio Receiver

Superheterodyne circuits are used in FM receivers as well as in AM receivers. The IF in FM receivers is much higher at 10.7 MHz, because the RF carrier frequencies for FM radio broadcasting are in the 88–108 MHz band. Figure 10–12 (page 313) shows a block diagram of an FM radio receiver.

The received signal is fed from the antenna into an RF tuner that includes the RF amplifier, local oscillator, and mixer stages. The RF amplifier improves

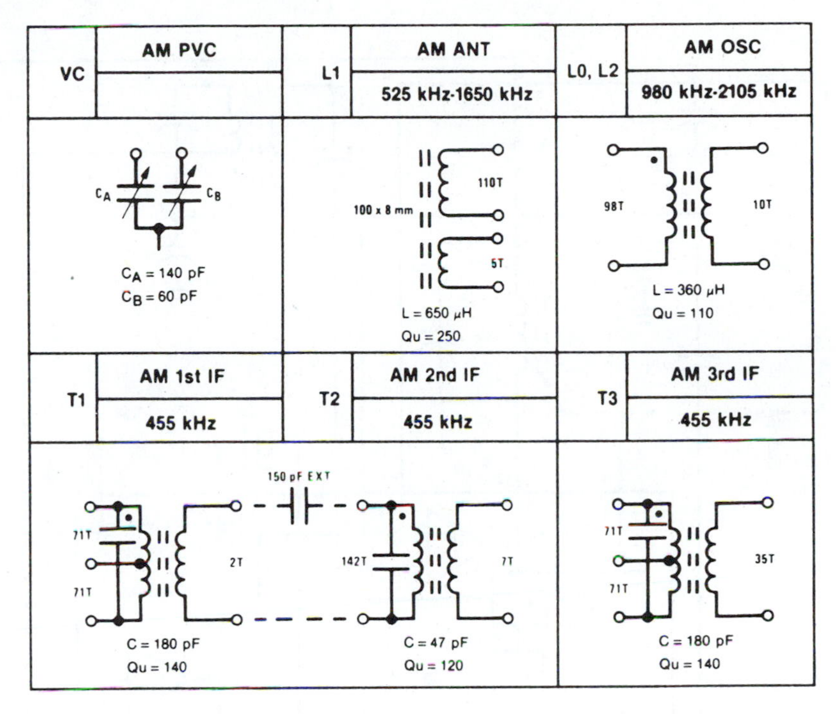


**Figure 10–10.** Coil Specifications for Figure 10–9 (Courtesy of National Semiconductor Corporation)

the signal-to-noise ratio ($S/N$), and the frequency-converter section converts the RF signal to an IF signal at 10.7 MHz. The mixer output is the input to the IF section.

### Limiter Stage

The *limiter stage* is an IF amplifier tuned to the 10.7 MHz IF. The limiter provides a relatively constant output level for different input levels. The stage operates between saturation and cutoff, and it usually has signal bias that automatically adjusts itself to the amount of signal.

### Detector Stage

The *detector stage* receives the amplified IF signal from the limiter. This circuit recovers the audio-modulating signal by allowing the frequency variations in the signal to provide equivalent variations in amplitude that can be rectified by a diode. This stage generally uses two diodes in a balanced detector circuit, such as the *Foster-Seeley discriminator* or the *ratio detector.* The Foster-Seeley discriminator and ratio detector circuits are not used in constructing LICs, therefore they will not be discussed further.

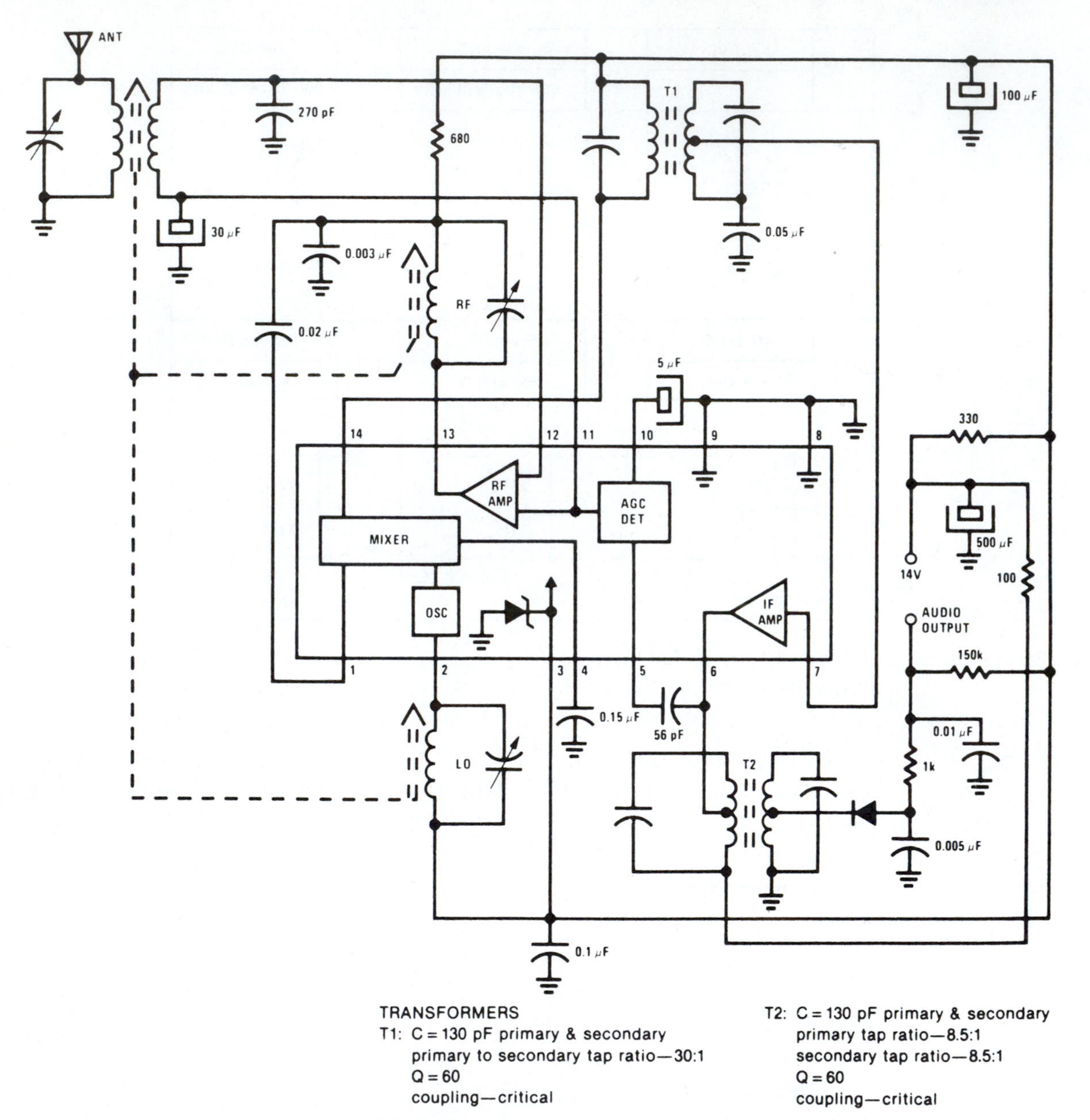


**Figure 10–11.** Slug-Tuned AM Car Radio Design Using the LM3820 (Courtesy of National Semiconductor Corporation)

### Quadrature Detector

For today's FM systems, a discriminator circuit called the *quadrature detector* has been designed. The quadrature detector circuit is used in most modern ICs that contain FM discriminators.

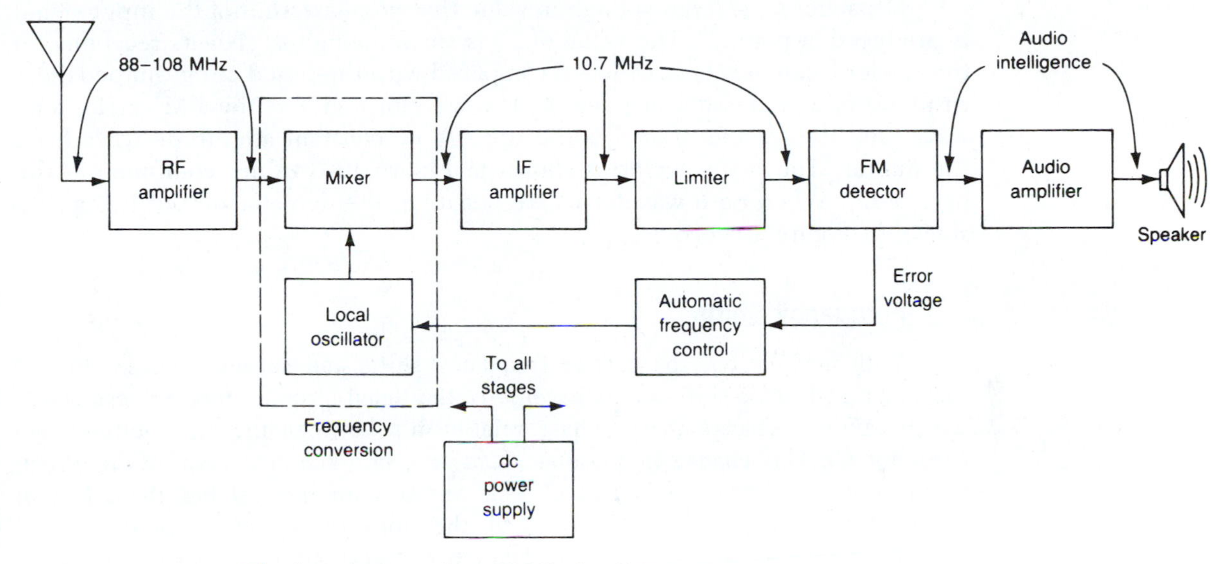


**Figure 10–12.** Block Diagram of an FM Radio Receiver

As shown in Figure 10–13, the quadrature detector circuit is basically a difference amplifier formed by $Q_1$ and $Q_2$, with the output taken at the collector of $Q_2$. With the current in $R_5$ being constant, any change in the current of $Q_1$ results in an equal but opposite change in $Q_2$. The quadrature circuit provides the audio output when an FM carrier is applied to the input.

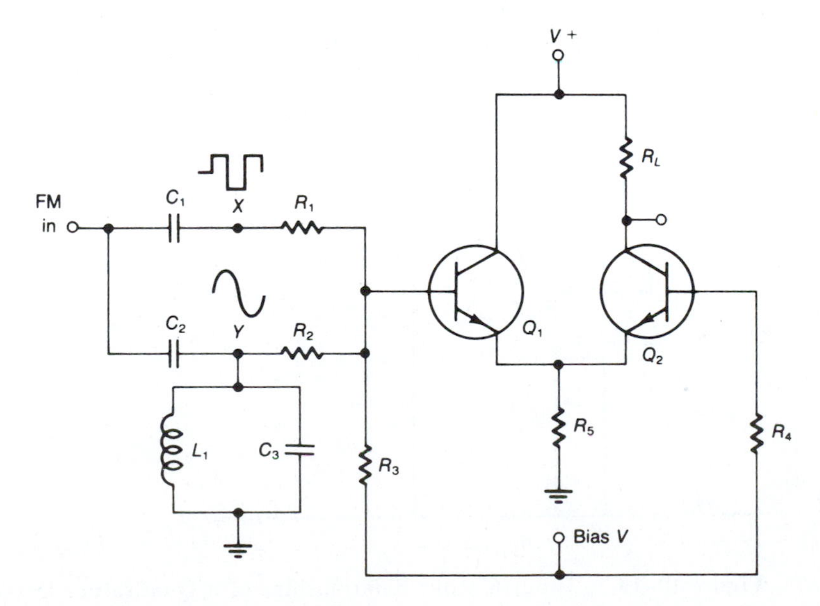


**Figure 10–13.** Quadrature Detector Circuit

Capacitor $C_1$ is large enough in value that no phase shift of the input signal is produced at point $X$. The value of $C_2$ is small enough so that its reactance to the carrier frequency is large when compared with the tuned circuit impedance, resulting in a phase shift at point $Y$. The waveform at $Y$ is the FM carrier sine wave, and the waveform at $X$ is the squared-off constant amplitude wave from the limiter. The two waveforms shown in Figure 10–14A are combined at the base of $Q_1$. When both waveforms are negative, the detector output is high, as shown in Figure 10–14B.

### Frequency Shifts

With modulation, the carrier frequency shifts and causes a phase shift of the signal at $Y$. This shift occurs because of the tuned circuit going off resonance, which causes a change in the phase relationship between the tuned circuit and capacitor $C_2$. This change in phase at $Y$ causes a variance in interval of the output when both waveforms are negative. This variance in interval has the effect of varying the output pulse width (PW), and, therefore, the average value of the output voltage. The change in the pulse average voltage is proportional to the amplitude of the original modulation, since it is the amplitude of the modulating signal voltage that causes the frequency to change. The audio can now be separated from the pulse output through an LP filter.

Using LIC devices that contain quadrature detectors requires a minimum of external components. One such device is the CA2111A shown in Figure 10–15. This circuit contains a limiter-amplifier, a quadrature detector, and an emitter-follower stage that provides a low-impedance output for driving an external audio amplifier. The detector circuit can be tuned easily because of the external coil.

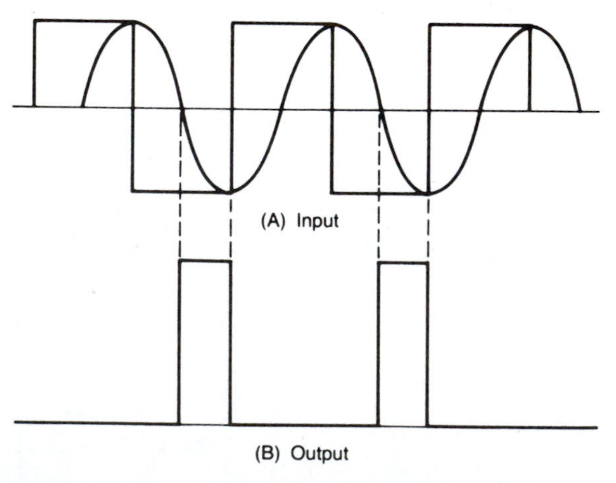


**Figure 10–14.** Input/Output Relationships of a Quadrature Detector

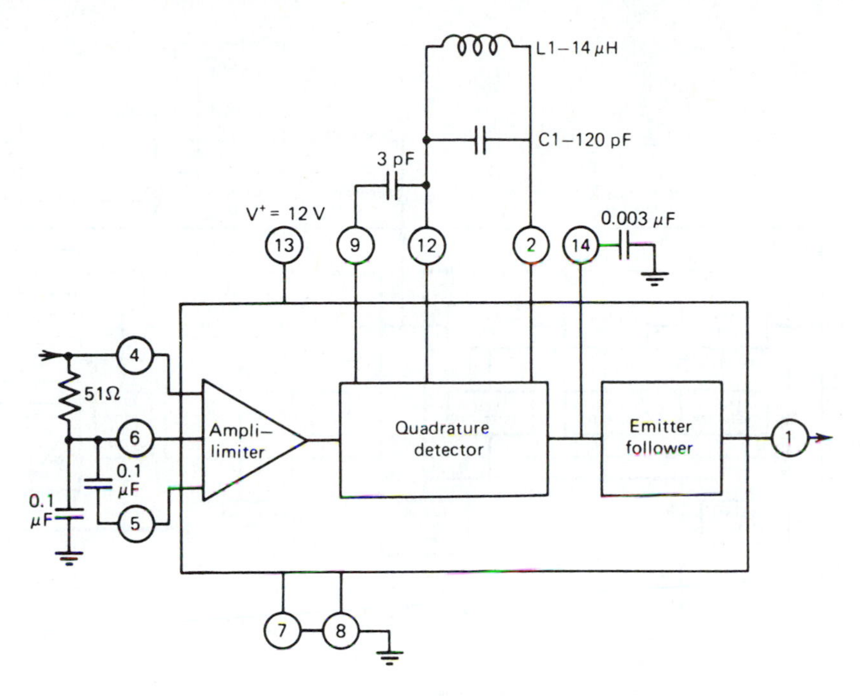


**Figure 10–15.** CA2111A Quadrature Detector IC (Courtesy of RCA Corporation)

## A Modern FM-IF System

Taking modern sophistication one step further results in such ICs as the LM3189N illustrated in Figure 10–16. This IC provides all the functions of a comprehensive FM-IF system. It includes a three-stage FM-IF amplifier/limiter configuration with level detectors for each stage, a doubly-balanced quadrature FM detector, and an audio amplifier that features the optional use of a *muting* (squelch) circuit.

The advanced design of this IC provides such desirable features as programmable delayed AGC for the RF tuner, an *automatic frequency control* (AFC) drive circuit, and an output signal to drive a tuning meter or provide stereo switching logic. In addition, internal power supply regulators maintain a nearly constant current drain over the voltage supply range of +8.5 - 16 V. This circuit is ideal for high-fidelity operation, providing single-coil tuning capability with the external detector coil.

# 10.5 A Balanced Modulator/Demodulator

The MC1496 is a single IC that incorporates both the modulation and demodulation functions. Figure 10–17 shows the pin configuration and equivalent schematic.

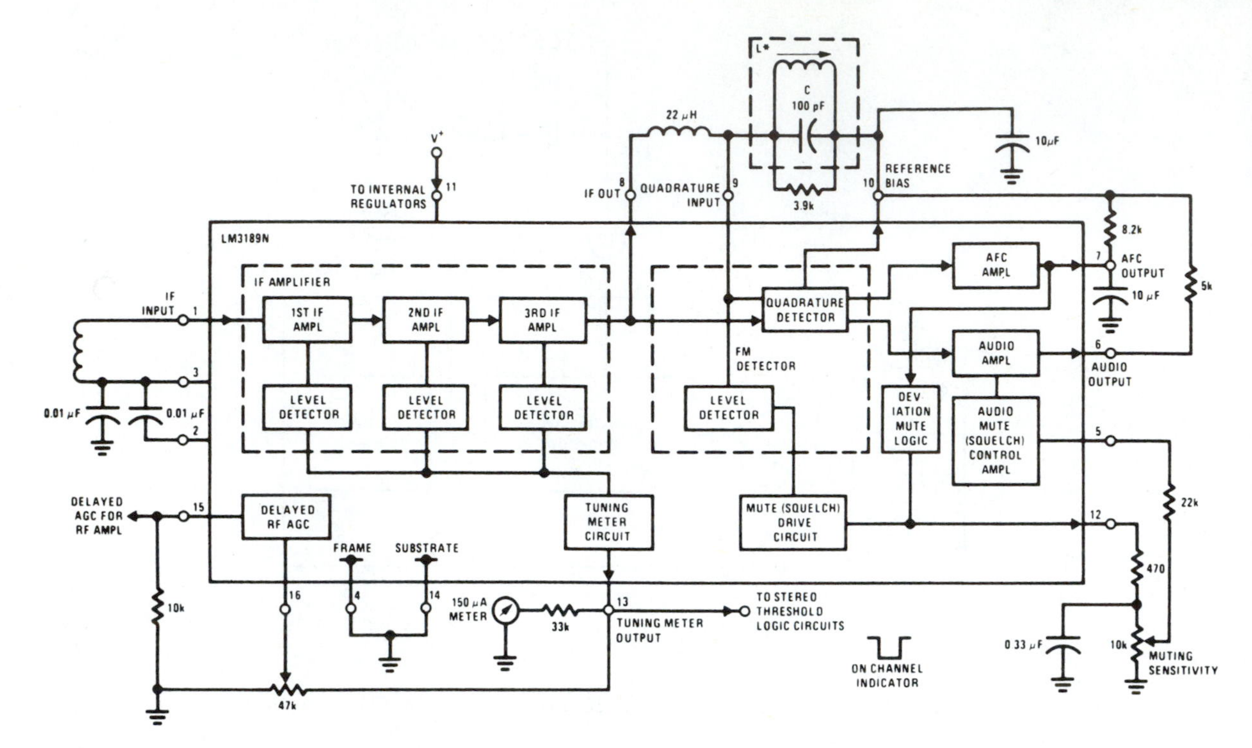


All resistance values are in ohms

* L tunes with 100 pF (C) at 10.7 MHz, $Q_0 \cong 75$
(Toko No. KACS K586HM or equivalent)

**Figure 10–16.** LM3189N FM-IF System (Courtesy of National Semiconductor Corporation)

This device is a monolithic, double-balanced modulator/demodulator designed for use where the output voltage is a product of an input voltage (signal) and a switched function (carrier). Its features include adjustable gain and signal handling, typical carrier suppression of 65 dB at 0.5 MHz and 50 dB at 10 MHz, balanced inputs and outputs, and high (85 dB) common-mode rejection.

Applications include suppressed carrier and amplitude modulation, synchronous detection, AM modulation and demodulation, FM detection, phase detection, sampling, and frequency doubling. It can operate from either single supply or dual supply systems.

For the theory of operation and application schematics, see application note AN189 in Appendix E.

# 10.6 Compandors

A *compandor* is a gain control device that is used for dynamic gain expansion or compression. The name is derived from its two functions—compression and

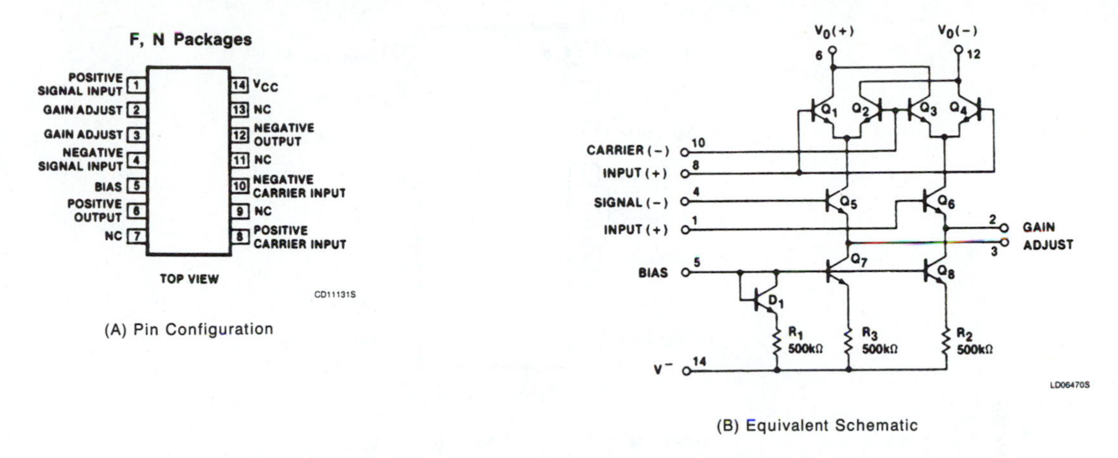


(A) Pin Configuration

(B) Equivalent Schematic

**Figure 10–17.** MC1496/1596 Balanced Modulator/Demodulator (Courtesy of Signetics Company)

expansion. Applications are in telephone subscriber and trunk carrier systems, communications systems, high-quality audio systems, and video recording systems.

## Compandor Building Blocks

The NE570/571 compandor is shown in Figure 10–18. The block diagram shows the internal building blocks for one-half of the dual-channel device. Each channel has a full-wave rectifier, a variable gain cell, an op amp, and a bias system. The arrangement of these blocks in the IC results in a circuit that can perform well with few external components, yet can be adapted to many diverse applications.

### Rectifier Stage

The full-wave rectifier rectifies the input current that flows from the rectifier input (pin 2) to an internal summing node that is biased at $V_{ref}$. The rectified current is averaged on an external filter capacitor tied to the $C_{rect}$ terminal (pin 1), and the average value of the input current controls the gain of the variable gain cell ($\Delta G$ cell). The gain will thus be proportional to the average value of the input signal for capacitively coupled inputs. Since capacitively coupled inputs have no error-producing offset voltage, any error will come from the internally supplied bias current of the rectifier, which is less than 0.1 μA. Expressed in equation form,

$$G \propto \frac{|V_{in} - V_{ref}|\ \text{av}}{R_1} \qquad \textbf{(10.2a)}$$

or

$$G \propto \frac{|V_{in}|\ \text{av}}{R_1} \qquad \textbf{(10.2b)}$$

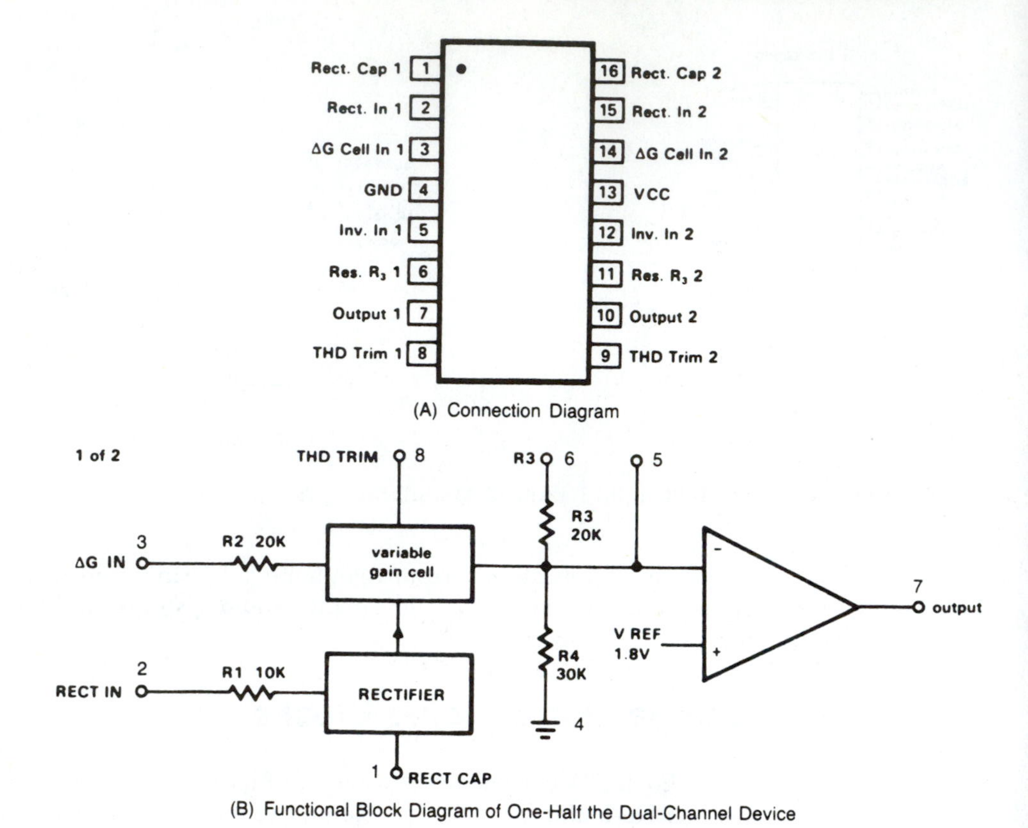


**Figure 10–18.** NE570/571 Dual-Channel Compandor (Courtesy of Signetics Company)

The speed with which gain changes to follow changes in input signal levels is determined by the rectifier filter capacitor at pin 1. A small value capacitor will yield rapid response but will not fully filter low-frequency signals. Any ripple on the gain control signal will modulate the signal passing through the variable gain cell. In an expandor or compressor application, this would lead to third harmonic distortion (THD), so there is a trade-off to be made between fast attack and decay times, and distortion. For step changes in amplitude, the change in gain with time is as follows:

$$G(t) = (G_{\text{initial}} - G_{\text{final}})\, e^{-t/\tau} + G_{\text{final}} \tag{10.3}$$

where

$$\tau = 10\ \text{k}\Omega \times C_{\text{rect}}$$

### Gain Cell Stage

The variable gain cell is a current in-current out device with the ratio $I_{out}/I_{in}$ controlled by the rectifier. $I_{in}$ is the current flowing from the $G$ input (pin 3) to an internal summing node biased to $V_{ref}$. For capacitively coupled inputs,

$$I_{in} = \frac{V_{in} - V_{ref}}{R_2} \quad \textbf{(10.4a)}$$

or

$$I_{in} = \frac{V_{in}}{R_2} \quad \textbf{(10.4b)}$$

The output current, $I_{out}$, is fed to the summing node of the op amp.

A compensation scheme built into the $\Delta G$ cell compensates for temperature and cancels out odd harmonic distortion. The only distortion that remains is even harmonics, and they exist only because of internal offset voltages. The THD trim terminal (pin 8) provides a means for nulling the internal offsets for low distortion operation.

### Op Amp Stage

The op amp (which is internally compensated) has the noninverting input tied to $V_{ref}$ and the inverting input connected to the $\Delta G$ cell output as well as brought out externally (pin 5). Resistor $R_3$ is brought out from the summing node (pin 6) and allows compressor or expandor gain to be determined only by internal components.

### Output Stage

The output stage is capable of ±20 mA output current. This allows a +13 dBm (3.5 $V_{rms}$) output (pin 7) into a 300 Ω load that, with a series resistor and proper transformer, can result in +13 dBm with a 600 Ω output impedance.

### Reference Voltage

A band gap reference provides the reference voltage for all summing nodes, a regulated supply voltage for the rectifier and $\Delta G$ cell, and a bias current for the $\Delta G$ cell. The low temperature coefficient (tempco) of this type of reference provides very stable biasing over a wide temperature range.

## Programmable Analog Compandor

A programmable analog compandor is the NE572 shown in Figure 10–19. This is a dual-channel high-performance device in which either channel may be used for dynamic gain compression or expansion. Each channel has a full-wave rectifier to detect the average value of input signal, a linearized temperature-compensated variable gain cell ($\Delta G$), and a dynamic time-constant buffer. The buffer permits independent control of dynamic attack and recovery time with minimum external components and improved low-frequency gain control ripple distortion.

The NE572 is designed to reduce noise in high-performance audio systems, but it can be used in a wide variety of such applications as a voltage control amplifier, a stereo expandor, an automatic level control, a high-level limiter, a low-noise gate, and a state-variable filter.

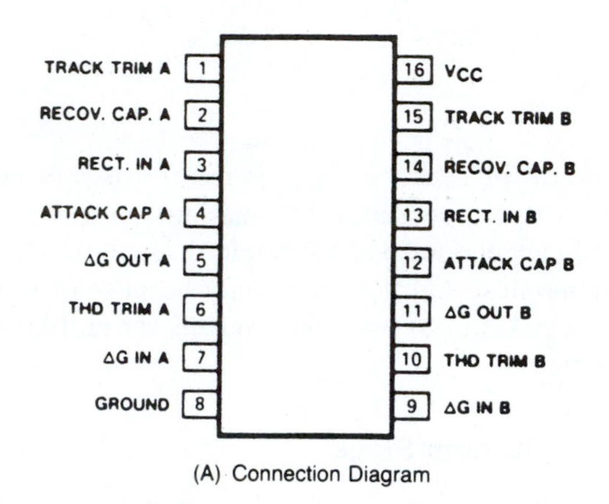


(A) Connection Diagram

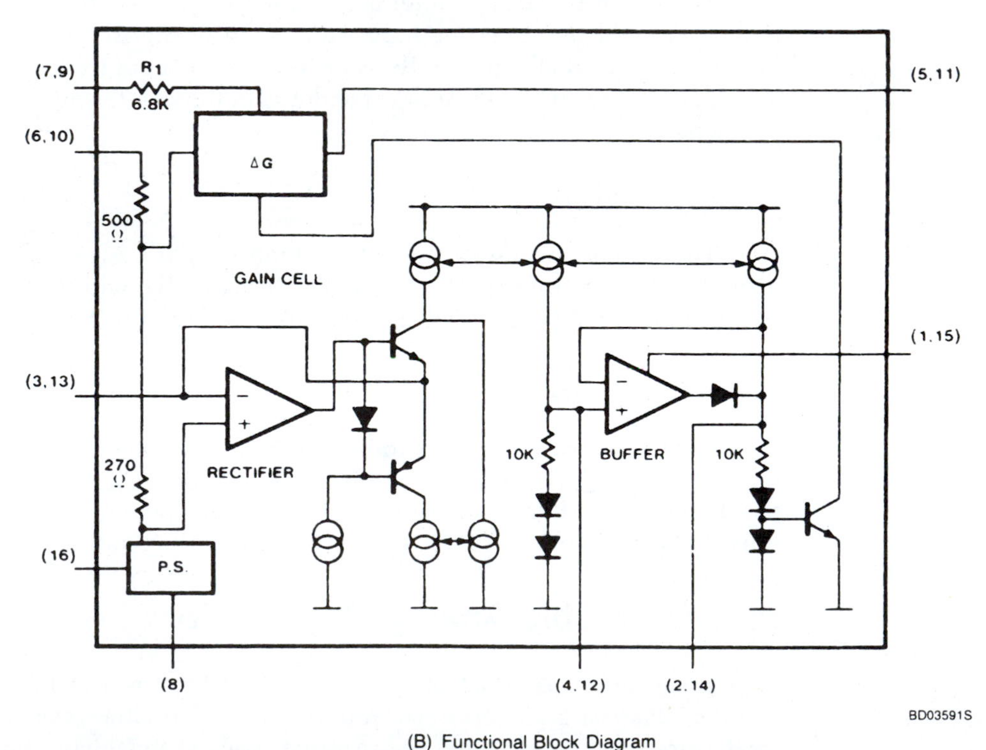


(B) Functional Block Diagram

Figure 10–19. NE572 Programmable Analog Compandor (Courtesy of Signetics Company)

# 10.7 Phase-Locked Loops

The *phase-locked loop* (PLL) is basically an electronic feedback loop that consists of a phase detector (or comparator), an LP filter, a dc amplifier, and a voltage-controlled oscillator (VCO). The basic block diagram of a PLL is shown in Figure 10–20. The components between the input and output are considered to be in the forward path of the loop, and the single connection between the VCO and the phase detector is the feedback path.

The purpose of the PLL circuit is to make a variable-frequency oscillator (VFO) lock in at the frequency and phase angle of a standard frequency ($f_s$) used as a reference. The oscillator will then have the same frequency accuracy as the referenced standard.

## PLL Functional Blocks

To better understand the operation of the PLL, we will examine each of the four functional blocks of the PLL—the phase detector, the LP filter, the dc amplifier, and the VCO.

### Phase Detector

All PLL systems, whether analog or digital, use a phase detector, or comparator, circuit to generate the dc control voltage. The basic difference between the analog and digital PLLs is the type of phase detector used. In general, digital systems use either an exclusive-OR gate, or some type of edge-triggered phase detector, and the analog system uses a double-balanced mixer. Basically, the analog phase detector uses two diodes in a balanced rectifier circuit. The phase detector output voltage, $V_{out}$, is proportional to the phase difference between the two inputs, and is determined as follows:

$$V_{out} = K_c \Delta\phi \qquad \textbf{(10.5)}$$

where

$K_c$ = phase detector conversion gain in volts/radian (V/rad)
$\Delta\phi$ = the input phase difference in radians (rad)
1 rad = 57.3°

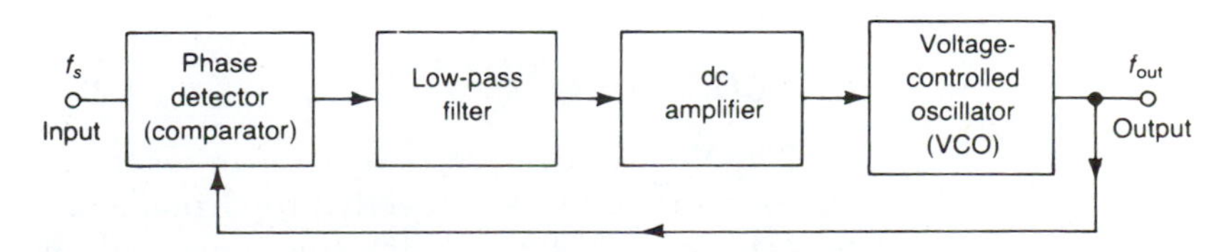


**Figure 10–20.** Block Diagram of a PLL

## Example 10.3

A phase detector has a conversion gain of 20 mV and an input phase difference of 0.15 rad. Calculate the phase detector output voltage.

*Solution*

Use Equation 10.5:

$$V_{out} = K_c\Delta\phi = 20 \text{ mV/rad} \times 0.15 \text{ rad} = 3 \text{ mV}$$

## Example 10.4

Calculate the phase detector output for a conversion gain of 15 mV and an input phase difference of 20°.

*Solution*

First, convert the phase difference from degrees to radians:

$$\frac{20°}{57.3°/\text{rad}} = 0.349 \text{ rad}$$

Next, use Equation 10.5:

$$V_{out} = K_c\Delta\phi = 15 \text{ mV/rad} \times 0.349 \text{ rad} = 5.2 \text{ mV}$$

### Low-Pass Filter

The LP filter network can be either passive or active. It serves two major functions: first, it removes the ac signal variations of the two oscillators, or traces of higher frequency noise, from the rectified dc output voltage of the phase detector; and second, it controls the lock, capture, BW, and transient response of the loop. That is, the filter determines the dynamic performance of the loop.

### dc Amplifier

The dc amplifier circuit amplifies the filtered dc control voltage to the desired level for better control and in the polarity needed for the varactor in the VCO.

### VCO

The VCO circuit uses a varactor to set the oscillator frequency. The output frequency of the VCO is directly proportional to its input control voltage, which keeps the VCO locked into the frequency and phase of the referenced standard oscillator. The VCO is also termed a *voltage-to-frequency converter*.

The VCO circuit is used for electronic tuning of the oscillator frequency. The *varactor,* also called a *varicap,* is a semiconductor capacitive diode that operates on the principle of a varying capacitance that is inversely proportional to the amount of reverse dc voltage applied. The reverse voltage applied can be either negative at the anode or positive at the cathode of the diode. The more reverse voltage that is applied, the wider the depletion area of the P-N junction. The effect of the widened depletion area is equivalent to an increased distance between capacitor plates, which produces less capacitance.

## PLL Operating Modes

There are three operating modes for a PLL. They are *free-running, capture,* and *phase-lock* (sometimes called lock-in or tracking). If the VCO output frequency ($f_{out}$) is too far from the standard frequency ($f_s$), the PLL cannot lock in the oscillator. Without such lock-in, the VCO is in the free-running mode. Once the control voltage from the dc amplifier starts to change the VCO frequency, the oscillator is in the capture mode. When $f_{out}$ is exactly the same as $f_s$, the VCO is in lock-in and the PLL is in the phase-locked mode. The PLL will remain in the phase-locked mode as long as the dc control voltage is applied.

### Lock Range and Capture Range

The frequency range over which the PLL can follow the incoming signal is called the *lock range.* The BW over which capture is possible is called the *capture range.* The capture range can never be wider than the lock range. To keep the capture range narrower than the lock range, the phase detector is used to compare the two frequencies. Any difference in phase causes an error signal at the output of the phase detector. This error signal is a dc voltage that is proportional to the difference in frequency and phase of the standard oscillator and the VCO. The dc error voltage is used to correct the VCO frequency by forcing it to change in a direction that reduces the frequency difference between the input oscillator and the VCO.

The LP filter circuit removes the ac variations of the two oscillators, or traces of higher frequency noise from the rectified dc output of the phase detector. The output of the filter is a filtered dc control voltage that is fed to the dc amplifier, where it is increased in amount to provide better control. The amplifier output provides the desired dc level and polarity for the control voltage needed for the varactor in the VCO.

The VCO uses the varactor to set the oscillator frequency. Input from the dc amplifier keeps the VCO locked into the frequency of the reference oscillator. Although locked in, there is always a finite phase difference between the input and output.

## A General-Purpose VCO

A popular monolithic IC VCO is the LM566C shown in Figure 10–21. This is a general-purpose VCO that may be used to generate square and triangle waves. The frequency of the square and triangle waves is a very linear function of a control voltage and of an external resistor and capacitor. The LM566C operates over a supply voltage range of 10–24 V, and provides very linear modulation characteristics, high-temperature stability, and excellent power supply rejection ratio (PSRR). It has a 10-to-1 frequency range with fixed capacitor, and frequency is programmable by means of current, voltage, resistor, or capacitor. It can be operated from either a single supply or a split (±) power supply. Its applications include FM modulation, signal generation, function generation, frequency shift keying (FSK), and tone generation.

## A Four-Block VCO

Another monolithic IC VCO is the XR-2207 shown in Figure 10–22. The circuit comprises four functional blocks: a VFO that generates the basic periodic waveforms; four current switches that are actuated by binary keying inputs; and two buffer amplifiers for simultaneous triangle and square wave outputs, available over a range of 0.01 Hz to 1 MHz. The internal switches transfer the oscillator current to any of four external timing resistors to produce four discrete frequencies that are selected according to the binary logic levels at the keying terminals. This device is

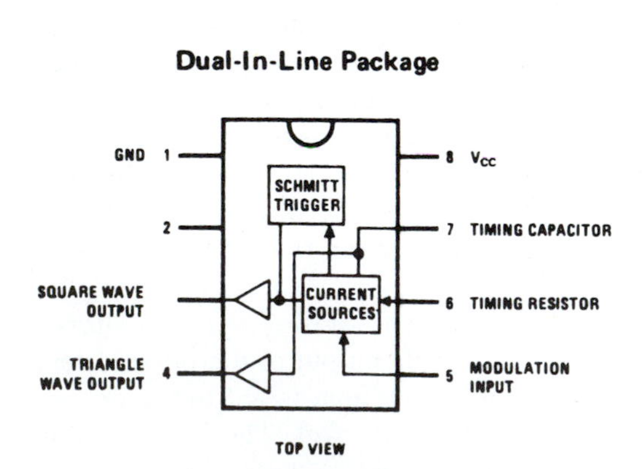


Figure 10–21. Block Diagram of the LM566C VCO (Courtesy of National Semiconductor Corporation)

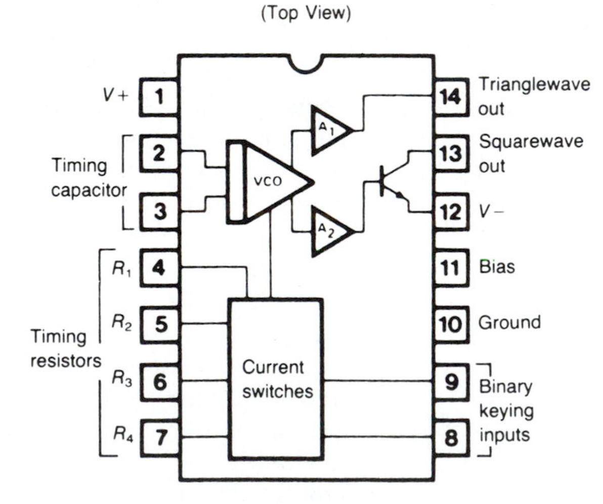


Figure 10–22. Connection and Block Diagram for the XR-2207 VCO (Courtesy of Raytheon Company, Semiconductor Division

ideally suited for FM, FSK, and sweep or tone generation, as well as for PLL applications.

# 10.8 PLL Applications

There are many applications for the PLL circuit. PLLs are ideally suited for routine applications as AM and FM detectors, FSK decoders, signal conditioners, prescalars for frequency counters, tone detectors, and touch-tone decoders. In general, when a PLL is used to control the frequency of an oscillator, such as in radio and TV receiver sound systems, it is referred to as *automatic frequency control* (AFC). In the RF tuner section of TV receivers the PLL is referred to as *automatic fine tuning* (AFT), and in the horizontal synchronizing circuit, it is called the horizontal AFC.

## Frequency Synthesis

An important application of the PLL is in communication systems, where a CCO, such as the MC4024 shown in Figure 10–23, is the standard source for a reference. The PLL circuit then provides an oscillator without a crystal with the same frequency stability as the crystal reference oscillator. This procedure is called *frequency synthesis,* that is, putting together, or mixing, two frequencies to provide the desired output.

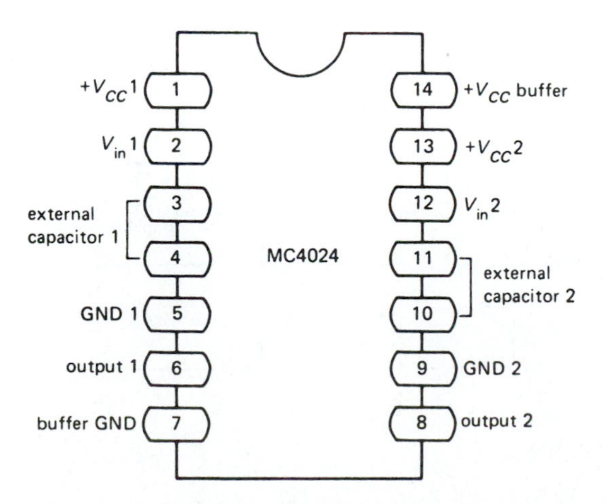


**Figure 10–23.** MC4024 VCO (Courtesy of Breton Publishers)

A basic frequency synthesizer is illustrated in Figure 10–24, where a divide-by-$N$ ($\div N$) counter is inserted in the feedback path of a PLL. The output frequency of the synthesizer is

$$f_{out} = Nf_{in} \tag{10.6}$$

where

$N$ = the divider value of the $\div N$ counter
$f_{in}$ = input frequency (generally from a CCO)

The synthesizer phase detector produces a dc control voltage that is proportional to the phase difference between the input frequency, $f_{in}$, and the $\div N$ counter output, $f_{out}/N$. The counter generates a single output pulse for every $N$ input pulse. The dc control voltage from the phase detector, after filtering and amplification, then controls the output frequency of the VCO. The output signal from the $\div N$ counter is equivalent to the reference input frequency, except for the following small phase difference:

$$f_{in} = \frac{f_{out}}{N} \tag{10.7}$$

## Example 10.5

A frequency synthesizer has an input frequency of 750 kHz. The $\div N$ counter is set at a count of 50. Calculate the synthesizer frequency output.

*Solution*

Use Equation 10.6:

$$f_{out} = Nf_{in} = 50 \times 750 \text{ kHz} = 37.5 \text{ MHz}$$

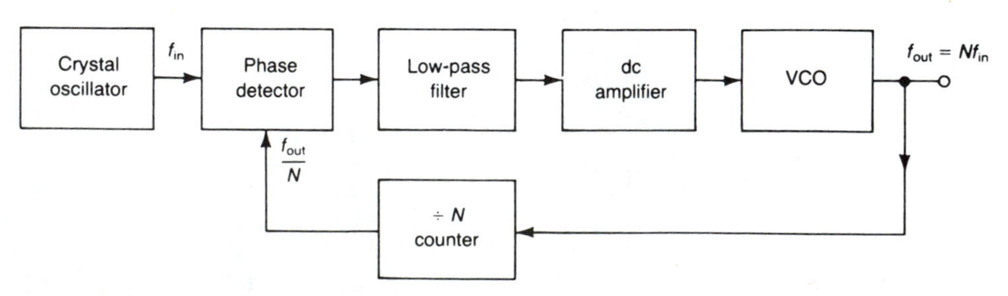


**Figure 10–24.** A Frequency Synthesizer

## Example 10.6

What is the input frequency of the phase detector if the output frequency of a synthesizer is 200 MHz and the $\div N$ counter is set at a count of 400?

*Solution*

Use Equation 10.7:

$$f_{\text{in}} = \frac{f_{\text{out}}}{N} = \frac{200\text{ MHz}}{400} = 500\text{ kHz}$$

## Data Communications

The XR-2211 shown in Figure 10–25 is a monolithic PLL system designed especially for use in data communications. It operates over a wide supply voltage range of 4.5–20 V and a wide frequency range of 0.01 Hz to 300 kHz. It can accommodate analog signals between 2 mV and 3 V, and can interface with conventional logic families. The circuit consists of a basic PLL for tracking an input signal frequency within the passband, a quadrature phase detector that provides carrier detection, and an FSK voltage comparator that provides FSK demodulation. External components are used to independently set carrier frequency, BW, and output delay. Design equations and typical applications of the XR-2211 are available in the data sheets in Appendix F.

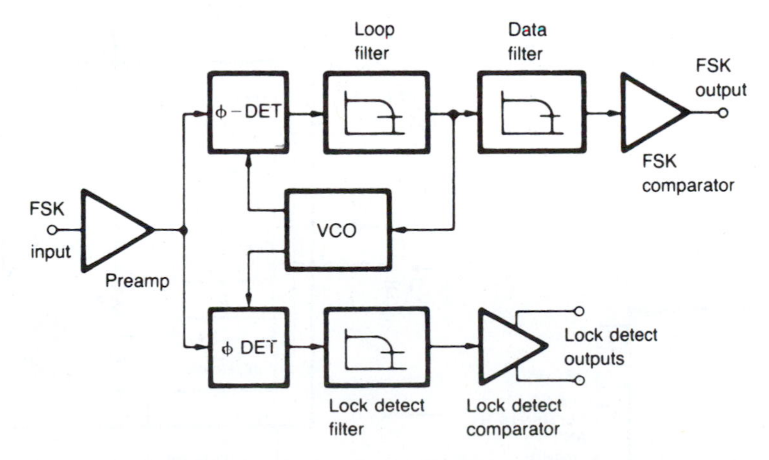


**Figure 10–25.** Functional Block Diagram of the XR-2211 PLL (Courtesy of Raytheon Company, Semiconductor Division)

## 560 Series PLL

A group of monolithic analog PLL devices are the 560 series, which use the double-balanced mixer. You will recall that the output of this circuit is an average dc voltage that is proportional to the phase difference between its two inputs. We will discuss the most popular of the series, the 565 and the 567, illustrated in Figure 10–26 and Figure 10–27, respectively.

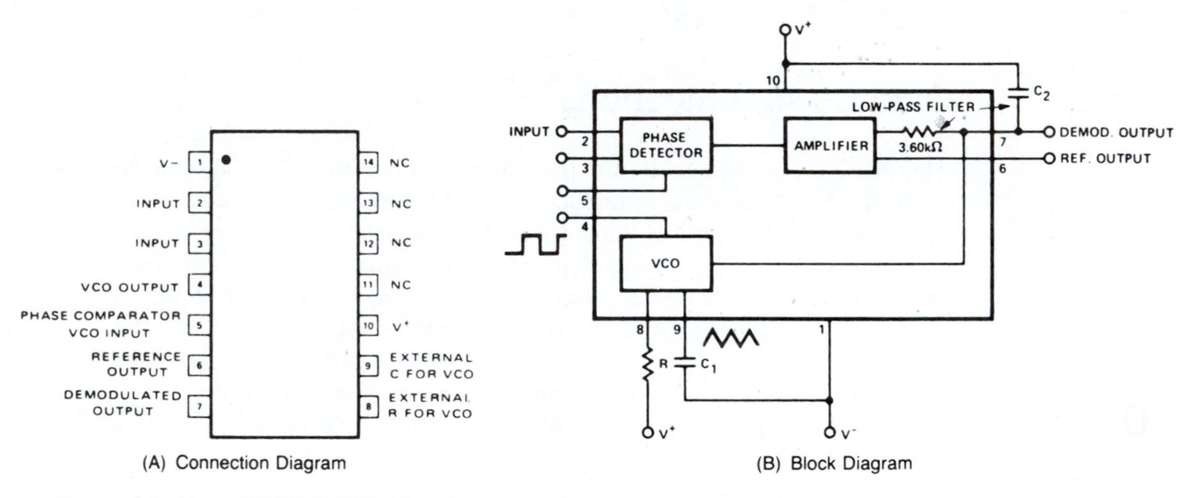


Figure 10–26. SE/NE565 PLL (Courtesy of Signetics Company)

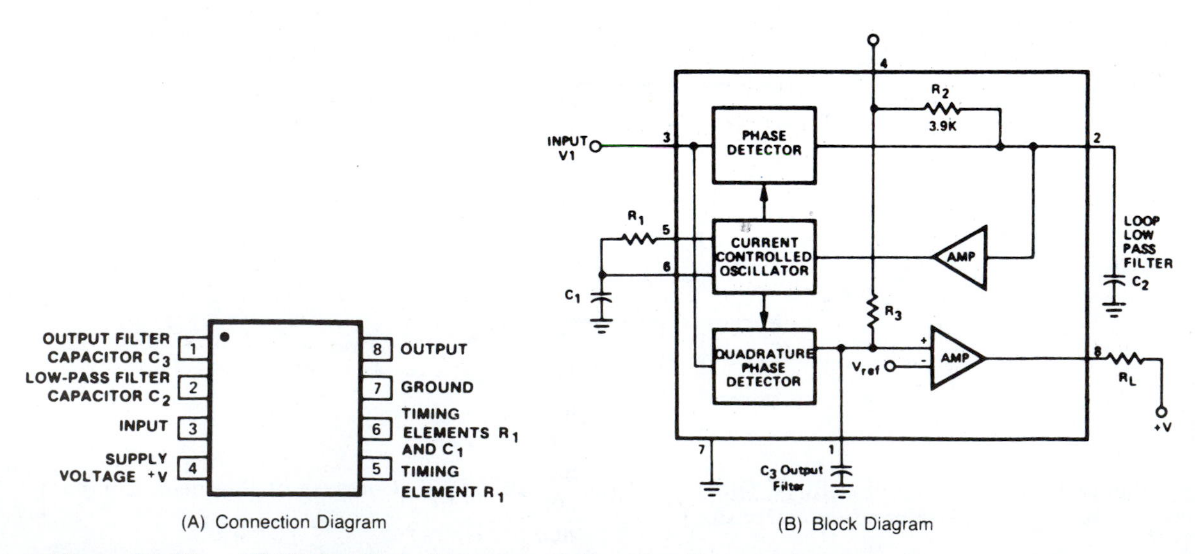


Figure 10–27. SE/NE567 PLL (Courtesy of Signetics Company)

The 565 is a general-purpose PLL that contains a stable, highly linear VCO for low distortion FM demodulation, and a double-balanced phase detector with good carrier suppression. The VCO frequency is set with an external resistor and capacitor, and a tuning range of 10:1 can be obtained with the same capacitor. The BW, response speed, capture, and pull-in range can be adjusted over a wide range with an external resistor and capacitor. Inserting a digital frequency divider between the VCO and the phase detector will provide frequency synthesis.

Other applications of the 565 are data and tape synchronization, FSK demodulation, FM demodulation, tone decoding, frequency multiplication and division, telemetry reception, and signal generation. For design considerations, see the data sheets in Appendix G.

The 567 is a general-purpose tone decoder designed to provide a saturated transistor switch to ground when an input signal is present within the passband. The BW, center frequency, and output delay are independently determined by external components. Typical applications of the 567 are in touch-tone decoding, precision oscillators, frequency monitoring and control, wideband FSK demodulation, ultrasonic controls, carrier current remote controls, communications paging decoders, and 0°–180° phase shifting. Diagrams for the applications are shown in the data sheets in Appendix H. Many additional applications for modulators, demodulators, and PLLs will be explored in Chapter 12.

## 10.9 Summary

☐ AM is the result of varying a carrier wave in amplitude at a certain rate. FM is the result of varying frequencies above and below the center frequency of a carrier wave, while maintaining a constant amplitude. PM is a modified form of FM.

☐ Demodulation, or detection, is the process of extracting the audio intelligence from the carrier wave and passing it on to a speaker or display device, such as a television screen. AM demodulators are called envelope detectors, and FM demodulators are called discriminators.

☐ Television signals are composite signals that consist of AM video information and FM audio information. The diode detector is a commonly used AM demodulator, and the quadrature detector circuit is commonly used in FM demodulation.

☐ The name *compandor* is a contraction of the two functions that the device performs—compression and expansion.

☐ PLLs are basically closed loops that provide an electronic feedback to lock in a desired frequency. A PLL consists of a phase detector, an LP filter, a dc amplifier, and a VCO. The phase detector in a PLL compares an input reference frequency and the output frequency of the VCO, and generates a dc control voltage proportional to the phase difference of the two input signals. The LP filter in a PLL can be either passive or active. It removes noise from the two oscillator signals, and it controls the lock, capture, BW, and transient response of the loop. The dc amplifier in a PLL provides the desired level of control voltage in the polarity needed for the VCO. The VCO in a PLL uses a varactor to set oscillator frequency. The varactor is a semiconductor capacitive diode that operates on the principle of a varying capacitance that is inversely proportional to the amount of applied dc reverse voltage. There are many monolithic IC PLL circuits available that require only a few external components to provide a multitude of applications.

☐ There are three operating modes for the PLL: free-running, capture, and phase-lock (sometimes referred to as lock-in or tracking). Lock range is the frequency range over which the PLL can follow incoming signals. Capture range is the BW over which capture is possible. Capture range is never wider than lock range.

☐ Changing the $\div N$ count in a frequency synthesizer can provide many different frequencies while using a single reference CCO.

☐ FSK means that the carrier frequency is changed from one value to another instead of keying the transmitter on and off.

# 10.10 Questions and Problems

**10.1** What is the purpose of modulation in a radio transmitter?

**10.2** What function does the detector in a radio receiver perform?

**10.3** State the major advantage of FM over AM.

**10.4** Name two FM demodulator circuits.

**10.5** Draw a block diagram of (a) an AM radio receiver and (b) an FM radio receiver.

**10.6** Briefly describe the operation of (a) an AM receiver and (b) an FM receiver.

**10.7** Draw a block diagram of a PLL and describe the operation of the four functional blocks.

**10.8** What is the purpose of the varactor in the VCO circuit? Explain its operation.

**10.9** What is the output frequency of a synthesizer if the input frequency is 1.5 MHz and the $\div N$ counter is set at a count of 20?

**10.10** What is the input frequency of the phase detector if the output frequency of a synthesizer is 350 MHz and the counter is set to a count of 240?

**10.11** Calculate the phase detector output voltage for a conversion gain of 0.02 V/rad and an input phase difference of 0.3 rad.

**10.12** Calculate the phase detector output voltage for a conversion gain of 30 mV/rad and an input phase difference of 23°. (HINT: 1 radian = 57.3°)

**10.13** What is the percent modulation of a signal with a carrier frequency of 3 A, a maximum current of 6 A, and a minimum current of 0 A?

**10.14** What is the percent modulation of a signal with a carrier frequency of 2 A, a maximum current of 6 A, and a minimum current of 4 A?

**10.15** What will be the result at the receiver if a transmitted signal exceeds 100% modulation? What is this situation called?

**10.16** List eight applications for the PLL circuit.

**10.17** List five applications for the programmable compandor.

Chapter 11

# Timers

## 11.1 Introduction

Among existing LICs, perhaps the most popular is the 555 IC timer. Introduced in 1972 by Signetics Company, the 555 is a versatile, reliable, low-cost device that is easy to use in a variety of applications and generally requires simple connections to a few external low-cost components. The device can operate from supply voltages ranging from 4.5 V to 15 V, making it compatible with both transistor-transistor logic (TTL) circuits and op amp circuits.

More sophisticated and wide-ranging versions of the simple IC timer have been developed. The 556 is a dual 555 timer package, the 7555 is a CMOS version of the basic 555, the 558 package contains four timer sections, and the 2240 is a programmable timer/counter combination that contains a 555 timer plus a programmable binary counter in a single 16-pin package. The LM322 is a precision timer that can operate from unregulated power supplies over a range of 4.5–40 V. The timing range for the single 555 timer is microseconds to hours, while the timing range for the timer/counter is microseconds to days. Cascading the devices can extend the timing range of both to months or even years.

Our study of IC timers will begin with the functional 555 and its modes of operation, followed by other timers and applications.

# 11.2 Objectives

When you complete this chapter, you should be able to:

- ☐ State the basic characteristics of the 555, 322, and 2240 IC timers.
- ☐ Explain the operation of a monostable timer circuit.
- ☐ Explain the operation of an astable timer circuit.
- ☐ Determine the output pulse width of a monostable timer circuit and the factors that affect the pulse width.
- ☐ Determine the output pulse width and frequency of an astable timer circuit and the factors that affect them.
- ☐ Determine duty cycle and how to change it.
- ☐ Design and construct both monostable and astable timer circuits using the 555, 322, and 2240 IC timers.
- ☐ Design and construct a variety of application circuits using the 555, 322, and 2240 IC timers.
- ☐ Program the outputs of the 2240 programmable timer/counter IC.
- ☐ List many applications for the different types of IC timers.

# 11.3 555 IC Timer

The 555 monolithic timing circuit shown in the block diagram of Figure 11–1 is a very stable controller for producing accurate time delays or oscillations. Terminals are provided for *triggering* (pin 2) and *reset* (pin 4) if desired. In the time delay mode, the delay time is precisely controlled by one external resistor and capacitor. For stable operation in the oscillator mode, the free-running frequency and the duty cycle are both accurately controlled with two external resistors and one capacitor. The output is capable of sinking or sourcing 200 mA, is compatible with TTL logic circuits, and can drive relays and indicator lamps.

## Modes of Operation

The 555 IC timer has two modes of operation. It can operate either as a monostable (one-shot) multivibrator or as an astable (free-running) multivibrator. It is available in two package styles, the eight-pin MINI DIP and the eight-pin TO-100, as shown in Figure 11–2. To understand how the 555 timer operates, we will briefly review the purpose of each of its terminals.

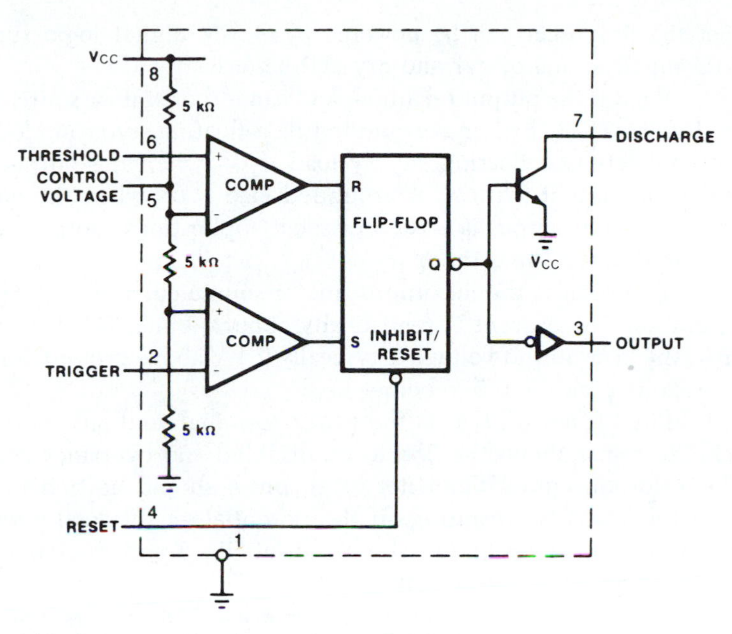


Figure 11–1. 555 IC Timer (Courtesy of Fairchild Camera & Instrument Corporation, Linear Division)

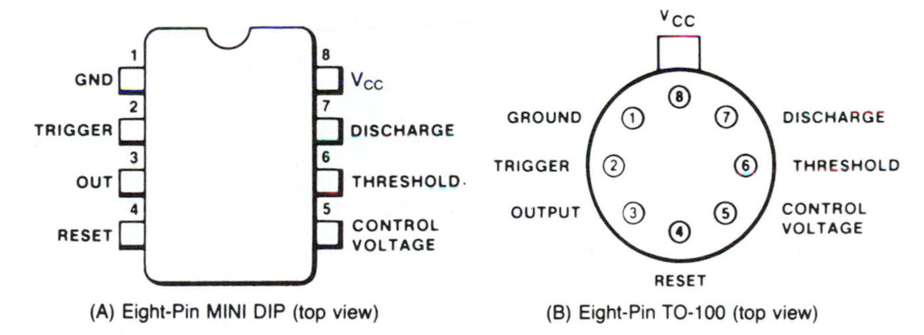


Figure 11–2. Connection Diagrams for the 555 IC Timer (Part A: Courtesy of Fairchild Camera & Instrument Corporation, Linear Division; Part B: Courtesy of National Semiconductor Corporation)

## Terminals and Their Purposes

Refer to Figure 11–1 as we identify the terminals and their functions. Pin 1 is the common, or ground, terminal, and pin 8 is the positive supply voltage terminal, $V_{CC}$, which can be any voltage between +4.5 V and +15 V. This means that the 555 timer can be powered by +5 V digital logic supplies, by +15 V LIC supplies, and by car and dry cell batteries.

Pin 3 is the output terminal, and can either sink or source current, as shown in Figure 11–3. Either a grounded or a floating (ungrounded) supply load can be connected. A floating supply load is ON when the output is LOW, and OFF when the output is HIGH. A grounded load is ON when the output is HIGH, and OFF when the output is LOW. For most applications, both types of loads are not required at the same time.

Technically, the maximum sink or source current is 200 mA, but in normal operation, the current is realistically about 40 mA. For load currents below 25 mA, the LOW output voltage is typically 0.1 V above ground, and the HIGH output voltage is typically 0.5 V below $V_{CC}$.

Pin 4 (Figure 11–1) is the RESET terminal, and has an overriding function. This terminal allows the 555 to be disabled and overrides command signals on the TRIGGER input. When not used, pin 4 should be tied to $V_{CC}$ to avoid any possibility of false resetting. If the potential on the RESET terminal is reduced below 0.4 V, or grounded, both output pin 3 and discharge pin 7 are forced LOW, holding the output LOW.

Pin 5 is the CONTROL VOLTAGE terminal. An external voltage applied to pin 5 will change both threshold and trigger voltages, since this terminal allows direct access to the upper comparator and indirect access to the lower comparator through a 2:1 voltage divider (Figure 11–1). Use of this terminal is optional, but it does offer great flexibility through modification of the timing period, and it can be used to modulate the output waveform. When not used, it is practical to connect a 0.01 μF bypass capacitor between pin 5 and ground. This bypasses

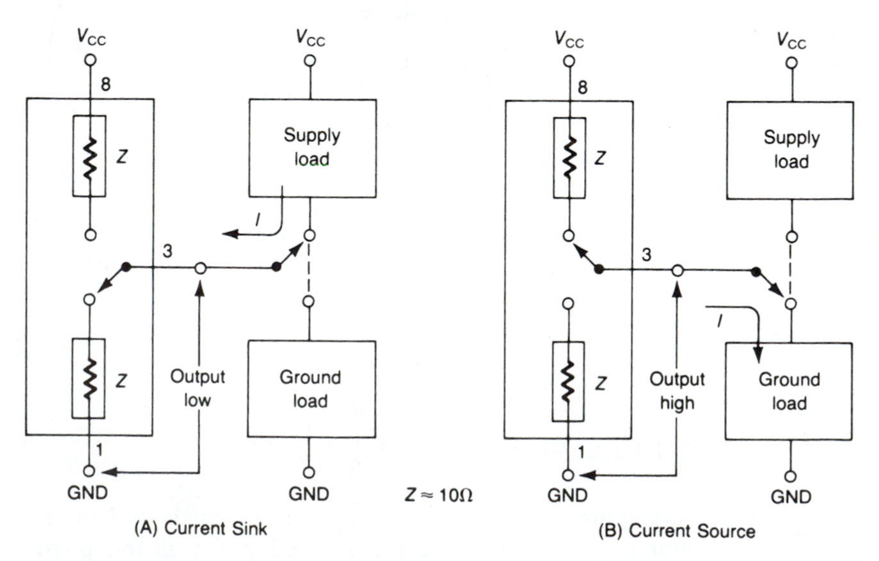


**Figure 11–3.** Terminal 3 Outputs for the 555 IC Timer

noise or ripple voltages from the power supply, thus minimizing their effect on threshold voltage.

Pin 7 is the DISCHARGE terminal, used to discharge an external timing capacitor during the LOW output period. When the output is HIGH, pin 7 acts as an open circuit and allows the timing capacitor to charge at a rate determined by the *RC* time constant of external resistors and capacitors. In certain applications, this terminal can be used as an auxiliary output terminal, with current sink capabilities similar to those of output pin 3.

Pins 2 and 6, TRIGGER and THRESHOLD terminals, respectively, will be considered together because the two possible operating states and two possible memory states of the 555 are determined by both pins. Pin 2 is the input to the lower comparator, where it is compared with a lower threshold voltage, $V_{lth}$, that is equal to 1/3 $V_{CC}$, determined by the voltage divider (Figure 11–1). This pin is used to set the latch, which in turn causes a HIGH output. Triggering is accomplished by bringing the input level on pin 2 from above to below $V_{lth}$. Pin 6 is one input (pin 5 is the other) to the upper comparator, where it is compared with a higher threshold voltage, $V_{uth}$, that is equal to 2/3 $V_{CC}$, determined by the voltage divider. This pin is used to reset the latch, which in turn causes a LOW output. Resetting is accomplished by bringing the input level on pin 6 from below to above $V_{uth}$. The voltage range that can safely be applied to pins 2 and 6 is between $V_{CC}$ and ground.

## 11.4 556 Timer

The 556 dual timing system shown in Figure 11–4 is simply two 555 timers in a single 14-pin DIP. The legends for the pins are the same as for the 555, but there are two separate circuits with different pin designations. Care should be exercised when using the 556 dual circuit, so that correct pin connections are made. For safety and accuracy, always refer to the data sheet when making connections.

Data sheets for the 555 and 556 timers are in Appendix I.

## 11.5 General Design Considerations for 555 and 556 Timers

The timers will operate over a guaranteed voltage range of 4.5 $V_{dc}$ to 15 $V_{dc}$, with absolute maximum ratings of 16 $V_{dc}$ (for NE/SA555, SE555C, NE/SA556/556–1, SE556C/556–1C) and 18 $V_{dc}$ (for SE555, SE556/556–1). Most of the devices, however, will operate at voltage levels as low as 3 $V_{dc}$. The timing interval is independent of supply voltage since the charge rate and threshold level of the comparator are both directly proportional to supply. Although the supply voltage may be

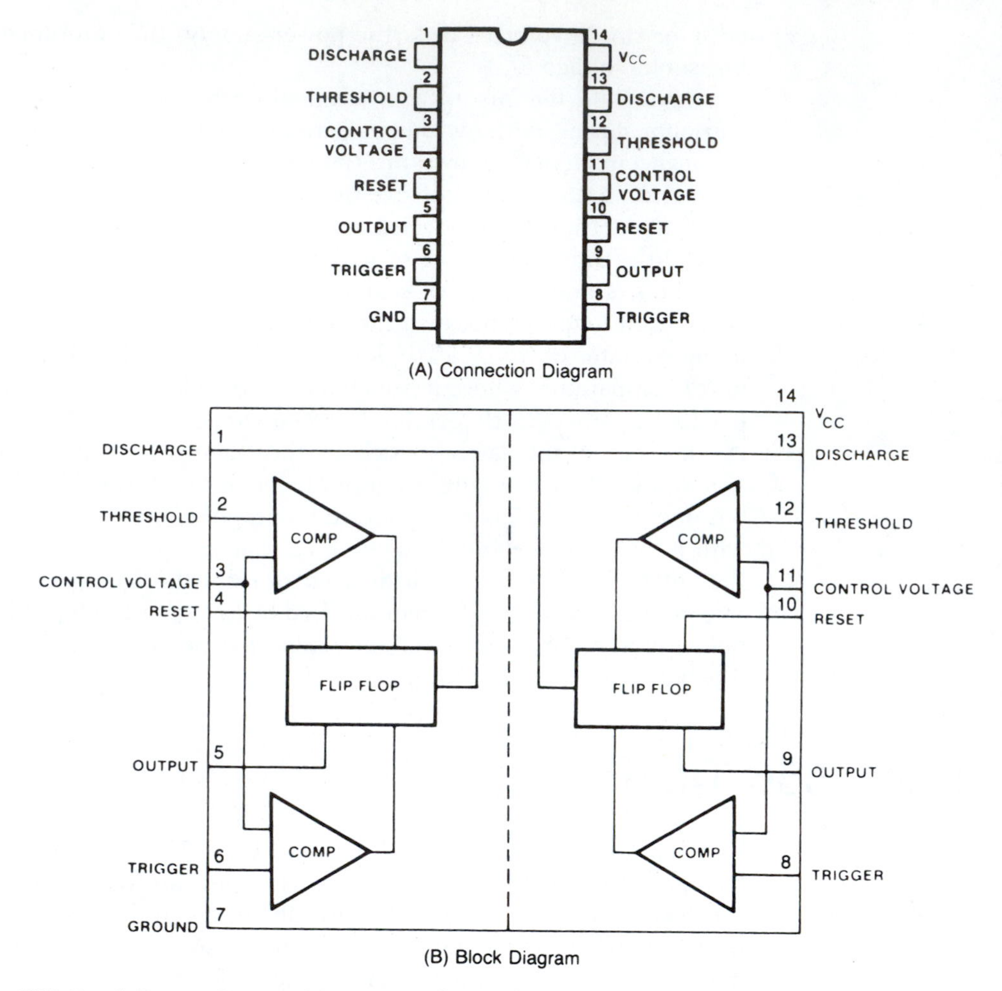


Figure 11–4. 556 Dual Timer Circuit (Courtesy of Fairchild Camera & Instrument Corporation, Linear Division)

provided by any number of sources, several precautions must be taken. The most important, the one that produces the most problems if not practiced, is good power supply filtering and adequate bypassing. Ripples on the supply line can cause loss of timing accuracy; the threshold level shifts, causing a change of charging current, thereby causing a timing error for that cycle.

Because of the high-power totem pole output structure, the output of the timer can exhibit large current spikes on the supply line. Bypassing is necessary to eliminate this phenomenon. A capacitor across $V_{CC}$ and ground, directly across the timer device, is necessary and ideal. The size of the capacitor will depend on the specific application. Values of capacitance from 0.01 μF to 10 μF are not uncommon. The capacitor should be placed as close as physically possible to the timer device.

## Selecting External Components

In selecting the external timing resistor and capacitor, several considerations must be taken into account. Stable external components are necessary for the *RC* network if good timing accuracy is to be maintained.

### Resistor Selection

Of the two timing components, the resistors are the easiest to select, because there is such a great number of standard values from which to choose, particularly for values up to 10 MΩ. For values above 10 MΩ, it may be necessary to procure special components at a higher cost.

The timing resistor(s) should be of the metal film variety if timing accuracy and repeatability are important design criteria. The timer exhibits a typical initial accuracy of 1%; that is, with any one *RC* network, from timer-to-timer, only 1% change is to be expected. Most of the initial timing error (deviation from the formula) is due to inaccuracies of external components. Resistors range from their rated values by 0.01% to 10% and 20%. (Capacitors may have a 5% to 10% deviation from rated capacity.) Therefore, in a system where timing is critical, an adjustable timing resistor or precision components are necessary. Normally, potentiometers in the timing circuit should be avoided because the reliability and accuracy of the circuit may be compromised. For best results, a good-quality trim pot placed in series with the largest feasible resistance will allow for best adjustability and performance.

### Capacitor Selection

Capacitor selection creates more problems than resistor selection for the designer. There are many different types of capacitors, but few types have tolerances below 5%. In general, capacitors are imprecise components with varying performance characteristics, and the range of available values is limited. The selected timing capacitor should be a high-quality, stable component.

Some desirable characteristics for timing circuit capacitors are *low leakage, low dielectric absorption,* and *low temperature coefficient.* For example, if a capacitor is to be charged from a 1 μA source, it must have a leakage much lower than that, and the leakage must remain low over varying conditions of voltage, temperature, and time. Dielectric absorption means that when a capacitor is charged and then discharged by short-circuiting, all the energy that was stored during charge is not given up by the dielectric, and a residual voltage is retained across the capacitor. This is a serious problem because the timing principle depends on starting from zero volts. Therefore, high dielectric absorption capacitor types such as paper, ceramic, and some mica ones should not be used.

NOTE: Under no circumstances should ceramic disc capacitors be used in the timing network! Ceramic disc capacitors are not sufficiently stable in capacitance to operate properly in an *RC* mode.

Several acceptable capacitor types are: silver mica, mylar, polycarbonate, polystyrene, tantalum, or similar types. The plastic-film capacitor family, which includes polystyrenes, polycarbonates, parylenes, and teflon, exhibits low dielectric absorption characteristics and is suitable for use in timing circuits.

A polystyrene capacitor is one of the best timing circuit choices in terms of cost and performance, but it is limited because it cannot be used above 85 °C, and it is available in values only up to about 1 μF. The next best choice is a polycarbonate capacitor, which is available from many suppliers at reasonable cost, in values ranging up to about 100 μF, many with close tolerances.

Another possible problem with capacitors is their relatively large size, particularly for the higher values. In general, the higher the value, the larger the physical size and the higher the cost. Therefore, it is wise to minimize the size of the capacitor by reducing its value and increasing the value of the resistor associated with it in the timing circuit.

Capacitor value has not proven to be a legitimate design criteria. Values ranging from picofarads (pF) to greater than 1000 μF have been used successfully. One precaution needs to be observed though. (It should be a cardinal rule that applies to the usage of all ICs.) *Make certain that the package power dissipation is not exceeded.* With extremely large capacitor values, a maximum duty cycle which allows some cooling for the discharge transistor may be necessary.

The most important characteristic of the capacitor should be as low a leakage as possible. Obviously, any leakage will subtract from the charge count, causing the calculated time to be longer than anticipated.

Except in very low-performance applications, electrolytic capacitors are not suitable for timing capacitor use, because of their characteristics of loose tolerances, high leakages, and poor stability over temperature changes.

## Other Considerations

The timer typically exhibits a small negative temperature coefficient (50ppm/°C). If timer accuracy over temperature is a consideration, timing components with a small positive temperature coefficient should be chosen. This combination will tend to cancel timing drift due to temperature.

In selecting the values for the timing resistors and capacitor, several points should be considered. A minimum value of threshold current (0.25 μA) is necessary to trip the threshold comparator. To calculate the maximum value of resistance, it is important to know that at the time the threshold current is required, the voltage potential on the threshold pin is two-thirds of supply. Therefore:

$$V_{\text{potential}} = V_{CC} - V_{\text{cap}} \tag{11.1a}$$

$$V_{\text{potential}} = V_{CC} - {}^2/_3\, V_{CC} \tag{11.1b}$$

Maximum resistance is then defined as:

$$R_{max} = \frac{V_{CC} - V_{cap}}{I_{thresh}} \tag{11.2a}$$

$$R_{max} = \frac{V_{CC} - 2/3\,V_{CC}}{I_{thresh}} \tag{11.2b}$$

**Example 11.1**

Calculate $R_{max}$ for (a) $V_{CC} = 15$ V and (b) $V_{CC} = 5$ V.

*Solution*

Use Equation 11.2b:

$$R_{max} = \frac{V_{CC} - 2/3\,V_{CC}}{I_{thresh}}$$

(a) $R_{max} = [15\text{ V} - 2/3(15\text{ V})]/0.25\ \mu\text{A}$
$= (15 - 10)/0.25 \times 10^{-6}$
$= 20\ \text{M}\Omega$

(b) $R_{max} = (5\text{ V} - 3.33\text{ V})/0.25 \times 10^{-6}$
$= 6.6\ \text{M}\Omega$

NOTE: If using a large value of timing resistor, be certain that the capacitor leakage is significantly lower than the charging current available to minimize timing error.

There are certain minimum values of resistance that should be observed. The discharge transistor of the timer is current-limited at 35 mA to 55 mA internally. Thus, at the current limiting values, the discharge transistor establishes high saturation voltages. When turned on, the transistor will be carrying two currents: the first is the constant current through timing resistor $R_A$; the second the varying discharge current from the timing capacitor. To provide best operation, the current contributed by the $R_A$ path should be minimized so that the majority of discharge current can be used to reset the capacitor voltage. Hence, it is recommended that a 5 kΩ value be the minimum feasible value for $R_A$. This does not mean that lower values cannot be used successfully in certain applications, but it does mean that extreme cases should be avoided, because the timer can be damaged or destroyed if $R_A$ is too low a value.

## 11.6 Monostable Circuits

The most basic mode of operation of timer devices is the *triggered monostable* circuit shown in Figure 11–5. Note the simplicity of the circuit; it consists of only the two timing components, $R_A$ and $C$, the timer itself, and bypass capacitor $C_1$, which is used only for noise immunity. When a negative-going input pulse applied to pin 2 reaches a level less than 1/3 $V_{CC}$, the timer is triggered and begins its timing cycle. The output goes HIGH and the voltage across $C$ begins to charge towards $V_{CC}$ at a rate determined by the $RC$ time constant of $R_A$ and $C$. When the capacitor voltage reaches 2/3 $V_{CC}$, the upper comparator (Figure 11–1) causes the output to go LOW, ending the timing cycle. The input and output voltage waveforms are shown in Figure 11–5.

The output pulse width ($PW$), expressed in seconds, is defined as

$$PW = 1.1 R_A C \quad \textbf{(11.3)}$$

There are relatively few restrictions on $PW$, and the timing components can have a wide range of values. Theoretically, there is no upper limit on $PW$, but practical limits are set by $R$ and $C$ values. The lower limit is set at 10 μs. It should be remembered that timer circuits can be cascaded to extend the pulse width to virtually any length, even periods of days, weeks, or months.

## Example 11.2

Assume the values of the timing components in Figure 11–5 are $R_A$ = 10 kΩ and $C$ = 0.001 μF. Calculate the $PW$ of the output pulse.

*Solution*

Use Equation 11.3:

$$PW = 1.1 R_A C = 1.1 \times 10\ \text{k}\Omega \times 0.001\ \mu\text{F} = 11\ \mu\text{s}$$

The practical lower value for $R_A$ is around 10 kΩ, and the practical lower value for $C$ is around 0.001 μF. For capacitances below this value, stray capacitance affects accuracy and predictability of the timer circuit. In practice, it is best to select $C$ first to minimize size and cost, then select $R_A$.

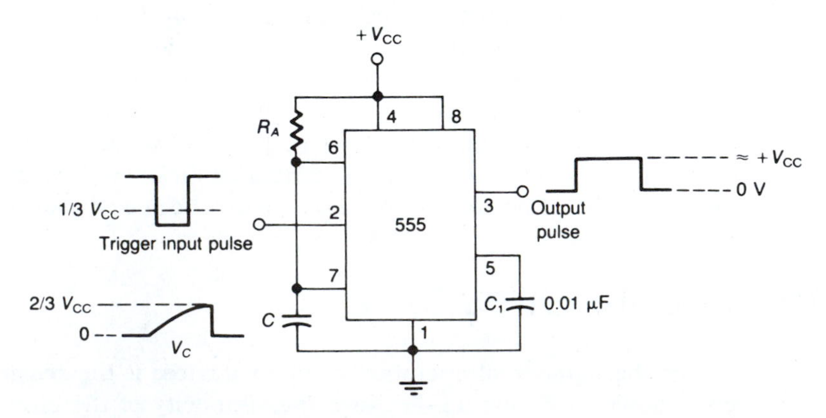


Figure 11–5. 555 Timer in the Triggered Monostable Mode

## Example 11.3

Assume a desired output *PW* of 100 ms. Select the values for the timing resistor if $C = 1.0\ \mu F$.

*Solution*

Rearrange Equation 11.3 to solve for $R_A$:

$$R_A = \frac{PW}{1.1C} = \frac{100\ \text{ms}}{1.1 \times 1.0\ \mu\text{F}} = 90.9\ \text{k}\Omega$$

## Example 11.4

If $R_A = 10\ \text{M}\Omega$, calculate $C$ for an output *PW* of 100 s.

*Solution*

Rearrange Equation 11.3 to solve for $C$:

$$C = \frac{PW}{1.1R_A} = \frac{100\ \text{s}}{1.1 \times 10\ \text{M}\Omega} = 9.1\ \mu\text{F}$$

Examples 11.3 and 11.4 demonstrate the wide range of values that may be used for the timing components. Figure 11–6 is a quick-reference design aid for selecting components and determining output *PW*.

## Trigger Input Conditioning

Note that Figure 11–5 shows no trigger input conditioning components, which implies that the driving source must be capable of satisfying the trigger voltage requirements. In general, the only restriction on the input pulse is that it have a width less than the output pulse. Because of the internal latching mechanism (the flip-flop in Figure 11–1), the timer will always time out once triggered, even if subsequent noise pulses, such as bounce, are present on the trigger input. This feature makes the 555 an excellent device to use when connecting to noisy sources.

To increase stability and accuracy, the input conditioning network shown in Figure 11–7 is added to the circuit. Input resistor $R_{in}$ insures that the output is LOW and that input capacitor $C_{in}$ is charged to $V_{CC}$ until the negative-going edge of the trigger pulse occurs. The time constant of $R_{in}$ and $C_{in}$ should be short compared with the output *PW*, established by the time constant of $R_A$ and $C$. Diode $D_1$ prevents the timer from triggering on the positive-going edge of the input pulse.

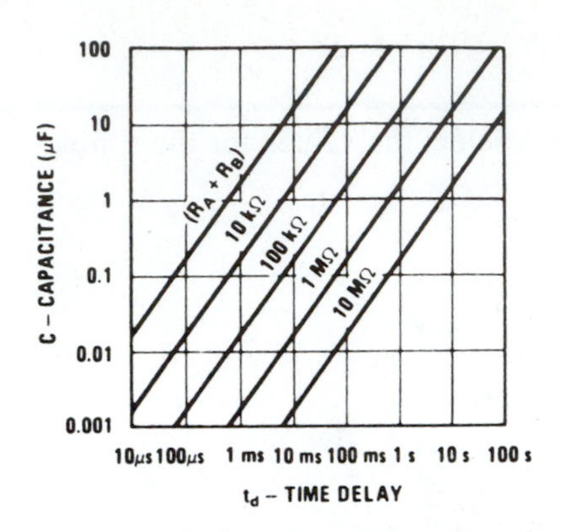


Figure 11–6. Quick-Reference Design Aid for Determining Output Pulse Width (Courtesy of National Semiconductor Corporation)

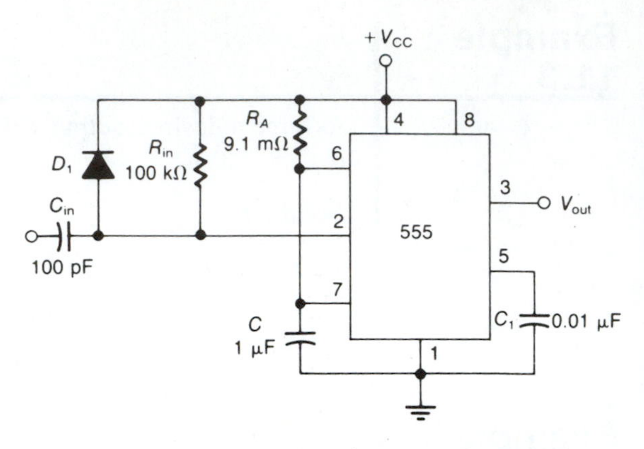


Figure 11–7. Input Conditioning Network for the 555 Timer

## Example 11.5

Calculate (a) the output *PW* and (b) the time constant ($\tau$) of the input circuit in Figure 11–7.

*Solutions*

**a.** Use Equation 11.3:
$PW = 1.1R_AC = 1.1 \times 9.1\ \text{M}\Omega \times 1\ \mu\text{F} = 10\ \text{s}$

**b.** $\tau = R_{in}C_{in} = 100\ \text{k}\Omega \times 100\ \text{pF} = 10\ \mu\text{s}$

### False Reset Prevention

Note that reset pin 4 is tied to $V_{CC}$ in both Figure 11–5 and Figure 11–7. This is to prevent false resetting, for if pin 4 is grounded at any time, both output pin 3 and discharge pin 7 go to ground potential, removing any accumulated charge on $C$ and pulling the output LOW. Those conditions will remain as long as the reset pin is grounded.

## 11.7 Astable Circuits

The 555 timer connected in the astable, or free-running, mode is shown in Figure 11–8A. This configuration uses three timing components—$R_A$, $R_B$, and $C$—the timer, and bypass capacitor $C_1$ for noise immunity. When voltage is first applied, the voltage across $C$ is LOW, causing triggering at pins 2 and 6, which drives output pin 3 HIGH and opens pin 7. Capacitor $C$ charges through resistors $R_A$

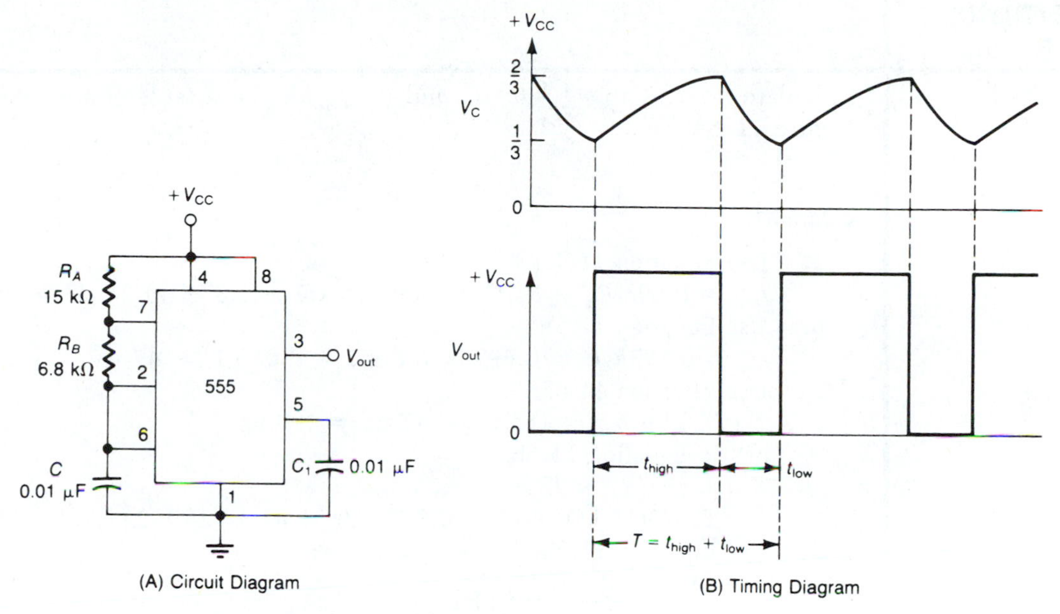


**Figure 11–8.** 555 Timer in the Astable Mode

and $R_B$ until $V_C$ reaches the upper threshold level of 2/3 $V_{CC}$, at which time output pin 3 goes LOW. Discharge pin 7 also goes LOW, providing a discharge path for $C$ through resistor $R_B$. When $V_C$ drops to the lower threshold level of 1/3 $V_{CC}$, the timer is triggered again, starting a new cycle.

The output remains HIGH for the time interval ($t$) that $C$ charges from 1/3 $V_{CC}$ to 2/3 $V_{CC}$, as shown in Figure 11–8B. This time interval is

$$t_{high} = 0.693(R_A + R_B)C \qquad \textbf{(11.4)}$$

For the time interval that $C$ discharges from 2/3 $V_{CC}$ to 1/3 $V_{CC}$, the output is LOW and is determined as follows:

$$t_{low} = 0.693R_BC \qquad \textbf{(11.5)}$$

The total period of oscillation ($T$) is the sum of the HIGH and LOW time intervals:

$$T = t_{high} + t_{low} \qquad \textbf{(11.6a)}$$

or

$$T = 0.693(R_A + 2R_B)C \qquad \textbf{(11.6b)}$$

The frequency of oscillation of the free-running oscillator configuration is

$$f_{op} = \frac{1}{T} \qquad \textbf{(11.7a)}$$

or

$$f_{op} = \frac{1.44}{(R_A + 2R_B)C} \qquad \textbf{(11.7b)}$$

## Example 11.6

Calculate (a) $t_{high}$, (b) $t_{low}$, (c) $T$, and (d) $f_{op}$ for the astable timer circuit in Figure 11–8A.

*Solutions*

a. Use Equation 11.4:
$t_{high} = 0.693(R_A + R_B)C = 0.693(15\ \text{k}\Omega + 6.8\ \text{k}\Omega)0.01\ \mu\text{F} \approx 152\ \mu\text{s}$

b. Use Equation 11.5:
$t_{low} = 0.693R_BC = 0.693 \times 6.8\ \text{k}\Omega \times 0.01\ \mu\text{F} \approx 47\ \mu\text{s}$

c. Use Equation 11.6a:
$T = t_{high} + t_{low} = 152\ \mu\text{s} + 47\ \mu\text{s} = 199\ \mu\text{s}$
or use Equation 11.6b:
$T = 0.693(R_A + 2R_B)C$
$= 0.693[15\ \text{k}\Omega + (2 \times 6.8\ \text{k}\Omega)]0.01\ \mu\text{F} \approx 199\ \mu\text{s}$

d. Use Equation 11.7a:
$$f_{op} = \frac{1}{T} = \frac{1}{199\ \mu\text{s}} = 5\ \text{kHz}$$
or use Equation 11.7b:
$$f_{op} = \frac{1.44}{(R_A + 2R_B)C} = \frac{1.44}{[15\ \text{k}\Omega + (2 \times 6.8\ \text{k}\Omega)]0.01\ \mu\text{F}} = 5\ \text{kHz}$$

## Duty Cycle

The duty cycle ($D$) is the ratio of the time interval of the LOW output ($t_{low}$) to the total period of oscillation ($T$), and is expressed as a percentage:

$$D = \frac{t_{low}}{T} \times 100\% \qquad \textbf{(11.8a)}$$

or

$$D = \frac{R_B}{R_A + 2R_B} \times 100\% \qquad \textbf{(11.8b)}$$

Within limits, the $D$ can be programmed by the ratios of $R_A$ and $R_B$. As you can see by observing Equation 11.8b, by making $R_B$ large with respect to $R_A$, the $D$ can approach 50%, or a square wave. On the other hand, if $R_A$ is large with respect to $R_B$, the $D$ approaches zero. However, since $R_B$ cannot be allowed to reach zero, the $D$ likewise can never reach zero. Therefore, the practical range for the $D$ is from around 1% to 50%. Later we will see how to design a variable duty cycle circuit that can range from 1% to 99%.

## Example 11.7

Determine $D$ for the circuit in Figure 11–8A.

*Solution*

Use Equation 11.8b:

$$D = \frac{R_B}{R_A + 2R_B} \times 100\% = \frac{6.8\text{ k}\Omega}{15\text{ k}\Omega + (2 \times 6.8\text{ k}\Omega)} \times 100\% = 23.8\%$$

## Example 11.8

Assume $R_B$ is 1000 times the value of $R_A$ in Figure 11–8A. Calculate $D$.

*Solution*

Since $R_A = 15\text{ k}\Omega$, $R_B = 1000 \times 15\text{ k}\Omega = 15\text{ M}\Omega$. Then,

$$D = \frac{15\text{ M}\Omega}{15\text{ k}\Omega + 30\text{ M}\Omega} \times 100\% = 49.9\%$$

## Example 11.9

Assume $R_A = 100R_B$ in Figure 11–8A. Calculate $D$.

*Solution*

Since $R_B = 6.8\text{ k}\Omega$, $R_A = 100 \times 6.8\text{ k}\Omega = 680\text{ k}\Omega$. Then,

$$D = \frac{6.8\text{ k}\Omega}{680\text{ k}\Omega + 13.6\text{ k}\Omega} \times 100\% = 0.998\%$$

### Fifty Percent Duty Cycle

One method of producing a 50% $D$ is shown in Figure 11–9A. In this network, $R_A$ is equal to $R_B$, and a diode is connected in parallel with $R_B$. The capacitor $C$ now charges through $R_A$ and the diode, but discharges through $R_B$, and the output waveform times are determined as follows:

$$t_{\text{high}} = 0.693R_AC \quad \textbf{(11.9a)}$$

$$t_{\text{low}} = 0.693R_BC \quad \textbf{(11.9b)}$$

$$T = 0.693(R_A + R_B)C \quad \textbf{(11.9c)}$$

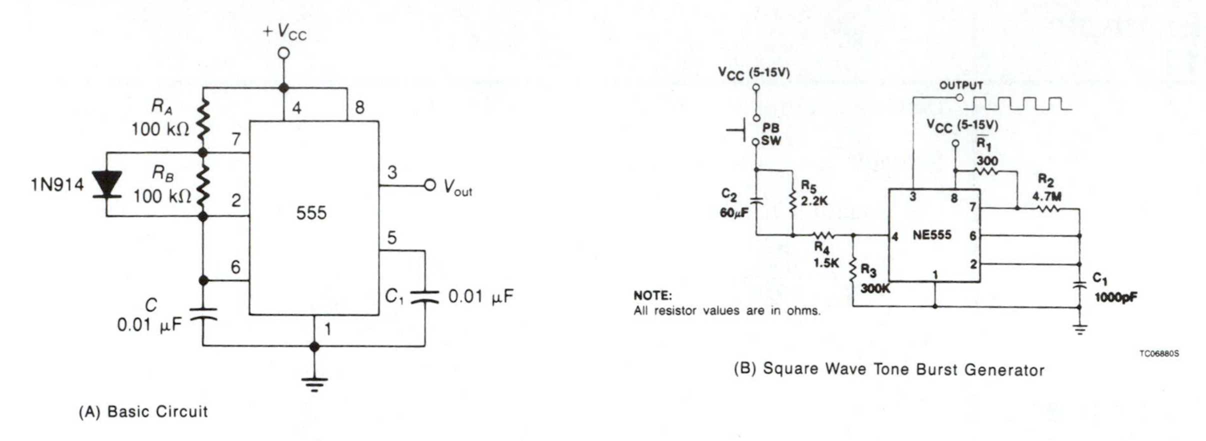


(A) Basic Circuit

(B) Square Wave Tone Burst Generator

**Figure 11-9.** Circuit for the 50 Percent Duty Cycle (Part B: Courtesy of Signetics Company)

Another method of producing a 50% *D* is shown in Figure 11-9B. Here, a 556 is connected as a square wave tone burst generator.

## 11.8 ICM7555 Timer

A direct replacement for the standard 555 timer in many applications, the ICM7555 general-purpose CMOS timer, shown in Figure 11-10, provides significantly improved performance. A stable controller capable of producing accurate time delays or frequencies, improved parameters of the 7555 include low supply current (80 μA typical), wide operating supply voltage range (2 V to 18 V), extremely low THRESHOLD, $\overline{\text{TRIGGER}}$, and $\overline{\text{RESET}}$ currents (20 pA typical), no crowbarring of the supply current during output transitions, higher frequency performance (500 kHz guaranteed), and no requirement to decouple CONTROL VOLTAGE for stable operation.

The 7555 timer can be set for times from μs through hours, operates in either astable or monostable modes, and can be used with higher-impedance timing elements for longer time constants.

Applications include precision timing, pulse generation, sequential timing, time delay generation, *PW* modulation, pulse position modulation (PPM), and missing pulse detection.

Figure 11-11 compares output transition supply current spikes between the standard 555 and the 7555. As shown, the 7555 produces current spikes of only 2 to 3 mA instead of the 300 to 400 mA of the 555. This means that supply decoupling is normally not required for the 7555. Also, in most instances, the CONTROL VOLTAGE decoupling capacitors are not required because of the very high input impedance of the on-chip CMOS comparators. Thus, for many applications, two capacitors can be saved by using the 7555.

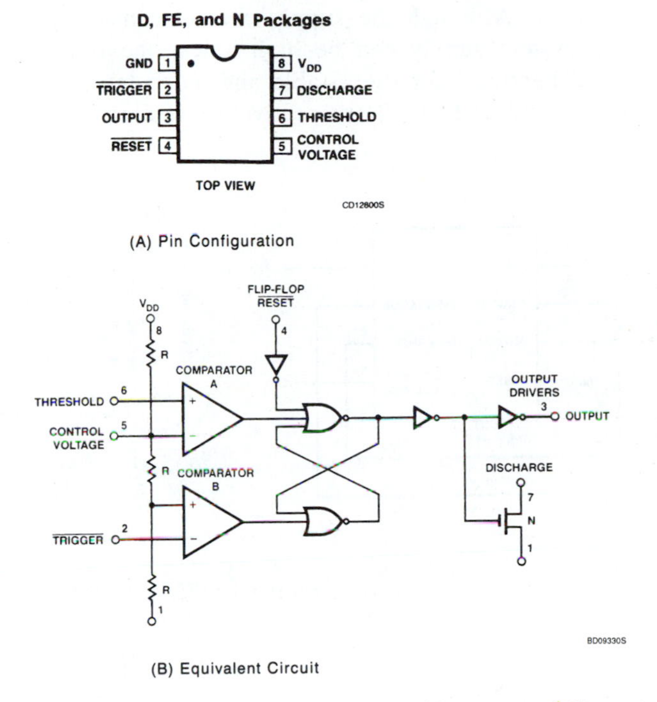


Figure 11–10. ICM7555 General-Purpose CMOS Timer (Courtesy of Signetics Company)

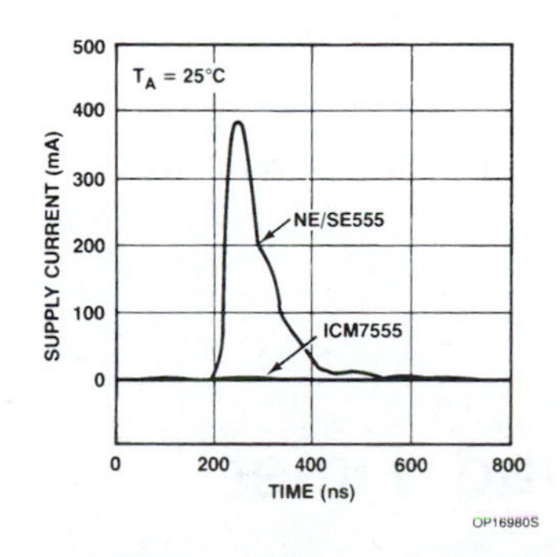


Figure 11–11. Supply Current Transient Compared with a Standard Bipolar 555 during an Output Transition (Courtesy of Signetics Company)

Although the supply current consumed by the 7555 is very low, the total system supply can be high unless the timing components are high impedance. Therefore, for the astable and monostable modes of operation shown in Figures 11–12 and 11–13, respectively, use high values for *R* and low values for *C*.

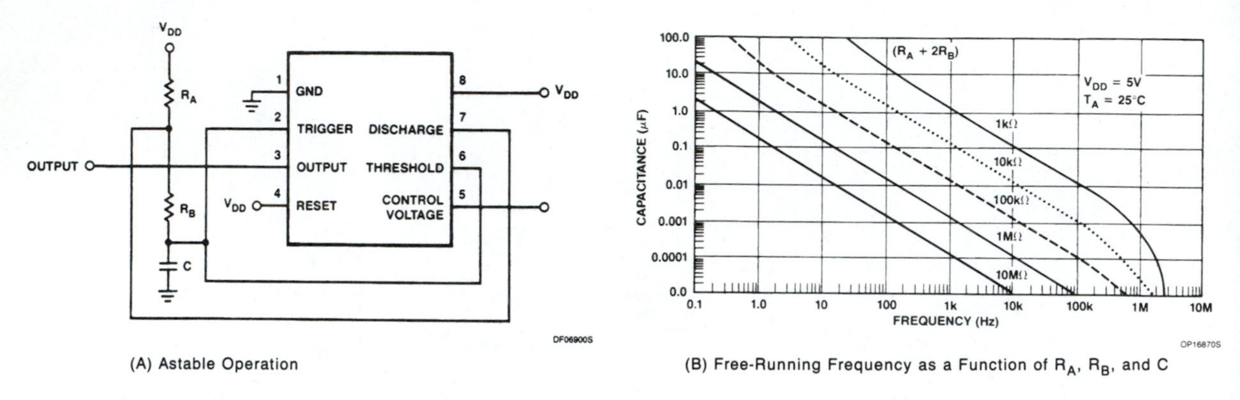


(A) Astable Operation

(B) Free-Running Frequency as a Function of $R_A$, $R_B$, and C

**Figure 11–12.** Astable Circuit with 7555 Timer (Courtesy of Signetics Company)

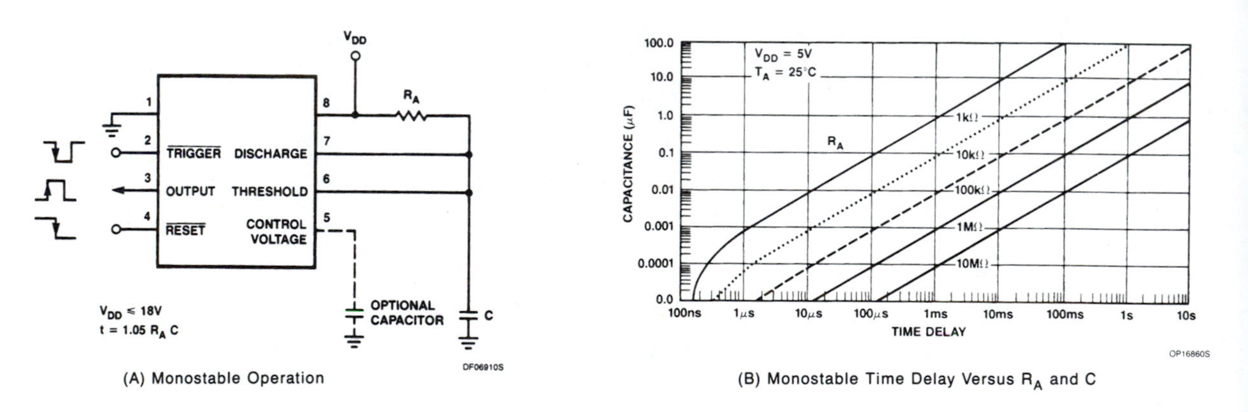


(A) Monostable Operation

(B) Monostable Time Delay Versus $R_A$ and C

**Figure 11–13.** Monostable Circuit with 7555 Timer (Courtesy of Signetics Company)

# 11.9 SE558 Quad Timer

The 558 quad timers, Figure 11–14, are monolithic timing devices. These highly stable, general-purpose controllers can be used in a monostable mode to produce accurate time delays from μs to hours. In the time delay mode of operation, four

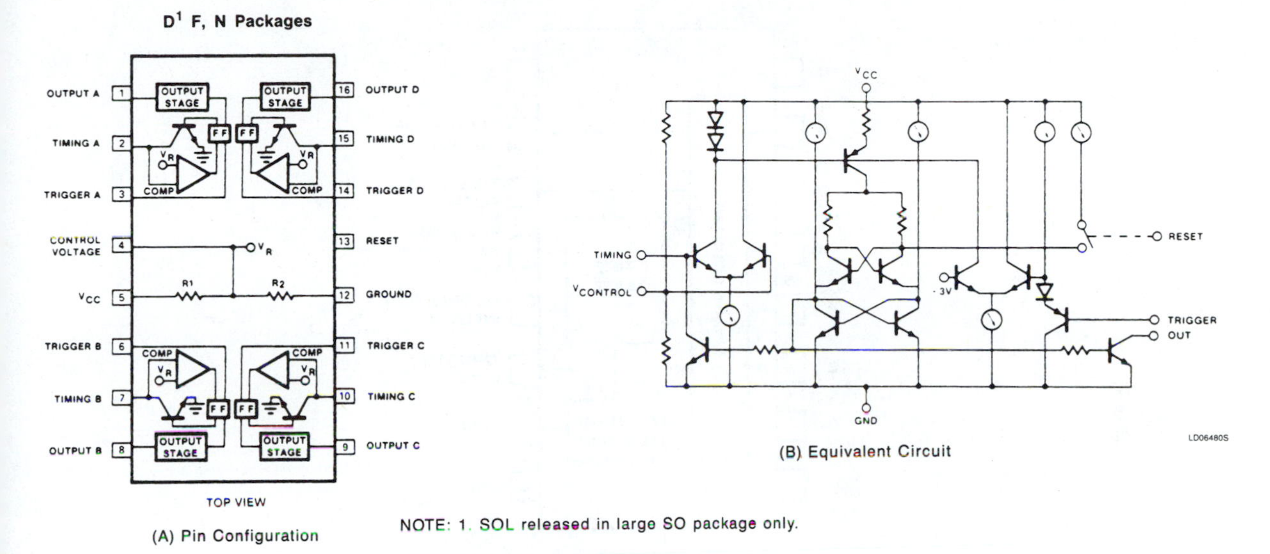


**Figure 11–14.** 558 Quad Timer (Courtesy of Signetics Company)

independent timing functions can be achieved. The time is precisely controlled by one external resistor and one capacitor. Two sections can be interconnected for astable operation, or all four sections can be used together, in tandem, for sequential timing applications up to several hours.

The four sections in the 558 are edge-triggered, which means that, when used in tandem for sequential timing applications, no coupling capacitors are required.

Application note AN171 in Appendix J shows several applications for the 558 quad timer.

# 11.10 2240 Programmable Timer/Counter

The 2240 programmable timer/counter shown in Figure 11–15 is a monolithic controller capable of producing accurate delays from µs to five days. By cascading two of these timers, delays up to three years can easily be generated. The timer consists of a time-base oscillator (TBO), a programmable eight-bit counter, and a control flip-flop. An external *RC* network sets the oscillator frequency and allows delay times from 1*RC* to 255*RC* to be selected. In the astable mode of operation, 255 frequencies or pulse patterns can be generated from a single *RC* network. These frequencies or pulse patterns can also easily be synchronized to an external signal. The trigger, reset, and outputs are all TTL- and diode-transistor logic (DTL)-compatible for easy interface with digital systems. The timer's high accuracy and versatility in producing a wide range of time delays makes it ideal as a direct replacement for mechanical or electromechanical devices.

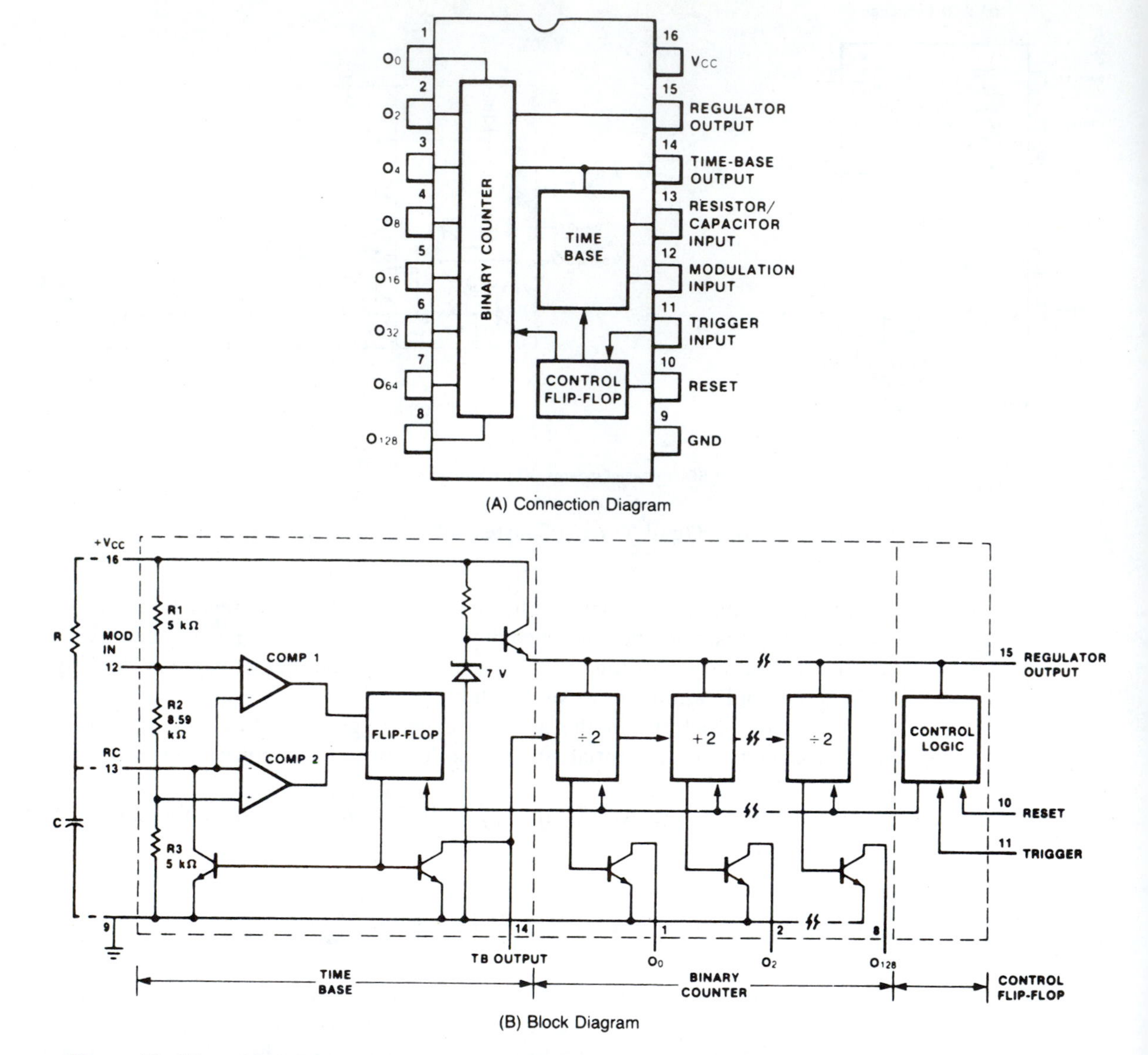


Figure 11–15. 2240 Programmable Timer/Counter (Courtesy of Fairchild Camera & Instrument Corporation)

## Operation

When power is applied to the 2240 with no trigger or reset inputs, the circuit starts with all outputs HIGH. The timing cycle is initiated with a positive-going trigger pulse at pin 11. This trigger pulse activates the TBO, enables the counter section, and sets the counter outputs LOW. The TBO generates timing pulses with a period of $T = 1RC$. These clock pulses are counted by the binary counter section. The timing sequence is completed when a positive-going reset pulse is applied to pin 10, which returns all outputs to HIGH.

Once triggered, the circuit is immune from additional trigger pulses until the timing cycle is completed or a reset input is applied. If both the reset and trigger are activated at the same time, the trigger takes precedence.

## Basic Circuit

The basic circuit connection for timing applications is shown in Figure 11–16. With switch $S_1$ closed, the device is operating in the monostable mode. With $S_1$ open, the device is operating in the astable mode.

## Monostable Operation

In precision timing applications the 2240 is used in its monostable mode. The output is normally HIGH and goes LOW following a trigger input. It remains in the LOW state for the time duration, $T_{out}$, and then returns to the HIGH state.

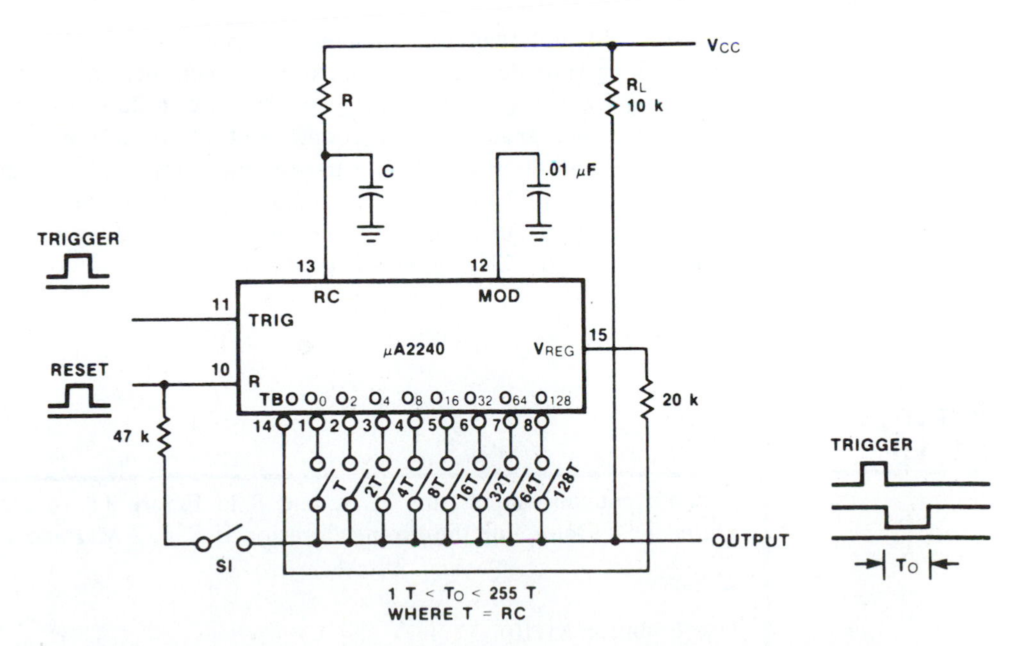


Figure 11–16. Basic Circuit Connection for Timing Applications (Courtesy of Fairchild Camera & Instrument Corporation, Linear Division)

The duration of the timing cycle is

$$T_{out} = N\tau \quad \textbf{(11.10)}$$

where

$\tau = RC$ = time base as set by the choice of timing components at $RC$ pin 13

$N$ = an integer between 1 and 255 as determined by the combination of counter outputs $0_0 \ldots 0_{128}$, pins 1 through 8, connected to the output bus.

The binary counter outputs, $0_0 \ldots 0_{128}$, pins 1 through 8, are open-collector type stages and can be shorted together to a common pull-up resistor to form a wired-OR connection; the combined output will be LOW as long as any one of the outputs is LOW. The time delays associated with each counter output can be added together by simply shorting the outputs together to form a common bus, as shown in Figure 11–16. For example, if only pin 6 is connected to the output and the rest are left open, the total timing duration $T_{out}$ is $32\tau$. However, if pins 1, 5, and 6 are shorted to the output bus, the total time delay is $T_{out} = (1 + 16 + 32)\tau = 49\tau$. In this manner, by proper choice of counter terminals connected to the output bus, the timing cycle can be programmed to be any duration between $1\tau$ and $255\tau$.

### Cascading Units

Two 2240 units can be cascaded, as shown in Figure 11–17, to generate extremely long time delays. The total timing cycle of two cascaded units can be programmed from $T_{out} = 256\tau$ to $T_{out} = 65{,}536\tau$ in 256 discrete steps by selectively shorting one or more of the counter outputs from unit 2 to the output bus. In this application the reset and the trigger input terminals of both units are tied together and the unit 2 time base is disabled. Normally, the output is HIGH when the system is reset. On triggering, the output goes LOW where it remains for a total of $(256)^2$, or 65,536, cycles of the TBO.

## Example 11.10

Assume counter output pins 2, 6, and 8 in Figure 11–16 are shorted to the output bus. Determine the timing duration if $R = 2\ \text{M}\Omega$ and $C = 1\ \mu\text{F}$.

*Solution*

Use Equation 11.10:

$$T_{out} = N\tau = (2 + 32 + 128)(2 \times 10^6\ \Omega \times 1 \times 10^{-6}\ \text{F}) = (162)(2)$$
$$= 324\ \text{s}$$

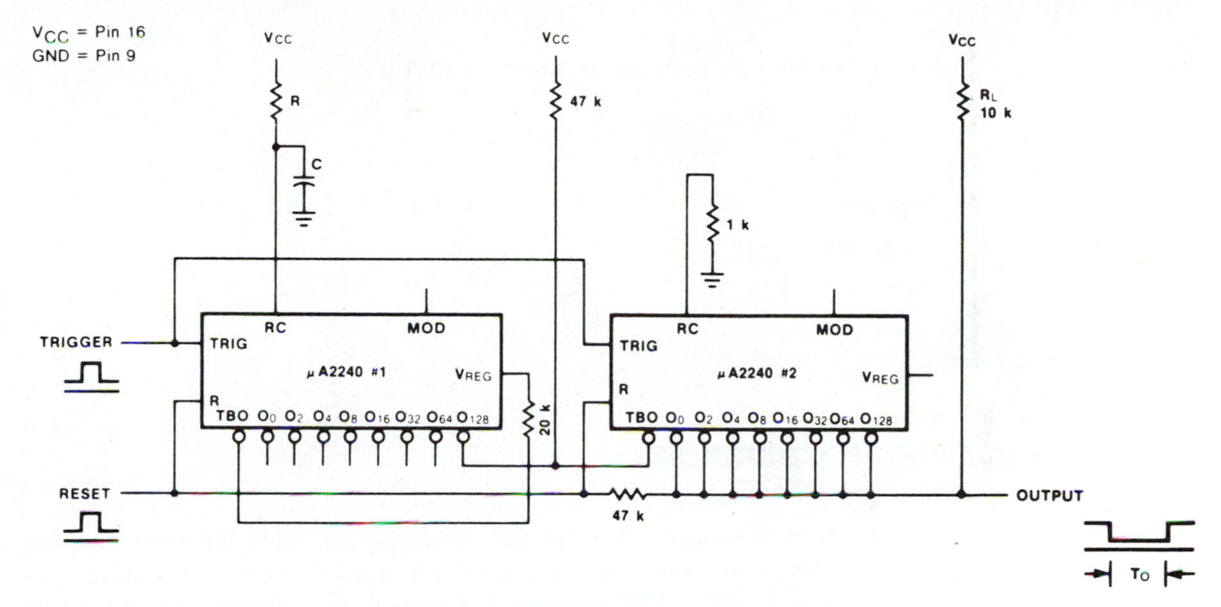


**Figure 11–17.** Circuit for a Long Time Delay (Courtesy of Fairchild Camera & Instrument Corporation, Linear Division)

## Example 11.11

Assume pins 1, 3, 5, and 7 of unit 2 in Figure 11–17 are shorted to the output bus. $R = 10\ \text{M}\Omega$ and $C = 100\ \mu\text{F}$. Determine $T_{out}$.

*Solution*

Use Equation 11.10:

$$\begin{aligned} T_{out} &= N\tau = ((256)(1 + 4 + 16 + 64))((10 \times 10^6\ \Omega)(100 \times 10^{-6}\ \text{F})) \\ &= (21760)(1000\ \text{s}) \\ &= 21.76 \times 10^6\ \text{s} = 361{,}670\ \text{min} = 6028\ \text{hr} \\ &= 251\ \text{days, } 3\ \text{hr, } 50\ \text{min} \end{aligned}$$

## Example 11.12

Design a long time delay system using two 2240 units with all outputs tied together, to provide a timing cycle, $T_{out}$, of one year (365 days).

*Solution*

First, calculate the timing cycle in seconds:

$$T_{out} = \frac{60\ \text{s}}{1\ \text{min}} \times \frac{60\ \text{min}}{1\ \text{hr}} \times \frac{24\ \text{hr}}{1\ \text{day}} \times 365\ \text{days} = 31.536 \times 10^6\ \text{s}$$

Next solve for $\tau$ by rearranging Equation 11.10:

$$\tau = \frac{T_{out}}{N} = \frac{31.536 \times 10^6 \text{ s}}{65{,}536} = 481.2 \text{ s}$$

Now select a suitable value for $C$, say 100 $\mu$F, and solve for $R$:

$$\tau = RC = 481.2 \text{ s}$$

and

$$R = \frac{\tau}{C} = \frac{481.2 \text{ s}}{100 \times 10^{-6} \text{ F}} = 4.8 \text{ M}\Omega$$

## Astable Operation

The 2240 can be operated in the astable mode by disconnecting reset terminal pin 10 from the counter outputs. The circuit in Figure 11–18 operates in its free-running mode with external trigger and reset signals. It starts counting and timing following a trigger input, continuing until an external reset pulse is applied. When a positive-going signal is applied to pin 10, the circuit reverts to its reset state, or HIGH outputs. This circuit is essentially the same as that in Figure 11–16 with switch $S_1$ open.

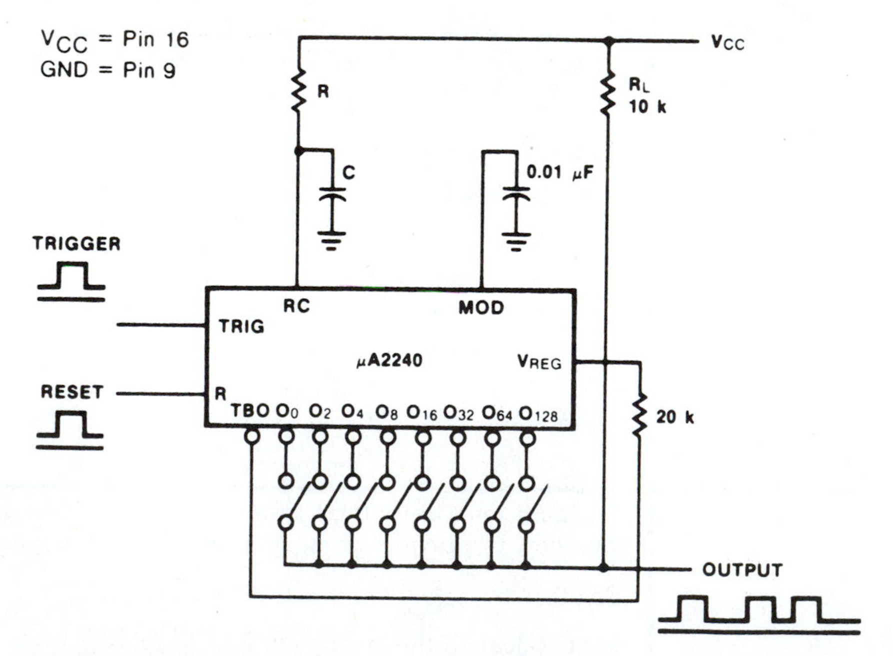


Figure 11–18. Astable Operation with External Trigger and Reset Inputs (Courtesy of Fairchild Camera & Instrument Corporation, Linear Division)

### Continuous Operation

The circuit in Figure 11–19 is designed for continuous operation. It self-triggers automatically when the power supply is turned on, and continues to operate in its free-running mode indefinitely.

### Synchronous Operation

When operating in the astable, or free-running, mode, the counter outputs can be used individually as synchronized oscillators, or they can be interconnected to generate complex pulse patterns, such as those shown in Figure 11–20. The pulse pattern repeats itself at a rate equal to the period of the highest counter bit connected to the common output bus. The minimum *PW* contained in the pulse train is determined by the lowest counter bit connected to the output.

### External Clock Operation

The 2240 can be operated with an external clock or time base when connected as shown in Figure 11–21. The internal TBO is disabled by connecting a 1 kΩ resistor from *RC* pin 13 to ground. The counters are triggered on the negative-going edges of the external clock pulse applied to the TBO pin 14. For proper operation, a minimum clock pulse amplitude of 3 V and a clock *PW* of more than 1 μs is required.

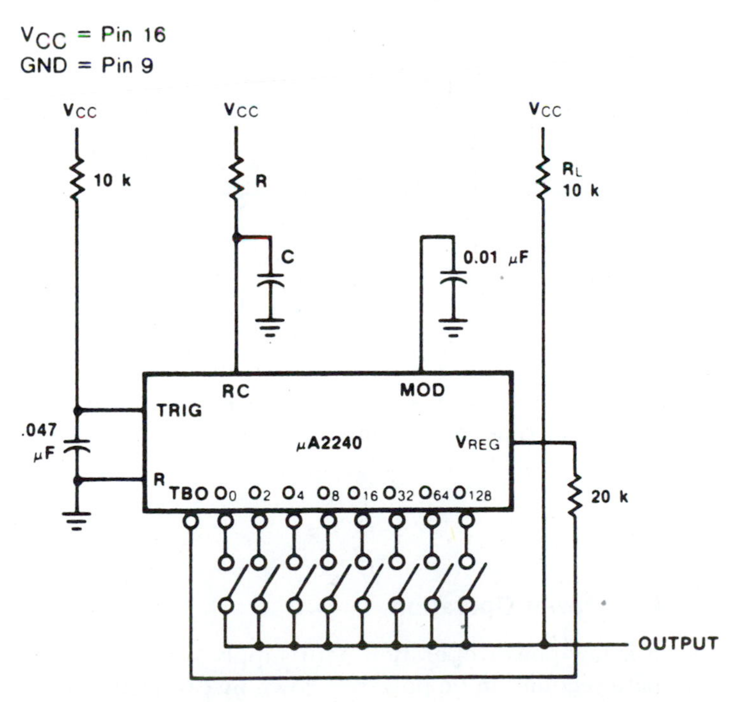


Figure 11–19. Free-Running Continuous Operation (Courtesy of Fairchild Camera & Instrument Corporation, Linear Division)

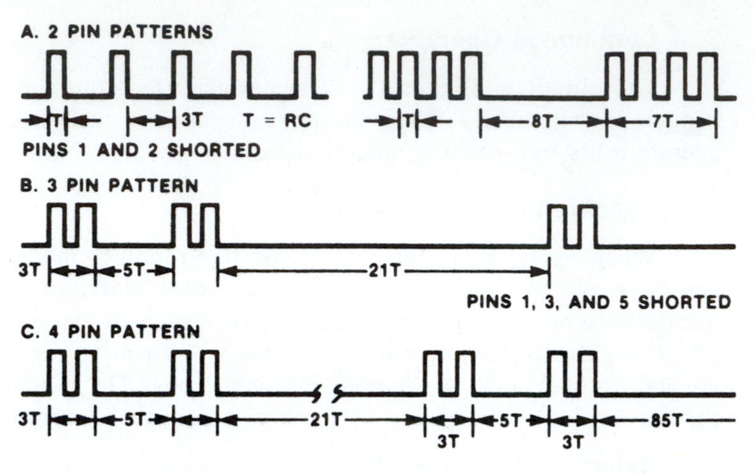


Figure 11–20. Binary Pulse Patterns (Courtesy of Fairchild Camera & Instrument Corporation, Linear Division)

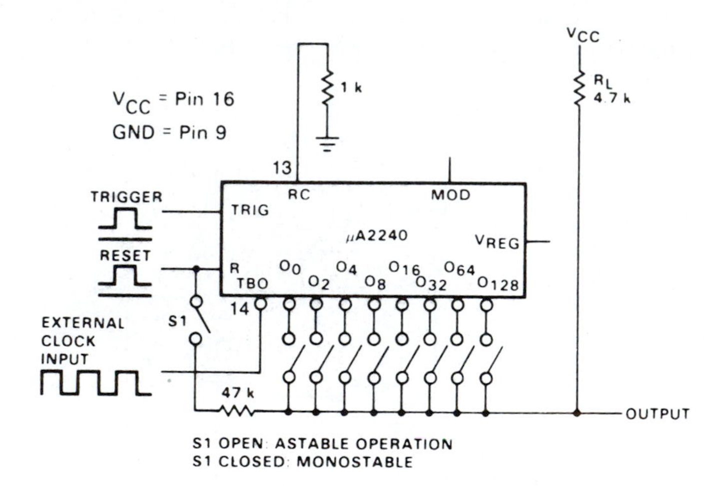


Figure 11–21. Operation with an External Clock (Courtesy of Fairchild Camera & Instrument Corporation, Linear Division)

### Low Power Operation

For low power operation with supply voltages of 6 V or less, the internal time-base section can be powered down by connecting $V_{CC}$ to pin 15 and leaving pin 16 open. In this configuration the internal time base does not draw any current, and the overall current drain is reduced by approximately 3 mA.

# 11.11 LM322 Precision Timer

The LM322 shown in Figure 11–22 is a precision timer of great versatility and high accuracy. It operates with unregulated supplies of 4.5–40 V while maintaining constant timing periods from μs to hours. Internal logic and regulator

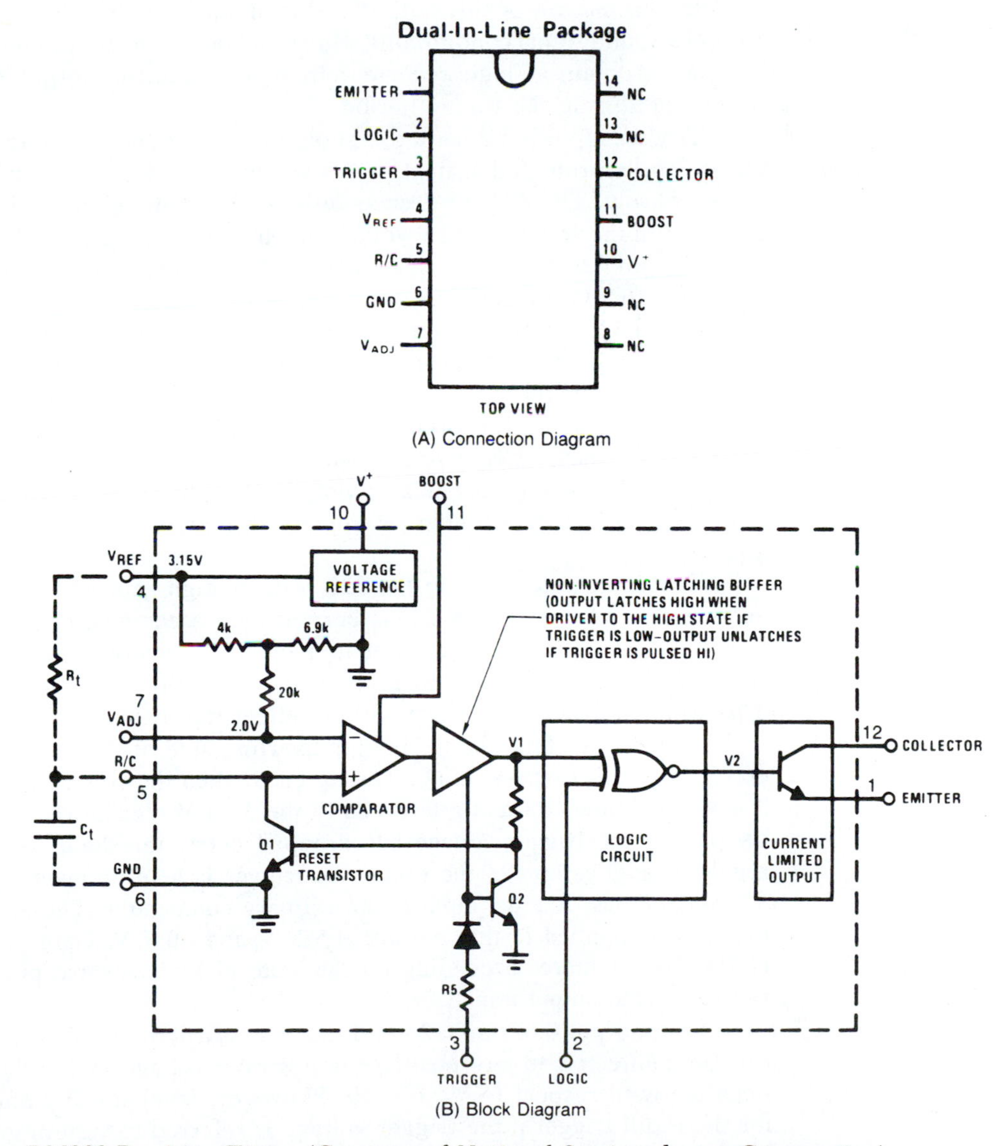

(A) Connection Diagram

(B) Block Diagram

**Figure 11–22.** LM322 Precision Timer (Courtesy of National Semiconductor Corporation)

circuits complement the basic timing function and enable the 322 to operate in many different applications with a minimum of external components.

## Operation

The output of the timer is a floating transistor with built-in current limiting. It can drive either ground-referred or supply-referred loads up to 40 V and 50 mA. The floating nature of this output makes it ideal for interfacing, driving lamps and relays, and signal conditioning where an open collector or emitter is required. You can program a "logic reverse" circuit to make the output transistor either ON or OFF during the timing period.

Trigger input pin 3 has a threshold of 1.6 V independent of supply voltage, but it is fully protected against inputs as high as ±40 V, even when using a 5 V power supply. The circuitry reacts only to the rising edge of the trigger pulse, and is immune to any trigger voltage during the timing periods.

$V_{ref}$ pin 4 is the output of an internal 3.15 V regulator referenced to ground pin 6. Up to 5.0 mA can be drawn from this regulator for driving external loads. An internal 2 V divider between the reference and ground sets the timing period to 1*RC*. The timing period can be voltage-controlled by driving this divider with an external source through $V_{ADJ}$ pin 7. Timing ratios of 50:1 can be easily achieved. In most applications the timing resistor is tied to $V_{ref}$, but it need not be if a more linear charging current is required. The regulated voltage is very useful in applications where the 322 is not used as a timer. Such applications are switching regulators, variable reference comparators, and temperature controllers.

The comparator used in the 322 utilizes high-gain PNP input transistors to achieve 300 pA typical input bias current over a common-mode range of 0–3 V. Boost terminal pin 11 allows you to increase comparator operating current for timing periods less than 1 millisecond (ms). This allows the timer to operate over a 3 μs to multihour timing range with excellent repeatability.

The *RC* pin 5 is tied to the noninverting side of the comparator and to the collector of reset transistor $Q_1$. Timing ends when the voltage on this pin reaches 2.0 V (1*RC* time-constant referenced to the 3.15 V regulator). Transistor $Q_1$ turns ON only if the trigger voltage has dropped below threshold. In comparator and regulator applications of the timer, the trigger is held permanently HIGH and the *RC* pin acts just like the input to an ordinary comparator. The maximum voltages that can be applied to this pin are +5.5 V and -0.7 V. Gain of the comparator is 200,000 or more, depending on the state of logic reverse pin 2 and the connection of the output transistor.

Ground pin 6 of the 322 need not necessarily be tied to system ground. It can be connected to any positive or negative voltage as long as the supply is negative with respect to $V^+$ pin 10. However, level shifting may be necessary for the input trigger if the trigger voltage is referred to system ground. This can be done by capacitive coupling, or by actual resistive or active level shifting.

### Output Variations

The emitter and collector outputs of the timer can be treated just as if they are an ordinary transistor with 40 V minimum collector-emitter breakdown voltage. Normally, the emitter is tied to ground and the signal is taken from the collector, or the collector is tied to $V^+$ and the signal is taken from the emitter. Variations of these basic connections are possible. The collector can be tied to any positive voltage up to 40 V when the signal is taken from the emitter. However, the emitter will not be pulled higher than the supply voltage on $V^+$ pin 10. Connecting the collector to a voltage less than $V^+$ is allowed, but the emitter should not be connected to a low impedance load other than that to which the ground pin is tied.

Logic pin 2 is used to reverse the signal that appears at the output transistor. An open or HIGH condition on pin 2 programs the output transistor to be OFF during the timing period and ON all other times. Grounding pin 2 reverses the sequence, making the transistor ON during the timing period. Minimum and maximum voltages that may appear on the logic pin are 0 V and +5.0 V, respectively.

## Basic Circuits

A basic timer using the 322 is shown in Figure 11–23A. The output is taken from the collector. $R_t$ and $C_t$ set the time interval with $R_L$ as the load. During the timing interval, the output may be either HIGH or LOW depending on the connection of the logic pin. In the timing waveforms shown, note that the trigger pulse can be either shorter or longer than the output *PW*.

Another basic timer using the 322 is shown in Figure 11–23B. In this circuit, the output is taken from the emitter of the output transistor. As with the collector output, the output taken from the emitter may be either HIGH or LOW during the timing interval. Also, the trigger input may be either longer or shorter than the output *PW*.

### Self-starting

The 322 can be made into a self-starting oscillator by feeding the output back to the trigger input through a capacitor, as shown in Figure 11–24A. The operating frequency is determined by

$$f_{op} = \frac{1}{(R_t + R_1)(C_t)} \tag{11.11}$$

The output is a narrow negative pulse whose width is approximately twice the product of resistor $R_2$ and feedback capacitor $C_f$:

$$PW = 2R_2C_f \tag{11.12}$$

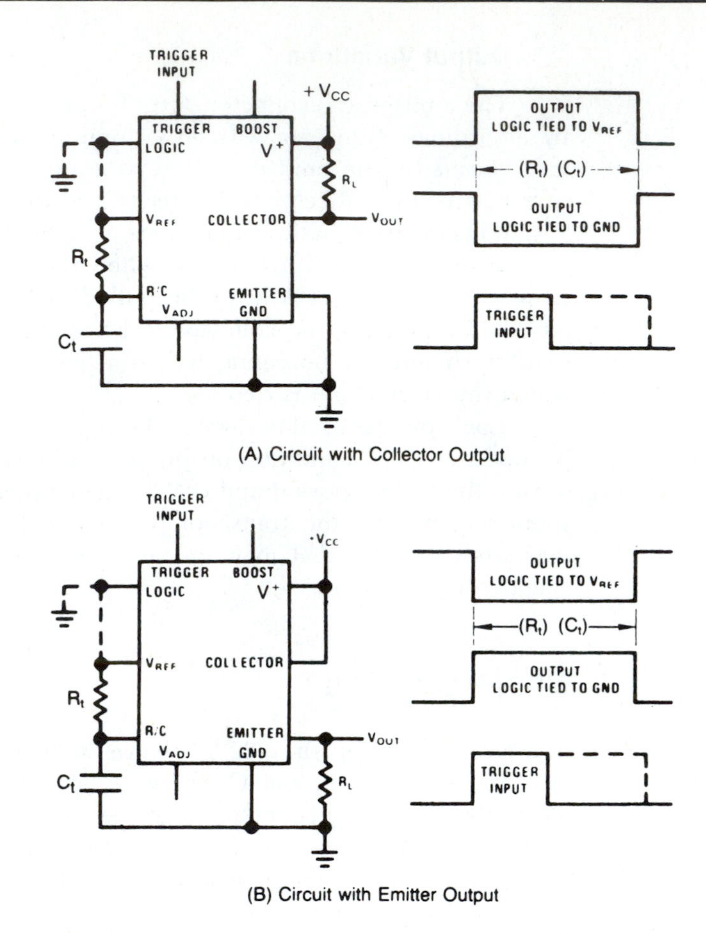


(A) Circuit with Collector Output

(B) Circuit with Emitter Output

**Figure 11–23.** Basic Timer Circuits (Courtesy of National Semiconductor Corporation)

For optimum frequency stability, $C_t$ should be as small as possible. The minimum value is determined by the time required to discharge $C_t$ through the internal discharge transistor. A conservative value for $C_t$ can be chosen from the graph shown in Figure 11–24B. For frequencies below 1 kHz, the frequency error introduced by $C_t$ is a few tenths of one percent or less for $R_t$ greater than 500 kΩ.

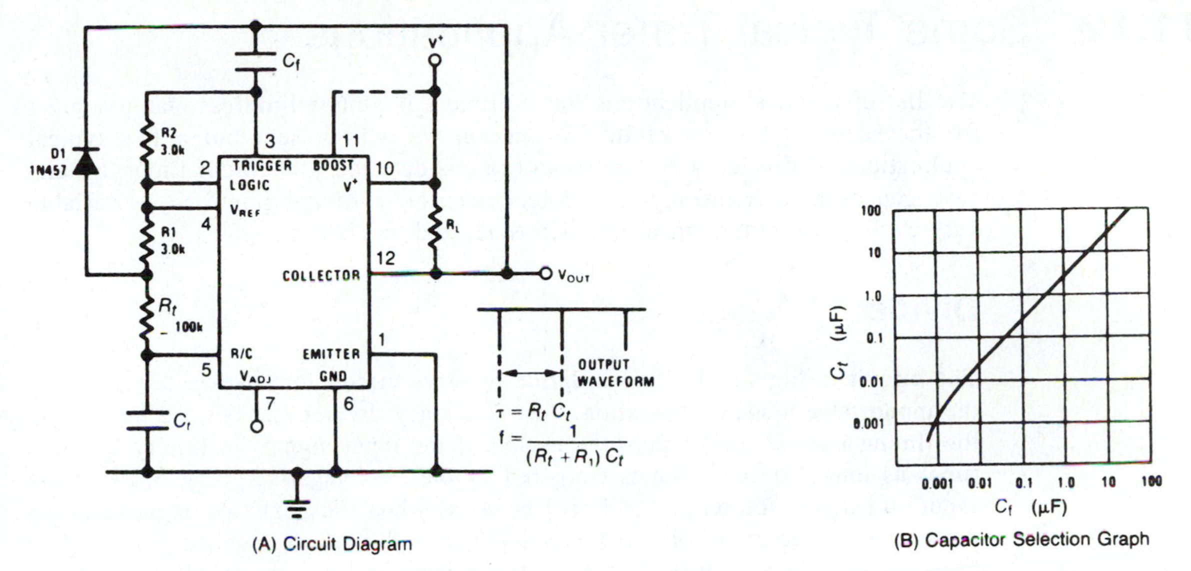


**Figure 11–24.** An Astable Oscillator (Courtesy of National Semiconductor Corporation)

## Example 11.13

Determine (a) the operating frequency of the self-starting oscillator in Figure 11–24A and (b) the $PW$ of the output. Let $C_t = 0.015\ \mu\text{F}$.

*Solutions*

a. Use Equation 11.11:

$$f_{\text{op}} = \frac{1}{(R_t + R_1)(C_t)} = \frac{1}{(100\ \text{k}\Omega + 3\ \text{k}\Omega)(0.015\ \mu\text{F})} = 647\ \text{Hz}$$

b. First determine $C_f$ from Figure 11–24B, which indicates a value of approximately 0.01 μF for $C_t = 0.015\ \mu\text{F}$. Then use Equation 11.10 to solve for $PW$:

$PW = 2R_2C_f = 2 \times 3\ \text{k}\Omega \times 0.01\ \mu\text{F} = 60\ \mu\text{s}$

# 11.12 Some Typical Timer Applications

The list of practical applications for IC timers is almost limitless, far too great for the scope of this book. In this section we will present but a few typical applications—a divider, a *PW* modulator, a *PW* detector, a one-hour timer, a staircase generator, a frequency-to-voltage converter, a digital interface, a variable duty-cycle pulse generator, a digital S & H, and an eight-bit ADC.

## Divider

The circuit in Figure 11–25 is a divide-by-three circuit that uses a 555 timer in the monostable mode of operation. The frequency divider is designed by making the timing interval longer than the period of the input signal, in this case, three times as long. The one-shot is triggered by the first negative-going pulse of the input signal, but the output will still be HIGH when the next two negative-going pulses occur. The one-shot will be retriggered on the fourth negative-going pulse. There is one output pulse for every three input pulses, therefore the input is divided by three.

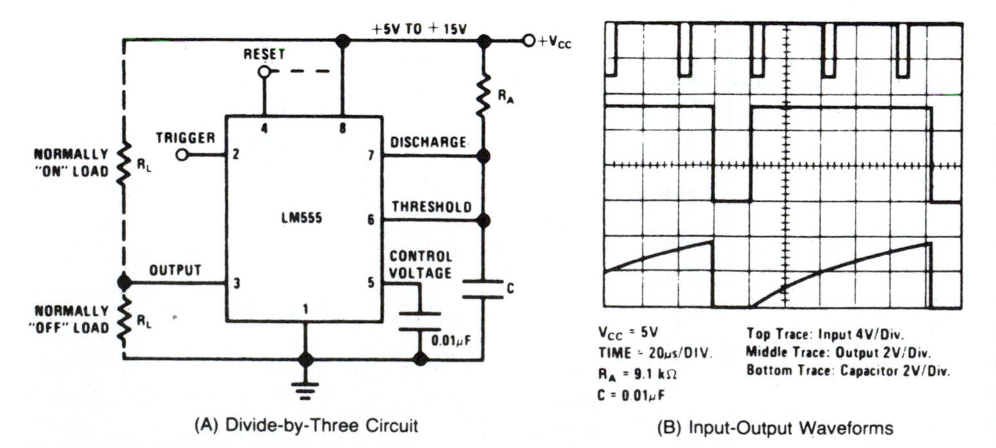


(A) Divide-by-Three Circuit

(B) Input-Output Waveforms

**Figure 11–25.** Frequency Divider Circuit (Courtesy of National Semiconductor Corporation)

## Pulse Width Modulator

When the timer is connected in the monostable mode and triggered with a continuous pulse train, the output *PW* can be modulated by a signal applied to pin 5. The 555 in Figure 11–26A is connected as a *PW* modulator, and Figure 11–26B shows some waveform examples.

## Pulse Width Detector

A simple, accurate *PW* detector using the LM322 is shown in Figure 11–27. In this application the logic terminal is normally held HIGH by resistor $R_3$. When a trigger pulse is received, transistor $Q_1$ is turned ON, driving the logic terminal to ground. The result of triggering the timer and reversing the logic at the same time is that the output does not change from its initial LOW condition. The only time the output will change states is when the trigger input stays HIGH longer than one time period set by $R_t$ and $C_t$. The output *PW* is equal to the input trigger width minus $R_tC_t$. $C_2$ insures no output pulse for short trigger pulses (those less than *RC*) by prematurely resetting the timing capacitor when the trigger pulse drops. $C_L$ filters the narrow spikes that would occur at the output due to propagation delays during switching.

## One-hour Timer

The LM322 connected as a one-hour timer with manual controls for start, reset, and cycle end is illustrated in Figure 11–28. Switch $S_1$ starts timing, but has no effect after timing has started. Switch $S_2$ is a center-OFF switch that can either end the cycle prematurely with the appropriate change in output state and discharging of $C_t$, or cause $C_t$ to be reset to 0 V without a change in output. In

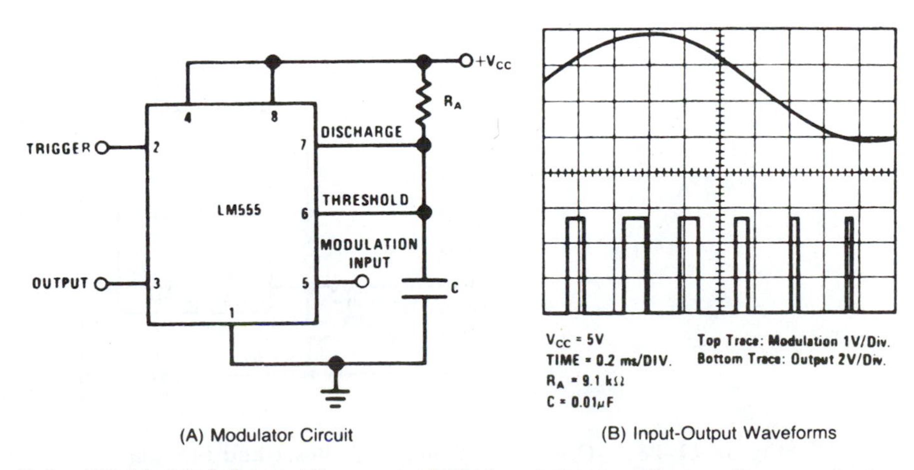


(A) Modulator Circuit (B) Input-Output Waveforms

**Figure 11–26.** Pulse Width Modulator (Courtesy of National Semiconductor Corporation)

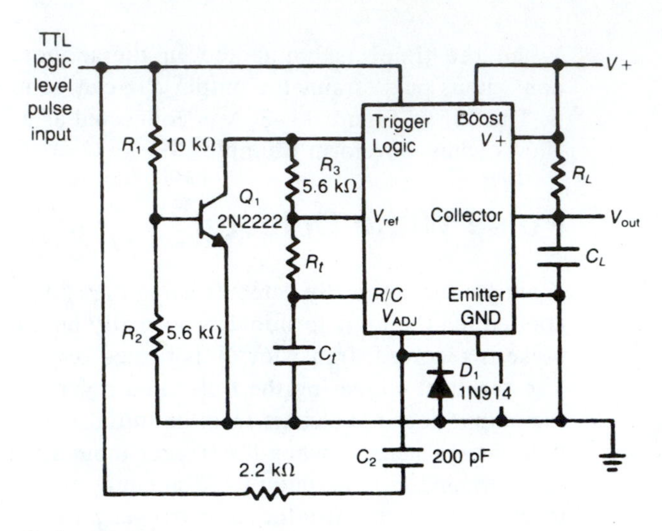


**Figure 11–27.** Pulse Width Detector (Courtesy of National Semiconductor Corporatio

the latter case, a new timing period starts as soon as $S_2$ is released. The average charging current through $R_t$ is about 30 nA, so some attention must be paid to parts layout to prevent stray leakage paths. The suggested timing capacitor has a typical self time constant of 300 hours and a guaranteed minimum of 25 hours at +25 °C. Other capacitor types may be used if sufficient information is available on their leakage characteristics.

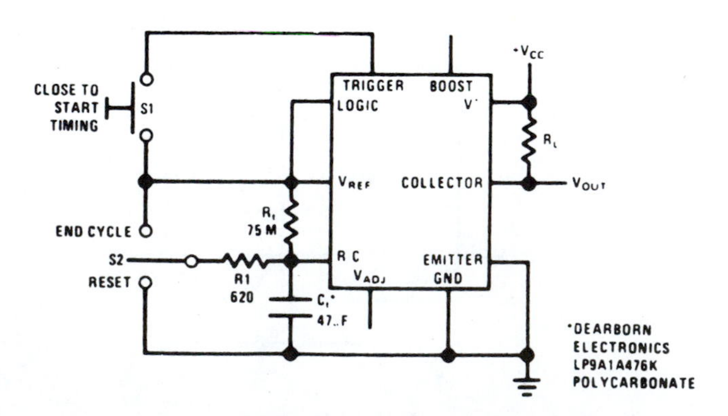


**Figure 11–28.** One-Hour Timer with Reset and Manual Cycle End (Courtesy of National Semiconductor Corporation)

## Staircase Generator

The 2240 timer/counter can be interconnected with an external op amp and a precision resistor ladder to form a staircase generator, as shown in Figure 11-29. Under reset conditions, the output is LOW. When a trigger is applied, the op amp output goes HIGH and generates a negative-going staircase of 256 equal steps. The time duration of each step is equal to the time-base period $T$. The staircase can be stopped at any level by applying a disable signal to pin 14 through a steering diode. The count is stopped when pin 14 is clamped at a voltage level not greater than 1.0 V.

## Frequency-to-Voltage Converter

An accurate frequency-to-voltage converter can be made with the LM322 by averaging output pulses with a simple filter, as illustrated in Figure 11-30. *PW* is adjusted with $R_2$ to provide initial calibration at 10 kHz. The collector of the output transistor is tied to $V_{ref}$, giving constant amplitude pulses equal to $V_{ref}$ at the emitter output. $R_4$ and $C_1$ filter the pulses to give a dc output equal to

$$V_{out(dc)} = R_t C_t V_{ref} f_{op} \tag{11.13}$$

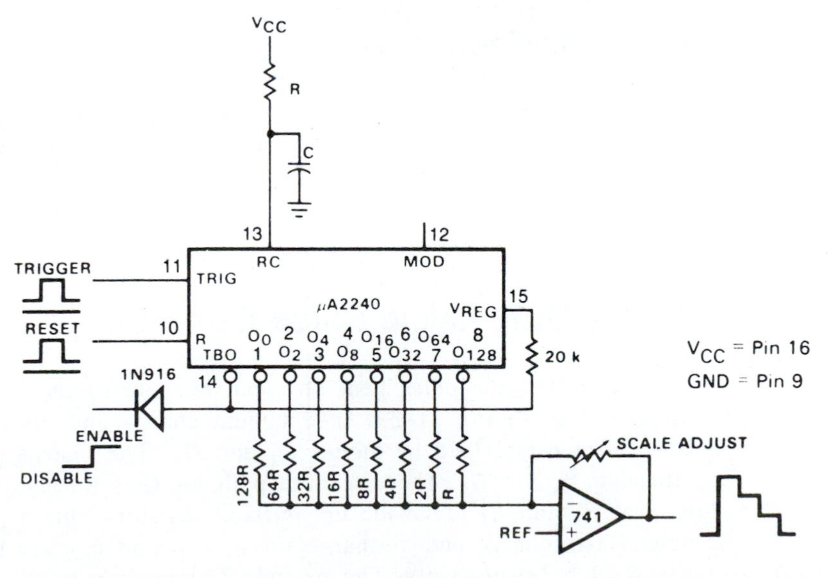


Figure 11-29. Staircase Generator (Courtesy of Fairchild Camera & Instrument Corporation, Linear Division)

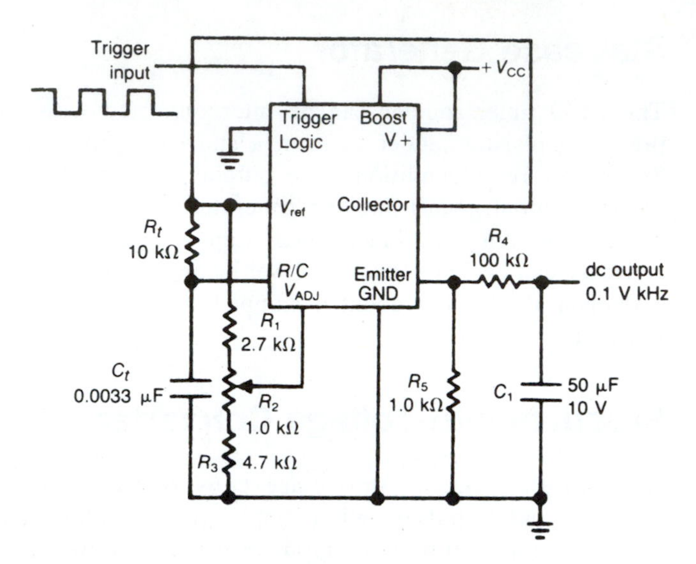


**Figure 11–30.** Frequency-to-Voltage Converter (Courtesy of National Semiconductor Corporation)

Linearity is about 0.2% for a 0–1 V output. If better linearity is desired, $R_5$ can be tied to the summing node of an op amp that has the filter in the feedback path. If a low output impedance is desired, a unity gain buffer can be tied to the output. An analog meter can be driven directly by placing it in series with $R_5$ to ground. A series *RC* network across the meter to provide damping will improve response at very low frequencies.

## Digital Interfacing

The ability of the timer to interface to digital logic when operating off a high supply voltage is demonstrated in Figure 11–31. $V_{out}$ swings between +5 V and ground with a minimum fanout of 5 for medium-speed TTL. If the logic is sensitive to rise/fall time of the trailing edge of the output pulse, the trigger pin should be LOW at that time.

## Variable Duty-Cycle Pulse Generator

A variable duty-cycle pulse generator can be constructed using the 555 timer, as illustrated in Figure 11–32. Independent charge and discharge paths for capacitor *C* are established by diodes $D_1$ and $D_2$. The charge path for *C* is from $V_{CC}$ through $R_A$ and $D_1$. The discharge path for *C* is through $D_2$, $R_B$, and pin 7. (Note that $R_A$ and $R_B$ are made up of fixed resistors plus a portion of a potentiometer.) The charge and discharge times, $t_{high}$ and $t_{low}$, are given by Equations 11.4 and 11.5, respectively. The period, *T*, and duty cycle, *D*, for this circuit are as follows:

$$T = 0.7(R_A + R_B)C \tag{11.14}$$

and

$$D = \frac{R_B}{R_A + R_B} \times 100\% \tag{11.15}$$

With this circuit configuration, duty cycles from 1% to 99% can be achieved.

## Example 11.14

In Figure 11-32 assume the potentiometer to be set so that $R_A = 20\ \text{k}\Omega$. Calculate (a) the period $T$ and (b) the duty cycle.

*Solutions*

a. Use Equation 11.14:
$T = 0.7(R_A + R_B)C = 0.7(20\ \text{k}\Omega + 100\ \text{k}\Omega)0.01\ \mu\text{F} = 840\ \mu\text{s}$

b. Use Equation 11.15:
$$D = \frac{R_B}{R_A + R_B} \times 100\% = \frac{100\ \text{k}\Omega}{20\ \text{k}\Omega + 100\ \text{k}\Omega} \times 100\% = 83.3\%$$

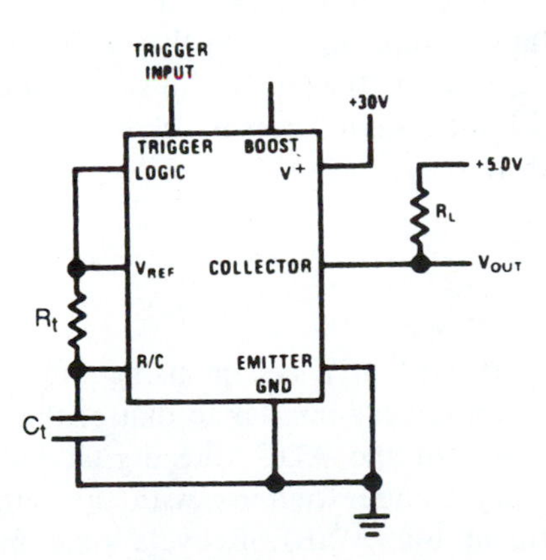


**Figure 11-31.** 30 V Supply Interface with 5 V Logic (Courtesy of National Semiconductor Corporation)

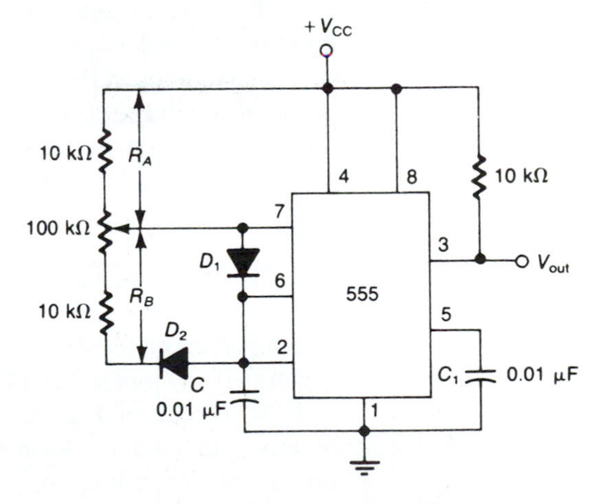


**Figure 11-32.** Variable Duty-Cycle Pulse Generator

**Example 11.15**

In Figure 11–32 assume the potentiometer to be set so that $R_A = 90\ \text{k}\Omega$. Calculate (a) the period $T$ and (b) the duty cycle.

*Solutions*

a. Use Equation 11.14:
$$T = 0.7(90\ \text{k}\Omega + 30\ \text{k}\Omega)0.01\ \mu\text{F} = 840\ \mu\text{s}$$
b. Use Equation 11.15:
$$D = \frac{30\ \text{k}\Omega}{90\ \text{k}\Omega + 30\ \text{k}\Omega} \times 100\% = 25\%$$

Close observation of Examples 11.14 and 11.15 will reveal that $R_A + R_B$ is constant, that $T$ is constant, and that placement of the wiper arm on the potentiometer determines the duty cycle, or the percentage of the total period ($T$) that the output signal is LOW.

## Digital Sample-and-Hold Circuit

A digital S & H circuit that uses the 2240 is shown in Figure 11–33. Circuit operation is similar to that of the staircase generator discussed earlier. When a strobe input is applied, the *RC* LP network between the reset and the trigger inputs resets the timer, then triggers it. This strobe input also sets the output of the bistable latch to a HIGH state and activates the counter.

The circuit generates a staircase voltage at the op amp output. When the level of the staircase reaches that of the analog input to be sampled, the comparator changes state, activates the bistable latch, and stops the count. At this point, the voltage level at the op amp output corresponds to the sampled analog input. Once the input is sampled, it is held until the next strobe signal. Minimum recycle time of the system is approximately 6 ms.

## Eight-Bit ADC

A simple eight-bit analog-to-digital converter (ADC) system using the 2240 is illustrated in Figure 11–34. Circuit operation is very similar to that of the digital S & H circuit in Figure 11–33. In the case of the ADC, the digital output is obtained in parallel format from the binary-counter outputs with the output at pin 8 corresponding to the most significant bit (MSB). Recycle time for this circuit is approximately 6 ms.

There are a great number of other applications for the timers discussed in this chapter, as well as for other available timers. Reference to manufacturers' data sheets and application notes is recommended for those interested in further investigation of timers and timer circuits.

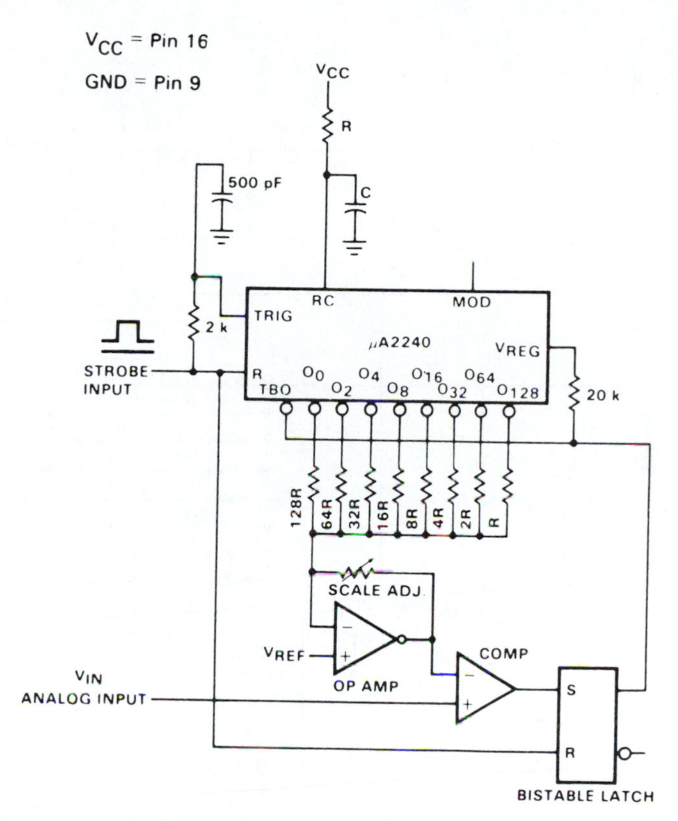


**Figure 11–33.** Digital Sample-and-Hold Circuit (Courtesy of Fairchild Camera & Instrument Corporation, Linear Division)

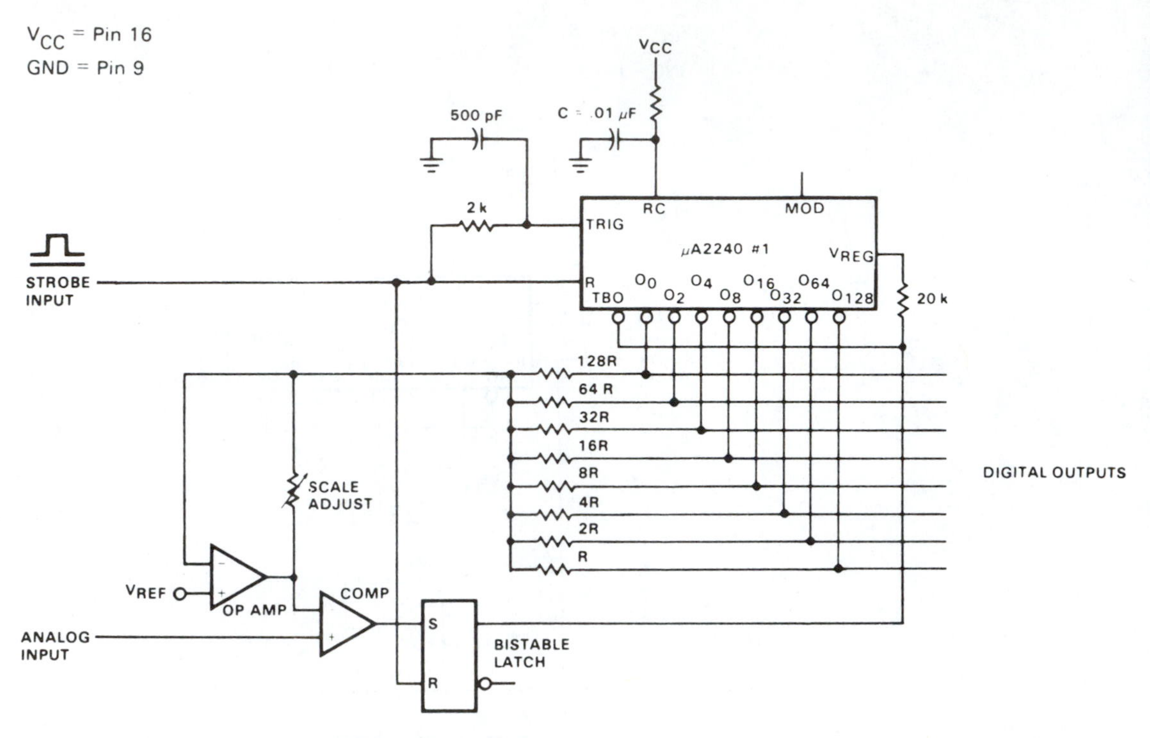


**Figure 11–34.** Analog-to-Digital Converter (Courtesy of Fairchild Camera & Instrument Corporation, Linear Division)

# 11.13 Summary

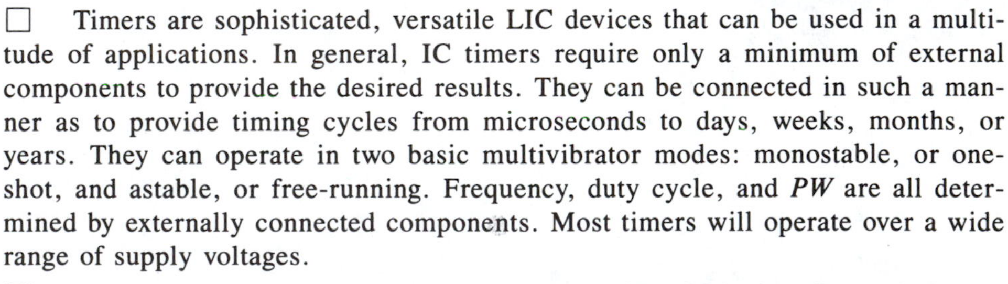

☐ Timers are sophisticated, versatile LIC devices that can be used in a multitude of applications. In general, IC timers require only a minimum of external components to provide the desired results. They can be connected in such a manner as to provide timing cycles from microseconds to days, weeks, months, or years. They can operate in two basic multivibrator modes: monostable, or one-shot, and astable, or free-running. Frequency, duty cycle, and *PW* are all determined by externally connected components. Most timers will operate over a wide range of supply voltages.

☐ One of the most important considerations when designing timer circuits is the careful selection of timing components. Capacitors are the most critical components because of the wide variance in their characteristics, while resistors are more easily selected.

☐ The 555 IC timer is an ideal timer for applications where noisy sources are used. The timer will always time out once triggered, providing immunity to noise pulses.

☐ The 556 is a dual timer consisting of two 555 timers in a single IC package.

☐ The 7555 general-purpose CMOS timer is an improved version of the basic 555 timer. In many applications, the 7555 can replace directly the 555.

☐ The 558 is a quad timer consisting of four timers in a single IC package. When connected in tandem for sequential timing applications, no coupling capacitors are required.

☐ The 2240 programmable timer/counter IC can deliver extremely long time delays when cascaded. Programming the outputs of the device is a simple matter of selecting the number of outputs to be tied to the output bus.

☐ The LM322 precision timer can operate with unregulated supply voltages over the range from 4.5 V to 40 V while maintaining accuracy in output timing periods from microseconds to hours. The 322 can provide outputs that are normally ON or normally OFF by use of a "logic reverse" circuit.

☐ Some of the applications of timers are as follows: frequency dividers, frequency multipliers, *PW* modulators, *PW* detectors, missing pulse detectors, window detectors, square-wave generators, staircase generators, digital S & H circuits, ADCs, frequency-to-voltage converters, voltage-to-frequency converters, dc-dc converters, and variable duty-cycle pulse generators. Many other applications are available through manufacturers' data sheets and application notes.

## 11.14 Questions and Problems

**11.1** What are the basic operating modes of IC timers?

**11.2** List the functional blocks inside a 555 IC timer.

**11.3** Explain the most basic mode of operation of the 555 timer.

**11.4** How many external components are required with a 555 timer operating in the time delay mode? in the free-running mode?

**11.5** Describe the characteristics of the 555 output pin 3.

**11.6** Is there such a thing as an override function on the 555 timer? If so, describe its operation.

**11.7** What is the purpose of a bypass capacitor in timer circuits?

**11.8** What is one of the most important considerations in timer circuit design?

**11.9** What is the best choice of resistor types for timer applications?

**11.10** Describe some of the desirable characteristics of timing capacitors.

**11.11** What is a good choice of timing capacitor types?

**11.12** Are electrolytic capacitors an acceptable choice for timing circuit applications? Explain your answer.

**11.13** Refer to Figure 11–5. Assume values of $R_A = 22\ \text{k}\Omega$ and $C = 0.01\ \mu\text{F}$. Calculate the *PW* of the output pulse.

**11.14** A circuit design calls for a timing capacitor value of 10 μF and a desired output *PW* of 100 seconds. Calculate the value for the timing resistor.

**11.15** A circuit design shows a timing resistor value of 2 MΩ and a desired output *PW* of 1 minute. Calculate the timing capacitor value.

**11.16** Refer to Figure 11–6. Determine the timing component values for an output signal width of 100 ms. (NOTE: There are several. List them all.)

**11.17** What is the purpose of an input-conditioning network?

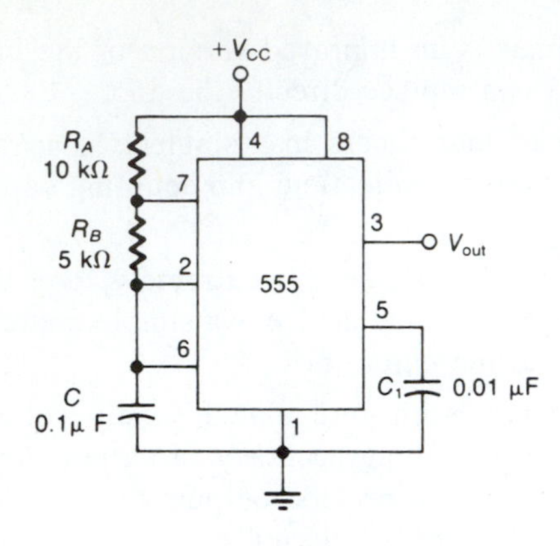


Figure 11-35. Circuit for Problem 11.21 and Problem 11.22

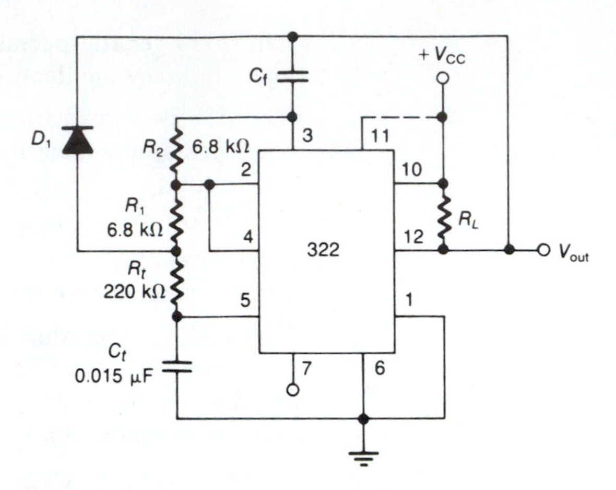


Figure 11-36. Circuit for Problem 11.26

**11.18** What function does the input resistor of the input-conditioning network serve?

**11.19** Refer to Figure 11-7. Assume $R_A$ = 10 MΩ and $C$ = 0.01 μF. Calculate the output *PW*.

**11.20** Refer to Figure 11-7. Assume $R_{in}$ = 10 kΩ and $C_{in}$ = 1000 pF. Calculate the input network time constant.

**11.21** Refer to Figure 11-35. Calculate (a) $t_{high}$, (b) $t_{low}$, (c) $T$, and (d) $f_{op}$.

**11.22** Calculate the duty cycle for the circuit in Figure 11.35.

**11.23** Refer to Figure 11.16. Assume $R$ = 20 MΩ and $C$ = 10 μF. Determine the timing duration for the following conditions: (a) output pins 2, 4, 6, and 8 are shorted to the output bus; (b) output pins 1, 3, 5, and 7 are shorted to the output bus; and (c) all pins are shorted to the output bus.

**11.24** Refer to Figure 11-17. Assume $R$ = 100 MΩ, $C$ = 100 μF, and the desired time delay is exactly 50 days. Determine which pins must be shorted to the output bus.

**11.25** Design a long time delay system with two 2240 units that will provide a timing cycle of 3 years. (HINT: Short all output pins to the output bus. This helps to keep the component values lower.)

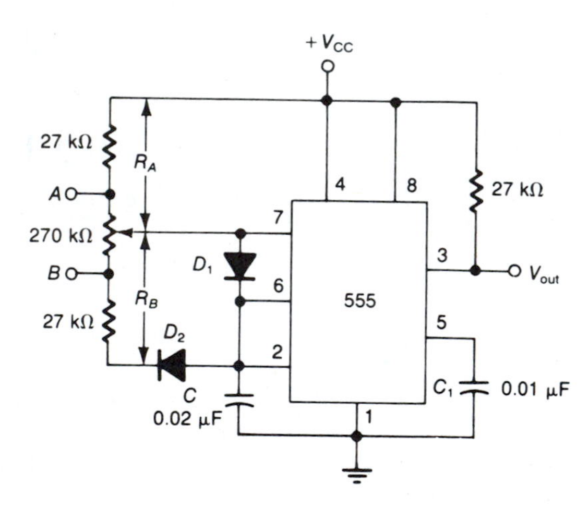


Figure 11-37. Circuit for Problem 11.27

**11.26** Refer to Figure 11-36. Determine the operating frequency and the PW of the output signal. (HINT: Refer to Figure 11-24B to determine the value of $C_f$.)

**11.27** Refer to Figure 11-37. Determine (a) the period $T$ and the duty cycle when the potentiometer is set at point $A$, and (b) the period $T$ and the duty cycle when the potentiometer is set at point $B$.

Chapter 12

# Special Analog Circuits and Devices

## 12.1 Introduction

The wide range of LIC devices that we have discussed up to this point is but a small sample of those available to the electronics technician. The versatility of such devices, and the applications for which they may be used, are limited only by the imagination of the designer. In this chapter we will present some special-purpose devices, as well as practical applications that use many circuits with which you are familiar. Some applications use individual ICs and some use a combination of ICs and discrete components.

## 12.2 Objectives

When you complete this chapter, you should be able to:

- ☐ Describe a variety of consumer-type LIC applications.
- ☐ Design a variety of useful circuits using op amps, monolithic ICs, discrete devices, and external components.

# 12.3 Audio, Radio, and Television Circuits

In this section we will present a wide selection of consumer-type applications of LIC devices. Many of the circuits discussed here can be constructed using either op amps with external components or monolithic ICs with the major circuit components contained within a single package.

## Dual Power Amplifier

The LM377 is a dual power amplifier that offers high-quality performance for stereo phonographs, tape players, recorders, and AM-FM stereo receivers. It will deliver two watts per channel into 8 Ω or 16 Ω loads. The device contains an internal bias regulator to bias each amplifier and contains both internal current limit and thermal shutdown for device protection.

The LM377 operates from a 10–26 V supply, and provides a typical open-loop output gain of 90 decibels (dB), with 70 dB channel separation and 70 dB ripple rejection. The connection diagram is shown in Figure 12–1.

A minimum of external components is required for this device, as shown in the simple stereo amplifier circuit of Figure 12–2. Addition of an *RC* network in the feedback path between the speakers and the inverting input of the two amplifiers, as shown in Figure 12–3, provides bass boost to the basic stereo amplifier.

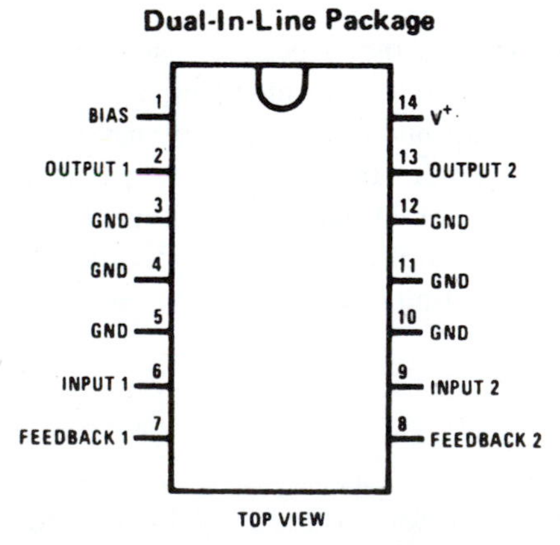


**Figure 12–1.** Connection Diagram for LM377 Dual 2-Watt Audio Amplifier (Courtesy of National Semiconductor Corporation)

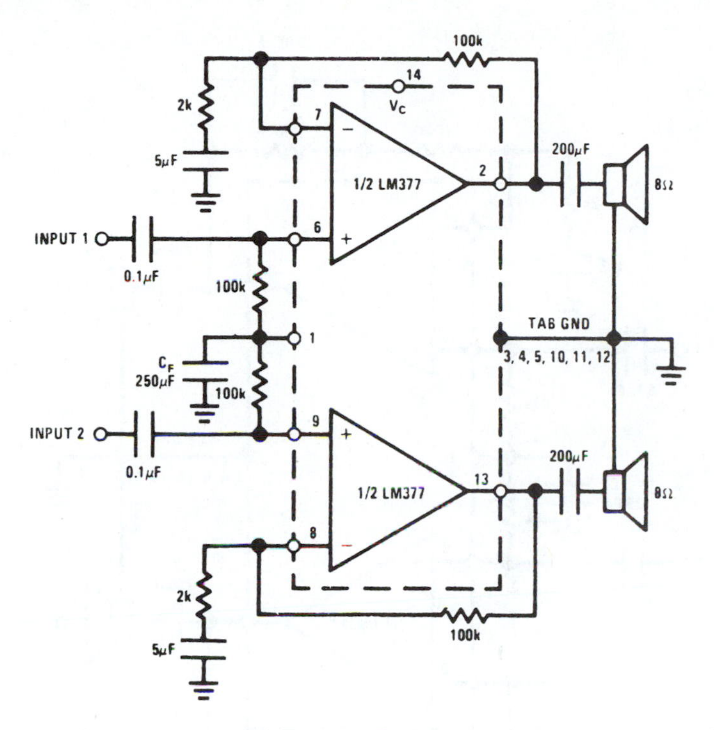


Figure 12–2. Simple Stereo Amplifier Using LM377 (Courtesy of National Semiconductor Corporation)

## Stereo Amplifiers

The LM1877 is a pin-for-pin replacement for the LM377. This device operates with a minimum of external components while still providing flexibility for use in stereo phonographs, tape recorders, and AM-FM stereo receivers. Other applications are servo amplifiers, intercom systems, and automotive products.

The LM1877 operates for a supply range of 6–24 V, has very low crossover distortion, internal current limiting, short circuit protection, and thermal shutdown. It provides 70 dB channel separation and 65 dB ripple rejection. Each amplifier is biased from a common internal regulator to provide high power supply rejection and output $Q$ point centering.

A stereo phonograph amplifier with bass tone control is shown in Figure 12–4A. The frequency response of the bass tone control of this circuit is essentially flat from 100 Hertz (Hz) to 20 kHz. The circuit voltage gain, $A_v$, equals 50 (34 dB) when feeding an 8 Ω load.

The circuit in Figure 12–4B is a stereo amplifier with output voltage gain $A_v$ of 200, which is about 46 dB, into an 8 Ω load.

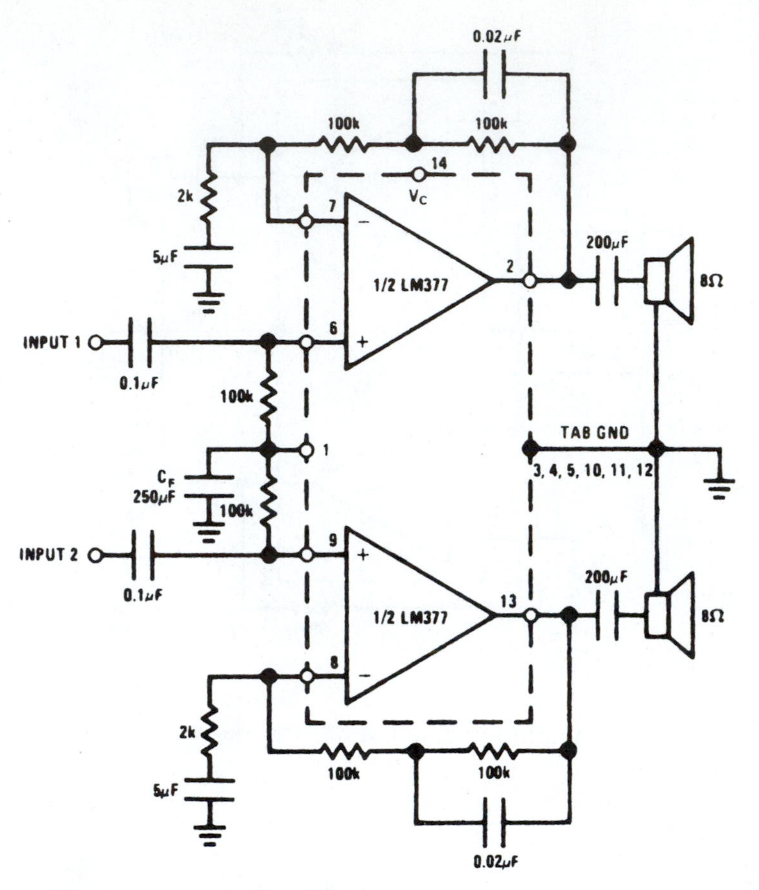


**Figure 12-3.** Stereo Amplifier with Bass Boost (Courtesy of National Semiconductor Corporation)

## Power Audio Amplifier

The LM380, illustrated in Figure 12-5 (page 379), is a power audio amplifier designed for low-cost consumer application. Gain is internally fixed at 34 dB, the unique configuration of the input stage allows inputs to be ground referenced, and the output is automatically self-centering to one-half the supply voltage. The output is short circuit-proof with internal thermal limiting.

Uses of the LM380 include simple phonograph amplifiers (Figure 12-6A), phase-shift oscillators (Figure 12-6B), intercoms (Figure 12-6C), and bridge amplifiers (Figure 12-6D) on page 380. Other uses are as line drivers, teaching machine outputs, alarms, ultrasonic drivers, TV sound systems, AM-FM radios, small servo drivers, and power converters.

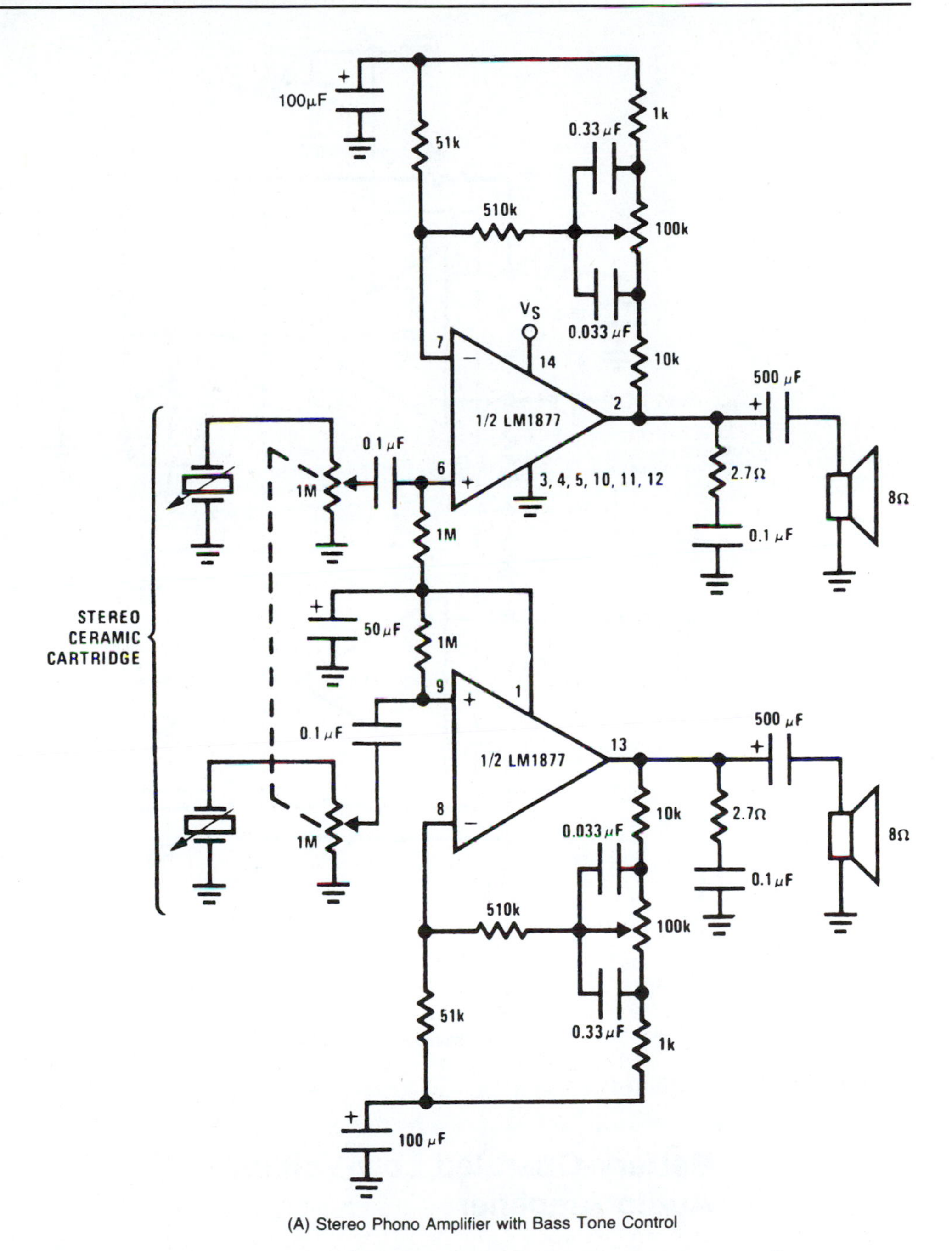


(A) Stereo Phono Amplifier with Bass Tone Control

**Figure 12–4.** Stereo Amplifiers (Courtesy of National Semiconductor Corporation) (Continued)

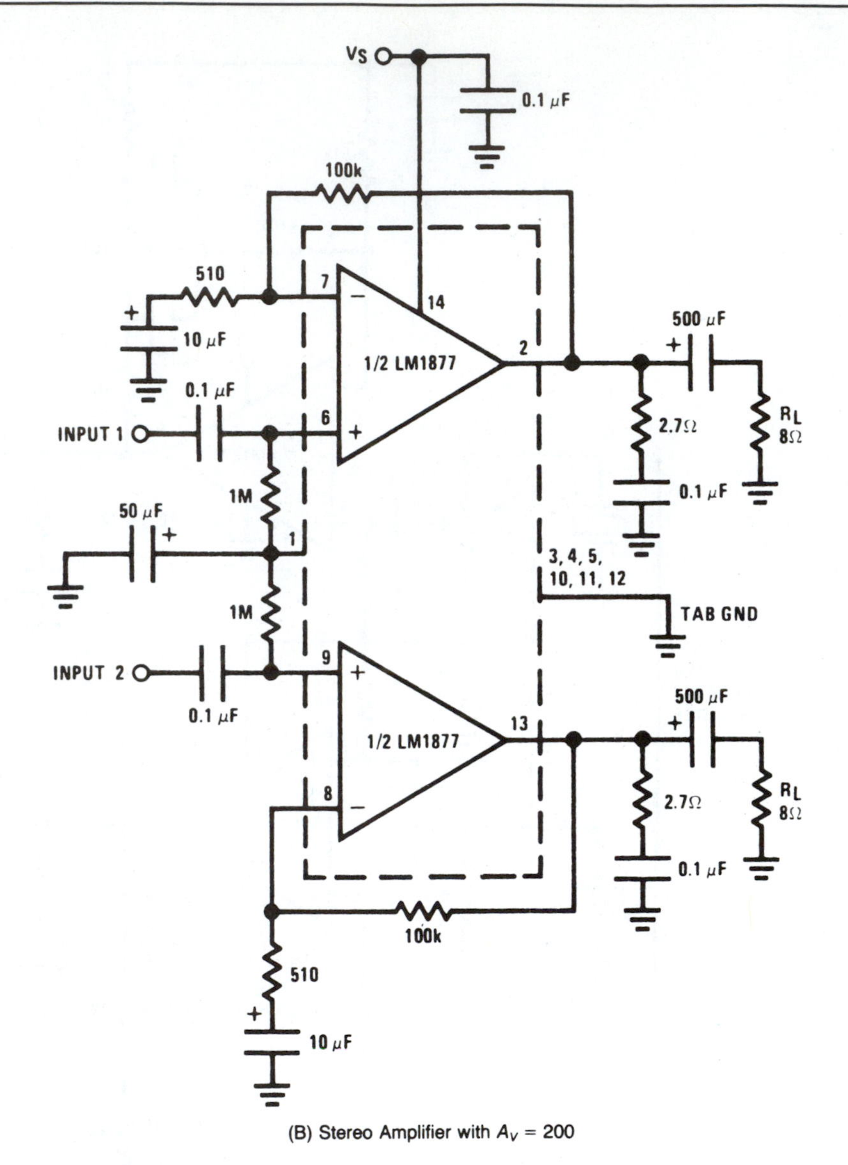


(B) Stereo Amplifier with $A_V = 200$

Figure 12–4. *Continued*

## Battery-Operated Low-Voltage Audio Amplifier

A battery-operated device is the LM389 low-voltage audio power amplifier with NPN transistor array, shown in Figure 12–7 on page 381. The amplifier inputs are ground referenced, and the output is automatically biased to one-half the supply voltage.

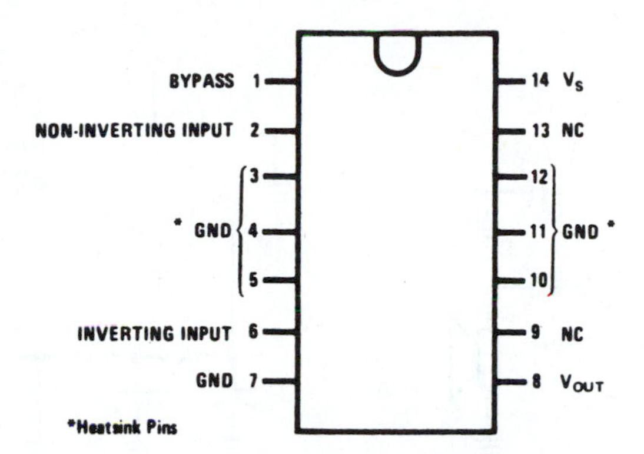


**Figure 12–5.** Connection Diagram for LM380 Audio Power Amplifier (Courtesy of National Semiconductor Corporation)

The three transistors have high gain and excellent matching characteristics. They are well suited to a wide variety of applications in dc through VHF systems. The transistors are general-purpose devices that can be used in the same manner as other small-signal transistors. As long as the currents and voltages are kept within the absolute maximum limitations, and the collectors are never allowed to go to ground potential with respect to pin 17, there is no limit to the way they can be used.

To make the LM389 a more versatile amplifier, two pins (4 and 12) are provided for gain control. With pins 4 and 12 open, the 1.35 kΩ resistor sets the gain at 20. However, bypassing the 1.35 kΩ resistor with a capacitor between pins 4 and 12 increases the gain to 200. If a resistor is placed in series with the bypass capacitor, gain can be set at any level between 20 and 200.

Additional external components can be placed in parallel with the internal feedback resistors to tailor the gain and frequency response for individual applications. For example, we can compensate for poor speaker bass response by placing a series *RC* network from pin 1 to pin 12, thus paralleling the internal 15 kΩ resistor.

The schematic of Figure 12–7B shows that both inputs are biased to ground with a 50 kΩ resistor. The base current of the input transistors is about 250 nA, so the inputs are at about 12.5 mV when left open. If the dc source resistance driving the LM389 is higher than 250 kΩ it will contribute very little additional offset (about 2.5 mV at the input, 50 mV at the output). If the dc source resistance is less than 10 kΩ, then shorting the unused input to ground will keep the offset low (about 2.5 mV at the input, 50 mV at the output). For dc source resistances between these values we can eliminate excess offset by putting a resistor of equal value to the dc source resistance, from the unused input to ground. Elimination

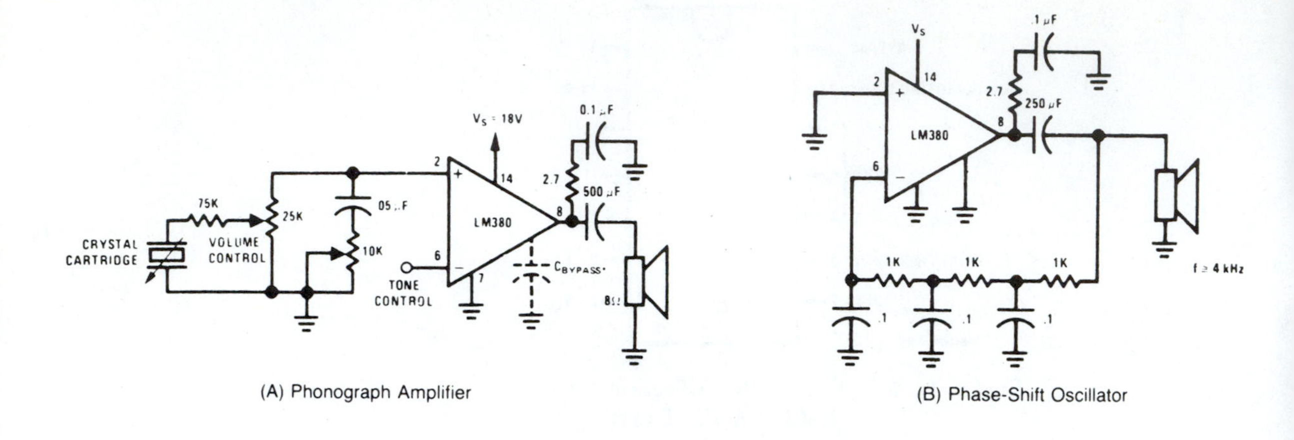


(A) Phonograph Amplifier

(B) Phase-Shift Oscillator

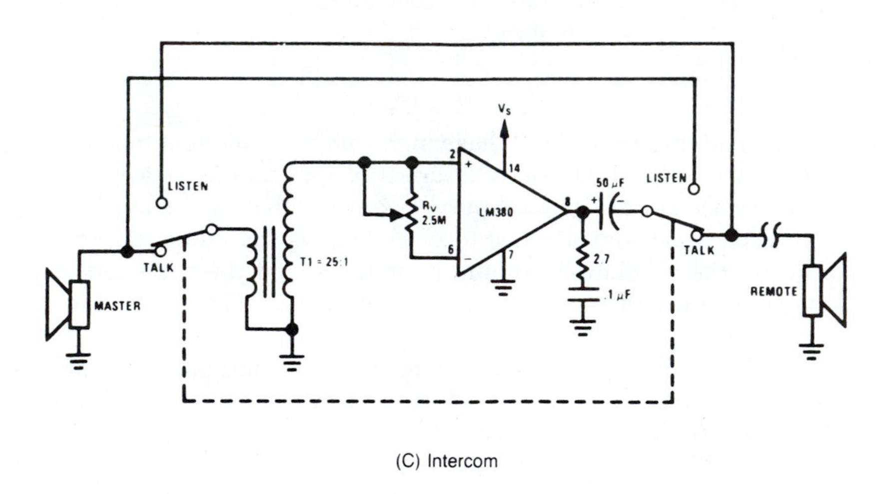


(C) Intercom

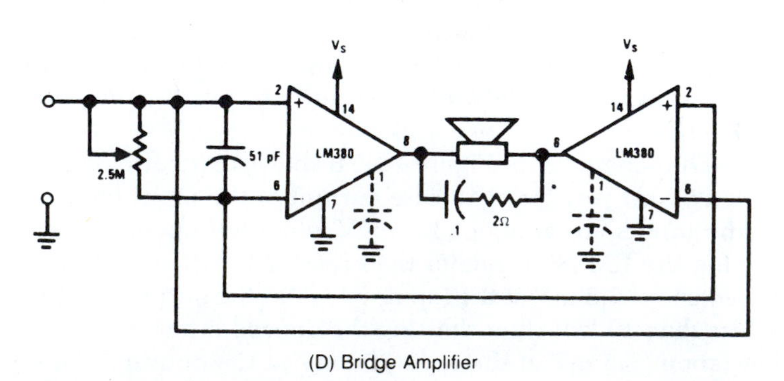


(D) Bridge Amplifier

Figure 12–6. LM380 Applications (Courtesy of National Semiconductor Corporation)

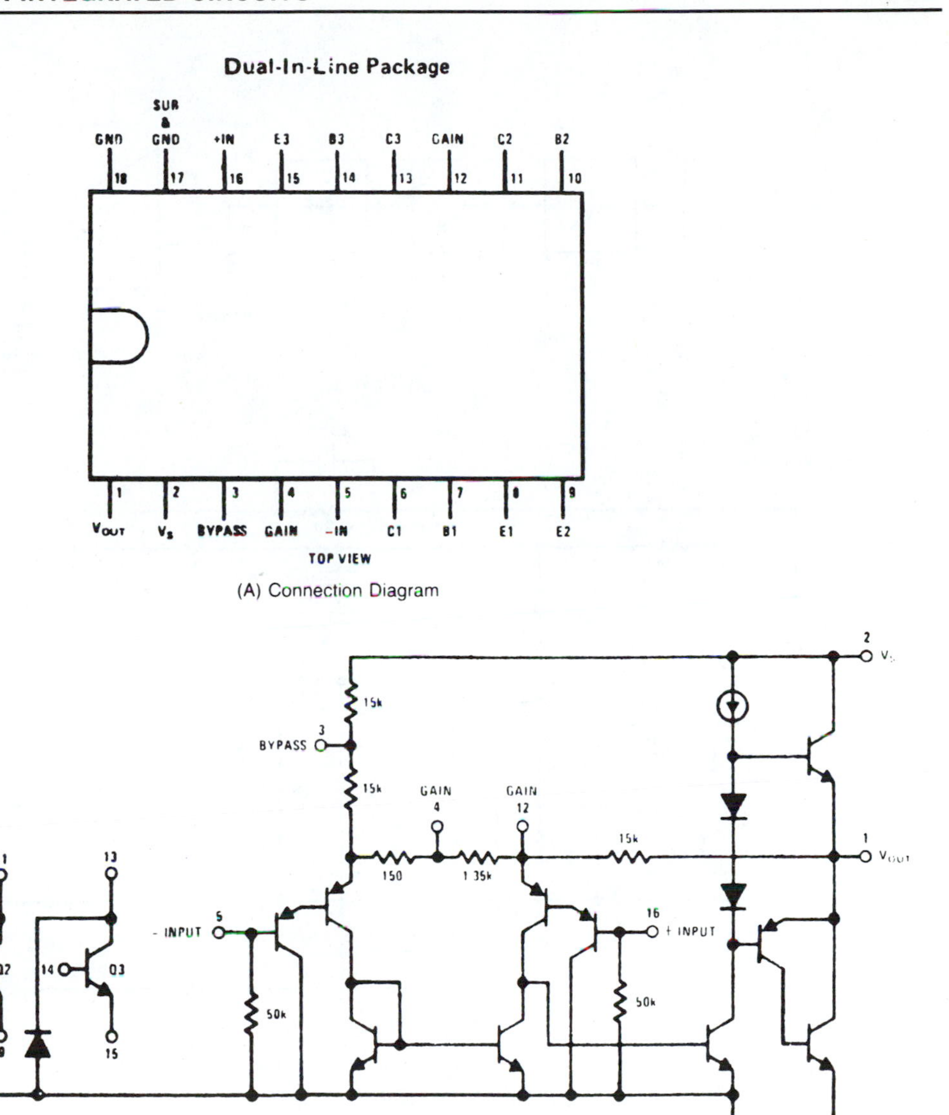


(A) Connection Diagram

(B) Equivalent Schematic

**Figure 12–7.** LM389 Low-Voltage Audio Power Amplifier with NPN Transistor Array (Courtesy of National Semiconductor Corporation)

of all offset problems can be accomplished by capacitively coupling the inputs.

When using the LM389 with higher gains (bypassing the internal 1.35 kΩ resistor), it is necessary to bypass the unused input to prevent gain degradation and possible instability. This is done with a 0.1 μF capacitor or a short to ground, depending on the dc source resistance.

Figure 12–8 shows four typical applications for the LM389.

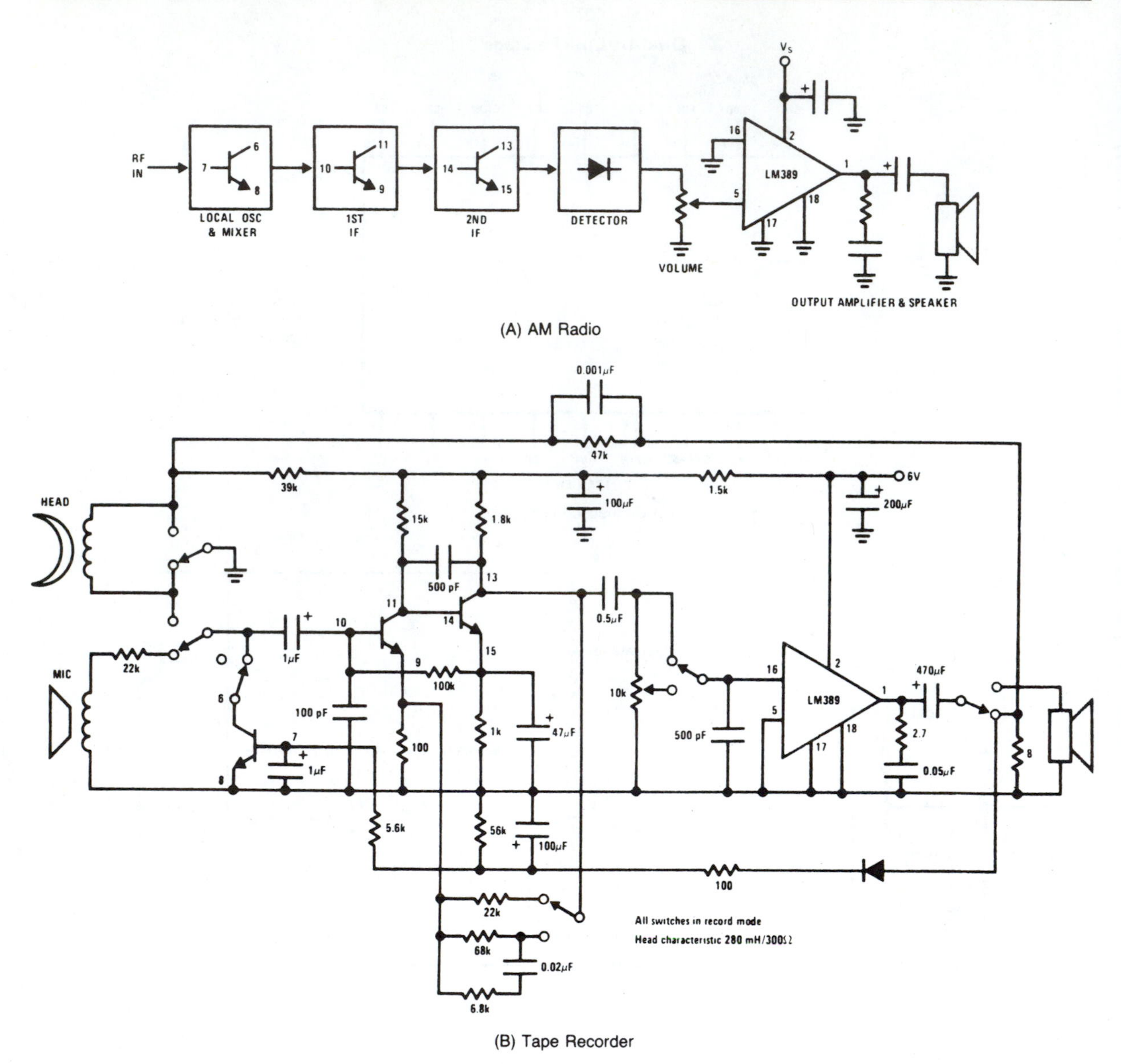


(A) AM Radio

(B) Tape Recorder

Figure 12–8. LM389 Applications (Courtesy of National Semiconductor Corporation) (Continued)

### Additional Audio Circuits

The NE5532, Figure 12–9, is an internally compensated dual low-noise op amp. Compared to most of the standard op amps, this device shows better noise performance, improved output drive capability, and considerably higher small signal and power bandwidths (BWs). This makes the 5532 especially suitable for

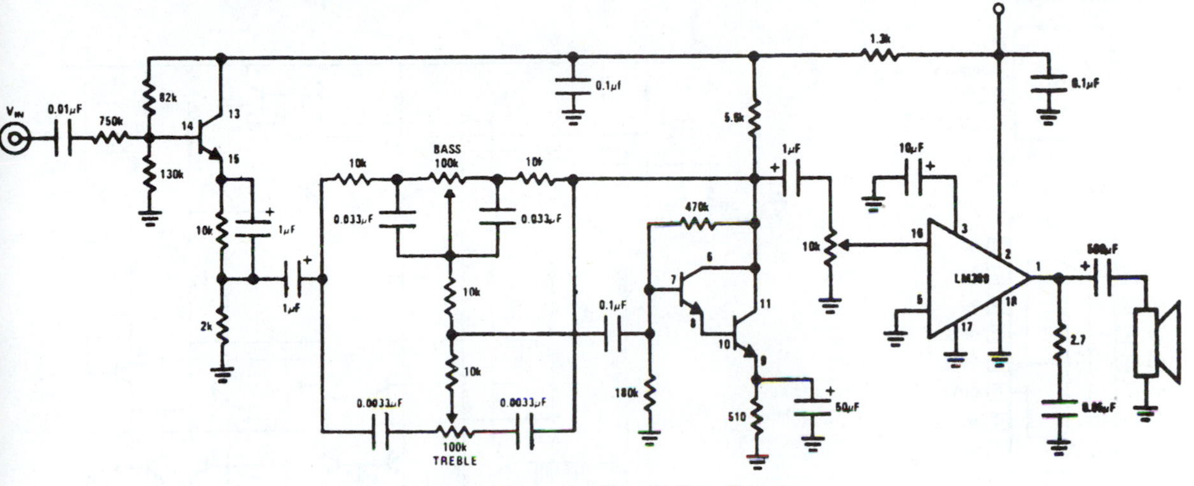

(C) Ceramic Phono Amplifier with Tone Controls

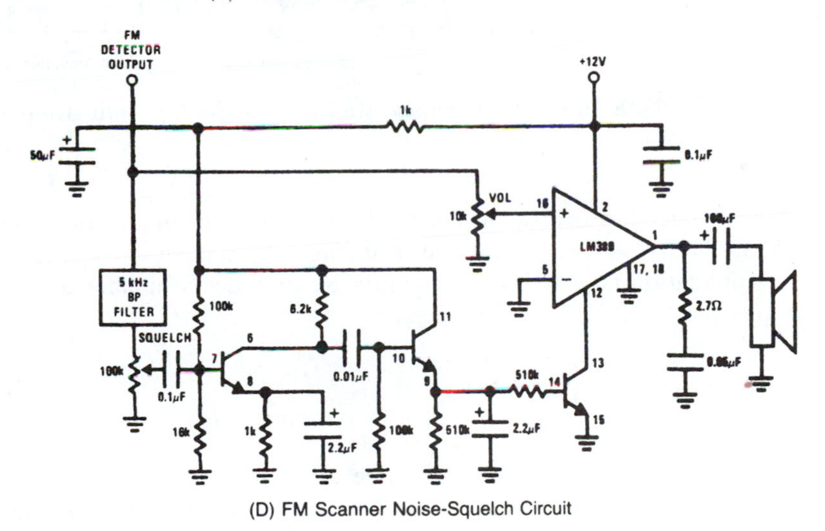

(D) FM Scanner Noise-Squelch Circuit

**Figure 12–8.** *Continued*

applications in high-quality and professional audio equipment, instrumentation and control circuits, and telephone channel amplifiers. The op amp is internally compensated for gains equal to one.

The NE5533/34, Figure 12–10, are dual and single high-performance low-noise op amps. The op amps are internally compensated for gain equal to, or higher than, three. The frequency response can be optimized with an external compensation capacitor for various applications (unity gain amplifier, capacitive

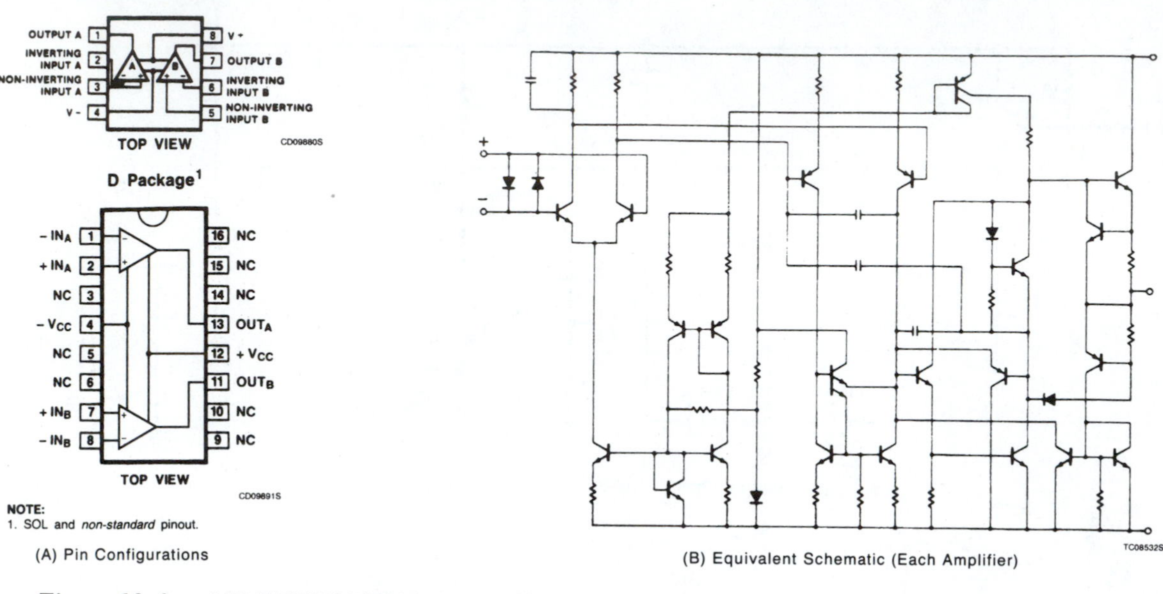


(A) Pin Configurations (B) Equivalent Schematic (Each Amplifier)

Figure 12–9. NE/SE5532/5532A Internally Compensated Low-Noise Op Amp (Courtesy of Signetics Company)

load, slew rate, low overshoot, etc.). In addition to the applications for the 5532, the 5533/34 are well suited for use in medical equipment.

Application note AN142, Appendix K, provides a number of audio applications for the NE5532/33/34 op amps.

## Analog Multiplier

The RC4200 IC multiplier features three on-board op amps designed specifically for use as multiplier logging circuits. The device can be used in a wide variety of applications with predictable accuracy. Applications include four-quadrant multiplication, two-quadrant division, squaring, square rooting, RMS-to-dc-conversion, automatic gain control (AGC), and modulation/demodulation. Data sheets and application notes for the RC4200 are in Appendix L.

### Example 12.1

Refer to Figure 2 in Appendix L. Assume $V_x = 1$ V, $R_1 = 2700\ \Omega$, $V_y = 1.25$ V, $R_2 = 3300\ \Omega$, $V_z = 2$ V, $R_4 = 4700\ \Omega$. Calculate (a) $I_1$, (b) $I_2$, (c) $I_4$, and (d) $I_3$.

*Solution*

(a) $I_1 = V_x/R_1 = 1\ \text{V}/2700\ \Omega = 370.4\ \mu\text{A}$

(b) $I_2 = V_y/R_2 = 1.25\ \text{V}/3300\ \Omega = 378.8\ \mu\text{A}$

(c) $I_4 = V_z/R_4 = 2\ \text{V}/4700\ \Omega = 425.5\ \mu\text{A}$

(d) $I_3 = (I_1 I_2)/I_4 = (370 \times 378.8)/425.5$

$= 329.75\ \mu\text{A}$

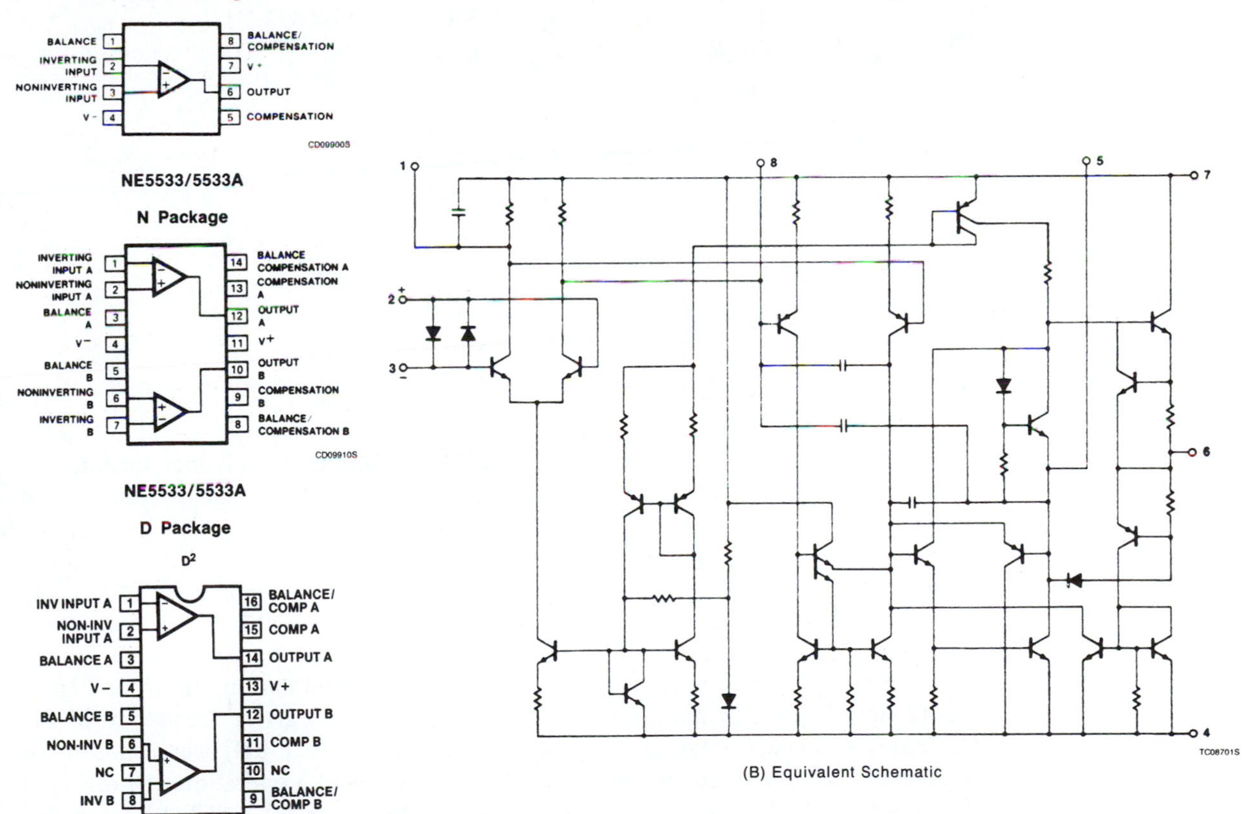


**Figure 12–10.** NE/SE5533/5534 Dual and Single Low-Noise Op Amp (Courtesy of Signetics Company)

## Example 12.2

Refer to Figure 4 in Appendix L. Assume the same voltage and resistor values as in Example 12.1, and $R_0 = 1500\ \Omega$. Calculate $V_o$ for (a) basic circuit, (b) multiplier, and (c) divider.

*Solution*

(a) $V_o = (V_x V_y R_0 R_4)/(V_z R_1 R_2)$
$= (1 \times 1.25 \times 1500 \times 4700)/(2 \times 2700 \times 3300)$
$= 494.53$ mV

(b) First solve for $K$:
$K = (R_0 R_4)/(VrR_1 R_2)$
$= (1500 \times 4700)/(2 \times 2700 \times 3300) = .0395623$
$V_o = V_x V_y K = 1 \times 1.25 \times .0395623 = 49.453$ mV

(c) First solve for $K$:
$K = (VrR_0 R_4)/(R_1 R_2)$
$= (2 \times 1500 \times 4700)/(2700 \times 3300) = 1.5825$
$V_o = K(V_x/V_z) = 1.5825(1/2) = 791.25$ mV

## TV Modulator Circuit

The MC1374 shown in Figure 12–11 is a TV modulator circuit that includes an FM audio modulator, sound carrier oscillator, RF oscillator, and RF dual input modulator. It is designed to generate a TV signal from audio and video inputs. The wide dynamic range and low distortion audio of the device make it particularly well-suited for applications such as video cassette recorders (VCRs) and video disc players (VDPs). It operates off a single supply, 5–12 V, and is designed for channel 3 or 4 operation.

The oscillator components shown in the typical application circuit of Figure 12–12 are selected to have a parallel resonance at the carrier frequency of the desired TV channel. The values of $C_2$ (56 pF) and $L_1$ (0.1 μH) were chosen for a channel 4 carrier frequency of 67.25 MHz. For channel 3 operation, the resonant frequency is 61.25 MHz, so the values would be $C_2 = 67.5$ pF and $L_1 = 0.1$ μH. Resistors $R_2$ and $R_3$ are chosen to provide an adequate amplitude of switching voltage, while $R_1$ is used to lower the maximum dc level of switching voltage below $V_{CC}$, thus preventing saturation within the IC.

The video modulator is a balanced modulator. Sound-carrier and video information are applied to pins 1 and 11. The other modulator inputs are internally connected to the RF oscillator. The modulator output appears at pin 9.

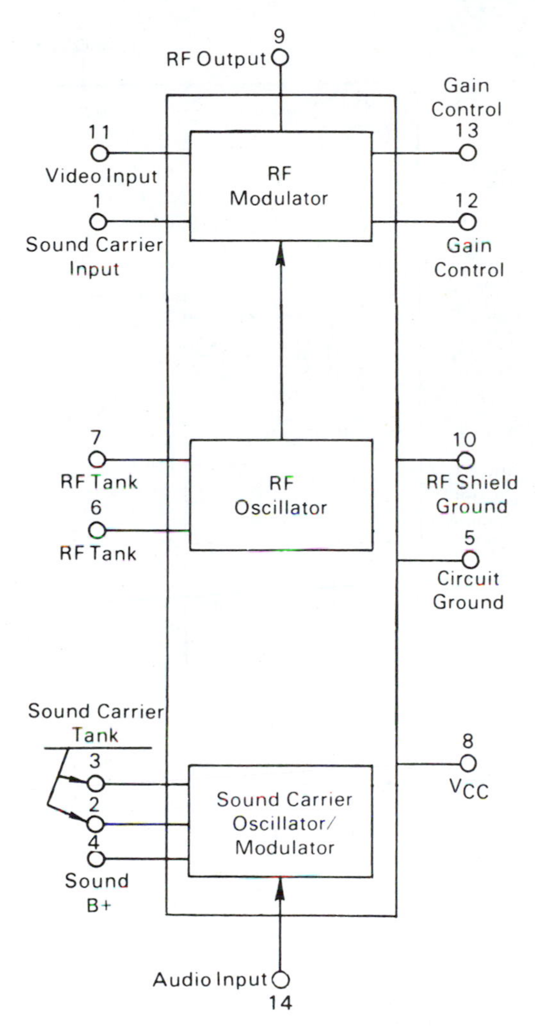


**Figure 12–11.** Block Diagram of MC1374 TV Modulator Circuit (Courtesy of Motorola Semiconductor Products, Inc.)

In a typical application where the composite video information is dc coupled to pin 11, the bias on pin 1 is set to give the desired modulation characteristics. Minimum carrier occurs when the voltage on pins 1 and 11 is equal, a desirable trait. The minimum permissible voltage on either input is 1.6 V. The maximum voltage should be 1.5 V below the dc voltage on pins 6 and 7. The value for gain-setting resistor $R_8$, between pins 12 and 13, is selected to give the proper modulation depth for the available composite video amplitude.

The modulated RF signal is presented as a current at the RF output pin 9. Since this pin represents a current source, any load impedance may be selected for matching purposes and gain selections, as long as the voltage on pin 9 is high

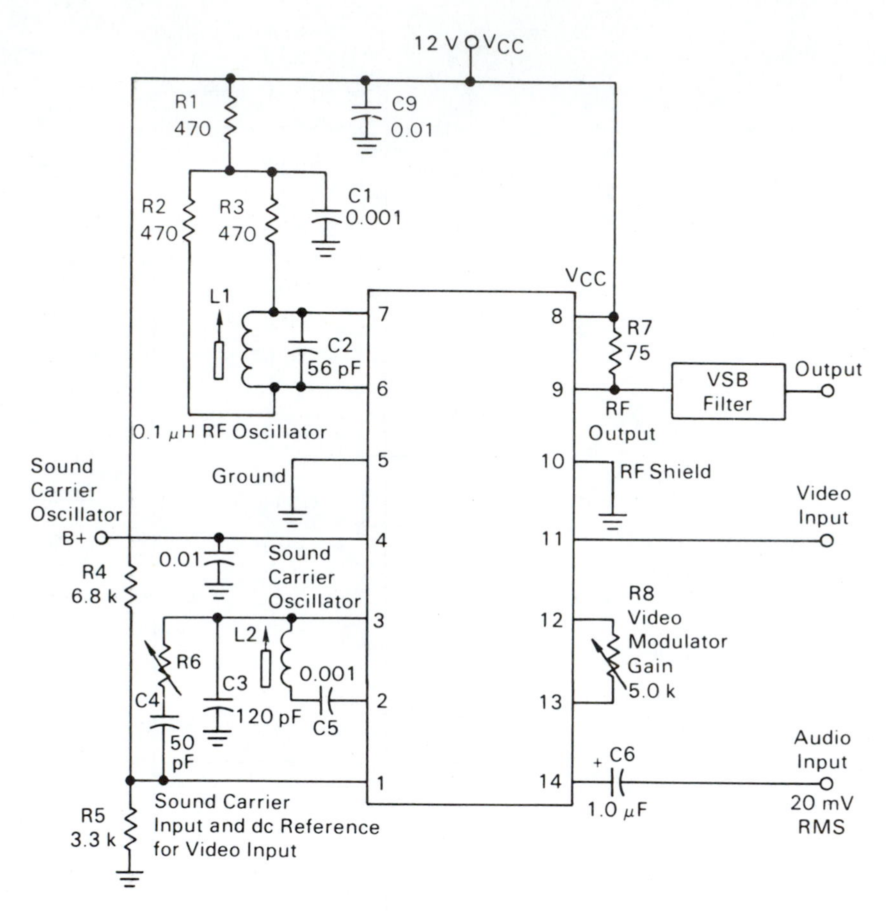


**Figure 12-12.** MC1374 Channel 4 Application (Courtesy of Motorola Semiconductor Products, Inc.)

enough to prevent output devices from reaching saturation. Lowering the dc voltage on pins 6 and 7 gives increased RF output capability at the expense of video input range.

The sound-carrier oscillator and audio modulator have internally set bias. A separate $B+$ is supplied to the oscillator through pin 4, so the oscillator may be easily disabled while tuning the RF tank. The sound-carrier frequency is determined by $L_2$ and $C_3$. The oscillator feedback is fed to $L_2$ through dc blocking capacitor $C_5$, and 4.5 MHz appears at the input to the oscillator, pin 3.

The sound carrier is coupled to the modulator input, pin 1, through variable resistor $R_6$ and capacitor $C_4$. The value for $R_6$ is chosen to give the desired sound-carrier amplitude and depends upon the $Q$ of $L_2$ and the values for $R_4$ and $R_5$, and the RF modulator gain is set by $R_8$.

Baseband audio (the frequency band of the audio modulating signal) is fed to the audio modulator on pin 14, where it directly modulates the sound-carrier oscillator for a flat characteristic and very low distortion. The input impedance on pin 14 is nominally 6 kΩ. If the audio available is much greater than necessary for proper deviation, a series resistor may be added to allow a low value for coupling capacitor $C_6$.

When the application calls for tight frequency stability, the sound carrier may be frequency controlled by supplying a dc current to pin 14 from a suitable automatic frequency control (AFC) circuit. The nominal voltage at pin 14 is approximately 3 V. Supplying current to pin 14 increases the frequency, and pulling current out of pin 14 reduces the frequency.

Two-channel operation is possible by switching in a second capacitor to tune the lower channel by means of a PIN diode (see glossary). Figure 12–13 shows the circuit for channel 3 or 4 operation.

## High-Gain, Low-Power FM-IF Circuit

The MC3359 illustrated in Figure 12–14 is a high-gain, low-power FM-IF circuit that includes oscillator, mixer, limiting amplifier, AFC, quadrature discriminator, op amp, squelch, scan control, and mute switch. It is designed primarily for use in voice communication scanning receivers.

The mixer-oscillator combination converts the 10.7 MHz input frequency down to 455 kHz, where, after external bandpass (BP) filtering (ceramic filter at pin 3), most of the amplification is done. The audio is recovered using a conventional quadrature FM detector (at pin 8). The absence of an input signal is indicated by the presence of noise after the desired audio frequencies. This "noise band" is monitored by an active filter and a detector. A squelch-trigger circuit (pin 14) indicates the presence of noise (or a tone) by an output (pin 15) that can be

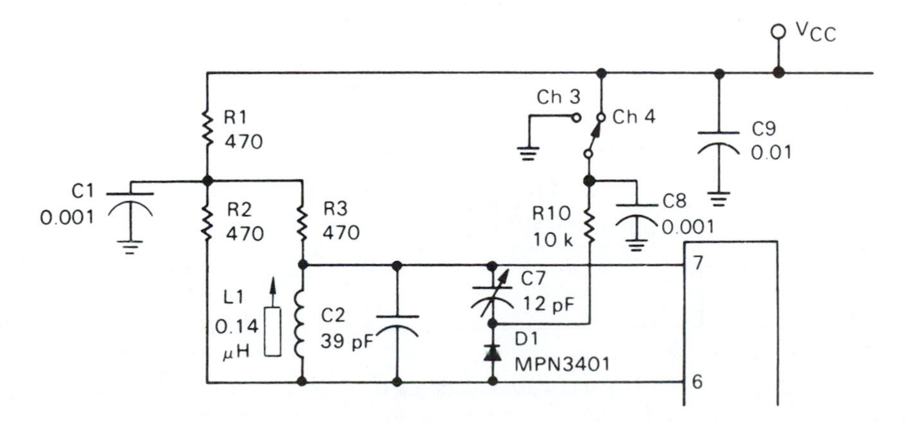


**Figure 12–13.** Oscillator Components Circuit for Channel 3 and Channel 4 Operation (Courtesy of Motorola Semiconductor Products, Inc.)

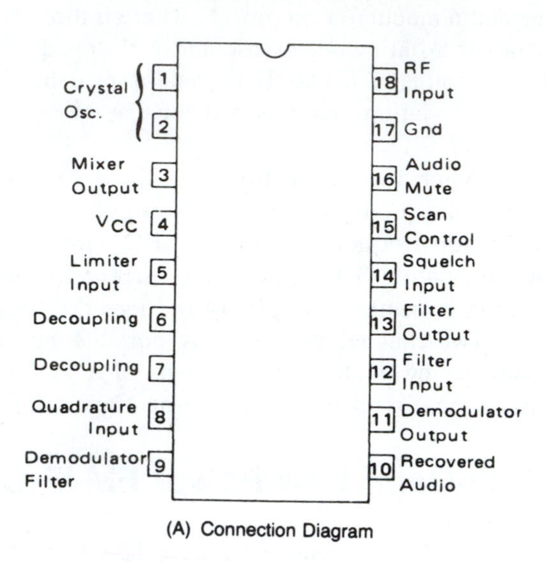


(A) Connection Diagram

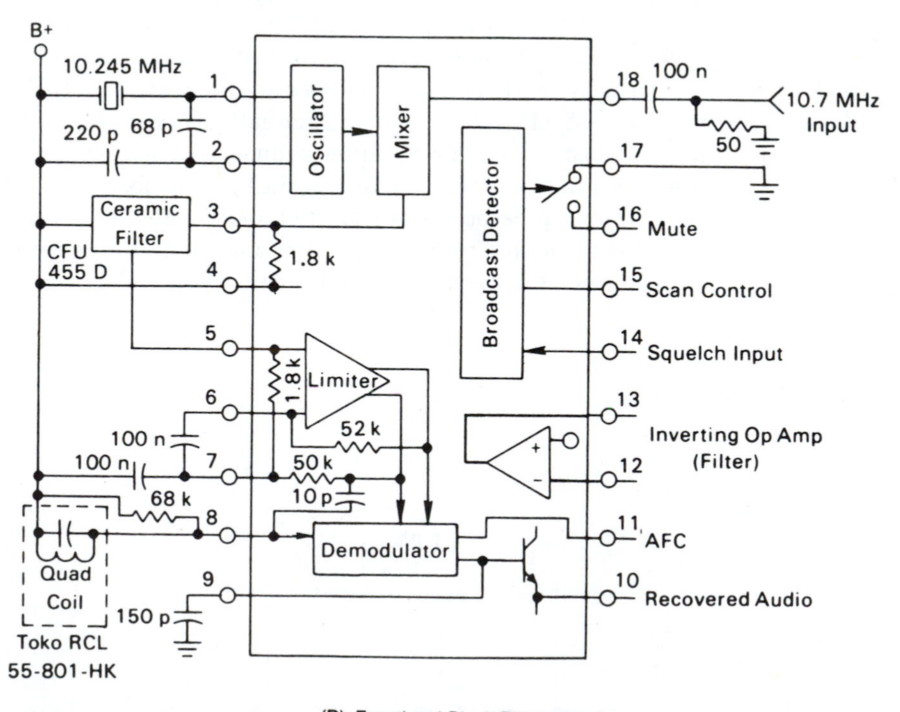


(B) Functional Block Diagram

Figure 12–14. MC3359 High-Gain, Low-Power FM-IF Circuit (Courtesy of Motorola Semiconductor Products, Inc.)

used to control scanning. At the same time, an internal switch (pins 16 and 17) is operated. This switch can be used to mute the audio.

The oscillator is an internally-biased Colpitts with the collector, base, and emitter connections at pins 4, 1, and 2, respectively. A 10.245 MHz crystal is used in place of the usual coil.

The mixer is doubly balanced to reduce spurious responses. The input impedance at pin 16 is set by a 3.6 kΩ internal biasing resistor and has low capacitance, which allows the circuit to be preceded by a crystal filter. The mixer output at pin 3 has a 1.8 kΩ impedance to match the external ceramic filter.

After BP filtering, the signal goes to the input of a six-stage limiter at pin 5 whose impedance is again 1.8 kΩ. The output of the limiter drives a multiplier, both directly and through the quadrature coil, to detect the FM.

The external capacitor at pin 9 can combine with the internal 50 kΩ resistor to form a low-pass (LP) filter for the audio.

The audio is delivered through the emitter follower to pin 10, which may require an external resistor-to-ground to prevent the signal from rectifying with some capacitive loads.

Pin 11 provides AFC. If AFC is not required, pin 11 should be grounded, or it can be tied to pin 9 to double the recovered audio amplitude.

A simple inverting op amp is provided, with an output at pin 13 providing dc bias externally to the input at pin 12, which is referred internally to 2.3 V. A filter can be made with external impedance elements to discriminate between frequencies. With an external AM detector, the filtered audio signal can be checked for the presence of either noise above the normal audio band or a tone signal. The result is applied to pin 14.

An external negative bias to pin 14 sets up the squelch-trigger circuit such that pin 15 is HIGH, at an impedance level of about 2.5 kΩ, and the audio mute (pin 16) is open-circuit. If pin 14 is raised to 0.7 V by the noise or tone detector, pin 15 will go open-circuit and pin 16 is internally short-circuited to ground. There is no hysteresis. Audio muting is accomplished by connecting pin 16 to a high-impedance ground-reference point in the audio path between pin 10 and the audio amplifier.

## VCO/Modulator

The MC1376 shown in Figure 12–15 is a voltage-controlled oscillator (VCO)/modulator that is ideally suited to cordless telephone and television intercarrier applications. It operates over a supply range of 5–12 V dc, has a useful frequency range of 1.4–14 MHz, has less than 1% distortion, and offers excellent oscillator stability.

This device was originally designed for the base station of a cordless telephone. It includes a separate, or auxiliary, transistor (pins 2, 3, and 4) suitable for service as an output buffer or amplifier for up to 50 mA. Though the oscillator contains internal phase-shift components that are not accessible, the device still

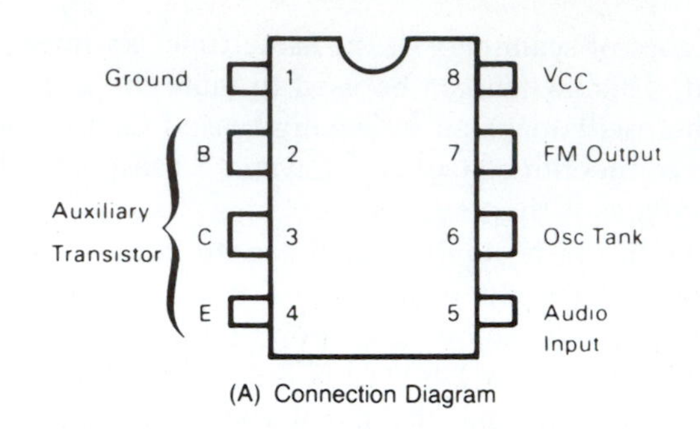


(A) Connection Diagram

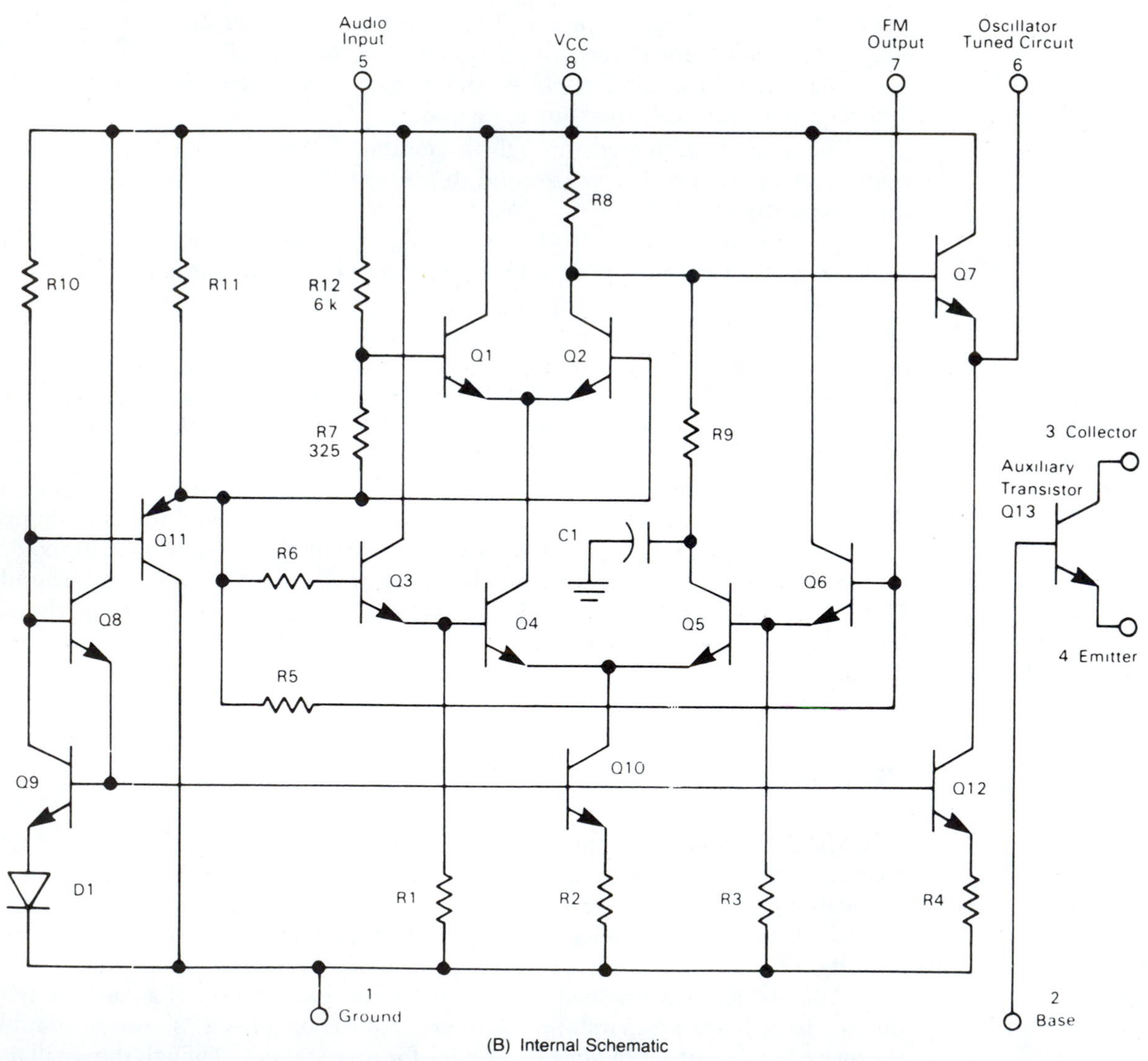


(B) Internal Schematic

**Figure 12–15.** MC1376 FM Modulator Circuit (Courtesy of Motorola Semiconductor Products, Inc.)

has an operating range of 1.4–14 MHz. This range makes the MC1376 a good companion to other devices, such as the MC1372 and MC1373, as a 4.5 MHz or 5.5 MHz intercarrier sound modulator for television signal generation. Also, the device can be used as a low-cost FM-IF (10.7 MHz) signal source. The modulator section of this device is identical to the FM portion of the MC1374 TV modulator discussed earlier.

## Cordless Telephone Base Station

A 1.76 MHz cordless telephone base station transmitter is shown in Figure 12–16. The oscillator center frequency is approximately the resonance of the inductor (pin 6 to pin 7) and the total capacitance from pin 7 to ground. If the internal capacitance of about 6 pF is included, the circuit strays in the resonant frequency calculations for the higher frequency applications. For overall oscillator stability, it is best to keep $X_L$ (inductive reactance) and $X_C$ (capacitive reactance) in the range of 300 Ω to 1 kΩ.

Most applications will require no dc connection at the audio input, pin 5. However, some performance improvements can be achieved by the addition of biasing circuitry. The unaided device will usually establish its own pin 5 bias at 2.9–3.0 V. This bias is a little high for optimum modulation linearity and results in some modulation distortion. This distortion can be significantly reduced by

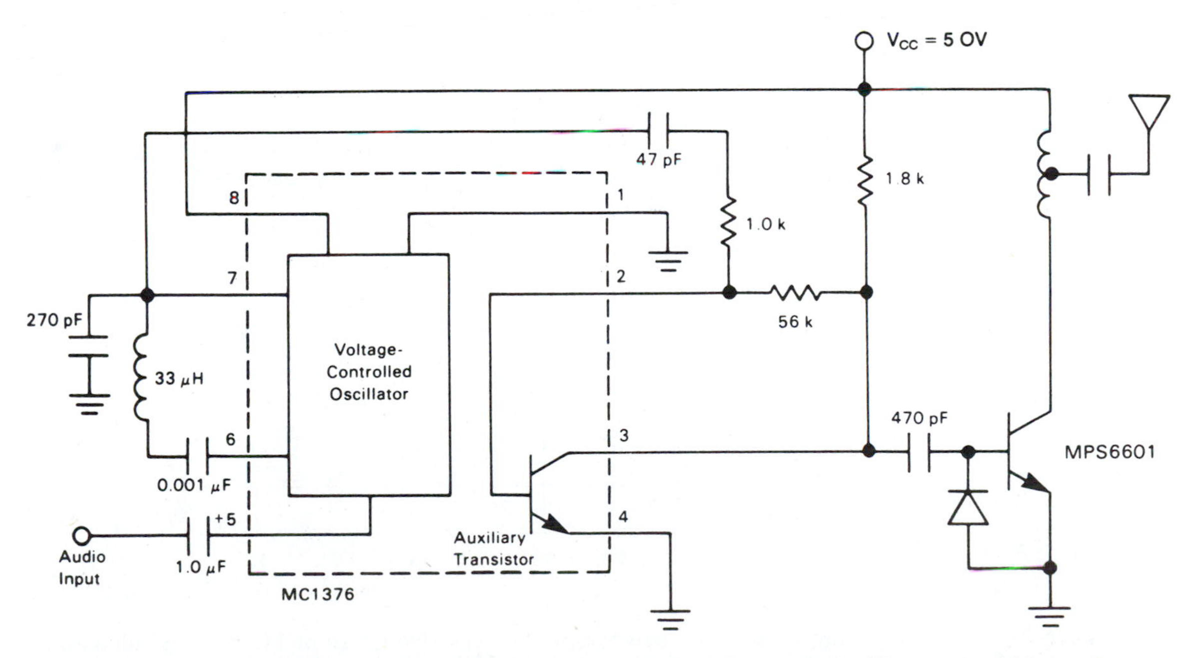


**Figure 12–16.** 1.76 MHz Cordless Telephone Base Station Transmitter (Courtesy of Motorola Semiconductor Products, Inc.)

pulling the pin bias down to 2.6–2.7 V. Temperature and supply voltage factors must also be considered when determining biasing.

Temperature stability can be improved by pulling pin 5 down to 2.6 V through a 27 kΩ resistor. If $V_{CC}$ is well regulated, then a simple 180 kΩ/30 kΩ resistor divider, as shown in Figure 12–17, provides optimum distortion-and-frequency stability versus temperature.

The FM output at pin 7 is usually about 600 $mV_{p\text{-}p}$ and has low harmonic content and high (2 kΩ) output impedance. The oscillator behavior is relatively unaffected by loading above 1.0 kΩ. If lower impedance must be driven, the capacitive divider, shown in Figure 12–18, can be used, or the auxiliary transistor can be used as a buffer.

## Electronically Switched Audio Tape System

The LM1818 electronically switched audio tape system shown in Figure 12–19 is an LIC that contains all of the active electronics necessary for building a tape recorder deck (excluding the bias oscillator). The electronic functions on the chip include a microphone and playback preamplifier, record and playback amplifiers, a meter driving circuit, and an automatic input level control circuit. The IC features complete internal electronic switching between the record and playback modes of operation.

A monaural application circuit using the LM1818 is shown in Figure 12–20 on page 397. Table 12–1 (pages 398–399) lists the external components and their functions.

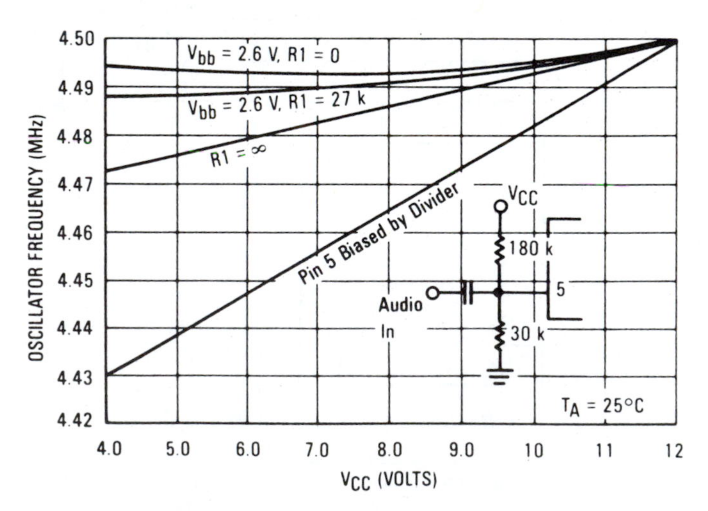


Figure 12–17. Frequency Stability Versus Supply Voltage (Courtesy of Motorola Semiconductor Products, Inc.)

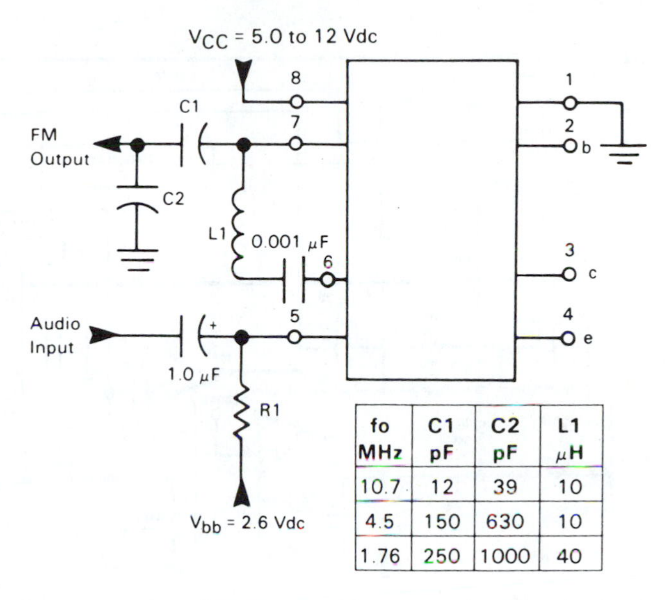


| fo MHz | C1 pF | C2 pF | L1 μH |
|---|---|---|---|
| 10.7 | 12 | 39 | 10 |
| 4.5 | 150 | 630 | 10 |
| 1.76 | 250 | 1000 | 40 |

**Figure 12–18.** Test Circuit for Capacitive Divider Output (Courtesy of Motorola Semiconductor Products, Inc.)

## No-Holds TV Circuit

The LM1880 no-holds vertical/horizontal circuit shown in Figure 12–21 (page 399) uses compatible linear/$I^2L$ (integrated injection logic) technology to produce a TV vertical and horizontal processing system that completely eliminates the hold controls. The heart of the system is a precision 32-times horizontal frequency VCO that is designed to use a low-cost resonator as a tuning element.

The VCO signal is divided in the horizontal section to produce a predriver output that is locked to negative synchronization (sync) by means of an on-chip phase detector. The vertical output ramp is injection-locked by vertical sync that is subject to a sync window derived from the vertical countdown section. A gate pulse centered on the chroma (color) burst is also provided.

A typical application circuit is illustrated in Figure 12–22 (page 400). Since the LM1880 uses a counter to derive the horizontal frequency, care must be taken to prevent extraneous signals from the horizontal driver and output stages from feeding back to the VCO where they could cause false counts and consequent severe phase jitter. To prevent this problem from occurring, keep the VCO feedback capacitor, $C_L$, as close as possible to device pins 6 and 7, and limit the lead length on the horizontal output pin 8. If a long line is required to the driver base, isolate it with a small series resistor (200–300 Ω) next to pin 8.

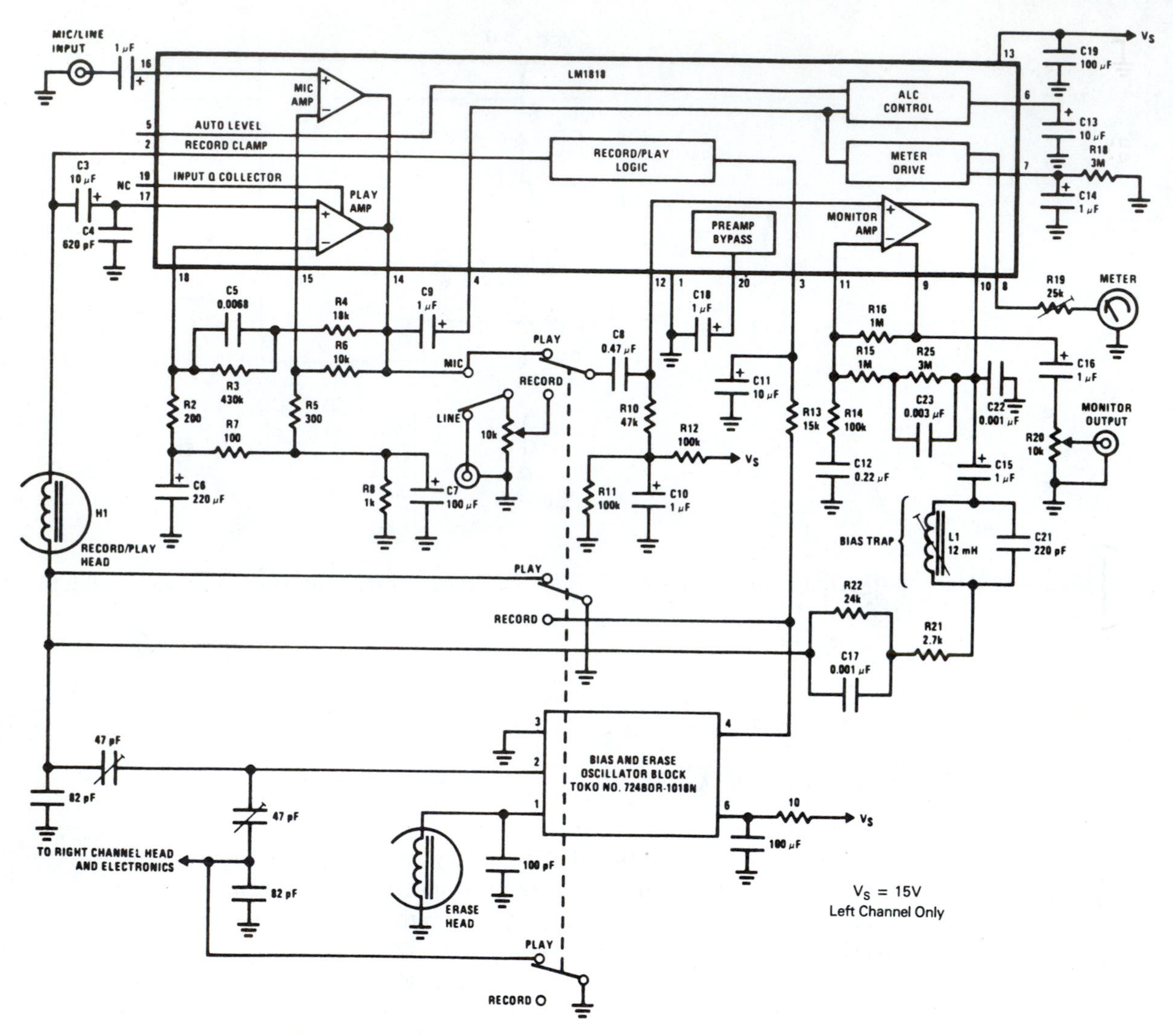


Figure 12–19. LM1818 Electronically Switched Audio Tape System Application (Courtesy of National Semiconductor Corporation)

## TV Signal Processing Circuit

The TBA950-2 TV signal processing circuit of Figure 12–23 (page 401) is designed for pulse separation and line synchronization in TV receivers with transistor output stages. It consists of a sync separator with noise suppression, a frame (a single, complete TV picture scanned in 1/30 second) pulse integrator, a phase comparator, a switching stage for automatic changeover of noise immunity, a line oscillator with frequency range limiter, a phase control circuit, and an output stage. It

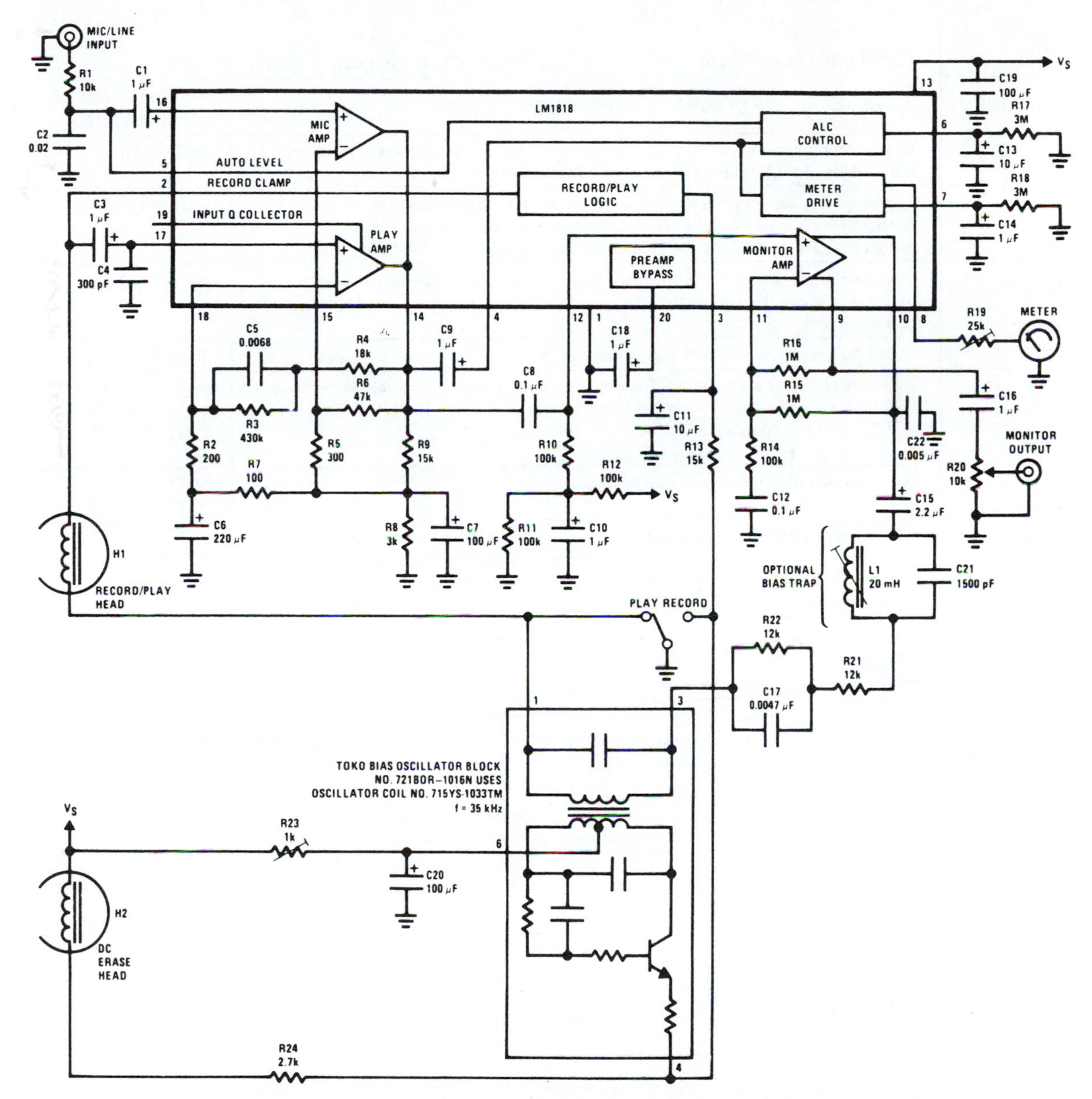


**Figure 12–20.** LM1818 Monaural Application Circuit (Courtesy of National Semiconductor Corporation)

delivers prepared frame sync pulses for triggering the frame oscillator. The phase comparator may be switched for video recording operation. Because of large-scale integration, few external components are needed.

The sync separator separates the synchronizing pulses from the composite video signal. The noise inverter circuit, which needs no external components,

**Table 12–1 External Components for Monaural Application Circuit, Figure 12–19**

| COMPONENT | EXTERNAL COMPONENT FUNCTION | NORMAL RANGE OF VALUE |
|---|---|---|
| R1 | Used in conjunction with varying impedance of pin 5, forming a resistor divider network to reduce input level in automatic level control circuit | 500 Ω–20 kΩ |
| C2 | Forms a noise reduction system by varying bandwidth as a function of the changing impedance on pin 5. With a small input signal, the bandwidth is reduced by R1 and C2. As the input level increases, so does the bandwidth. | 0.01 μF–0.5 μF |
| C1, C3 | Coupling capacitors. Because these are part of the source impedance, it is important to use the larger values to keep low frequency source impedance at a minimum. | 0.5 μF–10 μF |
| C4 | Radio frequency interference roll-off capacitor | 100 pF–300 pF |
| R2<br>R3<br>R4<br>C5 | Playback response equalization. C5 and R3 form a pole in the amplifier response at 50 Hz. C5 and R4 form a zero in the response at 1.3 kHz for 120 μs equalization and 2.3 kHz for 70 μs equalization. | 50 Ω–200 Ω<br>47 kΩ–3.3 MΩ<br>2 kΩ–200 kΩ |
| R5<br>R6 | Microphone preamplifier gain equalization | 50 Ω–200 Ω<br>5 kΩ–200 kΩ |
| R7<br>R8<br>R9<br>C6<br>C7 | DC feedback path. Provides a low impedance path to the negative input in order to sink the 50 μA negative input amplifier current. C6, R9, R7 and C7 provide isolation from the output so that adequate gain can be obtained at 20 Hz. This 2-pole technique also provides fast turn-ON settling time. | 0–2 kΩ<br>200Ω–5 kΩ<br>1 kΩ–30 kΩ<br>200 μF–1000 μF<br>0–100 μF |
| C8 | Preamplifier output to monitor amplifier input coupling | 0.05 μF–1 μF |
| C9 | ALC coupling capacitor. Note that ALC input impedance is 2 kΩ | 0.1 μF–5 μF |
| R10<br>R11<br>R12<br>C10 | These components bias the monitor amplifier output to half supply since the amplifier is unity gain at DC. This allows for maximum output swing on a varying supply. | 10 kΩ–100 kΩ<br>10 kΩ–100 kΩ<br>10 kΩ–100 kΩ<br>1 μF–100 μF |
| C11<br>R13 | Exponentially falling or rising signal on pin 3 determines sequencing, time delay, and operational mode of the record/play anti-pop circuitry. See anti-pop diagram. | 0–10 μF<br>0–50 kΩ |
| R14<br>R15<br>R16<br>C12 | R16, R14 and C12 determine monitor amplifier response in the play mode. R15, R14 and C12 determine monitor amplifier response in the record mode. | 1k–100k<br>30 kΩ–3 MΩ<br>30 kΩ–3 MΩ<br>0.1 μF–20 μF |
| C13<br>R17 | Determines decay response on ALC characteristic and reduces amplifier pop | 5 μF–20 μF<br>100k–∞ |
| C14<br>R18 | Determines time constant of meter driving circuitry | 0.1 μF–10 μF<br>100k–∞ |
| R19 | Meter sensitivity adjust | 10 kΩ–100 kΩ |
| C15 | Record output DC blocking capacitor | 1 μF–10 μF |
| C16 | Play output DC blocking capacitor | 0.1 μF–10 μF |
| C17<br>R21<br>R22 | Changes record output response to approximate a constant current output in conjunction with record head impedance resulting in proper recording equalization | 500 pF–0.1 μF<br>5 kΩ–100 kΩ<br>5 kΩ–100 kΩ |
| C18 | Preamplifier supply decoupling capacitor. Note that large value capacitor will increase turn-ON time | 0.1 μF–500 μF |
| C19 | Supply decoupling capacitor | 100 μF–1000 μF |

**Table 12–1** *(continued)*

| COMPONENT | EXTERNAL COMPONENT FUNCTION | NORMAL RANGE OF VALUE |
|---|---|---|
| C20 | Decouples bias oscillator supply | 10 μF–500 μF |
| R23 | Allows bias level adjustment | 0–1 kΩ |
| R24 | Adjusts DC erase current in DC erase machines (for AC erase, "Stereo Application Hook-up") | |
| L1<br>C21 | Optional bias trap | 1 mH–30 mH<br>100 pF–2000 pF |
| C22 | Bias Roll-Off | 0.001 μF–0.01 μF |
| H1 | Record/play head | 100 Ω–500 Ω; 70 mH–300 mH |
| H2 | Erase head (DC type, AC optional) | 10 Ω–300 Ω |

Source: (Courtesy of National Semiconductor Corporation)

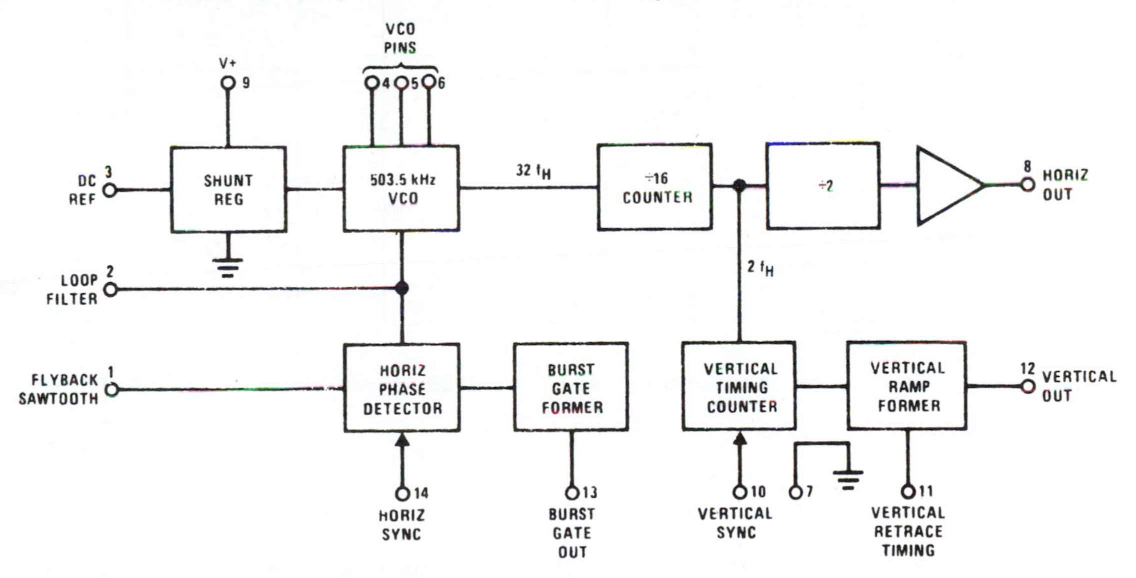


**Figure 12–21.** LM1880 No-Holds Vertical/Horizontal Circuit Block Diagram (Courtesy of National Semiconductor Corporation)

in connection with an integrating and differentiating network frees the synchronizing signal from distortion and noise.

The frame sync pulse is obtained by multiple integration and limitation of the synchronizing signal and is available at pin 7. It is recommended to use the leading edge of the frame sync pulse for triggering, because of possible pulse duration differences in production of the sync pulses.

The frequency of the line oscillator is determined by a 10 nF polystyrene capacitor at pin 13 that is charged and discharged periodically by two internal

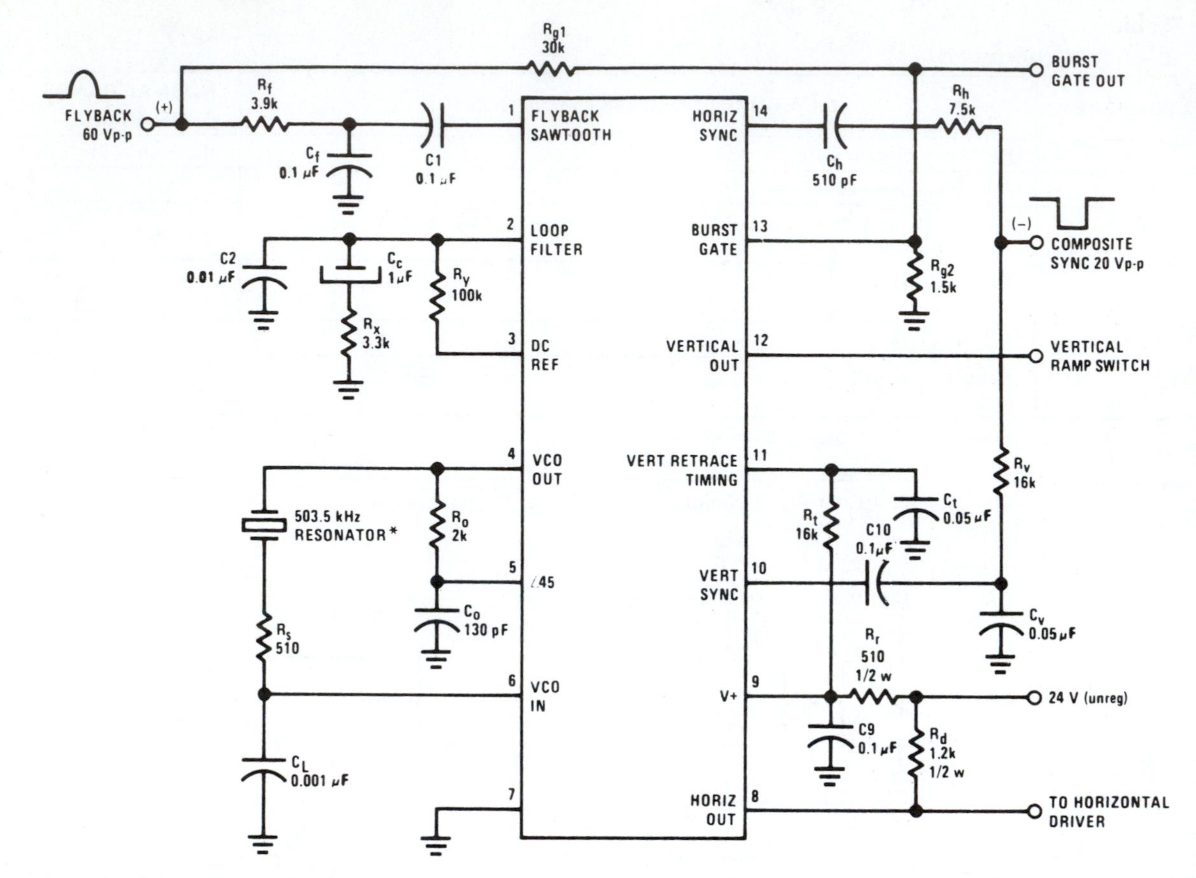


*MuRata Corporation of America, Part No. FX-1028, Vernitron Corp. VTFA3-01-503.5

**Figure 12–22.** LM1880 Application Circuit (Courtesy of National Semiconductor Corporation)

current sources. The external resistor at pin 14 defines the charging current and, consequently, in conjunction with the oscillator capacitor, the line frequency.

The phase comparator compares the sawtooth voltage of the oscillator with the line sync pulses. Simultaneously, an AFC voltage is generated that influences the oscillator frequency. A frequency range limiter restricts the frequency holding range.

The oscillator sawtooth voltage, which is in a fixed ratio to the line sync pulses, is compared with the flyback pulse in the phase control circuit to compensate all drift of delay times in the driver and line output stage. The correct phase position and, hence, the horizontal position of the picture, can be adjusted by the 10 kΩ potentiometer connected to pin 11. Within the adjustable range, the output pulse duration (pin 2) is constant. Any larger displacements of the

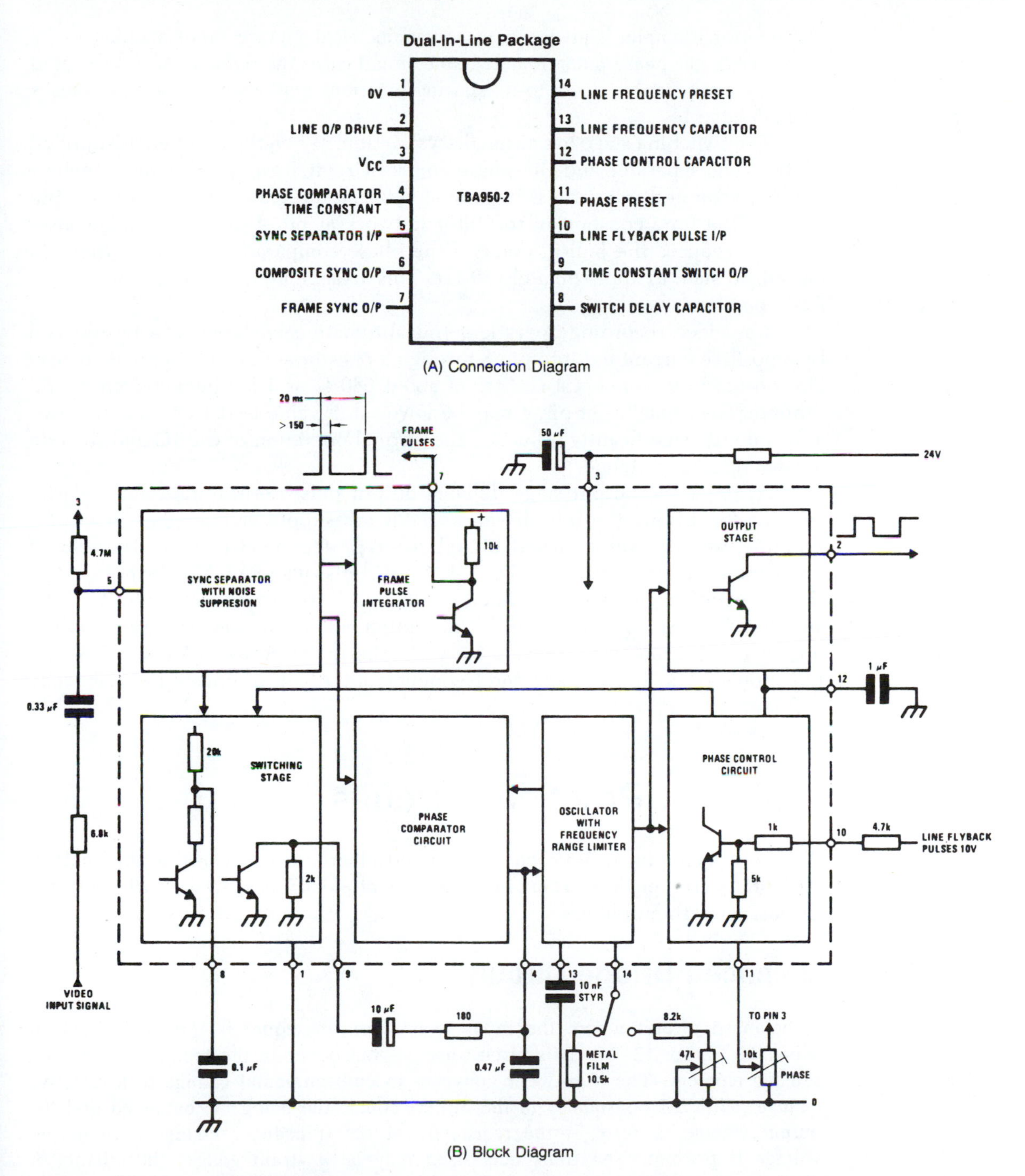


(A) Connection Diagram

(B) Block Diagram

**Figure 12–23.** TBA950-2 Television Signal Processing Circuit (Courtesy of National Semiconductor Corporation)

picture (for example, that due to a nonsymmetrical picture tube) should not be corrected by the phase potentiometer, since in all cases the flyback pulse (the rapid return of the scanning beam to its starting position) must overlap the sync pulses on both edges.

The switching stage has an auxiliary function. When the two signals supplied by the sync separator and the phase control circuit, respectively, are synchronized, a saturated transistor is in parallel with the integrated 2 kΩ resistor at pin 9. Thus the time constant of the filter network at pin 4 increases and, consequently, reduces the pull-in range of the phase comparator circuit for the synchronized state to approximately 50 Hz. This arrangement ensures disturbance-free operation.

For video recording operation, this automatic switchover can be blocked by a positive current fed into pin 8 through a resistor connected to pin 3. It may also be useful to connect a resistor of about 680 Ω or 1 kΩ between pin 9 and ground. The capacitor at pin 4 may be lowered in value to 0.1 μF. These alterations do not significantly influence the normal operation of the IC and thus do not need to be switched.

At pin 2 the output stage delivers output pulses of duration and polarity suitable for driving the line driver stage. If the supply voltage goes down (for example, when power is turned off), a built-in protection circuit ensures defined line frequency pulses down to $V_3 = 4$ V and shuts OFF when $V_3$ falls below 4 V, thus preventing pulses of undefined duration and frequency. Conversely, if the supply voltage rises, pulses defined in duration and frequency will appear at the output pin as soon as $V_3$ reaches 4.5 V. In the range between $V_3 = 4.5$ V and full supply voltage, the shape and frequency of the output pulses are practically constant.

## 12.4 Collection of Practical Circuits

A variety of simple, easy to construct, practical circuits is covered in this section. Amplifiers are very important in all aspects of electronics, so we will begin the discussion with amplifiers.

### Balanced Bridge Circuit

A balanced circuit where the input resistances are equal is the bridge circuit shown in Figure 12–24. This circuit uses a transducer in the feedback path as a sensing element. The transducer converts an environmental change to a resistive change. With all resistances in the circuit equal, the bridge is balanced and the output voltage is zero. If the resistance of the transducer changes, an output voltage is present. The transducer used may be a strain gauge, thermistor, or photodetector. Almost any type of op amp may be used. The output voltage $V_{out}$ is determined as follows:

$$V_{out} = -V_{ref}\left(\frac{R_T + \Delta R_T}{R_1 + R_T}\right) \qquad (12.1)$$

where

$R_T$ = transducer resistance
$\Delta R_T$ = a change in the transducer resistance

## Current Amplifier

An easily assembled current amplifier is shown in Figure 12–25A. A small current generated by the solar cell is amplified by the op amp, providing enough current to cause a visible indication from the light-emitting diode (LED). Taking this simple circuit one step further, as shown in Figure 12–25B, an optoisolator can be connected to the amplifier, with a resultant higher voltage output.

## dc and ac Controllers

An important and very effective amplifier circuit for controlling dc servomotors is illustrated in Figure 12–26. The speed and direction of the motor are determined by the amplitude and polarity of the voltage at the input to the circuit.

Using two op amps, as shown in Figure 12–27, produces an extremely effective ac servo amplifier. The speed and direction of the motor are again determined by the input voltage. The noninverting inputs are held at a predetermined dc reference voltage, usually slightly less than the level of the power supply for the op amps.

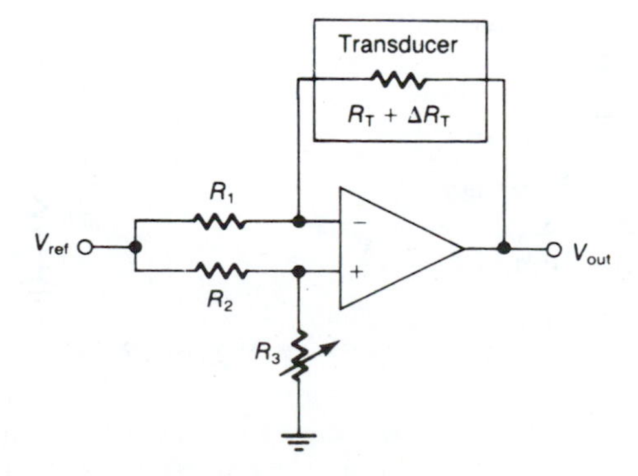


**Figure 12–24.** Bridge Amplifier with Transducer Sensing Element

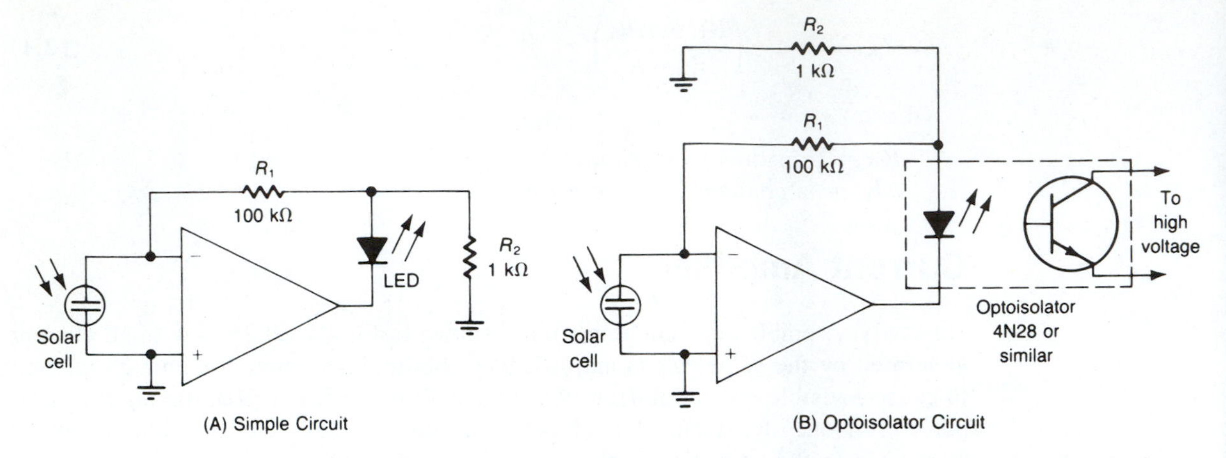


**Figure 12–25.** Current Amplifiers (Part B: Courtesy of National Semiconductor Corporation)

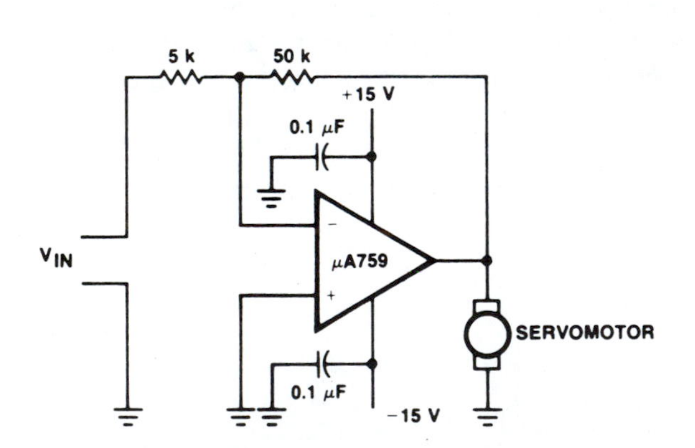


**Figure 12–26.** dc Servo Amplifier (Courtesy of Fairchild Camera & Instrument Corporation, Linear Division)

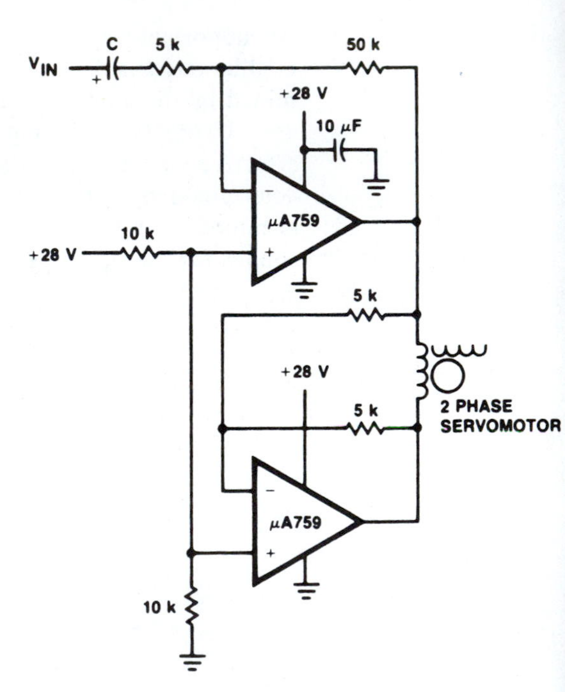


**Figure 12–27.** ac Servo Amplifier (Courtesy of Fairchild Camera & Instrument Corporation, Linear Division)

## Absolute Value Amplifier

An absolute value amplifier, illustrated in Figure 12–28, produces a positive voltage at the output, regardless of the polarity of the input signal. A positive input signal causes diode $D_2$ to conduct, and the circuit acts as a noninverting amplifier to produce a positive output voltage. A negative input signal causes diode $D_1$ to conduct, and the circuit acts as an inverting amplifier, again producing a positive output voltage. Input signals greater than 1 V provide the best accuracy for the circuit.

## More Timer Circuits

We have already established that IC timers are extremely versatile. We will now discuss some practical and easily constructed timer circuits.

### Tone-Burst Generator

A tone-burst generator constructed with one 556 dual timer is shown in Figure 12–29. With switch $S_1$ set to the *continuous* position, timer *B* functions as a free-running multivibrator. The oscillating frequency can be varied from about 1.3 kHz to 14 kHz by the 10 kΩ potentiometer. If the potentiometer is replaced by a thermistor or photoconductive cell, the oscillating frequency will be proportional to temperature or light intensity, respectively.

When $S_1$ is set to the *burst* position, timer *A* output pin 5 alternately places a LOW or HIGH voltage on reset pin 4 of timer *B*. When pin 4 is LOW, timer *B* cannot oscillate; but when pin 4 is HIGH, timer *B* does oscillate. The result of the alternately high and low voltage, then, is that timer *B* oscillates in bursts.

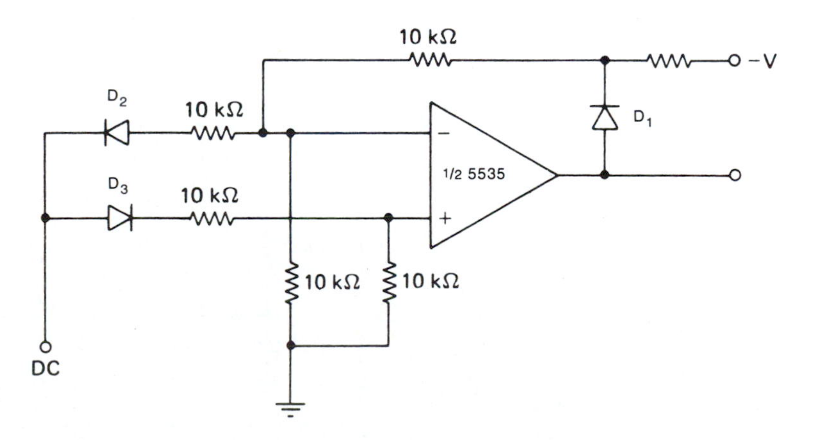


**Figure 12–28.** Absolute Value Amplifier (Courtesy of Signetics Company)

The output of the tone-burst generator is taken from pin 9 of timer $B$. This output can be used to drive either an audio amplifier or a stepdown transformer directly to a speaker.

### Water-Level Fill Control

A simple water-level fill control circuit that uses a 555 timer is shown in Figure 12–30. When switch $S_1$ is closed, the output of the timer is LOW. When $S_1$ is opened, the output goes HIGH and starts the pump. The time interval for the HIGH output sets the height of the water level, and is determined as follows:

$$t_{high} = 1.1 R_A C \qquad \text{(in seconds)} \tag{12.2}$$

To prevent overflow, at some predetermined water level, the overfill switch will place a LOW on reset pin 4, causing the output of the timer to go LOW and stop the pump.

### Missing-Pulse Detector

Adding a transistor to a 555 one-shot multivibrator, as illustrated in Figure 12–31, produces a missing-pulse detector. When $V_{in} = 0$ V, the emitter diode of $Q_1$ clamps capacitor voltage $V_C$ to a few tenths of a volt above ground, forcing the 555 timer into its idle state with a HIGH output voltage at pin 3. When $V_{in}$ goes HIGH, $Q_1$ is cut off and capacitor C begins to charge towards $V_{CC}$. If $V_{in}$ goes LOW before the timer completes its timing cycle, the voltage across capacitor C is reset to near 0 V. However, if $V_{in}$ does not go LOW before the timer completes its timing cycle, the timer will enter its normal state, and output pin 3 will go

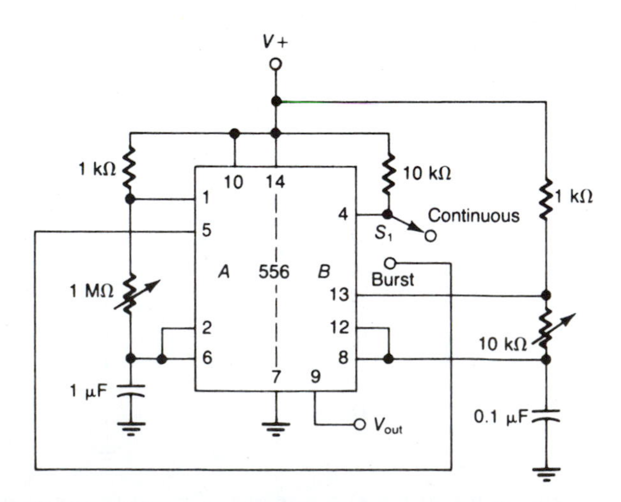


**Figure 12–29.** Tone-Burst Oscillator Using a Single 556 Timer

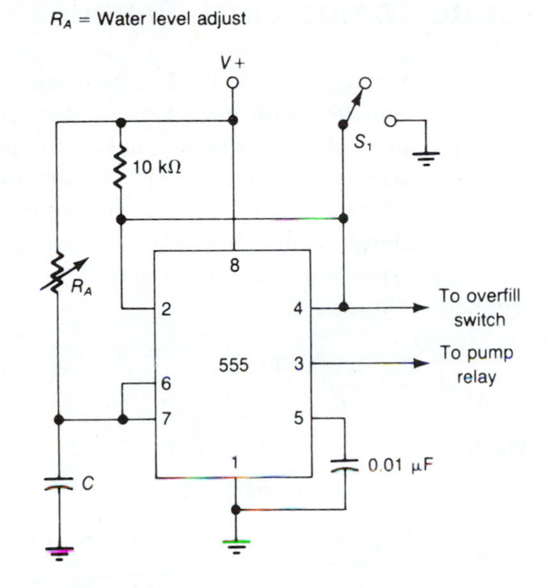


**Figure 12–30.** Water-Level Fill Control Using a 555 Timer

LOW. This is what happens if the timing interval is slightly longer than the input pulse at $V_{in}$ and $V_{in}$ misses a pulse. The missing-pulse detector is useful in medical electronics, where it can be used to detect a missing heartbeat.

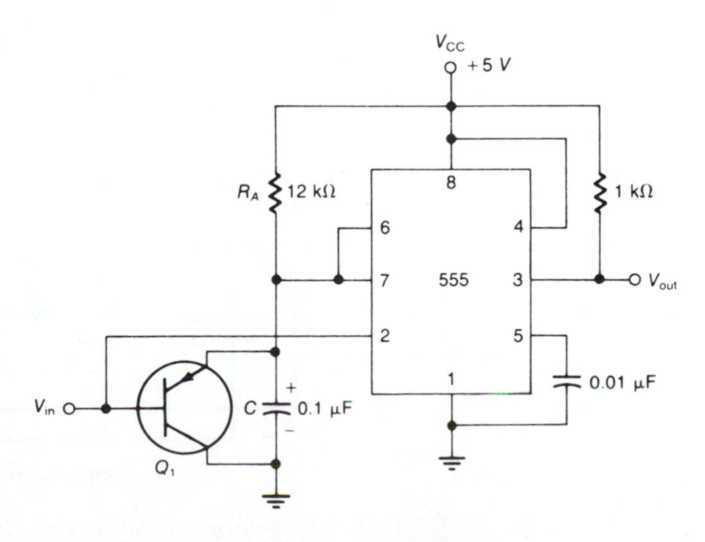


**Figure 12–31.** Missing-Pulse Detector Using a 555 Timer

## Remote Temperature Sensor

The LM134 shown in Figure 12–32 is a three-terminal adjustable current source. Current is established with one external resistor (*R* at pin 1) and no other components are needed. The device makes an ideal remote temperature sensor because its current-mode operation does not lose accuracy over long wire runs. Figure 12–33 shows two possible methods for constructing a low output impedance thermometer using the LM134 current source. Output current ($I_{out}$) is directly proportional to absolute temperature ($T$) in degrees Kelvin ($K$) and is determined as follows:

$$I_{out} = \frac{(227\ \mu V/K)(T)}{R} \quad \textbf{(12.3)}$$

where

$R$ = the external resistor connected between the $V^-$ and $R$ pins

To maintain high accuracy, a low temperature coefficient resistor must be used for $R$.

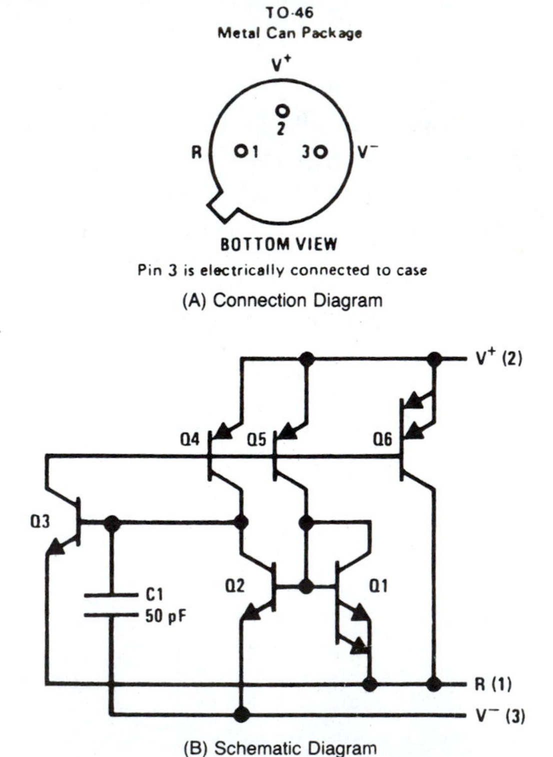


(A) Connection Diagram

(B) Schematic Diagram

Figure 12–32. LM134 Three-Terminal Adjustable Current Source (Courtesy of National Semiconductor Corporation)

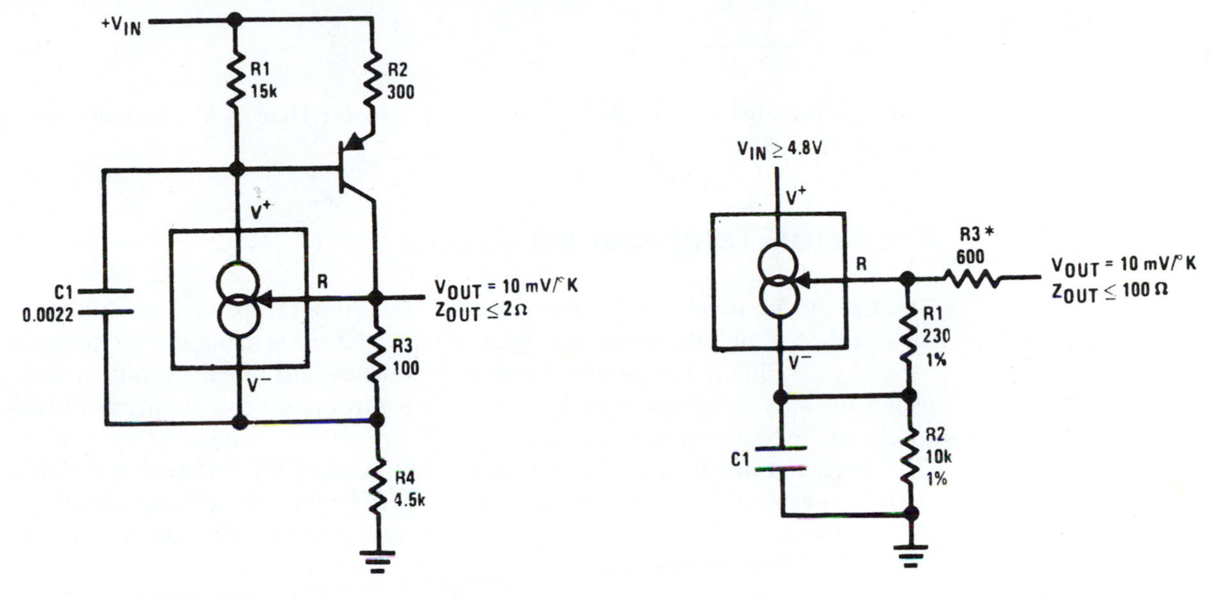


Figure 12–33. Two Methods for Constructing Low Output Impedance Thermometers (Courtesy of National Semiconductor Corporation)

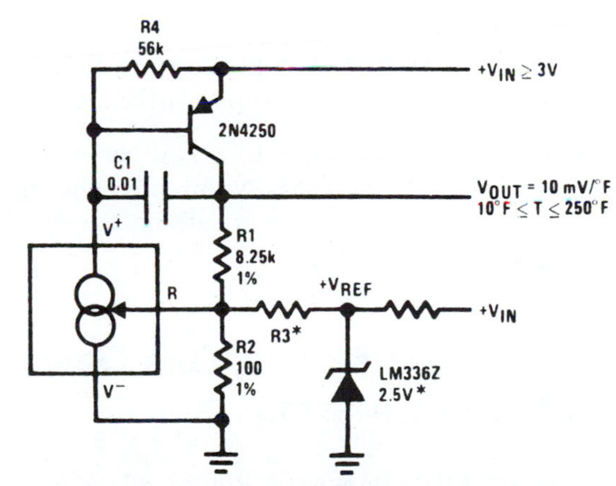


*Select R3 = VREF/583 μA. VREF may be any stable positive voltage ≥ 2V · Trim R3 to calibrate

Figure 12–34. Ground-Referred Fahrenheit Thermometer (Courtesy of National Semiconductor Corporation)

## Fahrenheit Thermometer

A ground-referred Fahrenheit thermometer that uses the LM134 device is shown in Figure 12–34. In this circuit, the value for $R_3$ is selected so that

$$R_3 = \frac{V_{ref}}{583\ \mu A} \tag{12.4}$$

and $V_{ref}$ may be any stable positive value greater than 2 V. To calibrate the circuit, trim $R_3$.

## Precision Temperature Sensor

The LM135 is a precision, easily calibrated, IC temperature sensor. Applications for this device include almost any type of temperature sensing over a range from −55°C to +150°C. The device has low impedance and linear output, making it especially easy to interface with readout or control circuitry. Its linear output is unlike other temperature sensors.

Included on the LM135 chip is an easy method for calibrating the device for higher accuracies. A potentiometer connected across the device with the arm tied to the adjustment terminal allows a single-point calibration of the sensor that corrects for inaccuracies over the full temperature range.

The output of the device (calibrated and uncalibrated) is expressed as

$$V_{out(T)} = V_{out(T_O)} \times \frac{T}{T_O} \tag{12.5}$$

where

$T$ = unknown temperature expressed in degrees Kelvin (K)
$T_O$ = reference temperature, also expressed in K

By calibrating the output to read correctly at one temperature, the output at all temperatures is correct. Nominally, the output is calibrated at 10 mV/K.

A wide range of applications for this temperature sensor are shown in the data sheets in Appendix M.

## Smoke Detectors, Gas Detectors, Intrusion Alarms

The LM1801 shown in Figure 12–35A is an ionization type smoke detector designed to operate off a 9 V alkaline battery, although provisions are made for operation off supplies up to 14 V and for line operation.

Low battery threshold, alarm threshold, hysteresis (for noise immunity), and stand-by current drain are externally programmed by resistors. The device includes a power transistor capable of directly driving a typical 85 dB horn. The ionization chamber requires an external field effect transistor (FET) buffer.

A parallel alarm output is provided to enable up to eight similar detectors to be connected in parallel. In this mode of operation, a fault in the line cannot prevent local operation. The low battery alarm signal is confined to the local unit.

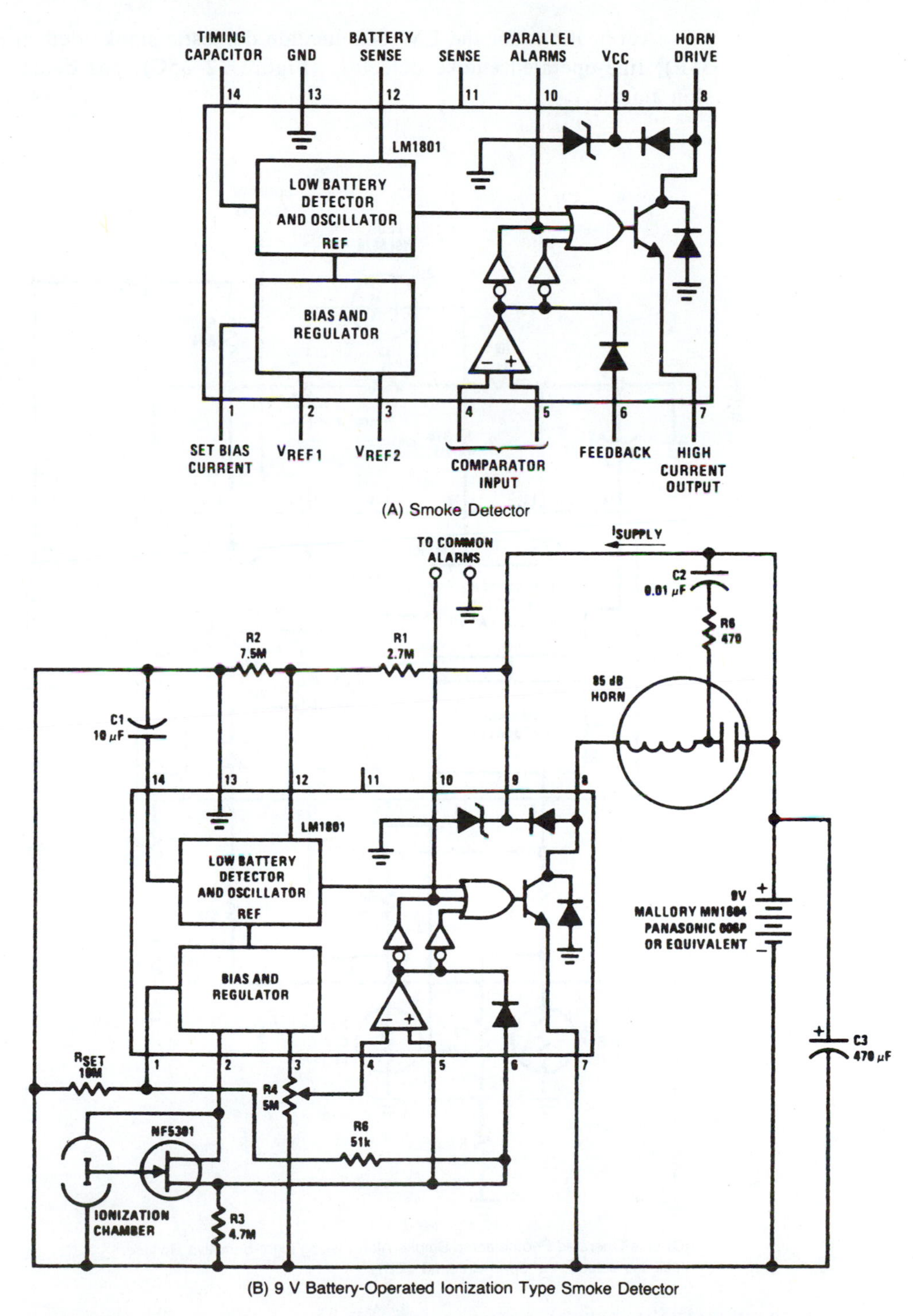


Figure 12–35. LM1801 Applications (Courtesy of National Semiconductor Corporation) (Continued)

Applications for the LM1801 include domestic smoke detectors (Figure 12–35B), line-operated smoke detectors (Figure 12–35C), gas detectors, and intrusion alarms.

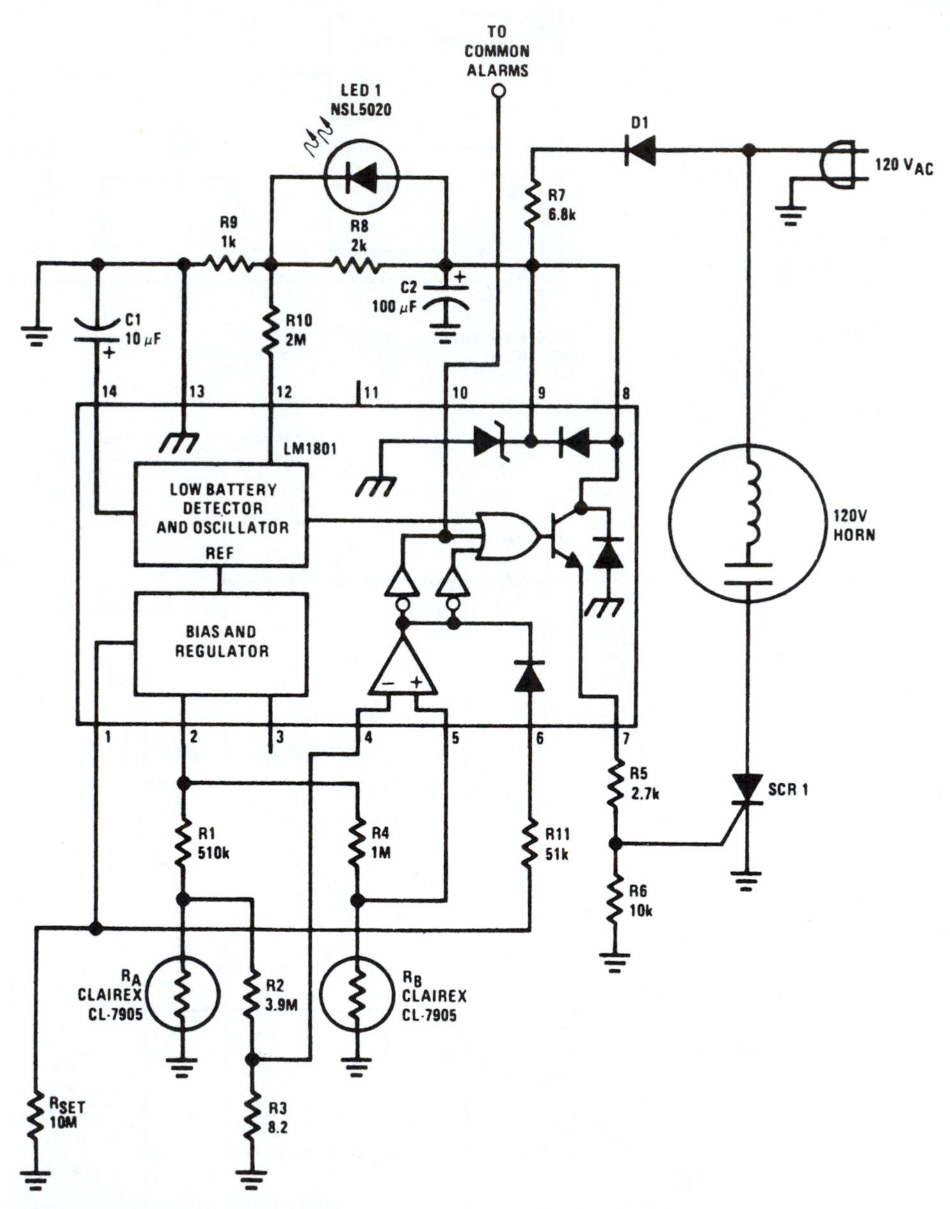


(C) Line-Operated Photoelectric Smoke Alarm Using Light-Sensitive Resistor (Includes detection of open-circuited LED)

**Figure 12–35.** *Continued*

In Figure 12–35B low battery threshold is set by resistors $R_1$ and $R_2$. The value for these resistors is selected so that the voltage at pin 12 is equal to the oscillator trip voltage when the battery voltage is at the low limit at which the low battery alarm is to operate. The values shown provide a warning at about 8.2 V.

Parallel operation of two or more units is easily achieved with a pair of wires connecting pin 10 of each unit and ground. In this mode, every alarm will sound if any single unit detects smoke.

## Fluid Detection System

An IC designed for use in fluid detection systems is the LM1830 shown in Figure 12–36. Applications include beverage dispensers, water softeners, irrigation pumps, sump pumps, radiators, reservoirs, and boilers. In typical applications, the output can be used to drive an LED, loud speaker, or a low-current relay.

An ac signal is passed through two probes within the fluid. A detector determines the presence or absence of the fluid by comparing the resistance of the fluid between the probes with the resistance internal to the IC. A pin is available for connecting an external resistance in cases where the fluid impedance is of a different magnitude than that of the internal resistance. When the probe resistance increases above the preset value, the oscillator signal is coupled to the base of the open-collector output transistor, which drives the external alarm indicator.

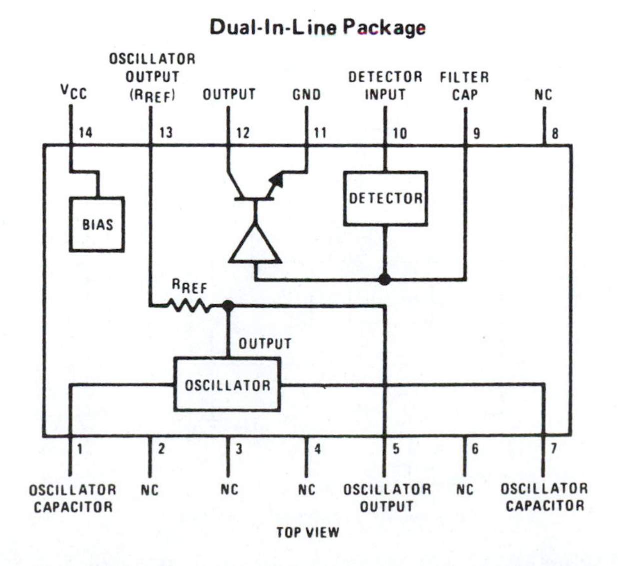


**Figure 12–36.** LM1830 Fluid Detector (Courtesy of National Semiconductor Corporation)

### Low- and High-Level Warning Devices

Typical circuits that use the LM1830 are the basic low-level warning device with an LED indicator in Figure 12–37A, the low-level warning device with audio output in Figure 12–37B, and the high-level warning device in Figure 12–37C, which is suitable for driving a sump pump or opening a drain valve. In all cases, note the simplicity of the circuit.

## LED Flasher/Oscillator

An extremely useful device is the LM3909 LED flasher/oscillator shown in Figure 12–38, designed to provide low power drain and operation from weak batteries so that continuous operation life exceeds that expected from battery rating. By using the timing capacitor for voltage boost, the device delivers pulses of 2 V or more to the LED while operating on a supply of 1.5 V or less.

Application is made simple by inclusion of internal timing resistors and an internal LED current limit resistor. Timing capacitors will generally be of the electrolytic type, and a small 3 V rated part will be suitable for an LED flasher that uses a supply up to 6 V. The circuit is inherently self-starting and requires the addition of only a battery and a capacitor to function as an LED flasher.

Typical applications of the LM3909 IC are the flashlight finder in Figure 12–39A, the emergency lantern/flasher in Figure 12–39B, and the 1 kHz square wave oscillator in Figure 12–39C. Other typical applications include devices for locating boat mooring floats in the dark, sales and advertising gimmicks, emergency locators for such things as fire extinguishers, and electronic applications such as trigger and sawtooth generators.

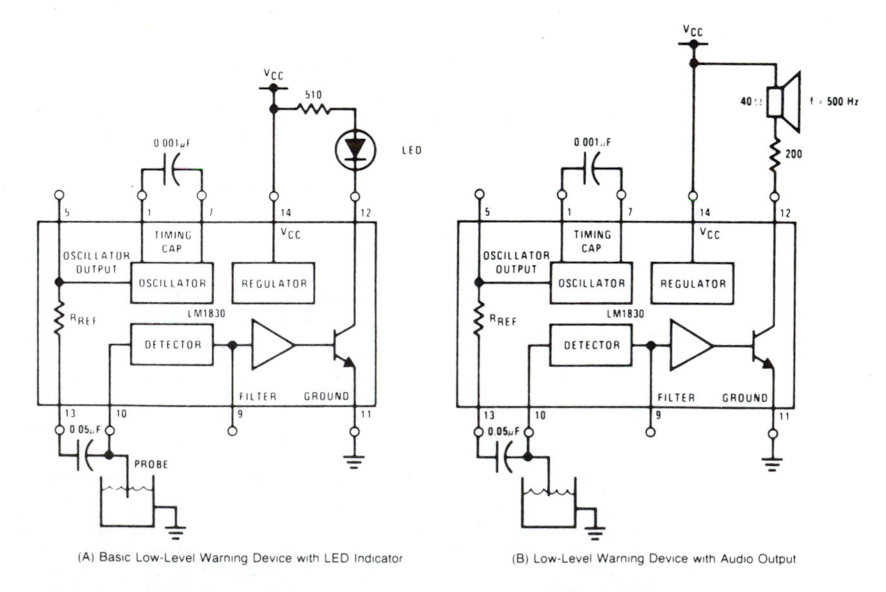


(A) Basic Low-Level Warning Device with LED Indicator

(B) Low-Level Warning Device with Audio Output

Figure 12–37. LM1830 Applications (Courtesy of National Semiconductor Corporation) (Continued)

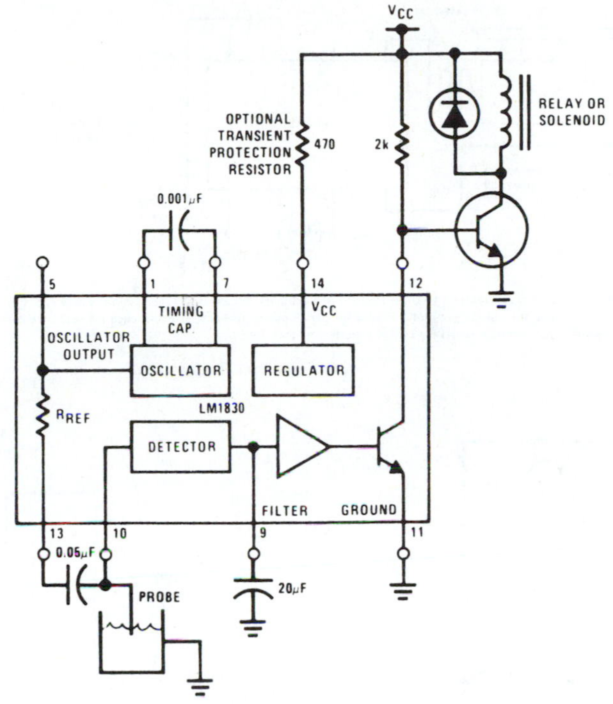


(C) High-Level Warning Device with Relay Output

**Figure 12-37.** *Continued*

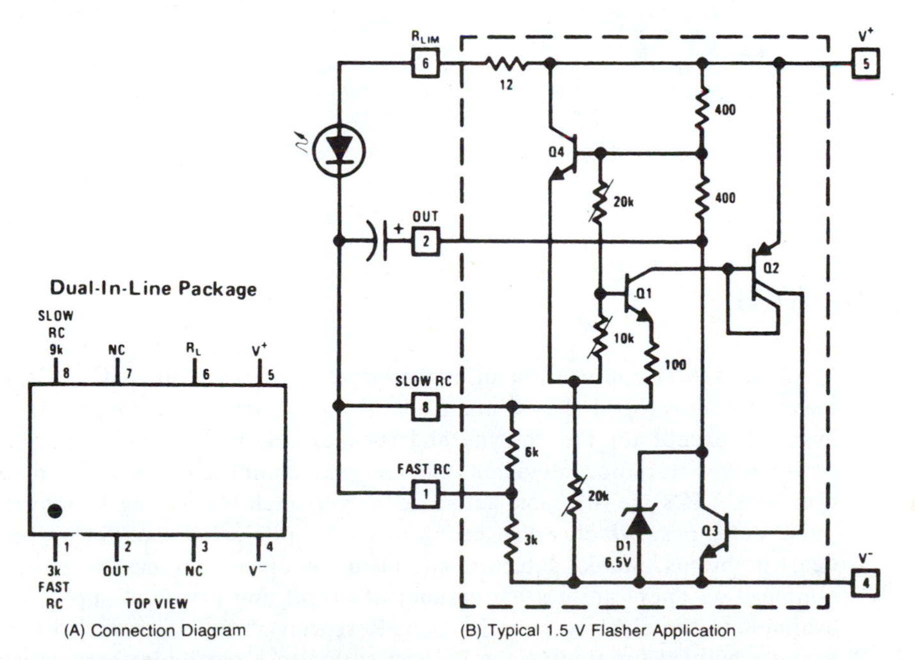


(A) Connection Diagram

(B) Typical 1.5 V Flasher Application

**Figure 12-38.** LM3909 LED Flasher/Oscillator (Courtesy of National Semiconductor Corporation)

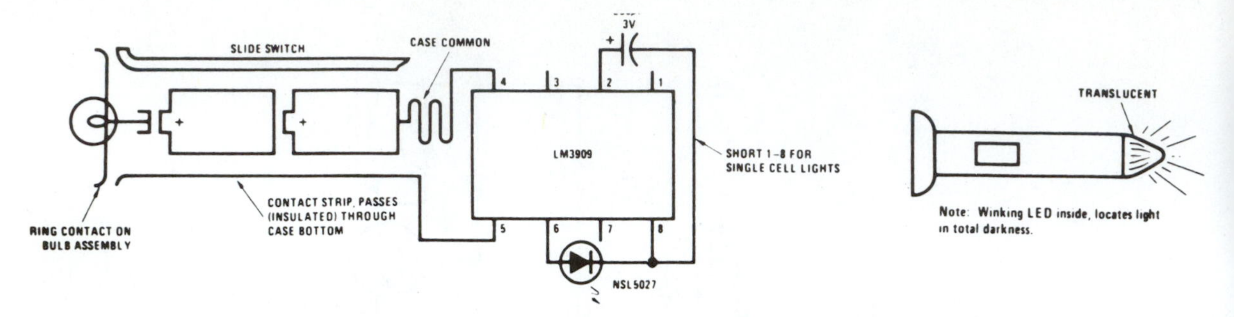


Note: LM3909, capacitor, and LED are installed in a white translucent cap on the flashlight's back end. Only one contact strip (in addition to the case connection) is needed for flasher power. Drawing current through the bulb simplifies wiring and causes negligible loss since bulb resistance cold is typically less than 2Ω.

(A) Flashlight Finder

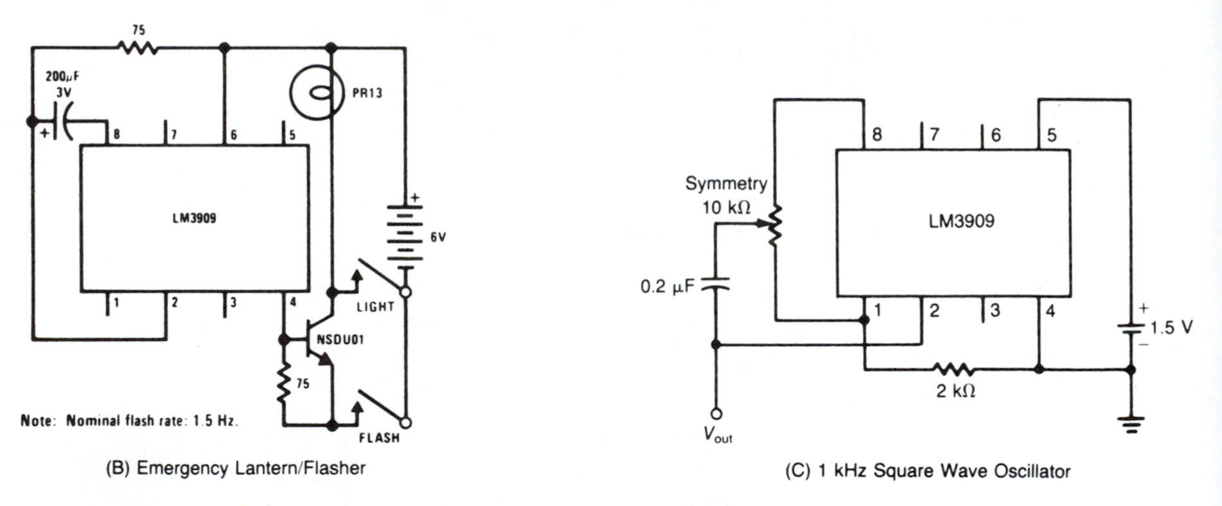


(B) Emergency Lantern/Flasher

(C) 1 kHz Square Wave Oscillator

**Figure 12–39.** LM3909 Applications (Courtesy of National Semiconductor Corporation)

# 12.5 Summary

☐ Consumer applications offer the largest single use for LICs. Single-chip ICs have been developed that contain all the necessary circuitry for complete radio systems, except for the antenna and speaker. Many single-chip ICs that replace entire stages in radio, television, and two-way communications systems are readily available. LICs are the basic active devices in such life-saving systems as missing-pulse detectors, which are used by many hospitals for monitoring patients with heart problems. Smoke detectors and intrusion alarms are easily constructed using simple ICs. There are a great number of useful and practical applications for ICs available to the electronics technician. Research of manufacturer's literature is extremely helpful for finding the IC best suited to a particular application.

# 12.6 Questions and Problems

**12.1** Of the devices discussed in this chapter, select one that would best suit the requirements for a stereo phono amplifier with bass tone control that operates on a 12 V system and provides at least 60 dB channel selection and 60 dB ripple rejection. You desire $A_v = 50$, and you are using 8 Ω speakers.

**12.2** You plan to build a simple intercom system with ground-referenced input having $A_v = 50$. The output will be centered to one-half the supply voltage. What will be a low-cost device to use for this application?

**12.3** You are using a battery-operated LM389 audio amplifier as an AM radio. You desire to increase the gain from 20 to 200. What modification to the circuit will you make to accomplish this increase?

**12.4** You are in an area in which TV channel 3 provides full-length feature movies. You wish to record these movies on a VCR. What device can you use for this purpose, and what oscillator frequency will you need to design? What are the component values for this frequency?

**12.5** If you also want to design the circuit in Problem 12.4 so that both channel 3 and channel 4 are available for recording, what frequencies must be used, and what are the component values for those frequencies?

**12.6** You are using an MC3359 FM-IF circuit in a voice communications system. You find that AFC is not required, but you desire to increase the audio output level. What action will you take to accomplish this increased output?

**12.7** You find that the circuit of Problem 12.6 sometimes rectifies signals when using capacitive loads. How can you correct this problem?

**12.8** You find that the cordless base station transmitter you operate is showing some modulation distortion. What action can you take to minimize this problem? What values are the components used to correct the problem?

**12.9** You are using a no-holds vertical/horizontal circuit to eliminate the hold controls on your TV receiver. State the actions necessary to prevent false counts and consequent severe phase jitter.

**12.10** A TBA950-2 TV signal processing circuit is to be used for video recording. What must be done to block the automatic switchover feature of the device?

**12.11** Explain the operation of a balanced bridge circuit using a transducer in the feedback path.

**12.12** What determines the speed and direction of a motor in a dc servo amplifier? in an ac servo amplifier?

**12.13** You desire to have a positive output signal regardless of the polarity of the input signal. What type of circuit satisfies this requirement?

**12.14** Explain the operation of the burst portion of a tone-burst generator constructed with a 556 timer.

**12.15** In a water-level fill control circuit that uses a 555 timer, calculate the value needed for $C$ if $R_A = 10$ MΩ and the HIGH time interval is 10 seconds.

**12.16** What value would you need for $R_A$ in the water-level fill control circuit if the HIGH time interval is 2 minutes and $C = 10$ μF?

**12.17** What would the HIGH time interval be if $R_A = 2$ MΩ and $C = 0.1$ μF?

**12.18** How might you use a 555 timer to detect a missing heartbeat in medical electronics?

**12.19** You have the task of monitoring the temperature in an environmentally controlled room that is located several hundred feet from your work station. What device can you use that will allow you to perform the monitoring task without leaving your work area?

**12.20** You have the task of designing a smoke alarm system for a large home. The customer wants all alarms to be activated if any one of them detects smoke, and wants each alarm to operate independently if anything goes wrong with any one in the system. What type of device can you select to satisfy this requirement?

**12.21** You are a technician with an industrial firm that uses large boilers in the plant. You are instructed to design a system that will monitor the water in the boilers and automatically cut off the burner for a boiler that loses its water. What is one type of device that can be used for this task? What type of load would be logically driven by the output of the device for this task?

**12.22** You are hired by a marina to design a system for marking boat mooring floats and fire extinguishers so that they may be located easily in the dark. The fire extinguishers should be identified with a red signal and the floats by green. What is a simple device selection for this problem? What type of output indicators will you use?

**For Problems 12.23 through 12.30, refer to Appendix L.**

**12.23** Explain how input and output voltage ranges can be extended.

**12.24** Explain the procedure for offset adjustment of a multiplying circuit.

**12.25** Explain the procedure for offset adjustment of a divider circuit.

**12.26** Explain the procedure for offset adjustment of a square root circuit.

**12.27** Explain the procedure for offset adjustment of a squaring circuit.

**12.28** Explain nonlinearity accuracy error and its cause.

**12.29** List and explain the cause of three other system accuracy errors.

**12.30** Refer to the first quadrant multiplier/divider in Figure 18 in Appendix L. Using resistor and voltage values as listed in Example 1 in Appendix L, calculate $V_O$.

# Chapter 13

# Troubleshooting

## 13.1 Introduction

One of the major problems that you face as a technician is to find the defect in a faulty system or circuit, correct it, and make the system or circuit perform as designed. When faced with such a problem, you must make use of any technique available to determine the fault, keeping in mind that there is no single approach to solving every problem. It takes a combination of experience, knowledge of the system, sound analytical ability, and persistence to reach any solution.

In this chapter we will look at several basic troubleshooting techniques. While these are standard procedures, you will, as a technician, develop your own method.

## 13.2 Objectives

When you complete this chapter, you should be able to:

- ☐ List several troubleshooting techniques.
- ☐ Realize the limitations of the quick-check method of troubleshooting.
- ☐ Recognize that a different approach is needed for troubleshooting circuits with feedback.
- ☐ List several precautions that must be taken when troubleshooting circuits containing ICs.

- ☐ Recognize the value of the information contained within systems manuals.
- ☐ List the important troubleshooting information contained within systems manuals.
- ☐ Select the right piece of test equipment to use for the troubleshooting task at hand.
- ☐ Realize the value of data sheets for setting up test circuits.
- ☐ List some PCB troubles and the steps that can be taken to correct or minimize those problems.
- ☐ List several typical LIC failure modes and steps that can be taken to minimize failures.

# 13.3 Troubleshooting Methods

It is not always possible to draw a fine line between good and bad troubleshooting procedures. An acceptable method of locating trouble in one system can result in the destruction of a device in another system. As more sophisticated devices and systems become available, troubleshooting methods have to be altered. However, in any modern electronic system, the power supply is most often at fault.

## Locating Troubles Using the Senses

Troubleshooting involves testing various branches of the defective circuit to find the specific trouble. After the problem is isolated, first perform a preliminary inspection using the senses of sight, smell, touch, and hearing. For example, burned resistors, charred capacitors, inductors, and transformers can be detected by visual observation or by smell. Overheated transistors or ICs can be located by touch. The sense of hearing can detect hum or high-voltage arcing caused by overloaded or overheated transformers.

The sensory procedures just discussed are most often referred to as *visual inspection*. When defective components are located by means of visual inspection, further inspection should be performed before replacement of components. The real cause of trouble may be another component or situation; for example, a short. If not detected at this point, this condition could result in secondary failure.

## Secondary Failure

When an electronic component malfunctions, it may cause other parts of a circuit or system to malfunction also. The failure of the other components is known as *secondary failure*.

Secondary failure often represents a much more serious problem than the original malfunction. For example, suppose that a regulated power supply, having no provision for overload protection, malfunctions and produces 25 V at the output instead of the intended 13.5 V. The single component that causes this failure (the pass transistor) might cost a few pennies to replace. But the damage to a radio transceiver, connected to the power supply when the overvoltage condition occurs, may result in a far greater repair cost.

In well-designed electronic equipment, the possibility of secondary failure is addressed and minimized. In the situation just discussed, for example, an overvoltage protection circuit would prevent the secondary damage to the transceiver.

In the troubleshooting and repair of any electronic apparatus, a technician must always be aware of the possibility of secondary failure. If a component has burned out because of secondary failure, merely replacing that component will not solve the problem and it will probably recur.

## Shotgun Method

In the vacuum-tube circuit days, some technicians resorted to the shotgun approach; that is, they replaced all the parts in the circuit without taking the time to determine which of the individual parts was faulty. This was considered to be a very poor practice, since it overlooked the value of logically determining the fault. In modern systems, however, with their integrated circuits and packaged subassemblies, or *modules,* it is an accepted practice to replace the IC or module after isolating the fault to that stage, because it is impossible to locate and replace the faulty component within such a device.

## Troubleshooting Procedures

When troubleshooting an electronic system, a good procedure is to start with general facts about the system's behavior and move to the specific device at fault, as outlined in the block diagram in Figure 13–1.

### Defining the Problem

The first step is to define the problem and record the failure mode. The operator of the system can often provide valuable information about its behavior, both prior to and after failure. Also, make a record of the failure mode to refer to when future failures of a similar nature occur in the same type of system. Such information should become a permanent part of the records for that particular system.

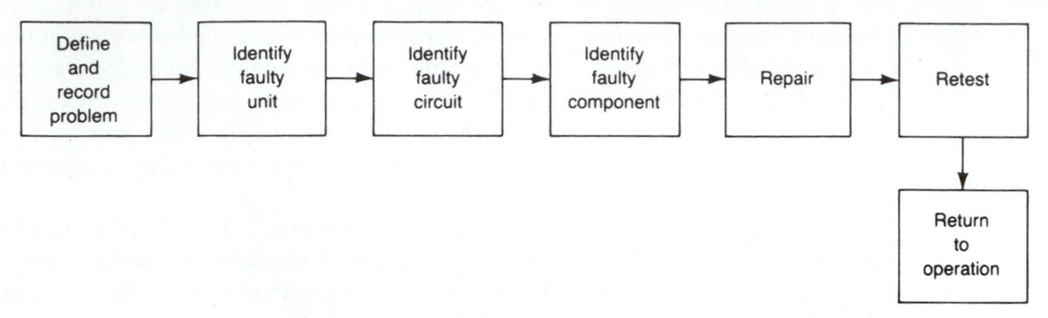


**Figure 13–1.** Block Diagram for General Troubleshooting Procedures

## Isolating the Faulty Unit

Many troubleshooting manuals for systems identify possible failure modes and possible causes. Such manuals also describe important checkpoints in the system, usually with waveforms and voltage levels. These manuals are extremely helpful for quickly isolating the faulty unit. If the faulty unit cannot be identified from the symptoms, *signal injection* or *signal tracing* can be used, as shown in Figure 13–2.

If the system has several units, the divide-and-conquer concept works well with signal injection. This method involves dividing the system units in half, injecting a signal at that halfway point, and checking for a normal output. If the

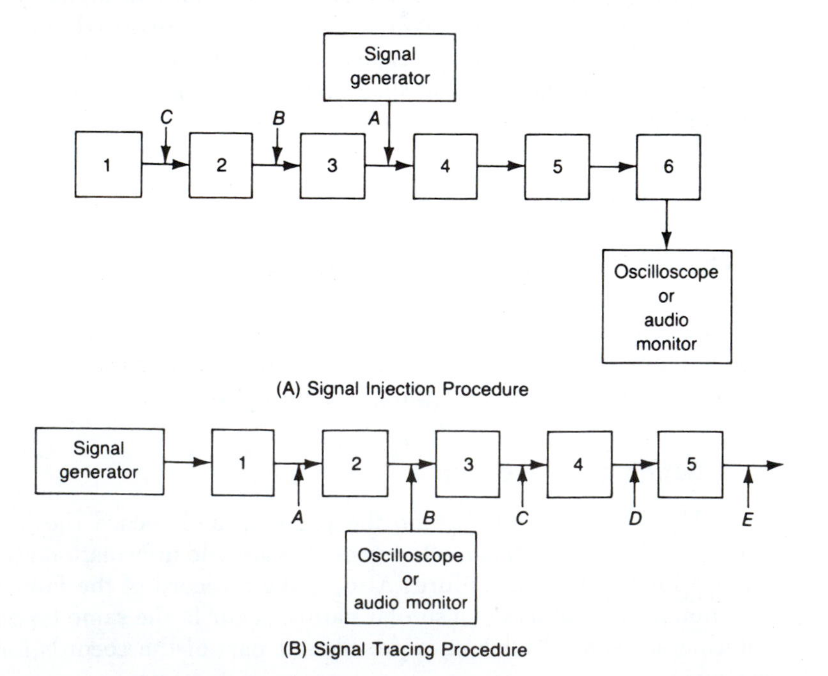


**Figure 13–2.** Two Standard Troubleshooting Techniques

output signal is normal, then you know that the last half of the system is working properly. The next step is to divide the remaining portion of the system in half and repeat the procedure as many times as necessary to locate the problem. In this manner, the number of tests required to isolate the faulty unit is reduced.

### Locating the Faulty Circuit

Once the faulty unit is located, the next step is to locate the faulty circuit within that unit. The same techniques of signal injection and signal tracing can be used to locate faulty circuits within the unit. The divide-and-conquer concept is useful, also, if there are several stages within the unit.

### Identifying the Faulty Component

After locating the faulty circuit, the next step is to identify the faulty device or component in that circuit. This is usually done by making voltage and resistance checks.

### Repairing and Retesting

Replacing the faulty device or component should complete the trouble-shooting procedures, but the system must be tested for proper operation before it is returned to its operator. *Retesting* is a step often overlooked by many technicians, yet it is as important as all the other procedures. It is advisable to let the system operate over an extended period of time to verify complete operational capability.

## Quick-Check Methods

Quick-check methods are useful only if you understand their limitations and take proper precautions to protect the equipment under test. For example, a popular quick-check for bipolar transistor circuits is to cause a temporary short between the base and emitter while measuring the collector voltage, as illustrated in Figure 13–3. The short circuit should cause the collector voltage to rise to the $V_{CC}$ voltage.

This procedure must be used with caution, because a momentary short between the base and collector instead of between the base and emitter can destroy the transistor. Also, if you are working with the direct-coupled amplifier circuit shown in Figure 13–4, a base-emitter short at $Q_1$ will destroy transistor $Q_2$. The following is a description of what happens. The collector current of $Q_1$ is conducted through resistor $R_2$. The base voltage of $Q_2$ is $V_{CC}$ minus the voltage drop across $R_2$ caused by the collector current of $Q_1$. The base current of $Q_2$ also flows through $R_2$, but is so small that it can be ignored. Shorting the base-emitter

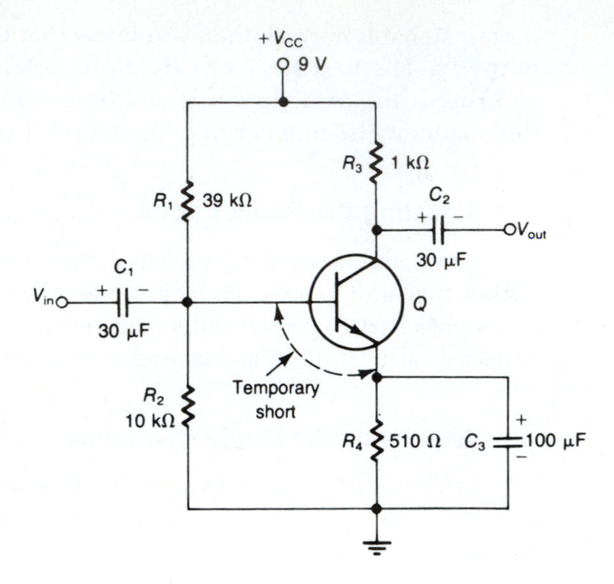


**Figure 13–3.** Base-Emitter Short Circuit Quick Test

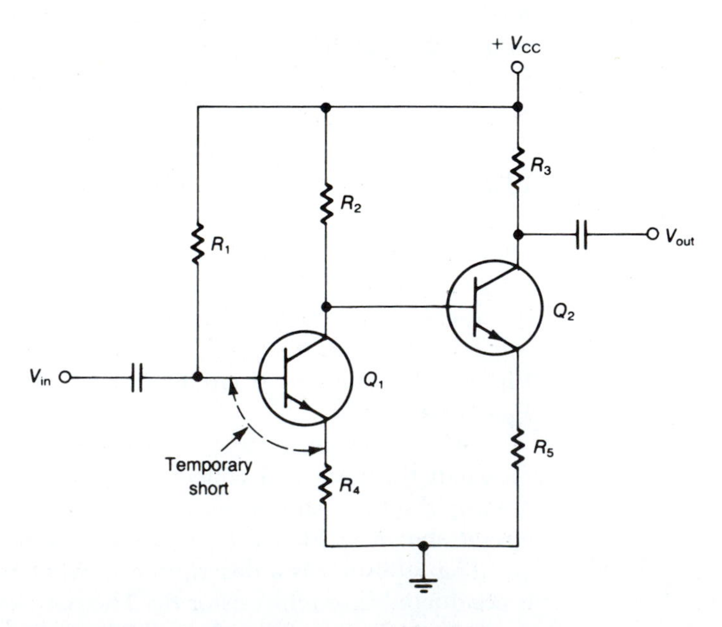


**Figure 13–4.** Direct-Coupled Amplifier

of $Q_1$ shuts off $Q_1$ and causes $Q_1$ collector current to stop flowing, forcing the base voltage of $Q_2$ to rise to the $V_{CC}$ voltage and destroying transistor $Q_2$.

### Bad-Habit Quick-Check

A quick-check that has been used by some technicians could prove to be dangerous to your health and well-being. That *bad-habit* quick-check is placing a finger on the grid, base, or gate of the amplifying device. A change in the output that results from the disturbance caused by the finger at the input generally indicates that the stage is operating correctly. However, this test is based on the false assumptions that the voltages present at these points are always too small to cause personal injury and that solid-state circuits always use low voltage supplies.

In some vacuum-tube circuits, direct coupling of amplifiers could place voltages as high as 300–500 V on the grids. Likewise, solid-state circuits may have voltages as high as 100 V or more, and accidentally placing your finger on the wrong terminal could cause severe shock.

A *safer* method for this type of quick-check is to use a screwdriver blade instead of your finger to disturb the input of the stage. However, this approach also must be used cautiously, because the screwdriver blade may cause an accidental short circuit and introduce problems into the circuit that did not exist before the tests began.

### Other Quick-Checks

Quick-checks using ohmmeters in transistor or semiconductor diode circuits also must be made with care. Since an ohmmeter delivers a voltage, it can forward bias the device into conduction, resulting in faulty resistance readings. Reversing the ohmmeter leads may eliminate the forward biasing problem (if the semiconductor device is operating properly), but you still have to interpret the readings in relationship to the circuit under test.

Some ohmmeters are designed to provide either *high ohms* or *low ohms* switchable positions. In the low ohms position, the ohmmeter supplies such a low voltage to the circuit that it cannot forward bias a P-N junction. Interpretation of readings in relationship to the circuit under test is still required.

## Troubleshooting Closed-Loop Circuits

Troubleshooting in closed-loop circuitry requires special methods and techniques, because each section in the circuit depends upon the others for its operation.

A block diagram of a regulated power supply that uses closed-loop circuitry is shown in Figure 13–5. Let's assume that this device has incorrect output voltage. The problem might be due to defects in the series-pass power amplifier,

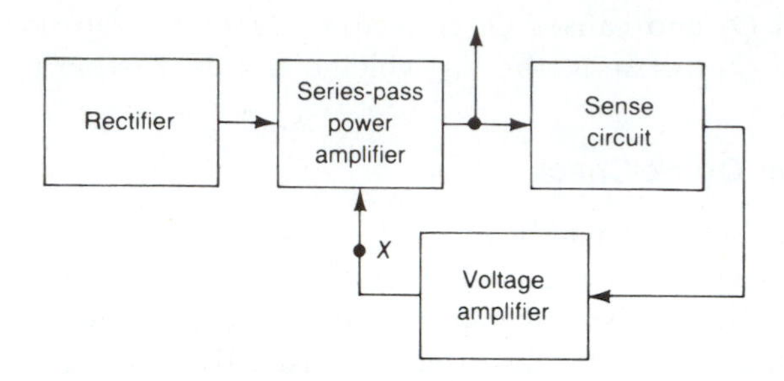


**Figure 13–5.** Closed-Loop Power Supply Regulator

the sense circuit, or the voltage amplifier. In the closed loop, the power amplifier gets its input voltage from the voltage amplifier, which receives its voltage from the sense circuit, which in turn receives its voltage from the power amplifier. A defect in any of these stages causes all stages in the closed loop to receive improper voltage and results in incorrect output voltage. The defective stage cannot be isolated by making voltage measurements around the loop.

The most popular method used to troubleshoot closed loops is to *open the feedback loop* at some convenient place, such as at point $X$ in Figure 13–5, and insert a dc power supply to provide the control voltage for the series-pass power amplifier. Measurements can then be taken in each section to determine the faulty stage.

Another procedure that works well in modular systems where the circuits can be replaced without much soldering is to simply replace each unit, one at a time, until the output voltage is correct. Keep in mind that the power to the circuit must be turned off before making such replacements, otherwise transient voltage spikes may destroy the unit you are working with.

All or part of the feedback signal in some closed-loop systems may be ac rather than dc. An example is the motor-speed control system shown in Figure 13–6. The voltage-controlled oscillator (VCO) develops the basic motor signal, which is amplified by the voltage and power amplifier and delivered to the motor.

In the feedback loop the VCO frequency, $f_1$, is compared in the discriminator with the countdown frequency, $f_2$, from the crystal-controlled oscillator (CCO). The output of the discriminator is a dc voltage that is amplified and used to set the VCO frequency. The gain control at the dc amplifier input allows setting the VCO frequency to produce the desired motor speed.

Two possible approaches, each equally effective, can be used to isolate the faulty unit. One method is to open the circuit at point $X$ and insert a dc power supply to set and control VCO frequency $f_1$. The other is to open the circuit at point $Y$ and use a signal generator to inject frequency $f_1$ into the discriminator. After the loop is opened, using either method, measurements are made to determine the faulty stage of the circuit.

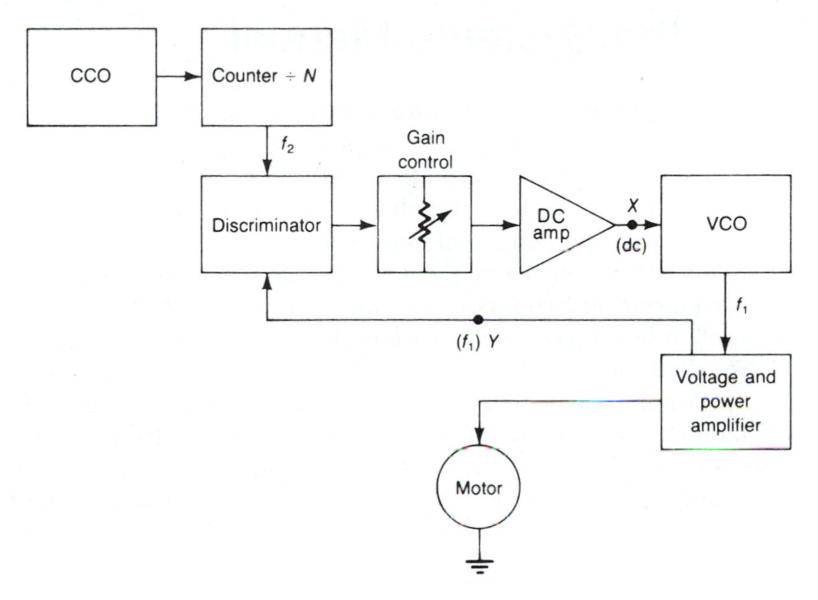


Figure 13-6. Closed-Loop Motor-Speed Control System

## Precautions

When troubleshooting circuits that contain ICs, take the following precautions. Do not short pins by using large test probes, do not use excessive heat when unsoldering a component, and never remove or insert an IC into its socket without first making sure the power to the circuit has been removed. With power on, excessive surge currents can occur and destroy the IC. Whenever possible, use available aids such as IC insertion tools, test clips, and so on, when servicing IC units.

A logical approach to troubleshooting IC units is as follows:

1. Check the power supply input at the IC pins. If it is within its rated value, and the ripple level is low, proceed to the next step.
2. Check for the proper input signal at the IC pin indicated as input on the IC diagram.
3. Check the output pin of the IC for correct output signal.
4. Visually check, and measure with an ohmmeter, for any open or short circuits in the copper track to the IC.
5. If the IC must be removed from the circuit by desoldering, take the extra time necessary to remove the solder completely from each pin with a good desoldering tool, so that the IC can be lifted out without placing strain on the copper tracks of the printed circuit board (PCB).

# 13.4 Using the Systems Manual

Most equipment manufacturers provide a manual that usually includes a schematic diagram for the particular piece of equipment. A good schematic provides much valuable information for the technician, such as how the components are electrically connected, the physical layout of the unit, typical voltage values for a properly operating unit, and expected waveforms at key points. With this information, you can go through the circuit, make voltage measurements, check waveforms, and compare your measurements with those of a properly operating unit. When discrepancies are found in a section, closer examination of that section is called for.

Some manufacturer's manuals also provide a *trouble/cause table*. These tables are extremely helpful for speeding up the troubleshooting process. However, you should not rely completely on such aids, because the manufacturer cannot list all possible faults and their causes. It is important that you develop and use good troubleshooting techniques. Use what aids are available to you, but rely on your own good judgment.

# 13.5 Selecting the Right Test Equipment

In order to locate and correct faults in a system, test instruments of some kind are essential. Apart from any specialized instrument that may be required for complex systems, a large majority of system faults can be located using only the following three standard test instruments:

1. **Multirange meter (multimeter)**—either analog or digital, for voltage, current, and resistance measurements
2. **Oscilloscope**—single trace or dual trace, for checking waveforms
3. **Function generator**—to provide sine wave, square wave, triangle wave, or pulse inputs at various points in a system (signal injection)

## Multimeter

The simplest and most useful of these instruments is the multimeter. The range and accuracy of the meter must be such that it can accurately measure all voltages, ac and dc, that you might be expected to encounter in the system under test. Perhaps the most important consideration is the input impedance of the instrument. It must be as high as possible to prevent loading effects on the circuit being measured. Most modern multimeters use FET input devices, which produce an input impedance of 10 MΩ or more.

As mentioned earlier in this chapter, the ohmmeter function of the multimeter is most valuable to you if it has the high-low ohms capability. The high ohms mode allows in-circuit resistance measurements when no active devices are connected to the resistance under test. This mode could cause a P-N junction to be forward biased. The low ohms mode offers such a low voltage that it cannot forward bias a P-N junction, so it can be used in circuits where active devices are connected to the resistance under test. However, you must still interpret the results of the measurements in relationship to the overall circuit. Also, remember that all power to the circuit under ohms test must be turned off.

## Oscilloscope

The oscilloscope is a versatile and extremely useful test instrument. It is possible to measure both dc values and ac waveforms with the oscilloscope. It offers high sensitivity, typically 10 mV/division, and low loading effect, with typical input impedances greater than 1 MΩ. The frequency, shape, and time period of a single waveform can be determined, or waveforms can be displayed in time and phase relationship to one another. Since many of the ICs we have discussed involve time-varying signals, the oscilloscope will prove to be very useful in any fault-finding effort in these circuits. For best results, the oscilloscope used should have a *minimum* bandwidth (BW) of 5 MHz.

The accuracy of both the $Y$ (vertical, or amplitude) input and the $X$ (horizontal, or time) input is limited to about $\pm 3\%$. At low frequencies, the voltage signal to be measured can be applied directly to the $Y$ input through suitable wires or a coaxial cable. However, when measuring high frequencies, to prevent signal degradation, a fully *shielded* cable should be used. For example, when measuring a rapidly changing signal, such as in comparators, or determining a slew-rate, you should make sure that the probe used is *compensated* to account for the loading of the oscilloscope amplifier input impedance. This compensation consists of an $RC$ parallel network in series with the input impedance of the oscilloscope. Figure 13–7 illustrates a typical compensated probe circuit, in which the actual input impedance components of the oscilloscope have typical values of 1 MΩ and 30 pF.

Such probes are commercially available. To verify full compensation, the probe is connected to a square wave generator (usually available on the oscilloscope as a calibration signal), and the waveform is observed on the cathode ray tube (CRT). The variable capacitor, normally a small screwdriver adjustment on the probe, is adjusted until the square wave seen on the CRT has no overshoot and shows sharp rising and falling edges.

If you are limited to an economy oscilloscope, Figure 13–8 shows how a triggered sweep can be constructed using a 555 timer. The circuit's input op amp triggers the timer, setting its flip-flop and cutting off its discharge transistor so that capacitor $C$ can charge. When capacitor $C$ voltage reaches the timer's control voltage (1/3 $V_{CC}$), the flip-flop resets and the transistor conducts, discharging the capacitor.

Greater linearity can be achieved by substituting a constant-current source for the frequency adjust resistor $R$.

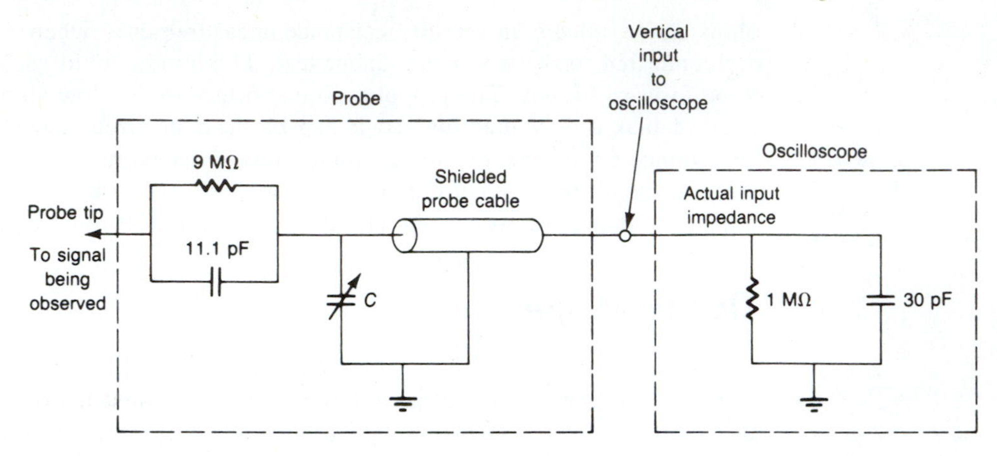


Figure 13–7. Equivalent Circuit for Compensated Oscilloscope Probe and Input Impedance

## Function Generator

Function generators are used in troubleshooting to inject a suitable test signal into the system. The resulting output signal is observed with an oscilloscope or voltmeter and then compared with the data listed in the manufacturer's manual.

The complexity and performance characteristics of this instrument are usually dictated by the system under test. Signal generators are available that provide sine, square, triangle, and pulse waveforms. In some sophisticated units, a sweep control is added, which allows you to automatically vary the frequency of the applied signal between two predetermined values. This feature allows examination of the frequency response of a device.

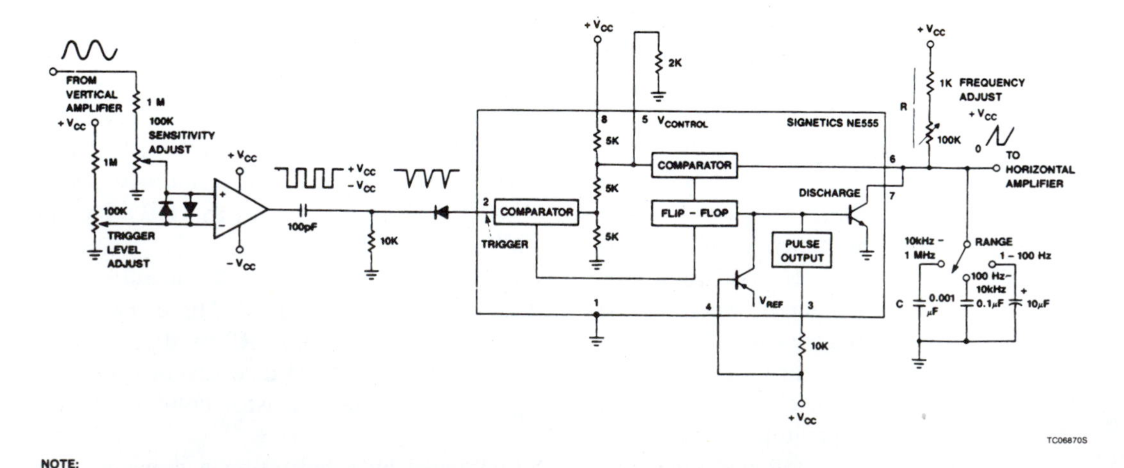


Figure 13–8. Schematic of Triggered Sweep (Courtesy of Signetics Company)

### Signal Injector and Continuity Tester

A more simple troubleshooting device for use in analog systems is a small *hand-held signal injector*. This is usually a fixed-frequency battery-powered oscillator that runs at about 1 kHz, with its output available from a metal prod and a lead with an alligator clip provided for a ground connection. An extension of this design is the *continuity tester*, a battery-powered 1 kHz oscillator with an audible output provided by a small loudspeaker. When the two output leads are shorted together, or connected by the low resistance of a cable wire, the output of the oscillator is fed to the loudspeaker. This type of tester is very useful for checking the continuity of cables, connecting wires, and PCB traces.

## 13.6 Setting Up Test Circuits

Most data sheets provide a test section and a diagram for testing the IC. This testing can be done before the IC is used in a circuit, or when it is removed from a circuit if it is suspected of being at fault. An example is the test circuit for the LM1830 fluid detector shown in Figure 13–9. In this test circuit, using the 0.001 μF capacitor between pins 1 and 7 produces an oscillator output frequency of approximately 6 kHz, available at pin 5. (In normal applications, the output is taken from pin 13 so that the internal 13 kΩ resistor, $R_{ref}$, can be used to compare with the probe resistance. Pin 13 is coupled to the probe by a blocking capacitor so that there is no net dc on the probe.)

The ac test circuit for the LM567 tone decoder is shown in Figure 13–10. In this circuit, a typical value for $R_L$ is 20 kΩ. The input signal to pin 3 should be 100 kHz at ±5 V, and capacitor $C_1$ should be adjusted for a center frequency, $f_{op}$, of 100 kHz.

Other examples of test circuits available in data sheets are shown in Figure 13–11, the LM1818 electronically switched audio tape system, and Figure 13–12 (page 434), the LM3075 FM detector/limiter and audio preamplifier.

These test circuits are but a few examples of the valuable information available to you in the manufacturers' data sheets. The information provides test circuits and component values that enable you to test the relevant operating characteristics of the device. You should keep in mind, however, that the device may test good in the test circuit but may not perform properly when placed in the circuit of the system. This can be a result of feedback paths and faults in other parts of the system.

## 13.7 Printed Circuit Boards

PCBs are used in almost all systems and circuits in which ICs are used. In some circuits the IC is soldered permanently to the board, and in others the IC is inserted into a mount soldered to the PCB.

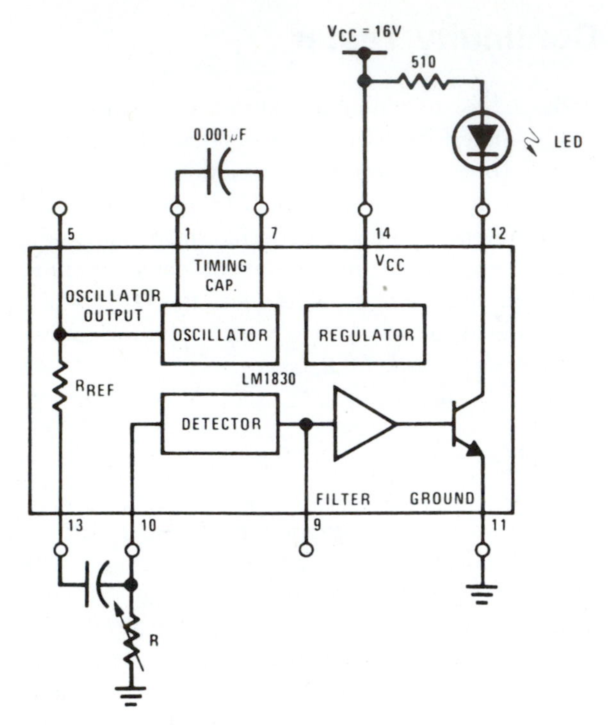


**Figure 13–9.** Test Circuit for LM1830 Fluid Detector (Courtesy of National Semiconductor Corporation)

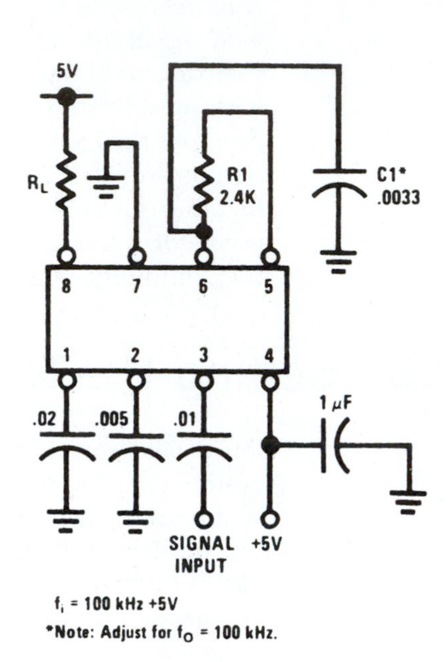


**Figure 13–10.** Test Circuit for LM567 Tone Decoder (Courtesy of National Semiconductor Corporation)

## Construction-Induced Problems

PCBs are constructed with strips of copper mounted to a rigid or flexible insulating material, which usually serves no other purpose than to support the copper strips and the components. Boards that are constructed with care and precision offer high reliability, but there are some problems that may arise in less well constructed boards. For example, if the original artwork or the etching process is poorly done, the copper strip may not have sufficient cross-sectional area, or it may be too narrow in spots. These deficiencies may cause high circuit resistances, or the copper strips may burn open due to heat generated at that portion of the board.

Another problem you may encounter is distortion of the shape of the PCB due to either improper insertion into the PCB holder or the heat generated in the operating unit. Such distortion may cause the copper strips to crack, resulting in open circuits or intermittent failures. This same problem may arise in PCBs used in systems that are subjected to serious vibrations.

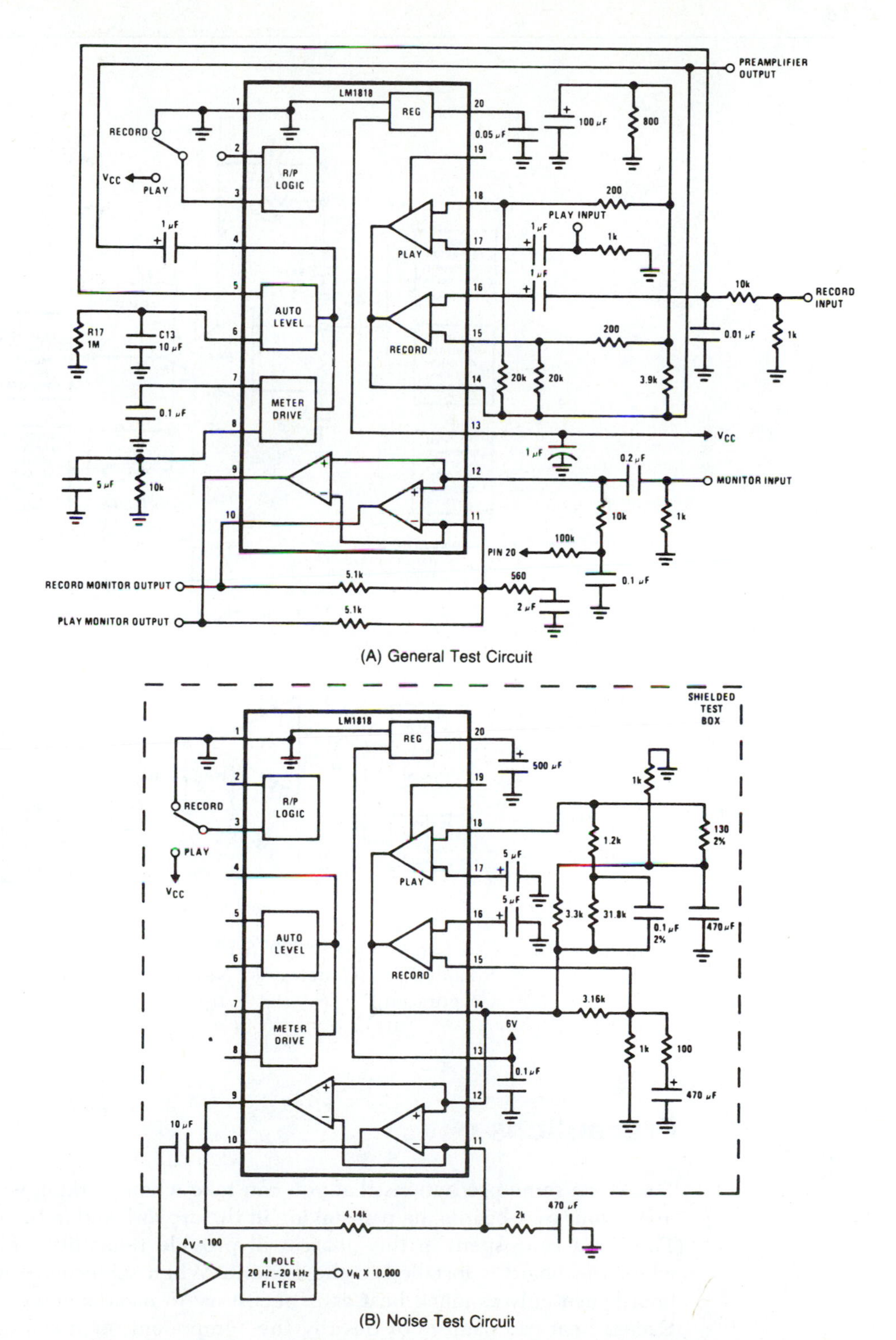


(A) General Test Circuit

(B) Noise Test Circuit

**Figure 13-11.** Test Circuits for LM1818 Electronically Switched Audio Tape System (Courtesy of National Semiconductor Corporation)

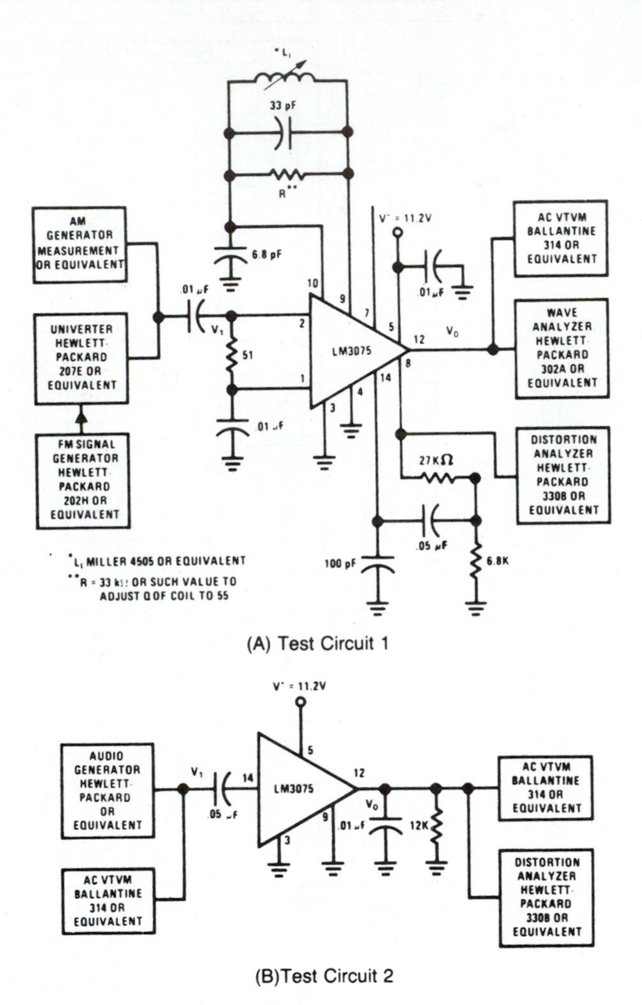


(A) Test Circuit 1

(B)Test Circuit 2

**Figure 13–12.** Test Circuits for LM3075 FM Detector/Limiter and Audio Preamplifier (Courtesy of National Semiconductor Corporation)

## Precautions

There are some precautions that you can take when working with PCBs. If you make your own boards, be painstaking in the artwork and in the etching process. The extra time spent in this phase will provide reliability and peace of mind when the board is installed in the system. When soldering components to the board, use only as much heat as is necessary to make a good solid connection. Excess heat can damage or destroy the component, or it can cause the copper strip to be lifted from the board. Be especially careful that excessive solder is not used, because it may flow between two strips and cause a bridge. After all soldering is completed, use a good solvent to clean the board of flux.

To prevent high resistance or open circuits caused by corrosion of the copper, hermetically seal the PCB. If this is not possible (or not desired), occasional cleaning of the board after it is put into operation (preventive maintenance) will provide longer life and longer trouble-free operation of the system.

## 13.8 Typical Linear IC Failures

### Catastrophic Failure

In general, LICs are extremely reliable devices, but they can and do fail. Sudden, or catastrophic, failures occur when the device is overloaded. This overload condition can result from a shorted output or some condition that causes excessive current flow over a long period of time, or when excessive voltages or voltages of the wrong polarity are applied to the device. Transient power supply spikes on the voltage supply line can also cause a device to be destroyed. Identifying and eliminating the cause of the catastrophic failure is essential prior to replacing the faulty device. The most difficult problem to identify is the power supply spike, because this occurs so infrequently. However, if the replacement device fails again soon after installation, and you are sure that all other possible causes of the failure have been corrected, look to the power supply.

### Intermittent Problems

Electronic equipment does not always malfunction catastrophically; sometimes the problem occurs intermittently. In order to find the source of an intermittent problem, the equipment must be evaluated while the malfunction can be observed. This process is sometimes time-consuming and frustrating.

Intermittents usually occur either because of a faulty connection, or as a result of temperature sensitivity. Some intermittents, however, occur for no apparent reason. A bad connection can often be located visually, or by wiggling the leads in a suspected area. Thermal intermittents require manual temperature control. Small heaters, along with "freeze mist," can be used to subject individual components to thermal shock until the source of thermal intermittent is found. In some stubborn cases, a technician must use intuition or perhaps even guesswork to solve an intermittent problem.

## Gradual Failure

Long periods of usage may cause degradation of the IC's operating parameters. This is particularly true if the device has been operating in a high temperature or high humidity environment. The change in the device's specifications will show up as a gradual deterioration in system performance. Replacing the device and altering the environmental conditions that caused the gradual failure should eliminate this type of failure.

## Poor Performance

Poor performance of a device in a circuit designed for a specific purpose may be the result of the external components or improper connections. If you assume that the device is being used properly, check that the external components are of the correct value. They could have changed value because of excessive heat used during soldering, leakage paths, or unwanted feedback paths. Solder bridges or leakage paths can very easily provide a path for current or signals to flow to the wrong location in the circuit, because the terminals of ICs are very close together. Care in mounting the IC and the external components, plus a thorough cleaning of the board should eliminate poor performance. If care and cleaning do not restore the IC to the proper condition, recheck the design specifications. You may decide to use a different design to achieve higher performance.

### Other Considerations

The effects of offset voltage, offset current, and temperature drift are often overlooked when evaluating poor performance, particularly in circuits using op amps. If circuitry that allows you to null the output or compensate for offsets is not included in the circuit design, you should look at the effect these parameters have on the overall performance of the circuit. A check of the manufacturer's data sheet will generally provide information on null and compensation circuits. Serious consideration should be given to adding the extra circuitry needed to overcome the influence of offset and drift.

# 13.9 Summary

☐ There are five basic methods used for troubleshooting: (a) shotgun, (b) signal injection, (c) signal tracing, (d) quick-check, and (e) divide-and-conquer. Many circuit troubles can be located by using the senses of sight, smell, touch, and hearing.

☐ The best procedure for troubleshooting a system is to work from general facts about the problem to the specific fault. There are six basic steps in the troubleshooting procedure: (a) define the problem, (b) isolate the faulty unit, (c) locate the faulty circuit, (d) identify the faulty component, (e) repair, and (f) retest. A failure of one component may cause secondary failure. Before replacing the faulty component, always check the entire system for possible secondary failure.

☐ A bad-habit quick-check to avoid is using your finger to cause a disturbance in a circuit. Using an ohmmeter in transistor or semiconductor diode circuits may forward bias the device into conduction. Troubleshooting closed-loop circuits requires special considerations.

☐ When troubleshooting circuits containing ICs, special care must be taken not to short-circuit the pins of the IC. Always make sure that the power to the circuit has been turned off before removing or replacing an IC. To assist in the troubleshooting process, use equipment manufacturers' manuals and schematics.

☐ It is essential to learn to select and correctly use the right test instruments for good troubleshooting practices. The three standard test instruments are (a) the multimeter, (b) the oscilloscope, and (c) the function generator. Simple devices that are helpful for troubleshooting analog systems are the signal injector and the continuity tester.

☐ Troubleshooting PCBs requires special skills.

☐ Some of the types of failure in ICs are (a) catastrophic failure, (b) intermittent failure, (c) gradual failure, (d) poor performance, and (e) the effects of offset voltage, offset current, and temperature drift.

☐ The best methods and procedures for you to use as a troubleshooter will come with experience, for which there is no substitute.

# 13.10 Questions and Problems

**13.1** List the steps that should be taken when troubleshooting a system or circuit.

**13.2** What type of information might you expect to find in a systems manual?

**13.3** Are quick-check methods useful in all circumstances? Explain your answer.

**13.4** What is meant by *divide-and-conquer?*

**13.5** Can the divide-and-conquer method be used for circuit isolation as well as unit isolation?

**13.6** Explain signal tracing.

**13.7** Explain signal injection.

**13.8** List the principal troubleshooting instruments and explain how they are used.

**13.9** Explain how to verify full compensation of a probe for measuring high frequencies.

**13.10** Where can you obtain information for testing ICs out of the circuit in which they operate?

**13.11** List some problems you may encounter with PCBs.

**13.12** List the precautions you may take when working with PCBs that may prevent troubles in future operations.

**13.13** List some possible failures of LICs.

**13.14** What is often overlooked when evaluating poor circuit performance in op amp circuits?

**13.15** What can you do to overcome the effects caused by the problem identified in Problem 13.14?

**13.16** Define visual inspection.

**13.17** Define secondary failure.

**13.18** Define intermittent failure and explain how these type failures may be found.

# Glossary

**Absolute accuracy** In ADC/DAC operations, a measure of the deviation of the analog output level from the ideal value under any input combination.

**Absolute maximum ratings** Electrical limitations that must not be exceeded; specified by the device manufacturer and listed on data sheets.

**Absolute value** The output of an absolute value amplifier, which is always positive regardless of the polarity of the input signal.

**ac** Alternating current, an electrical current which changes in both amplitude and direction.

**ac controller** A two op amp amplifier circuit for controlling ac servo-motors.

**Acquisition time** In sample-and-hold operations, the time required to acquire a new analog voltage with an output step of 10 V.

**ac stability** The capability of a power supply device to reject ac interference; improved by use of bypass capacitors with short leads.

**Active filter** A network of resistors and capacitors around an active device, usually an op amp.

**Adder** An amplifier circuit whose output is the sum of individual inputs (*see* Summing amplifier).

**Adjustable voltage regulator** A regulator circuit that provides adjustable output voltages; formed by adding a few external components to a standard fixed regulator.

**Alphanumeric code** A system used by manufacturers to identify their products; consists of a combination of alphabetic characters and decimal numbers.

**Ambient** Surrounding; for example, the temperature of the air about a component or a device.

**Amplitude modulation (AM)** An intelligence signal that is varied in amplitude (strength) while the frequency is held constant.

**Amplitude response** The gain of an active filter, represented by the relationship between the filter's output voltage and its input voltage at various frequencies, usually expressed in decibels.

**Analog mathematical circuit** A variation of the difference amplifier; input and feedback resistors are made to be of equal values.

**Analog multiplier** A device designed specifically for use as a multiplier of logging circuits.

**Analog signal** An ac or dc voltage or current that varies smoothly or continuously.

**Analog-to-digital converter (ADC)** A device used as an interface between analog and digital systems; converts an analog input signal to a digital output signal.

**Aperture time** In sample-and-hold operations, the delay required between "hold" command and an input analog transition, so that the transition does not affect the held output.

**Astable multivibrator** A free-running MV with a square wave output whose frequency is determined by the values of externally connected components; requires no input trigger pulse.

**Attenuation** The reduction of signal strength through a circuit or device.

**Attenuator** A circuit or device used to reduce the strength of a signal being passed through it.

**Automatic fine tuning (AFT)** A circuit, usually found in the RF tuner section of TV receivers, that automatically locks on to the constant frequency of a TV channel.

**Automatic frequency control (AFC)** A circuit used to vary the frequency of an FM amplifier in proportion to the input signal frequency so that its output remains at a constant frequency.

**Automatic gain control (AGC)** A circuit used to vary the gain of an AM amplifier in proportion to the input signal strength so that its output remains at a constant level.

**Averaging amplifier** A variation of the summing amplifier; all input resistors ($n$) are of equal value and the feedback resistor is equal to $1/n$ of that value.

**Band gap reference** A low temperature coefficient reference used for reference voltages in compandors.

**Bandpass (BP)** A filter that allows a band of frequencies between a designated high value and a designated low value to pass, rejecting frequencies above and below those values.

**Band reject** A filter that prevents a band of frequencies between two designated values from passing.

**Bandwidth (*BW*)** The difference between the upper cutoff frequency and the lower cutoff frequency of a filter network; the band of frequencies allowed for transmitting a modulated signal.

**Barkhausen Criterion** A condition that must be met for an oscillator to be self-starting and self-sustaining; occurs when the product of amplifier gain and a fractional feedback factor equals one, that is, $A_v B_v = 1$.

**Blanking** The method used in oscilloscopes and TV receiver screens to blank out, or make invisible, the sweep beam as it returns to its starting point at the left of the screen.

**Break point** One designation for the frequency at which the voltage gain of a filter circuit drops to 70.7 percent of its maximum value.

**Buffer** A device used to prevent interaction between stages; an isolation amplifier.

**Butterworth response** A filter response where the passband amplitude response is relatively constant, or flat, with no oscillations; the response of a highly damped filter.

**Capture mode** One of three operating modes for PLLs; occurs when control voltage from the dc amplifier starts to change the VCO frequency.

**Capture range** The bandwidth over which capture is possible; can never be wider than the lock range.

**Carrier** A modulated radio frequency wave used to carry a transmitted intelligence signal over great distances.

**Cascading** Connecting circuits such as filters or amplifiers in series to improve system operation; the output of one stage becomes the input to the next stage.

**Catastrophic failure** The sudden failure of a device, usually caused by overloading or transient power supply spikes.

**Cathode ray tube (CRT)** A vacuum tube in which electrons emitted from the cathode are shaped into a narrow beam and accelerated to a high velocity before striking a phosphor-coated viewing screen; used in oscilloscopes and TV receivers.

**Center operating frequency ($f_{op}$)** The highest point of a filter response curve, or the point at which maximum voltage gain is reached.

**Ceramic transducers** Transducers such as microphones and pickups that use a ceramic material to transform mechanical or sound vibrations into electrical impulses.

**Chebyshev response** A filter response where the passband amplitude response is not flat, but has oscillations; the response of a filter with low damping.

**Closed loop gain ($A_{CL}$)** The gain of an op amp when feedback is present.

**Colpitts oscillator** An inductor-capacitor oscillator whose identifying feature is a tapped capacitor arrangement.

**Common-mode gain ($A_{CM}$)** The ratio of a change in output voltage to a change in common-mode voltage.

**Common-mode rejection (CMR)** The ability of an op amp to reject common-mode signals while amplifying differential signals.

**Common-mode rejection ratio (CMRR)** The ratio of difference gain to common-mode gain, expressed in decibels.

**Common-mode signal** Signal voltages that are in phase, of equal amplitude and frequency, applied to both inputs of a differential amplifier.

**Compandor** A gain control device used for dynamic gain expansion or compression; name derived from the two functions, compressor and expander; primary use is in telephone subscriber and trunk carrier systems.

**Comparator** An op amp circuit used as a sensing device that compares an input voltage to a reference voltage; not used to amplify signals; typically operates in the open-loop mode.

**Comparator ADC** An analog-to-digital converter that uses comparators to convert analog signals to digital form.

**Compliance** In ADC/DAC operations, the maximum output voltage range that can be tolerated while maintaining the specified accuracy.

**Constant-current regulator** A three-terminal fixed voltage regulator arranged to provide a constant-current output.

**Constant-current source** An op amp circuit configuration that has a fixed, stable reference voltage and a predetermined value for the input resistor, which results in a constant-current flow regardless of the value of the load resistance.

**Continuity tester** A simple, easily constructed troubleshooting device for checking the continuity of cables, wires, and PCB traces.

**Control transistor** The transistor that controls the current flow to a regulator output; series-pass transistor.

**Control voltage** An external voltage applied to a timer circuit that can alter both threshold and trigger voltages; can modify the timing period and can be used to modulate the output waveform.

**Converter** A circuit that converts a dc input voltage to a higher or lower dc output voltage.

**Correlation** A mathematical expression for the relationship between two quantities.

**Crossover distortion** A nonlinearity in the output signal that occurs during the signal crossover from positive to negative in a push-pull output circuit.

**Crystal oscillator** An oscillator circuit that uses a crystal that is cut to a precise thickness and has a natural frequency of vibration; used for increased oscillator frequency stability.

**Current-compensating resistor** A resistor placed in series with the (+) input terminal of an op amp to correct for output offset voltage caused by input current imbalances.

**Current-differencing amplifier** Another designation for the Norton amplifier; derived from circuit negative feedback that is used to keep the two input currents equal.

**Current-driven device** The Norton amplifier; driven by input currents rather than input voltages.

**Current limit** The most commonly used regulator protection scheme; guards the output series-pass transistor against excessive output currents or short circuit conditions.

**Current-mirror (biasing)** The method of biasing of the Norton amplifier; sets the resting dc output voltage at one-half the $V^+$ supply level.

**Current regulator** A regulator circuit that maintains a constant current in a load, independent of changes in that load.

**Current sinking** Occurs in a timer circuit when the output terminal is LOW and causes a floating supply load to be ON, thus providing a path between the load supply and ground through the timer.

**Current sourcing** Occurs in a timer circuit when the output terminal is HIGH and causes a grounded load to be ON, thus providing a path between the timer supply and ground through the load.

**Current-to-voltage converter** An inverting amplifier without an input resistor; input current is applied directly to the inverting input terminal of an op amp.

**Cutoff frequency ($f_c$)** The frequency at which the voltage gain of a filter circuit drops to 70.7 percent of its maximum value; also called 0.707 point, 3 dB point, break point, and half-power point.

**Damping** The determining factor in the shaping of a filter's passband; represented by the symbol alpha $\alpha$.

**Darlington pairs** The second stage of an op amp; a very high gain voltage amplifier stage; cascaded Darlington pairs produce the extremely high open-loop gain of the op amp.

**Data acquisition** The gathering, or acquiring, of data in process control systems; usually involves ADC/DAC operations.

**Data communications** The transmission of data, such as computer information, from point to point.

**dc** Direct current, a current that flows in one direction only; its magnitude may change but its direction does not.

**dc beta ($\beta_{dc}$)** The dc current gain of a transistor.

**dc controller** An op amp circuit used to control dc servo-motors.

**dc-dc converter** A circuit that converts a dc input to a higher or lower dc output voltage.

**Decade** A tenfold increase or decrease in bandwidth or in frequency.

**Decibel (dB)** One-tenth of a bel; a unit used to express the relative increase or decrease in power or the gain or loss in a circuit.

**Degenerative feedback** Feedback that is 180° out of phase with the input signal so that it subtracts from input; negative feedback.

**Demodulation** The process of stripping the carrier wave from the intelligence information and amplifying the original intelligence signal to a usable level.

**Demodulator** The section of a radio or TV receiver that rectifies the incoming modulated signal and removes the intelligence; a detector; a discriminator.

**Deposition** The process of depositing impurities, or dopants, onto the surface of a wafer.

**Detector** A circuit used for demodulation; for AM signals, diode detector or envelope detector; for FM signals, ratio detector or quadrature detector.

**Dielectric absorption** The characteristic of a charged and discharged capacitor that retains a residual voltage across the capacitor; all the energy stored in the dielectric is not given up when the capacitor is discharged.

**Difference amplifier** An op amp circuit that has input signals simultaneously applied to both input terminals; output voltage is the algebraic sum of the inverting and noninverting outputs.

**Difference gain ($A_D$)** In a differential amplifier, the ratio of a change in the output voltage to a change in the differential input voltage.

**Differential amplifier** The input stage of an op amp; a circuit that amplifies the difference between two inputs; a difference amplifier.

**Differential nonlinearity** In ADC/DAC operations, a measure of the deviation between the actual output level change and the ideal output level change for a one-bit change in input code.

**Differentiator** A wave-forming circuit with a short time constant, the output of which is proportional to the rate of change of the input.

**Diffusion** The process of building electrical characteristics into silicon wafers, step by step; diffusing or driving dopants to the desired depth in a wafer by exposing the wafer to very high temperatures.

**Digital signal** A series of pulses or rapidly changing voltage levels that vary in discrete steps or increments.

**Digital-to-analog converter (DAC)** A device used for converting a digital input signal to an analog output signal; an interface between digital and analog systems.

**Diode detector** A type of detector (demodulator) used for AM signals.

**Diode-transistor logic (DTL)** A diode logic gate to which a transistor has been added for amplification purposes; encapsulated in a digital IC.

**Discharge** A timer terminal used to discharge an external timing capacitor during the LOW output period.

**Discrete component** Any component not a physical part of an IC; resistor, capacitor, transistor, diode, inductor.

**Discriminator** A type of detector (demodulator) used for FM signals.

**Dissipative voltage regulator circuit** A series or shunt regulator circuit in which the control transistor operates in its active region, resulting in power loss, or dissipation, in the transistor.

**Divide-and-conquer** A troubleshooting method of dividing system units or system stages in half to evaluate one-half at a time, repeating the division and evaluation until the problem is located.

**Dopant** An impurity, carrying either negative or positive charges, deposited onto wafer surfaces to establish either N-type or P-type sections.

**Double-ended limit detector** A circuit designed to monitor an input voltage and indicate predetermined upper and lower limits; a window detector.

**Drift** The change in offset current and offset voltage caused by temperature change.

**Droop** In sample-and-hold circuits, the variation, or drift, caused by the charge leaking out of the holding capacitor through the amplifier input terminals and the switch; expressed in millivolts per second.

**Dual in-line package (DIP)** A packaging design to improve IC mounting; internal connections are brought out to external legs on either side of the package; can be plugged into a mounting socket.

**Dual tracking regulator** A regulator circuit that provides both plus and minus regulated voltages, with the voltages tracking (changing) proportionally with line changes.

**Dual power supply** A power source that provides equal positive and negative voltages; can be two separate supplies.

**Duty cycle** A specific interval of ON time compared to the period of the signal; a ratio, normally expressed in percent.

**Dynamic regulation** *See* Load regulation

**Dynamic sampling error** In sample-and-hold circuits, the error introduced into the held output caused by a changing analog input at the time the "hold" command is given; expressed in millivolts.

**Effective series resistance (ESR)** The internal effective series resistance that must be considered in the design of a switching regulator.

**Envelope detector** A type of detector (demodulator) used for AM signals.

**Equal component filter** Any second-order VCVS filter circuit where all capacitors are equal in value and all input resistors are of equal value; the gain must be fixed at 1.59.

**Etching** The process of cutting openings or windows into oxide, metal, or glass surfaces of a wafer.

**External nulling** A method for cancelling dc offset voltage in an op amp; a network of resistors and a potentiometer at the input terminal.

**Fabrication** The process of manufacturing an IC; the step-by-step procedure from wafer to packaged device.

**Falloff** The rate of decrease (or increase) in amplitude with an increase (or decrease) of frequency in a filter circuit; rate determined by the order of filter.

**Feedback element** The resistor or capacitor used between the output of an op amp and its input terminal; the determining factor for op amp gain and bandwidth.

**Filter** A device that screens out certain frequencies or passes electric current of only certain frequencies; can be active or passive.

**Fixed voltage regulator** A small, three-terminal device that requires no external components; provides a single, fixed, well regulated voltage.

**Flash ADC** *See* Comparator ADC

**Flat-pack** An early packaging style for ICs; leads extend from two or four sides and are on the same plane as the package; used where space is a problem.

**Flip-flop (FF)** A nonoscillating circuit that has two conditions of equilibrium; an ON-OFF device.

**Floating load** A load that is ungrounded; it has no common connection.

**Flyback** The blanked out portion of a wave; the retrace portion of a wave; used in oscilloscopes and TV receivers.

**Foldback current limiting** A method of power supply protection; designed to cause the power supply voltage to decrease if load impedance becomes too low and draws excessive current from the power supply.

**Frame pulse** A synchronizing pulse used to trigger a frame oscillator in a TV signal processing circuit; there are 30 frames per second in a TV system.

**Free-running mode** The operating mode of a VCO before lock-in occurs; present when the VCO output frequency is too far from the standard frequency.

**Free-running multivibrator** An astable MV.

**Frequency compensation** A method used to prevent high-frequency output signals from being fed back to the input of an op amp and causing undesired oscillations; can be internally or externally connected.

**Frequency converter stage** The combined oscillator and mixer stages of a radio receiver.

**Frequency-determining network** The *RC* or *LC* configuration of a circuit that sets the operating frequency of the circuit.

**Frequency deviation** The maximum departure from center frequency at the peak of the modulating signal.

**Frequency modulation (FM)** An intelligence signal that is modulated by varying the frequency of the RF carrier wave at an audio rate.

**Frequency shift keying (FSK)** The process of varying the carrier frequency rather than keying a transmitter on and off.

**Frequency swing** The total frequency swing from maximum to minimum; it is equal to twice the deviation.

**Frequency synthesis** The process of putting together, or mixing, two frequencies to provide a desired output; an application for the PLL.

**Frequency-to-voltage converter** A device that converts an input pulse rate into an output analog value.

**Full-scale range (FSR)** In ADC/DAC operations, the maximum range of steps, number of bits, or maximum analog output.

**Function generator** Any signal generator that has two or more different output waveforms; usually has square, triangle, and sine wave outputs.

**Gain-bandwidth product (*GBP*)** Equal to the unity-gain frequency of an op amp; determined by multiplying the gain and bandwidth of any specific circuit; indicates the upper useful frequency of a circuit and provides a means of determining bandwidth.

**Gain drift** In ADC/DAC operations, a measure of the change in full-scale analog output, with all bits ones, over the specified temperature range; expressed in parts per million of full-scale range per °C.

**Gain error** In sample-and-hold circuits, the ratio of output voltage swing to input voltage swing in the sample mode; expressed as a percent difference.

**Gradual failure** The slow deterioration in performance of a device or system; caused by long periods of use or operation in a high temperature or high humidity environment.

**Grounded load** A load with a ground, or common connection, on one side.

**Half-power point** *See* Cutoff frequency

**Harmonic frequency** Any multiple of a fundamental frequency; used for extending the range of crystal oscillators.

**Hartley oscillator** An inductor-capacitor oscillator whose identifying feature is a tapped inductor arrangement.

**Heat sink** A device for or method of dissipating heat away from components in which current flow generates heat at the junction.

**High pass (HP)** A filter circuit that rejects or attenuates lower frequencies and allows only frequencies above a given value to pass.

**High *Q*** Quality factor of a narrow bandwidth, high selectivity, high output voltage bandpass filter circuit.

**Hold settling time** The time required for the output of an S & H circuit to settle with 1 mV of final value after the "hold" logic command.

**Hold step** The voltage step at the output of the S & H circuit when switching from sample mode to hold mode with a steady dc analog input voltage.

**Hold voltage drift** *See* Droop

**Hybrid IC** An IC made up of a combination of monolithic and thin-film, thick-film, or individual semiconductor component circuits in a single package.

**Hysteresis voltage ($V_H$)** The voltage difference between the upper threshold voltage and the lower threshold voltage in a Schmitt trigger circuit.

**Index of modulation (*m*)** The variation in an AM signal compared with the unmodulated carrier; modulation factor.

**In-line package** Pins extend out from, and along the side(s) of the IC package to improve mounting and soldering ease.

**Input bias current ($I_b$)** The average of the two input currents (+ and –) of an op amp, with no signal applied.

**Input capacitance** An op amp electrical characteristic, an important factor to be considered when the op amp is to be operated at high frequencies.

**Input conditioning** The process of adding an *RC* network to the TRIGGER input of a timer circuit; increases stability and accuracy of the device.

**Input element** The passive device (resistor or capacitor) between an input signal source and the input terminal of an op amp.

**Input impedance ($Z_{in}$)** An op amp electrical characteristic, usually designated as resistance, typically higher than 1 MΩ.

**Input offset current ($I_{oi}$)** The absolute value of the difference between the two input currents (+ and −) of an op amp, for which the output will be driven to change states.

**Input offset voltage ($V_{oi}$)** The absolute value of the voltage between the input terminals of an op amp required to make the output voltage greater than or less than some specified value.

**Input voltage range** The range of voltage on the input terminals (common-mode) over which the offset specifications apply.

**Instrumentation amplifier (IA)** An amplifier circuit with high-impedance differential inputs and high common-mode rejection; gain is set by one or two resistors that do not connect to the input terminals.

**Integrated circuit (IC)** A complete electronic circuit formed in semiconductor material; many types may be used with no externally connected discrete components.

**Integrated-injection logic ($I^2L$, or IIL)** A relatively new logic family; has high-speed low-power dissipation and high reliability.

**Integrator** A wave-forming circuit with a long time constant, the output of which represents the average energy content of the input signal; square wave input produces triangle wave output.

**Intelligence** The original signal, such as voice communication or music, being modulated and transmitted by a radio frequency signal; information.

**Intermediate frequency (IF)** The resultant frequency produced in a receiver by combining a local oscillator and a mixer to form a frequency-converter stage; 455 kHz for AM and 10.7 MHz for FM.

**Intermittent problem** A problem or failure that comes and goes in a circuit or system; to find such a problem, the equipment must be evaluated while the malfunction can be observed.

**Internal nulling** Internally connected circuits in an op amp that are used to cancel dc output offset voltage; controlled by an externally connected potentiometer across offset null or balance terminals.

**Intrinsic input impedance ($Z_{iin}$)** The input impedance (resistance) of an op amp under open-loop conditions.

**Intrinsic output impedance ($Z_{iout}$)** The output impedance (resistance) of an op amp under open-loop conditions.

**Inverter** A circuit that converts a dc input voltage to a higher or lower ac output voltage.

**Inverting amplifier** An op amp used with the input signal applied to the (−) input terminal; the output signal is 180° out of phase with the input signal.

**Inverting input terminal (−)** One of two input terminals of an op amp, designated by a minus (−) sign; the output signal will be 180° out of phase with the input signal at this terminal.

**Isolation amplifier** An amplifier that is electrically isolated between input and output in order to be able to amplify a differential signal superimposed on a high common-mode voltage.

**JAN part number** Joint Army-Navy numbering system used by manufacturers to designate devices developed to meet military standards.

**Junction temperature ($T_J$)** The heat generated at the junction of any semiconductor device that carries current.

**Latching** The effect of latch-up in the safe area protection circuit of a regulator.

**Latch-up** Occurs under heavy and high input-output conditions of a regulator; a safe area protection circuit action occurring after a momentary short.

**Lead-lag network** The frequency-selective network in a Wien-bridge oscillator in which both degenerative and regenerative feedback exists.

**Leakage** Current flow over the surface or through a path of high insulation value; the undesirable loss of charge in a capacitor.

**Least significant bit (LSB)** The rightmost bit in a data converter code; the analog output level shift associated with this bit, which is the smallest possible analog output step.

**Level detector** An op amp circuit that compares an input signal to a reference voltage; the output swing indicates which input signal is higher.

**Limiter** An IF amplifier tuned to the 10.7 MHz IF in an FM receiver; provides constant output level.

**Linear integrated circuit (LIC)** An IC used specifically for analog applications; some may be a combination of analog and digital circuits.

**Linear signal** Sometimes used interchangeably with analog signal.

**Linear system** Any unit or assembly of components that generates or processes analog signals.

**Line driver** An amplifier circuit for driving a signal over transmission lines; data communications line amplifier.

**Line regulation** Refers to changes in output as input ac (RMS) is varied slowly from its rated minimum value to its rated maximum value.

**Loading effect** In voltage regulators, the effect that occurs in the sensing divider circuit because of the low input impedance of the sensing transistor.

**Load protection** A circuit, normally built into a power supply, that automatically reduces the voltage or limits the current of the supply if anything goes wrong with the power supply.

**Load regulation** Refers to changes in output when load conditions are suddenly changed.

**Local oscillator** An oscillator circuit in a receiver, the frequency of which is mixed with an incoming signal frequency to produce a difference frequency called the intermediate frequency.

**Local regulation** A method of physically locating a small regulator, usually in the form of a single IC, at any point in a system where it is needed; sometimes referred to as on-board regulation.

**Lock-in mode** The phase-locked operating mode of a PLL; occurs when the VCO output frequency is exactly the same as the standard frequency.

**Lock range** The frequency range over which a PLL can follow the incoming signal.

**Logic gates** Digital devices with two or more inputs and a single output; used in many combined analog-digital systems.

**Logic reverse** A characteristic of a timer circuit that allows the user to choose either an ON or OFF condition of the output transistor during the timing period.

**Logic threshold voltage** The voltage at the output of a comparator at which the driven logic circuitry changes its digital state; relates to the operation of digital devices.

**Long-term stability** The ability of a power supply to provide a constant output voltage over a long period of operation under constant load, temperature, line voltage, and circuit output control adjustment conditions.

**Loop gain ($A_L$)** The ratio of the open-loop gain to the closed-loop gain in an op amp circuit.

**Lower frequency ($f_{low}$)** In a bandpass filter, the frequency below center operating frequency at which the voltage is 0.707 times maximum.

**Low pass (LP)** A filter that allows frequencies below a given value to pass, rejecting frequencies above that value.

**Low *Q*** Quality factor of a wide bandwidth, low selectivity, low output voltage bandpass filter circuit.

**Masking** A series of steps that selectively cut openings or windows into the oxide, metal, or glass surfaces of a wafer.

**Metal package** One of several packaging types for ICs; leads connected to the chip are brought out from the base of the package, like transistors.

**Missing-pulse detector** A timing circuit in which the output will go low if the timing interval is longer than the input pulse; useful in medical electronics for monitoring heartbeats.

**Mixer** A circuit in a receiver that mixes an incoming RF signal with a local oscillator to produce an intermediate frequency.

**Mode control signal** A signal pulse that allows FETs to turn on; usually a clock pulse of predetermined frequency.

**Modulation** The process of varying or shaping an intelligence signal for transmission by an RF carrier wave.

**Modulation envelope** The outline around a modulated signal, usually symmetrical.

**Modulation factor (*m*)** *See* Index of modulation

**Monolithic IC** An IC built entirely on a single base of semiconductor material.

**Monostable multivibrator** A one-shot MV with a single output pulse, the width of which is determined by the values of external components; circuit must be triggered.

**Monotonicity** A characteristic of a DAC that requires a nonnegative output step for an increasing input digital code.

**Most significant bit (MSB)** The digital bit that carries the highest numerical weight; an analog output level shift associated with this bit.

**Multifunction waveform generator** A signal generator that provides multiple output waveforms; normally provides square, triangle, and sine waves.

**Multiple feedback** A filter circuit that has more than one feedback path; gain must be unity for a Butterworth response.

**Multivibrator (MV)** An IC circuit in which the output is either a square wave or a rectangular wave; *see also* Astable MV and Monostable MV.

**Narrow bandpass** A high $Q$, high selectivity, high voltage output bandpass filter; has sharp rolloff from center operating frequency.

**Natural frequency of vibration ($f_n$)** The frequency of operation of a crystal; determined by the thickness of the crystal.

**Noninverting amplifier** An op amp used with the input signal applied to the (+) input terminal; the output signal is in phase with the input signal.

**Noninverting input terminal (+)** One of two input terminals of an op amp, designated by a plus (+) sign; the output signal will be in phase with the input signal at this terminal.

**Nonlinearity** A measure of the deviation of an analog output level from an ideal straight-line transfer curve drawn between zero and full scale; commonly referred to as endpoint linearity.

**Nonsinusoidal wave** All waveforms other than sine waves (square, triangle, sawtooth, staircase, rectangular).

**Normalized curve** A curve in which the maximum amplitude response is set to unity, or 0 dB.

**Norton amplifier** An op amp designed for use in ac amplifier circuits, operating from a single power supply; a current-driven device.

**Notch** A filter that prevents a band of frequencies between two designated values from passing; also band-reject or band-stop.

**Null frequency ($f_{null}$)** The frequency at which maximum attenuation or rejection is obtained in a notch filter circuit.

**Octave** A doubling or halving of frequency; the interval between frequencies that have a two-to-one ratio.

**Offset drift** A measure of the change in analog output, with all bits zero, over a specified temperature range.

**Offset null** The cancelling of dc output offset voltage in an op amp used in dc operations.

**One-shot multivibrator** A monostable MV; has one stable state; requires trigger signal and produces a single output pulse before returning to its stable state.

**Open-loop gain ($A_{OL}$)** The gain of an op amp at 0 Hz or dc input, without feedback; can be considered as infinite gain.

**Operational amplifier (op amp)** An amplifier circuit with very high gain and differential inputs.

**Optoisolator** A device that contains both an infrared LED and a photodetector; used for electrical isolation between two stages, such as electronic monitoring devices and a hospital patient.

**Order** For a filter, it defines the rate of decrease or rolloff from the operating frequency; the higher the order, the sharper the rolloff.

**Oscillator** A circuit that generates a continuously repetitive output signal; used for timing or synchronizing operations.

**Oscillograph** An instrument that provides a permanent visual trace of a waveform or shape; a chart recorder.

**Output impedance ($Z_{out}$)** The output resistance of an op amp under loaded conditions; the ratio of the intrinsic output impedance of the op amp to the loop gain.

**Output offset voltage ($V_{os}$)** A small dc output voltage inherent in op amps with zero input; not present when operating with ac signals.

**Overmodulation** A condition that exists when the modulating wave exceeds the amplitude of the continuous carrier wave, thereby reducing the carrier wave power to zero.

**Overtone crystals** Crystals designed especially to operate at a harmonic frequency; almost always used at frequencies above 20 MHz.

**Overvoltage crowbar** A load-protection circuit, built into many power supplies, that uses an SCR to automatically reduce voltage to protect the load if something goes wrong with the power supply.

**Overvoltage protection** A load protection system that uses zener diodes and current-limiting resistors to protect op amps from an overvoltage condition of a power supply.

**Oxidation** The process of growing a very thin layer of silicon dioxide on the surface of a wafer during IC fabrication.

**Packaging** The method used for encapsulating a chip after the circuit is formed, resulting in a completed IC.

**Passband** The range of frequencies passing through a filter with maximum gain or minimum attenuation.

**Passive filter** A filter constructed with resistors, capacitors, and inductors; a filter that contains no active devices.

**Peak detector** A circuit that detects and remembers the peak value of an input signal.

**Phase detector** In analog systems, a double-balanced mixer using two diodes in a balanced rectifier circuit; a comparator circuit.

**Phase difference detection** An ability of a comparator that uses the common-mode rejection characteristic to detect phase differences of out-of-phase input signals.

**Phase-locked loop (PLL)** An electronic feedback loop consisting of a phase detector, a low-pass filter, a dc amplifier, and a voltage-controlled oscillator; a closed-loop circuit.

**Phase-lock mode** Mode of operation of a PLL when the VCO output frequency is exactly the same as the standard frequency; lock-in.

**Phase modulation (PM)** A process of changing the instantaneous frequency of RF energy already generated by a constant frequency; a variation of FM.

**Phase-shifting network** An *RC* or *LC* network in the feedback path of an oscillator; designed to shift the feedback signal by 180° to cause it to be in phase with input to the amplifier.

**Phase-shift oscillator** An oscillator circuit that uses an *RC* feedback network (usually three sections) to produce the required 180° phase shift.

**Photocell** *See* Photoelectric transducer

**Photoconductive cell** A device containing materials that increase in conductivity when exposed to light.

**Photodiode** A diode that exhibits a variable resistance depending on the intensity of visible light that strikes its P-N junction.

**Photoelectric transducer** Converts luminous (light) energy into electrical impulses.

**Photoresistor** A device that decreases in resistance depending on the amount of light that strikes it.

**Photovoltaic cell** A semiconductor device that generates a direct current when it is exposed to visible light.

**Pierce oscillator** A variation of the Colpitts crystal-controlled oscillator; the crystal replaces the inductor of a standard Colpitts oscillator and appears as an inductor.

**Piezoelectricity** The quality of a crystal to generate a difference of potential across its face when subjected to mechanical pressure, and to compress when a difference of potential is applied across its face.

**PIN diode** A photodiode of sandwichlike construction. A P-material anode and an N-material cathode are separated by a thin layer of *intrinsic* (undoped) semiconductor material. The PIN is operated in the reverse bias mode without breaking down.

**Power amplifier (PA)** The amplifier in a transmitter that supplies the antenna with the amount of current needed for the desired power output; the higher the current, the stronger the radiated signal.

**Power supply rejection ratio (PSRR)** The ability of an op amp to reject power-supply-induced noise and drift.

**Power supply sensitivity** A measure of the effect of power supply changes on a DAC full-scale output.

**Precision level amplifier (PLA)** An amplifier placed between each digital input and its corresponding input resistor to a summing amplifier DAC; provides precise output levels of 0 V or 5 V.

**Precision reference** A monolithic IC developed to provide precise reference voltages or currents for signal processing and conditioning applications; is temperature stabilized and contains an active reference zener diode.

**Prescalar** An electronic device that produces an output pulse for a specified number of input pulses; used in frequency counters.

**Printed circuit board (PCB)** A circuit in which electrical conductors are printed on an insulating base; constructed for use of insertion and removal of ICs.

**Process control** A method used to measure such physical quantities as temperature, fluid flow rates, speed, pressure, light intensity, strain, and vibration.

**Pro-Electron** An organization located in Belgium that has attempted to standardize an identification system for linear ICs; an IC numbering system.

**Protective circuits** Circuits added on regular chips to improve reliability and to protect the regulator against certain types of overloads.

**Pull-in range** The range of frequencies over which the phase comparator circuit in a TV signal processing circuit will react for synchronization.

**Pulse width (PW)** The time duration of a pulse; pulse may be either positive or negative.

**Pulse width detector** A timer circuit used to determine when the trigger input stays HIGH longer than one time period set by the timing components.

**Pulse width modulator** A timer circuit connected in the monostable mode, triggered with a continuous pulse train, and modulated by a signal applied to the CONTROL VOLTAGE terminal.

**Push-pull amplifier** Two transistors used to amplify a signal in such a manner that each transistor amplifies one-half cycle of the signal; the output stage of an op amp.

**Quad in-line package (QIP)** Pins extend from two sides of the package but are bent so that mounting is in four lines.

**Quadrature detector** An FM discriminator circuit; a difference amplifier used in modern ICs for FM systems.

**Quadrature oscillator** An oscillator circuit that produces two sine wave output signals, 180° out of phase with one another; a sine-cosine oscillator.

**Quality factor (*Q*)** The relationship between the bandwidth and the center operating frequency in a bandpass filter circuit.

**Quantization uncertainty** A direct consequence of the resolution of a DAC or ADC; can only be reduced by increasing resolution.

**Quick-check** A troubleshooting method involving a minimum of test equipment; should be used with caution to prevent further problems in a system or a circuit.

**Quiescent** At rest, or static condition; inactive.

**Radiated signal** The radio frequency signal transmitted (radiated) from a transmitter; includes the carrier and intelligence.

**Radio frequency (RF)** The frequency that is used as a carrier for the intelligence transmitted.

**Ramp generator** A triangle wave generator.

**Ratiometric conversion** An ADC conversion method in which the magnitude of the reference voltage is a factor in both the output of the source transducer and the output of the ADC.

**Reference voltage ($V_{ref}$)** A small voltage applied to one of the inputs of a comparator to compare with the input signal at the other input.

**Regenerative feedback** Positive feedback; a small feedback signal that provides an in-phase signal from the output to the input.

**Reset** A control that allows an operator to disable and override command signals of a timer circuit, driving the output LOW.

**Resistive temperature coefficient** The tendency of a component to change in value with temperature.

**Resistor-transistor logic (RTL)** An early type of IC logic; easy to interface with discrete component circuits.

**Resolution** An indication of the number of possible analog output levels a DAC will produce, normally expressed as the number of input bits.

**Response time** The interval between the application of an input step function and the time when the output crosses the logic threshold voltage; similar to propagation delay of standard op amps.

**Retest** A troubleshooting step; the final step to verify proper operation after repairs are made.

**Retrace** The period of time when a display sweep in a CRT returns to a starting point after each line is scanned; also called flyback.

**Ripple rejection** The ability of a regulator to reject ripple from a power supply.

**Rolloff** *See* Falloff

**R-2*R* ladder** A resistive network in ladder form used in summing amplifier DACs.

**Safe area limit** In regulators, the limits placed on input-output differential conditions.

**Safe-area protection** Protection built into IC regulators to protect the series-pass transistor against excessive input-output differential conditions.

**Safe operating area (SOA)** A protective operating range for regulators that prevents excessive output current when excessive input-output differential conditions occur.

**Sallen & Key (S & K)** A VCVS second-order active filter.

**Sample-and-hold (S & H)** A circuit that periodically samples a signal and then holds it constant; used as an input to an ADC.

**Saturation voltage ($V_{sat}$)** In comparator circuits, normally one or two volts less than the applied power supply voltage of the op amp.

**Sawtooth wave** A nonsymmetrical triangle wave; the time period of the negative ramp is extremely short compared to that of the positive ramp.

**Scaling** A method of design convenient for use in second-order active filters.

**Schmitt trigger** A comparator circuit with feedback that provides a reference voltage dependent upon the output voltage level; helps eliminate false output switching caused by noise at the input.

**Scribing** Cutting into individual chips the hundreds of identical circuits fabricated on a wafer.

**Secondary failure** The failure of components in a circuit or system caused by an original component malfunction.

**Selectivity** The relative ability of the bandpass filter circuit to select the desired frequency while rejecting all others.

**Semiconductor** A material with resistivity somewhere in the range between conductors and insulators.

**Semiconductor substrate** The base material, such as silicon, into which other semiconductor materials are diffused to form an IC.

**Sensing device** A device used to determine if a voltage is greater or less than a given reference voltage; a comparator.

**Series pass transistor** The variable control element in a regulator; control transistor.

**Series regulator** A type of regulator circuit in which the variable control element is in series with the load.

**Servo amplifier** An amplifier circuit for controlling either dc or ac servo-motors; speed and direction of the motor is determined by the input voltage to the ICs.

**Settling time** The total time measured from a digital input change to the time the analog output reaches its new value within a specified error band.

**Shotgun approach** A troubleshooting method in which all parts in a circuit are replaced without verifying the faulty part.

**Shunt regulator** A type of regulator circuit in which the variable control element is in parallel (shunt) with the load.

**Signal conditioning** The process of shaping or changing a signal to be used as a control signal; signal processing.

**Signal injection** A troubleshooting method in which a signal of known value is injected into a circuit or system while the output is monitored.

**Signal processing** *See* Signal conditioning

**Signal-to-noise ratio (S/N)** The ratio of signal level to noise level expressed in similar units; the higher the ratio, the less noise interference.

**Signal tracing** A troubleshooting method in which a signal of known value is applied to the input of a system and the signal is traced through the system.

**Sine-cosine oscillator** An oscillator circuit that produces two sine wave output signals, 180° out of phase with one another; a quadrature oscillator.

**Sine wave** A wave form of a single frequency ac, constantly changing smoothly in both amplitude and direction.

**Single in-line bent to dual in-line package** (SIP-Bent-to-DIP) Pins extend from one side of the package but are bent in a manner that allows mounting in a dual line.

**Single in-line package** (SIP) Pins extend from one side of the package in a single line.

**Single-voltage power supply** A power supply that provides only a positive or a negative output voltage; care must be taken when using such a voltage source with op amps.

**Sink current** Current received by the output of an op amp or timer for an ungrounded (floating) load.

**Sinusoidal** A wave varying in proportion to the sine of an angle.

**Slew rate *(SR)*** An indication of how fast the output voltage of an op amp can change, normally designated in volts per microsecond.

**Snap action** The effect of comparator output signal switching action when hysteresis voltage is present.

**Source current** Current supplied to a grounded load by the output of an op amp or a timer.

**Source follower** A circuit in which the output signal is an exact reproduction of the input (source) signal; a voltage follower, unity-gain amplifier, buffer amplifier, isolation amplifier.

**Square wave** A wave that alternatively assumes two fixed values for equal lengths of time; symmetrical pulses.

**Squaring circuit** A comparator in which the (+) terminal is grounded and the output signal swings to the opposite polarity every time the sine wave input signal crosses the zero point, resulting in a square wave output; a zero-crossing detector.

**Squelch** To automatically quiet a radio receiver by reducing its gain when its input signal is below a specified level; a muting system.

**Staircase wave** A multisegmented wave produced by the addition of a number of square or rectangular waves.

**State-variable** A multiple feedback filter that uses three or four op amps, providing three different outputs from a single input.

**Static regulation** *See* Line regulation

**Step size** The amount that output voltage of a DAC will change as the digital input goes from one step to the next; always equal to the weight of the LSB.

**Stopband** The band of frequencies that is rejected by a filter; all frequencies other than those in the passband.

**Strain gauge** A type of transducer that converts a physical change to a resistance change.

**Stress** Any static force exerted against a body; normally measured in pounds or kilograms per unit of cross-sectional area.

**Strobing** A method used to enable or to disable a comparator.

**Subtractor** A variation of the difference amplifier; the output voltage is the difference between the two input voltages; all resistors are of equal values.

**Summing amplifier** An amplifier circuit whose output voltage is the sum of the individual input voltages applied through two or more resistors to the (−) input terminal of the op amp.

**Summing point** The point at which all summing amplifier inputs and the feedback resistor join; the virtual ground point.

**Superheterodyne** A radio receiver in which the incoming signal is converted to a fixed intermediate frequency before detecting the audio signal component.

**Surface-mount devices (SMDs)** Smaller than conventional components, SMDs are placed on the surface of the substrate, not through it.

**Switching regulator** A regulator in which the control transistor is switched between the cutoff and saturation modes; results in low power dissipation and high circuit efficiency.

**Synchronizing signal** A signal transmitted with a TV signal to lock in the received image in the same manner as the original image.

**Sync pulse** The horizontal (15,750 Hz) and the vertical (60 Hz) pulses used in TV signals for synchronization.

**Sync separator** A circuit in a TV receiver used to separate the horizontal and vertical pulses from the synchronizing signal.

**Temperature coefficient** A measure of the change in voltage or resistance of a device with temperature.

**Thermal conductivity** The extent to which a substance can transfer heat efficiently from one place to another.

**Thermal limit** The maximum allowable temperature limit of a regulator IC.

**Thermal overload** A condition of excessive temperature in a regulator IC that activates the thermal shutdown transistor to protect the regulator circuit.

**Thermal regulation** Refers to changes caused by ambient variations or thermal drift.

**Thermal resistance ($\theta_{JA}$)** The opposition to the flow of heat from the junction of a device to the surrounding air.

**Thermal shutdown** The effect of removing the base drive from a regulator transistor to prevent chip damage from excessive temperatures.

**Thermistor** A type of transducer that converts a temperature change to a resistance change.

**Thermocouple** A device that generates a voltage by the heating of a junction between two metal electrodes, consisting of dissimilar metals placed in physical contact.

**Thermodynamics** A branch of physics concerned with the interaction between thermal energy and mechanical energy.

**Thermopile** Two or more thermocouples connected in series; used to obtain higher voltages than a single thermocouple delivers.

**Thermostat** A temperature-sensitive transducer used to actuate some apparatus for regulation or protection.

**Thick-film** A circuit formed with passive elements (resistors, capacitors, conductors) on an insulating substrate, using a silk-screen process.

**Thin-film** A circuit with passive elements formed from thin layers of metals and oxides deposited on an insulating substrate.

**3 dB point** *See* Frequency cutoff

**Threshold** One of two terminals of a timer that determines the operating state of the device.

**Threshold voltage ($V_{th}$)** The input voltage to a device at which the device is turned on or off, or the output is switched to the opposite polarity.

**Time-base oscillator (TBO)** One stage of a programmable timer/counter; generates timing pulses with a period of 1 *RC* time constant.

**Time delay** The length of a timing cycle; may be extended by cascading timers.

**Timing state** The unstable or nonnormal state of a monostable MV.

**Tracking mode** The phase-locked or lock-in mode of a PLL.

**Transducer** A device that produces an electrical output signal proportional to an applied physical stimulus, or that converts an environmental change to a resistance change.

**Transfer function** The gain or amplitude response of an active filter, represented by the ratio of the filter's output voltage to its input voltage at various frequencies.

**Transistor-transistor logic (TTL or $T^2L$)** The most widely used IC logic; characterized by low propagation delay and good noise immunity.

**Transmission** The result of sending (transmitting) an intelligence signal; may be voice transmission over a telephone line, or other intelligence via radio frequency signal.

**Transmittance amplifier** Voltage-to-current converter circuits similar in form to inverting and noninverting amplifiers, used to drive relays and analog meters.

**Triangle wave** A wave consisting of symmetrical positive- and negative-going ramp voltages.

**Triggered monostable circuit** A timer circuit that requires only two external timing components and an input trigger pulse.

**Triggering** Applying a pulse to the input of a monostable timer to cause the timing cycle to begin.

**Trouble-cause table** A manufacturer-prepared table of possible causes of specific equipment troubles; found in operating and/or maintenance manuals.

**Troubleshooting** The logical step-by-step process of locating and repairing faulty systems; may be down to component level.

**Twin-T notch** An active filter that consists of a passive twin-T filter connected to an op amp voltage follower.

**Unity gain** In filters, $V_{out} = V_{in}$, and the transfer function has a maximum amplitude of 0 dB.

**Unity gain amplifier** *See* Source follower

**Unity gain analog subtractor** *See* Subtractor

**Unity gain frequency** The gain-bandwidth product of an op amp circuit; the upper useful frequency of a circuit; the maximum frequency of an op amp operating at unity gain.

**Unity gain inverter** An op amp inverting amplifier in which both input and feedback resistors are of equal value.

**Upper frequency ($f_{high}$)** In a bandpass filter, the frequency above center operating frequency at which the voltage is 0.707 times maximum.

**Varactor** A semiconductor diode that operates on the principle of a varying capacitance inversely proportional to the amount of reverse dc voltage applied; a varicap.

**Variable gain cell ($\Delta G$)** A current-in–current-out device used in compandors.

**Varicap** *See* Varactor

**Virtual ground** A point whose voltage is zero with respect to ground, yet is isolated from ground; the (−) input terminal of an op amp inverting amplifier circuit.

**Voltage-controlled oscillator (VCO)** An oscillator in which the frequency is controlled by a signal voltage; the amount of change in frequency is directly proportional to the input voltage level.

**Voltage-controlled voltage source (VCVS)** A second-order active filter; also called Sallen and Key (S & K).

**Voltage follower** *See* Source follower

**Voltage level detector** A comparator circuit in which a reference voltage is applied to one of the input terminals and the input voltage is applied to the other; the output indicates which voltage level is higher.

**Voltage subtractor** *See* Subtractor

**Voltage-to-current converter** *See* Transmittance amplifier

**Voltage-to-frequency converter** A reference sometimes used for a VCO.

**Wafer** The starting piece of semiconductor material, a few inches in diameter, into which many identical circuits are formed in some process.

**Wideband amplifier** An amplifier circuit capable of amplifying a wide range of frequencies, extending from dc to several hundred megahertz.

**Wide bandpass** A bandpass filter that, in general, has a bandwidth greater than 10 percent of its center operating frequency.

**Wien-bridge oscillator** An oscillator in which *RC* networks are part of a bridge circuit that provides both degenerative and regenerative feedback, applied to both inputs of an op amp.

**Window detector** An op amp circuit designed to monitor an input voltage and indicate predetermined upper and lower limits; a double-ended limit detector.

**Zero-crossing detector** A comparator circuit in which the output swings to the opposite polarity every time the sine wave input crosses the zero point; a squaring circuit; the (+) input terminal is grounded.

**0.707 point** *See* Cutoff frequency

# Appendixes

# Appendix A Package Outlines

**2-Pin Metal Package**
**Similar to JEDEC TO-3**

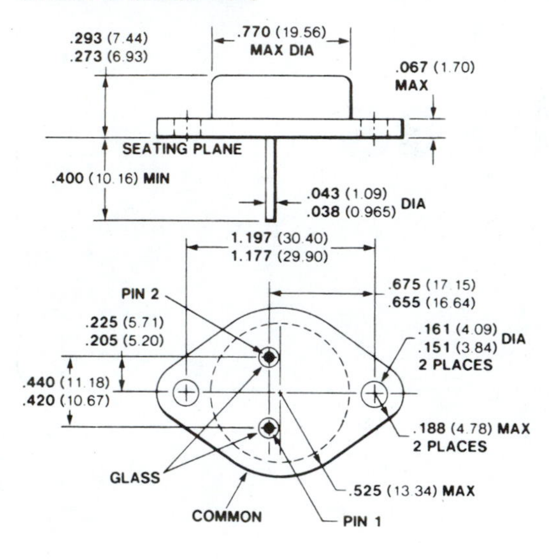


**2-Pin Metal Package**
**Similar to JEDEC TO-39**

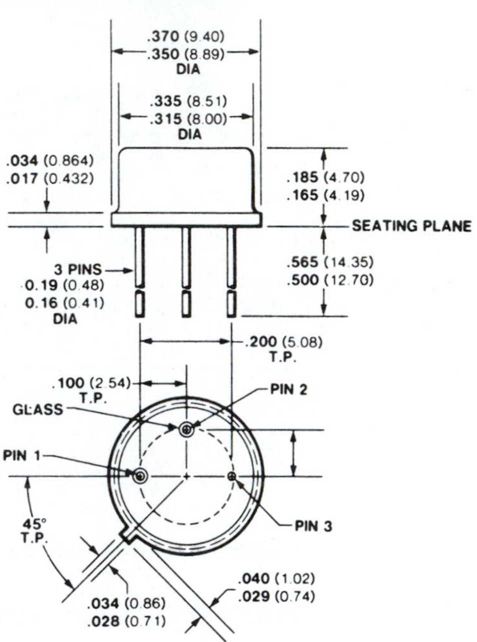


**3-Pin Molded Package**
**Similar to JEDEC TO-220**

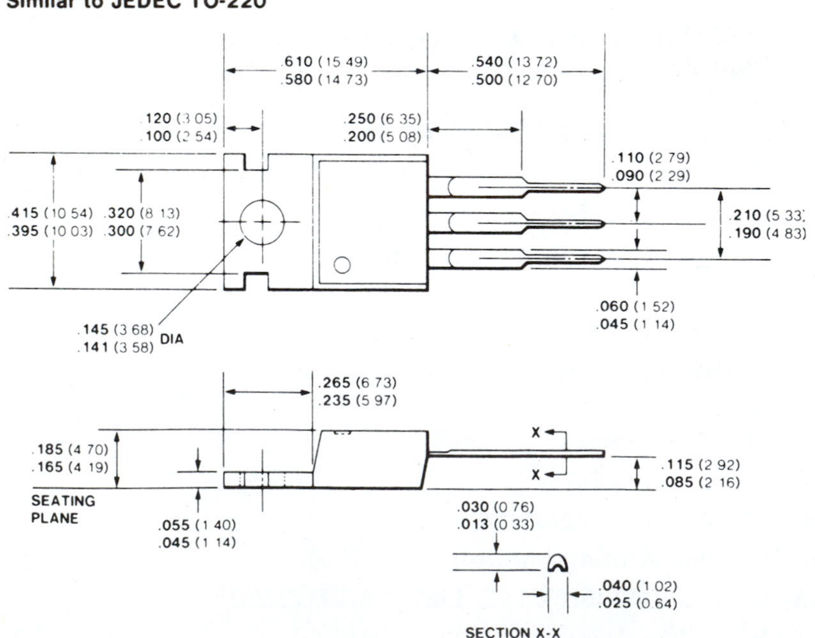


(Courtesy of Fairchild Camera & Instrument Corporation, Linear Division)

### 3-Pin Molded Package Similar to JEDEC TO-92

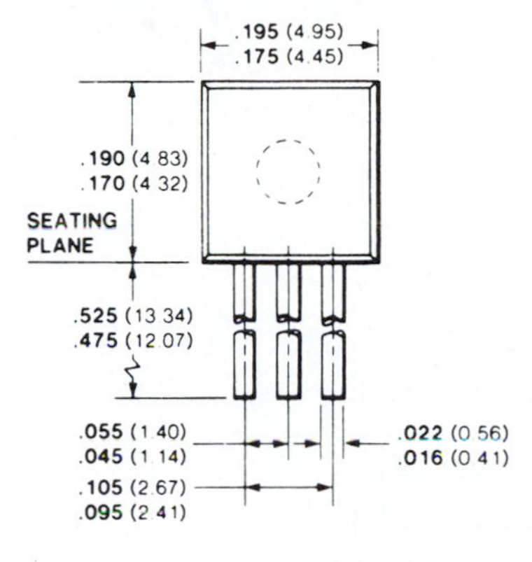


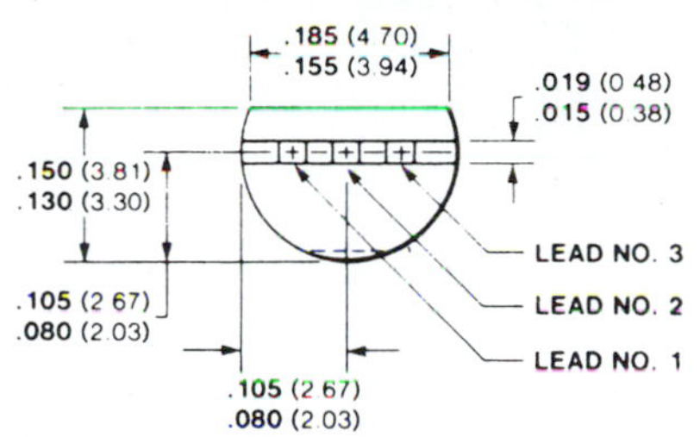


### 4-Pin Molded Single Wing

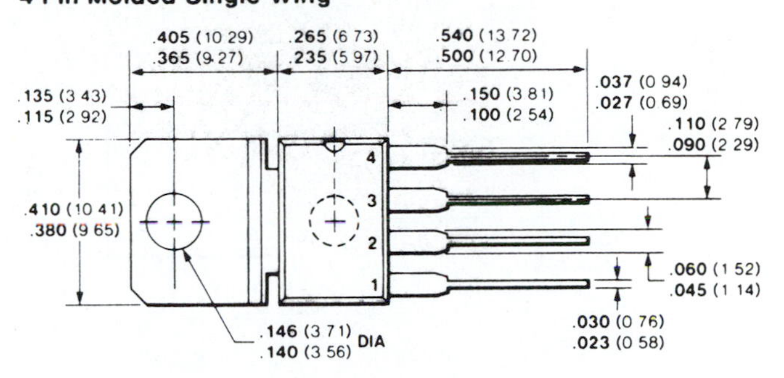


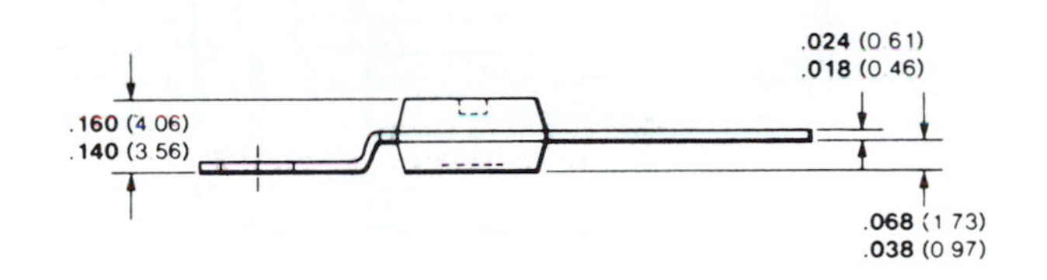


### 4-Pin Metal Package Similar to JEDEC TO-3

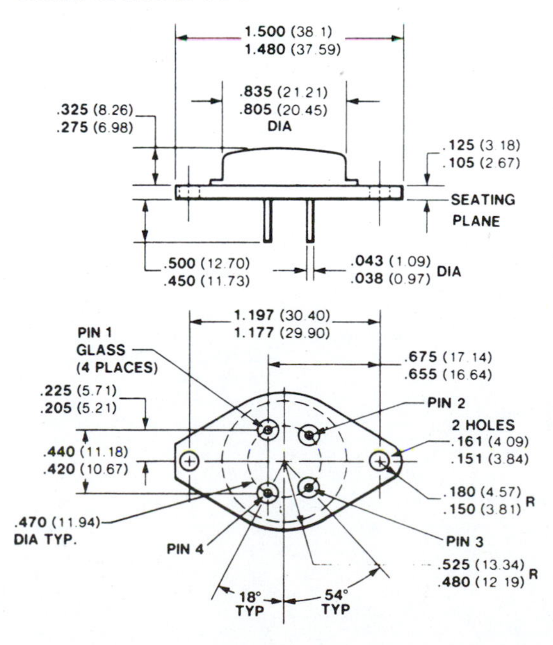


### 8-Pin Metal Package Similar to JEDEC TO-3

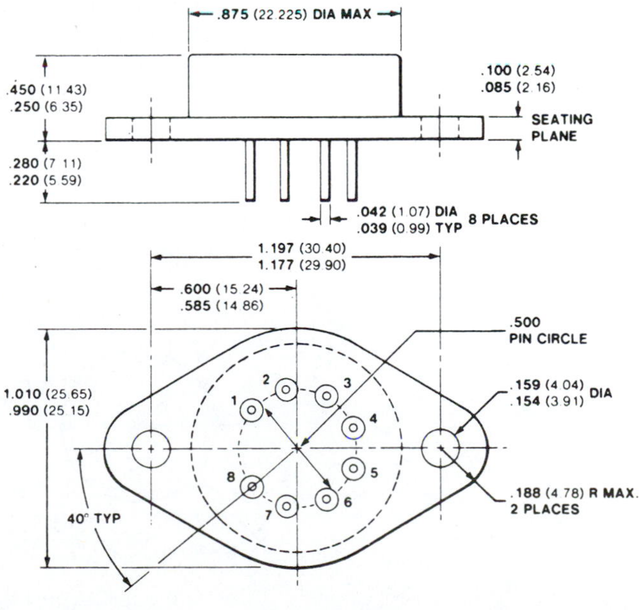

8-Pin Molded Dual In-Line

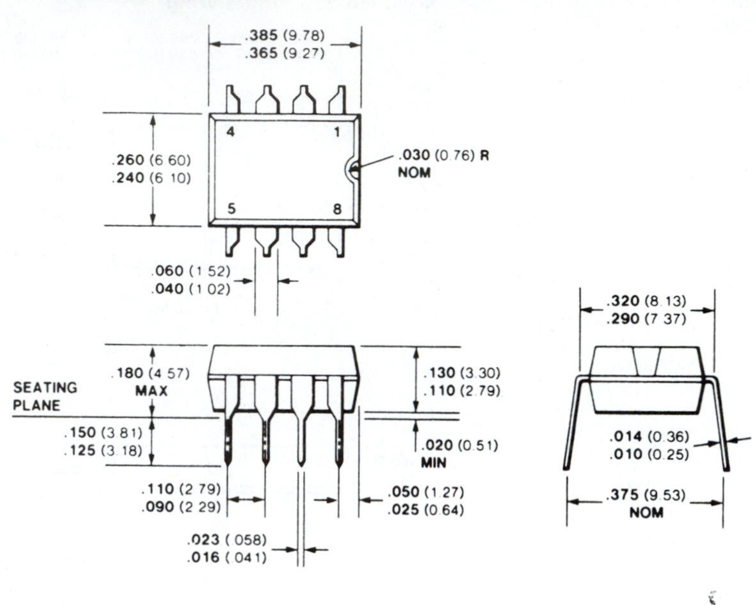


8-Pin Metal Package
In Accordance with JEDEC TO-99

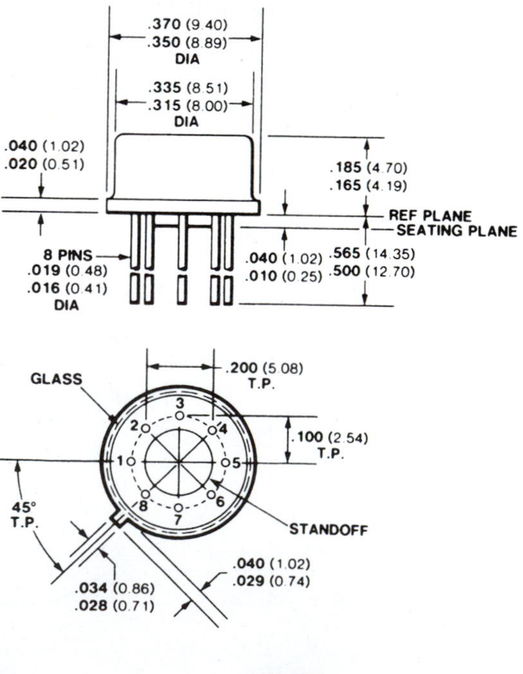

**8-Pin Ceramic Dual In-Line**

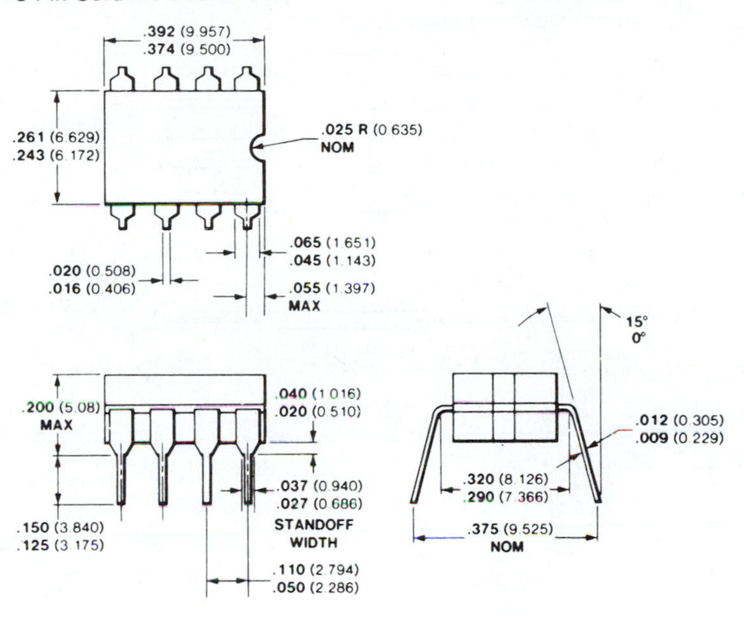


**10-Pin Metal Package**
**In Accordance with JEDEC TO-100**

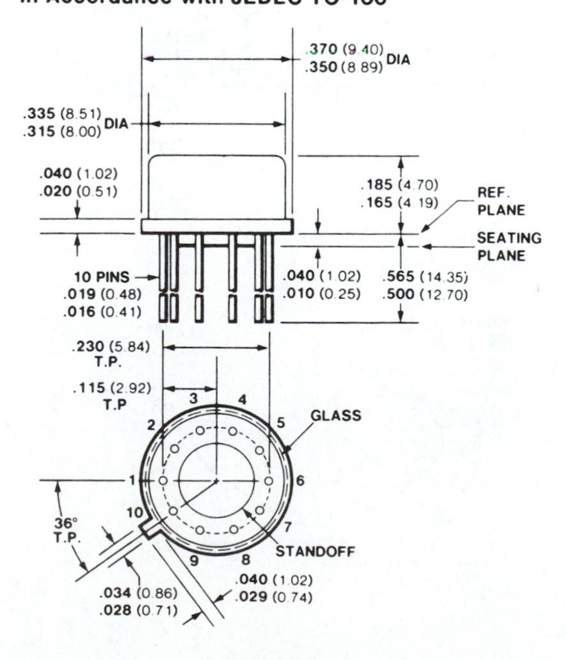

10-Pin Flatpak
In Accordance with JEDEC TO-91

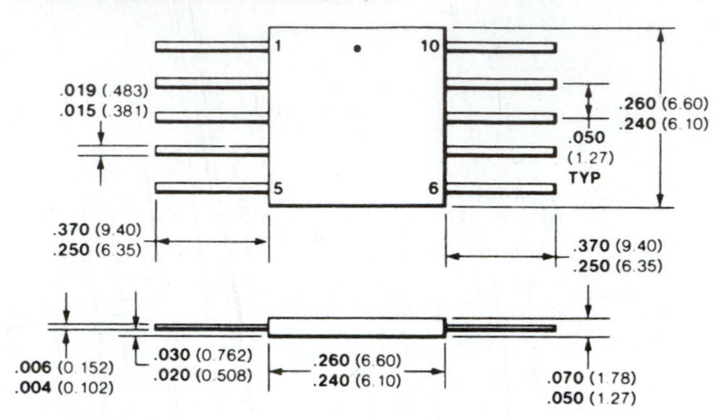


10-Pin Metal Package
Similar to JEDEC TO-3

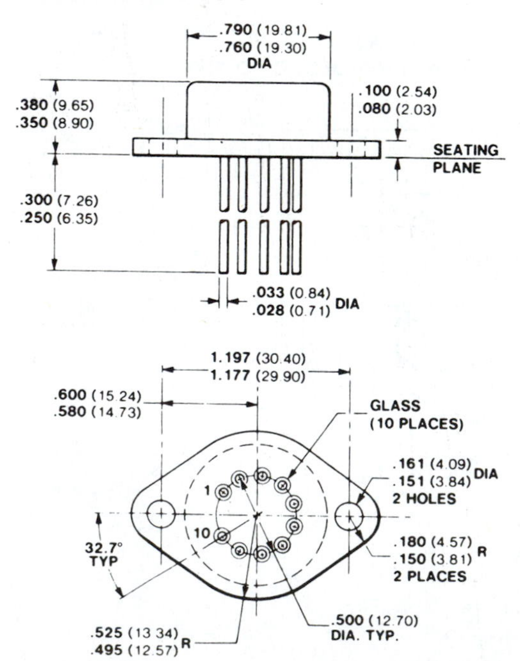

**14-Pin Ceramic Dual In-Line**
**In Accordance with JEDEC TO-116**

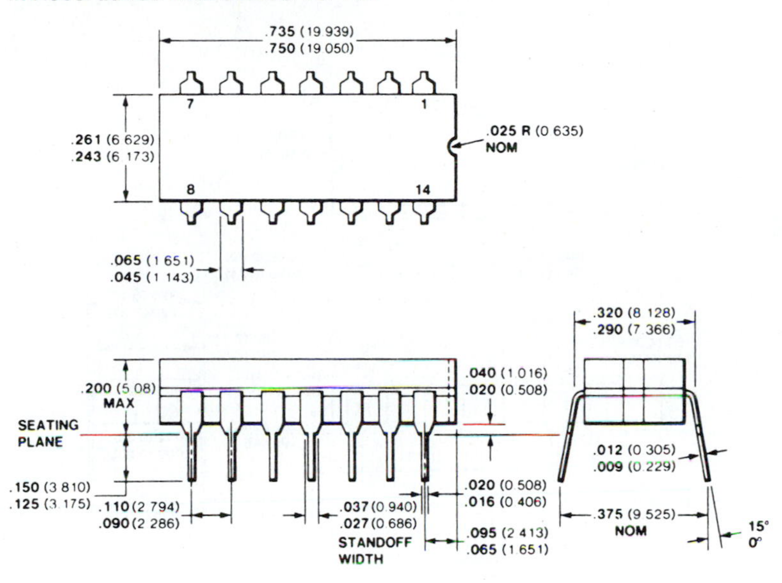


**14-Pin Molded Dual In-Line**
**In Accordance with JEDEC TO-116**

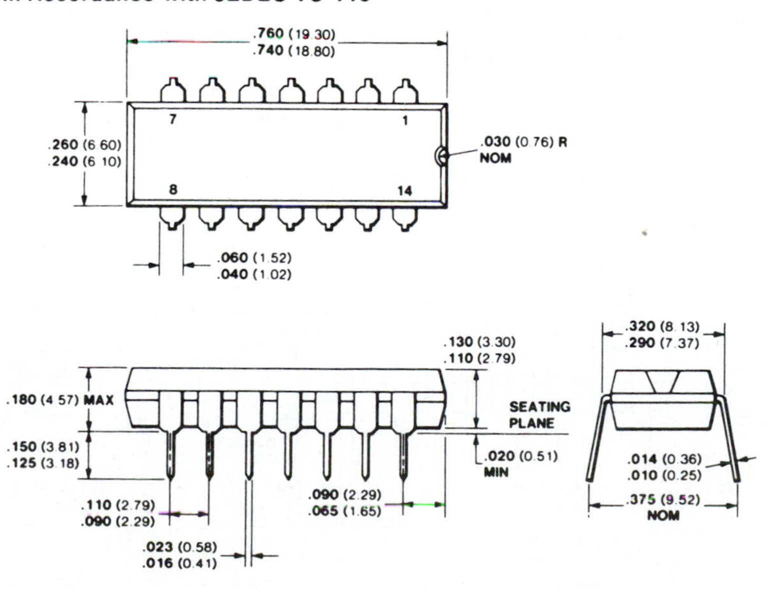

**14-Pin Ceramic Dual In-Line**

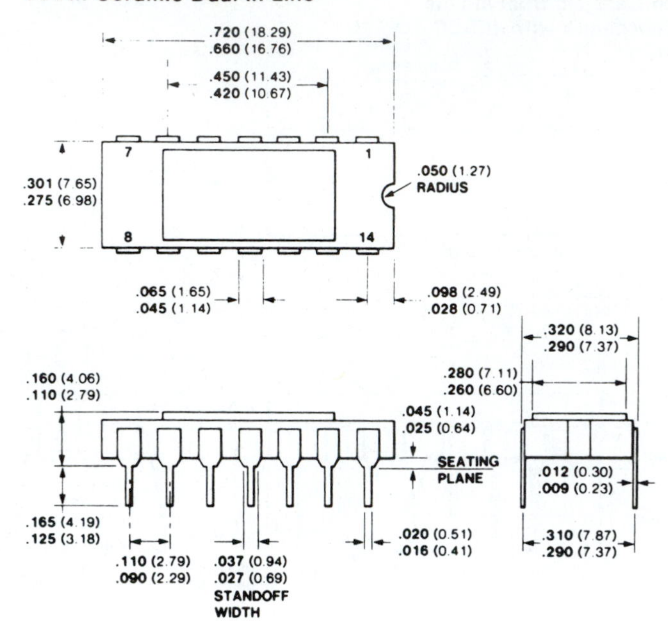


**16-Pin Molded Dual In-Line**

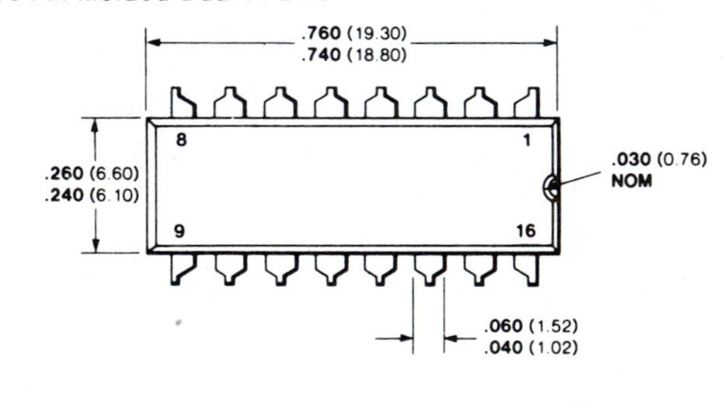


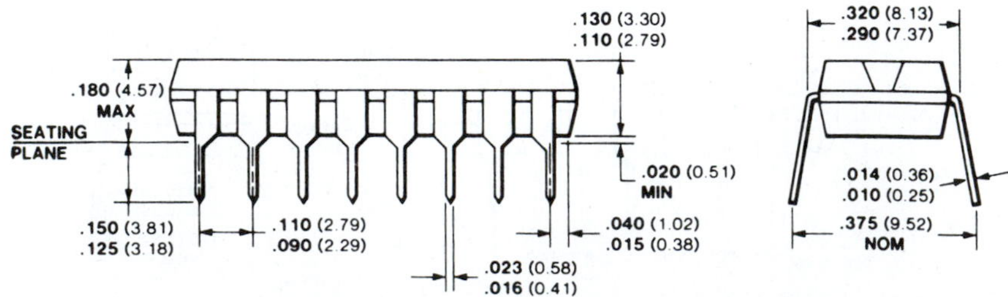

### 16-Pin Ceramic Dual In-Line

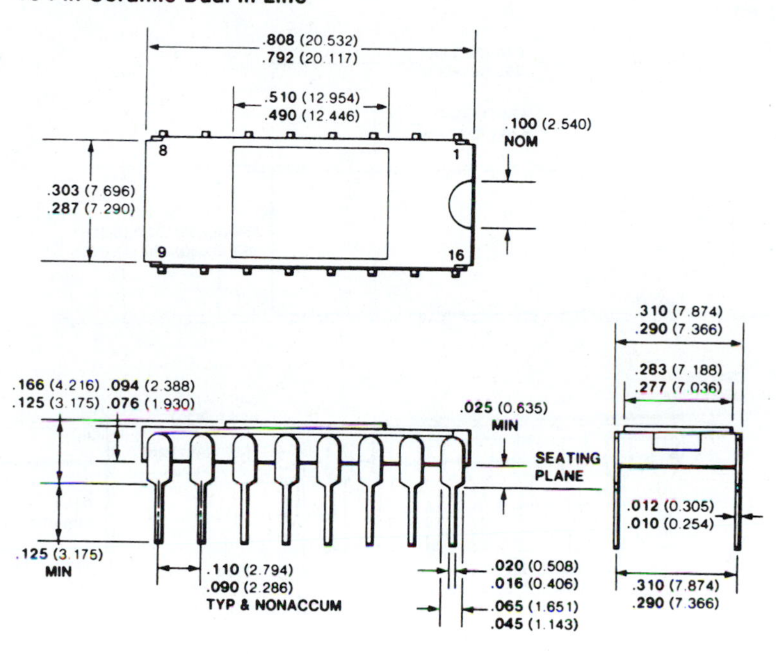


### 16-Pin Ceramic Dual In-Line

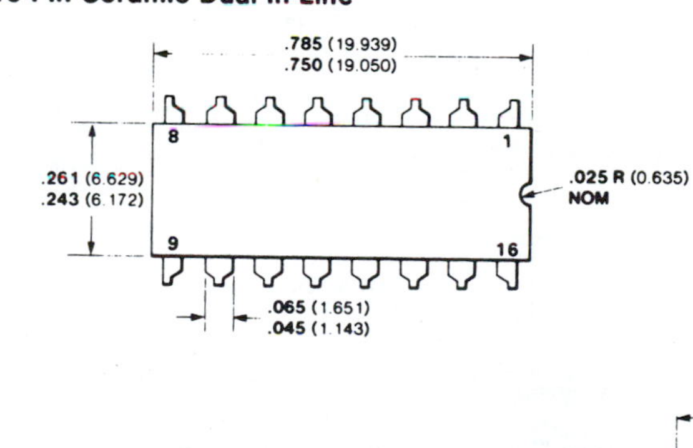


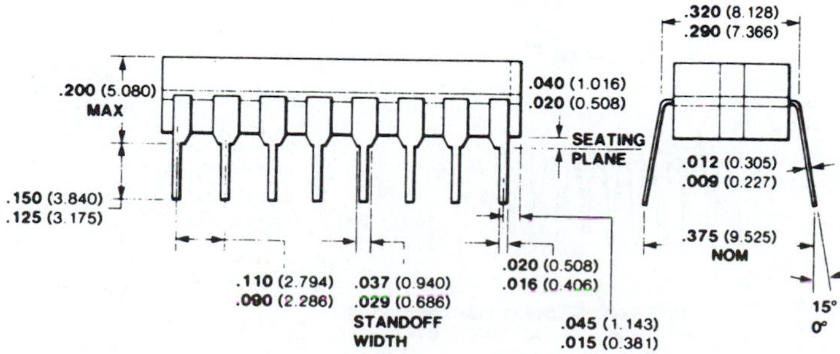

18-Pin Ceramic DIP
Side Brazed

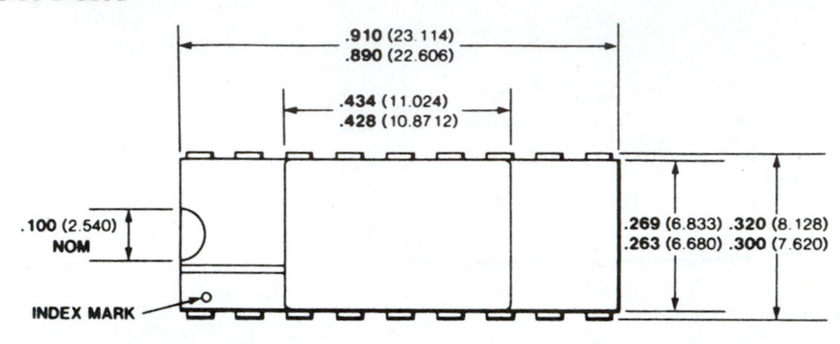


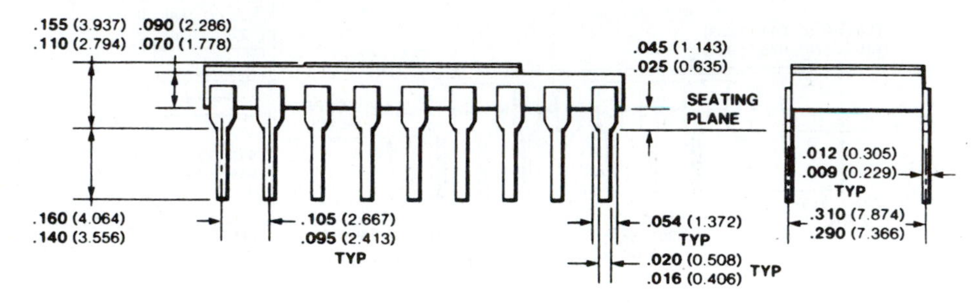


24-Pin Ceramic Dual In-Line
Side Brazed

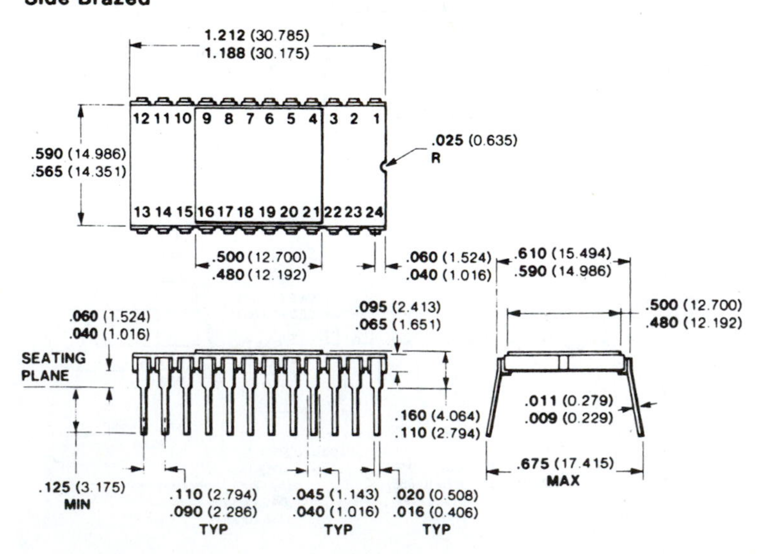

### 24-Pin Ceramic Dual In-Line

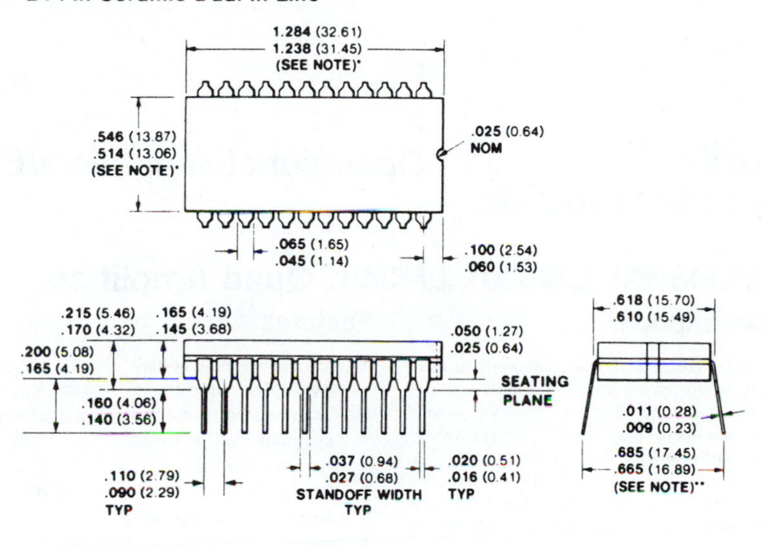


### 40-Pin Ceramic Dual In-Line Side Brazed

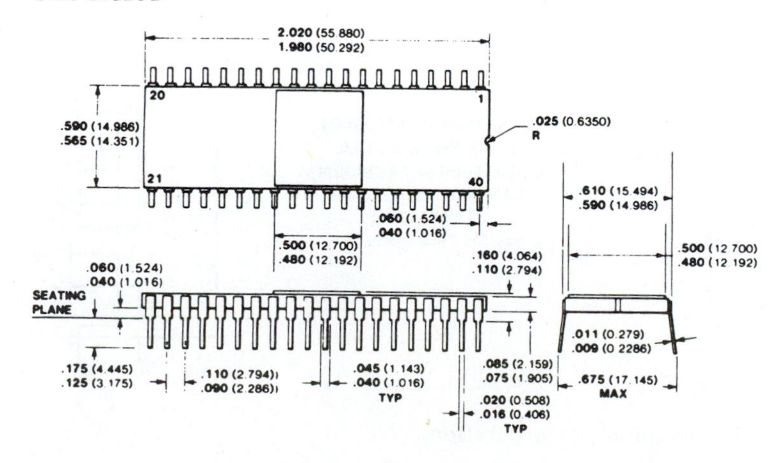

# Appendix B LM3900 Quad Amplifier

## Operational Amplifiers/Buffers

## LM2900/LM3900, LM3301, LM3401 Quad Amplifiers

### General Description

The LM2900 series consists of four independent, dual input, internally compensated amplifiers which were designed specifically to operate off of a single power supply voltage and to provide a large output voltage swing. These amplifiers make use of a current mirror to achieve the non-inverting input function. Application areas include: ac amplifiers, RC active filters, low frequency triangle, squarewave and pulse waveform generation circuits, tachometers and low speed, high voltage digital logic gates.

### Features

- Wide single supply voltage range or dual supplies — $4\ V_{DC}$ to $36\ V_{DC}$; $\pm 2\ V_{DC}$ to $\pm 18\ V_{DC}$
- Supply current drain independent of supply voltage
- Low input biasing current — 30 nA
- High open-loop gain — 70 dB
- Wide bandwidth — 2.5 MHz (Unity Gain)
- Large output voltage swing — $(V^+ - 1)$ Vp-p
- Internally frequency compensated for unity gain
- Output short-circuit protection

### Schematic and Connection Diagrams

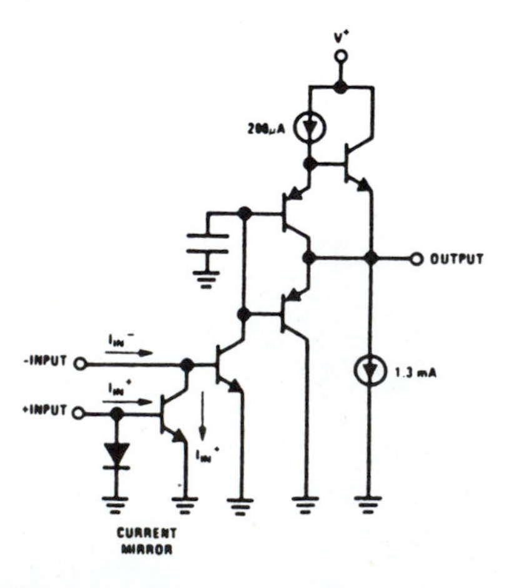


**Order Number LM2900J**
**See NS Package J14A**
**Order Number LM2900N, LM3900N, LM3301N or LM3401N**
**See NS Package N14A**

**Dual-In-Line and Flat Package**

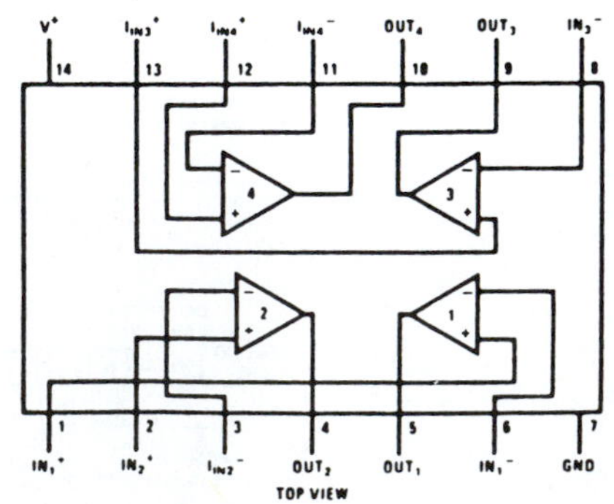


(Courtesy of National Semiconductor Corporation)

## Typical Applications $(V^{+} = 15\ V_{DC})$

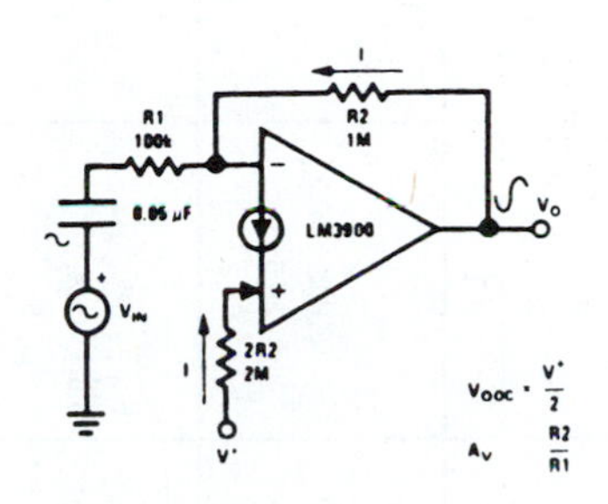


Inverting Amplifier

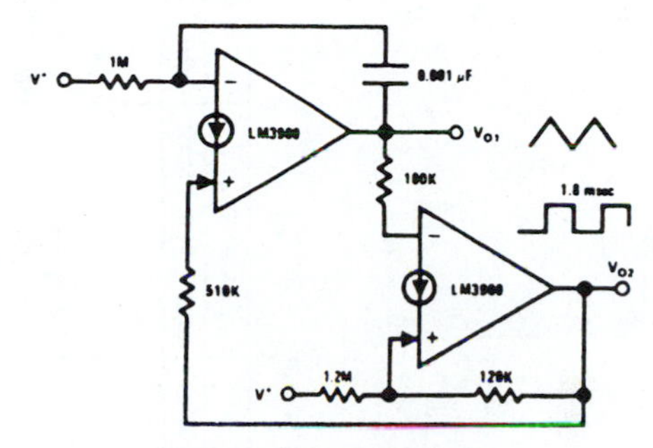


Triangle/Square Generator

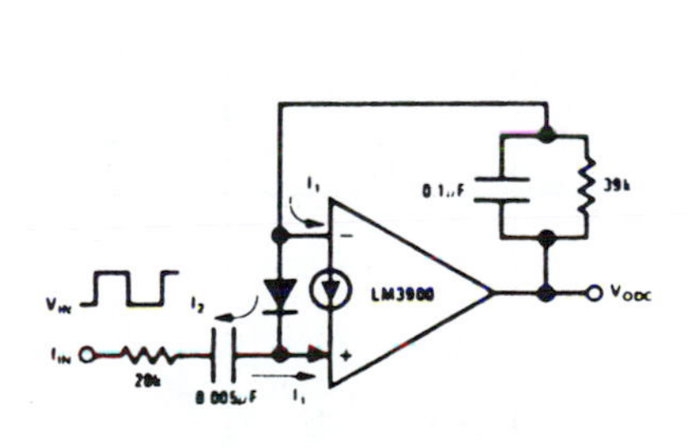


Frequency-Doubling Tachometer

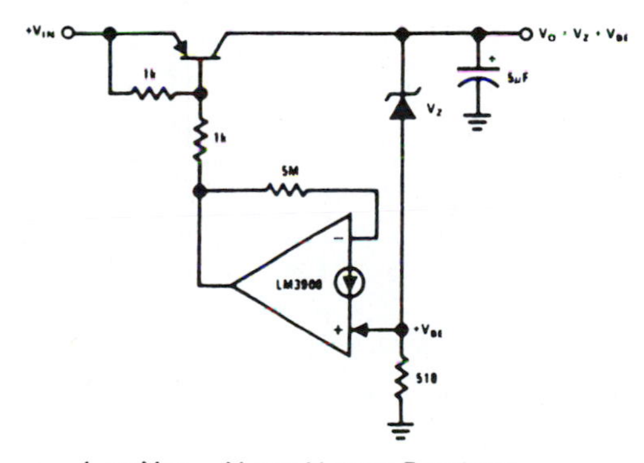


Low $V_{IN} - V_{OUT}$ Voltage Regulator

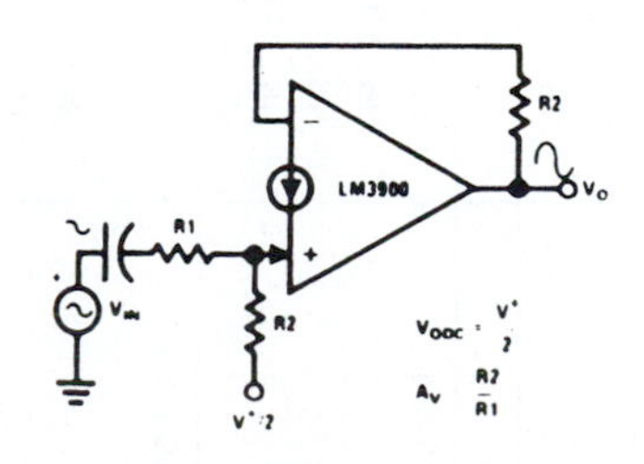


Non-Inverting Amplifier

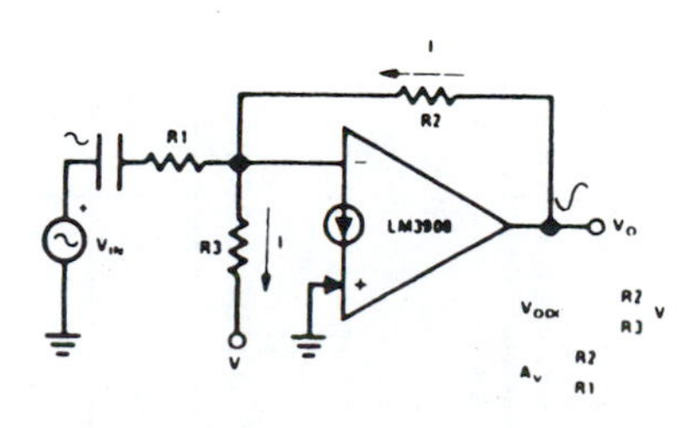


Negative Supply Biasing

## Absolute Maximum Ratings

| | LM2900/LM3900 | LM3301 | LM3401 |
|---|---|---|---|
| Supply Voltage | 32 $V_{DC}$<br>±16 $V_{DC}$ | 28 $V_{DC}$<br>±14 $V_{DC}$ | 18 $V_{DC}$<br>±9 $V_{DC}$ |
| Power Dissipation ($T_A$ = 25°C) (Note 1) | | | |
| Cavity DIP | 900 mW | | |
| Flat Pack | 800 mW | | |
| Molded DIP | 570 mW | 570 mW | 570 mW |
| Input Currents, $I_{IN}^+$ or $I_{IN}^-$ | 20 $mA_{DC}$ | 20 $mA_{DC}$ | 20 $mA_{DC}$ |
| Output Short-Circuit Duration – One Amplifier $T_A$ = 25°C (See Application Hints) | Continuous | Continuous | Continuous |
| Operating Temperature Range | | −40°C to +85°C | 0°C to +75°C |
| LM2900 | −40°C to +85°C | | |
| LM3900 | 0°C to +70°C | | |
| Storage Temperature Range | −65°C to +150°C | −65°C to +150°C | −65°C to +150°C |
| Lead Temperature (Soldering, 10 seconds) | 300°C | 300°C | 300°C |

## Electrical Characteristics (Note 6)

| PARAMETER | CONDITIONS | LM2900 | | | LM3900 | | | LM3301 | | | LM3401 | | | UNITS |
|---|---|---|---|---|---|---|---|---|---|---|---|---|---|---|
| | | MIN | TYP | MAX | MIN | TYP | MAX | MIN | TYP | MAX | MIN | TYP | MAX | |
| Open Loop | | | | | | | | | | | | | | |
| Voltage Gain | | | | | | | | | | | 800 | | | V/mV |
| Voltage Gain | $T_A$ = 25°C, f = 100 Hz | 1.2 | 2.8 | | 1.2 | 2.8 | | 1.2 | 2.8 | | 1.2 | 2.8 | | V/mV |
| Input Resistance | $T_A$ = 25°C, Inverting Input | | 1 | | | 1 | | | 1 | | 0.1 | 1 | | MΩ |
| Output Resistance | | | 8 | | | 8 | | | 8 | | | 8 | | kΩ |
| Unity Gain Bandwidth | $T_A$ = 25°C, Inverting Input | | 2.5 | | | 2.5 | | | 2.5 | | | 2.5 | | MHz |
| Input Bias Current | $T_A$ = 25°C, Inverting Input | | 30 | 200 | | 30 | 200 | | 30 | 300 | | 30 | 300 | nA |
| | Inverting Input | | | | | | | | | | | | 500 | nA |
| Slew Rate | $T_A$ = 25°C, Positive Output Swing | | 0.5 | | | 0.5 | | | 0.5 | | | 0.5 | | V/μs |
| | $T_A$ = 25°C, Negative Output Swing | | 20 | | | 20 | | | 20 | | | 20 | | V/μs |
| Supply Current | $T_A$ = 25°C, $R_L$ = ∞ On All Amplifiers | | 6.2 | 10 | | 6.2 | 10 | | 6.2 | 10 | | 6.2 | 10 | $mA_{DC}$ |
| Output Voltage Swing | $T_A$ = 25°C, $R_L$ = 2k, $V_{CC}$ = 15.0 $V_{DC}$ | | | | | | | | | | | | | |
| $V_{OUT}$ High | $I_{IN}^-$ = 0, $I_{IN}^+$ = 0 | 13.5 | | | 13.5 | | | 13.5 | | | 13.5 | | | $V_{DC}$ |
| $V_{OUT}$ Low | $I_{IN}^-$ = 10μA, $I_{IN}^+$ = 0 | | 0.09 | 0.2 | | 0.09 | 0.2 | | 0.09 | 0.2 | | 0.09 | 0.2 | $V_{DC}$ |
| $V_{OUT}$ High | $I_{IN}^-$ = 0, $I_{IN}^+$ = 0 $R_L$ = ∞, $V_{CC}$ = Absolute Maximum Ratings | | 29.5 | | | 29.5 | | | 25.5 | | | 15.5 | | $V_{DC}$ |
| Output Current Capability | $T_A$ = 25°C | | | | | | | | | | | | | |
| Source | | 6 | 18 | | 6 | 10 | | 5 | 18 | | 5 | 10 | | $mA_{DC}$ |
| Sink | (Note 2) | 0.5 | 1.3 | | 0.5 | 1.3 | | 0.5 | 1.3 | | 0.5 | 1.3 | | $mA_{DC}$ |
| $I_{SINK}$ | $V_{OL}$ = 1V, $I_{IN}$ = 5μA | | 5 | | | 5 | | | 5 | | | 5 | | $mA_{DC}$ |

## Electrical Characteristics (Continued) (Note 6)

| PARAMETER | CONDITIONS | LM2900 | | | LM3900 | | | LM3301 | | | LM3401 | | | UNITS |
|---|---|---|---|---|---|---|---|---|---|---|---|---|---|---|
| | | MIN | TYP | MAX | MIN | TYP | MAX | MIN | TYP | MAX | MIN | TYP | MAX | |
| Power Supply Rejection | $T_A = 25°C$, f = 100 Hz | | 70 | | | 70 | | | 70 | | | 70 | | dB |
| Mirror Gain | @ 20μA (Note 3) | 0.90 | 1.0 | 1.1 | 0.90 | 1.0 | 1.1 | 0.90 | 1 | 1.10 | 0.90 | 1 | 1.10 | μA/μA |
| | @ 200μA (Note 3) | 0.90 | 1.0 | 1.1 | 0.90 | 1.0 | 1.1 | 0.90 | 1 | 1.10 | 0.90 | 1 | 1.10 | μA/μA |
| ΔMirror Gain | @ 20μA To 200μA (Note 3) | | 2 | 5 | | 2 | 5 | | 2 | 5 | | 2 | 5 | % |
| Mirror Current | (Note 4) | | 10 | 500 | | 10 | 500 | | 10 | 500 | | 10 | 500 | $\mu A_{DC}$ |
| Negative Input Current | $T_A = 25°C$ (Note 5) | | 1.0 | | | 1.0 | | | 1.0 | | | 1.0 | | $mA_{DC}$ |
| Input Bias Current | Inverting Input | | 300 | | | 300 | | | | | | | | nA |

**Note 1:** For operating at high temperatures, the device must be derated based on a 125°C maximum junction temperature and a thermal resistance of 175°C/W which applies for the device soldered in a printed circuit board, operating in a still air ambient.

**Note 2:** The output current sink capability can be increased for large signal conditions by overdriving the inverting input. This is shown in the section on Typical Characteristics.

**Note 3:** This spec indicates the current gain of the current mirror which is used as the non-inverting input.

**Note 4:** Input $V_{BE}$ match between the non-inverting and the inverting inputs occurs for a mirror current (non-inverting input current) of approximately 10μA. This is therefore a typical design center for many of the application circuits.

**Note 5:** Clamp transistors are included on the IC to prevent the input voltages from swinging below ground more than approximately $-0.3\ V_{DS}$. The negative input currents which may result from large signal overdrive with capacitance input coupling need to be externally limited to values of approximately 1 mA. Negative input currents in excess of 4 mA will cause the output voltage to drop to a low voltage. This maximum current applies to any one of the input terminals. If more than one of the input terminals are simultaneously driven negative smaller maximum currents are allowed. Common-mode current biasing can be used to prevent negative input voltages; see for example, the "Differentiator Circuit" in the applications section.

**Note 6:** These specs apply for $-55°C \leq T_A \leq +125°C$, unless otherwise stated.

## Typical Applications (Continued)

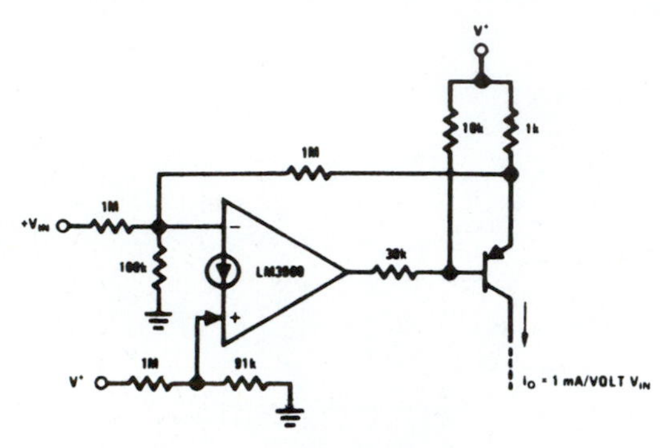


Voltage-Controlled Current Source
(Transconductance Amplifier)

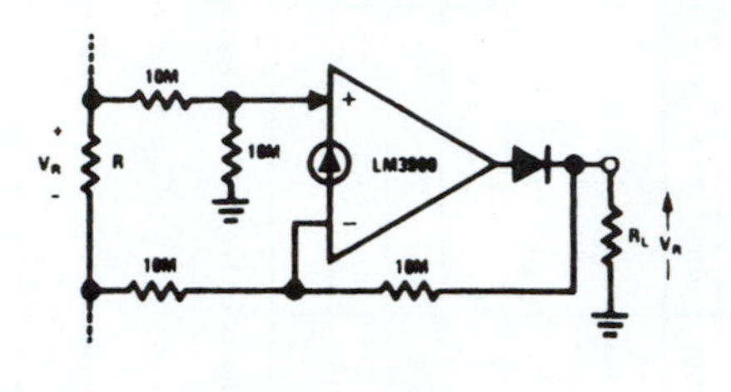


Ground-Referencing a
Differential Input Signal

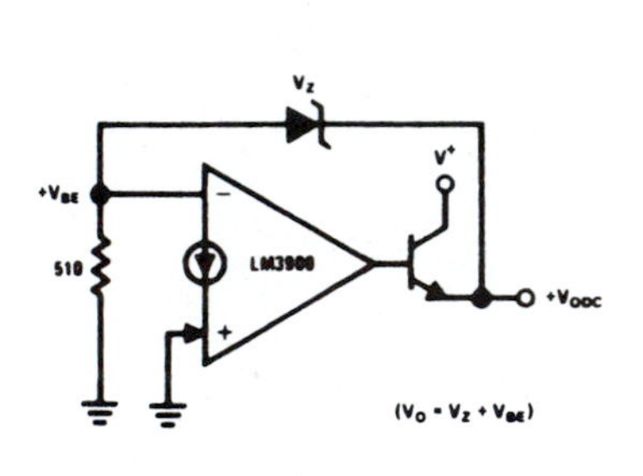


Voltage Regulator

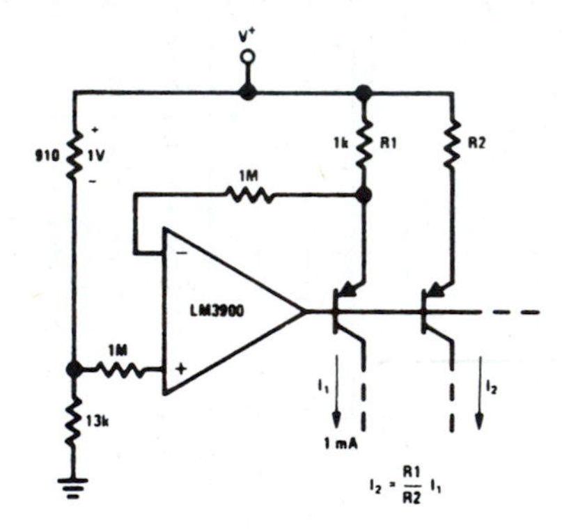


Fixed Current Sources

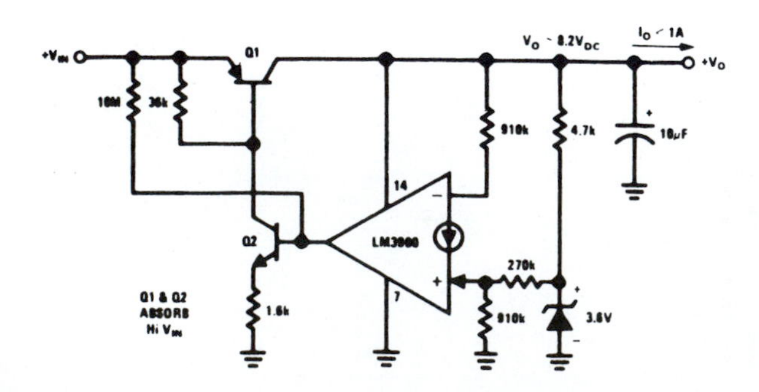


Hi $V_{IN}$, Lo $(V_{IN} - V_O)$ Self-Regulator

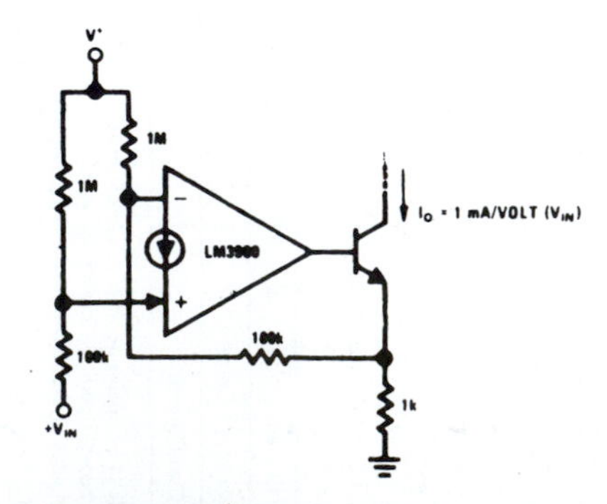


Voltage-Controlled Current Sink
(Transconductance Amplifier)

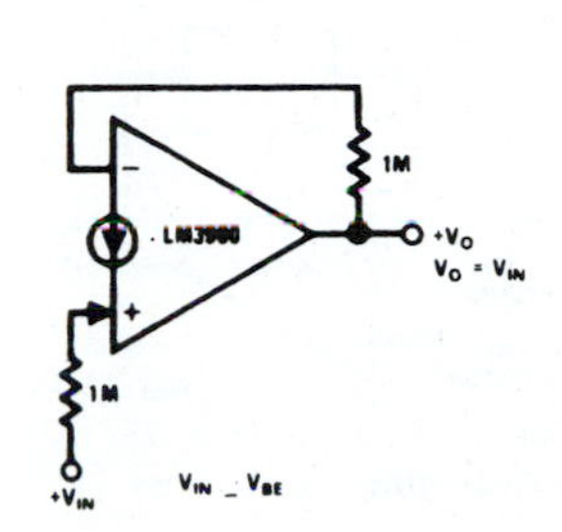


Buffer Amplifier

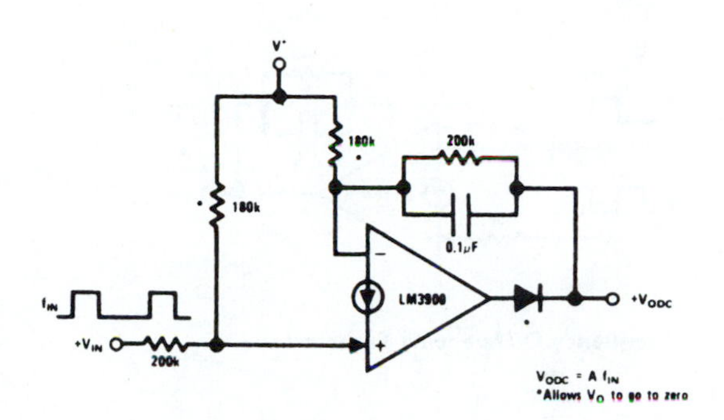


Tachometer

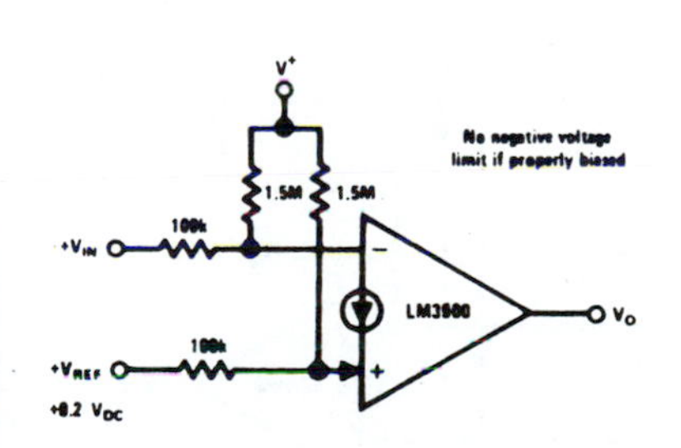


Low-Voltage Comparator

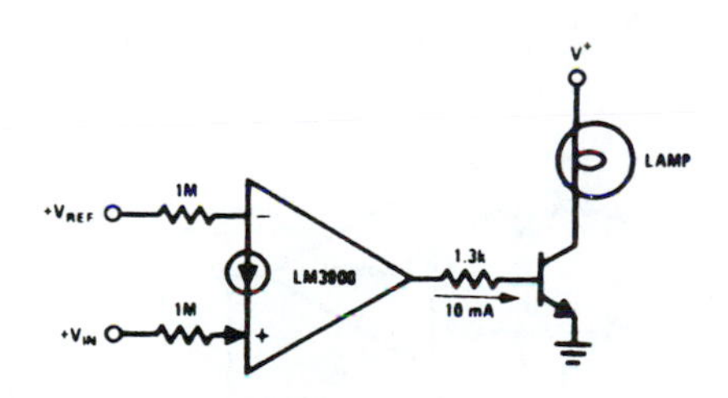


Power Comparator

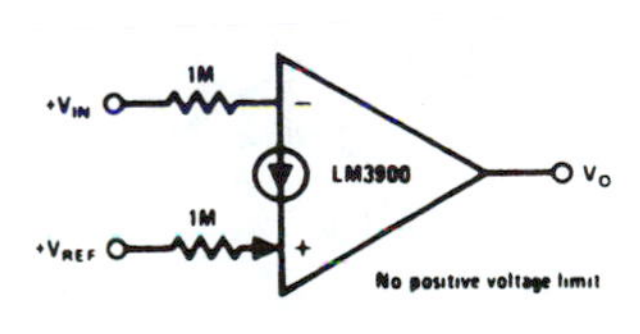


Comparator

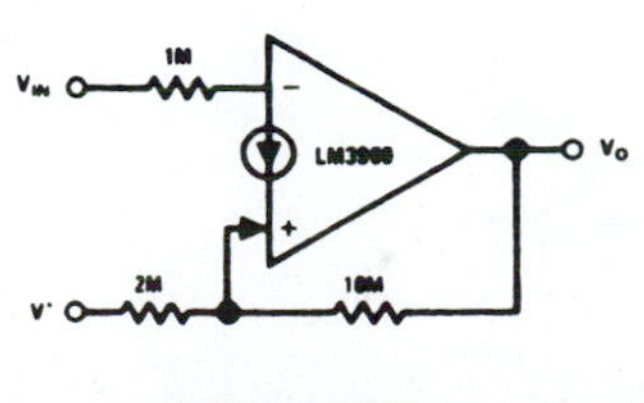


Schmitt-Trigger

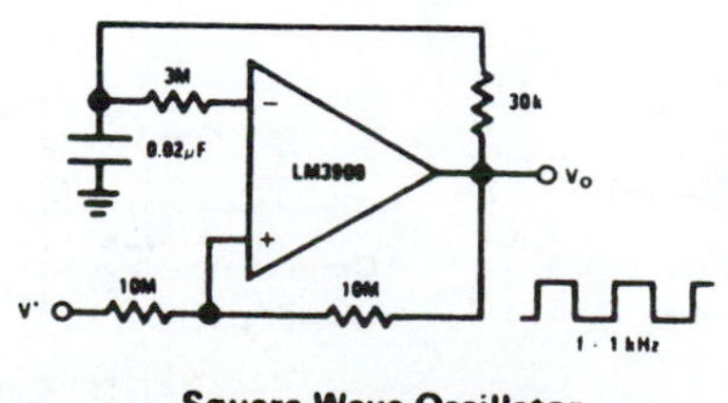


Square-Wave Oscillator

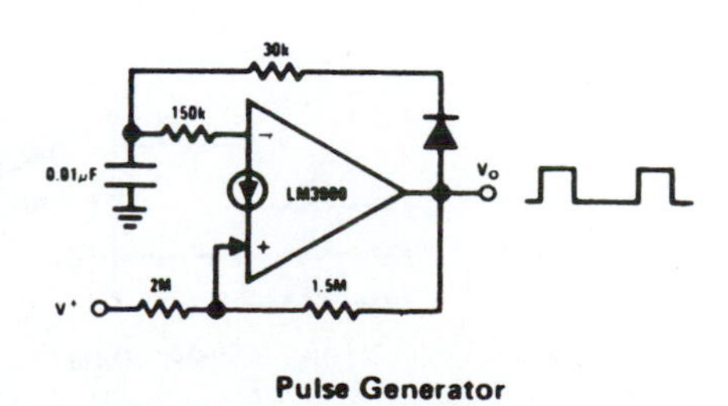


Pulse Generator

## Typical Applications (Continued)

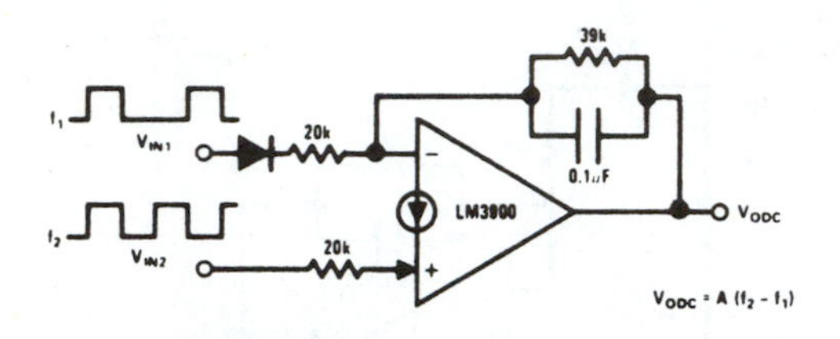


Frequency Differencing Tachometer

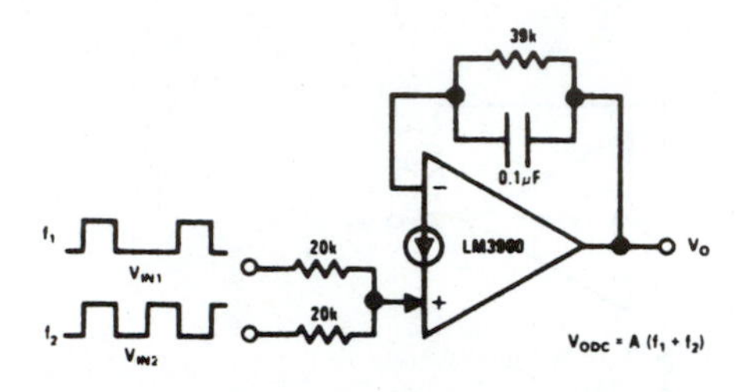


Frequency Averaging Tachometer

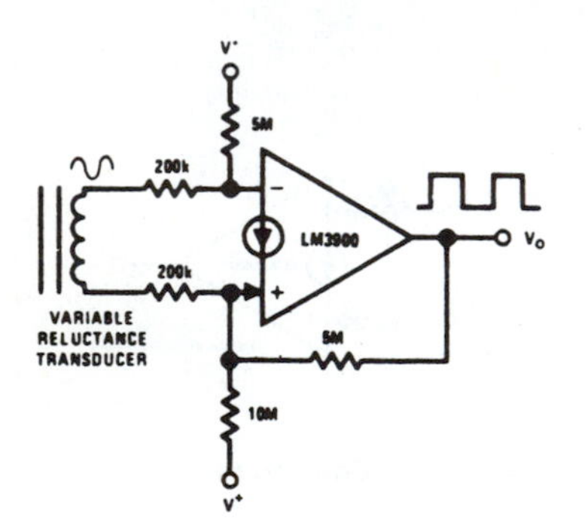


Squaring Amplifier (W/Hysteresis)

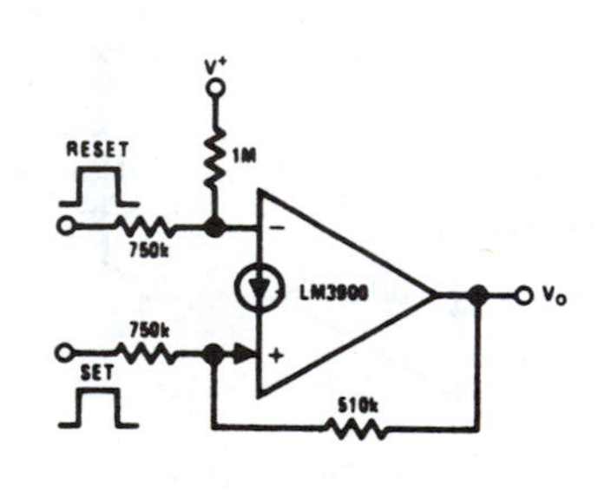


Bi-Stable Multivibrator

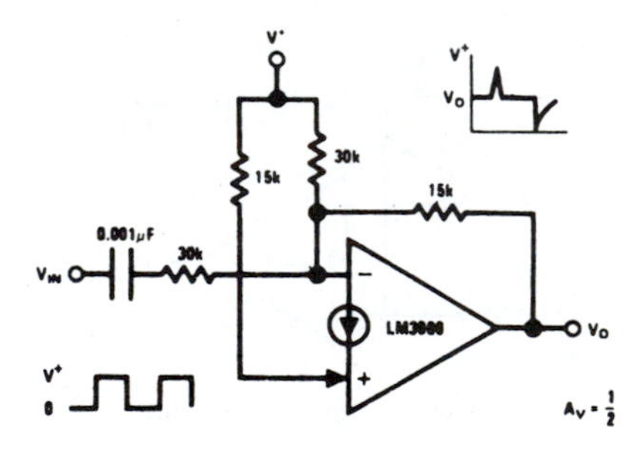


Differentiator (Common-Mode Biasing Keeps Input at $+V_{BE}$)

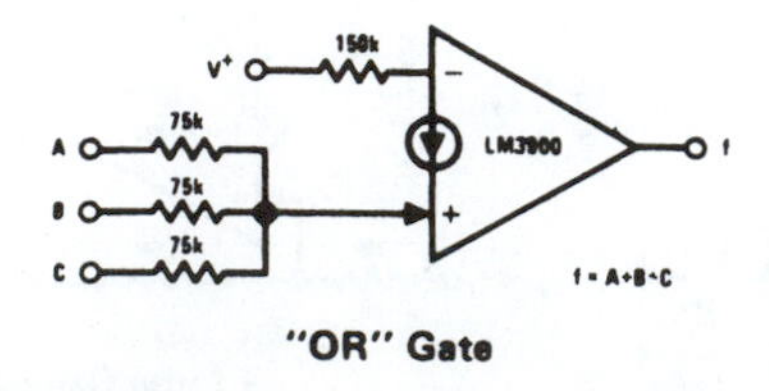


"OR" Gate

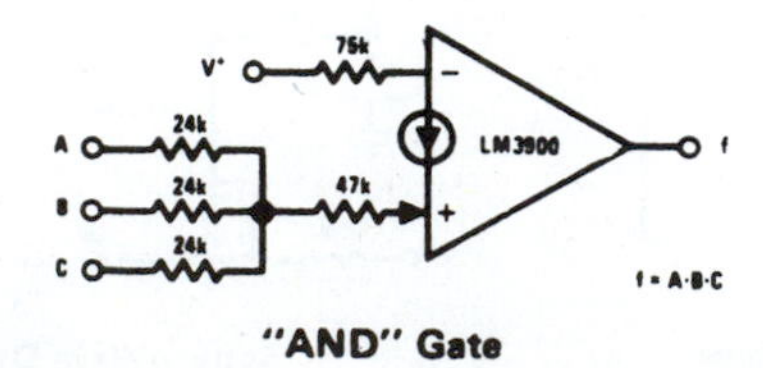


"AND" Gate

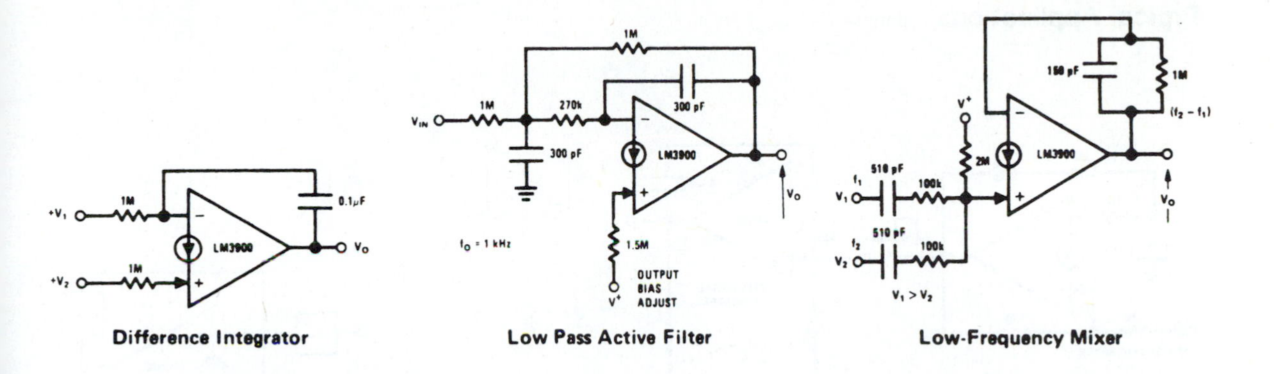


Difference Integrator Low Pass Active Filter Low-Frequency Mixer

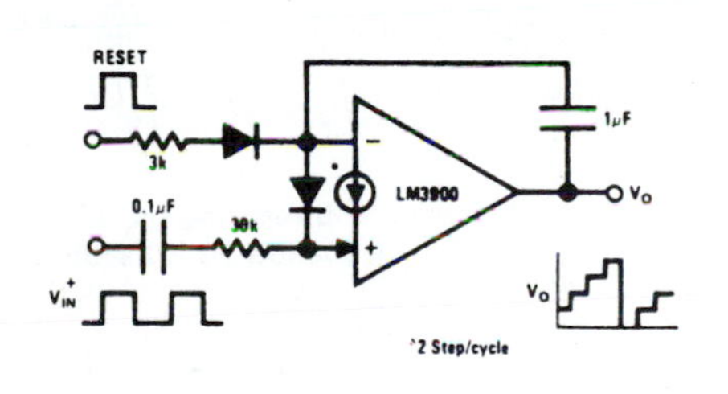

Staircase Generator

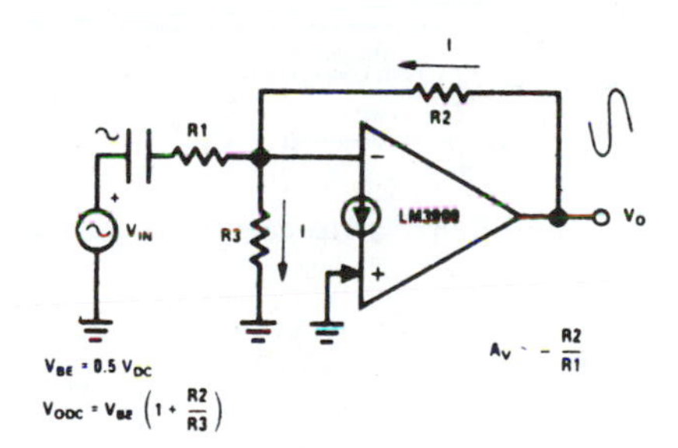


$V_{BE} = 0.5\ V_{DC}$

$V_{ODC} = V_{BE}\left(1 + \frac{R2}{R3}\right)$

$A_V = -\frac{R2}{R1}$

$V_{BE}$ Biasing

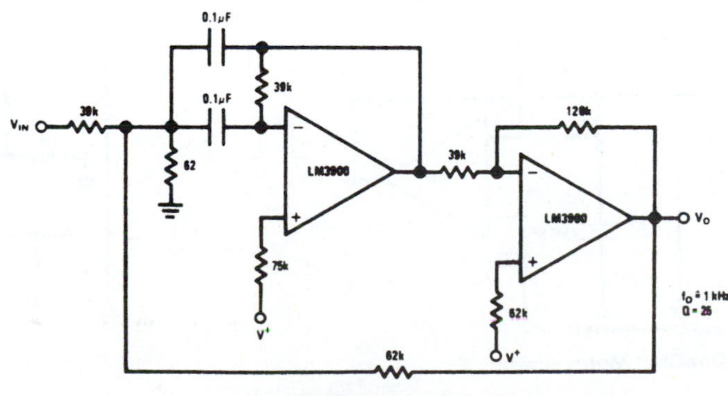


$f_O = 1$ kHz
$Q = 25$

Bandpass Active Filter

(B-8)

## Typical Applications (Continued)

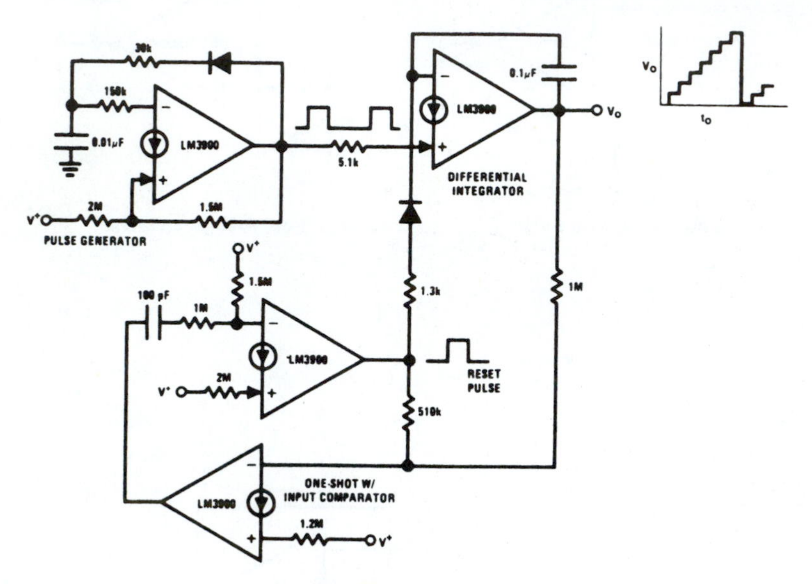


Free-Running Staircase Generator/Pulse Counter

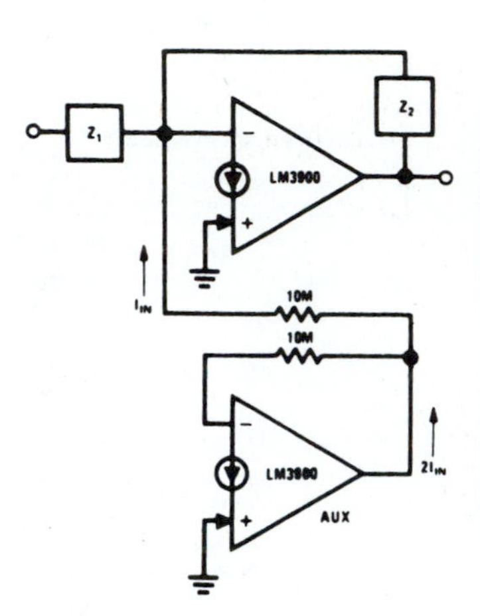


Supplying $I_{IN}$ with Aux. Amp (to Allow Hi-Z Feedback Networks)

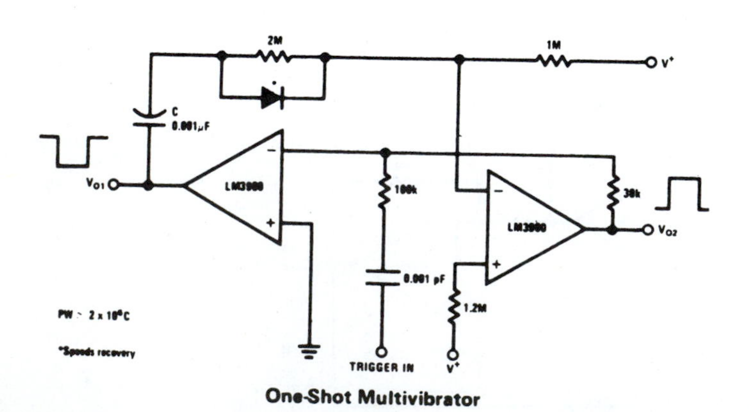


One-Shot Multivibrator

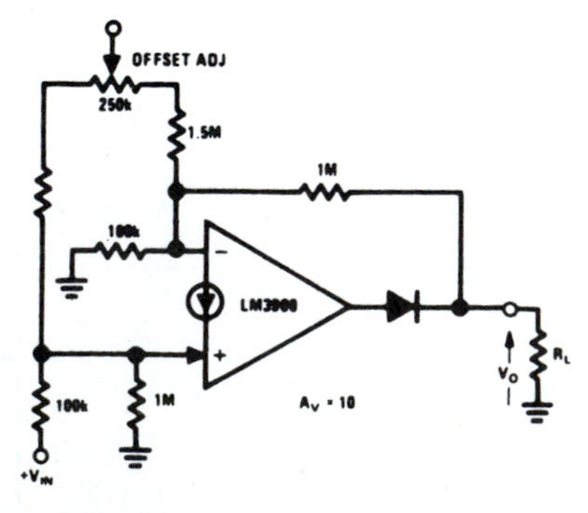


Non-Inverting DC Gain to (0,0)

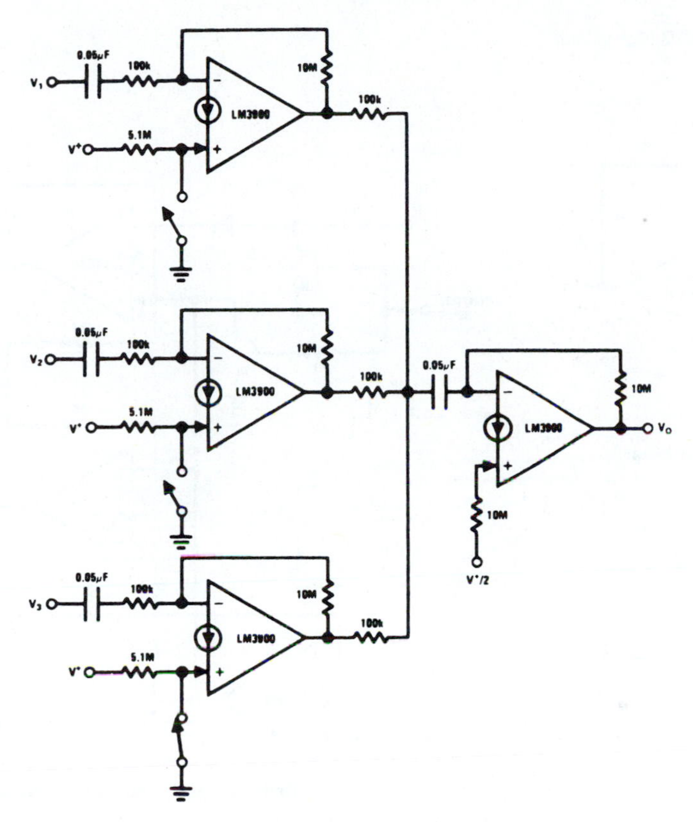


Channel Selection by DC Control (or Audio Mixer)

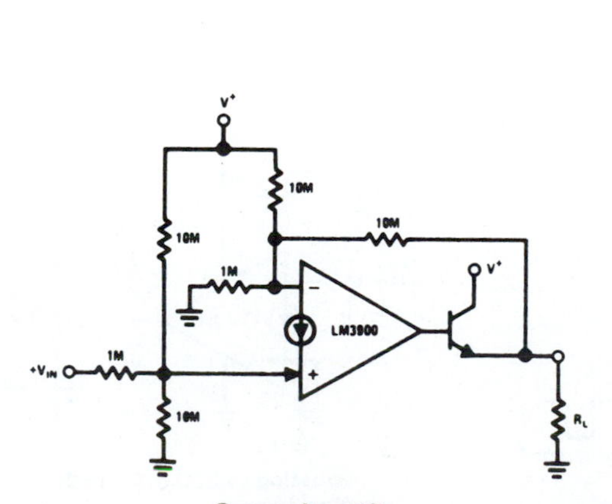


Power Amplifier

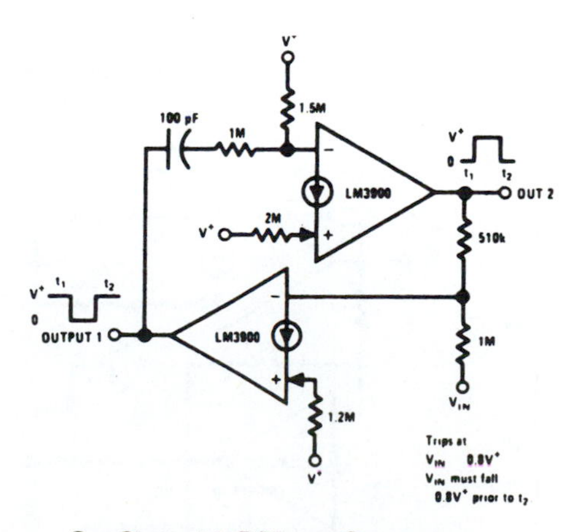


One-Shot with DC Input Comparator

## Typical Applications (Continued)

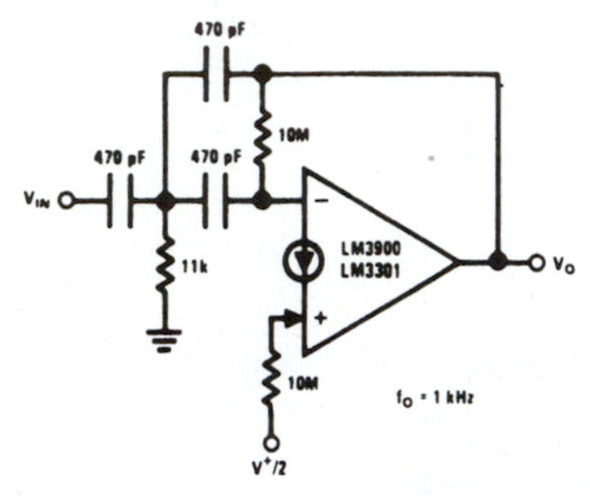


High Pass Active Filter

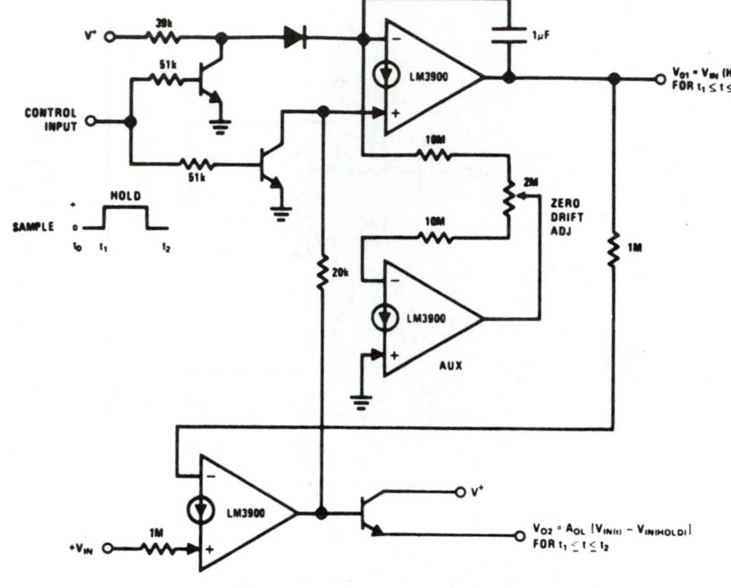


Sample-Hold and Compare with New $+V_{IN}$

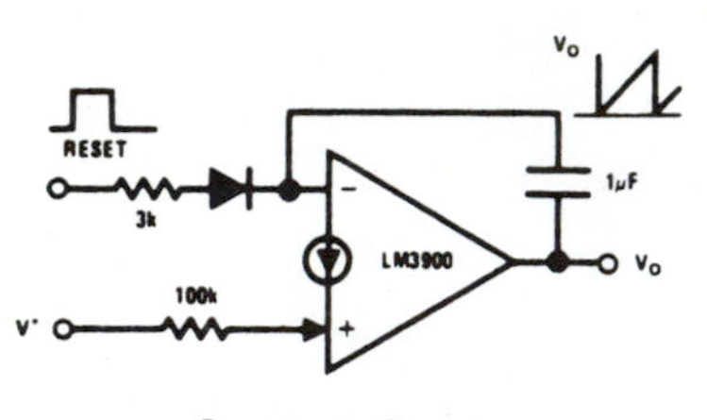


Sawtooth Generator

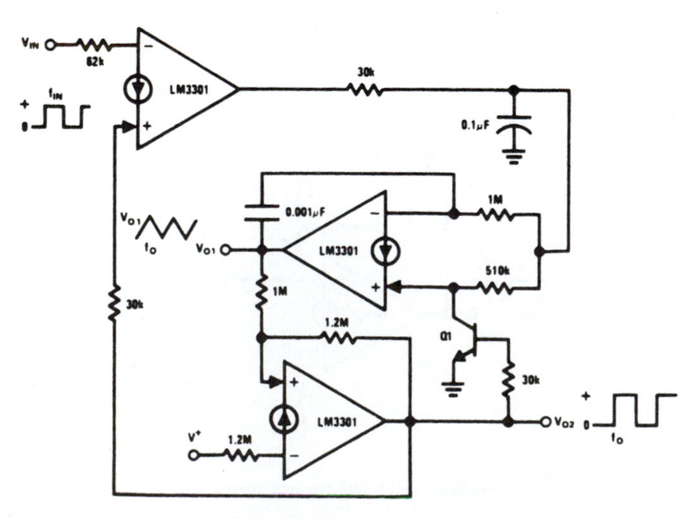


Phase-locked Loop

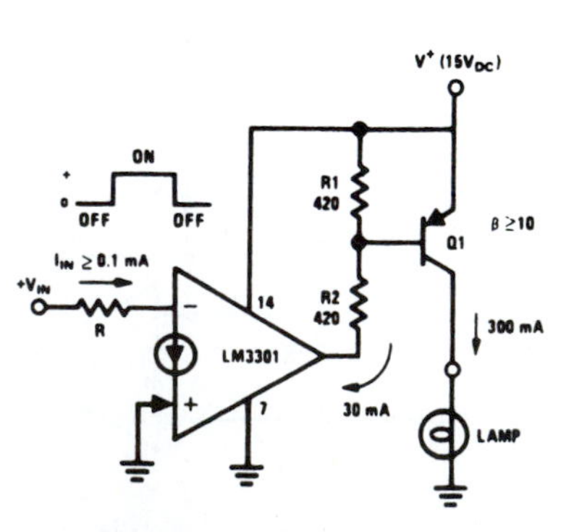


Boosting to 300 mA Loads

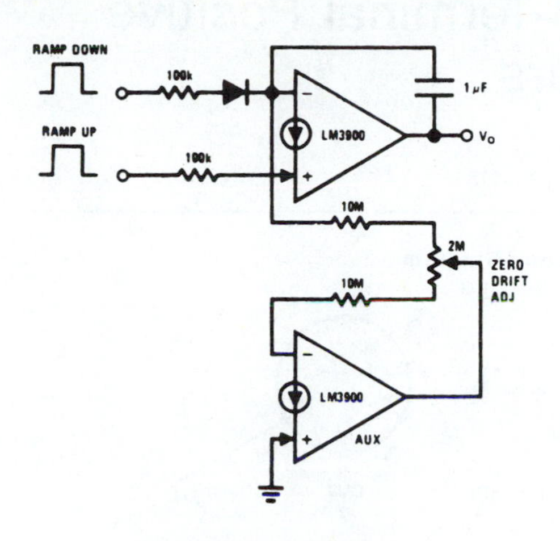


Low-Drift Ramp and Hold Circuit

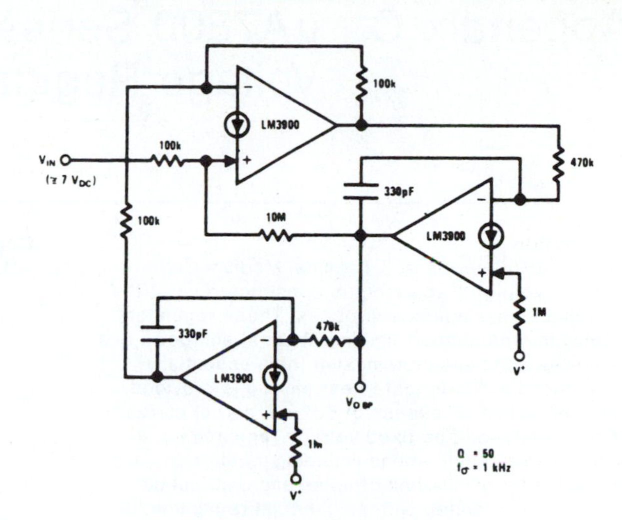


Bi-Quad Active Filter
(2nd Degree State-Variable Network)

## Split-Supply Applications ($V^+ = +15\ V_{DC}$ & $V^- = -15\ V_{DC}$)

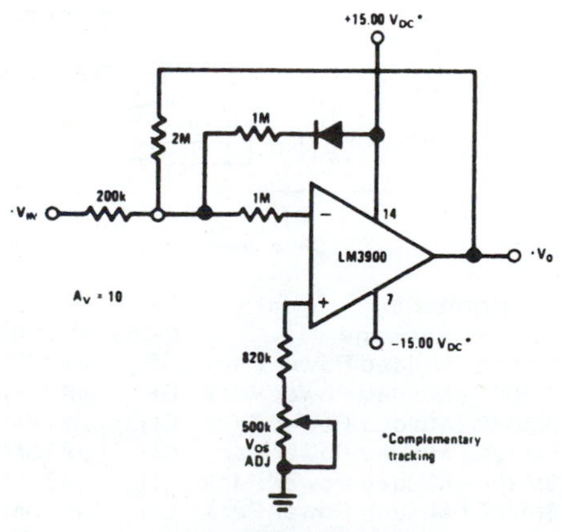


Non-Inverting DC Gain

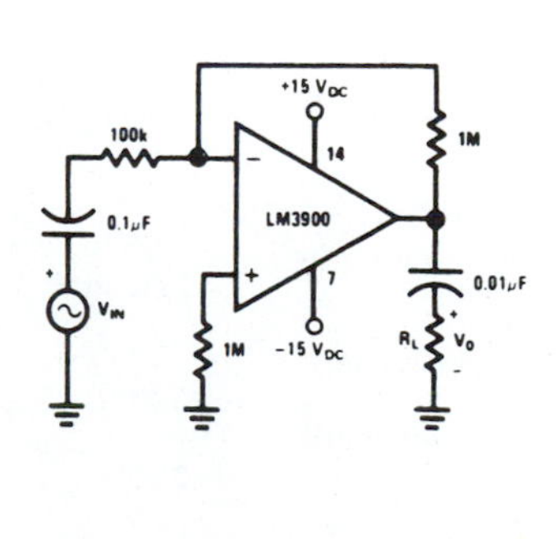


AC Amplifier

# Appendix C μA7800 Series 3-Terminal Positive Voltage Regulators

Linear Products

**Description**

The μA78M00 series of 3-Terminal Medium Current Positive Voltage Regulators is constructed using the Fairchild Planar epitaxial process. These regulators employ internal current-limiting, thermal-shutdown and safe-area compensation making them essentially indestructible. If adequate heat sinking is provided, they can deliver in excess of 500 mA output current. They are intended as fixed voltage regulators in a wide range of applications including local or on-card regulation for elimination of noise and distribution problems associated with single point regulation. In addition to use as fixed voltage regulators, these devices can be used with external components to obtain adjustable output voltages and currents.

- **OUTPUT CURRENT IN EXCESS OF 0.5 A**
- **NO EXTERNAL COMPONENTS**
- **INTERNAL THERMAL-OVERLOAD PROTECTION**
- **INTERNAL SHORT-CIRCUIT CURRENT LIMITING**
- **OUTPUT TRANSISTOR SAFE-AREA COMPENSATION**
- **AVAILABLE IN JEDEC TO-220 AND TO-39 PACKAGES**
- **OUTPUT VOLTAGES OF 5 V, 6 V, 8 V, 12 V, 15 V, AND 24 V**
- **MILITARY AND COMMERCIAL TEMPERATURE RANGE**

**Absolute Maximum Ratings**

| | |
|---|---|
| Input Voltage | |
| (5 V through 15 V) | 35 V |
| (20 V, 24 V) | 40 V |
| Internal Power Dissipation | Internally Limited |
| Storage Temperature Range | |
| TO-39 | −65°C to + 150°C |
| TO-220 | −55°C to + 150°C |
| Operating Junction Temperature Range | |
| μA78M00 | −55°C to + 150°C |
| μA78M00C | 0°C to + 125°C |
| Pin Temperatures | |
| (Soldering, 60 s time limit) | |
| TO-39 | 300°C |
| (Soldering, 10 s time limit) | |
| TO-220 | 230°C |

**Connection Diagram**
**TO-39 Package**

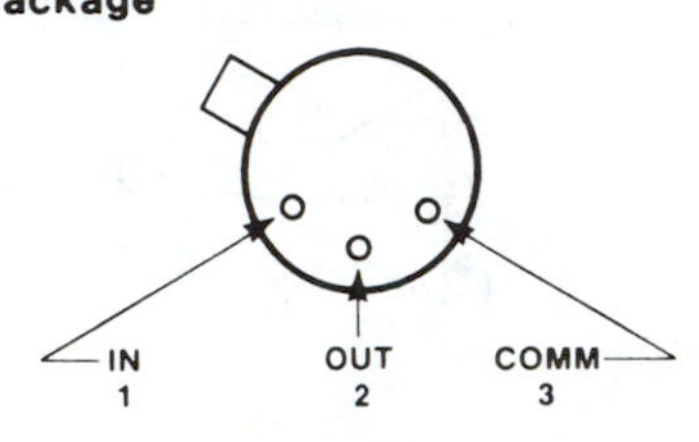


**(Top View)**

**Order Information**

| Type | Package | Code | Part No. |
|---|---|---|---|
| μA78M05 | Metal | FC | μA78M05HM |
| μA78M06 | Metal | FC | μA78M06HM |
| μA78M08 | Metal | FC | μA78M08HM |
| μA78M12 | Metal | FC | μA78M12HM |
| μA78M15 | Metal | FC | μA78M15HM |
| μA78M24 | Metal | FC | μA78M24HM |
| μA78M05C | Metal | FC | μA78M05HC |
| μA78M06C | Metal | FC | μA78M06HC |
| μA78M08C | Metal | FC | μA78M08HC |
| μA78M12C | Metal | FC | μA78M12HC |
| μA78M15C | Metal | FC | μA78M15HC |

**Connection Diagram**
**TO-220 Package**

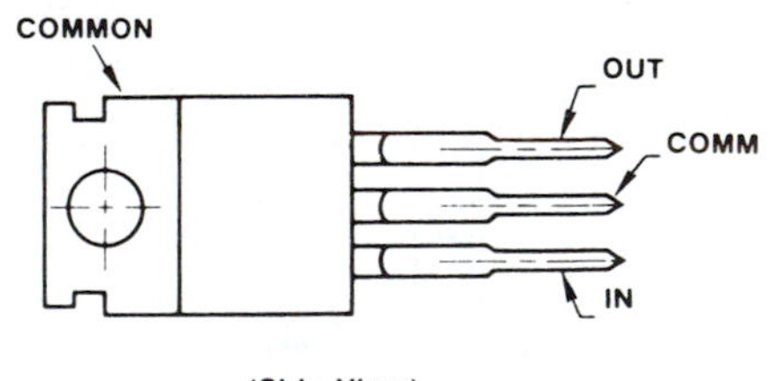


**(Side View)**

**Order Information**

| Type | Package | Code | Part No. |
|---|---|---|---|
| μA78M05C | Molded Power Pack | GH | μA78M05UC |
| μA78M06C | Molded Power Pack | GH | μA78M06UC |
| μA78M08C | Molded Power Pack | GH | μA78M08UC |
| μA78M12C | Molded Power Pack | GH | μA78M12UC |
| μA78M15C | Molded Power Pack | GH | μA78M15UC |
| μA78M24C | Molded Power Pack | GH | μA78M24UC |

(Courtesy of Fairchild Camera & Instrument Corporation, Linear Division)

**Equivalent Circuit**

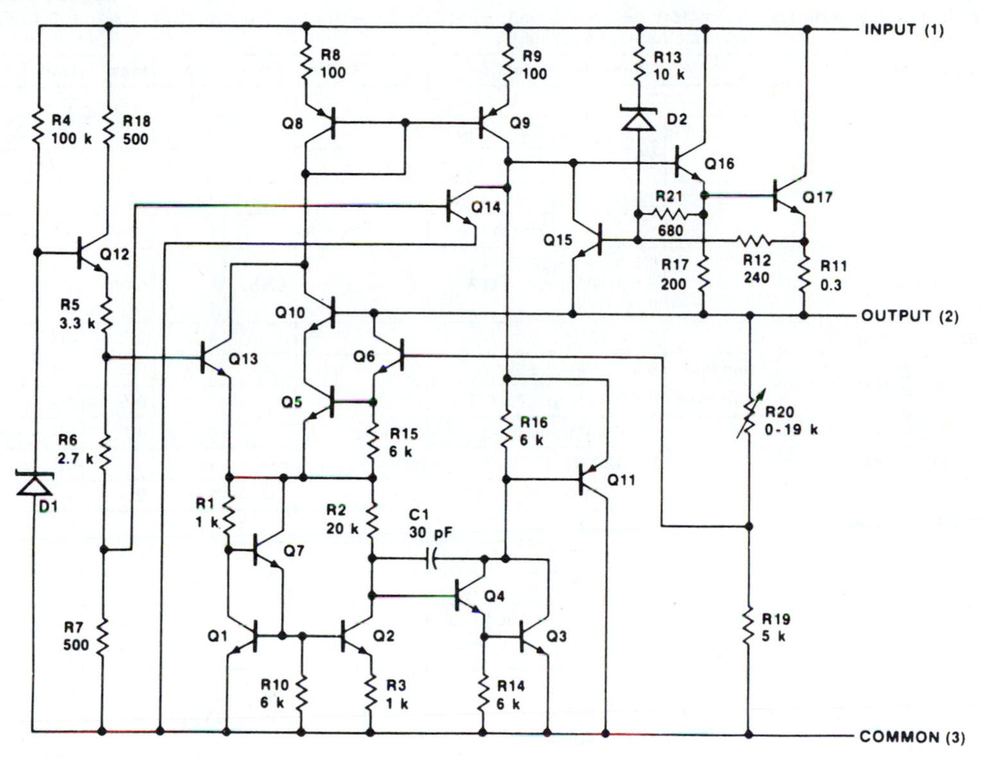

**μA7805**

**Electrical Characteristics** $V_{IN}$ = 10 V, $I_{OUT}$ = 500 mA, −55°C ≤ $T_J$ ≤ 150°C, $C_{IN}$ = 0.33 μF, $C_{OUT}$ = 0.1 μF, unless otherwise specified.

| Characteristic | | Condition (Note) | | Min | Typ | Max | Unit |
|---|---|---|---|---|---|---|---|
| Output Voltage | | $T_J$ = 25°C | | 4.8 | 5.0 | 5.2 | V |
| Line Regulation | | $T_J$ = 25°C | 7 V ≤ $V_{IN}$ ≤ 25 V | | 3 | 50 | mV |
| | | | 8 V ≤ $V_{IN}$ ≤ 12 V | | 1 | 25 | mV |
| Load Regulation | | $T_J$ = 25°C | 5 mA ≤ $I_{OUT}$ ≤ 1.5 A | | 15 | 100 | mV |
| | | | 250 mA ≤ $I_{OUT}$ ≤ 750 mA | | 5 | 25 | mV |
| Output Voltage | | 8.0 V ≤ $V_{IN}$ ≤ 20 V<br>5 mA ≤ $I_{OUT}$ ≤ 1.0 A<br>P ≤ 15 W | | 4.65 | | 5.35 | V |
| Quiescent Current | | $T_J$ = 25°C | | | 4.2 | 6.0 | mA |
| Quiescent Current Change | with line | 8 V ≤ $V_{IN}$ ≤ 25 V | | | | 0.8 | mA |
| | with load | 5 mA ≤ $I_{OUT}$ ≤ 1.0 A | | | | 0.5 | mA |
| Output Noise Voltage | | $T_A$ = 25°C, 10 Hz ≤ f ≤ 100 kHz | | | 8 | 40 | μV/$V_{OUT}$ |
| Ripple Rejection | | f = 120 Hz, 8 V ≤ $V_{IN}$ ≤ 18 V | | 68 | 78 | | dB |
| Dropout Voltage | | $I_{OUT}$ = 1.0 A, $T_J$ = 25°C | | | 2.0 | 2.5 | V |
| Output Resistance | | f = 1 kHz | | | 17 | | mΩ |
| Short-Circuit Current | | $T_J$ = 25°C, $V_{IN}$ = 35 V | | | 0.75 | 1.2 | A |
| Peak Output Current | | $T_J$ = 25°C | | 1.3 | 2.2 | 3.3 | A |
| Average Temperature Coefficient of Output Voltage | | $I_{OUT}$ = 5 mA | −55° C ≤ $T_J$ ≤ +25°C | | | 0.4 | mV/°C/ |
| | | | +25°C ≤ $T_J$ ≤ +150°C | | | 0.3 | $V_{OUT}$ |

**μA7805C**

**Electrical Characteristics** $V_{IN}$ = 10 V, $I_{OUT}$ = 500 mA, 0°C ≤ $T_J$ ≤ 125°C, $C_{IN}$ = 0.33 μF, $C_{OUT}$ = 0.1 μF, unless otherwise specified.

| Characteristic | | Condition (Note) | | Min | Typ | Max | Unit |
|---|---|---|---|---|---|---|---|
| Output Voltage | | $T_J$ = 25°C | | 4.8 | 5.0 | 5.2 | V |
| Line Regulation | | $T_J$ = 25°C | 7 V ≤ $V_{IN}$ ≤ 25 V | | 3 | 100 | mV |
| | | | 8 V ≤ $V_{IN}$ ≤ 12 V | | 1 | 50 | mV |
| Load Regulation | | $T_J$ = 25°C | 5 mA ≤ $I_{OUT}$ ≤ 1.5 A | | 15 | 100 | mV |
| | | | 250 mA ≤ $I_{OUT}$ ≤ 750 mA | | 5 | 50 | mV |
| Output Voltage | | 7 V ≤ $V_{IN}$ ≤ 20 V<br>5 mA ≤ $I_{OUT}$ ≤ 1.0 A<br>P ≤ 15 W | | 4.75 | | 5.25 | V |
| Quiescent Current | | $T_J$ = 25°C | | | 4.2 | 8.0 | mA |
| Quiescent Current Change | with line | 7 V ≤ $V_{IN}$ ≤ 25 V | | | | 1.3 | mA |
| | with load | 5 mA ≤ $I_{OUT}$ ≤ 1.0 A | | | | 0.5 | mA |
| Output Noise Voltage | | $T_A$ = 25°C, 10 Hz ≤ f ≤ 100 kHz | | | 40 | | μV |
| Ripple Rejection | | f = 120 Hz, 8 V ≤ $V_{IN}$ ≤ 18 V | | 62 | 78 | | dB |
| Dropout Voltage | | $I_{OUT}$ = 1.0 A, $T_J$ = 25°C | | | 2.0 | | V |
| Output Resistance | | f = 1 kHz | | | 17 | | mΩ |
| Short-Circuit Current | | $T_J$ = 25°C, $V_{IN}$ = 35 V | | | 750 | | mA |
| Peak Output Current | | $T_J$ = 25°C | | | 2.2 | | A |
| Average Temperature Coefficient of Output Voltage | | $I_{OUT}$ = 5 mA, 0°C ≤ $T_J$ ≤ 125°C | | | 1.1 | | mV/°C |

**$\mu$A7806C**

**Electrical Characteristics** $V_{IN} = 11$ V, $I_{OUT} = 500$ mA, $0°C \leq T_J \leq 125°C$, $C_{IN} = 0.33$ $\mu$F, $C_{OUT} = 0.1$ $\mu$F, unless otherwise specified.

| Characteristic | | Condition (Note) | | Min | Typ | Max | Unit |
|---|---|---|---|---|---|---|---|
| Output Voltage | | $T_J = 25°C$ | | 5.75 | 6.0 | 6.25 | V |
| Line Regulation | | $T_J = 25°C$ | $8 V \leq V_{IN} \leq 25 V$ | | 5 | 120 | mV |
| | | | $9 V \leq V_{IN} \leq 13 V$ | | 1.5 | 60 | mV |
| Load Regulation | | $T_J = 25°C$ | $5 mA \leq I_{OUT} \leq 1.5 A$ | | 14 | 120 | mV |
| | | | $250 mA \leq I_{OUT} \leq 750 mA$ | | 4 | 60 | mV |
| Output Voltage | | $8 V \leq V_{IN} \leq 21 V$<br>$5 mA \leq I_{OUT} \leq 1.0 A$<br>$P \leq 15 W$ | | 5.7 | | 6.3 | V |
| Quiescent Current | | $T_J = 25°C$ | | | 4.3 | 8.0 | mA |
| Quiescent Current Change | with line | $8 V \leq V_{IN} \leq 25 V$ | | | | 1.3 | mA |
| | with load | $5 mA \leq I_{OUT} \leq 1.0 A$ | | | | 0.5 | mA |
| Output Noise Voltage | | $T_A = 25°C$, $10 Hz \leq f \leq 100 kHz$ | | | 45 | | $\mu$V |
| Ripple Rejection | | $f = 120 Hz$, $9 V \leq V_{IN} \leq 19 V$ | | 59 | 75 | | dB |
| Dropout Voltage | | $I_{OUT} = 1.0 A$, $T_J = 25°C$ | | | 2.0 | | V |
| Output Resistance | | $f = 1 kHz$ | | | 19 | | m$\Omega$ |
| Short-Circuit Current | | $T_J = 25°C$, $V_{IN} = 35 V$ | | | 550 | | mA |
| Peak Output Current | | $T_J = 25°C$ | | | 2.2 | | A |
| Average Temperature Coefficient of Output Voltage | | $I_{OUT} = 5 mA$, $0°C \leq T_J \leq 125°C$ | | | 0.8 | | mV/°C |

**Note**

1. For all tables, all characteristics except noise voltage and ripple rejection ratio are measured using pulse techniques ($t_w \leq 10$ ms, duty cycle $\leq 5\%$). Output voltage changes due to changes in internal temperature must be taken into account separately.

**μA7815**

**Electrical Characteristics** $V_{IN}$ = 23 V, $I_{OUT}$ = 500 mA, −55°C ≤ $T_J$ ≤ 150°C, $C_{IN}$ = 0.33 μF, $C_{OUT}$ = 0.1 μF, unless otherwise specified

| Characteristic | | Condition (Note) | | Min | Typ | Max | Unit |
|---|---|---|---|---|---|---|---|
| Output Voltage | | $T_J$ = 25°C | | 14.4 | 15.0 | 15.6 | V |
| Line Regulation | | $T_J$ = 25°C | 17.5 V ≤ $V_{IN}$ ≤ 30 V | | 11 | 150 | mV |
| | | | 20 V ≤ $V_{IN}$ ≤ 26 V | | 3 | 75 | mV |
| Load Regulation | | $T_J$ = 25°C | 5 mA ≤ $I_{OUT}$ ≤ 1.5 A | | 12 | 150 | mV |
| | | | 250 mA ≤ $I_{OUT}$ ≤ 750 mA | | 4 | 75 | mV |
| Output Voltage | | 18.5 V ≤ $V_{IN}$ ≤ 30 V<br>5 mA ≤ $I_{OUT}$ ≤ 1.0 A<br>P ≤ 15 W | | 14.25 | | 15.75 | V |
| Quiescent Current | | $T_J$ = 25°C | | | 4.4 | 6.0 | mA |
| Quiescent Current Change | with line | 18.5 V ≤ $V_{IN}$ ≤ 30 V | | | | 0.8 | mA |
| | with load | 5 mA ≤ $I_{OUT}$ ≤ 1.0 A | | | | 0.5 | mA |
| Output Noise Voltage | | $T_A$ = 25°C, 10 Hz ≤ f ≤ 100 kHz | | | 8 | 40 | μV/$V_{OUT}$ |
| Ripple Rejection | | f = 120 Hz, 18.5 V ≤ $V_{IN}$ ≤ 28.5 V | | 60 | 70 | | dB |
| Dropout Voltage | | $I_{OUT}$ = 1.0 A, $T_J$ = 25°C | | | 2.0 | 2.5 | V |
| Output Resistance | | f = 1 kHz | | | 19 | | mΩ |
| Short-Circuit Current | | $T_J$ = 25°C, $V_{IN}$ = 35 V | | | 0.75 | | A |
| Peak Output Current | | $T_J$ = 25°C | | 1.3 | 2.2 | 3.3 | A |
| Average Temperature Coefficient of Output Voltage | | $I_{OUT}$ = 5 mA | −55° C ≤ $T_J$ ≤ +25°C | | | 0.4 | mV/°C/$V_{OUT}$ |
| | | | +25°C ≤ $T_J$ ≤ +150°C | | | 0.3 | |

**μA7815C**

**Electrical Characteristics** $V_{IN}$ = 23 V, $I_{OUT}$ = 500 mA, 0°C ≤ $T_J$ ≤ 125°C, $C_{IN}$ = 0.33 μF, $C_{OUT}$ = 0.1 μF, unless otherwise specified.

| Characteristic | | Condition (Note) | | Min | Typ | Max | Unit |
|---|---|---|---|---|---|---|---|
| Output Voltage | | $T_J$ = 25°C | | 14.4 | 15.0 | 15.6 | V |
| Line Regulation | | $T_J$ = 25°C | 17.5 V ≤ $V_{IN}$ ≤ 30 V | | 11 | 300 | mV |
| | | | 20 V ≤ $V_{IN}$ ≤ 26 V | | 3 | 150 | mV |
| Load Regulation | | $T_J$ = 25°C | 5 mA ≤ $I_{OUT}$ ≤ 1.5 A | | 12 | 300 | mV |
| | | | 250 mA ≤ $I_{OUT}$ ≤ 750 mA | | 4 | 150 | mV |
| Output Voltage | | 17.5 V ≤ $V_{IN}$ ≤ 30 V<br>5 mA ≤ $I_{OUT}$ ≤ 1.0 A<br>P ≤ 15 W | | 14.25 | | 15.75 | V |
| Quiescent Current | | $T_J$ = 25°C | | | 4.4 | 8.0 | mA |
| Quiescent Current Change | with line | 17.5 V ≤ $V_{IN}$ ≤ 30 V | | | | 1.0 | mA |
| | with load | 5 mA ≤ $I_{OUT}$ ≤ 1.0 A | | | | 0.5 | mA |
| Output Noise Voltage | | $T_A$ = 25°C, 10 Hz ≤ f ≤ 100 kHz | | | 90 | | μV |
| Ripple Rejection | | f = 120 Hz, 18.5 V ≤ $V_{IN}$ ≤ 28.5 V | | 54 | 70 | | dB |
| Dropout Voltage | | $I_{OUT}$ = 1.0 A, $T_J$ = 25°C | | | 2.0 | | V |
| Output Resistance | | f = 1 kHz | | | 19 | | mΩ |
| Short-Circuit Current | | $T_J$ = 25°C, $V_{IN}$ = 35 V | | | 230 | | A |
| Peak Output Current | | $T_J$ = 25°C | | | 2.1 | | A |
| Average Temperature Coefficient of Output Voltage | | $I_{OUT}$ = 5 mA, 0°C ≤ $T_J$ ≤ 125°C | | | 1.0 | | mV/°C |

## Typical Performance Curves

### Worst Case Power Dissipation Versus Ambient Temperature (TO-3)

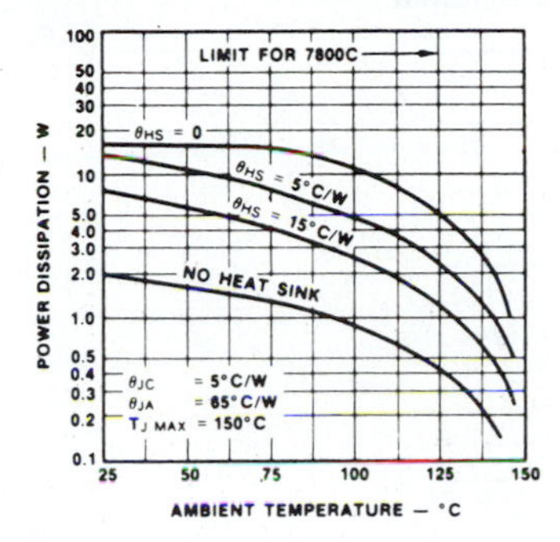


### Worst Case Power Dissipation Versus Ambient Temperature (TO-220)

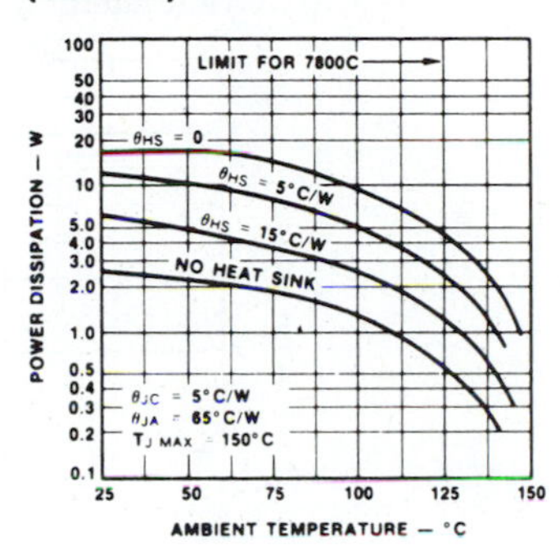


### Dropout Voltage as a Function of Junction Temperature

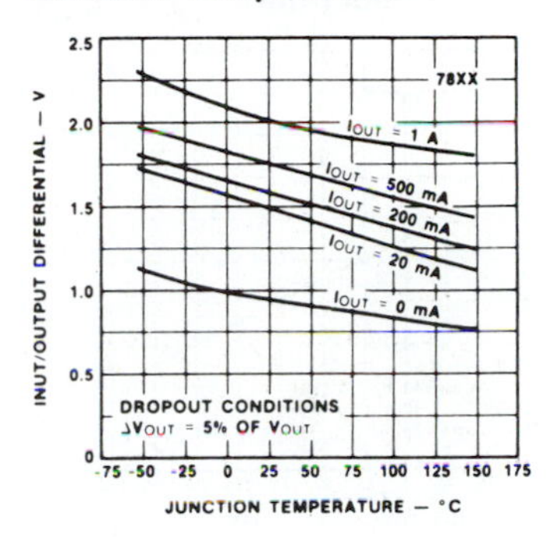


### Ripple Rejection as a Function of Frequency

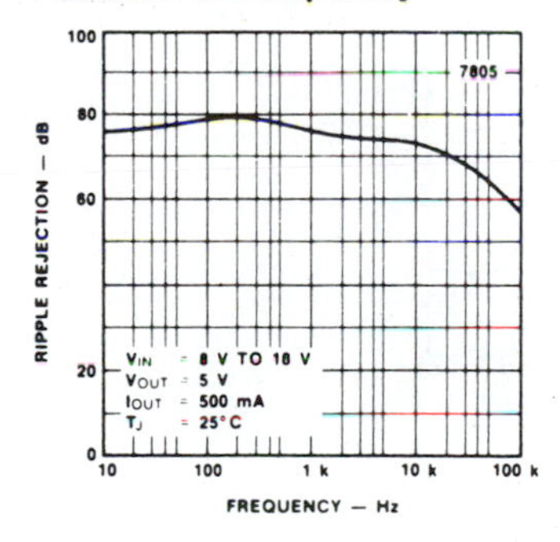


### Dropout Characteristics

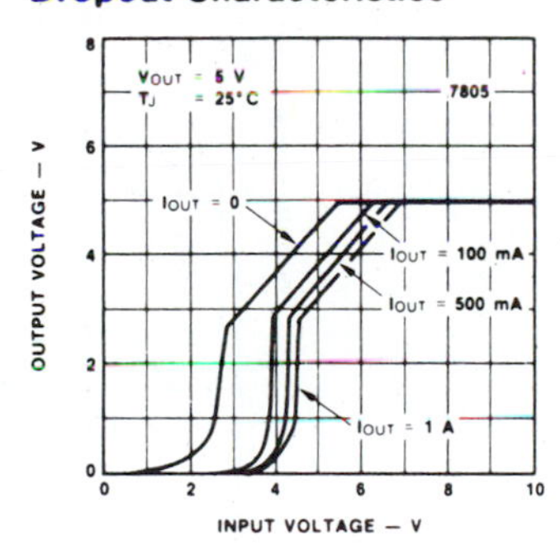


### Line Transient Response

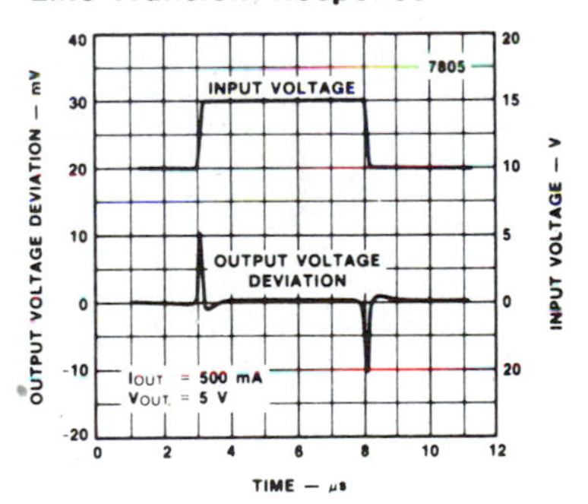


### Peak Output Current as a Function of Input/Output Differential Voltage

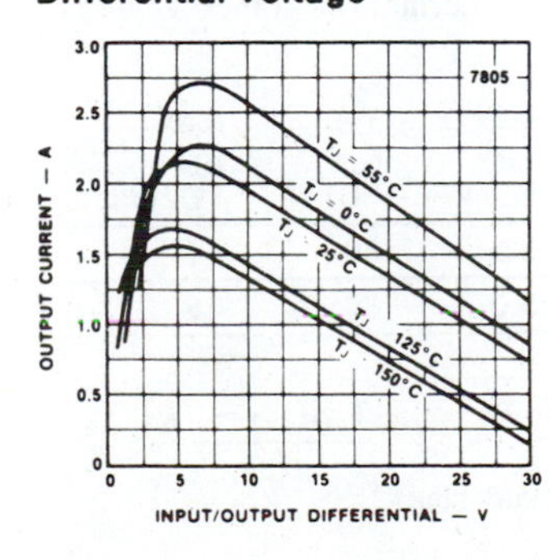


### Load Transient Response

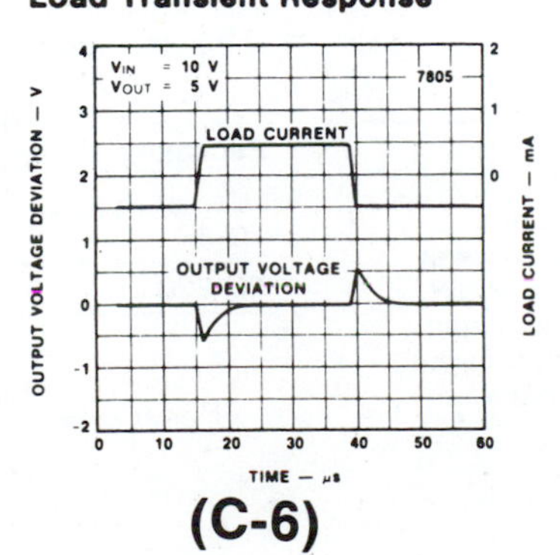


### Quiescent Current as a Function of Input Voltage

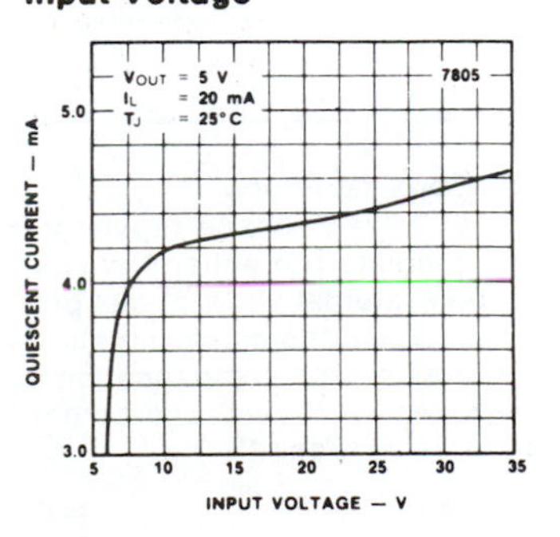

**Typical Performance Curves** (Cont.)

### Output Voltage as a Function of Junction Temperature

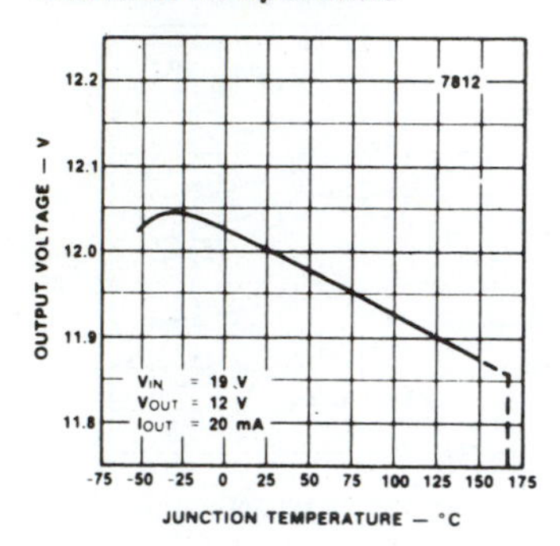


### Current Limiting Characteristics

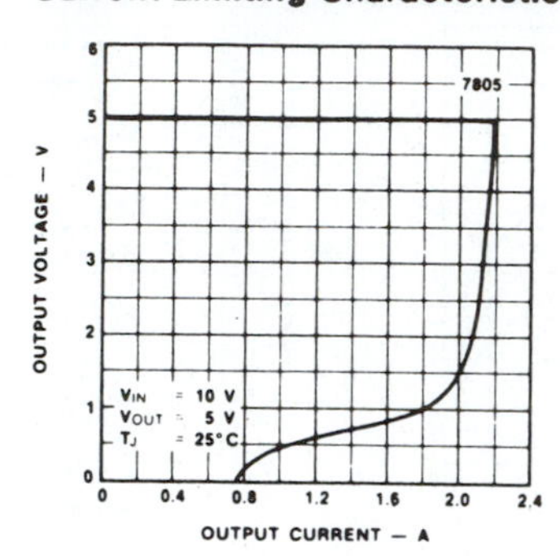


### Output Impedance as a Function of Frequency

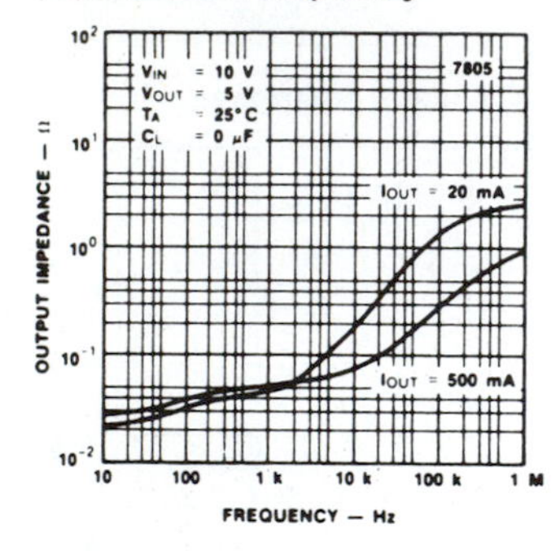


### Quiescent Current as a Function of Temperature

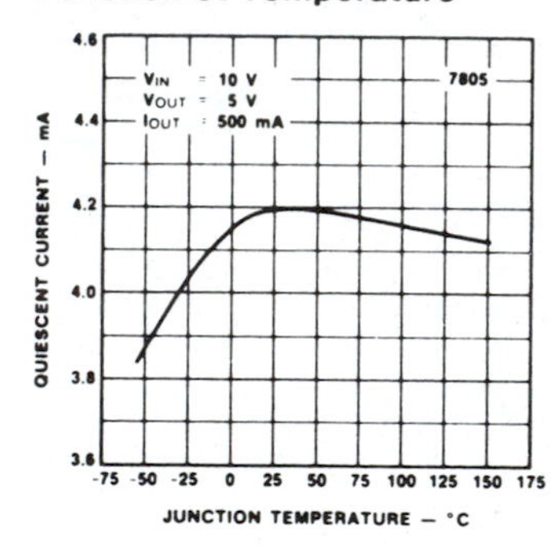


**Note**
The other μA7800 series devices have similar curves.

### DC Parameter Test Circuit

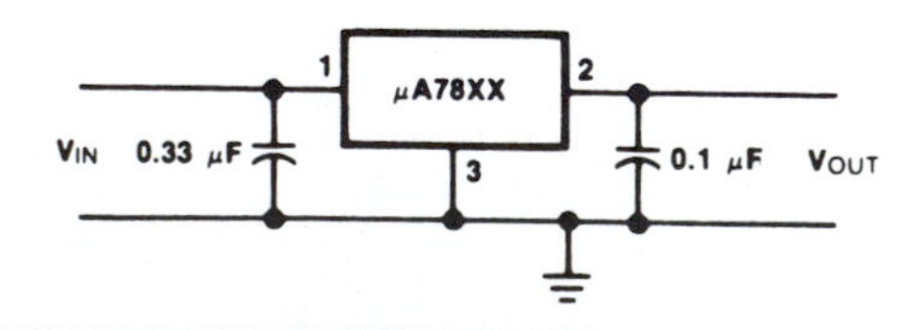


### Design Considerations

The μA7800 fixed voltage regulator series has thermal-overload protection from excessive power dissipation, internal short circuit protection which limits the regulator's maximum current, and output transistor safe area-compensation for reducing the output current as the voltage across the pass transistor is increased.

Although the internal power dissipation is limited, the junction temperature must be kept below the maximum specified temperature (150°C for 7800, 125°C for 7800C) in order to meet data sheet specifications. To calculate the maximum junction temperature or heat sink required, the following thermal resistance values should be used:

| Package | Typ $\theta_{JC}$ °C/W | Max $\theta_{JC}$ °C/W | Typ $\theta_{JA}$ °C/W | Max $\theta_{JA}$ °C/W |
|---|---|---|---|---|
| TO-3 | 3.5 | 5.5 | 40 | 45 |
| TO-220 | 3.0 | 5.0 | 60 | 65 |

$$P_{D(MAX)} = \frac{T_{J(Max)} - T_A}{\theta_{JC} + \theta_{CA}} \text{ or } \frac{T_{J(Max)} - T_A}{\theta_{JA}}$$

(Without heat sink)

$$\theta_{CA} = \theta_{CS} + \theta_{SA}$$

solving for $T_J$: $T_J = T_A + P_D(\theta_{JC} + \theta_{CA})$
or $T_A + P_D\theta_{JA}$ (Without heat sink)

where $T_J$ = Junction Temperature
$T_A$ = Ambient Temperature
$P_D$ = Power Dissipation
$\theta_{JC}$ = Junction-to-case-thermal resistance
$\theta_{CA}$ = Case-to-ambient thermal resistance
$\theta_{CS}$ = Case-to-heat sink to thermal resistance
$\theta_{SA}$ = Heat sink-to-ambient thermal resistance
$\theta_{JA}$ = Junction-to-ambient thermal resistance

## Typical Applications

### Fixed Output Regulator

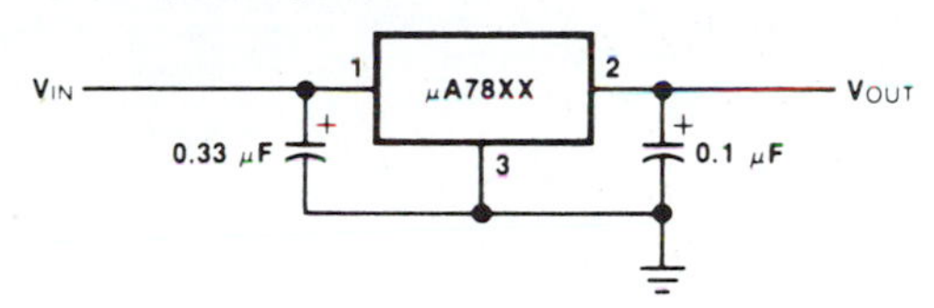


**Notes**

1. To specify an output voltage, substitute voltage value for "XX."
2. Bypass capacitors are recommended for optimum stability and transient response, and should be located as close as possible to the regulator.

### High Input Voltage Circuits

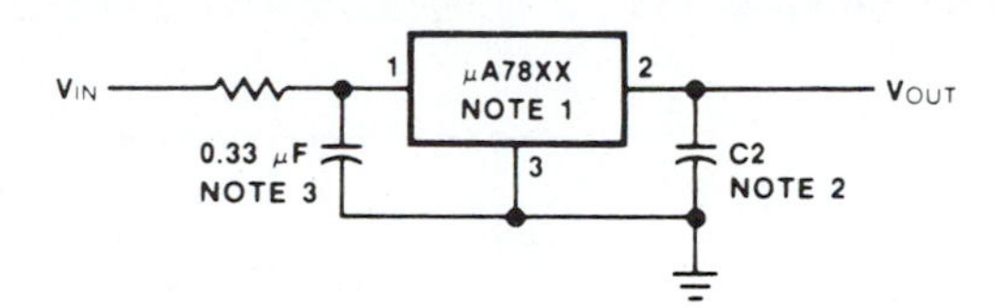


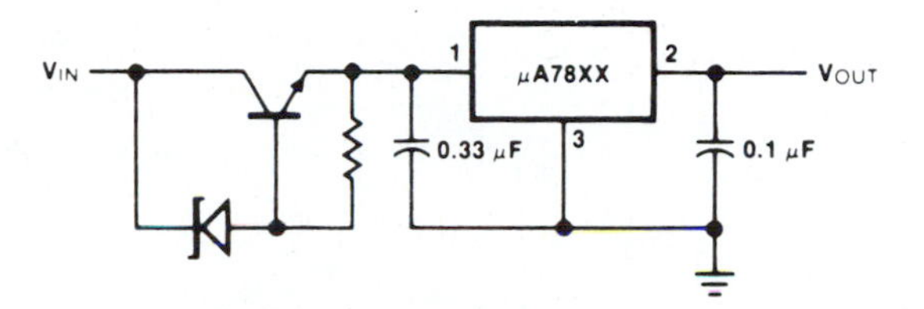


### Positive and Negative Regulator

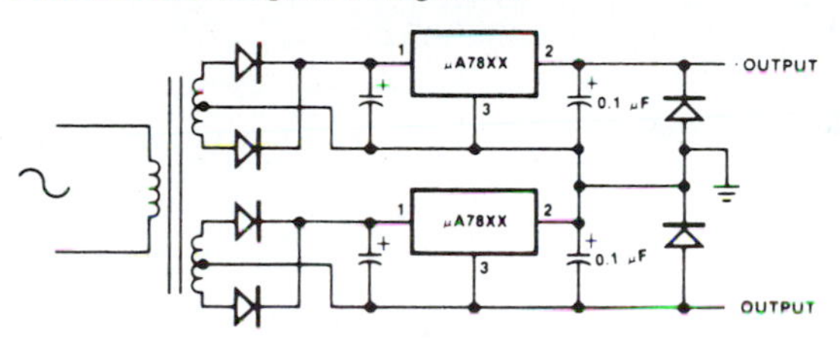


### High Current Voltage Regulator

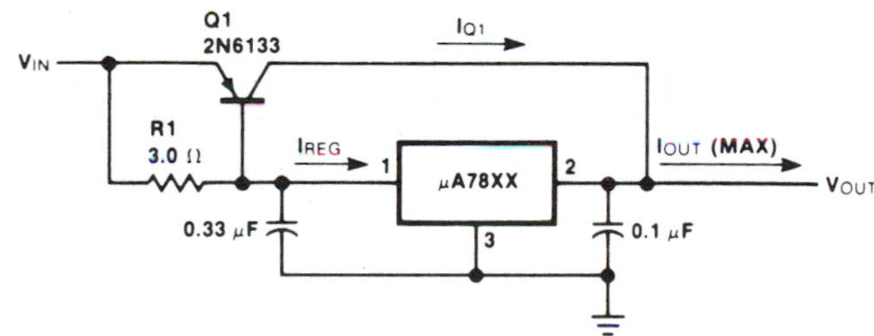


$$\beta(Q1) \geq \frac{I_{OUT(Max)}}{I_{REG(Max)}}$$

$$R1 = \frac{0.9}{I_{REG}} = \frac{\beta(Q1)\,V_{BE(Q1)}}{I_{REG(Max)}(\beta + 1) - I_{OUT(Max)}}$$

### Dual Supply
### Operational Amplifier Supply (± 15 V @ 1.0 A)

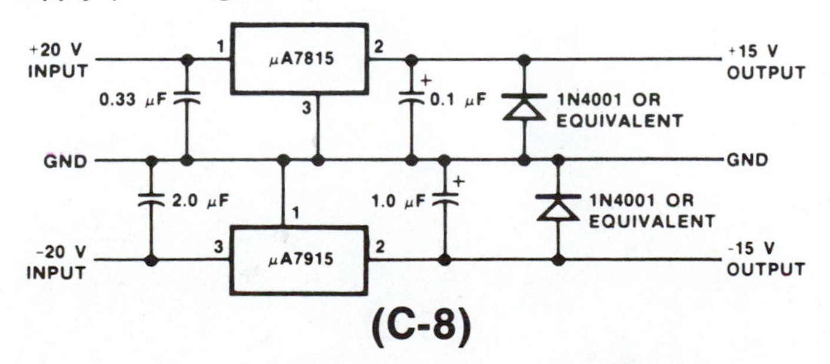

**High Output Current, Short Circuit Protected**

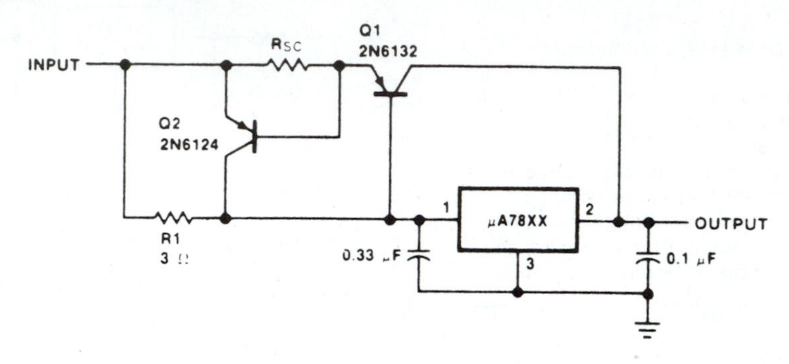


$$R_{SC} = \frac{0.8}{I_{SC}}$$

$$R1 = \frac{\beta V_{BE(Q1)}}{I_{REG(Max)}(\beta + 1) - I_{OUT(Max)}}$$

**Positive and Negative Regulator**

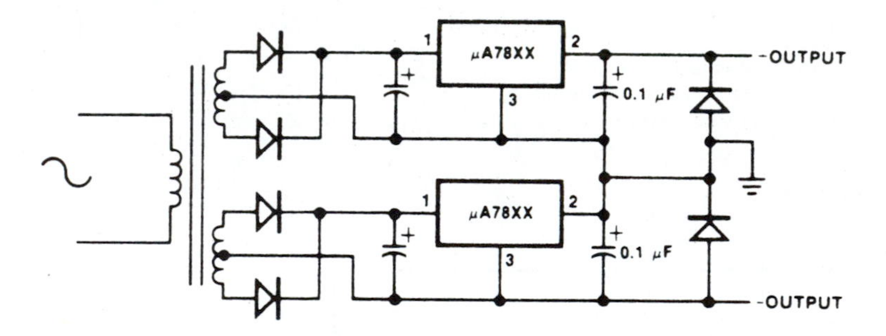

# Appendix D μA78S40 Universal Switching Regulator Subsystem

**Description**

The μA78S40 is a Monolithic Regulator Subsystem consisting of all the active building blocks necessary for switching regulator systems. The device consists of a temperature-compensated voltage reference, a duty-cycle controllable oscillator with an active current limit circuit, an error amplifier, high-current, high-voltage output switch, a power diode and an uncommitted operational amplifier. The device can drive external npn or pnp transistors when currents in excess of 1.5 A or voltages in excess of 40 V are required. The device can be used for step-down, step-up or inverting switching regulators as well as for series pass regulators. It features wide supply voltage range, low standby power dissipation, high efficiency and low drift. It is useful for any stand-alone, low part count switching system and works extremely well in battery operated systems.

- **STEP-UP, STEP DOWN OR INVERTING SWITCHING REGULATORS**
- **OUTPUT ADJUSTABLE FROM 1.3 to 40 V**
- **PEAK CURRENTS TO 1.5 A WITHOUT EXTERNAL TRANSISTORS**
- **OPERATION FROM 2.5 to 40 V INPUT**
- **LOW STANDBY CURRENT DRAIN**
- **80 dB LINE AND LOAD REGULATION**
- **HIGH GAIN, HIGH CURRENT, INDEPENDENT OP AMP**
- **PULSE WIDTH MODULATION WITH NO DOUBLE PULSING**

**Connection Diagram**
**16-Pin DIP**

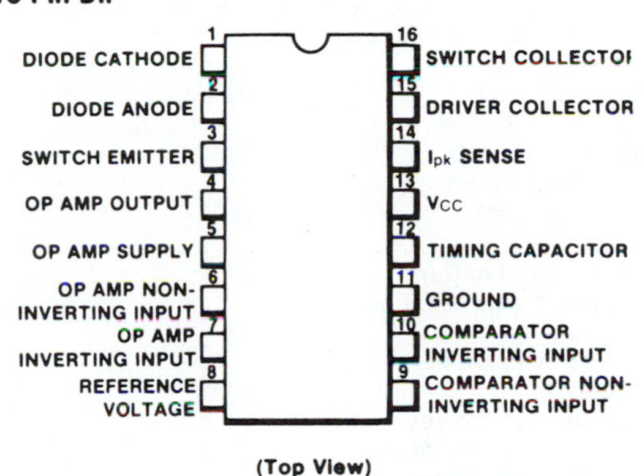


(Top View)

**Order Information**

| Type | Package | Code | Part No. |
|---|---|---|---|
| μA78S40 | Ceramic DIP | 6B | μA78S40DM |
| μA78S40 | Ceramic DIP | 6B | μA78S40DC |
| μA78S40 | Molded DIP | 9B | μA78S40PC |

**Block Diagram**

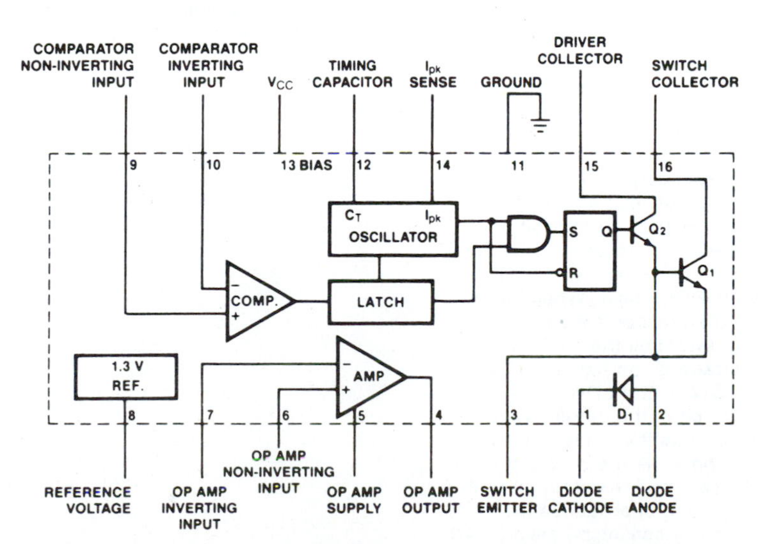


(Courtesy of Fairchild Camera & Instrument Corporation, Linear Division)

**Absolute Maximum Ratings**

| | |
|---|---|
| Input Voltage from V+ to V− | 40 V |
| Input Voltage from V+ Op Amp to V− | 40 V |
| Common Mode Input Range (Error Amplifier and Op Amp) | −0.3 to V+ |
| Differential Input Voltage (Note 1) | ±30 V |
| Output-Short Circuit Duration (Op Amp) | continuous |
| Current from $V_{REF}$ | 10 mA |
| Voltage from Switch Collectors to GND | 40 V |
| Voltage from Switch Emitters to GND | 40 V |
| Voltage from Switch Collectors to Emitter | 40 V |
| Voltage from Power Diode to GND | 40 V |
| Reverse Power Diode Voltage | 40 V |
| Current through Power Switch | 1.5 A |
| Current through Power Diode | 1.5 A |
| Internal Power Dissipation (Note 2) | |
| Molded DIP | 1500 mW |
| Ceramic DIP | 1000 mW |
| Storage Temperature Range | −65°C to +150°C |
| Operating Temperature Range | |
| Military (μA78S40M) | −55°C to 125°C |
| Commercial (μA78S40C) | 0°C to 70°C |
| Pin Temperature | |
| Ceramic DIP (Soldering, 60 s) | 300°C |
| Molded DIP (Soldering, 10 s) | 260°C |

**Notes**

1. For supply voltages less than 30 V, the absolute maximum voltage is equal to the supply voltage.
2. Ratings apply to 25°C ambient, derate ceramic DIP at 8 mW/°C and plastic DIP at 14 mW/°C.

### Functional Description

The μA78S40 is a variable frequency, variable duty cycle device. The initial switching frequency is set by the timing capacitor. The initial duty cycle is 6:1. This switching frequency and duty cycle can be modified by two mechanisms—the current limit circuitry ($I_{pk}$ sense) and the comparator.

The comparator modifies the OFF time. When the output voltage is correct, the comparator output is in the HIGH state and has no effect on the circuit operation. If the output voltage is too high then the comparator output goes LOW. In the LOW state the comparator inhibits the turn on of the output stage switching transistors. As long as the comparator is LOW the system is in OFF time. As the output current rises the OFF time decreases. As the output current nears its maximum the OFF time approaches its minimum value. The comparator can inhibit several ON cycles, one ON cycle or any portion of an ON cycle. Once the ON cycle has begun the comparator cannot inhibit until the beginning of the next ON cycle.

The current limit modifies the ON time. The current limit is activated when a 300 mV potential appears between pin 13 ($V_{CC}$) and pin 14 (I pk). This potential is intended to result when designed for peak current flows through $R_{SC}$. When the peak current is reached the current limit is turned on. The current limit circuitry provides for a quick end to ON time and the immediate start of OFF time. Generally the oscillator is free running but the current limit action tends to reset the timing cycle.

Increasing load results in more current limited ON time and less OFF time. The switching frequency increases with load current.

$V_D$ is the forward voltage drop across the internal power diode. It is listed on the data sheet as 1.25 V typical, 1.5 V maximum. If an external diode is used, then its own forward voltage drop must be used for $V_D$.

$V_S$ is the voltage across the switch element (output transistors Q1 and Q2) when the switch is closed or on. This is listed on the data sheet as output saturation voltage.

Output saturation voltage 1 — defined as the switching element voltage for Q2 and Q1 in the Darlington configuration with collectors tied together. On the data sheet this applies to *Figure 1,* the step down mode.

Output saturation voltage 2 — switching element voltage for just Q1 used as a transistor switch. This applies to *Figure 2* of the data sheet, the step-up mode.

For the inverting mode, *Figure 3,* the saturation voltage of the external transistor should be used for $V_S$.

**Electrical Characteristics** $V_{IN}$ = 5.0 V, $V_{Op\ Amp}$ = 5.0 V, $T_A$ = Operating temperature range, unless otherwise specified.

| Characteristic | Condition | | Min | Typ | Max | Unit |
|---|---|---|---|---|---|---|
| **General Characteristics** | | | | | | |
| Supply Voltage | | | 2.5 | | 40 | V |
| Supply Current (Op Amp Disconnected) | $V_{IN}$ = 5.0 V<br>$V_{IN}$ = 40 V | | | 1.8<br>2.3 | 3.5<br>5.0 | mA<br>mA |
| Supply Current Op Amp Connected | $V_{IN}$ = 5.0<br>$V_{IN}$ = 40 V | | | | 4.0<br>5.5 | mA<br>mA |
| **Reference Section** | | | | | | |
| Reference Voltage | $I_{REF}$ = 1.0 mA | $0 < T_A < 70°C$ μA78S40C<br>$-55°C < T_A < 125°C$ μA78S40M | 1.180 | 1.245 | 1.310 | V |
| Reference Voltage Line Regulation | $V_{IN}$ = 3.0 V to $V_{IN}$ = 40 V, $I_{REF}$ = 1.0 mA, $T_A$ = 25°C | | | 0.04 | 0.2 | mV/V |
| Reference Voltage Load Regulation | $I_{REF}$ = 1.0 mA to $I_{REF}$ = 10 mA, $T_A$ = 25°C | | | 0.2 | 0.5 | mV/mA |
| **Oscillator Section** | | | | | | |
| Charging Current | $V_{IN}$ = 5.0 V, $T_A$ = 25°C | | 20 | | 50 | μA |
| Charging Current | $V_{IN}$ = 40 V, $T_A$ = 25°C | | 20 | | 70 | μA |
| Discharge Current | $V_{IN}$ = 5.0 V, $T_A$ = 25°C | | 150 | | 250 | μA |
| Discharge Current | $V_{IN}$ = 40 V, $T_A$ = 25°C | | 150 | | 350 | μA |
| Oscillator Voltage Swing | $V_{IN}$ = 5 V, $T_A$ = 25°C | | | 0.5 | | V |
| $t_{on}/t_{off}$ | | | | 6.0 | | μs/μs |
| **Current Limit Section** | | | | | | |
| Current Limit Sense Voltage | $T_A$ = 25°C | | 250 | | 350 | mV |
| **Output Switch Section** | | | | | | |
| Output Saturation Voltage 1 | $I_{SW}$ = 1.0 A, *Figure 1* | | | 1.1 | 1.3 | V |
| Output Saturation Voltage 2 | $I_{SW}$ = 1.0 A, *Figure 2* | | | 0.45 | 0.7 | V |
| Output Transistor $h_{FE}$ | $I_C$ = 1.0 A, $V_{CE}$ = 5.0 V, $T_A$ = 25°C | | | 70 | | |
| Output Leakage Current | $V_{OUT}$ = 40 V, $T_A$ = 25°C | | | 10 | | nA |
| **Power Diode** | | | | | | |
| Forward Voltage Drop | $I_D$ = 1.0 A | | | 1.25 | 1.5 | V |
| Diode Leakage Current | $V_D$ = 40 V, $T_A$ = 25°C | | | 10 | | nA |
| **Comparator** | | | | | | |
| Input Offset Voltage | $V_{CM} = V_{REF}$ | | | 1.5 | 15 | mV |
| Input Bias Current | $V_{CM} = V_{REF}$ | | | 35 | 200 | nA |
| Input Offset Current | $V_{CM} = V_{REF}$ | | | 5.0 | 75 | nA |
| Common Mode Voltage Range | $T_A$ = 25°C | | 0 | | V+ −2 | V |
| Power Supply Rejection Ratio | $V_{IN}$ = 3.0 V to 40 V, $T_A$ = 25°C | | 70 | 96 | | dB |

**Electrical Characteristics** $V_{IN}$ = 5.0 V, $V_{Op\ Amp}$ = 5.0 V, $T_A$ = Operating temperature range, unless otherwise specified.

| Characteristic | Condition | Min | Typ | Max | Unit |
|---|---|---|---|---|---|
| **Output Operational Amplifier** | | | | | |
| Input Offset Voltage | $V_{CM}$ = 2.5 V | | 4.0 | 15 | mV |
| Input Bias Current | $V_{CM}$ = 2.5 V | | 30 | 200 | nA |
| Input Offset Current | $V_{CM}$ = 2.5 V | | 5.0 | 75 | nA |
| Voltage Gain + | $R_L$ = 2.0 k to GND; $V_O$ = 1.0 to 2.5 V, $T_A$ = 25°C | 25 k | 250 k | | V/V |
| Voltage Gain − | $R_L$ = 2.0 k to V+ Op Amp; $V_O$ = 1.0 to 2.5 V, $T_A$ = 25°C | 25 k | 250 k | | V/V |
| Common Mode Voltage Range | $T_A$ = 25°C | 0 | | V+ −2 | V |
| Common Mode Rejection Ratio | $V_{CM}$ = 0 to 3.0 V, $T_A$ = 25°C | 76 | 100 | | dB |
| Power Supply Rejection Ratio | V+ Op Amp = 3.0 to 40 V, $T_A$ = 25°C | 76 | 100 | | dB |
| Output Source Current | $T_A$ = 25°C | 75 | 150 | | mA |
| Output Sink Current | $T_A$ = 25°C | 10 | 35 | | mA |
| Slew Rate | $T_A$ = 25°C | | 0.6 | | V/µs |
| Output LOW Voltage | $I_L$ = −5.0 mA, $T_A$ = 25°C | | | 1.0 | V |
| Output HIGH Voltage | $I_L$ = 50 mA, $T_A$ = 25°C | V+OP Amp −3.0 V | | | V |

**Design Formulas**

| Characteristic | Step Down | Step Up | Inverting | Unit |
|---|---|---|---|---|
| $I_{pk}$ | $2\,I_{OUT(Max)}$ | $2\,I_{OUT(Max)} \bullet \frac{V_{OUT}+V_D-V_S}{V_{IN}-V_S}$ | $2\,I_{OUT(Max)} \bullet \frac{V_{IN}+\lvert V_{OUT}\rvert+V_D-V_S}{V_{IN}-V_S}$ | A |
| $R_{SC}$ | $0.33/I_{pk}$ | $0.33\ I_{pk}$ | $0.33\ I_{pk}$ | Ω |
| $\frac{t_{on}}{t_{off}}$ | $\frac{V_{OUT}+V_D}{V_{IN}-V_S-V_{OUT}}$ | $\frac{V_{OUT}+V_D-V_{IN}}{V_{IN}-V_S}$ | $\frac{\lvert V_{OUT}\rvert+V_D}{V_{IN}-V_S}$ | |
| L | $\frac{V_{OUT}+V_D}{I_{pk}} \bullet t_{off}$ | $\frac{V_{OUT}+V_D-V_{IN}}{I_{pk}} \bullet t_{off}$ | $\frac{\lvert V_{OUT}\rvert+V_D}{I_{pk}} \bullet t_{off}$ | µH |
| $t_{off}$ | $\frac{I_{pk} \bullet L}{V_{OUT}+V_D}$ | $\frac{I_{pk} \bullet L}{V_{OUT}+V_D-V_{IN}}$ | $\frac{I_{pk} \bullet L}{\lvert V_{OUT}\rvert+V_D}$ | µs |
| $C_T$ (µF) | $45 \times 10^{-5}\ t_{off}(\mu s)$ | $45 \times 10^{-5}\ t_{off}(\mu s)$ | $45 \times 10^{-5}\ t_{off}(\mu s)$ | µF |
| $C_O$ | $\frac{I_{pk} \bullet (t_{on}+t_{off})}{8\,V_{ripple}}$ | $\frac{(I_{pk}-I_{OUT})^2 \bullet t_{off}}{2\,I_{pk} \bullet V_{ripple}}$ | $\frac{(I_{pk}-I_{OUT})^2 \bullet t_{off}}{2\,I_{pk} \bullet V_{ripple}}$ | µF |
| Efficiency | $\frac{V_{IN}-V_S+V_D}{V_{IN}} \bullet \frac{V_{OUT}}{V_{OUT}+V_D}$ | $\frac{V_{IN}-V_S}{V_{IN}} \bullet \frac{V_{OUT}}{V_{OUT}+V_D-V_S}$ | $\frac{V_{IN}-V_S}{V_{IN}} \bullet \frac{\lvert V_{OUT}\rvert}{V_{OUT}+V_D}$ | |
| $I_{IN(Avg)}$ (Max load Condition) | $\frac{I_{pk}}{2} \bullet \frac{V_{OUT}+V_D}{V_{IN}-V_S+V_D}$ | $\frac{I_{pk}}{2}$ | $\frac{I_{pk}}{2} \bullet \frac{\lvert V_{OUT}\rvert+V_D}{V_{IN}+\lvert V_{OUT}\rvert+V_D-V_S}$ | A |

**Typical Step-Down Performance**
**$T_A = 25°C$**

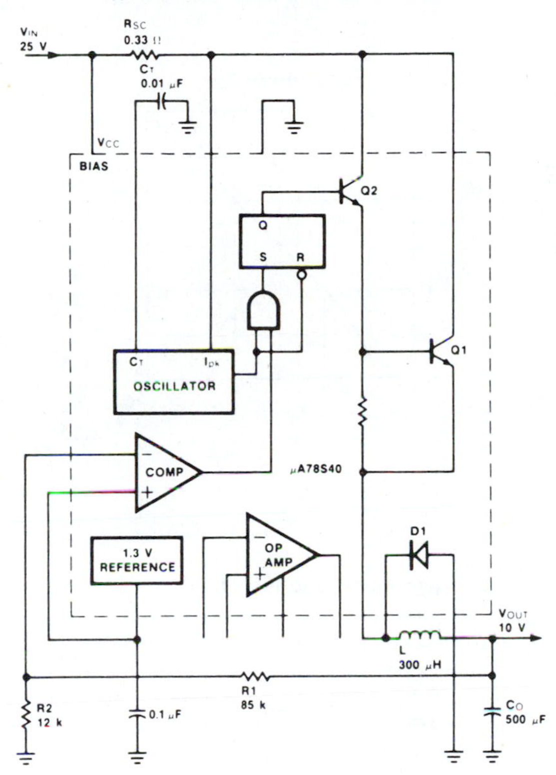


| Characteristic | Condition | Typical Value |
|---|---|---|
| Output Voltage | $I_{OUT}$ = 200 mA | 10 V |
| Line Regulation | 20 ≤ $V_{IN}$ ≤ 30 V | 1.5 mV |
| Load Regulation | 5 mA ≤ $I_{OUT}$ ≤ 300 mA | 3.0 mV |
| Max Output Current | $V_{OUT}$ = 9.5 V | 500 mA |
| Output Ripple | $I_{OUT}$ = 200 mA | 50 mV |
| Efficiency | $I_{OUT}$ = 200 mA | 74% |
| Standby Current | $I_{OUT}$ = 200 mA | 2.8 mA |

**Notes**

1. For $I_{OUT}$ ≥ 200 mA use external diode to limit on chip power dissipation.
2. It is recommended that the internal reference (pin 8) be bypassed by a 0.1 μF capacitor directly to (pin 11) the ground point of the μA78S40.

**Typical Step-Up Operational Performance**
**$T_A = 25°C$**

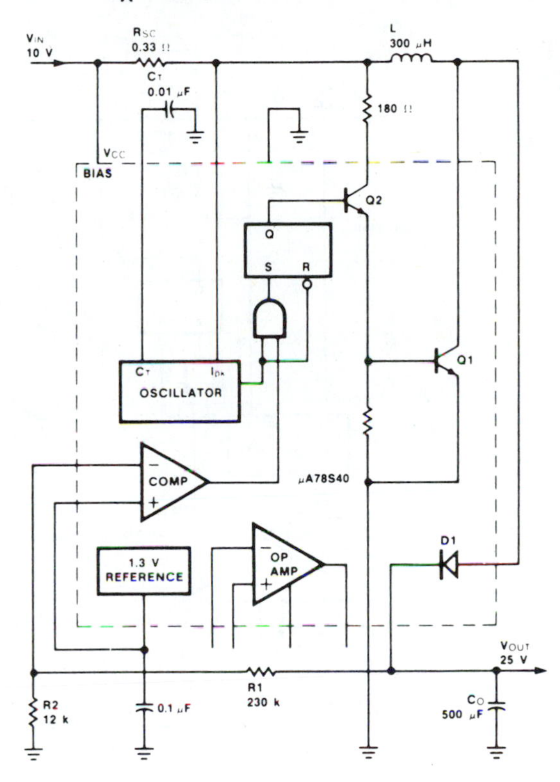


| Characteristic | Condition | Typical Value |
|---|---|---|
| Output Voltage | $I_{OUT}$ = 50 mA | 25 V |
| Line Regulation | 5 V ≤ $V_{IN}$ ≤ 15 V | 4.0 mV |
| Load Regulation | 5 mA ≤ $I_{OUT}$ ≤ 100 mA | 2.0 mV |
| Max Output Current | $V_{OUT}$ = 23.75 V | 160 mA |
| Output Ripple | $I_{OUT}$ = 50 mA | 30 mV |
| Efficiency | $I_{OUT}$ = 50 mA | 79% |
| Standby Current | $I_{OUT}$ = 50 mA | 2.6 mA |

### Typical Inversion Operational Performance $T_A = 25°C$

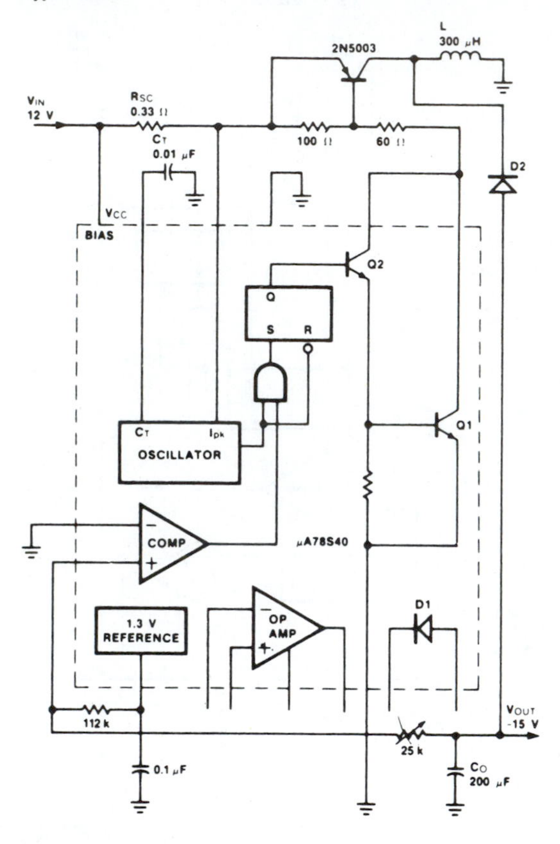


| Characteristic | Condition | Typical Value |
|---|---|---|
| Output Voltage | $I_{OUT}$ = 100 mA | −15 V |
| Line Regulation | 8 V ≤ $V_{IN}$ ≤ 18 V | 5.0 mV |
| Load Regulation | 5 mA ≤ $I_{OUT}$<br>$I_{OUT}$ ≤ 150 mA | 3.0 mV |
| Max Output Current | $V_{OUT}$ = 14.25 V | 160 mA |
| Output Ripple | $I_{OUT}$ = 100 mA | 20 mV |
| Efficiency | $I_{OUT}$ = 100 mA | 70% |
| Standby Current | $I_{OUT}$ = 100 mA | 2.3 mA |

### Typical Performance Curves

#### $C_T$ as a Function of $t_{off}$

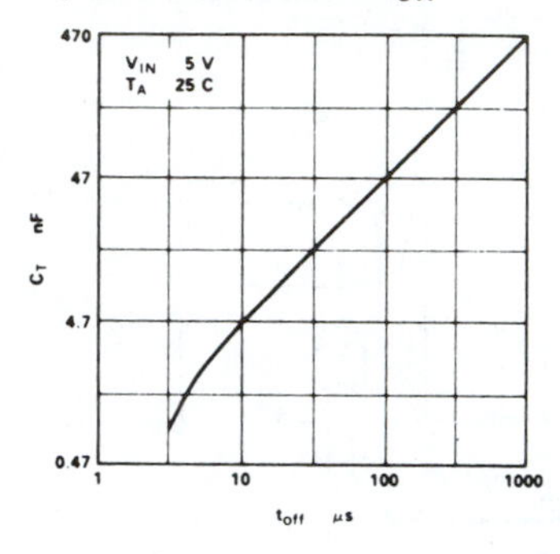


#### $V_{REF}$ as a Function of $T_J$

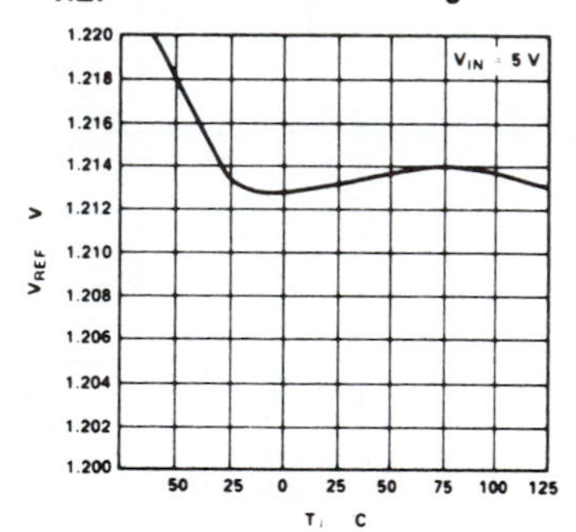


#### $I_{discharge}$ as a Function of $V_{IN}$

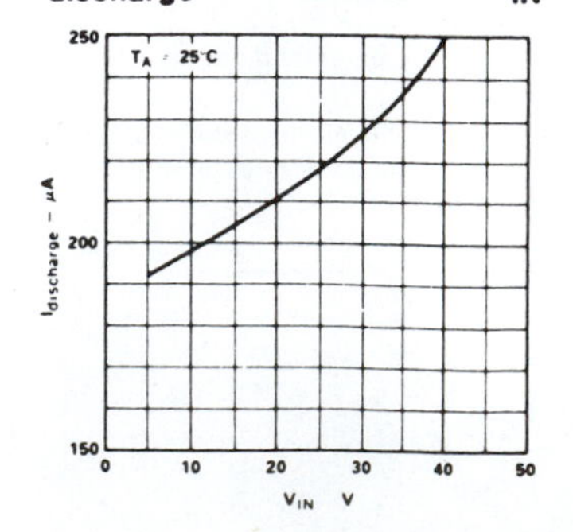

**Typical Performance Curves** (Cont.)

### $V_{Sense}$ as a Function of $V_{IN}$

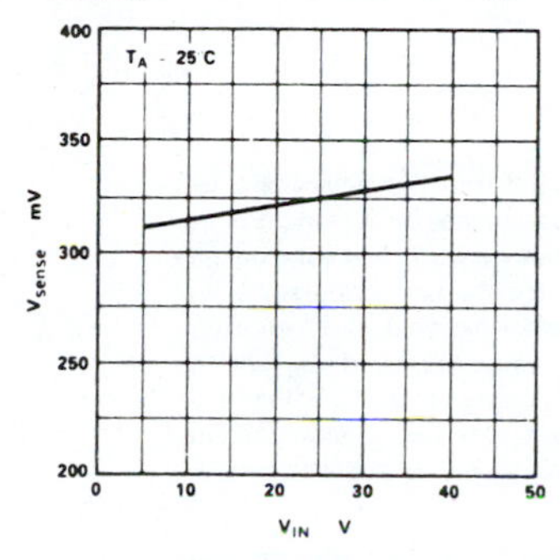


### Typical Pulse Width Modulator Application

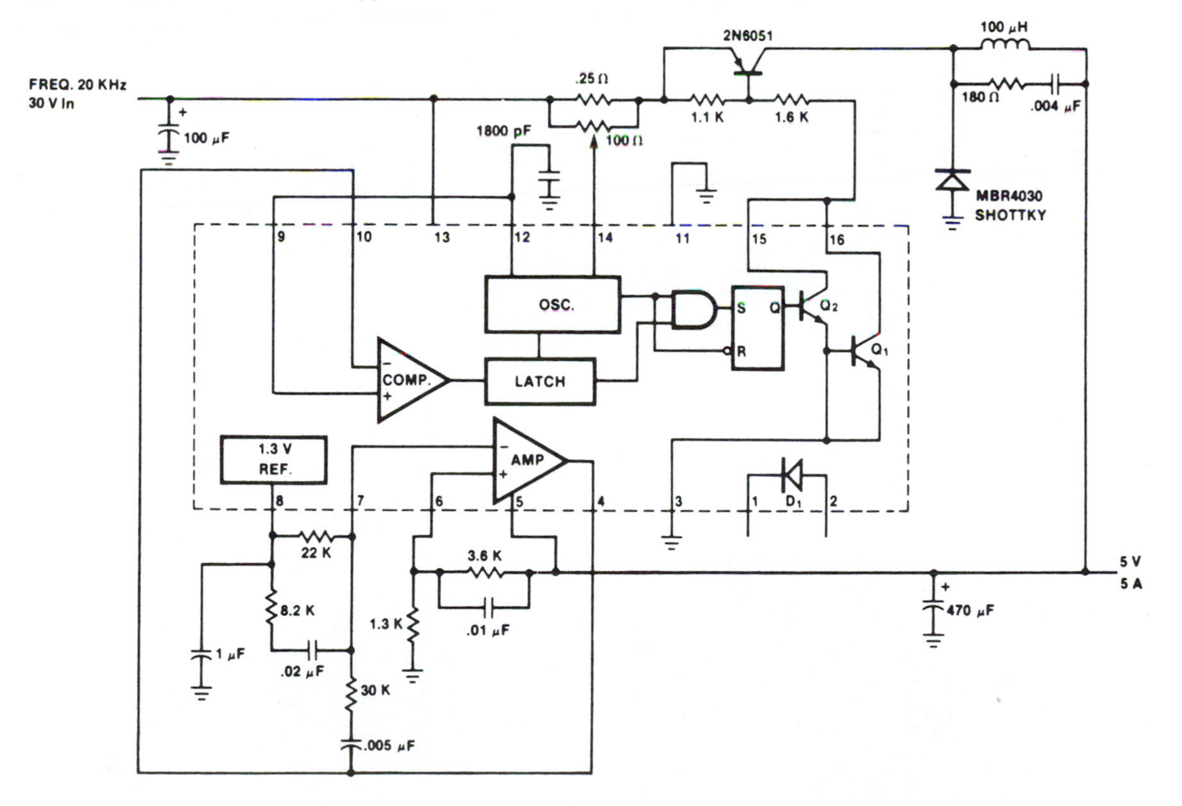

# Appendix E AN189 Balanced Modulator/ Demodulator Applications Using the MC1496/MC1596

**Linear Products**

## BALANCED MODULATOR/ DEMODULATOR APPLICATIONS USING MC1496/MC1596

The MC1496 is a monolithic transistor array arranged as a balanced modulator-demodulator. The device takes advantage of the excellent matching qualities of monolithic devices to provide superior carrier and signal rejection. Carrier suppressions of 50dB at 10MHz are typical with no external balancing networks required.

Applications include AM and suppressed carrier modulators, AM and FM demodulators, and phase detectors.

## THEORY OF OPERATION

As Figure 1 suggests, the topography includes three differential amplifiers. Internal connections are made such that the output becomes a product of the two input signals $V_C$ and $V_S$.

To accomplish this the differential pairs Q1 – Q2 and Q3 – Q4, with their cross-coupled collectors, are driven into saturation by the zero crossings of the carrier signal $V_C$. With a low level signal, $V_S$ driving the third differential amplifier Q5 – Q6, the output voltage will be full wave multiplication of $V_C$ and $V_S$. Thus for sine wave signals, $V_{OUT}$ becomes:

$$V_{OUT} = E_x E_y \left[ \cos(\omega x + \omega y)t + \cos(\omega x - \omega y)t \right] \quad (1)$$

As seen by equation (1) the output voltage will contain the sum and difference frequencies of the two original signals. In addition, with the carrier input ports being driven into saturation, the output will contain the odd harmonics of the carrier signals. (See Figure 4.)

## BIASING

Since the MC1496 was intended for a multitude of different functions as well as a myriad of supply voltages, the biasing techniques are specified by the individual application. This allows the user complete freedom to choose gain, current levels, and power supplies. The device can be operated with single-ended or dual supplies.

Internally provided with the device are two current sources driven by a temperature-compensated bias network. Since the transistor geometries are the same and since $V_{BE}$ matching in monolithic devices is excellent, the currents through $Q_7$ and $Q_8$ will be identical to the current set at Pin 5. Figures 2 and 3 illustrate typical biasing arrangements from split and single-ended supplies, respectively.

Of primary interest in beginning the bias circuitry design is relating available power supplies and desired output voltages to device requirements with a minimum of external components.

The transistors are connected in a cascode fashion. Therefore, sufficient collector voltage must be supplied to avoid saturation if linear operation is to be achieved. Voltages greater than 2V are sufficient in most applications.

Biasing is achieved with simple resistor divider networks as shown in Figure 3. This configuration assumes the presence of symmetrical supplies. Explaining the DC biasing technique is probably best accomplished by

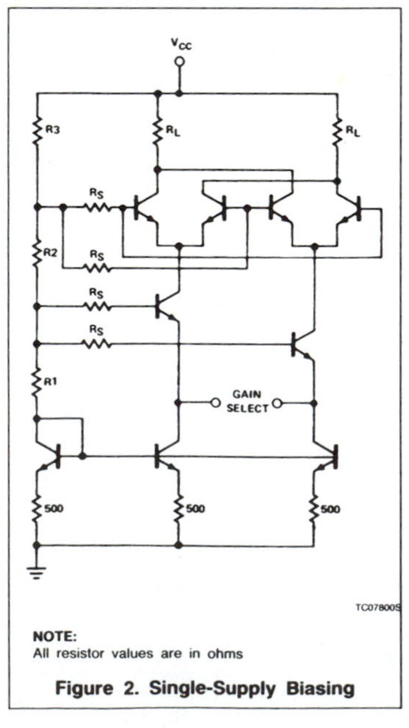


**NOTE:**
All resistor values are in ohms

**Figure 2. Single-Supply Biasing**

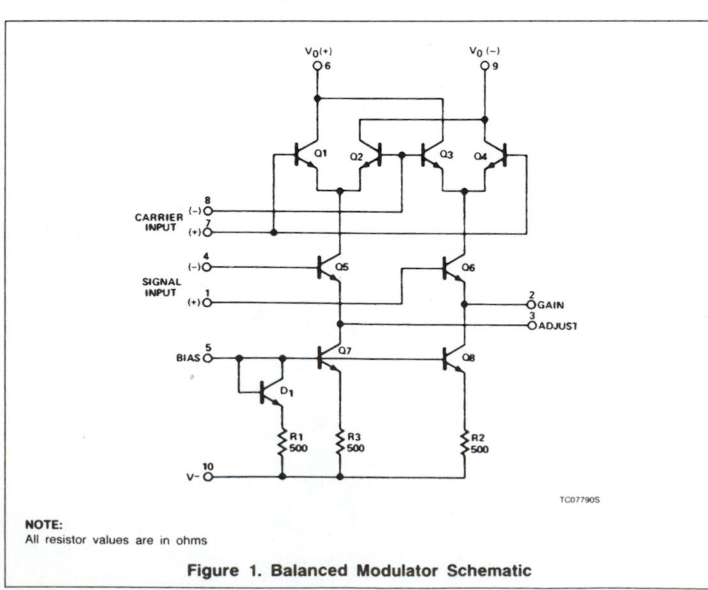


**NOTE:**
All resistor values are in ohms

**Figure 1. Balanced Modulator Schematic**

(Courtesy of Signetics Company)

an example. Thus, the initial assumptions and criteria are set forth:

1. Output swing greater than $4V_{P\text{-}P}$.
2. Positive and negative supplies of 6V are available.
3. Collector current is 2mA. It should be noted here that the collector output current is equal to the current set in the current sources.

As a matter of convenience, the carrier signal ports are referenced to ground. If desired, the modulation signal ports could be ground referenced with slight changes in the bias arrangement. With the carrier inputs at DC ground, the quiescent operating point of the outputs should be at one-half the total positive voltage or 3V for this case. Thus, a collector load resistor is selected which drops 3V at 2mA or 1.5kΩ. A quick check at this point reveals that with these loads and current levels the peak-to-peak output swing will be greater than 4V. It remains to set the current source level and proper biasing of the signal ports.

The voltage at Pin 5 is expressed by

$$V_{BIAS} = V_{BE} = 500 \times I_S$$

where $I_S$ is the current set in the current sources.

For the example $V_{BE}$ is 700mV at room temperature and the bias voltage at Pin 5 becomes 1.7V. Because of the cascode configuration, both the collectors of the current sources and the collectors of the signal transistors must have some voltage to operate properly. Hence, the remaining voltage of the negative supply (−6V + 1.7V = −4.3V) is split between these transistors by biasing the signal transistor bases at −2.15V.

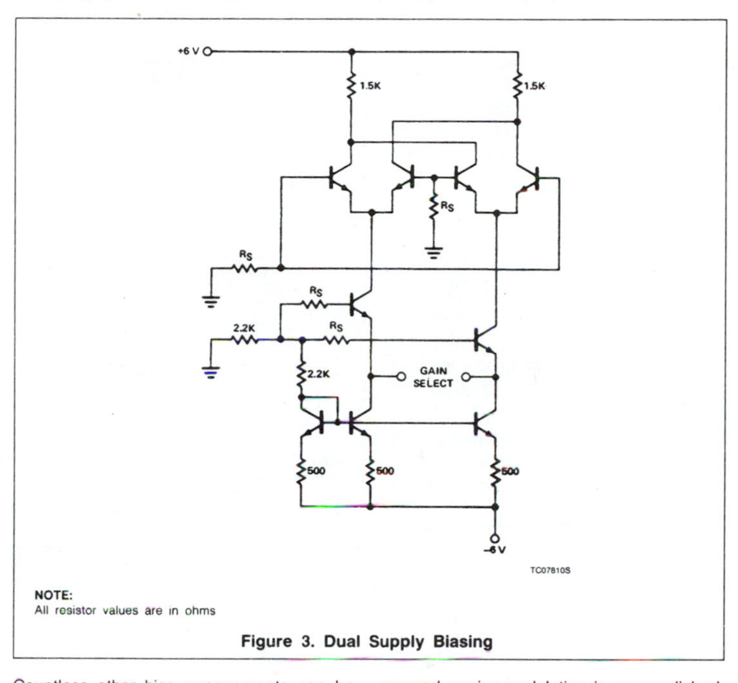


Figure 3. Dual Supply Biasing

Countless other bias arrangements can be used with other power supply voltages. The important thing to remember is that sufficient DC voltage is applied to each bias point to avoid collector saturation over the expected signal wings.

## BALANCED MODULATOR

In the primary application of balanced modulation, generation of double sideband suppressed carrier modulation is accomplished. Due to the balance of both modulation and carrier inputs, the output, as mentioned, contains the sum and difference frequencies while attenuating the fundamentals. Upper and lower sideband signals are the strongest signals present with harmonic sidebands being of diminishing amplitudes as characterized by Figure 4.

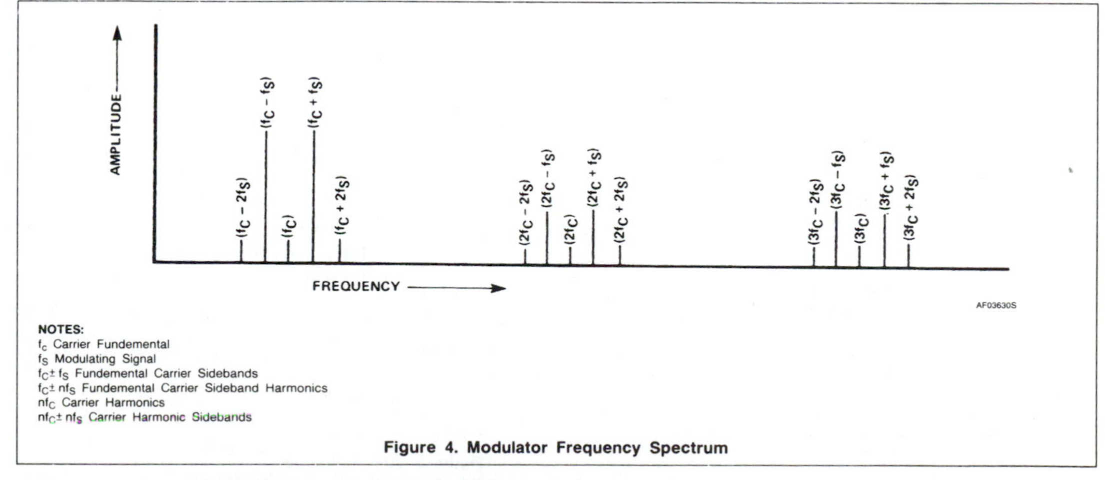


Figure 4. Modulator Frequency Spectrum

## Using the MC1496/MC1596 *Continued*

Gain of the 1496 is set by including emitter degeneration resistance located as $R_E$ in Figure 5. Degeneration also allows the maximum signal level of the modulation to be increased. In general, linear response defines the maximum input signal as

$$V_S \leqslant 15 \cdot R_E(\text{Peak})$$

and the gain is given by

$$A_{VS} = \frac{R_L}{R_E + 2r_e} \tag{2}$$

This approximation is good for high levels of carrier signals. Table 1 summarizes the gain for different carrier signals.

As seen from Table 1, the output spectrum suffers an amplitude increase of undesired sideband signals when either the modulation or carrier signals are high. Indeed, the modulation level can be increased if $R_E$ is increased without significant consequence. However, large carrier signals cause odd harmonic sidebands (Figure 4) to increase. At the same time, due to imperfections of the carrier waveforms and small imbalances of the device, the second harmonic rejection will be seriously degraded. Output filtering is often used with high carrier levels to remove all but the desired sideband. The filter removes unwanted signals while the high carrier level guards against amplitude variations and maximizes gain. Broadband modulators, without benefit of filters, are implemented using low carrier and modulation signals to maximize linearity and minimize spurious sidebands.

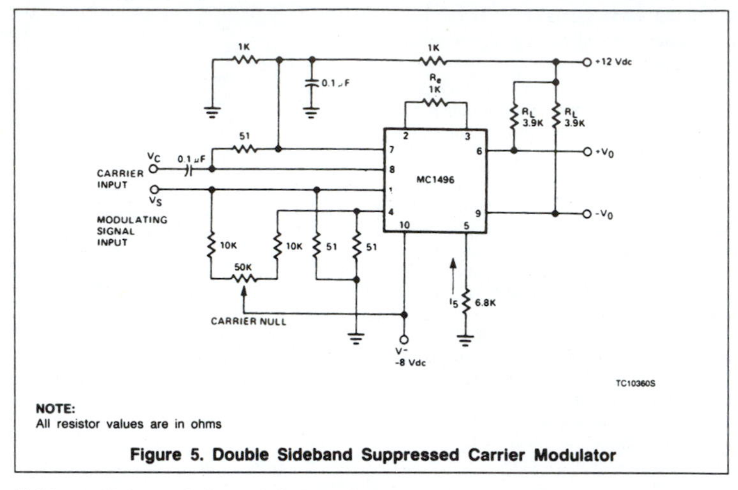


NOTE:
All resistor values are in ohms

**Figure 5. Double Sideband Suppressed Carrier Modulator**

**Table 1. Voltage Gain and Output Spectrum vs Input Signal**

| CARRIER INPUT SIGNAL ($V_C$) | APPROXIMATE VOLTAGE GAIN | OUTPUT SIGNAL FREQUENCY(S) |
|---|---|---|
| Low-level DC | $\dfrac{R_L V_C}{2(R_E + 2r_E)\left(\frac{KT}{q}\right)}$ | $f_M$ |
| High-level DC | $\dfrac{R_L}{R + 2r_e}$ | $f_M$ |
| Low-level AC | $\dfrac{R_L V_C(\text{rms})}{2\sqrt{2}\left(\frac{KT}{q}\right)(R_E + 2r_e)}$ | $f_C \pm f_M$ |
| High-level AC | $\dfrac{0.637R_L}{R_E + 2r_e}$ | $f_C \pm f_M$, $3f_C \pm f_M$, $5f_C \pm f_M$... |

### AM MODULATOR

The basic current of Figure 5 allows no carrier to be present in the output. By adding offset to the carrier differential pairs, controlled amounts of carrier appear at the output whose amplitude becomes a function of the modulation signal or AM modulation. As shown, the carrier null circuit is changed from Figure 5 to have a wider range so that wider control is achieved. All connections are shown in Figure 6.

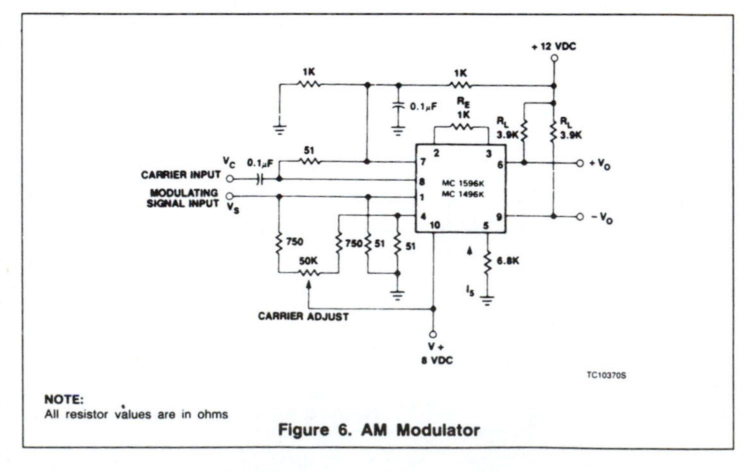


NOTE:
All resistor values are in ohms

**Figure 6. AM Modulator**

### AM DEMODULATION

As pointed out in Equation 1, the output of the balanced mixer is a cosine function of the angle between signal and carrier inputs. Further, if the carrier input is driven hard enough to provide a switching action, the output becomes a function of the input amplitude. Thus the output amplitude is maximum when there is 0° phase difference as shown in Figure 7.

Amplifying and limiting of the AM carrier is accomplished by IF gain block providing 55dB

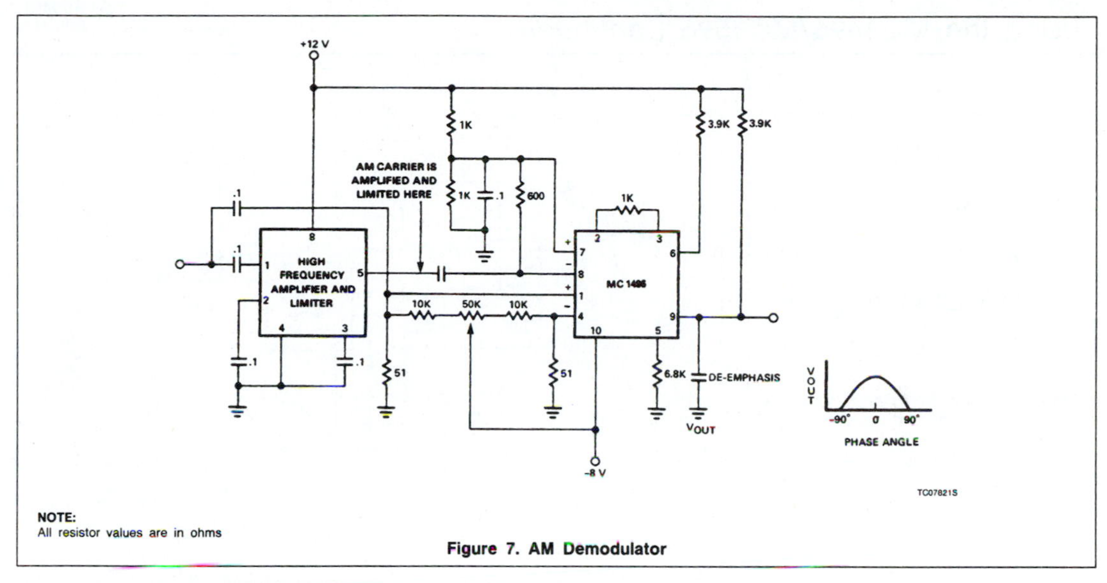


NOTE:
All resistor values are in ohms

Figure 7. AM Demodulator

of gain or higher with limiting of 400$\mu$V. The limited carrier is then applied to the detector at the carrier ports to provide the desired switching function. The signal is then demodulated by the synchronous AM demodulator (1496) where the carrier frequency is attenuated due to the balanced nature of the device. Care must be taken not to overdrive the signal input so that distortion does not appear in the recovered audio. Maximum conversion gain is reached when the carrier signals are in phase as indicated by the phase-gain relationship drawn in Figure 7.

Output filtering will also be necessary to remove high frequency sum components of the carrier from the audio signal.

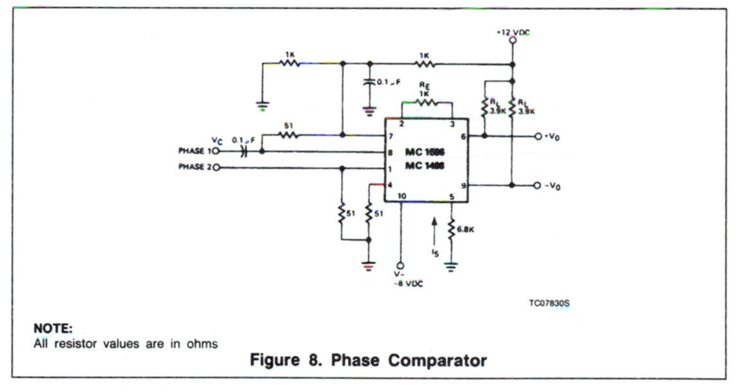


NOTE:
All resistor values are in ohms

Figure 8. Phase Comparator

## PHASE DETECTOR

The versatility of the balanced modulator or multiplier also allows the device to be used as a phase detector. As mentioned, the output of the detector contains a term related to the cosine of the phase angle. Two signals of equal frequency are applied to the inputs as per Figure 8. The frequencies are multiplied together producing the sum and difference frequencies. Equal frequencies cause the difference component to become DC while the undesired sum component is filtered out. The DC component is related to the phase angle by the graph of Figure 9. At 90° the cosine becomes zero, while being at maximum positive or maximum negative at 0° and 180°, respectively.

The advantage of using the balanced modulator over other types of phase comparators is the excellent linearity of conversion. This configuration also provides a conversion gain rather than a loss for greater resolution. Used in conjunction with a phase-locked loop, for instance, the balanced modulator provides a very low distortion FM demodulator.

## FREQUENCY DOUBLER

Very similar to the phase detector of Figure 8, a frequency doubler schematic is shown in Figure 10. Departure from Figure 8 is primarily the removal of the low-pass filter. The output then contains the sum component which is twice the frequency of the input, since both input signals are the same frequency.

## Using the MC1496/MC1596 *Continued*

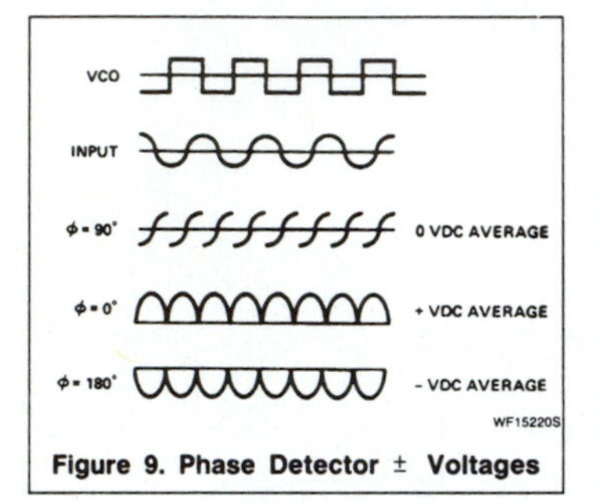


**Figure 9. Phase Detector ± Voltages**

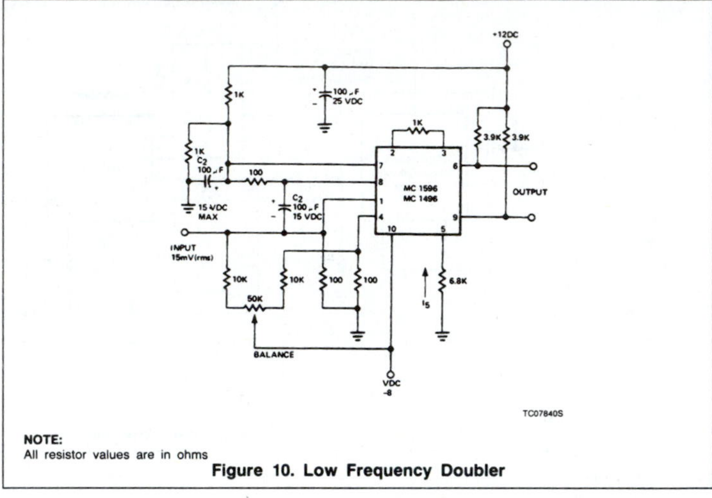


**NOTE:**
All resistor values are in ohms

**Figure 10. Low Frequency Doubler**

# Appendix F XR-2211 FSK Demodulator/Tone Decoder

## Features

- Wide Frequency Range — 0.01Hz to 300kHz
- Wide Supply Voltage Range — 4.5V to 20V
- DTL/TTL/ECL Logic Compatibility
- FSK Demodulation with Carrier-Detector
- Wide Dynamic Range — 2mV to $3V_{RMS}$
- Adjustable Tracking Range — ±1% to ±80%
- Excellent Temperature Stability — 20ppm/°C Typical

## Applications

- FSK Demodulation
- Data Synchronization
- Tone Decoding
- FM Detection
- Carrier Detection

## Description

The XR-2211 is a monolithic phase-locked loop (PLL) system especially designed for data communications. It is particularly well suited for FSK modem applications, and operates over a wide frequency range of 0.01Hz to 300kHz. It can accommodate analog signals between 2mV and 3V, and can interface with conventional DTL, TTL and ECL logic families. The circuit consists of a basic PLL for tracking an input signal frequency within the passband, a quadrature phase detector which provides carrier detection, and an FSK voltage comparator which provides FSK demodulation. External components are used to independently set carrier frequency, bandwidth, and output delay.

## Schematic Diagram

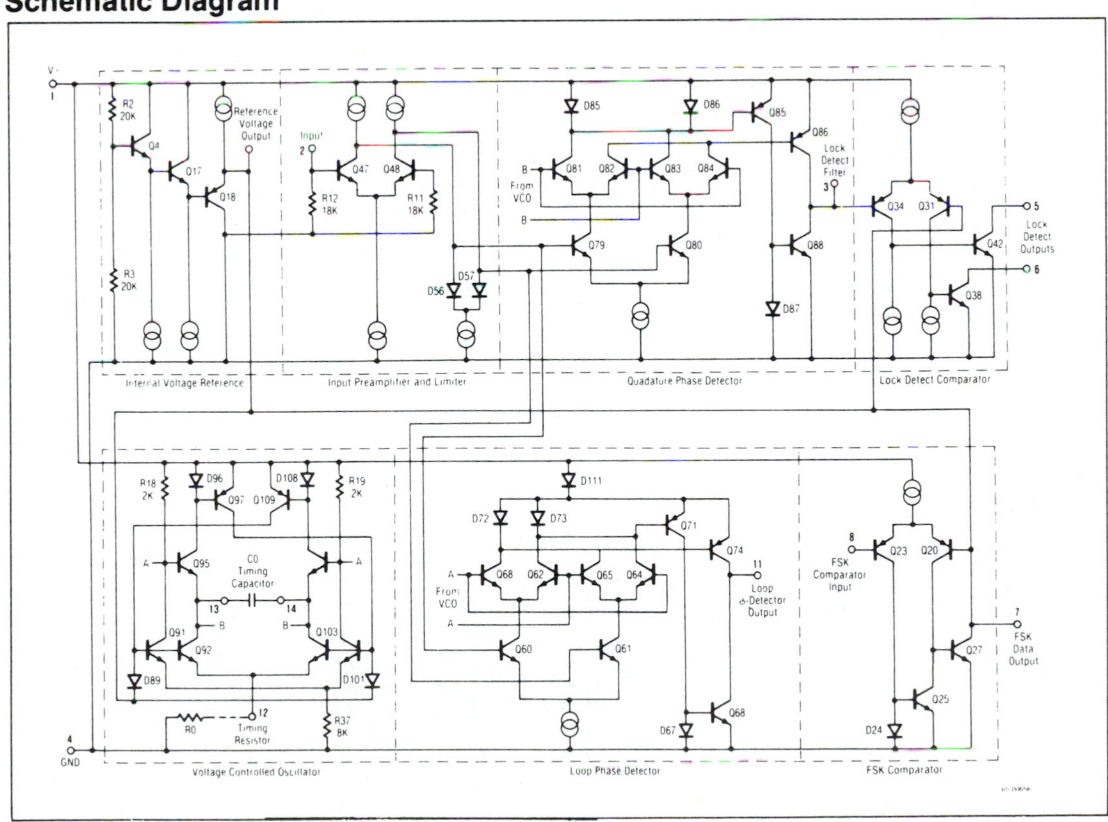


(Courtesy of Raytheon Company, Semiconductor Division)

## Connection Information

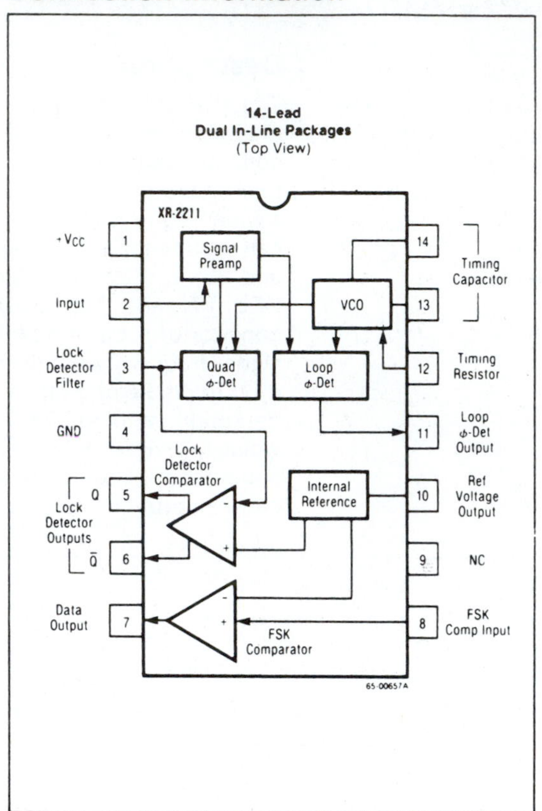


## Functional Block Diagram

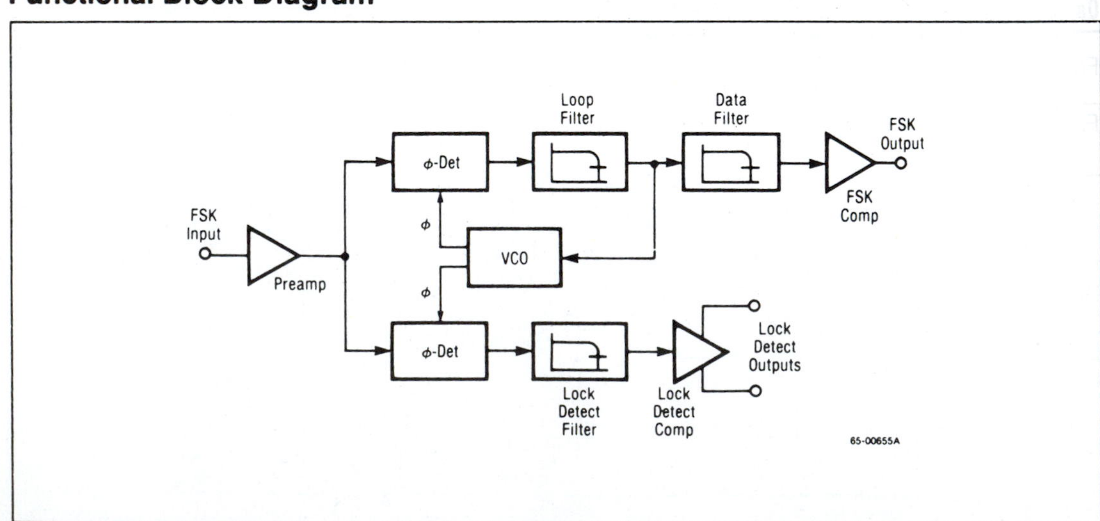

## Thermal Characteristics

| | 14-Lead Plastic DIP | 14-Lead Ceramic DIP |
|---|---|---|
| Max. Junction Temp. | 125°C | 175°C |
| Max. $P_D$ $T_A < 50°C$ | 468mW | 1042mW |
| Therm. Res. $\theta_{JC}$ | — | 50°C/W |
| Therm. Res. $\theta_{JA}$ | 160°C/W | 120°C/W |
| For $T_A > 50°C$ Derate at | 6.25mW per °C | 8.33mW per °C |

## Ordering Information

| Part Number | Package | Operating Temperature Range |
|---|---|---|
| XR-2211M | Ceramic | -55°C to +125°C |
| XR-2211N | Ceramic | -40°C to +85°C |
| XR-2211P | Plastic | -40°C to +85°C |
| XR-2211CN | Ceramic | 0°C to +75°C |
| XR-2211CP | Plastic | 0°C to +75°C |

## Absolute Maximum Ratings

Supply Voltage ...................... +20V
Input Signal Level .................. $3V_{RMS}$
Storage Temperature
Range .................. -65°C to +150°C
Operating Temperature Range
XR-2211CN/CP ............ 0°C to +75°C
XR-2211N/P ............ -40°C to +85°C
XR-2211M ............. -55°C to +125°C
Lead Soldering
Temperature (60 Sec) ............ 300°C

## Electrical Characteristics

Test Conditions $V^+ = +12V$, $T_A = +25°C$, $R0 = 30k\Omega$, $C0 = 0.033\mu F$.
See Figure 1 for component designations.

| Parameters | Conditions | XR-2211/M | | | XR-2211C | | | Units |
|---|---|---|---|---|---|---|---|---|
| | | Min | Typ | Max | Min | Typ | Max | |
| **General** | | | | | | | | |
| Supply Voltage | | 4.5 | | 20 | 4.5 | | 20 | V |
| Supply Current | $R0 \geq 10k\Omega$ | | 4.0 | 9.0 | | 5.0 | 11.0 | mA |
| **Oscillator** | | | | | | | | |
| Frequency Accuracy | Deviation from $f_0 = 1/R0C0$ | | ±1.0 | ±3.0 | | ±1.0 | | % |
| Frequency Stability Temperature Coeffient | $R1 = \infty$ | | ±20 | ±50 | | ±20 | | ppm/°C |
| Power Supply Rejection | $V^+ = 12 \pm 1V$ | | 0.05 | 0.5 | | 0.05 | | %/V |
| | $V^+ = 5 \pm 0.5V$ | | 0.2 | | | 0.2 | | %/V |
| Upper Frequency Limit | $R0 = 8.2k\Omega$, $C0 = 400pF$ | 100 | 300 | | | 300 | | kHz |
| Lowest Practical Operating Frequency | $R0 = 2M\Omega$ $C0 = 50\ \mu F$ | | | 0.01 | | 0.01 | | Hz |
| Timing Resistor, R0 Operating Range | | 5.0 | | 2000 | 5.0 | | 2000 | kΩ |
| Recommended Range | | 15 | | 100 | 15 | | 100 | kΩ |

## Electrical Characteristics (Cont'd)

Test Conditions $V^+ = +12V$, $T_A = +25°C$, $R0 = 30k\Omega$, $C0 = 0.033\mu F$.
See Figure 1 for component designations.

| Parameters | Conditions | XR-2211/M | | | XR-2211C | | | Units |
|---|---|---|---|---|---|---|---|---|
| | | Min | Typ | Max | Min | Typ | Max | |
| **Loop Phase Detector** | | | | | | | | |
| Peak Output Current | Meas. at Pin 11 | ±150 | ±200 | ±300 | ±100 | ±200 | ±300 | μA |
| Output Offset Current | | | ±1.0 | | | ±2.0 | | μA |
| Output Impedance | | | 1.0 | | | 1.0 | | MΩ |
| Maximum Swing | Ref. to Pin 10 | ±4.0 | ±5.0 | | ±4.0 | ±5.0 | | V |
| **Quadrature Phase Detector** | | | | | | | | |
| Peak Output Current | Meas. at Pin 3 | 100 | 150 | | | 150 | | μA |
| Output Impedance | | | 1.0 | | | 1.0 | | MΩ |
| Maximum Swing | | | 11 | | | 11 | | Vp-p |
| **Input Preamp** | | | | | | | | |
| Input Impedance | Meas. at Pin 2 | | 20 | | | 20 | | kΩ |
| Input Signal Voltage Required to Cause Limiting | | | 2.0 | 10 | | 2.0 | | $mV_{RMS}$ |
| **Voltage Comparator** | | | | | | | | |
| Input Impedance | Meas. at Pins 3 & 8 | | 2.0 | | | 2.0 | | MΩ |
| Input Bias Current | | | 100 | | | 100 | | nA |
| Voltage Gain | $R_L = 5.1k\Omega$ | 55 | 70 | | 55 | 70 | | dB |
| Output Voltage Low | $I_C = 3mA$ | | 300 | | | 300 | | mV |
| Output Leakage Current | $V_O = 12V$ | | 0.01 | | | 0.01 | | μA |
| **Internal Reference** | | | | | | | | |
| Voltage Level | Meas. at Pin 10 | 4.9 | 5.3 | 5.7 | 4.75 | 5.3 | 5.85 | V |
| Output Impedance | | | 100 | | | 100 | | Ω |

The information contained in this data sheet has been carefully compiled; However, it shall not by implication or otherwise become part of the terms and conditions of any subsequent sale. Raytheon's liability shall be determined solely by its standard terms and conditions of sale. No representation as to application or use or that the circuits are either licensed or free from patent infringement is intended or implied. Raytheon reserves the right to change the circuitry and other data at any time without notice and assumes no liability for inadvertent errors.

## Description of Circuit Controls

**Signal Input** (Pin 2)
Signal is ac coupled to this terminal. The internal impedance at pin 2 is 20kΩ. Recommended input signal level is in the range of $10mV_{RMS}$ to $3V_{RMS}$.

**Quadrature Phase Detector Output** (Pin 3)
This is the high-impedance output of quadrature phase detector, and is internally connected to the input of lock-detect voltage-comparator. In tone-detection applications, pin 3 is connected to ground through a parallel combination of $R_D$ and $C_D$ (see Figure 1) to elimate the chatter at lock-detect outputs. If this tone-detect section is not used, pin 3 can be left open circuited.

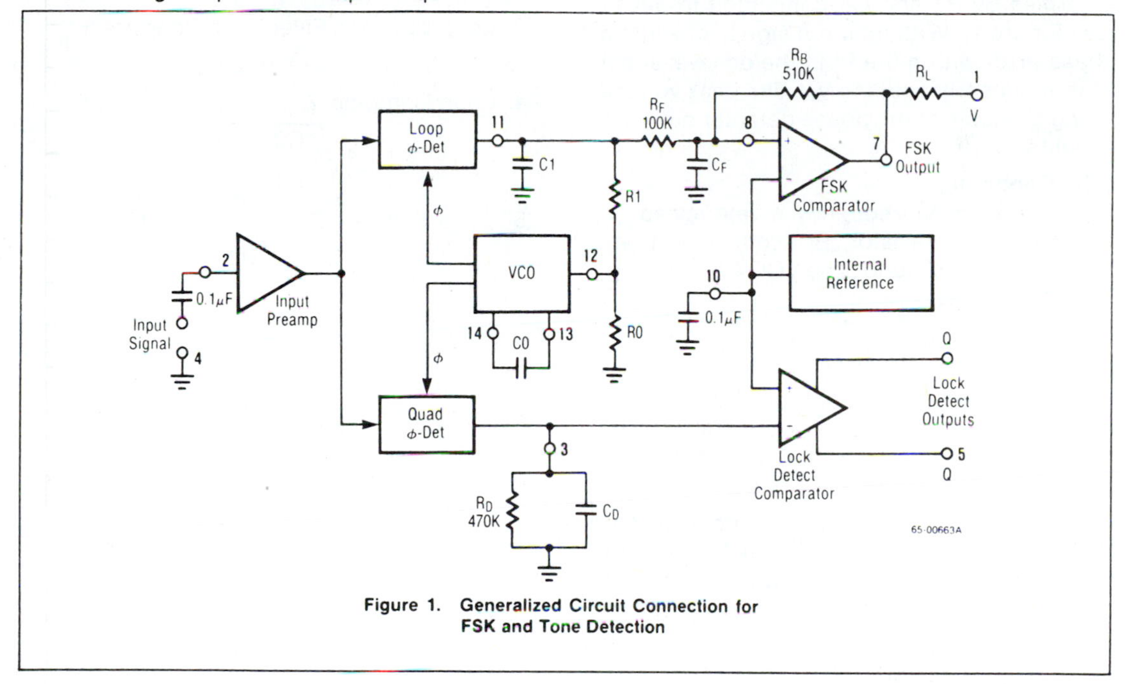


**Figure 1. Generalized Circuit Connection for FSK and Tone Detection**

**Lock-Detect Output, Q** (Pin 5)
The output at pin 5 is at "high" state when the PLL is out of lock and goes to "low" or conducting state when the PLL is locked. It is an open-collector type output and requires a pull-up resistor, $R_L$, to $V^+$ for proper operation. At "low" state, it can sink up to 5mA of load current.

**Lock-Detect Complement, $\overline{Q}$** (Pin 6)
The output at pin 6 is the logic complement of the lock-detect output at pin 5. This output is also an open-collector type stage which can sink 5mA of load current at low or "on" state.

**FSK Data Output** (Pin 7)
This output is an open-collector logic stage which requires a pull-up resistor, $R_L$, to $V^+$ for proper operation. It can sink 5mA of load current. When decoding FSK signals, FSK data output is at "high" or off state for low input frequency; and at "low" or on state for high input frequency. If no input signal is present, the logic state at pin 7 is indeterminate.

**FSK Comparator Input** (Pin 8)
This is the high-impedance input to the FSK voltage comparator. Normally, an FSK post-detection or data filter is connected between this terminal and the PLL phase-detector output (pin 11). This data filter is formed by $R_F$ and $C_F$ of Figure 1. The threshold voltage of the comparator is set by the internal reference voltage, $V_R$, available at pin 10.

**Reference Voltage, $V_R$** (Pin 10)
This pin is internally biased at the reference voltage level, $V_R$; $V_R = V^+/2 - 650mV$. The dc

voltage level at this pin forms an internal reference for the voltage levels at pins 3, 8, 11 and 12. Pin 10 must be bypassed to ground with a 0.1μF capacitor, for proper operation of the circuit.

**Loop Phase Detector Output** (Pin 11)
This terminal provides a high-impedance output for the loop phase-detector. The PLL loop filter is formed by R1 and C1 connected to pin 11 (see Figure 1). With no input signal, or with no phase-error within the PLL, the dc level at pin 11 is very nearly equal to $V_R$. The peak voltage swing available at the phase detector output is equal to $\pm V_R$.

**VCO Control Input** (Pin 12)
VCO free-running frequency is determined by external timing resistor, R0, connected from this terminal to ground. The VCO free-running frequency, $f_O$, is:

$$f_O = \frac{1}{R0C0} \text{ Hz}$$

where C0 is the timing capacitor across pins 13 and 14. For optimum temperature stability, R0 must be in the range of 10kΩ to 100kΩ (see Typical Electrical Characteristics).

This terminal is a low-impedance point, and is internally biased at a dc level equal to $V_R$. The maximum timing current drawn from pin 12 must be limited to ≤ 3mA for proper operation of the circuit.

**VCO Timing Capacitor** (Pins 13 and 14)
VCO frequency is inversely proportional to the external timing capacitor, C0, connected across these terminals. C0 must non-polar, and in the range of 200pF to 10μF.

**VCO Frequency Adjustment**
VCO can be fine-tuned by connecting a potentiometer, $R_X$, in series with R0 at pin 12 (see Figure 2).

**VCO Free-Running Frequency, $f_O$**
The XR-2211 does not have a separate VCO output terminal. Instead, the VCO outputs are internally connected to the phase-detector sections of the circuit. However, for set-up or adjustment purposes, VCO free-running frequency can be measured at pin 3 (with $C_D$ disconnected), with no input and with pin 2 shorted to pin 10.

## Design Equations

See Figure 1 for Definitions of Components.

1. VCO Center Frequency, $f_O$:

$$f_O = 1/R0C0\text{Hz}$$

2. Internal Reference Voltage, $V_R$ (measured at pin 10):

$$V_R = V^+/2 - 650\text{mV}$$

3. Loop Lowpass Filter Time Constant, $\tau$:

$$\tau = R1C1$$

4. Loop Damping, $\zeta$:

$$\zeta = 1/4\sqrt{\frac{C0}{C1}}$$

5. Loop Tracking Bandwidth, $\pm\Delta f/f_O$:

$$\Delta f/f_O = R0/R1$$

$\Delta f/f_0 = R0/R1$

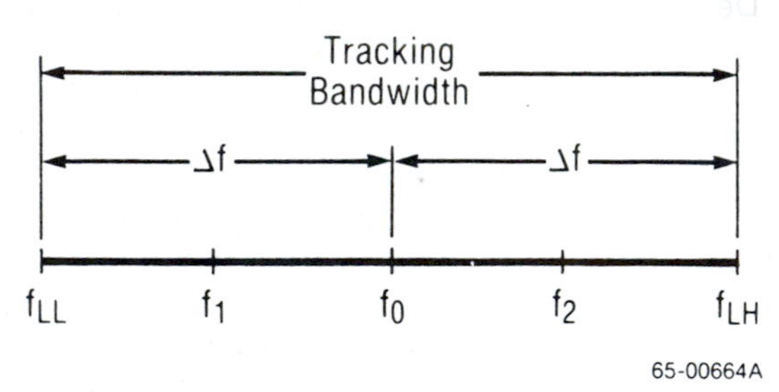


6. FSK Data Filter Time Constant, $\tau_F$:

$$\tau_F = R_F C_F$$

7. Loop Phase Detector Conversion Gain, $K_\phi$: ($K_\phi$ is the differential dc voltage across pins 10 and 11, per unit of phase error at phase-detector input):

$$K_\phi = -2V_R/\pi\text{Volts/Radian}$$

8. VCO Conversion Gain, K0 is the amount of change in VCO frequency, per unit of dc voltage change at pin 11):

$$K0 = -1/V_R\, C0R1 \text{ Hz/Volt}$$

9. Total Loop Gain, $K_T$:

$$K_T = 2\pi K_\phi K0 = 4/C0R1\text{Rad/Sec/Volt}$$

10. Peak Phase-Detector Current, $I_A$:

$$I_A = V_R \text{ (Volts)}/25\text{mA}$$

**Linear FM Detection**

The XR-2211 can be used as a linear FM detector for a wide range of analog communications and telemetry applications. The recommended circuit connection for the application is shown in Figure 5. The demodulated output is taken from the loop phase detector output (pin 11), through a post detection filter made up of $R_F$ and $C_F$, and an external buffer amplifier. This buffer amplifier is necessary because of the high impedance output at pin 11. Normally, a non-inverting unity gain op amp can be used as a buffer amplifier, as shown in Figure 5.

The FM detector gain, i.e., the output voltage change per unit of FM deviation, can be given as:

$$V_{OUT} = R1\ V_R/100\ R0 \text{ Volts/\% deviation}$$

where $V_R$ is the internal reference voltage. ($V_R = V^+/2 - 650mV$). For the choice of external components R1, R0, $C_D$, C1 and $C_F$, see section on Design Equations.

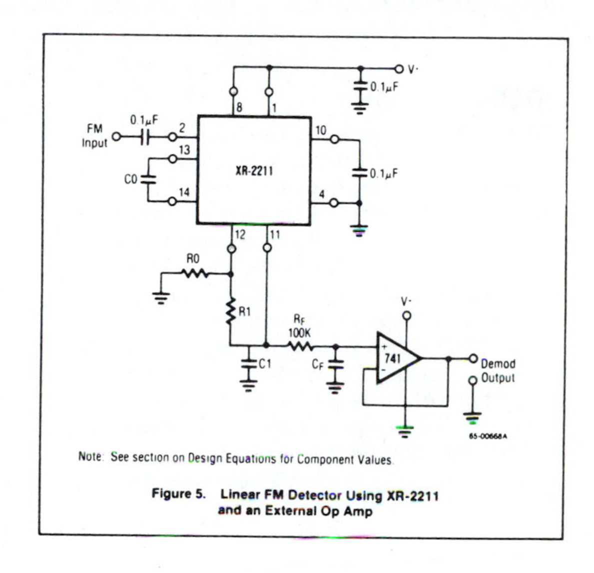


**Figure 5. Linear FM Detector Using XR-2211 and an External Op Amp**

# Appendix G NE/SE 565 Phase-Locked Loop

## DESCRIPTION

The SE/NE565 Phase-Locked Loop (PLL) is a self-contained, adaptable filter and demodulator for the frequency range from 0.001Hz to 500kHz. The circuit comprises a voltage-controlled oscillator of exceptional stability and linearity, a phase comparator, an amplifier and a low-pass filter as shown in the block diagram. The center frequency of the PLL is determined by the free-running frequency of the VCO; this frequency can be adjusted externally with a resistor or a capacitor. The low-pass filter, which determines the capture characteristics of the loop, is formed by an internal resistor and an external capacitor.

## APPLICATIONS

- **Frequency shift keying**
- **Modems**
- **Telemetry receivers**
- **Tone decoders**
- **SCA receivers**
- **Wideband FM discriminators**
- **Data synchronizers**
- **Tracking filters**
- **Signal restoration**
- **Frequency multiplication & division**

## FEATURES

- **Highly stable center frequency (200ppm/°C typ.)**
- **Wide operating voltage range (±6 to ±12 volts)**
- **Highly linear demodulated output (0.2% typ.)**
- **Center frequency programming by means of a resistor or capacitor, voltage or current**
- **TTL and DTL compatible square-wave output; loop can be opened to insert digital frequency divider**
- **Highly linear triangle wave output**
- **Reference output for connection of comparator in frequency discriminator**
- **Bandwidth adjustable from < ±1% to > ±60%**
- **Frequency adjustable over 10 to 1 range with same capacitor**

## PIN CONFIGURATIONS

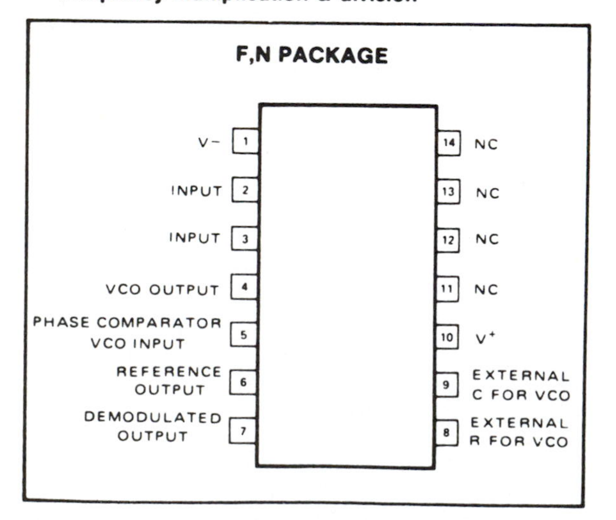


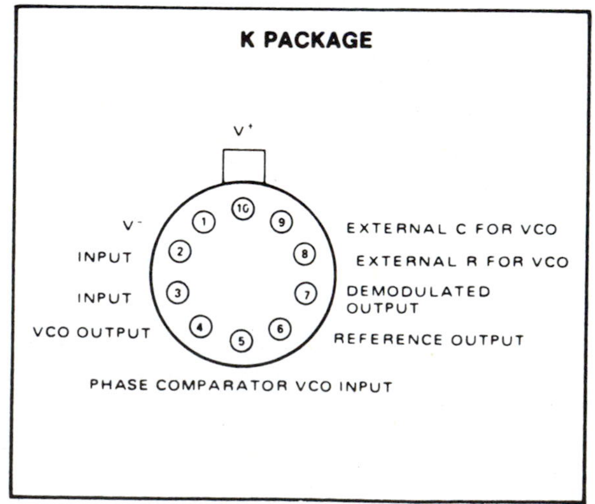


(Courtesy of Signetics Company)

## BLOCK DIAGRAM

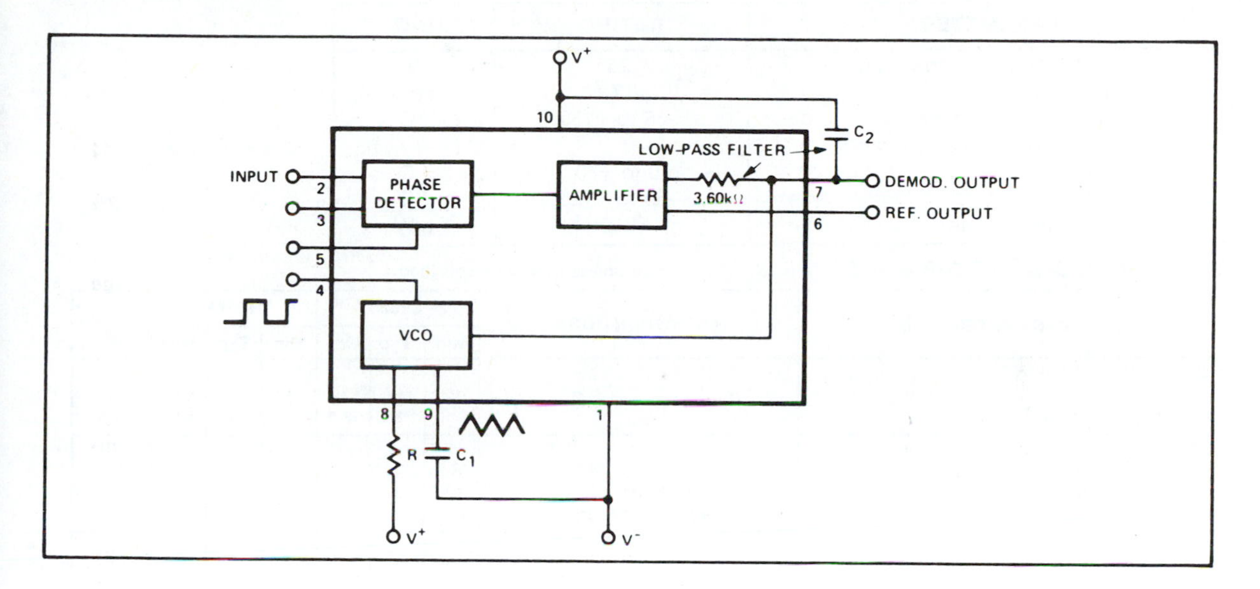


## EQUIVALENT SCHEMATIC

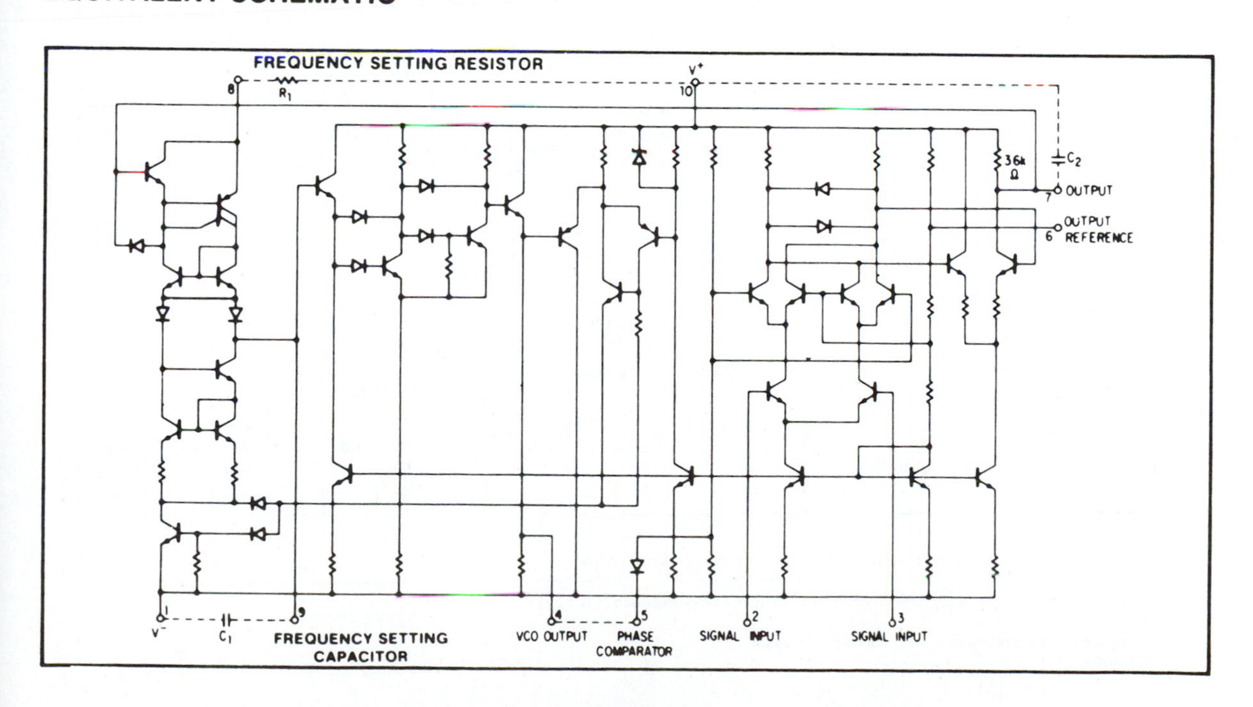

## ABSOLUTE MAXIMUM RATINGS $T_A = 25°C$ unless otherwise specified.

| PARAMETER | RATING | UNIT |
|---|---|---|
| Maximum operating voltage | 26 | V |
| Input voltage | 3 | Vp-p |
| Storage temperature | -65 to +150 | °C |
| Operating temperature range | | |
| NE565 | 0 to +70 | °C |
| SE565 | -55 to +125 | °C |
| Power dissipation | 300 | mW |

## ELECTRICAL CHARACTERISTICS $T_A = 25°C$, $V_{CC} = \pm 6V$ unless otherwise specified.

| PARAMETER | TEST CONDITIONS | SE565 | | | NE565 | | | UNIT |
|---|---|---|---|---|---|---|---|---|
| | | Min | Typ | Max | Min | Typ | Max | |
| SUPPLY REQUIREMENTS | | | | | | | | |
| Supply voltage | | 12 | | ±12 | ±6 | | ±12 | V |
| Supply current | | | 8 | 12.5 | | 8 | 12.5 | mA |
| INPUT CHARACTERISTICS | | | | | | | | |
| Input impedance[1] | | 7 | 10 | | 5 | 10 | | kΩ |
| Input level required for tracking | $f_o$ = 50kHz, ±10% frequency deviation | 10 | 1 | | 10 | 1 | | mVrms |
| VCO CHARACTERISTICS | | | | | | | | |
| Center frequency | | | | | | | | |
| Maximum value | $C_1$ = 2.7pF | 300 | 500 | | | 500 | | kHz |
| Distribution[2] | Distribution taken about $f_o$ = 50kHz, $R_1$ = 5.0kΩ, $C_1$ = 1200pF | -10 | 0 | +10 | -30 | 0 | +30 | % |
| Drift with temperature | $f_o$ = 50kHz | | 200 | | | 300 | | ppm/°C |
| Drift with supply voltage | $f_o$ = 50kHz, $V_{CC}$ = ±6 to ±7 volts | | 0.1 | 1.0 | | 0.2 | 1.5 | %/V |
| Triangle wave | | | | | | | | |
| Output voltage level | | 1.9 | 0 | | 1.9 | 0 | | V |
| Amplitude | | | 2.4 | 3 | | 2.4 | 3 | Vp-p |
| Linearity | | | 0.2 | | | 0.5 | | % |
| Square wave | | | | | | | | |
| Logical "1" output voltage | $f_o$ = 50kHz | +4.9 | +5.2 | | +4.9 | +5.2 | | V |
| Logical "0" output voltage | $f_o$ = 50kHz | | -0.2 | +0.2 | | -0.2 | +0.2 | V |
| Duty cycle | $f_o$ = 50kHz | 45 | 50 | 55 | 40 | 50 | 60 | % |
| Rise time | | | 20 | 100 | | 20 | | ns |
| Fall time | | | 50 | 200 | | 50 | | ns |
| Output current (sink) | | 0.6 | 1 | | 0 6 | 1 | | mA |
| Output current (source) | | 5 | 10 | | 5 | 10 | | mA |
| DEMODULATED OUTPUT CHARACTERISTICS | | | | | | | | |
| Output voltage level | Measured at pin 7 | 4.25 | 4.5 | 4.75 | 4.0 | 4.5 | 5.0 | V |
| Maximum voltage swing[3] | | | 2 | | | 2 | | Vp-p |
| Output voltage swing | ±10% frequency deviation | 250 | 300 | | 200 | 300 | | mVp-p |
| Total harmonic distortion | | | 0.2 | 0.75 | | 0.4 | 1.5 | % |
| Output impedance[4] | | | 3.6 | | | 3.6 | | kΩ |
| Offset voltage (V6-V7) | | | 30 | 100 | | 50 | 200 | mV |
| Offset voltage vs temperature (drift) | | | 50 | | | 100 | | μV/°C |
| AM rejection | | 30 | 40 | | | 40 | | dB |

NOTES

1. Both input terminals (pins 2 and 3) must receive identical dc bias. This bias may range from 0 volts to -4 volts.
2. The external resistance for frequency adjustment (R1) must have a value between 2kΩ and 20kΩ.
3. Output voltage swings negative as input frequency increases.
4. Output not buffered.

# Appendix H NE/SE 567 Tone Decoder/Phase-Locked Loop

## DESCRIPTION

The SE/NE567 tone and frequency decoder is a highly stable phase-locked loop with synchronous AM lock detection and power output circuitry. Its primary function is to drive a load whenever a sustained frequency within its detection band is present at the self-biased input. The bandwidth center frequency, and output delay are independently determined by means of four external components.

## FEATURES

- **Wide frequency range (.01Hz to 500kHz)**
- **High stability of center frequency**
- **Independently controllable bandwidth (up to 14 percent)**
- **High out-band signal and noise rejection**
- **Logic-compatible output with 100mA current sinking capability**
- **Inherent immunity to false signals**
- **Frequency adjustment over a 20 to 1 range with an external resistor**
- **Military processing available**

## APPLICATIONS

- **Touch Tone® decoding**
- **Carrier current remote controls**
- **Ultrasonic controls (remote TV, etc.)**
- **Communications paging**
- **Frequency monitoring and control**
- **Wireless intercom**
- **Precision oscillator**

## PIN CONFIGURATIONS

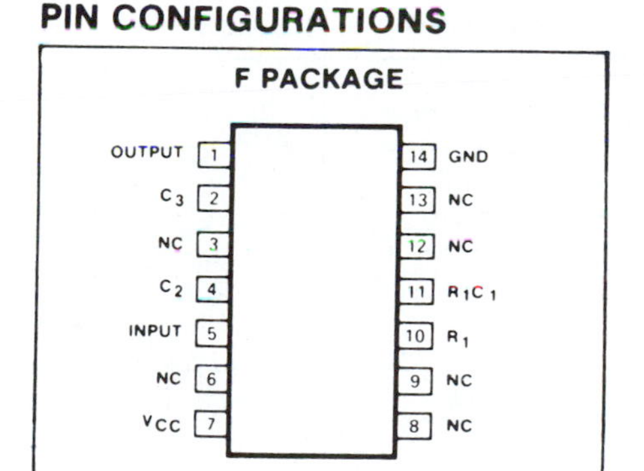


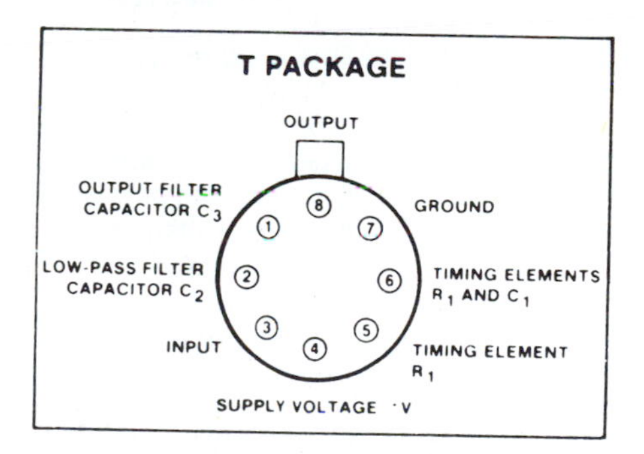


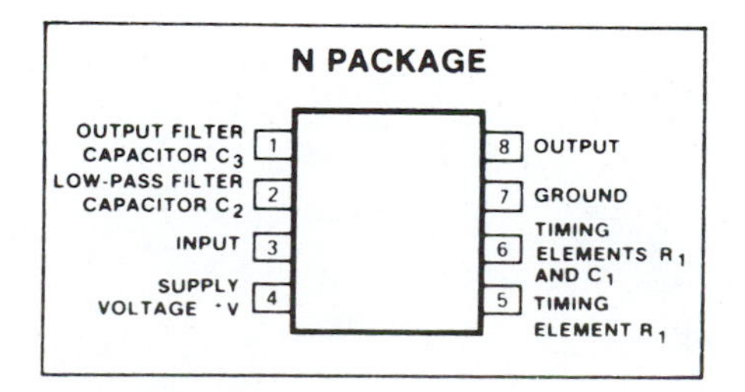


(Courtesy of Signetics Company)

## BLOCK DIAGRAM

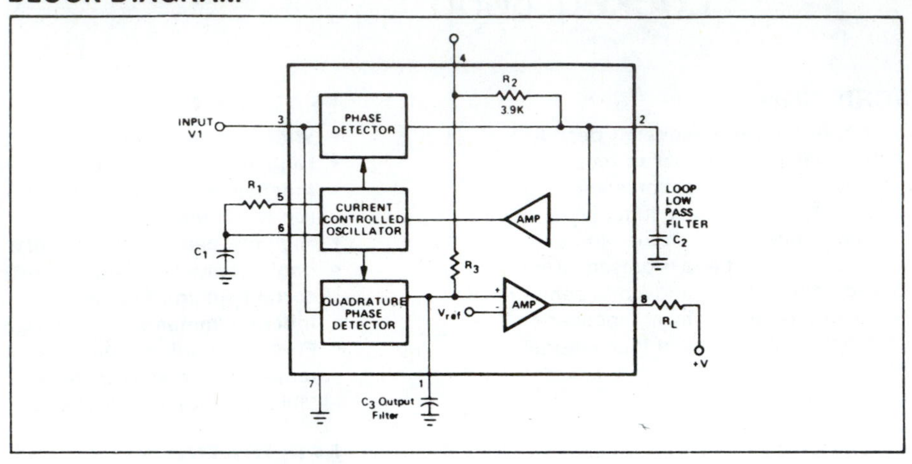


## ABSOLUTE MAXIMUM RATINGS

| PARAMETER | RATING | UNIT |
|---|---|---|
| Operating temperature | | |
| NE567 | 0 to +70 | °C |
| SE567 | -55 to +125 | °C |
| Operating voltage | 10 | V |
| Positive voltage at input | 0.5 + $V_S$ | V |
| Negative voltage at input | -10 | Vdc |
| Output voltage (collector of output transistor) | 15 | Vdc |
| Storage temperature | -65 to +150 | °C |
| Power dissipation | 300 | mW |

## EQUIVALENT SCHEMATIC

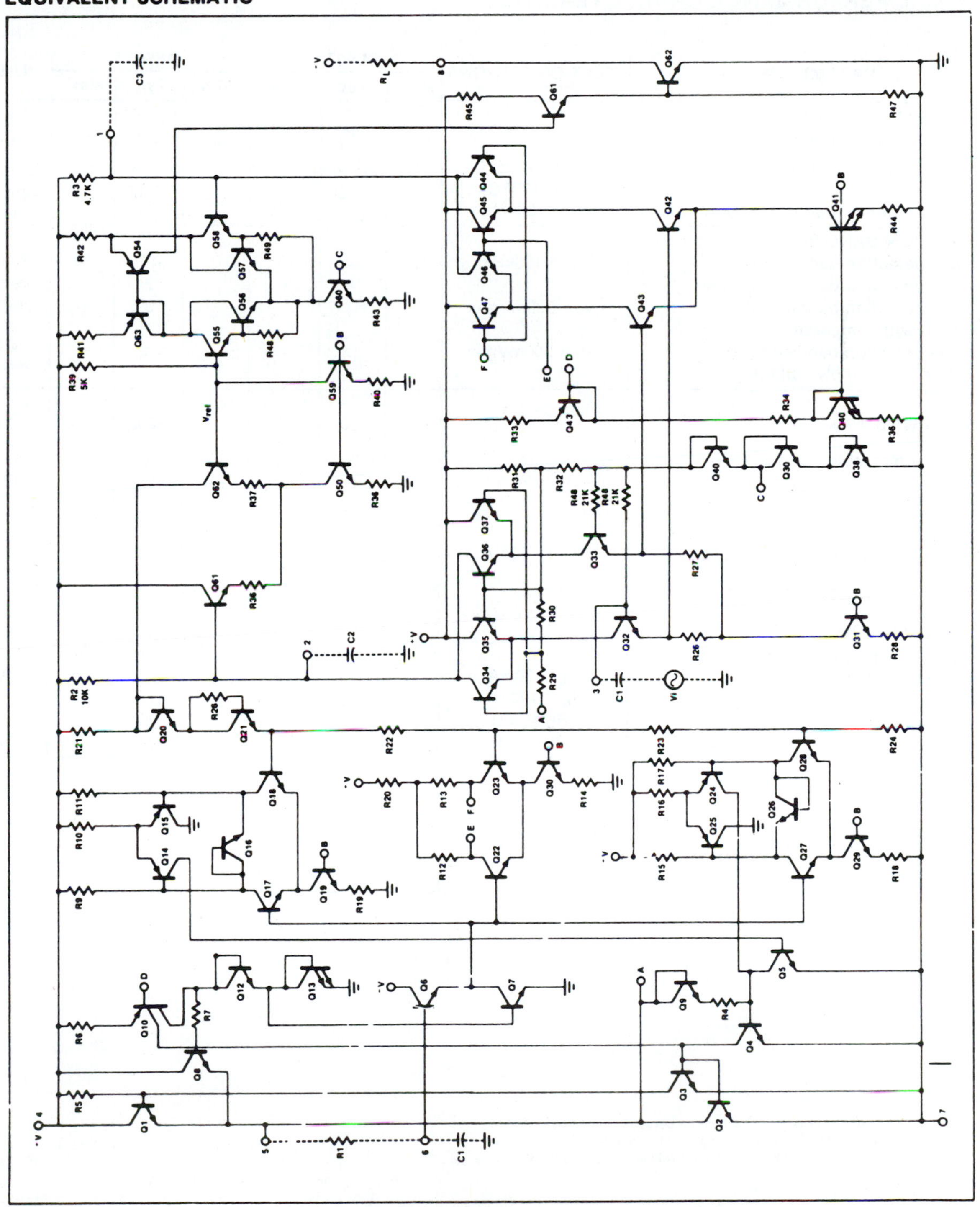

(H-3)

## TYPICAL PERFORMANCE CHARACTERISTICS

| PARAMETER | TEST CONDITIONS | SE567 | | | NE567 | | | UNIT |
|---|---|---|---|---|---|---|---|---|
| | | Min | Typ | Max | Min | Typ | Max | |
| CENTER FREQUENCY[1] | | | | | | | | |
| Highest center frequency ($f_o$) | | 100 | 500 | | 100 | 500 | | kHz |
| Center frequency stability[2] | -55 to +125°C | | 35±140 | | | 35±140 | | ppm/°C |
| | 0 to +70°C | | 35±60 | | | 35±60 | | ppm/°C |
| Center frequency shift with supply voltage | $f_o$ = 100kHz | | 0.5 | 1 | | 0.7 | 2 | %/V |
| DETECTION BANDWIDTH | | | | | | | | |
| Largest detection bandwidth | $f_o$ = 100kHz | 12 | 14 | 16 | 10 | 14 | 18 | % of $f_o$ |
| Largest detection bandwidth skew | | | 1 | 2 | | 2 | 3 | % of $f_o$ |
| Largest detection bandwidth—variation with temperature | $V_i$ = 300mVrms | | ±0.1 | | | ±0.1 | | %/°C |
| Largest detection bandwidth—variation with supply voltage | $V_i$ = 300mVrms | | ±2 | | | ±2 | | %/V |
| INPUT | | | | | | | | |
| Input resistance | | | 20 | | | 20 | | kΩ |
| Smallest detectable input voltage ($V_i$) | $I_L$ = 100mA, $f_i = f_o$ | | 20 | 25 | | 20 | 25 | mVrms |
| Largest no-output input voltage | $I_L$ = 100mA, $f_i = f_o$ | 10 | 15 | | 10 | 15 | | mVrms |
| Greatest simultaneous outband signal to inband signal ratio | | | +6 | | | +6 | | dB |
| Minimum input signal to wideband noise ratio | $B_n$ = 140kHz | | -6 | | | -6 | | dB |
| OUTPUT | | | | | | | | |
| Fastest on-off cycling rate | | | $f_o$/20 | | | $f_o$/20 | | |
| "1" output leakage current | | | 0.01 | 25 | | 0.01 | 25 | A |
| "0" output voltage | $I_L$ = 30mA | | 0.2 | 0.4 | | 0.2 | 0.4 | V |
| | $I_L$ = 100mA | | 0.6 | 1.0 | | 0.6 | 1.0 | V |
| Output fall time[3] | $R_L$ = 50Ω | | 30 | | | 30 | | ns |
| Output rise time[3] | $R_L$ = 50Ω | | 150 | | | 150 | | ns |
| GENERAL | | | | | | | | |
| Operating voltage range | | 4.75 | | 9.0 | 4.75 | | 9.0 | V |
| Supply current quiescent | | | 6 | 8 | | 7 | 10 | mA |
| Supply current—activated | $R_L$ = 20kΩ | | 11 | 13 | | 12 | 15 | mA |
| Quiescent power dissipation | | | 30 | | | 35 | | mW |

NOTES

1. Frequency determining resistor $R_1$ should be between 1 and 20kΩ.
2. Applicable over 4.75 to 5.75 volts. See graphs for more detailed information.
3. Pin 8 to Pin 1 feedback $R_L$ network selected to eliminate pulsing during turn-on and turn-off.

# Appendix I NE/SA/SE555/SE555C Timer

**Linear Products**

## DESCRIPTION

The 555 monolithic timing circuit is a highly stable controller capable of producing accurate time delays, or oscillation. In the time delay mode of operation, the time is precisely controlled by one external resistor and capacitor. For a stable operation as an oscillator, the free running frequency and the duty cycle are both accurately controlled with two external resistors and one capacitor. The circuit may be triggered and reset on falling waveforms, and the output structure can source or sink up to 200mA.

## BLOCK DIAGRAM

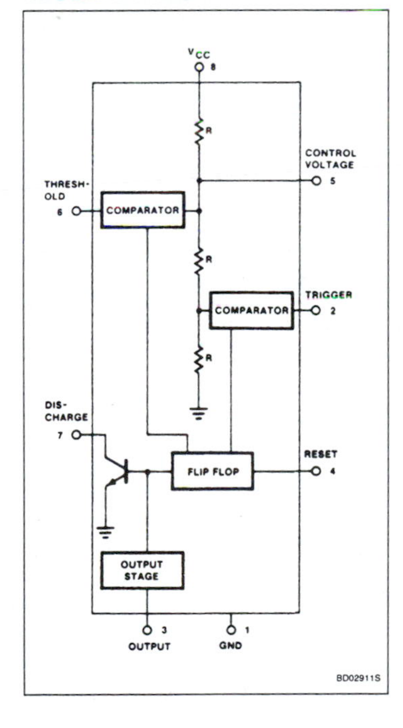


## FEATURES

- Turn-off time less than 2μs
- Max. operating frequency greater than 500kHz
- Timing from microseconds to hours
- Operates in both astable and monostable modes
- High output current
- Adjustable duty cycle
- TTL compatible
- Temperature stability of 0.005% per °C

## APPLICATIONS

- Precision timing
- Pulse generation
- Sequential timing
- Time delay generation
- Pulse width modulation
- Pulse position modulation
- Missing pulse detector

## PIN CONFIGURATIONS

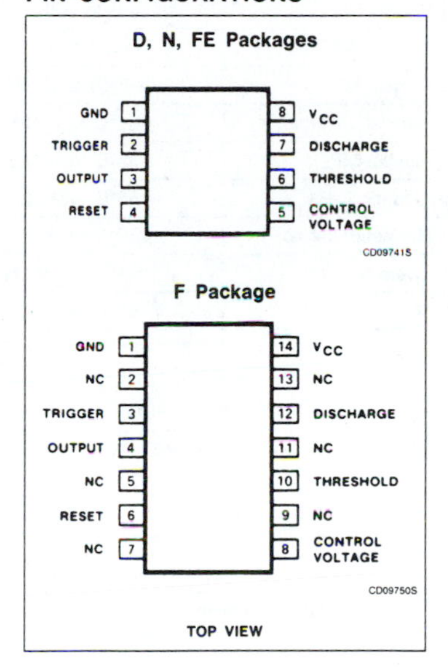


## EQUIVALENT SCHEMATIC

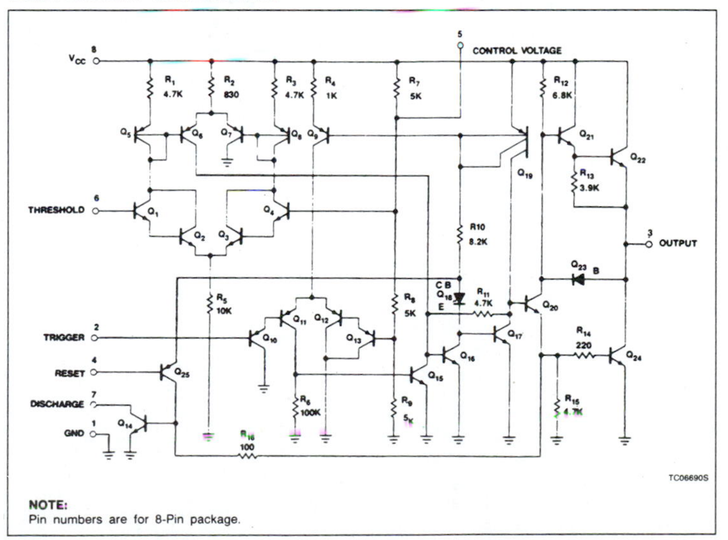


(Courtesy of Signetics Company)

## ORDERING INFORMATION

| DESCRIPTION | TEMPERATURE RANGE | ORDER CODE |
|---|---|---|
| 8-Pin Hermetic Cerdip | 0 to +70°C | NE555FE |
| 8-Pin Plastic SO | 0 to +70°C | NE555D |
| 8-Pin Plastic DIP | 0 to +70°C | NE555N |
| 8-Pin Plastic DIP | -40°C to +85°C | SA555N |
| 8-Pin Plastic SO | -40°C to +85°C | SA555D |
| 8-Pin Hermetic Cerdip | -55°C to +125°C | SE555CFE |
| 8-Pin Plastic DIP | -55°C to +125°C | SE555CN |
| 14-Pin Plastic DIP | -55°C to +125°C | SE555N |
| 8-Pin Hermetic Cerdip | -55°C to +125°C | SE555FE |
| 14-Pin Ceramic DIP | 0 to +70°C | NE555F |
| 14-Pin Ceramic DIP | -55°C to +125°C | SE555F |
| 14-Pin Ceramic DIP | -55°C to +125°C | SE555CF |

## ABSOLUTE MAXIMUM RATINGS

| SYMBOL | PARAMETER | RATING | UNIT |
|---|---|---|---|
| $V_{CC}$ | Supply voltage<br>SE555<br>NE555, SE555C, SA555 | <br>+18<br>+16 | <br>V<br>V |
| $P_D$ | Maximum allowable power dissipation[1] | 600 | mW |
| $T_A$ | Operating ambient temperature range<br>NE555<br>SA555<br>SE555, SE555C | <br>0 to +70<br>-40 to +85<br>-55 to +125 | <br>°C<br>°C<br>°C |
| $T_{STG}$ | Storage temperature range | -65 to +150 | °C |
| $T_{SOLD}$ | Lead soldering temperature (10sec max) | +300 | °C |

**NOTE:**

1. The junction temperature must be kept below 125°C for the D package and below 150°C for the FE, N and F packages. At ambient temperatures above 25°C, where this limit would be derated by the following factors:

   D package 160 °C/W
   FE package 150 °C/W
   N package 100 °C/W
   F package 105 °C/W

**DC AND AC ELECTRICAL CHARACTERISTICS** $T_A = 25°C$, $V_{CC} = +5V$ to $+15$ unless otherwise specified.

| SYMBOL | PARAMETER | TEST CONDITIONS | SE555 | | | NE555/SE555C | | | UNIT |
|---|---|---|---|---|---|---|---|---|---|
| | | | Min | Typ | Max | Min | Typ | Max | |
| $V_{CC}$ | Supply voltage | | 4.5 | | 18 | 4.5 | | 16 | V |
| $I_{CC}$ | Supply current (low state)[1] | $V_{CC} = 5V$, $R_L = \infty$<br>$V_{CC} = 15V$, $R_L = \infty$ | | 3<br>10 | 5<br>12 | | 3<br>10 | 6<br>15 | mA<br>mA |
| $t_M$<br>$\Delta t_M/\Delta T$<br>$\Delta t_M/\Delta V_S$ | Timing error (monostable)<br>Initial accuracy[2]<br>Drift with temperature<br>Drift with supply voltage | $R_A = 2k\Omega$ to $100k\Omega$<br>$C = 0.1\mu F$ | | 0.5<br>30<br>0.05 | 2.0<br>100<br>0.2 | | 1.0<br>50<br>0.1 | 3.0<br>150<br>0.5 | %<br>ppm/°C<br>%/V |
| $t_A$<br>$\Delta t_A/\Delta T$<br>$\Delta t_A/\Delta V_S$ | Timing error (astable)<br>Initial accuracy[2]<br>Drift with temperature<br>Drift with supply voltage | $R_A$, $R_B = 1k\Omega$ to $100k\Omega$<br>$C = 0.1\mu F$<br>$V_{CC} = 15V$ | | 4<br><br>0.15 | 6<br>500<br>0.6 | | 5<br><br>0.3 | 13<br>500<br>1 | %<br>ppm/°C<br>%/V |
| $V_C$ | Control voltage level | $V_{CC} = 15V$<br>$V_{CC} = 5V$ | 9.6<br>2.9 | 10.0<br>3.33 | 10.4<br>3.8 | 9.0<br>2.6 | 10.0<br>3.33 | 11.0<br>4.0 | V<br>V |
| $V_{TH}$ | Threshold voltage | $V_{CC} = 15V$<br>$V_{CC} = 5V$ | 9.4<br>2.7 | 10.0<br>3.33 | 10.6<br>4.0 | 8.8<br>2.4 | 10.0<br>3.33 | 11.2<br>4.2 | V<br>V |
| $I_{TH}$ | Threshold current[3] | | | 0.1 | 0.25 | | 0.1 | 0.25 | μA |
| $V_{TRIG}$ | Trigger voltage | $V_{CC} = 15V$<br>$V_{CC} = 5V$ | 4.8<br>1.45 | 5.0<br>1.67 | 5.2<br>1.9 | 4.5<br>1.1 | 5.0<br>1.67 | 5.6<br>2.2 | V<br>V |
| $I_{TRIG}$ | Trigger current | $V_{TRIG} = 0V$ | | 0.5 | 0.9 | | 0.5 | 2.0 | μA |
| $V_{RESET}$ | Reset voltage[4] | | 0.3 | | 1.0 | 0.3 | | 1.0 | V |
| $I_{RESET}$ | Reset current<br>Reset current | $V_{RESET} = 0.4V$<br>$V_{RESET} = 0V$ | | 0.1<br>0.4 | 0.4<br>1.0 | | 0.1<br>0.4 | 0.4<br>1.5 | mA<br>mA |
| $V_{OL}$ | Output voltage (low) | $V_{CC} = 15V$<br>$I_{SINK} = 10mA$<br>$I_{SINK} = 50mA$<br>$I_{SINK} = 100mA$<br>$I_{SINK} = 200mA$<br>$V_{CC} = 5V$<br>$I_{SINK} = 8mA$<br>$I_{SINK} = 5mA$ | | <br>0.1<br>0.4<br>2.0<br>2.5<br><br>0.1<br>0.05 | <br>0.15<br>0.5<br>2.2<br><br><br>0.25<br>0.2 | | <br>0.1<br>0.4<br>2.0<br>2.5<br><br>0.3<br>0.25 | <br>0.25<br>0.75<br>2.5<br><br><br>0.4<br>0.35 | <br>V<br>V<br>V<br>V<br><br>V<br>V |
| $V_{OH}$ | Output voltage (high) | $V_{CC} = 15V$<br>$I_{SOURCE} = 200mA$<br>$I_{SOURCE} = 100mA$<br>$V_{CC} = 5V$<br>$I_{SOURCE} = 100mA$ | <br><br>13.0<br><br>3.0 | <br>12.5<br>13.3<br><br>3.3 | | <br><br>12.75<br><br>2.75 | <br>12.5<br>13.3<br><br>3.3 | | <br>V<br>V<br><br>V |
| $t_{OFF}$ | Turn-off time[5] | $V_{RESET} = V_{CC}$ | | 0.5 | 2.0 | | 0.5 | 2.0 | μs |
| $t_R$ | Rise time of output | | | 100 | 200 | | 100 | 300 | ns |
| $t_F$ | Fall time of output | | | 100 | 200 | | 100 | 300 | ns |
| | Discharge leakage current | | | 20 | 100 | | 20 | 100 | ns |

**NOTES:**
1. Supply current when output high typically 1mA less.
2. Tested at $V_{CC} = 5V$ and $V_{CC} = 15V$.
3. This will determine the max value of $R_A + R_B$, for 15V operation, the max total $R = 10M\Omega$, and for 5V operation, the max. total $R = 3.4M\Omega$.
4. Specified with trigger input high.
5. Time measured from a positive going input pulse from 0 to $0.8 \times V_{CC}$ into the threshold to the drop from high to low of the output. Trigger is tied to threshold.

Timer *Continued* NE/SA/SE555/SE555C

## TYPICAL PERFORMANCE CHARACTERISTICS

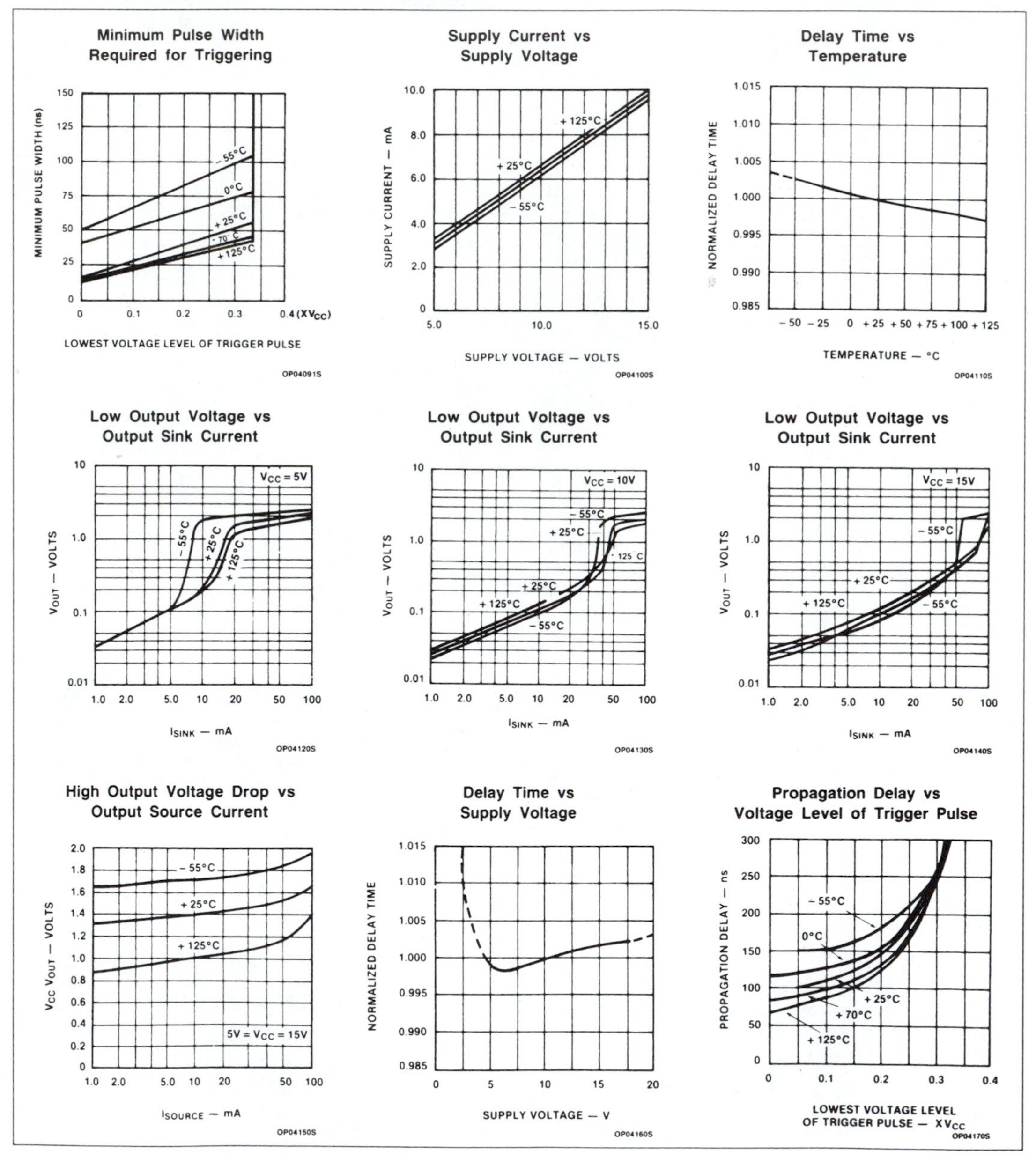

## TYPICAL APPLICATIONS

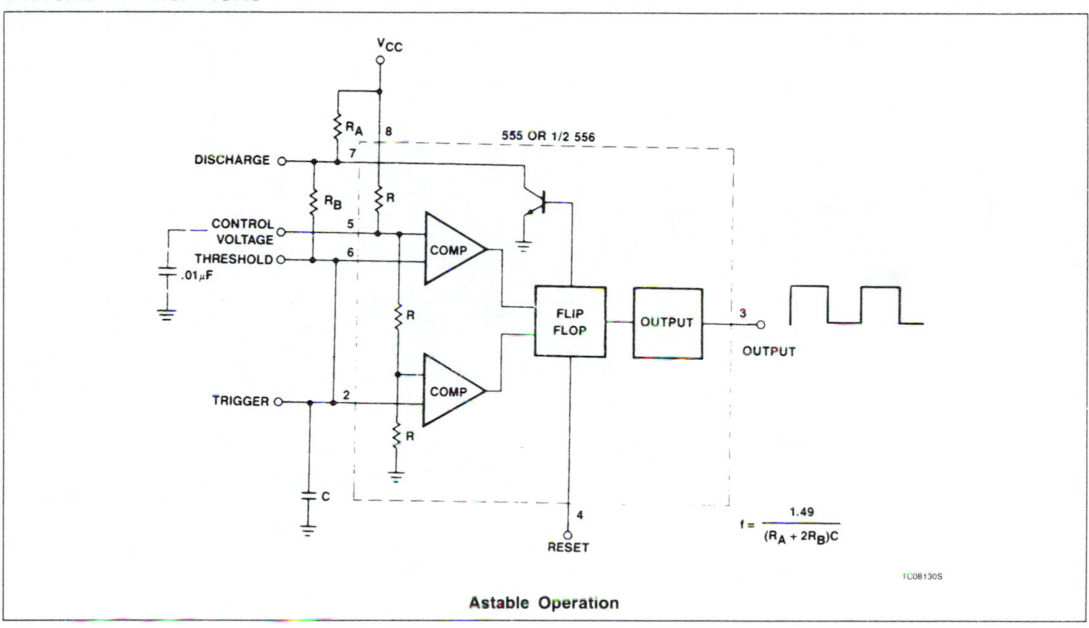


Astable Operation

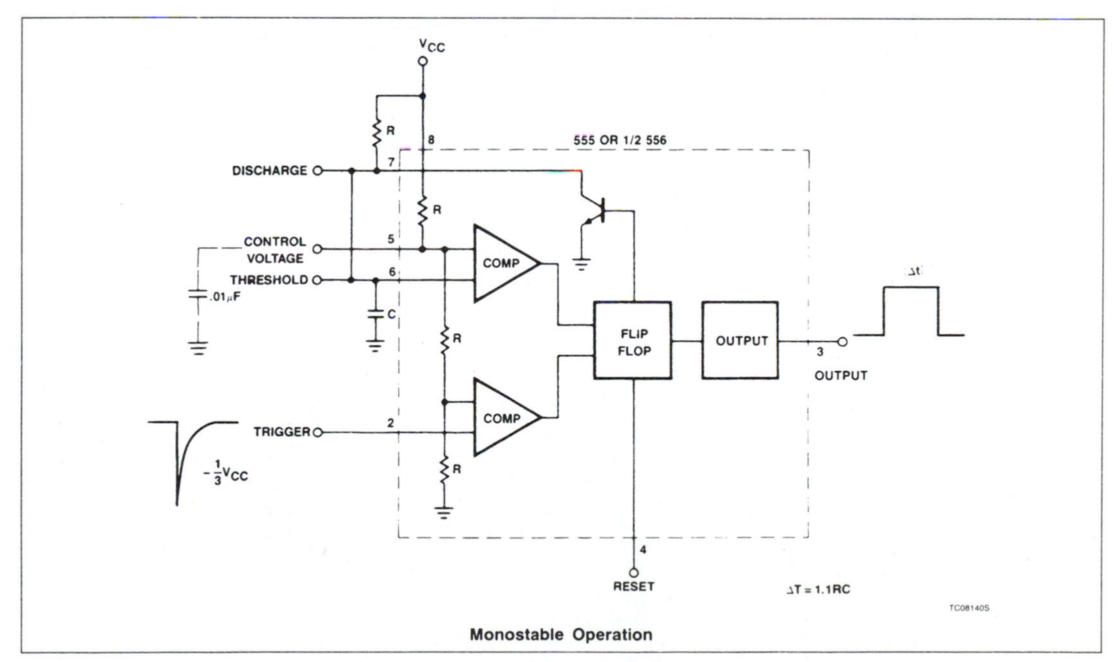


Monostable Operation

## TYPICAL APPLICATIONS

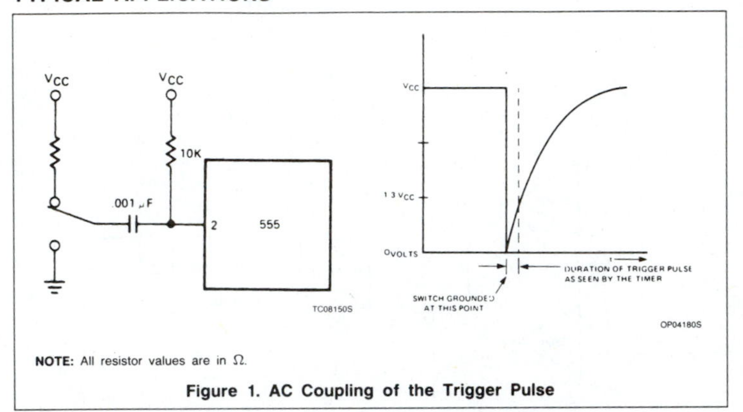


**NOTE:** All resistor values are in Ω.

**Figure 1. AC Coupling of the Trigger Pulse**

### Trigger Pulse Width Requirements and Time Delays

Due to the nature of the trigger circuitry, the timer will trigger on the negative going edge of the input pulse. For the device to time out properly, it is necessary that the trigger voltage level be returned to some voltage greater than one third of the supply before the time out period. This can be achieved by making either the trigger pulse sufficiently short or by AC coupling into the trigger. By AC coupling the trigger, see Figure 1, a short negative going pulse is achieved when the trigger signal goes to ground. AC coupling is most frequently used in conjunction with a switch or a signal that goes to ground which initiates the timing cycle. Should the trigger be held low, without AC coupling, for a longer duration than the timing cycle the output will remain in a high state for the duration of the low trigger signal, without regard to the threshold comparator state. This is due to the predominance of $Q_{15}$ on the base of $Q_{16}$, controlling the state of the bistable flip-flop. When the trigger signal then returns to a high level, the output will fall immediately. Thus, the output signal will follow the trigger signal in this case.

Another consideration is the "turn-off time". This is the measurement of the amount of time required after the threshold reaches 2/3 $V_{CC}$ to turn the output low. To explain further, $Q_1$ at the threshold input turns on after reaching 2/3 $V_{CC}$, which then turns on $Q_5$, which turns on $Q_6$. Current from $Q_6$ turns on $Q_{16}$ which turns $Q_{17}$ off. This allows current from $Q_{19}$ to turn on $Q_{20}$ and $Q_{24}$ to given an output low. These steps cause the 2μs max. delay as stated in the data sheet.

Also, a delay comparable to the turn-off time is the trigger release time. When the trigger is low, $Q_{10}$ is on and turns on $Q_{11}$ which turns on $Q_{15}$. $Q_{15}$ turns off $Q_{16}$ and allows $Q_{17}$ to turn on. This turns off current to $Q_{20}$ and $Q_{24}$, which results in output high. When the trigger is released, $Q_{10}$ and $Q_{11}$ shut off, $Q_{15}$ turns off, $Q_{16}$ turns on and the circuit then follows the same path and time delay explained as "turn off time". This trigger release time is very important in designing the trigger pulse width so as not to interfere with the output signal as explained previously.

Signetics

Linear Products

# NE/SA/SE556/NE/SA/SE556-1/SE556-1C Dual Timer

*Product Specification*

## DESCRIPTION

Both the 556 and 556-1 Dual Monolithic timing circuits are highly stable controllers capable of producing accurate time delays or oscillation. The 556 and 556-1 are a dual 555. Timing is provided by an external resistor and capacitor for each timing function. The two timers operate independently of each other, sharing only $V_{CC}$ and ground. The circuits may be triggered and reset on falling waveforms. The output structures may sink or source 200mA.

## FEATURES

- **Turn-off time less than 2μs (556-1, 1C)**
- **Maximum operating frequency >500kHz (556-1, 1C)**
- **Timing from microseconds to hours**
- **Replaces two 555 timers**
- **Operates in both astable and monostable modes**
- **High output current**
- **Adjustable duty cycle**
- **TTL compatible**
- **Temperature stability of 0.005%/°C**
- **SE556-1 compliant to MIL-STD or JAN available from Signetics' Military Division**

## APPLICATIONS

- **Precision timing**
- **Sequential timing**
- **Pulse shaping**
- **Pulse generator**
- **Missing pulse detector**
- **Tone burst generator**
- **Pulse width modulation**
- **Time delay generator**
- **Frequency division**
- **Industrial controls**
- **Pulse position modulation**
- **Appliance timing**
- **Traffic light control**
- **Touch-Tone® encoder**

## PIN CONFIGURATION

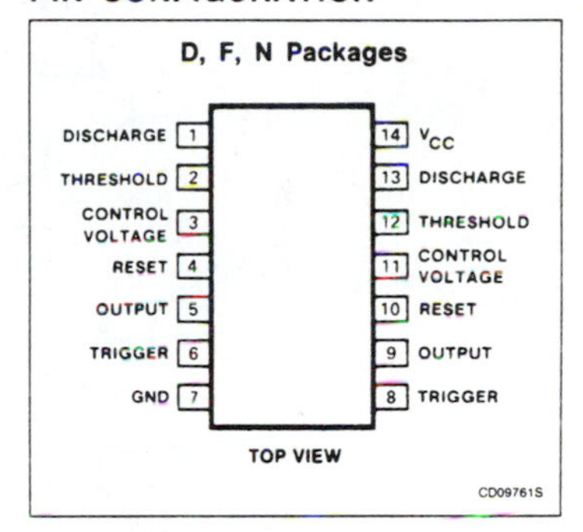


## BLOCK DIAGRAM

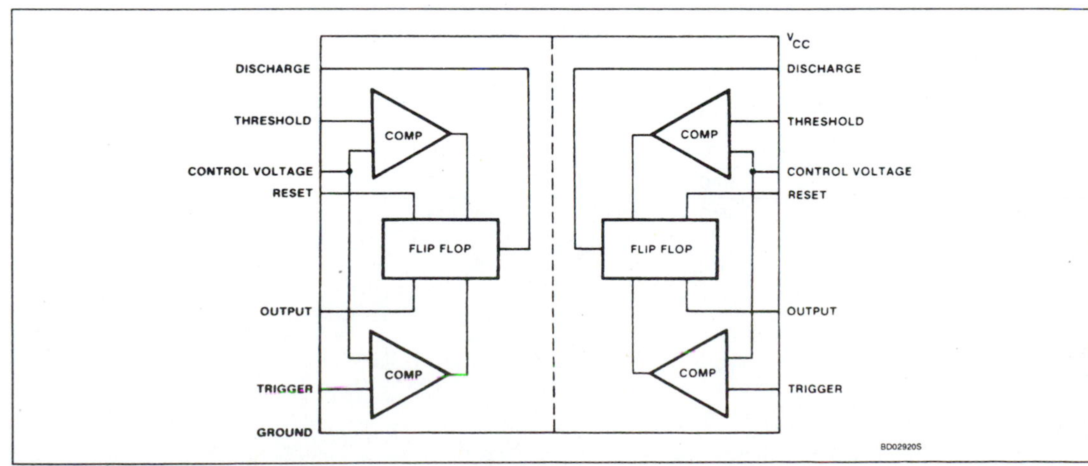


®Touch-Tone is a registered trademark of AT&T

Dual Timer *Continued* NE/SA/SE556/NE/SA/SE556-1/SE556-1C

**EQUIVALENT SCHEMATIC** (Shown for one circuit only)

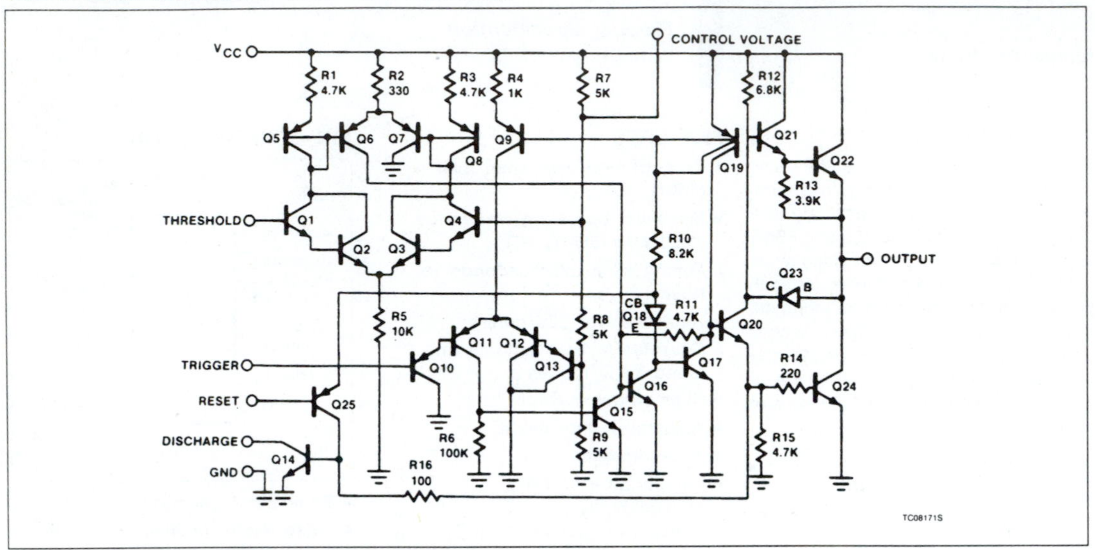


## ORDERING INFORMATION

| DESCRIPTION | TEMPERATURE RANGE | ORDER CODE |
|---|---|---|
| 14-Pin Plastic SO | 0 to +70°C | NE556D |
| 14-Pin Cerdip | 0 to +70°C | NE556F |
| 14-Pin Plastic DIP | 0 to +70°C | NE556N |
| 14-Pin Cerdip | 0 to +70°C | NE556-1F |
| 14-Pin Plastic DIP | 0 to +70°C | NE556-1N |
| 14-Pin Plastic DIP | -40°C to +85°C | SA556N |
| 14-Pin Cerdip | -40°C to +85°C | SA556-1F |
| 14-Pin Plastic DIP | -40°C to +85°C | SA556-1N |
| 14-Pin Cerdip | -55°C to +125°C | SE556F |
| 14-Pin Plastic DIP | -55°C to +125°C | SE556N |
| 14-Pin Plastic DIP | -55°C to +125°C | SE556CN |
| 14-Pin Cerdip | -55°C to +125°C | SE556-1F |
| 14-Pin Plastic DIP | -55°C to +125°C | SE556-1N |
| 14-Pin Cerdip | -55°C to +125°C | SE556-1CF |
| 14-Pin Plastic DIP | -55°C to +125°C | SE556-1CN |

Dual Timer NE/SA/SE556/NE/SA/SE556-1/SE556-1C

## ABSOLUTE MAXIMUM RATINGS

| SYMBOL | PARAMETER | RATING | UNIT |
|---|---|---|---|
| $V_{CC}$ | Supply voltage<br>NE/SA556, 556-1, SE556C, SE556-1C<br>SE556-1, SE556 | <br>+16<br>+18 | <br>V<br>V |
| $P_D$ | Maximum allowable power dissipation[1] | 800 | mW |
| $T_A$ | Operating temperature range<br>NE556-1, NE556<br>SA556-1, SA556<br>SE556-1, SE556-1C, SE556, SE556C | <br>0 to +70<br>-40 to +85<br>-55 to +125 | <br>°C<br>°C<br>°C |
| $T_{STG}$ | Storage temperature range | -65 to +150 | °C |
| $T_{SOLD}$ | Lead soldering temperature (10sec max) | +300 | °C |

**NOTE:**

1. The junction temperature must be kept below 125°C for the D package and below 150°C for the N and F packages. At ambient temperatures above 25°C, where this limit would be exceeded, the Maximum Allowable Power Dissipation must be derated by the following:
   D package 115 °C/W
   N package 80 °C/W
   F package 100 °C/W

## ELECTRICAL CHARACTERISTICS $T_A = 25°C$, $V_{CC} = +5V$ to $+15V$, unless otherwise specified.

| SYMBOL | PARAMETER | TEST CONDITIONS | SE556/556-1 | | | NE/SA556/SE556C NE556-1/SE556-1C | | | UNIT |
|---|---|---|---|---|---|---|---|---|---|
| | | | Min | Typ | Max | Min | Typ | Max | |
| $V_{CC}$ | Supply voltage | | 4.5 | | 18 | 4.5 | | 16 | V |
| $I_{CC}$ | Supply current (low state)[1] | $V_{CC} = 5V$, $R_L = \infty$<br>$V_{CC} = 15V$, $R_L = \infty$ | | 6<br>20 | 10<br>24 | | 6<br>20 | 12<br>30 | mA<br>mA |
| <br>$t_M$<br>$\Delta t_M/\Delta T$<br>$\Delta t_M/\Delta V_S$ | Timing error (monostable)<br>Initial accuracy[2]<br>Drift with temperature<br>Drift with supply voltage | $R_A = 2k\Omega$ to $100k\Omega$<br>$C = 0.1\mu F$<br>$T = 1.1$ RC | | <br>0.5<br>30<br>0.05 | <br><br>100<br>0.2 | | <br>0.75<br>50<br>0.1 | <br>3.0<br>150<br>0.5 | <br>%<br>ppm/°C<br>%/V |
| <br>$t_A$<br>$\Delta t_A/\Delta T$<br>$\Delta t_A/\Delta V_S$ | Timing error (astable)<br>Initial accuracy[2]<br>Drift with temperature<br>Drift with supply voltage | $R_A$, $R_B = 1k\Omega$ to $100k\Omega$<br>$C = 0.\mu F$<br>$V_{CC} = 15V$ | | <br>4<br>400<br>0.15 | <br>6<br>500<br>0.6 | | <br>5<br>400<br>0.3 | <br>13<br>500<br>1 | <br>%<br>ppm/°C<br>%/V |
| $V_C$ | Control voltage level | $V_{CC} = 15V$<br>$V_{CC} = 5V$ | 9.6<br>2.9 | 10.0<br>3.33 | 10.4<br>3.8 | 9.0<br>2.6 | 10.0<br>3.33 | 11.0<br>4.0 | V<br>V |
| $V_{TH}$ | Threshold voltage | $V_{CC} = 15V$<br>$V_{CC} = 5V$ | 9.4<br>2.7 | 10.0<br>3.33 | 10.6<br>4.0 | 8.8<br>2.4 | 10.0<br>3.33 | 11.2<br>4.2 | V<br>V |
| $I_{TH}$ | Threshold current[3] | | | 30 | 250 | | 30 | 250 | nA |
| $V_{TRIG}$ | Trigger voltage | $V_{CC} = 15V$<br>$V_{CC} = 5V$ | 4.8<br>1.45 | 5.0<br>1.67 | 5.2<br>1.9 | 4.5<br>1.1 | 5.0<br>1.67 | 5.6<br>2.2 | V<br>V |
| $I_{TRIG}$ | Trigger current | $V_{TRIG} = 0V$ | | 0.5 | 0.9 | | 0.5 | 2.0 | μA |
| $V_{RESET}$ | Reset voltage[5] | | 0.4 | 0.7 | 1.0 | 0.4 | 0.7 | 1.0 | V |
| $I_{RESET}$ | Reset current<br>Reset current | $V_{RESET} = 0.4V$<br>$V_{RESET} = 0V$ | 0.4 | 0.1<br>0.4 | 0.4<br>1.0 | 0.4 | 0.1<br>0.4 | 0.6<br>1.5 | mA<br>mA |
| $V_{OL}$ | Output voltage (low) | $V_{CC} = 15V$<br>$I_{SINK} = 10mA$<br>$I_{SINK} = 50mA$ | | <br>0.1<br>0.4 | <br>0.15<br>0.5 | | <br>0.1<br>0.4 | <br>0.25<br>0.75 | <br>V<br>V |
| | SE556<br>SE556-1<br>NE/SA556/SE556C<br>NE556-1/SE556-1C | $I_{SINK} = 100mA$ | | 2.0<br>0.8 | 2.25<br>1.2 | | <br><br>2.0<br>2.0 | <br><br>3.2<br>2.5 | V<br>V<br>V<br>V |

## Dual Timer *Continued* NE/SA/SE556/NE/SA/SE556-1/SE556-1C

**ELECTRICAL CHARACTERISTICS** (Continued) $T_A = 25°C$, $V_{CC} = +5V$ to $+15V$, unless otherwise specified.

| SYMBOL | PARAMETER | TEST CONDITIONS | SE556/556-1 | | | NE/SA556/SE556C NE556-1/SE556-1C | | | UNIT |
|---|---|---|---|---|---|---|---|---|---|
| | | | Min | Typ | Max | Min | Typ | Max | |
| | | $I_{SINK} = 200mA$ | | 2.5 | | | 2.5 | | V |
| | | $V_{CC} = 5V$ | | | | | | | |
| | | $I_{SINK} = 8mA$ | | 0.1 | 0.2 | | 0.25 | 0.3 | V |
| | | $I_{SINK} = 5mA$ | | 0.05 | 0.15 | | 0.15 | 0.25 | V |
| $V_{OH}$ | Output voltage (high) | $V_{CC} = 15V$ | | | | | | | |
| | | $I_{SOURCE} = 200mA$ | | 12.5 | | | 12.5 | | V |
| | | $I_{SOURCE} = 100mA$ | 13.0 | 13.3 | | 12.75 | 13.3 | | V |
| | | $V_{CC} = 5V$ | | | | | | | |
| | | $I_{SOURCE} = 100mA$ | 3.0 | 3.3 | | 2.75 | 3.3 | | V |
| $t_{OFF}$ | Turn-off time[6] NE556-1/SE556-1/SE556-1C | $V_{RESET} = V_{CC}$ | | 0.5 | 2.0 | | 0.5 | | µs |
| $t_R$ | Rise time of output | | | 100 | 200 | | 100 | 300 | ns |
| $t_F$ | Fall time of output | | | 100 | 200 | | 100 | 300 | ns |
| | Discharge leakage current | | | 20 | 100 | | 20 | 100 | nA |
| | Matching characteristics[4] | | | | | | | | |
| | Initial accuracy[2] | | | 0.5 | 1.0 | | 1.0 | 2.0 | % |
| | Drift with temperature | | | 10 | | | ± 10 | | ppm/°C |
| | Drift with supply voltage | | | 0.1 | 0.2 | | 0.2 | 0.5 | %/V |

**NOTES:**

1. Supply current when output is high is typically 1.0mA less.
2. Tested at $V_{CC} = 5V$ and $V_{CC} = 15V$.
3. This will determine maximum value of $R_A + R_B$. For 15V operation, the max total $R = 10M\Omega$, and for 5V operation, the maximum total $R = 3.4M\Omega$.
4. Matching characteristics refer to the difference between performance characteristics for each timer section in the monostable mode.
5. Specified with trigger input high. In order to guarantee reset the voltage at reset pin must be less than or equal to 0.4V. To disable reset function, the voltage at reset pin has to be greater than 1V.
6. Time measured from a positive-going input pulse from 0 to 0.4 $V_{CC}$ into the threshold to the drop from high to low of the output. Trigger is tied to threshold.

### TYPICAL APPLICATIONS

One feature of the dual timer is that by utilizing both halves it is possible to obtain sequential timing. By connecting the output of the first half to the input of the second half via a 0.001μF coupling capacitor sequential timing may be obtained. Delay $t_1$ is determined by the first half and $t_2$ by the second half delay.

The first half of the timer is started by momentarily connecting Pin 6 to ground. When it is timed out (determined by $1.1R_1C_1$) the second half begins. Its duration is determined by $1.1R_2C_2$.

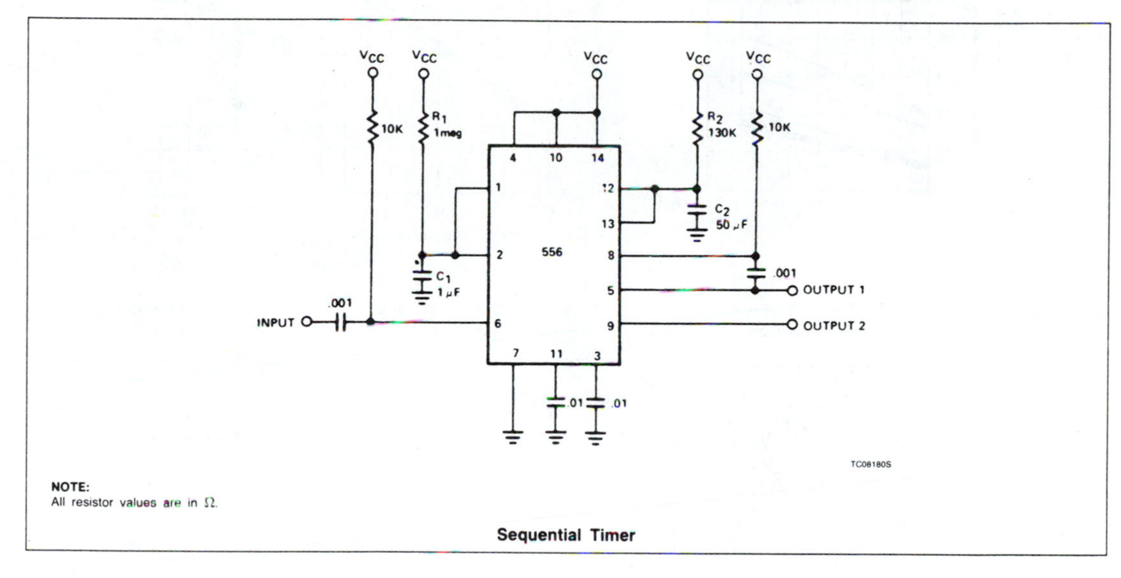


NOTE:
All resistor values are in Ω.

Sequential Timer

Dual Timer *Continued* NE/SA/SE556/NE/SA/SE556-1/SE556-1C

## TYPICAL PERFORMANCE CHARACTERISTICS

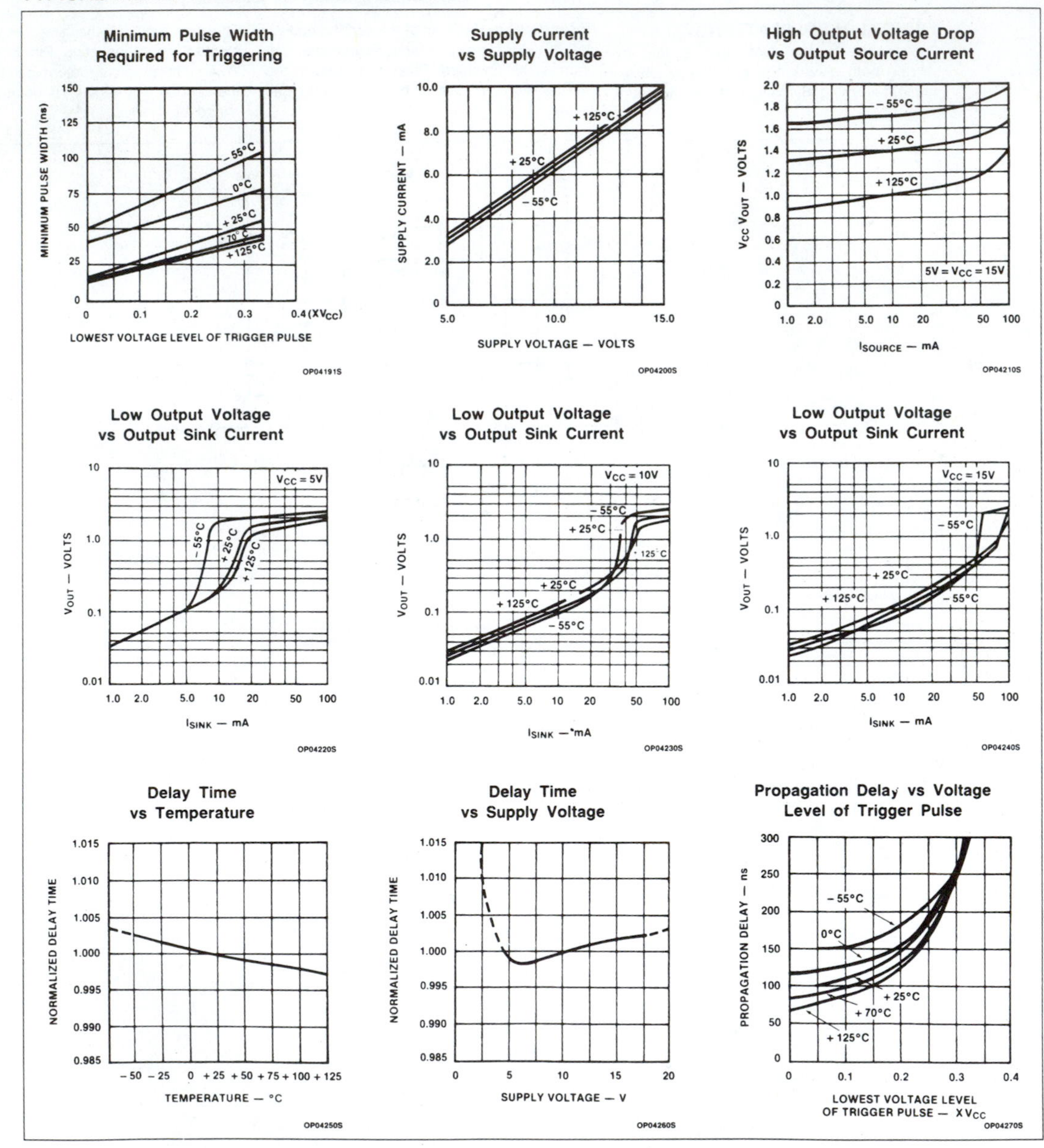

# Appendix J AN171 NE558 Applications

***Application Note***

**Linear Products**

## INTRODUCTION

The 558 is a monolithic Quad Timer designed to be used in the timing range from a few microseconds to a few hours. Four entirely independent timing functions can be achieved using a timing resistor and capacitor for each section. Two sections of the quad may be interconnected for astable operation. All four sections may be used together, in tandem, for sequential timing applications up to several hours. No coupling capacitors are required when connecting the output of one timer section to the input of the next.

## FEATURES

- **100mA output current per section**
- **Edge-triggered (no coupling capacitor)**
- **Output independent of trigger conditions**
- **Wide supply voltage range 4.5V to 16V**
- **Timer intervals from microseconds to hours**
- **Time period equals RC**

## CIRCUIT OPERATIONS

In the one-shot mode of operation, it is necessary to supply a minimum of two external components (the resistor and capacitor) for timing. The time period is equal to the product of R and C. An output load must be present to complete the circuit due to the output structure of the 558.

For astable operation, it is desirable to cross-couple two devices from the 558 Quad. The outputs are direct-coupled to the opposite trigger input. The duty cycle can be set by the ratio of $R_1C_1$ to $R_2C_2$, from close to zero to almost 100%. An astable circuit using one timer is shown in Figure 5b.

## OUTPUT STRUCTURE 558

The 558 structure is open-collector which requires a pull-up resistor to $V_{CC}$ and is capable of sinking 100mA per unit, but not to exceed the power dissipation and junction temperature rating of the die and package. The output is normally low and is switched high when triggered.

## RESET

A reset function has been made available to reset all sections simultaneously to an output low state. During reset the trigger is disabled. After reset is finished, the trigger voltage must be taken high and then low to implement triggering.

The reset voltage must be brought below 0.8V to insure reset.

## THE CONTROL VOLTAGE

The control voltage is also made available on the 558 timer. This allows the threshold voltage to be modulated, therefore controlling the output pulse width and duty cycle with an external control voltage. The range of this control voltage is from about 0.5V to $V_{CC}$ minus 1V. This will give a cycle time variation of about 50:1. In a sequential timer with voltage-controlled cycle time, the timing periods remain proportional over the adjustment range.

## TEST BOARD FOR 558

The circuit layout can be used to test and characterize the 558 timer. $S_2$ is used to connect the loads to either $V_{CC}$ or ground. The main precaution, in layout of the 558 circuit, is the path of the discharge current from the timing capacitor to ground (Pin 12). The path must be direct to Pin 12 and not on the ground bus. This is to prevent voltage spikes on the ground bus return due to current switching transient. It is also wise to use good power supply bypassing when large currents are being switched.

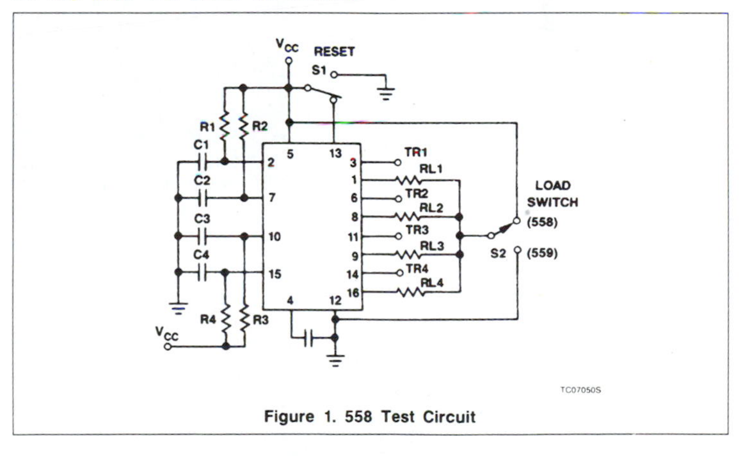


**Figure 1. 558 Test Circuit**

(Courtesy of Signetics Company)

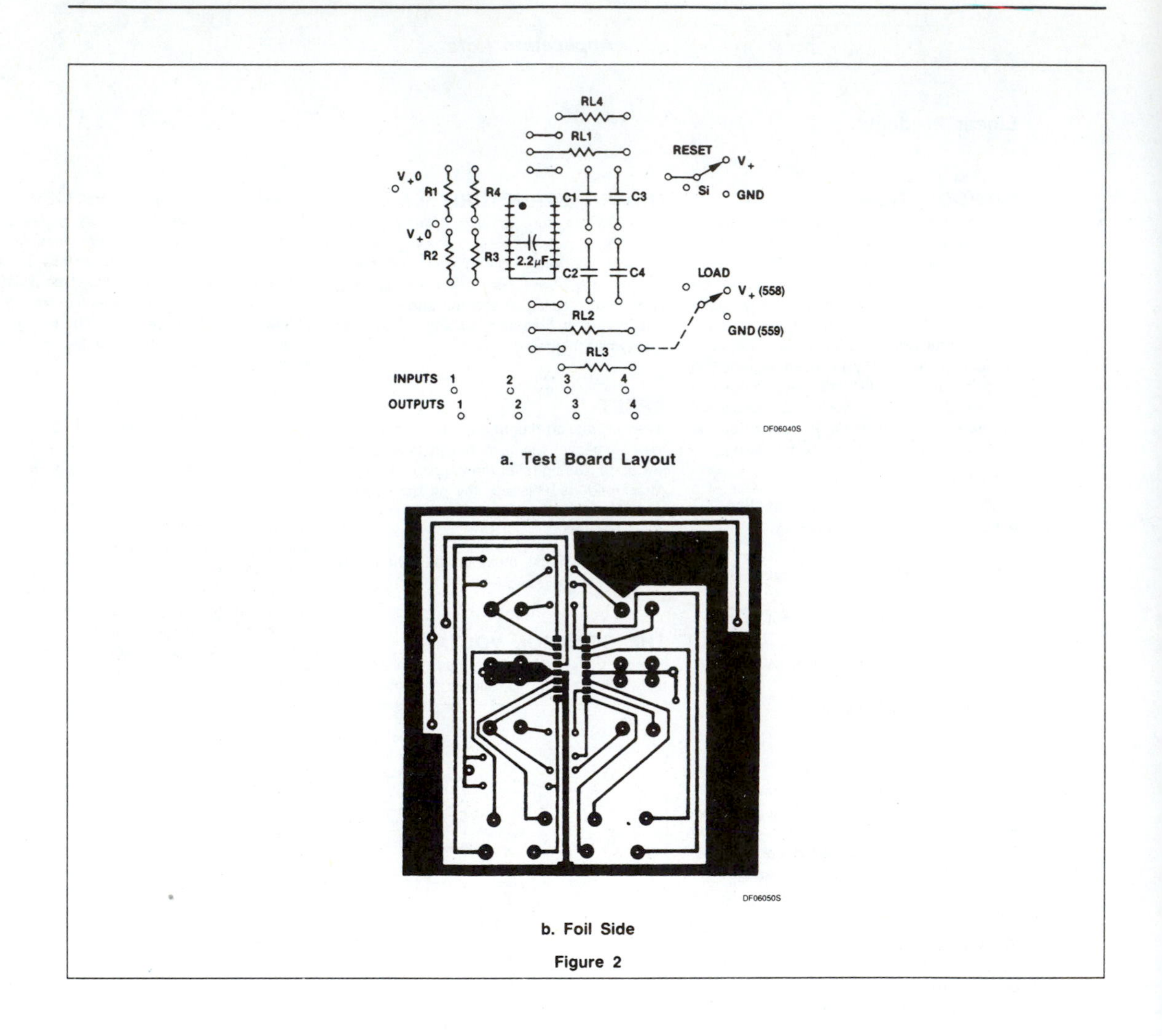


a. Test Board Layout

b. Foil Side

Figure 2

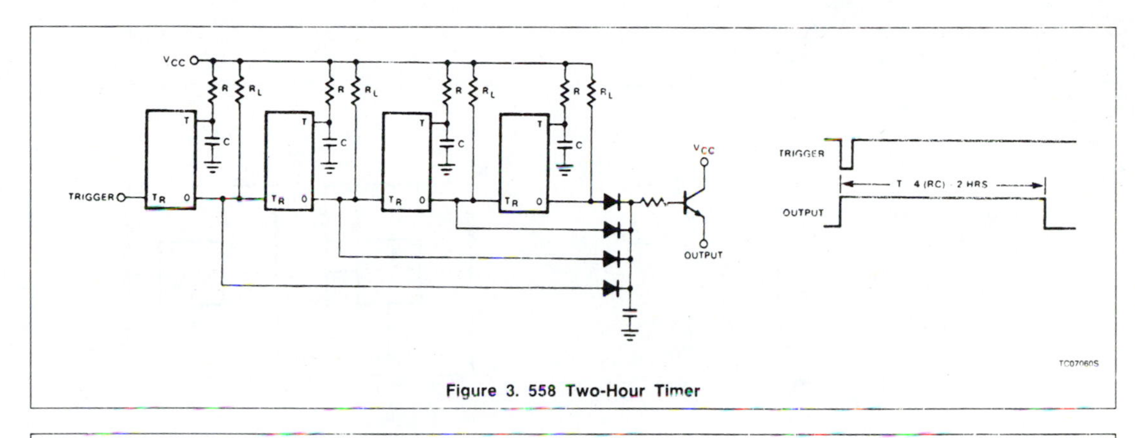


Figure 3. 558 Two-Hour Timer

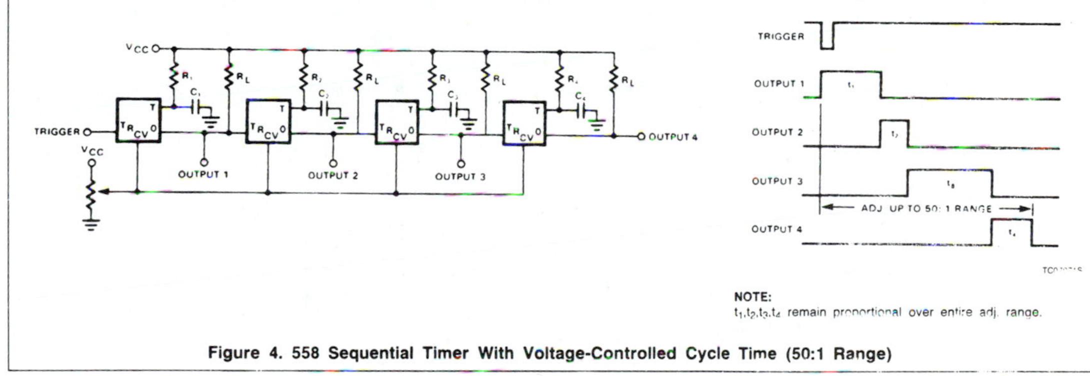


**NOTE:**
$t_1, t_2, t_3, t_4$ remain proportional over entire adj. range.

Figure 4. 558 Sequential Timer With Voltage-Controlled Cycle Time (50:1 Range)

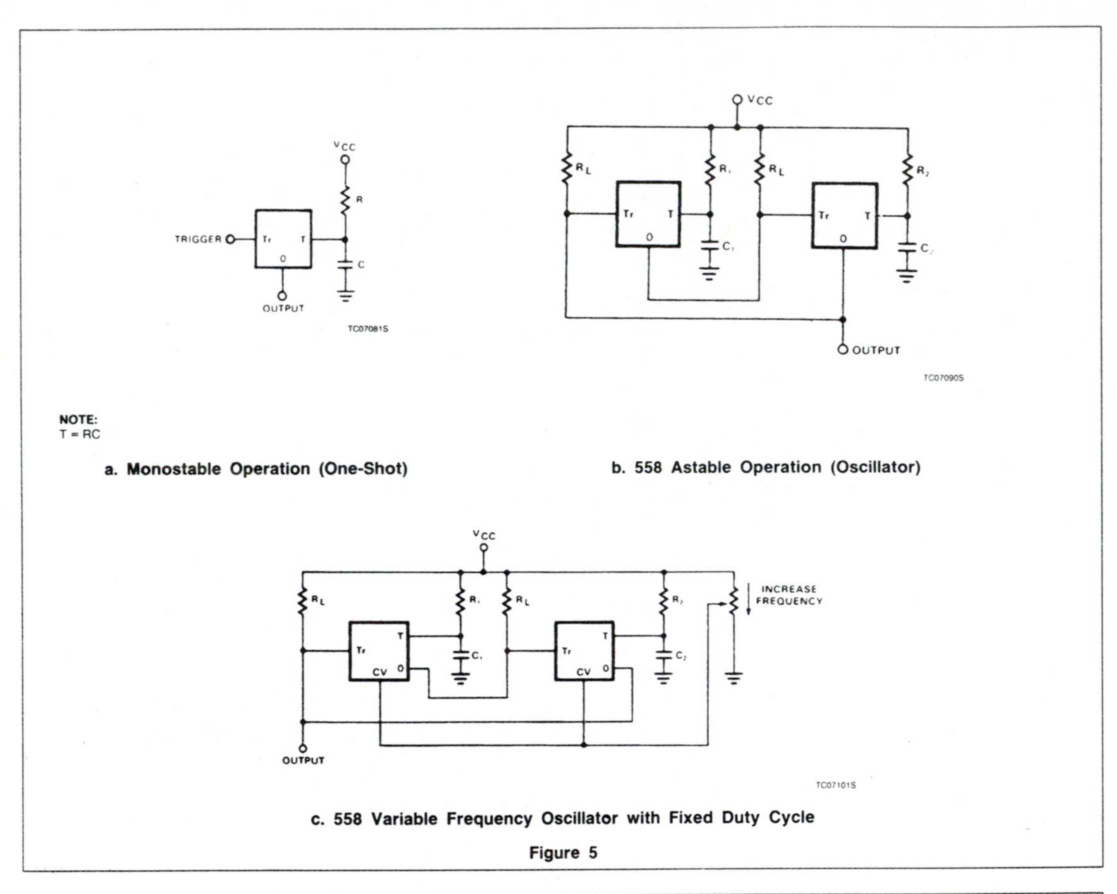


NOTE:
T = RC

a. Monostable Operation (One-Shot)

b. 558 Astable Operation (Oscillator)

c. 558 Variable Frequency Oscillator with Fixed Duty Cycle

Figure 5

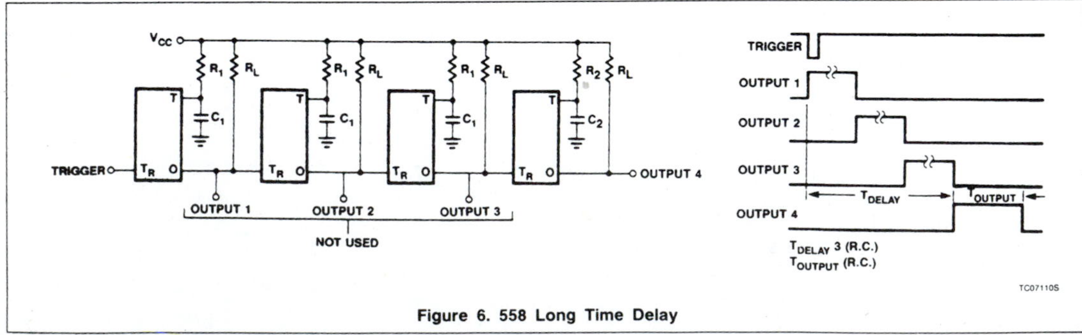


Figure 6. 558 Long Time Delay

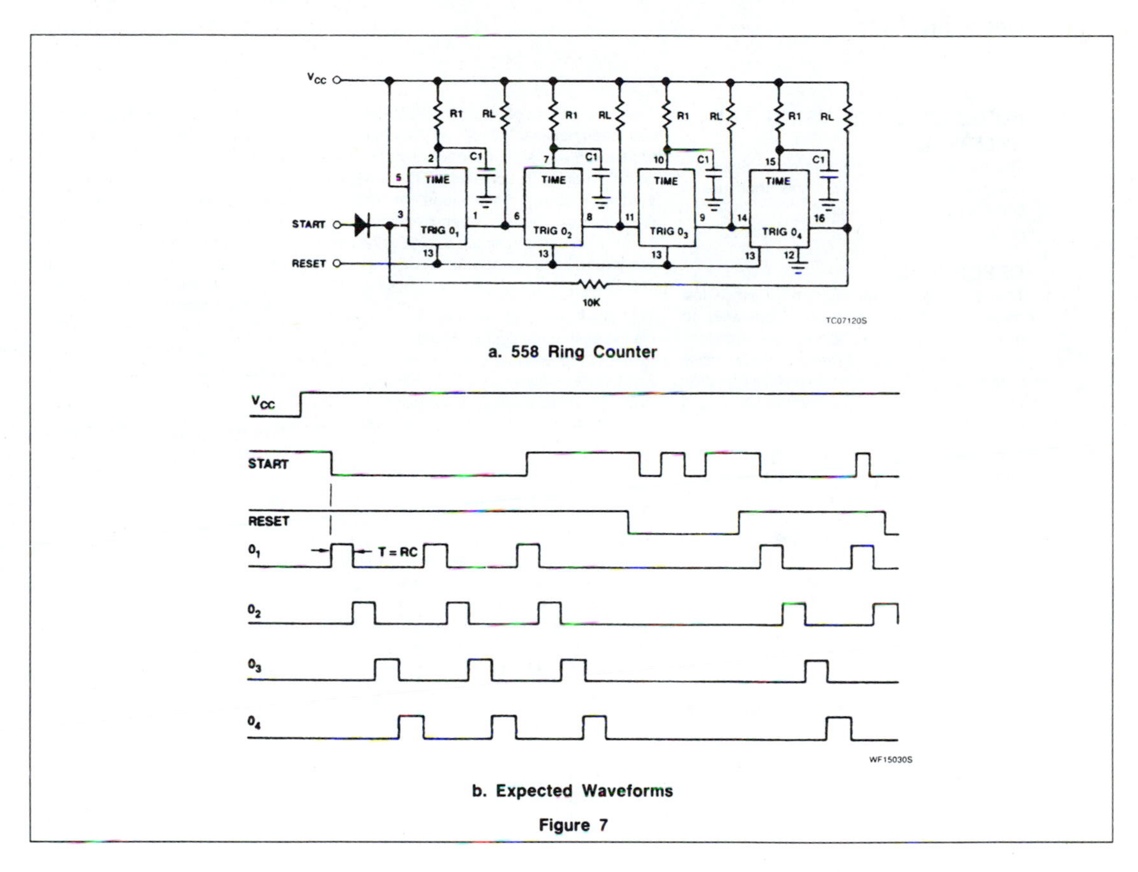


a. 558 Ring Counter

b. Expected Waveforms

Figure 7

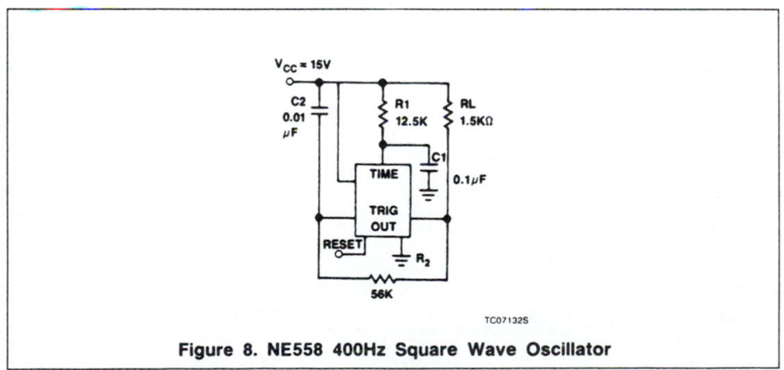


Figure 8. NE558 400Hz Square Wave Oscillator

A single section of the quad time may be used as a non-precision oscillator. The values given are for oscillation at about 400Hz. $T_1 \approx R_1C_1$ and $T_2 \approx 2.25\ R_2C_2$ for $V_{CC}$ of 15V. The frequency of oscillation is subject to the changes in $V_{CC}$.

# Appendix K AN142 Audio Circuits Using the NE5532/33/34

***Application Note***

**Linear Products**

## AUDIO CIRCUITS USING THE NE5532/33/34

The following will explain some of Signetics' low noise op amps and show their use in some audio applications.

## DESCRIPTION

The 5532 is a dual high-performance low noise operational amplifier. Compared to most of the standard operational amplifiers, such as the 1458, it shows better noise performance, improved output drive capability and considerably higher small-signal and power bandwidths.

This makes the device especially suitable for application in high quality and professional audio equipment, instrumentation and control circuits, and telephone channel amplifiers. The op amp is internally-compensated for gains equal to one. If very low noise is of prime importance, it is recommended that the 5532A version be used which has guaranteed noise voltage specifications.

## APPLICATIONS

The Signetics 5532 High-Performance Op Amp is an ideal amplifier for use in high quality and professional audio equipment which requires low noise and low distortion.

The circuit included in this application note has been assembled on a PC board, and tested with actual audio input devices (Tuner and Turntable). It consists of an RIAA (Recording Industry Association of America) preamp, input buffer, 5-band equalizer, and mixer. Although the circuit design is not new, its performance using the 5532 has been improved.

The RIAA preamp section is a standard compensation configuration with low frequency boost provided by the Magnetic cartridge and the RC network in the op amp feedback loop. Cartridge loading is accomplished via R1. 47k was chosen as a typical value, and may differ from cartridge to cartridge.

The Equalizer section consists of an input buffer, 5 active variable band pass/notch (depending on R9's setting) filters, and an output summing amplifier. The input buffer is a standard unity gain design providing impedance matching between the preamplifier and the equalizer section. Because the 5532 is internally-compensated, no external compensation is required. The 5-band active filter section is actually five individual active filters with the same feedback design for all five. The main difference in all five stages is the values of C5 and C6, which are responsible for setting the center frequency of each stage. Linear pots are recommended for R9. To simplify use of this circuit, a component value table is provided, which lists center frequencies and their associated capacitor values. Notice that C5 equals (10) C6, and that the Value of R8 and R10 are related to R9 by a factor of 10 as well. The values listed in the table are common and easily found standard values.

## RIAA EQUALIZATION AUDIO PREAMPLIFIER USING NE5532A

With the onset of new recording techniques with sophisticated playback equipment, a new breed of low noise operational amplifiers was developed to complement the state-of-the-art in audio reproduction. The first ultra-low noise op amp introduced by Signetics was called the NE5534A. This is a single operational amplifier with less than $4nV/\sqrt{Hz}$ input noise voltage. The NE5534A is internally-compensated at a gain of three. This device has been used in many audio preamp and equalizer (active filter) applications since its introduction early last year.

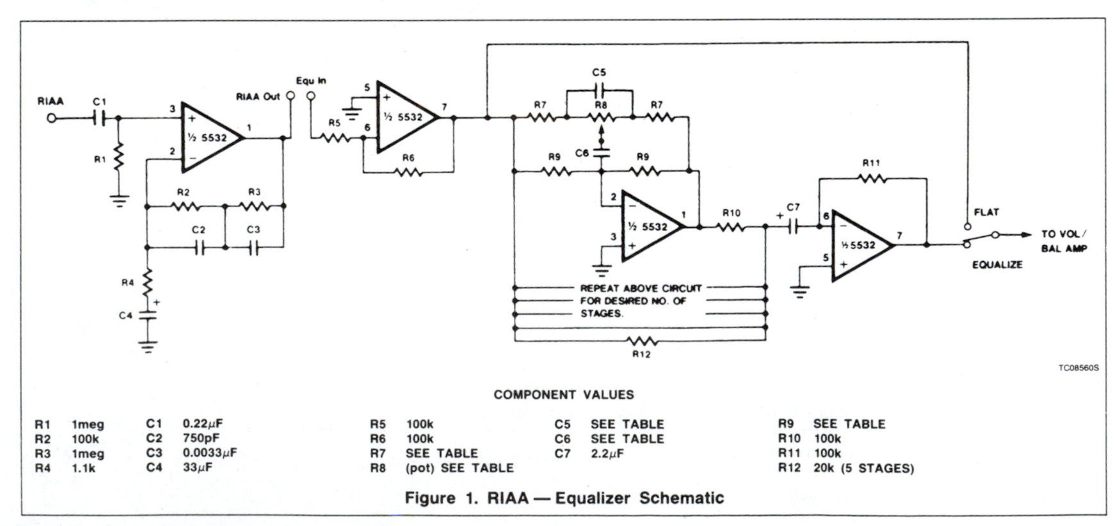


COMPONENT VALUES

| | | | | | | | | | | | |
|---|---|---|---|---|---|---|---|---|---|---|---|
| R1 | 1meg | C1 | 0.22μF | R5 | 100k | C5 | SEE TABLE | R9 | SEE TABLE | | |
| R2 | 100k | C2 | 750pF | R6 | 100k | C6 | SEE TABLE | R10 | 100k | | |
| R3 | 1meg | C3 | 0.0033μF | R7 | SEE TABLE | C7 | 2.2μF | R11 | 100k | | |
| R4 | 1.1k | C4 | 33μF | R8 | (pot) SEE TABLE | | | R12 | 20k (5 STAGES) | | |

**Figure 1. RIAA — Equalizer Schematic**

(Courtesy of Signetics Company)

### COMPONENT VALUES FOR FIGURE 1

| R8 = 25k R7 = 2.4k R9 = 240k | | | R8 = 50k R7 = 5.1k R9 = 510k | | | R8 = 100k R7 = 10k R9 = 1meg | | |
|---|---|---|---|---|---|---|---|---|
| $f_O$ | C5 | C6 | $f_O$ | C5 | C6 | $f_O$ | C5 | C6 |
| 23Hz | 1μF | 0.1μF | 25Hz | 0.47μF | 0.047μF | 12Hz | 0.47μF | 0.047μF |
| 50Hz | 0.47μF | 0.047μF | 36Hz | 0.33μF | 0.033μF | 18Hz | 0.33μF | 0.033μF |
| 72Hz | 0.33μF | 0.033μF | 54Hz | 0.22μF | 0.022μF | 27Hz | 0.22μF | 0.022μF |
| 108Hz | 0.22μF | 0.022μF | 79Hz | 0.15μF | 0.015μF | 39Hz | 0.15μF | 0.015μF |
| 158Hz | 0.15μF | 0.015μF | 119Hz | 0.1μF | 0.01μF | 59Hz | 0.1μF | 0.01μF |
| 238Hz | 0.1μF | 0.01μF | 145Hz | 0.082μF | 0.0082μF | 72Hz | 0.082μF | 0.0082μF |
| 290Hz | 0.082μF | 0.0082μF | 175Hz | 0.068μF | 0.0068μF | 87Hz | 0.068μF | 0.0068μF |
| 350Hz | 0.068μF | 0.0068μF | 212Hz | 0.056μF | 0.0056 μF | 106Hz | 0.056μF | 0.0056μF |
| 425Hz | 0.056μF | 0.0056μF | 253Hz | 0.047μF | 0.0047μF | 126Hz | 0.047μF | 0.0047μF |
| 506Hz | 0.047μF | 0.0047μF | 360Hz | 0.033μF | 0.0033μF | 180Hz | 0.033μF | 0.0033μF |
| 721Hz | 0.033μF | 0.0033μF | 541Hz | 0.022μF | 0.0022μF | 270Hz | 0.022μF | 0.0022μF |
| 1082Hz | 0.022μF | 0.0022μF | 794Hz | 0.015μF | 0.0015μF | 397Hz | 0.015μF | 0.0015μF |
| 1588Hz | 0.015μF | 0.0015μF | 1191Hz | 0.01μF | 0.001μF | 595Hz | 0.01μF | 0.001μF |
| 2382Hz | 0.01μF | 0.001μF | 1452Hz | 0.0082μF | 820pF | 726Hz | 0.0082μF | 820pF |
| 2904Hz | 0.0082μF | 820pF | 1751Hz | 0.0068μF | 680pF | 875Hz | 0.0068μF | 680pF |
| 3502Hz | 0.0068μF | 680pF | 2126Hz | 0.0056μF | 560pF | 1063Hz | 0.0056μF | 560pF |
| 4253Hz | 0.0056μF | 560pF | 2534Hz | 0.0047μF | 470pF | 1267Hz | 0.0047μF | 470pF |
| 5068Hz | 0.0047μF | 470pF | 3609Hz | 0.0033μF | 330pF | 1804Hz | 0.0033μF | 330pF |
| 7218Hz | 0.0033μF | 330pF | 5413Hz | 0.0022μF | 220pF | 2706Hz | 0.0022μF | 220pF |
| 10827Hz | 0.0022μF | 220pF | 7940Hz | 0.0015μF | 150pF | 3970Hz | 0.0015μF | 150pF |
| 15880Hz | 0.0015μF | 150pF | 11910Hz | 0.001μF | 100pF | 5955Hz | 0.001μF | 100pF |
| 23820Hz | 0.001μF | 100pF | 14524Hz | 820pF | 82pF | 7262Hz | 820pF | 82pF |
| | | | 17514Hz | 680pF | 68pF | 8757Hz | 680pF | 68pF |
| | | | 21267Hz | 560pF | 56pF | 10633Hz | 560pF | 56pF |
| | | | | | | 12670Hz | 470pF | 47pF |
| | | | | | | 18045Hz | 330pF | 33pF |

Many of the amplifiers that are being designed today are DC-coupled. This means that very low frequencies (2 – 15Hz) are being amplified. These low frequencies are common to turntables because of rumble and tone arm resonancies. Since the amplifiers can reproduce these sub-audible tones, they become quite objectionable because the speakers try to reproduce these tones. This causes non-linearities when the actual recorded material is amplified and converted to sound waves.

The RIAA has proposed a change in its standard playback response curve in order to alleviate some of the problems that were previously discussed. The changes occur primarily at the low frequency range with a slight modification to the high frequency range (See Figure 2). Note that the response peak for the bass section of the playback curve now occurs at 31.5Hz and begins to roll off below that frequency. The roll-off occurs by introducing a fourth RC network with a 7950μs time constant to the three existing networks that make up the equalization circuit. The high end of the equalization curve is extended to 20kHz, because recordings at these frequencies are achievable on many current discs.

## NE5533/34 DESCRIPTION

the 5533/5534 are dual and single high-performance low noise operational amplifiers. Compared to other operational amplifiers, such as TL083, they show better noise performance, improved output drive capability and considerably higher small-signal and power bandwidths.

This makes the devices especially suitable for application in high quality and professional audio equipment, instrumentation and control circuits, and telephone channel amplifiers.

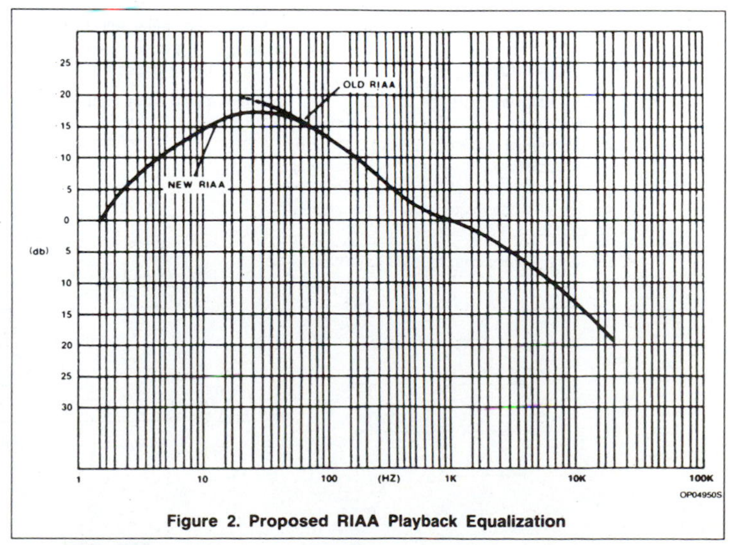


**Figure 2. Proposed RIAA Playback Equalization**

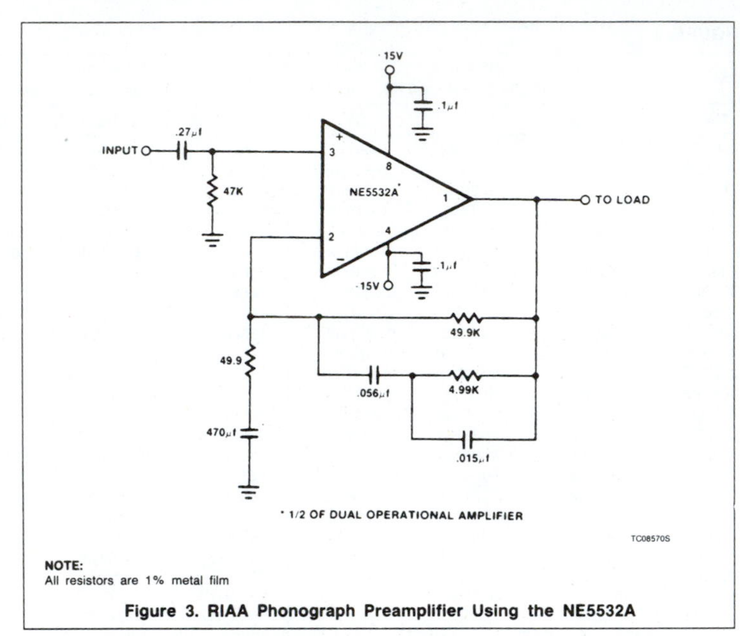


NOTE:
All resistors are 1% metal film

**Figure 3. RIAA Phonograph Preamplifier Using the NE5532A**

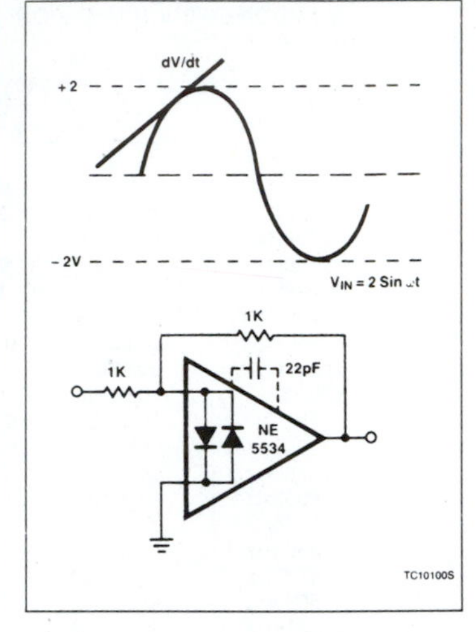


The op amps are internally-compensated for gain equal to, or higher than, three. The frequency response can be optimized with an external compensation capacitor for various applications (unity gain amplifier, capacitive load, slew rate, low overshoot, etc.) If very low noise is of prime importance, it is recommended that the 5533A/5534A version be used which has guaranteed noise specifications.

## APPLICATIONS

### Diode Protection of Input

The input leads of the device are protected from differential transients above ±0.6V by internal back-to-back diodes. Their presence imposes certain limitations on the amplifier dynamic characteristics related to closed-loop gain and slew rate.

Consider the unity gain follower as an example:

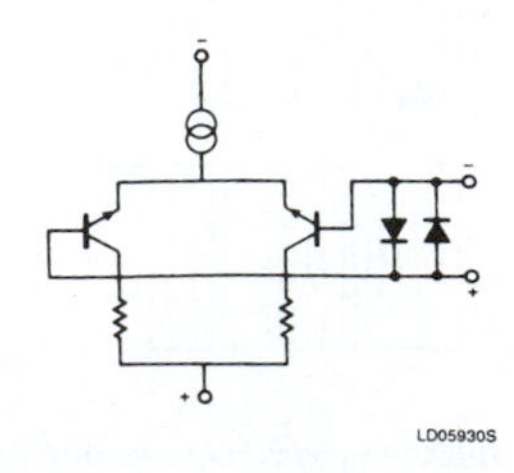


Assume a signal input square wave with dV/dt of 250V/µs and 2V peak amplitude as shown. If a 22pF compensation capacitor is inserted and the $R_1$ $C_1$ circuit deleted, the device slew rate falls to approximately 7V/µs. The input waveform will reach 2V/250V/µs or 8ns, while the output will have changed ($8 \times 10^{-3}$) only 56mV. The differential input signal is then ($V_{IN} - V_O$) $R_I/R_I + R_F$ or approximately 1V.

The diode limiter will definitely be active and output distortion will occur; therefore, $V_{IN} < 1V$ as indicated.

Next, a sine wave input is used with a similar circuit.

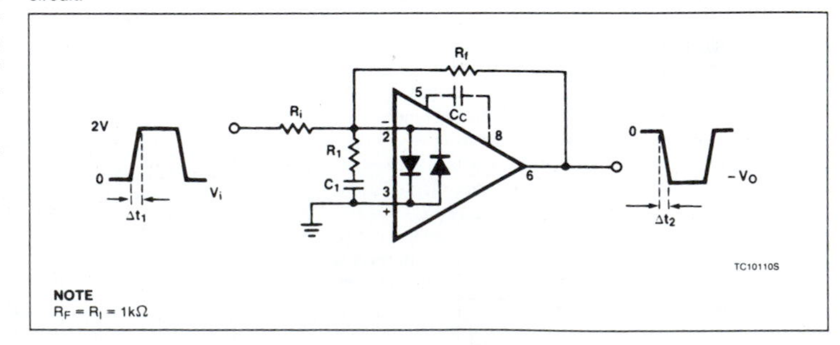


NOTE
$R_F = R_I = 1k\Omega$

The slew rate of the input waveform now depends on frequency and the exact expression is

$$\frac{dv}{dt} = 2\omega \cos \omega t$$

The upper limit before slew rate distortion occurs for *small-signal* ($V_{IN} < 100mV$) conditions is found by setting the slew rate to 7V/µs. That is:

$$7 \times 10^6 V/\mu s = 2\omega \cos \omega t$$

at $\omega t = 0$

$$\omega_{LIMIT} = \frac{7 \times 10^6}{2} = 3.5 \times 10^6 \text{ rad/s}$$

$$f_{LIMIT} = \frac{3.5 \times 10^6}{2\pi} \cong 560kHz$$

### External Compensation Network Improves Bandwidth

By using an external lead-lag network, the follower circuit slew rate and small-signal bandwidth can be increased. This may be useful in situations where a closed-loop gain less than 3 to 5 is indicated. A number of examples are shown in subsequent figures. The principle benefit of using the network approach is that the full slew rate and bandwidth of the device is retained, while impulse-related parameters such as damping and phase margin are controlled by choosing the appropriate circuit constants. For example, consider the following configuration:

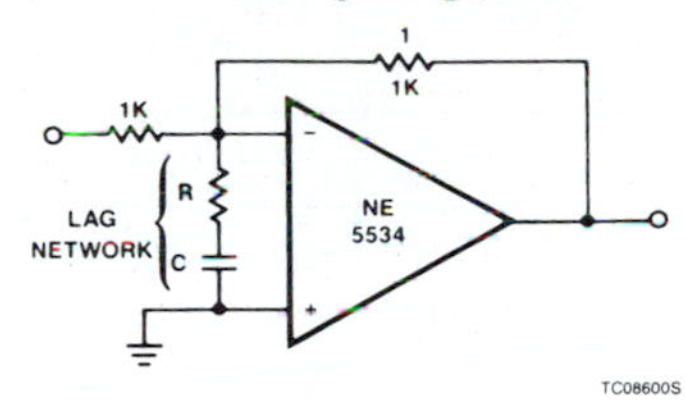


The major problem to be overcome is poor phase margin leading to instability.

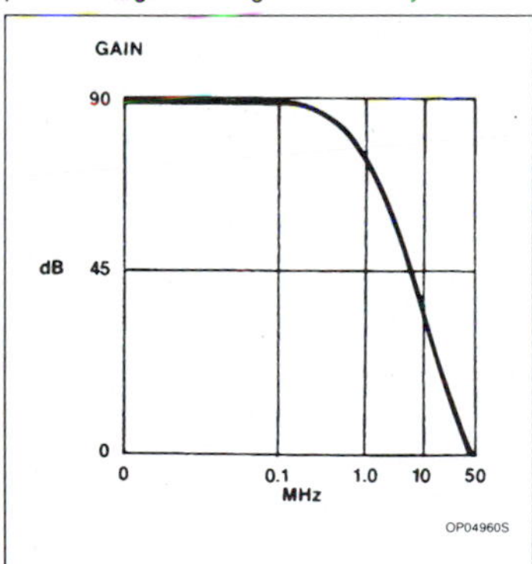


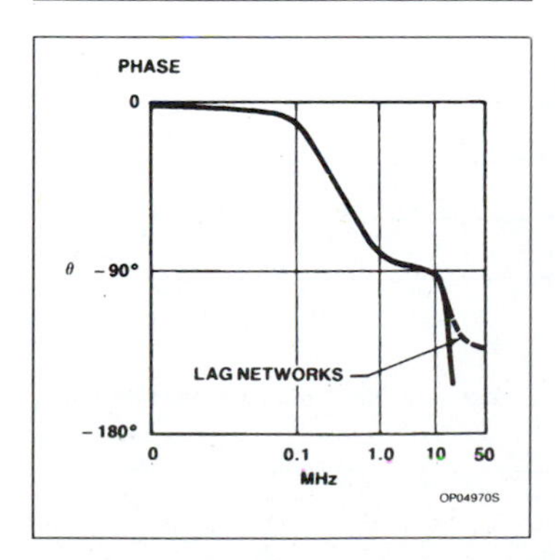


By choosing the lag network break frequency one decade below the unity gain crossover frequency (30 – 50MHz), the phase and gain margin are improved. An appropriate value for R is 270Ω. Setting the lag network break frequency at 5MHz, C may be calculated

$$C = \frac{1}{2\pi \cdot 270 \cdot 5 \times 10^6}$$
$$= 118\text{pF}$$

## RULES AND EXAMPLES

### Compensation Using Pins 5 and 8 (Limited Bandwidth and Slew Rate)

A single-pole and zero inserted in the transfer function will give an added 45° of phase margin, depending on the network values.

**Calculating the Lead-Lag Network**

$$C_1 = \frac{1}{2\pi F_1 R_1} \quad \text{Let } R_1 = \frac{R_{IN}}{10}$$

where

$$F_1 = \frac{1}{10}\,(\text{UGBW})$$

$$\text{UGBW} = 30\text{MHz}$$

### External Compensation for Wide-Band Voltage-Follower

**Shunt Capacitance Compensation**

$$C_F = \frac{1}{2\pi F_F R_F}, \quad F_F \cong 30\text{MHz}$$

or

$$C_F \cong \frac{C_{DIST}}{A_{CL}}$$

$C_{DIST} \cong$ Distributed Capacitance $\cong$ 2 – 3pF

Many audio circuits involve carefully-tailored frequency responses. Pre-emphasis is used in all recording mediums to reduce noise and produce flat frequency response. The most often used de-emphasis curves for broadcast and home entertainment systems are shown in Figure 7. Operational amplifiers are well suited to these applications because of their high gain and easily-tailored frequency response.

## RIAA PREAMP USING THE NE5534

The preamplifier for phono equalization is shown in Figure 8 with the theoretical and actual circuit response.

Low frequency boost is provided by the inductance of the magnetic cartridge with the

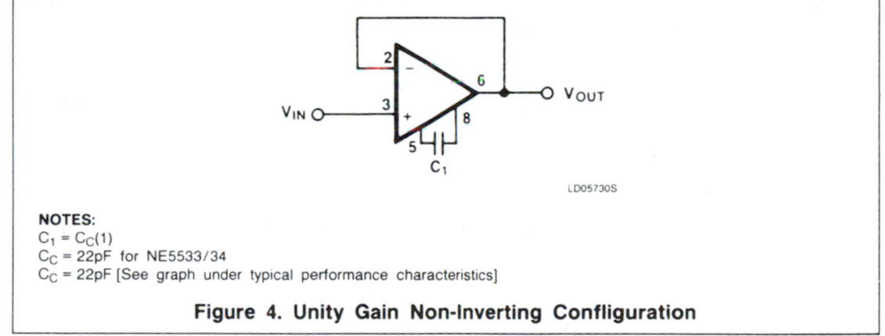


NOTES:
$C_1 = C_C(1)$
$C_C$ = 22pF for NE5533/34
$C_C$ = 22pF [See graph under typical performance characteristics]

**Figure 4. Unity Gain Non-Inverting Configuration**

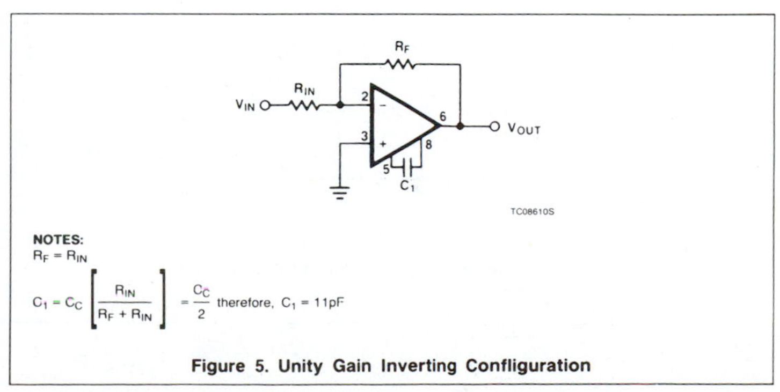


NOTES:
$R_F = R_{IN}$

$$C_1 = C_C\left[\frac{R_{IN}}{R_F + R_{IN}}\right] = \frac{C_C}{2} \text{ therefore, } C_1 = 11\text{pF}$$

**Figure 5. Unity Gain Inverting Configuration**

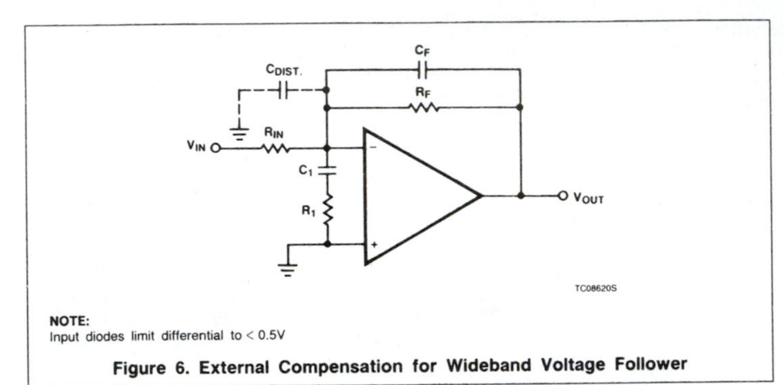


**NOTE:**
Input diodes limit differential to < 0.5V

**Figure 6. External Compensation for Wideband Voltage Follower**

RC network providing the necessary break points to approximate the theoretical RIAA curve.

## RUMBLE FILTER

Following the amplifier stage, rumble and scratch filters are often used to improve overall quality. Such a filter designed with op amps uses the 2-pole Butterworth approach and features switchable break points. With the circuit of Figure 9, any degree of filtering from fairly sharp to none at all is switch-selectable.

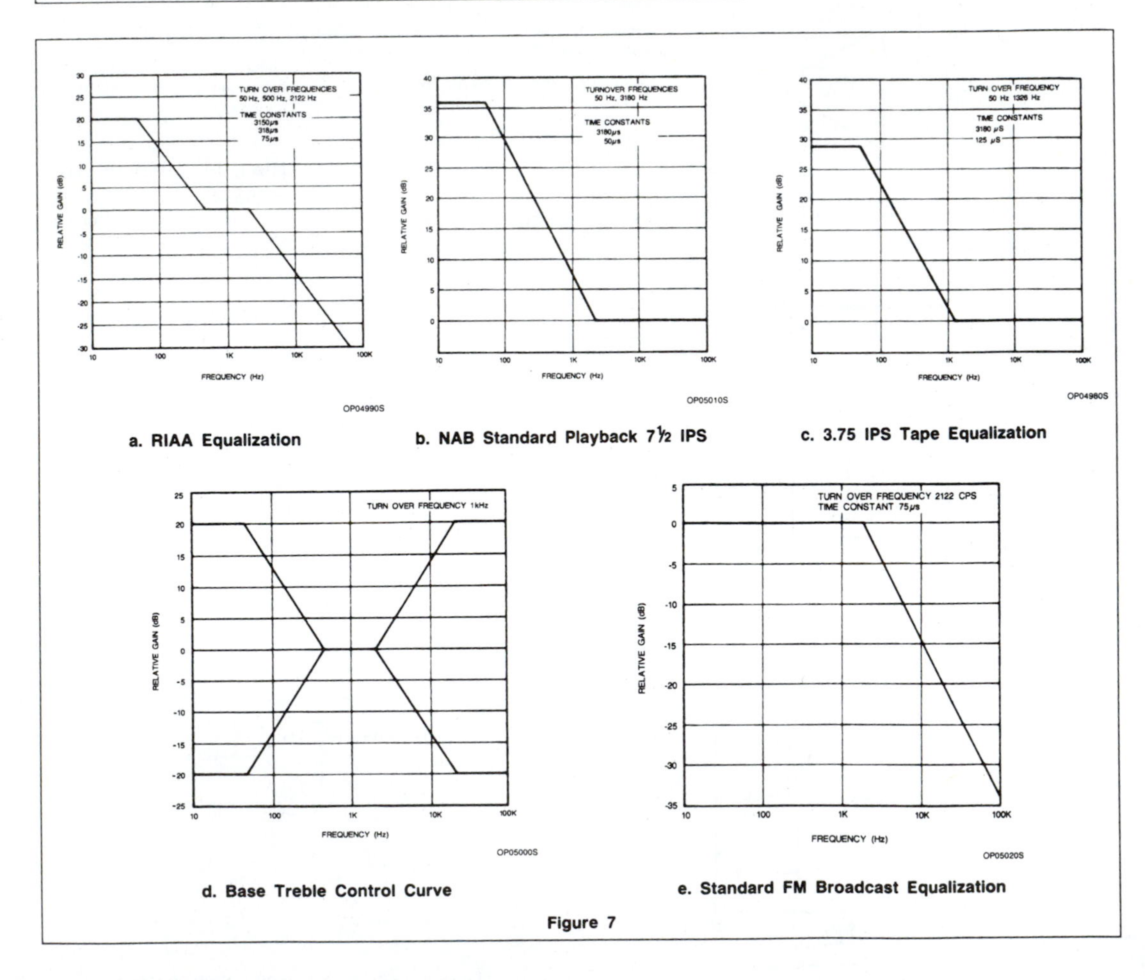


a. RIAA Equalization

b. NAB Standard Playback 7½ IPS

c. 3.75 IPS Tape Equalization

d. Base Treble Control Curve

e. Standard FM Broadcast Equalization

**Figure 7**

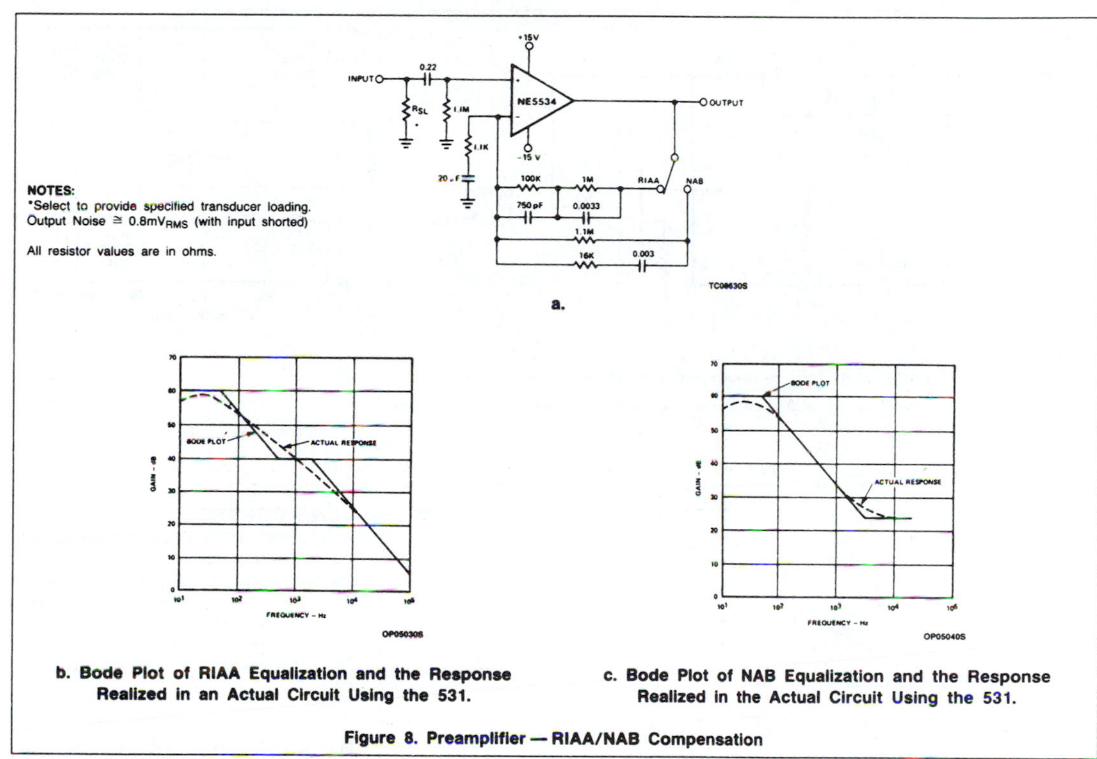


**NOTES:**
*Select to provide specified transducer loading.
Output Noise ≅ 0.8mV$_{RMS}$ (with input shorted)

All resistor values are in ohms.

**b. Bode Plot of RIAA Equalization and the Response Realized in an Actual Circuit Using the 531.**

**c. Bode Plot of NAB Equalization and the Response Realized in the Actual Circuit Using the 531.**

**Figure 8. Preamplifier — RIAA/NAB Compensation**

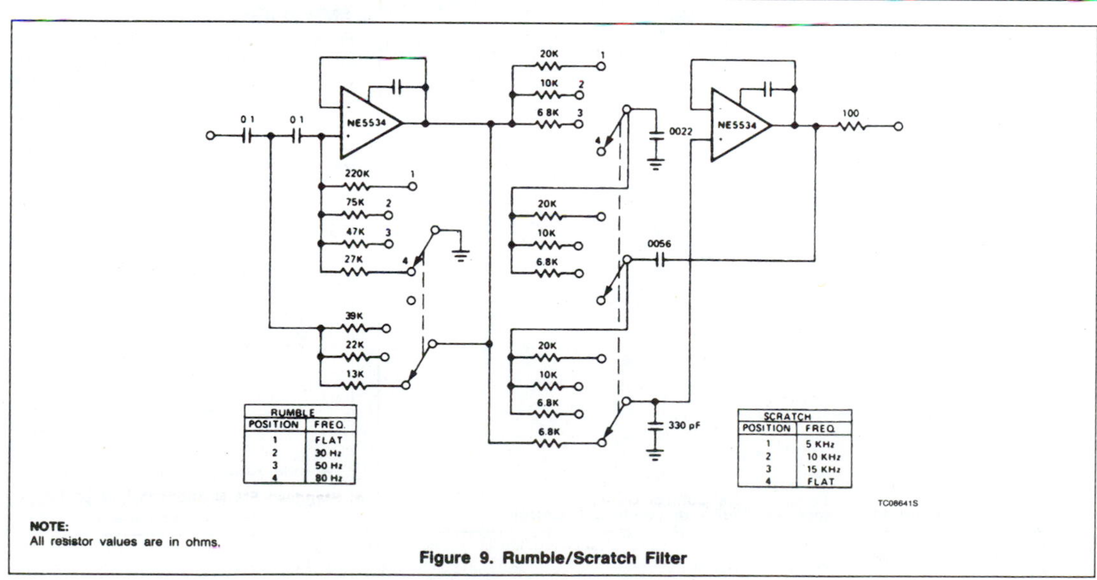


| RUMBLE | |
|---|---|
| POSITION | FREQ. |
| 1 | FLAT |
| 2 | 30 Hz |
| 3 | 50 Hz |
| 4 | 80 Hz |

| SCRATCH | |
|---|---|
| POSITION | FREQ. |
| 1 | 5 KHz |
| 2 | 10 KHz |
| 3 | 15 KHz |
| 4 | FLAT |

**NOTE:**
All resistor values are in ohms.

**Figure 9. Rumble/Scratch Filter**

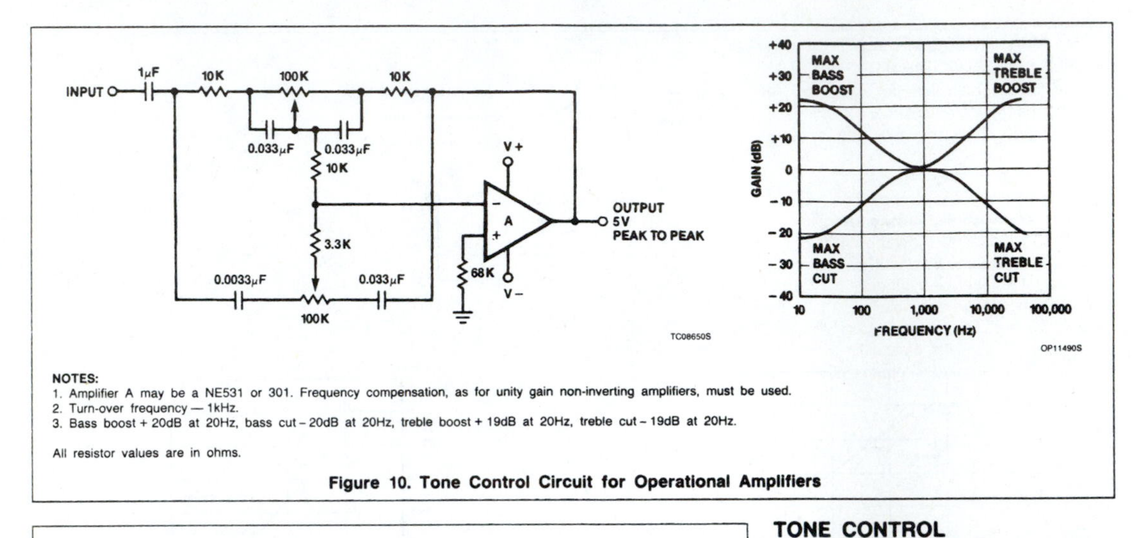


NOTES:
1. Amplifier A may be a NE531 or 301. Frequency compensation, as for unity gain non-inverting amplifiers, must be used.
2. Turn-over frequency — 1kHz.
3. Bass boost + 20dB at 20Hz, bass cut − 20dB at 20Hz, treble boost + 19dB at 20Hz, treble cut − 19dB at 20Hz.

All resistor values are in ohms.

**Figure 10. Tone Control Circuit for Operational Amplifiers**

## TONE CONTROL

Tone control of audio systems involves altering the flat response in order to attain more low frequencies or more high ones, dependent upon listener preference. The circuit of Figure 10 provides 20dB of bass or treble boost or cut as set by the variable resistance. The actual response of the circuit is shown also.

## BALANCE AND LOUDNESS AMPLIFIER

Figure 11 shows a combination of balance and loudness controls. Due to the non-linearity of the human hearing system, the low frequencies must be boosted at low listening levels. Balance, level, and loudness controls provide all the listening controls to produce the desired music response.

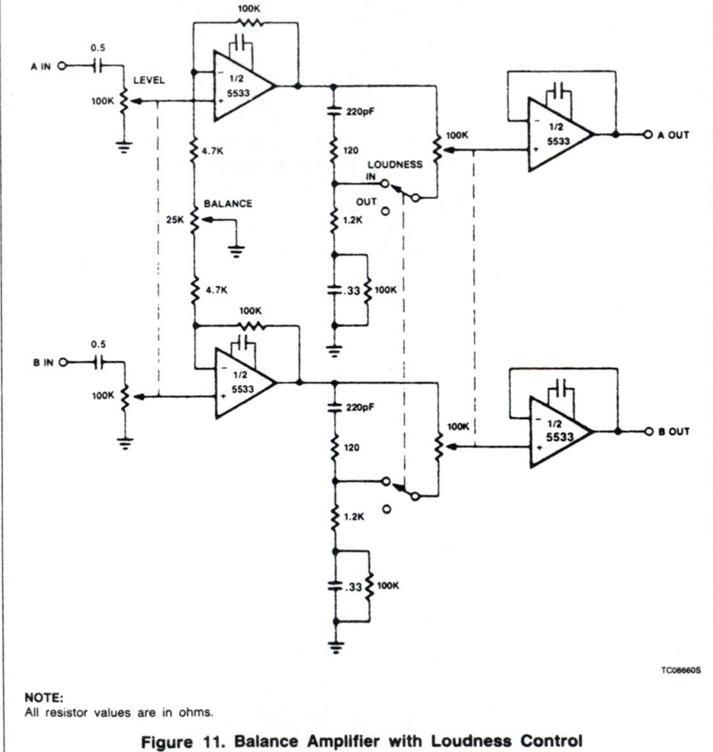


NOTE:
All resistor values are in ohms.

**Figure 11. Balance Amplifier with Loudness Control**

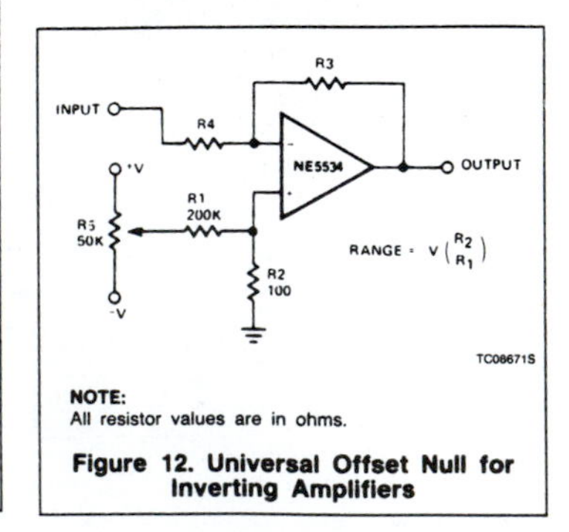


NOTE:
All resistor values are in ohms.

**Figure 12. Universal Offset Null for Inverting Amplifiers**

## VOLTAGE AND CURRENT OFFSET ADJUSTMENTS

Many IC amplifiers include the necessary pin connections to provide external offset adjustments. Many times, however, it becomes necessary to select a device not possessing external adjustments. Figures 12, 13, and 14 suggest some possible arrangements for offset voltage adjust and bias current nulling circuitry. The circuitry of Figure 14 provides sufficient current into the input to cancel the bias current requirement. Although more simplified arrangements are possible, the addition of Q2 and Q3 provide a fixed current level to Q1, thus, bias cancellation can be provided without regard to input voltage level.

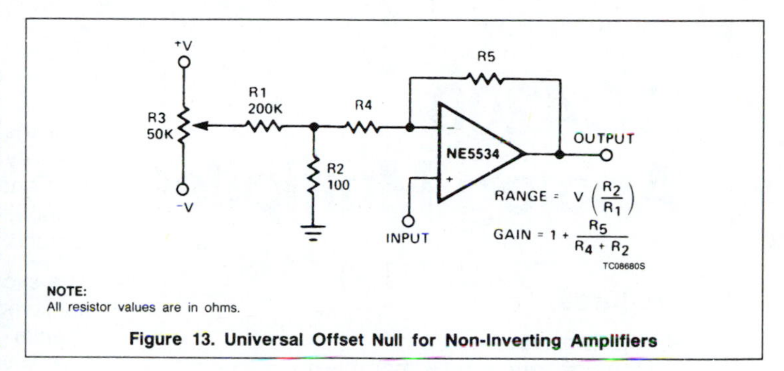


NOTE:
All resistor values are in ohms.

**Figure 13. Universal Offset Null for Non-Inverting Amplifiers**

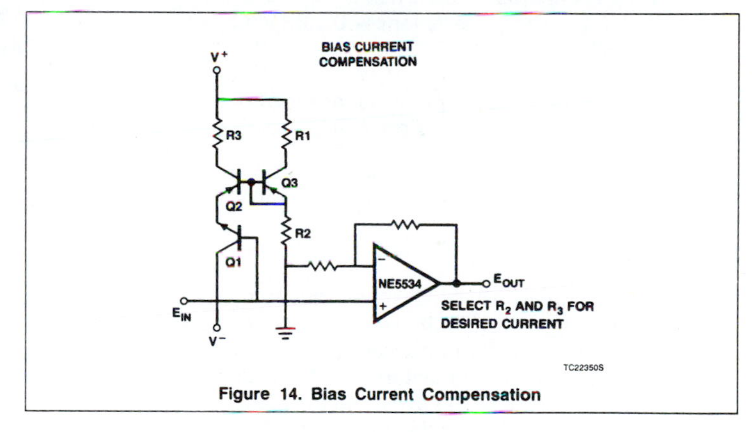


**Figure 14. Bias Current Compensation**

# Appendix L RC4200 Analog Multiplier

# RC4200 Analog Multiplier

## Features

- High accuracy
  Non-linearity — 0.1% maximum
  Temperature coefficient — 0.005%/°C typical
- Multiple functions
  Multiply, divide square, square root, RMS-to-DC conversion, AGC, and modulate/demodulate
- Wide bandwidth — 4MHz
- Signal-to-noise ratio — 94dB

## Description

The Raytheon RC4200 is the industry's first integrated circuit multiplier to have complete compensation for nonlinearity, the primary source of error and distortion. This is also the first IC multiplier to have three on-board operational amplifiers designed specifically for use in multiplier logging circuits. These specially designed amplifiers are frequency compensated for optimum AC response in a logging circuit, the heart of a multiplier, and can therefore provide superior AC response in comparison to other analog multipliers.

Versatility is unprecedented; this is the first IC multiplier that can be used in a wide variety of applications without sacrificing accuracy. Four-quadrant multiplication, two-quadrant division, square-rooting, squaring and RMS conversion can all be easily implemented with predictable accuracy. The nonlinearity compensation is not just trimmed at a single temperature, it is designed to provide compensation over the full temperature range. This nonlinearity compensation combined with the low gain and offset drift inherent in a well designed monolithic chip provides a very low accuracy tempco.

The excellent linearity and versatility were achieved through circuit design rather than special grading or trimming, and therefore unit cost is very low. Analog multipliers can now be used in applications where price was previously an inhibiting factor.

The Raytheon RC4200 is ideal for use in low distortion audio modulation circuits, voltage-controlled active filters, and precision oscillators.

## Connection Information

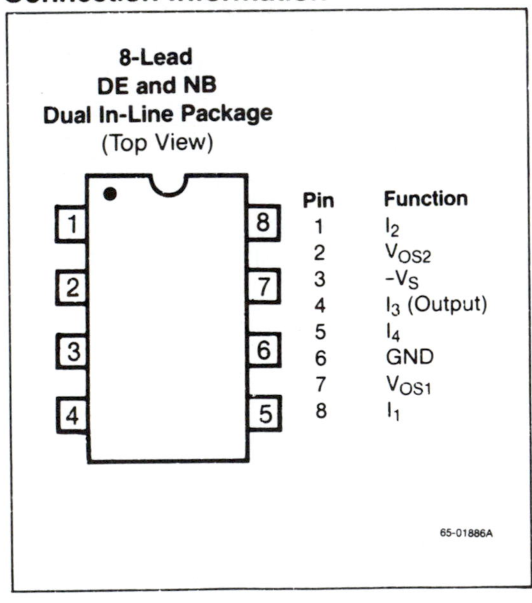


| Pin | Function |
|---|---|
| 1 | $I_2$ |
| 2 | $V_{OS2}$ |
| 3 | $-V_S$ |
| 4 | $I_3$ (Output) |
| 5 | $I_4$ |
| 6 | GND |
| 7 | $V_{OS1}$ |
| 8 | $I_1$ |

(Courtesy of Raytheon Company, Semiconductor Division)

## Absolute Maximum Ratings

Supply Voltage .......................................... -22V
Internal Power Dissipation** .................. 500 mV
Input Current .......................................... -5mA
Storage Temperature Range
RM4200/4200A .................. -65°C to +150°C
RV4200/4200A .................. -55°C to +125°C
RC4200/4200A .................. -55°C to +125°C
Operating Temperature Range
RM4200/4200A .................. -55°C to +125°C
RV4200/4200A .................. -25°C to +85°C
RC4200/4200A .................. 0°C to +70°C

*For supply voltages less than ±22V, the absolute maximum input voltage is equal to the supply voltage.
**Observe package thermal characteristics.

## Ordering Information

| Part Number | Package | Operating Temperature Range |
|---|---|---|
| RC4200N<br>RC4200AN | N<br>N | 0°C to +70°C<br>0°C to +70°C |
| RV4200D<br>RV4200AD | D<br>D | -25°C to +85°C<br>-25°C to +85°C |
| RM4200D<br>RM4200D/883B<br>RM4200AD<br>RM4200AD/883B | D<br>D<br>D<br>D | -55°C to +125°C<br>-55°C to +125°C<br>-55°C to +125°C<br>-55°C to +125°C |

Notes:
/883B suffix denotes Mil-Std-883, Level B processing
N = 8-lead plastic DIP
D = 8-lead ceramic DIP
Contact a Raytheon sales office or representative for ordering information on special package/temperature range combinations.

## Thermal Characteristics

| | 8-Lead Plastic DIP | 8-Lead Ceramic DIP |
|---|---|---|
| Max. Junction Temp. | 125°C | 175°C |
| Max. $P_D$ $T_A < 50°C$ | 468mW | 833mW |
| Therm. Res. $\theta_{JC}$ | — | 45°C/W |
| Therm. Res. $\theta_{JA}$ | 160°C/W | 150°C/W |
| For $T_A > 50°C$ Derate at | 6.25mW per °C | 8.33mW per °C |

## Mask Pattern

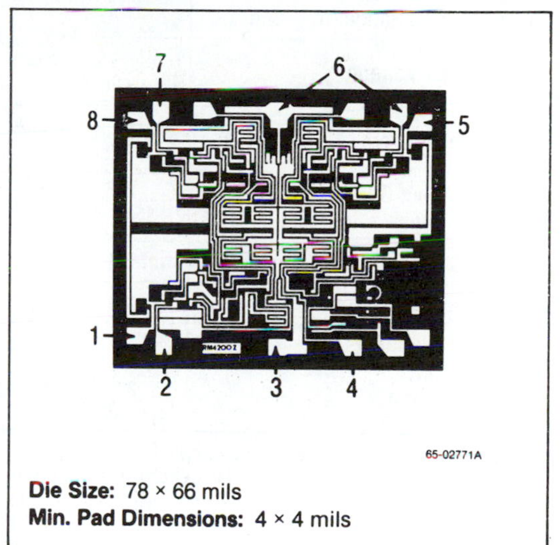


**Die Size:** 78 × 66 mils
**Min. Pad Dimensions:** 4 × 4 mils

# RC4200 Analog Multiplier *Continued*

## Electrical Characteristics

(Over Operating Temperature Range, $V_S$ = −15V unless otherwise noted)

| Parameters | Test Conditions | 4200A | | | 4200 | | | Units |
|---|---|---|---|---|---|---|---|---|
| | | Min | Typ | Max | Min | Typ | Max | |
| Total Error as Multiplier Untrimmed | $T_A$ = +25°C (Note 1) | | | ±2.0 | | | ±3.0 | % |
| With External Trim | | | ±0.2 | | | ±0.2 | | % |
| Versus Temperature | | | ±0.005 | | | ±0.005 | | %/°C |
| Versus Supply (−9 to −18V) | | | ±0.1 | | | ±0.1 | | %/V |
| Nonlinearity | $50\mu A \leq I_{1,2,4} \leq 250\mu A$, $T_A$ = +25°C (Note 2) | | | ±0.1 | | | ±0.3 | % |
| Input Current Range ($I_1$, $I_2$ and $I_4$) | | 1.0 | | 1000 | 1.0 | | 1000 | μA |
| Input Offset Voltage | $I_1 = I_2 = I_4 = 150\mu A$, $T_A$ = +25°C | | | ±5.0 | | | ±10 | mV |
| Input Bias Current | $I_1 = I_2 = I_4 = 150\mu A$, $T_A$ = +25°C | | | 300 | | | 500 | nA |
| Average Input Offset Voltage Drift | $I_1 = I_2 = I_4 = 150\mu A$ | | | ±50 | | | ±100 | μV/°C |
| Output Current Range ($I_3$) | (Note 3) | 1.0 | | 1000 | 1.0 | | 1000 | μA |
| Frequency Response, −3dB point | | | 4.0 | | | 4.0 | | MHz |
| Supply Voltage | | −18 | −15 | −9.0 | −18 | −15 | −9.0 | V |
| Supply Current | $I_1 = I_2 = I_4 = 150\mu A$, $T_A$ = +25°C | | | 4.0 | | | 4.0 | mA |

Notes:
1. Refer to Figure 6 for example.
2. The input circuits tend to become unstable at $I_1, I_2, I_4 < 50\mu A$ and linearity decreases when $I_1, I_2, I_4 > 250\mu A$ (eq. @ $I_1 = I_2 = 500\mu A$, nonlinearity error ≈ 0.5%).
3. These specifications apply with output ($I_3$) connected to an op amp summing junction. If desired, the output ($I_3$) at pin (4) can be used to drive a resistive load directly. The resistive load should be less than 700Ω and must be pulled up to a positive supply such that the voltage on pin (4) stays within a range of 0 to +5V.

## Functional Description

The RC4200 multiplier is designed to multiply two input currents ($I_1$ and $I_2$) and to divide by a third input current ($I_4$). The output is also in the form of a current ($I_3$). A simplified circuit diagram is shown in Figure 1. The nominal relationship between the three inputs and the output is:

$$I_3 = \frac{I_1 I_2}{I_4} \qquad (1)$$

The three input currents must be positive and restricted to a range of 1μA to 1mA. These currents go into the multiplier chip at op-amp summing junctions which are nominally at zero volts. Therefore, an input voltage can be easily converted to an input current by a series resistor. Any number of currents may be summed at the inputs. Depending on the application, the output current can be converted to a voltage by an external op amp or used directly. This capability of combining input currents and voltages in various combinations provides great versatility in application.

Inside the multiplier chip, the three op amps make the collector currents of transistors Q1, Q2, and Q4 equal to their respective input currents ($I_1$, $I_2$, and $I_4$). These op amps are designed with current-source outputs and are phase-compensated for optimum frequency response as a multiplier. Power drain of the op amps was minimized to prevent the introduction of undesired thermal gradients on the chip. The three op amps operate on a single supply voltage (nominally −15V) and total quiescent current drain is less than 4mA. These special op amps provide significantly improved performance in comparison to 741-type op amps.

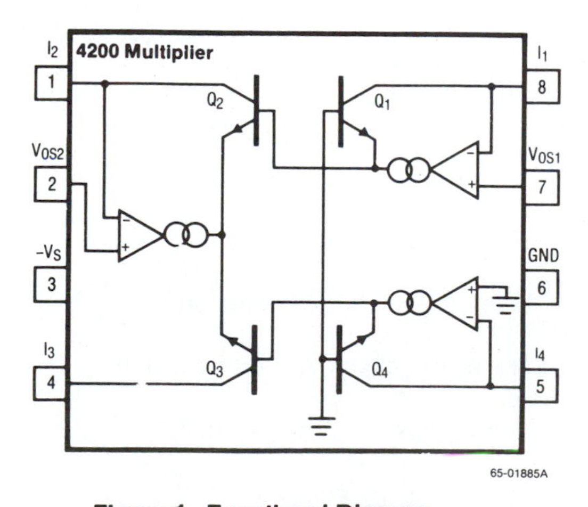


**Figure 1. Functional Diagram**

The actual multiplication is done within the log-antilog configuration of the Q1–Q4 transistor array. These four transistors, with associated proprietary circuitry, were specially designed to precisely implement the relationship

$$V_{BEN} = \frac{kT}{q} \ln \frac{I_{CN}}{I_{SN}} \qquad (2)$$

Previous multiplier designs have suffered from an additional undesired linear term in the above equation; the collector current times the emitter resistance. This $I_C r_E$ term introduces a parabolic nonlinearity even with matched transistors. Raytheon has developed a unique and proprietary means of inherently compensating for this undesired $I_C r_E$ term. Furthermore, this Raytheon-developed circuit technique compensates linearity error over temperature changes. The nonlinearity versus temperature is significantly improved over earlier designs.

From equation (2) and by assuming equal transistor junction temperatures, summing base-to-emitter voltage drops around the transistor array yields:

$$\frac{kT}{q}\left[\ln \frac{I_1}{I_{S1}} + \ln \frac{I_2}{I_{S2}} - \ln \frac{I_3}{I_{S3}} - \ln \frac{I_4}{I_{S4}}\right] = 0 \qquad (3)$$

This equation reduces to:

$$\frac{I_1 I_2}{I_3 I_4} = \frac{I_{S1} I_{S2}}{I_{S3} I_{S4}} \qquad (4)$$

The ratio of reverse saturation currents, $I_{S1} I_{S2} / I_{S3} I_{S4}$, depends on the transistor matching. In a monolithic multiplier this matching is easily achieved and the ratio is very close to unity, typically 1.0 ±1%. The final result is the desired relationship:

# RC4200 Analog Multiplier *Continued*

$$I_3 = \frac{I_1 I_1}{I_4} \qquad (5)$$

The inherent linearity and gain stability combined with low cost and versatility makes this new circuit ideal for a wide range of nonlinear functions.

## Basic Circuits

### Current Multiplier/Divider

The basic design criteria for all circuit configurations using the 4200 multiplier is contained in equation (1):

i.e., $$I_3 = \frac{I_1 I_2}{I_4}$$

The current-product-balance equation restates this as:

$$I_1 I_2 = I_3 I_4 \qquad (6)$$

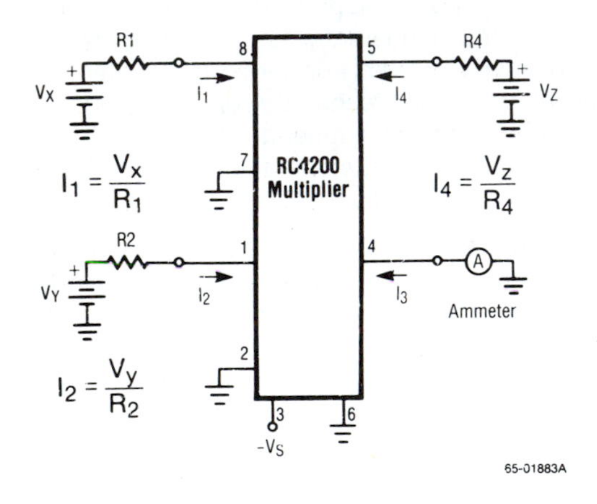


**Figure 2**

### Dynamic Range and Stability

The precision dynamic range for the 4200 is from +50μA to +250μA inputs for $I_1$, $I_2$ and $I_4$. Stability and accuracy degrade if this range is exceeded.

To improve the stability for input currents less than 50μA, filter circuits ($R_S C_S$) are added to each input (see Figure 3).

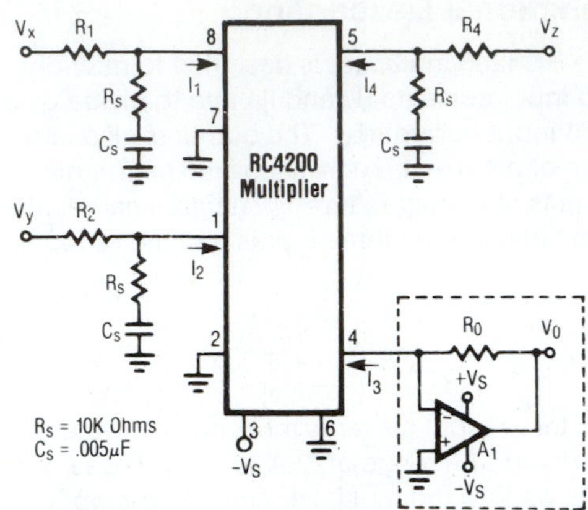


Amplifier $A_1$ is used to convert the $I_3$ current to an output voltage.

Multiplier: $V_z$ = constant $\neq 0$

Divider: $V_y$ = constant $\neq 0$

65-01882A

**Figure 3**

### Voltage Multiplier/Divider

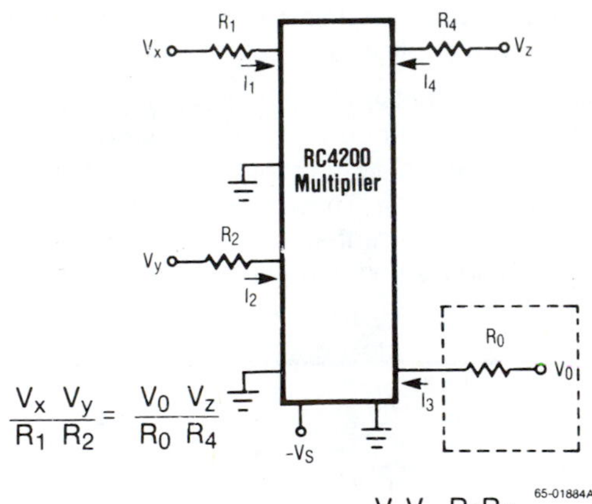


$$\frac{V_x}{R_1}\frac{V_y}{R_2} = \frac{V_0}{R_0}\frac{V_z}{R_4}$$

65-01884A

Solving for $V_0$: $$V_0 = \frac{V_x V_y}{V_z}\frac{R_0 R_4}{R_1 R_2}$$

For a multiplier circuit $V_z = V_R$ = constant

Therefore: $V_0 = V_x V_y K$ where $K = \frac{R_0 R_4}{V_R R_1 R_2}$

For a divider circuit $V_y = V_R$ = constant

Therefore: $V_0 = \frac{V_x}{V_z} K$ where $K = \frac{V_R R_0 R_4}{R_1 R_2}$

**Figure 4**

### Extended Range

The input and output voltage ranges can be extended to include 0 and negative voltage signals by adding bias currents. The $R_SC_S$ filter circuits are eliminated when the input and biasing resistors are selected to limit the respective currents to 50μA min. and 250μA max.

### Extended Range Multiplier

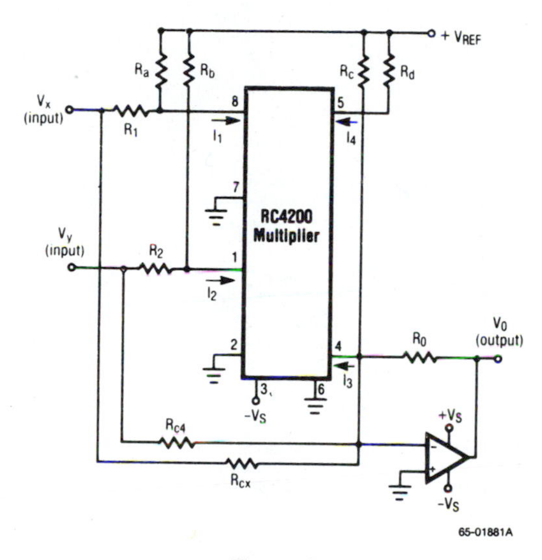


**Figure 5**

Resistors $R_a$ and $R_b$ extend the range of the $V_x$ and $V_y$ inputs by picking values such that:

$$I_1(\text{min.}) = \frac{V_x(\text{min.})}{R_1} + \frac{V_{REF}}{R_a} = 50\mu A,$$

$$\text{and } I_1(\text{max.}) = \frac{V_x(\text{max.})}{R_1} + \frac{V_{REF}}{R_a} = 250\mu A;$$

$$\text{also } I_2(\text{min.}) = \frac{V_y(\text{min.})}{R_2} + \frac{V_{REF}}{R_b} = 50\mu A,$$

$$\text{and } I_2(\text{max.}) = \frac{V_y(\text{max.})}{R_2} + \frac{V_{REF}}{R_b} = 250\mu A.$$

Resistor $R_c$ supplies bias current for $I_3$ which allows the output to go negative.

Resistors $R_{cx}$ and $R_{cy}$ permit equation (6) to balance, i.e.:

$$\left(\frac{V_x}{R_1} + \frac{V_{REF}}{R_a}\right)\left(\frac{V_y}{R_2} + \frac{V_{REF}}{R_b}\right) = \left(\frac{V_0}{R_0} + \frac{V_{REF}}{R_c} + \frac{V_x}{R_{cx}} + \frac{V_y}{R_{cy}}\right)\left(\frac{V_{REF}}{R_d}\right)$$

$$\frac{V_xV_y}{R_1R_2} + \frac{V_xV_{REF}}{R_1R_b} + \frac{V_yV_{REF}}{R_2R_a} + \frac{V_{REF}^2}{R_aR_b} =$$

$$\frac{V_0V_{REF}}{R_0R_d} + \frac{V_xV_{REF}}{R_{cx}R_d} + \frac{V_yV_{REF}}{R_{cy}R_d} + \frac{V_{REF}^2}{R_cR_d}$$

### Cross-Product Cancellation

Cross-products are a result of the $V_xV_R$ and $V_yV_R$ terms. To the extent that: $R_1R_b = R_{cx}R_d$ and $R_2R_a = R_{cy}R_d$, cross-product cancellation will occur.

### Arithmetic Offset Cancellation

The offset caused by the $V_{REF}^2$ term will cancel to the extent that: $R_aR_b = R_cR_d$, and the result is:

$$\frac{V_xV_y}{R_1R_2} = \frac{V_0V_{REF}}{R_0R_d} \quad \text{or } V_0 = V_xV_y K$$

$$\text{where } K = \frac{R_0R_d}{V_{REF}R_1R_2}$$

### Resistor Values

Inputs:

$$V_x(\text{min.}) \le V_x \le V_x(\text{max.})$$

$$\Delta V_x = V_x(\text{max.}) - V_x(\text{min.})$$

$$V_y(\text{min.}) \le V_y \le V_y(\text{max.})$$

$$\Delta V_y = V_y(\text{max.}) - V_y(\text{min.})$$

$$V_{REF} = \text{Constant } (+7V \text{ to } +18V)$$

$$K = \frac{V_0}{V_xV_y} \text{ (Design Requirement)}$$

$$R_1 = \frac{\Delta V_x}{200\mu A},\ R_2 = \frac{\Delta V_y}{200\mu A},\ R_d = \frac{V_{REF}}{250\mu A}$$

$$R_a = \frac{\Delta V_x V_{REF}}{250\mu A\ \Delta V_x - 200\mu A\ V_x(\text{max.})}$$

$$R_b = \frac{\Delta V_y V_{REF}}{250\mu A\ \Delta V_y - 200\mu A\ V_y(\text{max.})}$$

$$R_c = \frac{R_aR_b}{R_d},\ R_{cx} = \frac{R_1R_b}{R_d},\ R_{cy} = \frac{R_2R_a}{R_d}$$

$$R_0 = \frac{\Delta V_x\ \Delta V_y\ K}{160\mu A}$$

## RC4200 Analog Multiplier *Continued*

**Multiplying Circuit Offset Adjust**

$$10K \leq R_5 = R_9 = R_{16} \leq 50K$$

$$R_7 = R_{11} = R_{14} = 100\Omega$$

$$R_6 = R_{10} = 100\Omega \frac{V_S}{.05}$$

$$R_{15} = 100\Omega \frac{V_S}{.10}$$

$$R_8 = R_1 || R_a$$

$$R_{12} = R_2 || R_b$$

$$R_{13} = R_0 || R_c || R_{cx} || R_{cy}$$

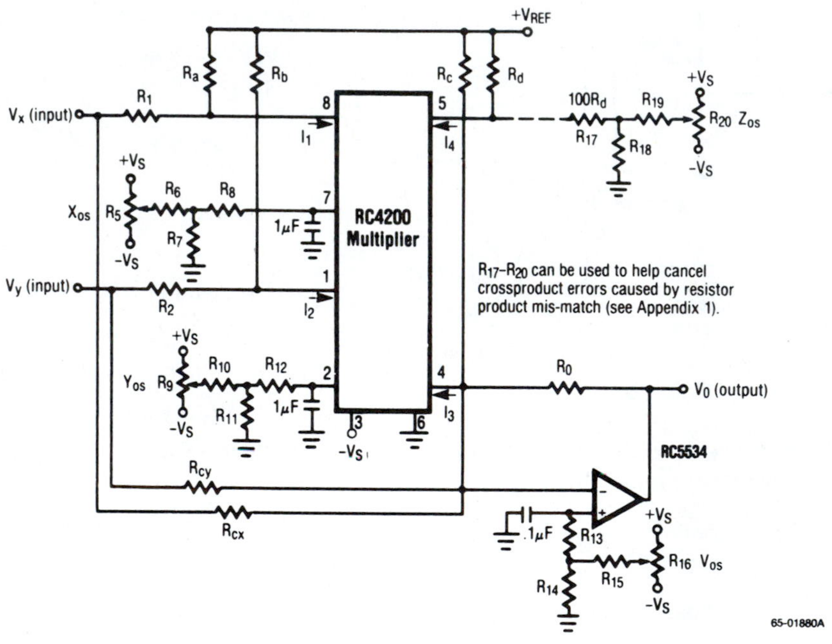


**Procedure:**

1. Set all trimmer pots to 0V on the wiper.
2. Connect $V_x$ input to ground. Put in a full scale square wave on $V_y$ input. Adjust $X_{OS}(R_5)$ for no square wave on $V_0$ output (adjust for 0 feedthrough).
3. Connect $V_y$ input to ground. Put in a full scale square wave on $V_x$ input. Adjust $Y_{OS}(R_9)$ for no square wave on $V_0$ output (adjust for 0 feedthrough).
4. Connect $V_x$ and $V_y$ to ground. Adjust $V_{OS}(R_{16})$ for 0V on $V_0$ output.

**Figure 6**

**Extended Range Divider**

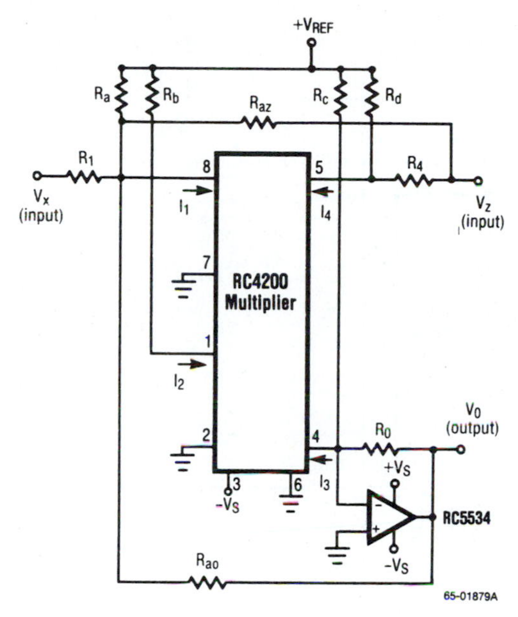


**Figure 7**

As with the extended range multiplier, resistors $R_{az}$ and $R_{ao}$ are added to cancel the cross-product error caused by the biasing resistors, i.e.,

$$\left(\frac{V_x}{R_1}+\frac{V_0}{R_{ao}}+\frac{V_z}{R_{az}}+\frac{V_{REF}}{R_a}\right)\left(\frac{V_{REF}}{R_b}\right)=\left(\frac{V_0}{R_0}+\frac{V_{REF}}{R_c}\right)\left(\frac{V_z}{R_4}+\frac{V_{REF}}{R_d}\right)$$

$$\frac{V_xV_{REF}}{R_1R_b}+\frac{V_0V_{REF}}{R_{ao}R_b}+\frac{V_zV_{REF}}{R_{az}R_b}+\frac{V_{REF}^2}{R_aR_b}=$$

$$\frac{V_0V_z}{R_0R_4}+\frac{V_0V_{REF}}{R_0R_d}+\frac{V_zV_{REF}}{R_4R_c}+\frac{V_{REF}^2}{R_cR_d}$$

To cancel cross-product and arithmetic offset:

$R_{ao}R_b = R_0R_d$, $R_{az}R_b = R_4R_c$ and $R_aR_b = R_cR_d$

and the result is:

$$\frac{V_xV_{REF}}{R_1R_b}=\frac{V_0V_z}{R_0R_4} \quad \text{or } V_0 = V_x/V_zK$$

$$\text{where } K = \frac{V_{REF}R_0R_4}{R_1R_b}$$

NOTE: It is necessary to match the resistor cross-products above to within the amount of error tolerable in the output offset, i.e., with a 10V F.S. output, 0.1% resistor cross-product match will give 0.1% x 10V = 10mV untrimmable output offset voltage.

**Resistor Values**

Inputs:

$$V_x(\text{min.}) \leq V_x \leq V_x(\text{max.})$$

$$\Delta V_x = V_x(\text{max.}) - V_x(\text{min.})$$

$$V_z(\text{min.}) \leq V_z \leq V_z(\text{max.})$$

$$\Delta V_z = V_z(\text{max.}) - V_z(\text{min.})$$

$$V_{REF} = \text{Constant } (+7V \text{ to } +18V)$$

Outputs:

$$V_0(\text{min.}) \leq V_0 \leq V_0(\text{max.})$$

$$\Delta V_0 = V_0(\text{max.}) - V_0(\text{min.})$$

$$K = \frac{V_0V_z}{V_x} \text{ (Design Requirement)}$$

$$R_0 = \frac{\Delta V_0}{750\mu A},\ R_b = \frac{V_{REF}}{250\mu A},\ R_4 = \frac{\Delta V_z}{200\mu A}$$

$$R_c = \frac{\Delta V_0 V_{REF}}{750\mu A\ \Delta V_0 - 700\mu A\ V_0(\text{max.})}$$

$$R_d = \frac{\Delta V_z V_{REF}}{250\mu A\ \Delta V_z - 200\mu A\ V_z(\text{max.})}$$

$$R_a = \frac{R_cR_d}{R_b},\ R_{az} = \frac{R_cR_4}{R_b},\ R_{ao} = \frac{R_0R_d}{R_b}$$

$$R_1 = \frac{\Delta V_0\ \Delta V_z}{600\mu AK}$$

# RC4200 Analog Multiplier *Continued*

**Divider Circuit with Offset Adjustment**

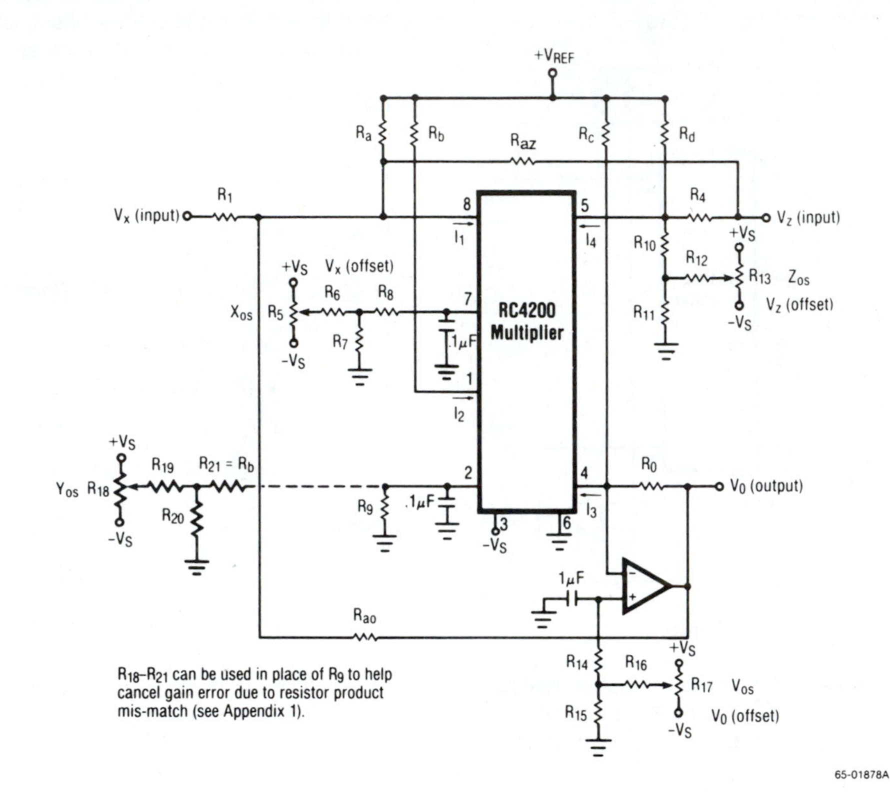


**General**

$10K \leq R_5 = R_{13} = R_{17} \leq 50K$

$R_7 + R_8 \approx R_1 || R_a || R_{az} || R_{ao}$

$R_6 \approx \frac{V_S}{.05} R_7$

$R_9 = R_b$

$R_{10} \approx 100 \times R_4$

$R_{11} = 20K$

$R_{12} = 100K$

$R_{14} + R_{15} \approx R_o || R_c$

$R_{16} \approx \frac{V_S}{.10} R_{15}$

**Example: Two-Quad Divider**

$V_O = K\frac{V_X}{V_Z}$, $K = k$, $V_{REF} = +V_S = +15V$

$-10 \leq V_x \leq +10$, therefore $\Delta V_x = 20$
$0 \leq V_z \leq +10$, therefore $\Delta V_z = 10$
$-10 \leq V_O \leq +10$, therefore $\Delta V_O = 20$

$R_O = 26.7K$
$R_b = 60K$
$R_4 = 50K$
$R_c = 37.5K$
$R_d = 300K$
$R_a = 187.5K$
$R_{az} = 31.25K$
$R_{ao} = 133K$

$R_1 = 333K$
$R_5$, $R_{13}$ $R_{17} = 10K$
$R_7$, $R_{15} = 1K$
$R_8$, $R_{11} = 20K$
$R_6$, $R_9$, $R_{16} = 300K$
$R_{10} = 4.7M$
$R_{12} = 100K$

**Figure 8**

### Divider Circuit Offset Adjustment Procedure

1. Set each trimmer pot to 0V on the wiper.

2. Connect $V_X$ (input) to ground. Put a DC voltage of approximately $\frac{1}{2}V_Z$ (max.) DC on the $V_Z$ (input) with an AC (squarewave is easiest) voltage of $\frac{1}{2}V_Z$ (max.) peak-to-peak superimposed on it. Adjust $X_{os}(R_5)$ for zero feedthrough. (No AC at $V_0$)

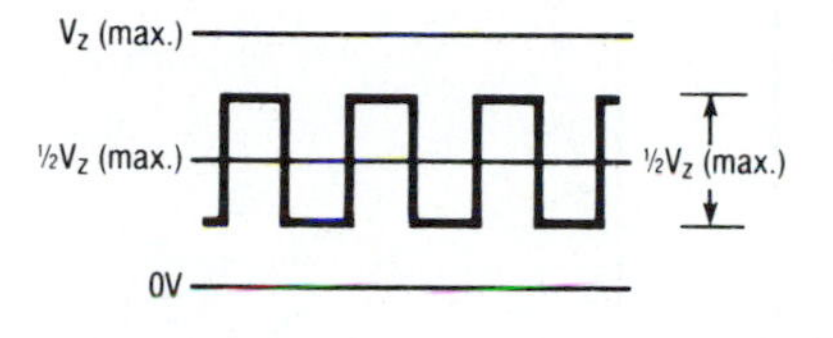


3. Connect $V_X$ (input to $V_Z$ (input) and put in the $\frac{1}{2}V_Z$ (max.) DC with an AC of approximately 20mV less than $V_Z$ (max.).

   Adjust $Z_{os}(R_{13})$ for zero feedthrough.

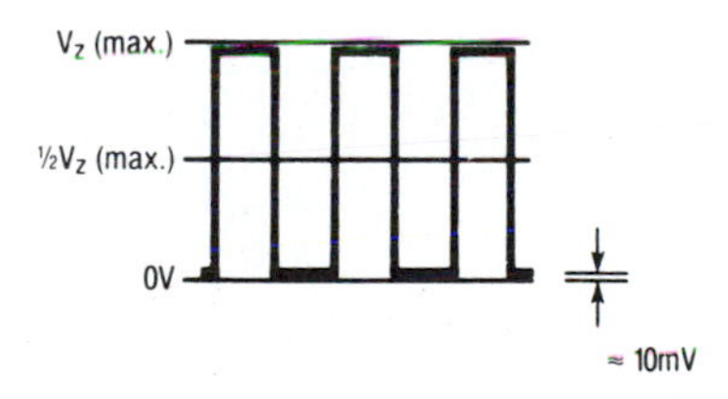


4. Return $V_X$ (input) to ground and connect $V_Z$ (max.) DC on $V_Z$ (input). Adjust output $V_{OS}(R_{17})$ for $V_0$ =.

5. Connect $V_X$ (input) to $V_Z$ (input) and put in $V_Z$ (max.) DC. (The output will equal K.) Decrease the input slowly until the output ($V_0$ = K) deviates beyond the desired accuracy. Adjust $Z_{os}$ to bring it back into tolerance and return to Step 4. Continue Steps 4 and 5 until $V_Z$ reduces to the lowest value desired.

NOTE: As the input to $V_X$ and $V_Z$ gets closer to zero (an illegal state) the system noise will predominate so much that an integrating voltmeter will be very helpful.

### Square Root Circuit $V_0 = N\sqrt{V_x}$

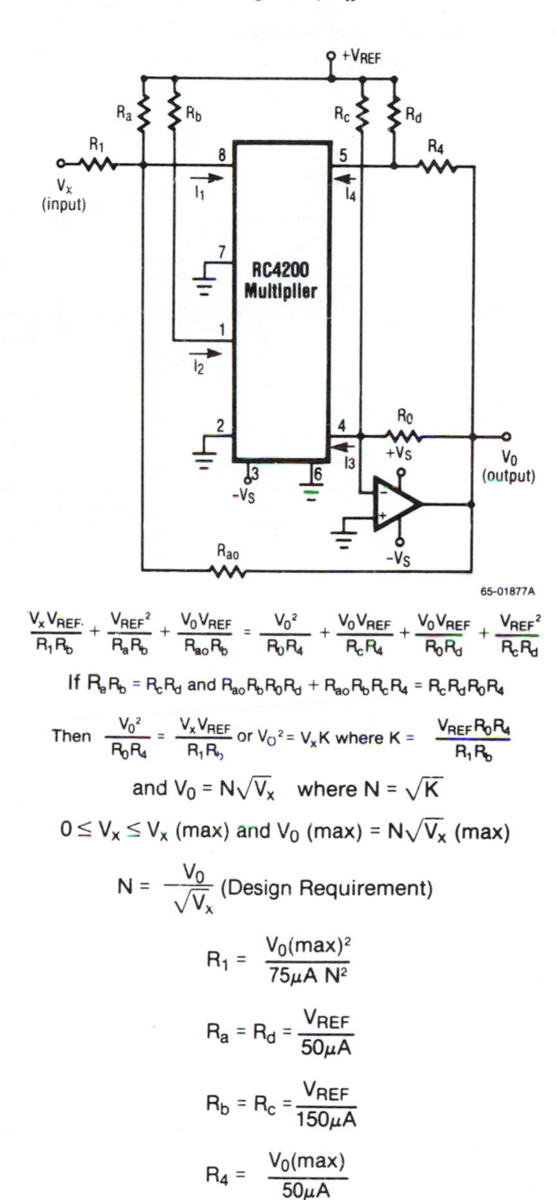


$$\frac{V_x V_{REF}}{R_1 R_b} + \frac{V_{REF}^2}{R_a R_b} + \frac{V_0 V_{REF}}{R_{ao} R_b} = \frac{V_0^2}{R_0 R_4} + \frac{V_0 V_{REF}}{R_c R_4} + \frac{V_0 V_{REF}}{R_0 R_d} + \frac{V_{REF}^2}{R_c R_d}$$

If $R_a R_b = R_c R_d$ and $R_{ao} R_b R_0 R_d + R_{ao} R_b R_c R_4 = R_c R_d R_0 R_4$

Then $\frac{V_0^2}{R_0 R_4} = \frac{V_x V_{REF}}{R_1 R_b}$ or $V_0^2 = V_x K$ where $K = \frac{V_{REF} R_0 R_4}{R_1 R_b}$

and $V_0 = N\sqrt{V_x}$ where $N = \sqrt{K}$

$0 \le V_x \le V_x$ (max) and $V_0$ (max) $= N\sqrt{V_x}$ (max)

$$N = \frac{V_0}{\sqrt{V_x}} \text{ (Design Requirement)}$$

$$R_1 = \frac{V_0(\text{max})^2}{75\mu A\ N^2}$$

$$R_a = R_d = \frac{V_{REF}}{50\mu A}$$

$$R_b = R_c = \frac{V_{REF}}{150\mu A}$$

$$R_4 = \frac{V_0(\text{max})}{50\mu A}$$

$$R_{ao} = \frac{V_0(\text{max})}{125\mu A}$$

$$R_0 = \frac{V_0(\text{max})}{225\mu A}$$

**Figure 9**

## RC4200 Analog Multiplier *Continued*

### Square Root Circuit Offset Adjust

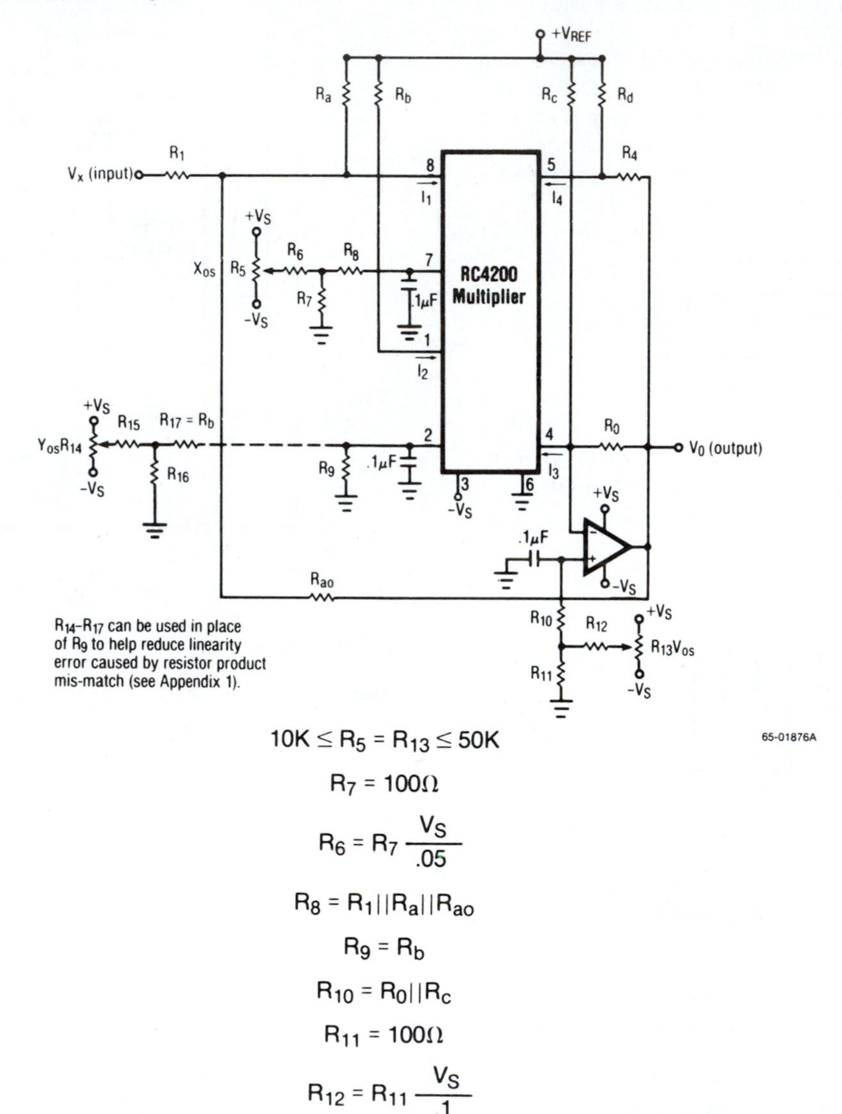


$$10K \leq R_5 = R_{13} \leq 50K$$

$$R_7 = 100\Omega$$

$$R_6 = R_7 \frac{V_S}{.05}$$

$$R_8 = R_1 || R_a || R_{ao}$$

$$R_9 = R_b$$

$$R_{10} = R_0 || R_c$$

$$R_{11} = 100\Omega$$

$$R_{12} = R_{11} \frac{V_S}{.1}$$

**Procedure**

1. Set both trimmer pots to 0V on the wiper.
2. Put in a full scale (0 to $V_x$(max.)) squarewave on $V_x$ input. Adjust $X_{os}(R_5)$ for proper peak-to-peak amplitude on $V_0$ output. (Scaling adjust)
3. Connect $V_x$ input to ground. Adjust $V_{os}(R_{13})$ for 0V on $V_0$ output.

**Figure 10**

**(L-11)**

**Squaring Circuits $V_0 = K\,V_x^2$**

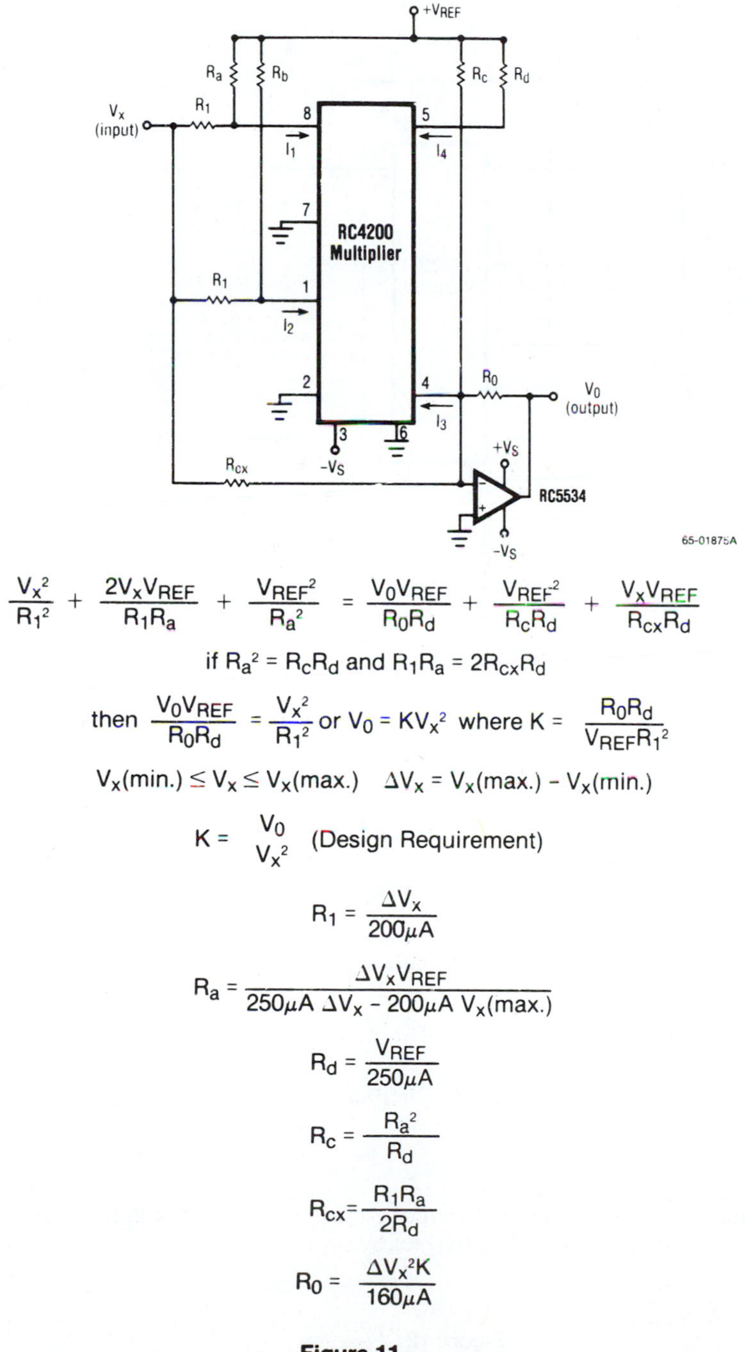


$$\frac{V_x^2}{R_1^2} + \frac{2V_xV_{REF}}{R_1R_a} + \frac{V_{REF}^2}{R_a^2} = \frac{V_0V_{REF}}{R_0R_d} + \frac{V_{REF}^2}{R_cR_d} + \frac{V_xV_{REF}}{R_{cx}R_d}$$

if $R_a^2 = R_cR_d$ and $R_1R_a = 2R_{cx}R_d$

then $\frac{V_0V_{REF}}{R_0R_d} = \frac{V_x^2}{R_1^2}$ or $V_0 = KV_x^2$ where $K = \frac{R_0R_d}{V_{REF}R_1^2}$

$V_x(\text{min.}) \leq V_x \leq V_x(\text{max.})$ $\quad \Delta V_x = V_x(\text{max.}) - V_x(\text{min.})$

$K = \frac{V_0}{V_x^2}$ (Design Requirement)

$$R_1 = \frac{\Delta V_x}{200\mu A}$$

$$R_a = \frac{\Delta V_x V_{REF}}{250\mu A\ \Delta V_x - 200\mu A\ V_x(\text{max.})}$$

$$R_d = \frac{V_{REF}}{250\mu A}$$

$$R_c = \frac{R_a^2}{R_d}$$

$$R_{cx} = \frac{R_1R_a}{2R_d}$$

$$R_0 = \frac{\Delta V_x^2 K}{160\mu A}$$

**Figure 11**

**(L-12)**

# RC4200 Analog Multiplier *Continued*

### Squaring Circuits Offset Adjust

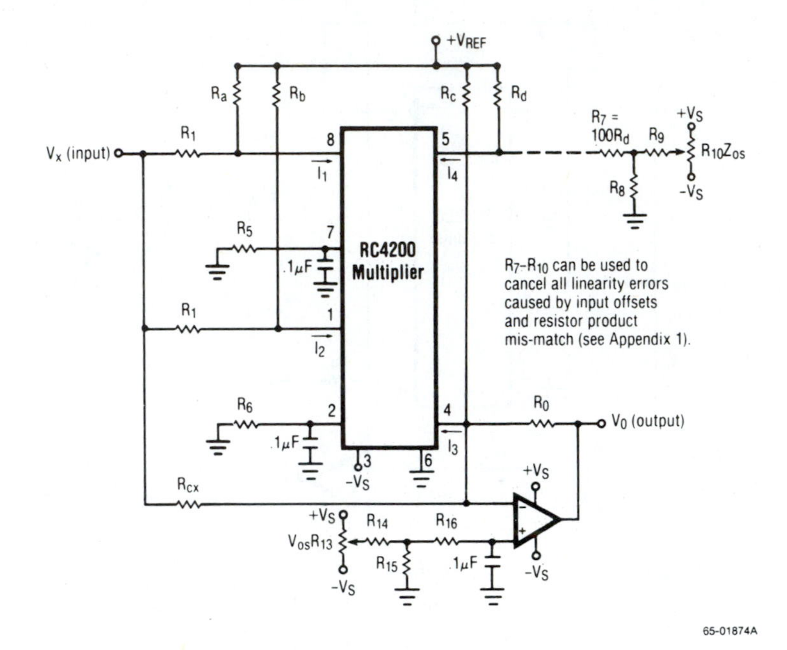


$$10K \leq R_{10}, R_{11} \leq 50K$$

$$R_8, R_{15} = 100\Omega$$

$$R_9, R_{14} = 100\Omega \frac{V_S}{.1}$$

$$R_5, R_6 = R_1 || R_a$$

$$R_{16} = R_O || R_c || R_{cx}$$

### Procedure

1. Set both trimmer pots to 0V on the wiper.
2. Put in a full scale ($\pm V_x$) squarewave on $V_x$ input. Adjust $Z_{OS}(R_{10})$ for uniform output.
3. Connect $V_x$ input to ground. Adjust $V_{OS}(R_{11})$ for 0V on $V_O$ output.

**Figure 12**

## Appendix 1 — System Errors

There are four types of accuracy errors which effect overall system performance. They are:

1. Nonlinearity — Incremental deviation from absolute accuracy.[1]
2. Scaling Error — Linear deviation from absolute accuracy.
3. Output Offset — Constant deviation from absolute accuracy.
4. Feedthrough[2] — Crossproduct errors caused by input offsets and external circuit limitations.

The nonlinearity error in the transfer function of the 4200 is ±0.1% max. (±0.03% max. for 4200A).

$$\text{i.e., } I_3 = \frac{I_1 I_2}{I_4} \pm 0.1\% \text{ F.S.}^{(4)}$$

The other system errors are caused by voltage offsets on the inputs of the 4200 and can be as high as ±3.0% (±2.0% for 4200A).

$$\text{i.e., } V_0 = \frac{V_x V_y}{V_z}\frac{R_0 R_4}{R_1 R_2} \pm 3.0\% \text{ F.S.}^{(3)(4)}$$

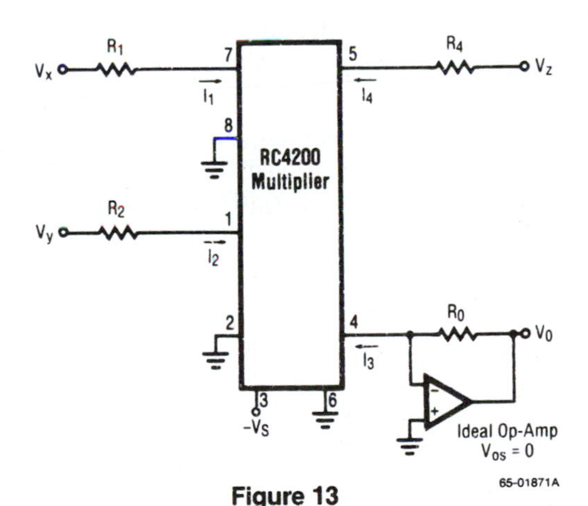


**Figure 13**

Notes:
1. The input circuits tend to become unstable at $I_1$, $I_2$, $I_4$ < 50μA and linearity decreases when $I_1$, $I_2$, $I_4$ > 250μA (e.g., @ $I_1 = I_2 = 500\mu A$ nonlinearity error ≈ 0.5%).
2. This section will not deal with feedthrough which is proportional to frequency of operation and caused by stray capacitance and/or bandwidth limitations. (Refer to Figure 21.)
3. Not including resistor tolerance or output offset on the op amp.
4. For $50\mu A \leq I_1, I_2, I_4 \leq 250\mu A$.

**Errors caused by input offsets.**

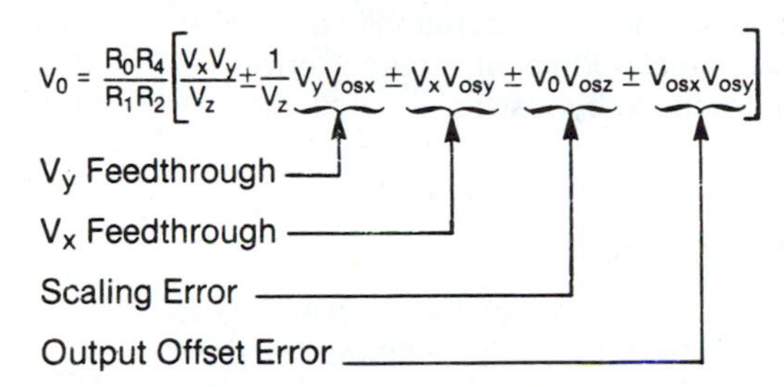


Systems errors can be greatly reduced by externally trimming the input offset voltages of the 4200. (±0.3% F.S. for 4200 and ±0.1% F.S. for 4200A.)

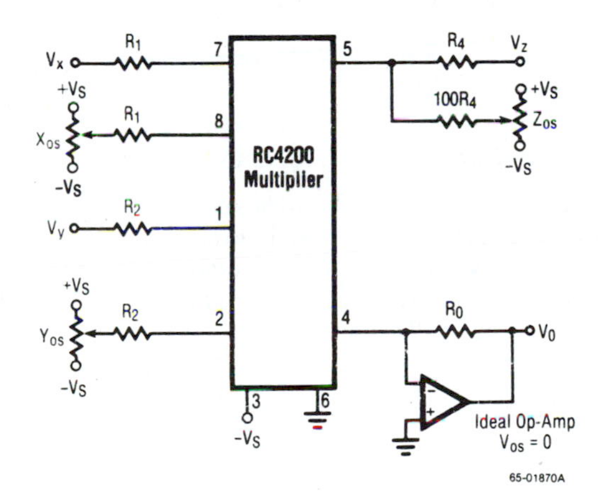


if; $X_{OS} = V_{OSX}$, $Y_{OS} = V_{OSY}$, $Z_{OS} = -V_{OSZ}$,

$$\text{then } V_0 = \frac{V_x V_y}{V_z}\frac{R_0 R_4}{R_1 R_2} \pm 0.3\% \text{ F.S.}^{(3)}$$

65-01870A

**Figure 14. 4200 With Input Offset Adjustment**

## RC4200 Analog Multiplier *Continued*

**Extended Range Circuit Errors**

The extended range configurations have a disadvantage in that additional accuracy errors may be introduced by resistor product mismatching.

**Multiplier (Figure 6)**

An error in resistor product matching will cause an equivalent feedthrough or output offset error:

1. $R_1R_b = R_{cx}R_d \pm \alpha$,
   $V_x$ feedthrough ($V_y = 0$) = $\pm \alpha V_x$
2. $R_2R_a = R_{cy}R_d \pm \beta$,
   $V_y$ feedthrough ($V_x = 0$) = $\pm \beta V_y$
3. $R_aR_b = R_cR_d \pm \gamma$,
   $V_0$ offset ($V_x = V_y = 0$) = $\pm \gamma V_{REF}$*

*Output offset errors can always be trimmed out with the output op amp offset adjust, $V_{os}(R_{16})$.

**Reducing Mis-Match Errors (Figure 6)**

You need not run out and buy .01% resistors to reduce resistor product mis-match errors. Here are a couple of ways to squeeze maximum accuracy out of the extended range multiplier (see Figure 6) using 1% resistors.

**Method #1**

$V_x$ feedthrough, for example, occurs when $V_y$ = 0 and $V_{osy} \neq 0$. This $V_x$ feedthrough will equal $\pm V_xV_{osy}$. Also, if $V_{osz} \neq 0$, there is a $V_x$ feedthrough equal to $\pm V_xV_{osz}$. A resistor-product error of $\alpha$ will cause a $V_x$ feedthrough of $\pm \alpha V_x$. Likewise, $V_y$ feedthrough errors are: $\pm V_yV_{osx}$, $\pm V_yV_{osz}$ and $\pm \beta V_y$.

Total feedthrough =
$\pm V_xV_{osy} \pm V_yV_{osx} \pm \alpha V_x \pm \beta V_y \pm (V_x + V_y)\, V_{osz}$

By carefully adjusting $X_{os}(R_5)$, $Y_{os}$(R9) and $Z_{os}(R_{20})$ this equation can be made to very nearly equal zero and the feedthrough error will practically disappear.

A residual offset will probably remain which can be trimmed out with $V_{os}(R_{16})$ at the output op amp.

**Method #2**

Notice that the ratios of $R_1R_b : R_{cx}R_d$ and $R_2R_a : R_{cy}R_d$ are both dependent on $R_d$, also that $R_1$, $R_2$, $R_a$ and $R_b$ are all functions of the maximum input requirements. By designing a multiplier for the same input ranges on both $V_x$ and $V_y$ then $R_1 = R_2$, $R_{cx} = R_{cy}$ and $R_a = R_b$. (Note: It is acceptable to design a four quadrant multiplier and use only two quadrants of it.)

Select $R_d$ to be 1% or 2% below (or above) the calculated value. This will cause $\alpha$ and $\beta$ to both be positive (or negative) by nearly the same amount. Now the effective value of $R_d$ can be trimmed with an offset adjustment $Z_{os}(R_{20})$ on pin 5.

This technique will cause: 1) a slight gain error which can be compensated for with the $R_0$ value, and 2) an output offset error that can be trimmed out with $V_{os}(R_{16})$ on the output op amp.

**Extended Range Divider (Figure 8)**

The only crossproduct error of interest is the $V_z$ feedthrough ($V_x = 0$ and $V_{osx} \neq 0$) which is easily adjusted with $X_{os}(R_5)$.

Resistor product mis-match will cause scaling errors (gain) that could be a problem for very low values of $V_z$. Adjustments to $Y_{os}(R_{18})$ can be made to improve the high gain accuracy.

**Square Root and Squaring (Figures 10 and 12)**

These circuits are functions of single variables so feedthrough, as such, is not a consideration. Crossproduct errors will effect incremental accuracy that can be corrected with $Y_{os}(R_{14})$ or $Z_{os}(R_{10})$.

## Appendix 2 — Applications

### Design Considerations for RMS-to-DC Circuits

#### Average Value

Consider $V_{in}$ = Asinωt. By definition,

$$V_{AVG} \equiv \frac{2}{T}\int_0^{\frac{T}{2}} V_{in} dt$$

Where T = Period
$\omega = 2\pi f$
$= \frac{2\pi}{T}$

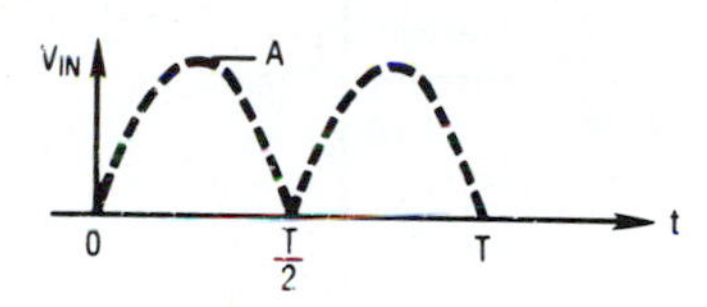


$$V_{AVG} \equiv \frac{2}{T}\int_0^{\frac{T}{2}} A\sin\omega t\, dt$$
$$= \frac{2A}{T}\left[-\frac{1}{\omega}\cos \omega t\right]_0^{\frac{T}{2}}$$
$$= \frac{2A}{2\pi}\left[-\cos(\pi) + \cos(0)\right]$$
$$= \frac{2}{\pi}A$$

Avg. Value of Asinωt is $\frac{2}{\pi}A$

#### RMS Value

Again consider $V_{in}$ = Asinωt

$$V_{rms} = \sqrt{V_{AVG}} = \sqrt{\frac{1}{T}\int_0^T [V_{in}]^2\, dt}$$

$V_{rms}$ for Asinωt:

$$V_{rms} = \sqrt{\frac{1}{T}\int_0^T A^2\sin^2\omega t\, dt}$$
$$= \sqrt{\frac{A^2}{T}\int_0^T\left(\frac{1}{2} - \frac{1}{2}\cos 2\,\omega t\right)dt}$$
$$= \sqrt{\frac{A^2}{2}\left(\frac{T}{2} - \frac{1}{4\omega}\sin 2\,\omega t\right)_0^T}$$
$$= \sqrt{\frac{A^2}{T}\left(\frac{T}{2}\right)}$$
$$= \sqrt{\frac{A^2}{2}}$$

therefore the rms value of Asinωt becomes:

$$V_{rms} = \frac{A}{\sqrt{2}}$$

#### RMS Value for Rectified Sine Wave

Consider $V_{in}$ = |A sin ωt|, a rectified wave. To solve, integrate over each half cycle.

$$\text{i.e. } \frac{1}{T}\int_0^T V_{in}^2\, dt =$$

$$\frac{1}{T}\left[\int_0^{\frac{T}{2}} A^2\sin^2\omega t\, dt + \int_{\frac{T}{2}}^T (-A\sin\omega t)^2 dt\right]$$

This is the same as $\frac{1}{1}\int_0^T A^2\sin^2\omega t\, dt$

so, $|A\sin\omega t|_{rms} = A\sin\omega t_{rms}$

Practical Consideration: |Asinωt| has high-order harmonics; Asinωt does not. Therefore, non-ideal integrators may cause different errors for two approaches:

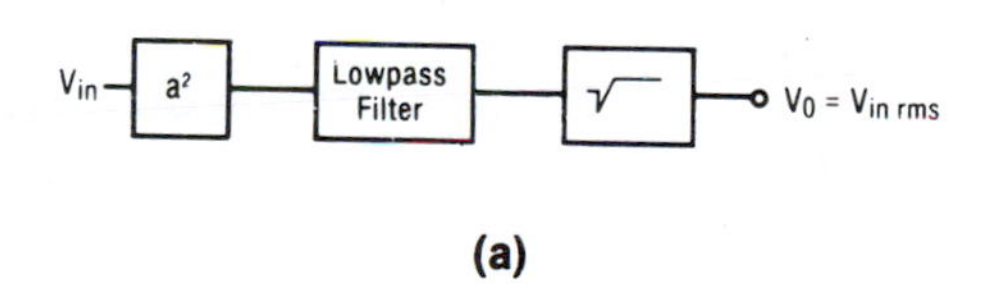


(a)

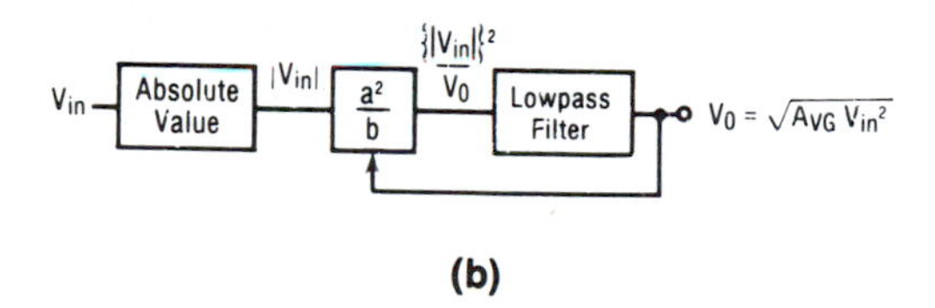


(b)

$$\text{Avg}\left\{\frac{V_{in}^2}{V_0}\right\} = V_0$$

implies $V_0 = \sqrt{\text{Avg}\ \{|V_{in}|^2\}}$

$$V_0 = \sqrt{\text{Avg}\ V_{in}^2}$$

65-01872A

**Figure 15**

RC4200 Analog Multiplier *Continued*

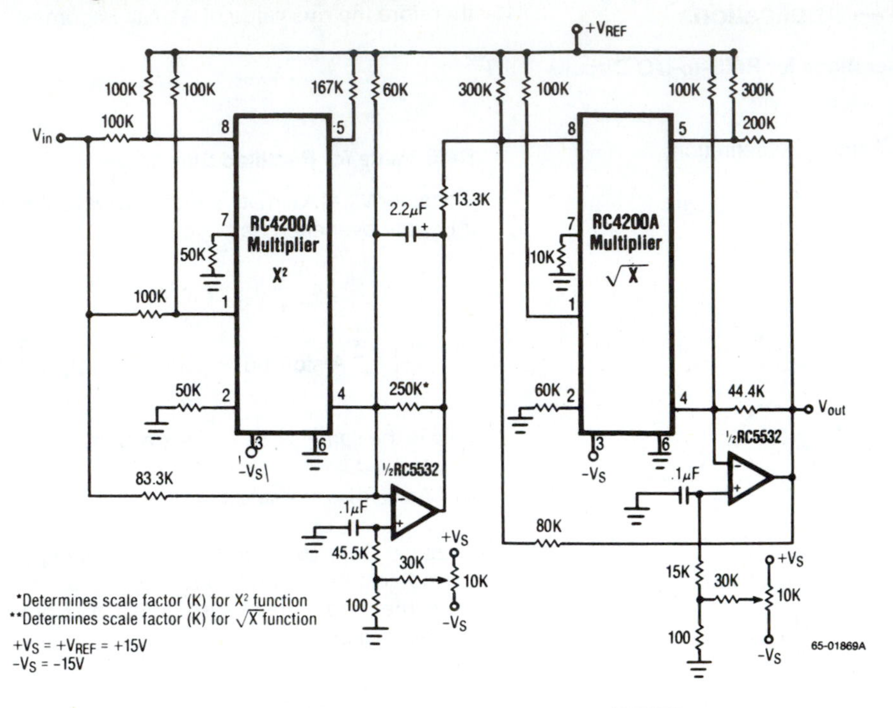


**Figure 16. RMS to DC Converter $V_{out} = \sqrt{\int V_{in}^2}$**

**Amplitude Modulator with A.G.C. (Figure 17)**

In many AC modulator applications unwanted output modulation is caused by variations in carrier input amplitude. The versatility of the RC4200 multiplier can be utilized to eliminate this undesired fluctuation. The extended range multiplier circuit (Figure 5) shows an output amplitude inversely proportional to the reference voltage $V_{REF}$.

$$\text{i.e., } V_0 = \frac{V_x V_y}{V_{REF}} \frac{R_0 R_d}{R_1 R_2}$$

By making $V_{REF}$ proportional to $V_y$ (where $V_y$ is the carrier input) such that:

$$V_{REF} = V_H = f(|V_y|),$$

Then the denominator becomes a variable value that automatically provides constant gain, such that the modulating input ($V_x$) modulates the carrier ($V_y$) with a fixed scale factor even though the carrier varies in amplitude.

If $V_H$ is made proportional to the average value of $A\sin\omega t$ (i.e., $2A/\pi$) and scaled by a value of $\pi/2$ then:

$$V_H = A$$

and if: $V_x$ = Modulating input ($V_M$)
and: $V_y$ = Carrier input ($A\sin\omega t$)

then: $V_0 = K\, V_M \sin\omega t$ where $K = \frac{R_0 R_d}{R_1 R_2}$

The resistor scaling is determined by the dynamic range of the carrier variation and modulating input.

The resistor values are solved, as with the other extended range circuits, in terms of the input voltages.

Input voltages:
- Modulation Voltage ($V_M$): $0 \le V_M \le V_x$(max.)
- Carrier ($V_y$: $V_y = A\sin\omega t$
- Carrier amplitude fluctuation ($\Delta A$):
  - $A(\text{min.}) \sin\omega t \le V_y \le A(\text{max.}) \sin\Omega\omega t$
- Dynamic Range (N): A(max.)/A(min.)
  - A(max.) = $V_H$(max.) and A(min.) = $V_H$(min.)

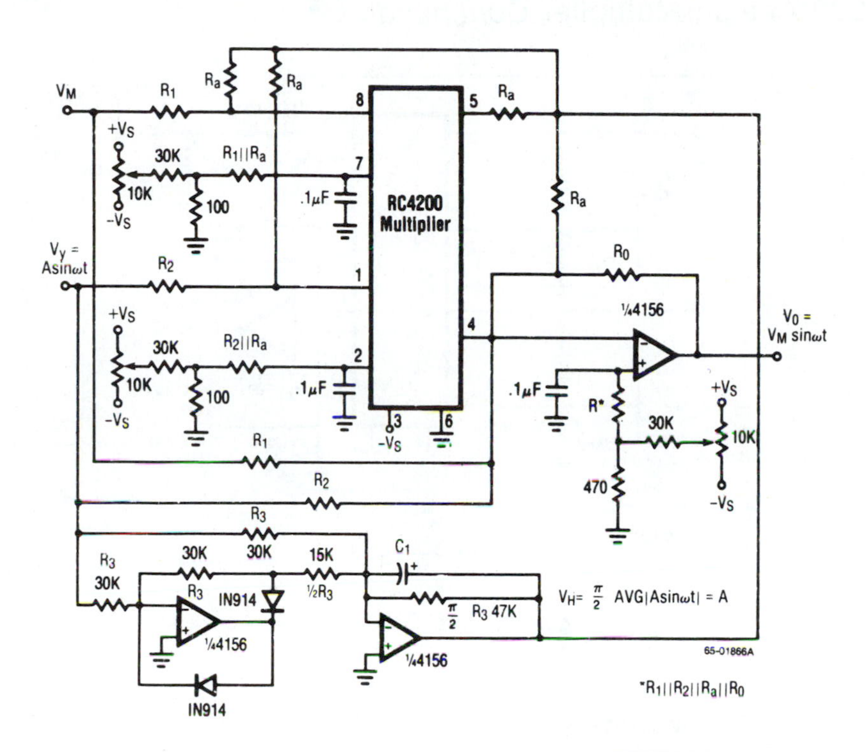


**Figure 17. Amplitude Modulator with A.G.C.**

The maximum and minimum values for $I_1$ and $I_2$ lead to:

$$I_1(\text{max.}) = \frac{V_x(\text{max.})}{R_1} + \frac{V_H(\text{max.})}{R_a} = 250\mu A$$

$$I_1(\text{min.}) = \frac{V_H(\text{min.})}{R_a} = 50\mu A \quad V_M(\text{min.}) = 0$$

$$I_2(\text{max.}) = \frac{A(\text{max.})}{R_2} + \frac{V_H(\text{max.})}{R_a} = 250\mu A$$

$$I_2(\text{min.}) = \frac{V_H(\text{min.})}{R_a} = 50\mu A$$

For a dynamic range of N, where

$$N = \frac{A(\text{max.})}{A(\text{min.})} < 5,$$

These equations combine to yield:

$$R_1 = \frac{V_x(\text{max.})}{(5 - N)50\mu A},\ R_2\ \frac{A(\text{max.})}{(5 - N)50\mu A},$$

$$R_a = \frac{A(\text{min.})}{50\mu A} \text{ and } R_0 = K\ \frac{R_1 R_2}{R_a}$$

**Example #1**

$V_y$ = Asinωt $2.5V \leq A \leq 10V$, therefore N = 4
$0V \leq V_M \leq 10V$, therefore $V_x(\text{max.}) = 10V$
K = 1, therefore $V_0 = V_M \sin\omega t$

$$R_1 = \frac{V_x(\text{max.})}{50\mu A} = \frac{10V}{50\mu A} = 200K$$

$$R_2 = \frac{A(\text{max.})}{50\mu A} = \frac{10V}{50\mu A} = 200K$$

$$R_a = \frac{A(\text{min.})}{50\mu A} = \frac{2.5V}{50\mu A} = 50K$$

$$R_0 = K\ \frac{R_1 R_2}{R_a} = 1\ \frac{200K \times 200K}{50K} = 800K$$

**Example #2**

$V_y$ = Asinωt $3 \leq A \leq 6$, therefore N = 2
$0V \leq V_M \leq 8V$, therefore $V_x(\text{max.}) = 8V$
K = .2, therefore $V_0 = .2V_M \sin\omega t$
so:
$R_1$ = 53.3K, $R_2$ = 40K
$R_a$ = 60K and $R_0$ = 7.11K

# RC4200 Analog Multiplier *Continued*

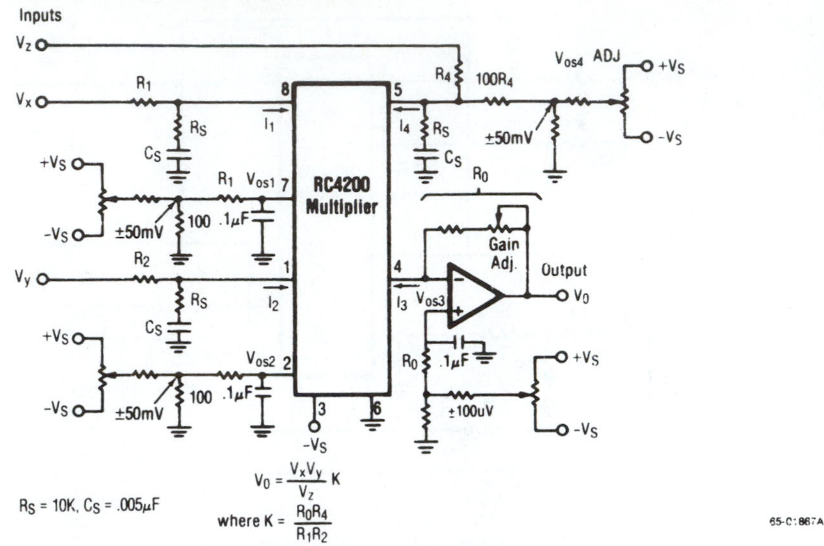


**Figure 18. First Quadrant Multiplier/Divider**

### Limited Range, First Quadrant Applications

The following circuit has the advantage that cross-product errors are due only to input offsets and nonlinearity error is slightly less for lower input currents.

The circuit also has no standby current to add to the noise content although the signal-to-noise ratio worsens at very low input currents (1–5μA) due to the noise current of the input stages.

The $R_sC_s$ filter circuits are added to each input to improve the stability for input currents below 50μA.

#### Caution

The bandpass drops off significantly for lower currents (<50μA) and nonsymmetrical rise and fall times can cause second harmonic distortion.

### Thermal Symmetry

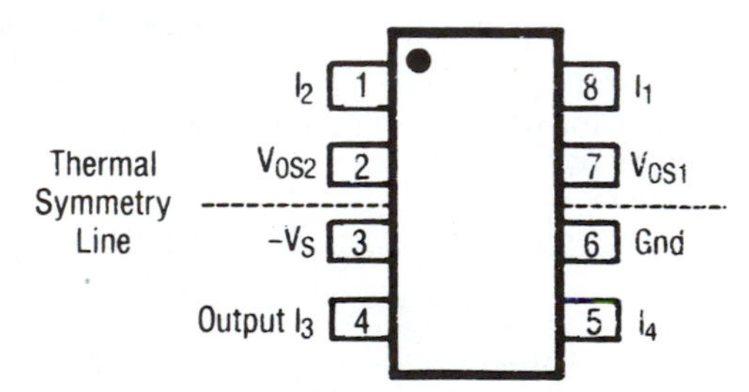


The scale factor is sensitive to temperature gradients across the chip in the lateral direction. Where possible, the package should be oriented such that sources generating temperature gradients are located physically on the line of thermal symmetry. This will minimize scale-factor error due to thermal gradients.

65-02775A

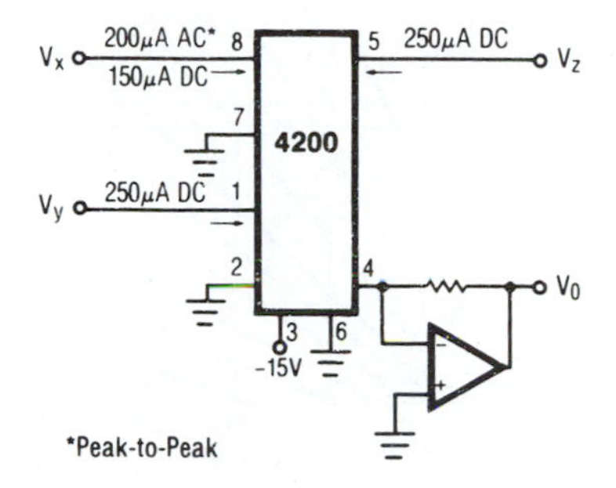

$V_x$
200μA AC*
150μA DC
8
5
250μA DC
$V_z$
7
4200
$V_y$
250μA DC
1
2
4
$V_0$
3
6
-15V
*Peak-to-Peak


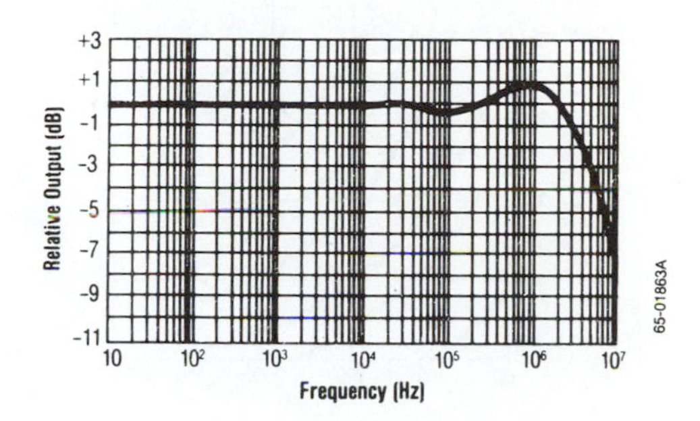

Relative Output (dB)
Frequency (Hz)
65-01863A


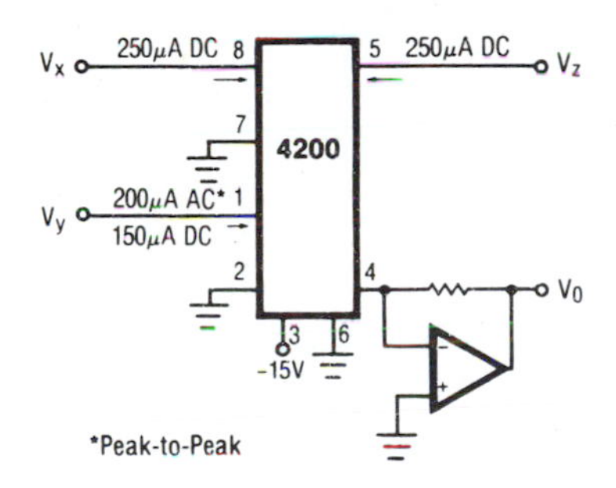

$V_x$
250μA DC
8
5
250μA DC
$V_z$
7
4200
$V_y$
200μA AC*
150μA DC
1
2
4
$V_0$
3
6
-15V
*Peak-to-Peak


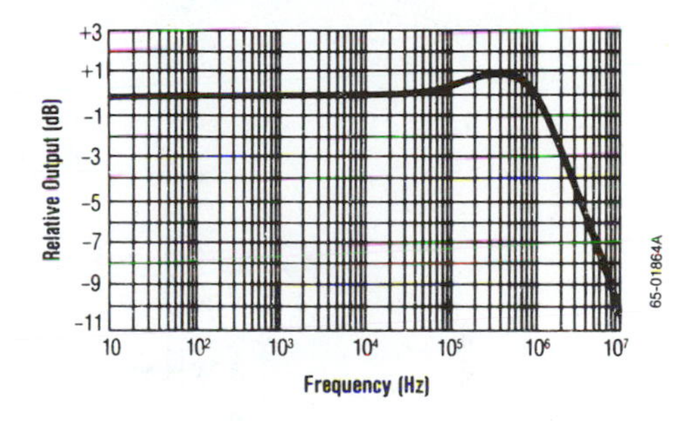

Relative Output (dB)
Frequency (Hz)
65-01864A


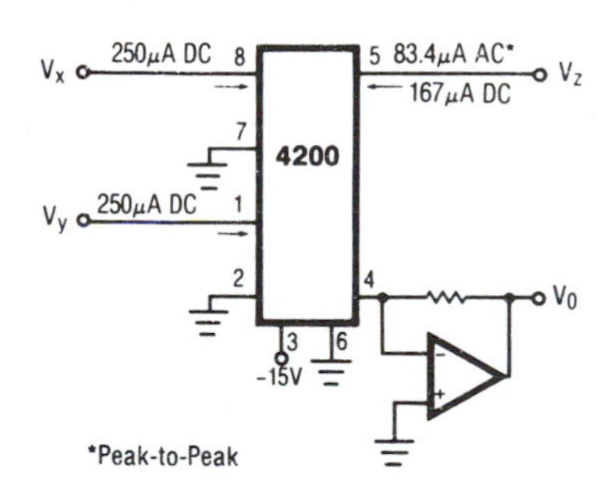

$V_x$
250μA DC
8
5
83.4μA AC*
167μA DC
$V_z$
7
4200
$V_y$
250μA DC
1
2
4
$V_0$
3
6
-15V
*Peak-to-Peak


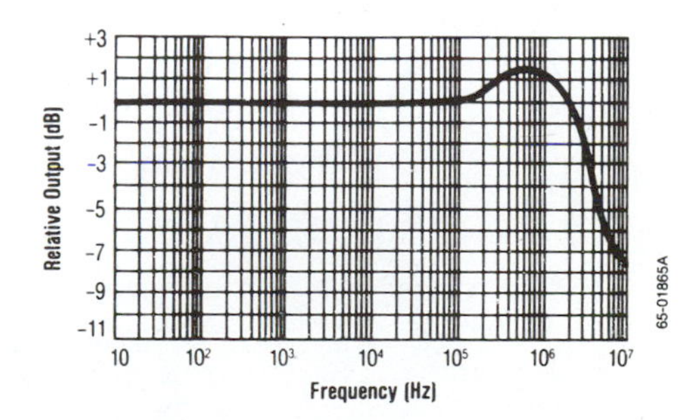

Relative Output (dB)
Frequency (Hz)
65-01865A


**Figure 19.**

## RC4200 Analog Multiplier *Continued*

250
200
150
100
50
I1 (μA)
50 100 150 200 250
I2 (μA)
5.0nA
4.0nA
3.0nA
2.5nA
2.0nA
1.5nA
1.2nA
1.0nA
65-01861A

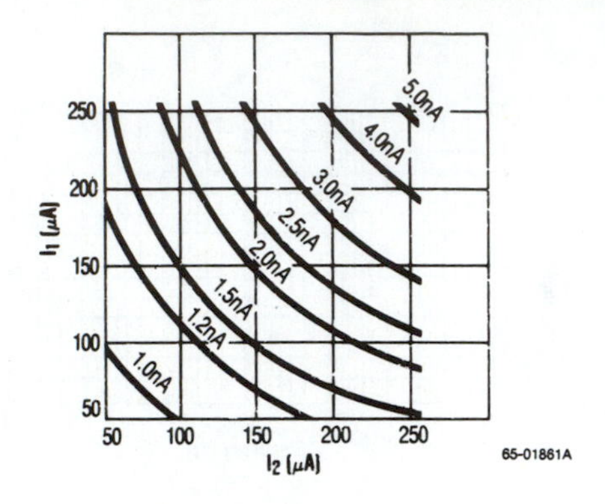

**Figure 20a. Output Noise Current vs. Input Current ($I_4$ = 250μA)**

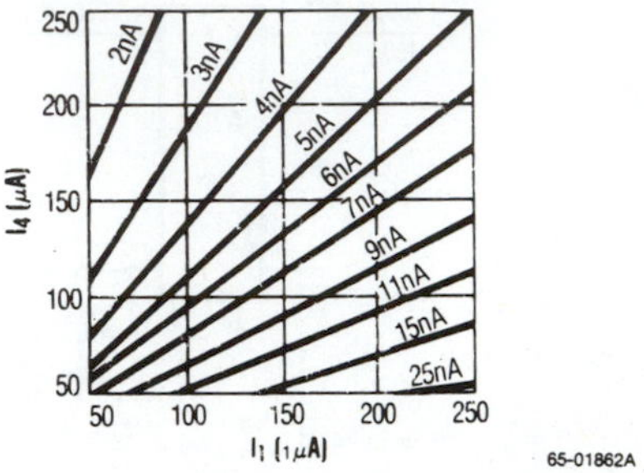

250
200
150
100
50
I4 (μA)
50 100 150 200 250
I1 (μA)
2nA
3nA
4nA
5nA
6nA
7nA
9nA
11nA
15nA
25nA
65-01862A

**Figure 20b. Output Noise Current vs. Input Current ($I_2$ = 250μA)**

250
200
150
100
50
Feedthrough (mVp-p)
Multiplier Configuration
V0 = VxVy / 10
Vy = 0
Vx = 10sinωt
Vx = 0
Vy = 10sinωt
1.0 10 100 1K 10K 100K
Frequency (Hz)
65-01860A

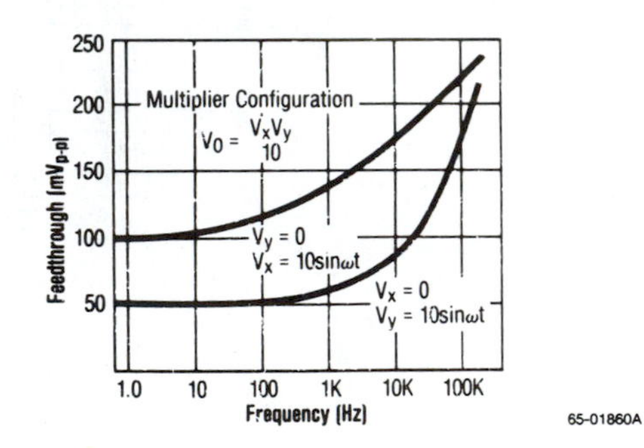

**Figure 21. AC Feedthrough vs. Frequency**

## Schematic Diagram

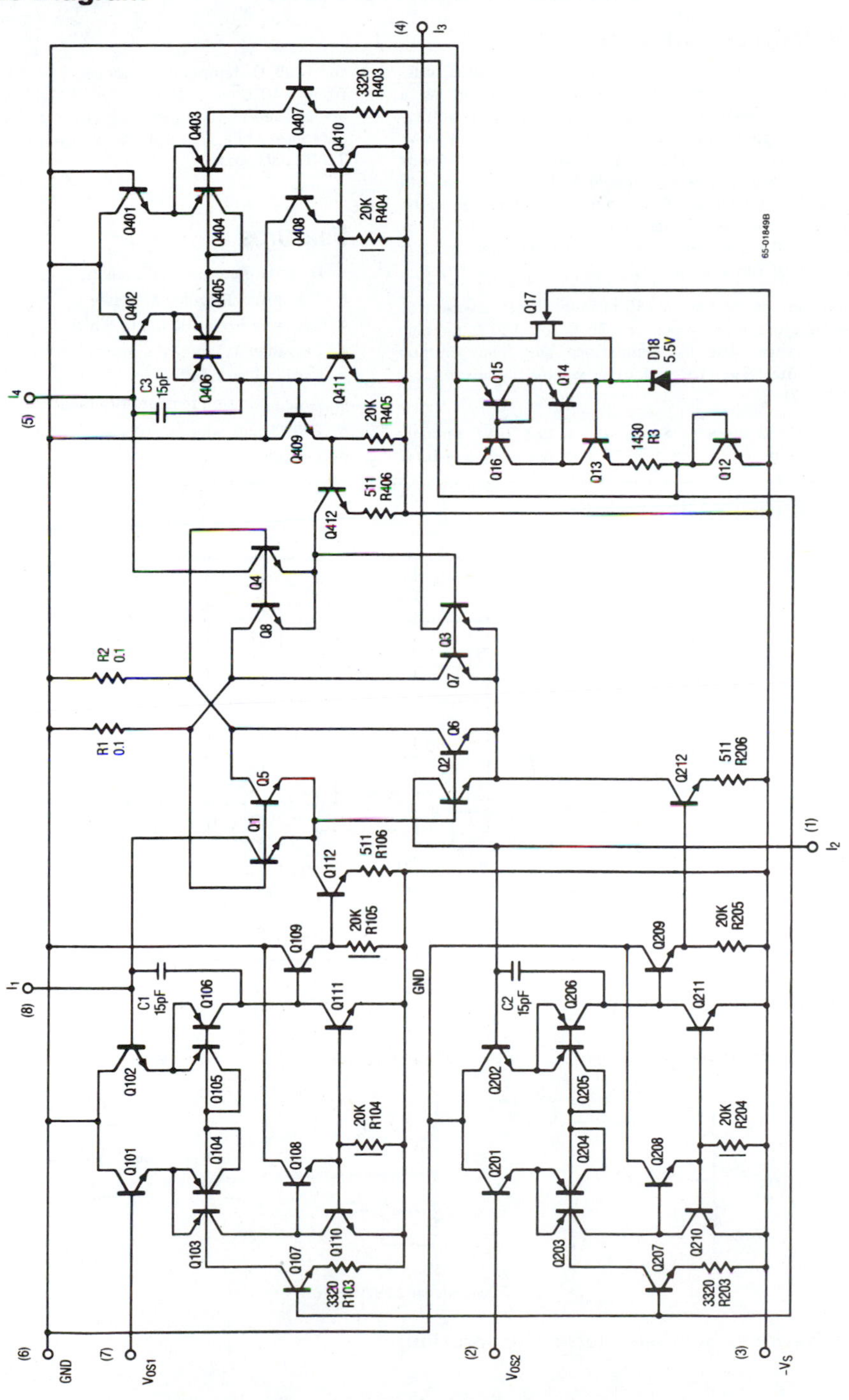

# Appendix M LM135/LM235/LM335, LM135A/ LM235A/LM335A

## General Description

The LM135 series are precision, easily-calibrated, integrated circuit temperature sensors. Operating as a 2-terminal zener, the LM135 has a breakdown voltage directly proportional to absolute temperature at +10 mV/°K. With less than 1Ω dynamic impedance the device operates over a current range of 400 μA to 5 mA with virtually no change in performance. When calibrated at 25°C the LM135 has typically less than 1°C error over a 100°C temperature range. Unlike other sensors the LM135 has a linear output.

Applications for the LM135 include almost any type of temperature sensing over a −55°C to +150°C temperature range. The low impedance and linear output make interfacing to readout or control circuitry especially easy.

The LM135 operates over a −55°C to +150°C temperature range while the LM235 operates over a −40°C to +125°C temperature range. The LM335 operates from −40°C to +100°C. The LM135/LM235/LM335 are available packaged in hermetic TO-46 transistor packages while the LM335 is also available in plastic TO-92 packages.

## Features

- Directly calibrated in °Kelvin
- 1°C initial accuracy available
- Operates from 400 μA to 5 mA
- Less than 1Ω dynamic impedance
- Easily calibrated
- Wide operating temperature range
- 200°C overrange
- Low cost

## Schematic Diagram

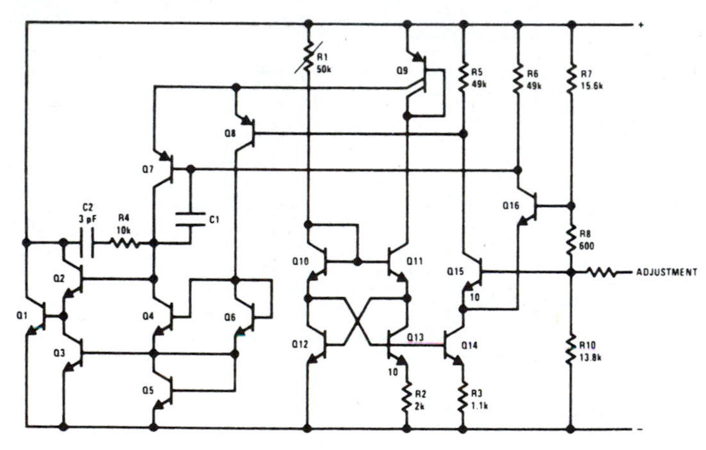


## Typical Applications

**Basic Temperature Sensor**

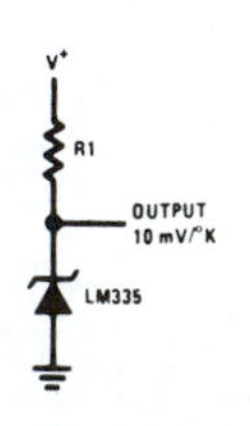


**Calibrated Sensor**

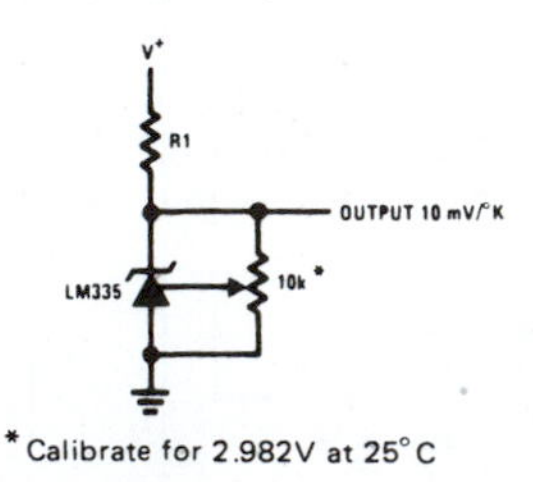


*Calibrate for 2.982V at 25°C

**Wide Operating Supply**

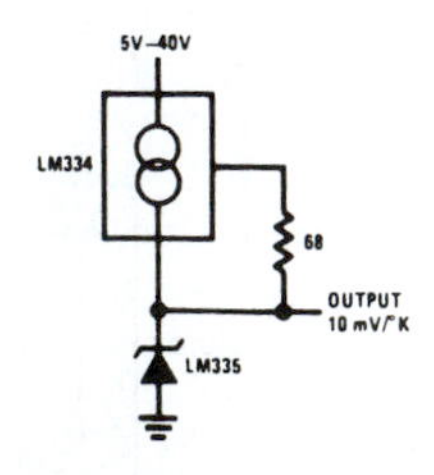


(Courtesy of National Semiconductor Corporation)

## Absolute Maximum Ratings

| | | |
|---|---|---|
| Reverse Current | | 15 mA |
| Forward Current | | 10 mA |
| Storage Temperature | | |
| TO-46 Package | | −60°C to +180°C |
| TO-92 Package | | −60°C to +150°C |
| Specified Operating Temperature Range | | |
| | Continuous | Intermittent (Note 2) |
| LM135, LM135A | −55°C to +150°C | 150°C to 200°C |
| LM235, LM235A | −40°C to +125°C | 125°C to 150°C |
| LM335, LM335A | −40°C to +100°C | 100°C to 125°C |
| Lead Temperature (Soldering, 10 seconds) | | 300°C |

## Temperature Accuracy LM135/LM235, LM135A/LM235A (Note 1)

| PARAMETER | CONDITIONS | LM135A/LM235A | | | LM135/LM235 | | | UNITS |
|---|---|---|---|---|---|---|---|---|
| | | MIN | TYP | MAX | MIN | TYP | MAX | |
| Operating Output Voltage | $T_C = 25°C$, $I_R = 1$ mA | 2.97 | 2.98 | 2.99 | 2.95 | 2.98 | 3.01 | V |
| Uncalibrated Temperature Error | $T_C = 25°C$, $I_R = 1$ mA | | 0.5 | 1 | | 1 | 3 | °C |
| Uncalibrated Temperature Error | $T_{MIN} < T_C < T_{MAX}$, $I_R = 1$ mA | | 1.3 | 2.7 | | 2 | 5 | °C |
| Temperature Error with 25°C Calibration | $T_{MIN} < T_C < T_{MAX}$, $I_R = 1$ mA | | 0.3 | 1 | | 0.5 | 1.5 | °C |
| Calibrated Error at Extended Temperatures | $T_C = T_{MAX}$ (Intermittent) | | 2 | | | 2 | | °C |
| Non-Linearity | $I_R = 1$ mA | | 0.3 | 0.5 | | 0.3 | 1 | °C |

## Temperature Accuracy LM335, LM335A (Note 1)

| PARAMETER | CONDITIONS | LM335A | | | LM335 | | | UNITS |
|---|---|---|---|---|---|---|---|---|
| | | MIN | TYP | MAX | MIN | TYP | MAX | |
| Operating Output Voltage | $T_C = 25°C$, $I_R = 1$ mA | 2.95 | 2.98 | 3.01 | 2.92 | 2.98 | 3.04 | V |
| Uncalibrated Temperature Error | $T_C = 25°C$, $I_R = 1$ mA | | 1 | 3 | | 2 | 6 | °C |
| Uncalibrated Temperature Error | $T_{MIN} < T_C < T_{MAX}$, $I_R = 1$ mA | | 2 | 5 | | 4 | 9 | °C |
| Temperature Error with 25°C Calibration | $T_{MIN} < T_C < T_{MAX}$, $I_R = 1$ mA | | 0.5 | 1 | | 1 | 2 | °C |
| Calibrated Error at Extended Temperatures | $T_C = T_{MAX}$ (Intermittent) | | 2 | | | 2 | | °C |
| Non-Linearity | $I_R = 1$ mA | | 0.3 | 1.5 | | 0.3 | 1.5 | °C |

## Electrical Characteristics (Note 1)

| PARAMETER | CONDITIONS | LM135/LM235 LM135A/LM235A | | | LM335 LM335A | | | UNITS |
|---|---|---|---|---|---|---|---|---|
| | | MIN | TYP | MAX | MIN | TYP | MAX | |
| Operating Output Voltage Change with Current | 400 $\mu$A $< I_R <$ 5 mA At Constant Temperature | | 2.5 | 10 | | 3 | 14 | mV |
| Dynamic Impedance | $I_R = 1$ mA | | 0.5 | | | 0.6 | | Ω |
| Output Voltage Temperature Drift | | | +10 | | | +10 | | mV/°C |
| Time Constant | Still Air | | 80 | | | 80 | | sec |
| | 100 ft/Min Air | | 10 | | | 10 | | sec |
| | Stirred Oil | | 1 | | | 1 | | sec |
| Time Stability | $T_C = 125°C$ | | 0.2 | | | 0.2 | | °C/khr |

**Note 1:** Accuracy measurements are made in a well-stirred oil bath. For other conditions, self heating must be considered.

**Note 2:** Continuous operation at these temperatures for 10,000 hours for H package and 5,000 hours for Z package may decrease life expectancy of the device.

## Application Hints

### CALIBRATING THE LM135

Included on the LM135 chip is an easy method of calibrating the device for higher accuracies. A pot connected across the LM135 with the arm tied to the adjustment terminal allows a 1-point calibration of the sensor that corrects for inaccuracy over the full temperature range.

This single point calibration works because the output of the LM135 is proportional to absolute temperature with the extrapolated output of sensor going to 0V output at 0°K (−273.15°C). Errors in output voltage versus temperature are only slope (or scale factor) so a slope calibration at one temperature corrects at all temperatures.

The output of the device (calibrated or uncalibrated) can be expressed as:

$$V_{OUT_T} = V_{OUT_{T_o}} \times \frac{T}{T_o}$$

where T is the unknown temperature and $T_o$ is a reference temperature, both expressed in degrees Kelvin. By calibrating the output to read correctly at one temperature the output at all temperatures is correct. Nominally the output is calibrated at 10 mV/°K.

To insure good sensing accuracy several precautions must be taken. Like any temperature sensing device, self heating can reduce accuracy. The LM135 should be operated at the lowest current suitable for the application. Sufficient current, of course, must be available to drive both the sensor and the calibration pot at the maximum operating temperature.

If the sensor is used in an ambient where the thermal resistance is constant, self heating errors can be calibrated out. This is possible if the device is run with a temperature stable current. Heating will then be proportional to zener voltage and therefore temperature. This makes the self heating error proportional to absolute temperature the same as scale factor errors.

### WATERPROOFING SENSORS

Meltable inner core heat shrinkable tubing such as manufactured by Raychem can be used to make low-cost waterproof sensors. The LM335 is inserted into the tubing about 1/2″ from the end and the tubing heated above the melting point of the core. The unfilled 1/2″ end melts and provides a seal over the device.

## Typical Applications

Simple Temperature Control

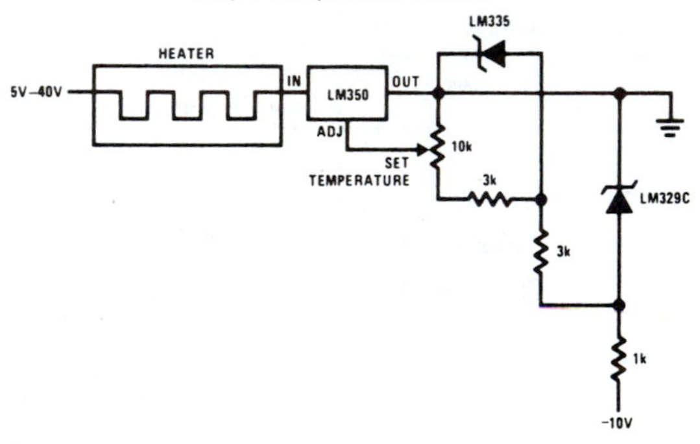


Simple Temperature Controller

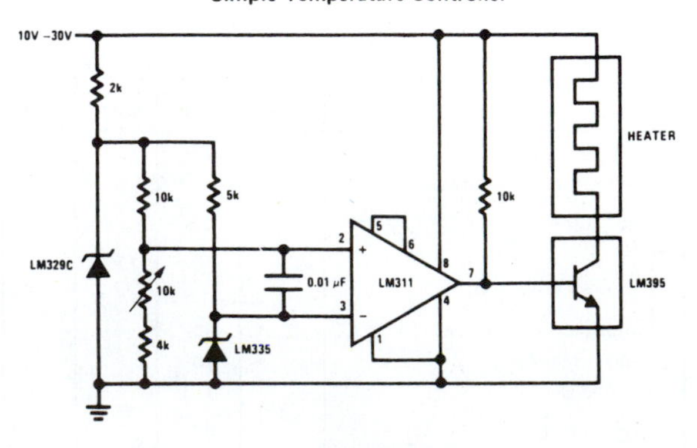

## Fast Charger for Nickel-Cadmium Batteries

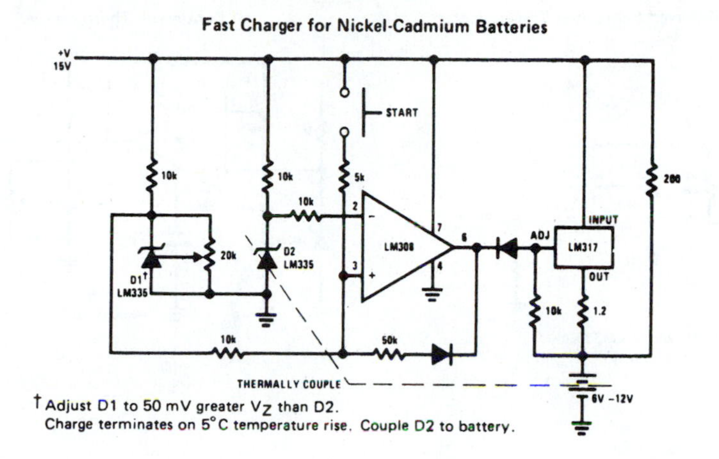


† Adjust D1 to 50 mV greater $V_Z$ than D2.
Charge terminates on 5°C temperature rise. Couple D2 to battery.

## Air Flow Detector*

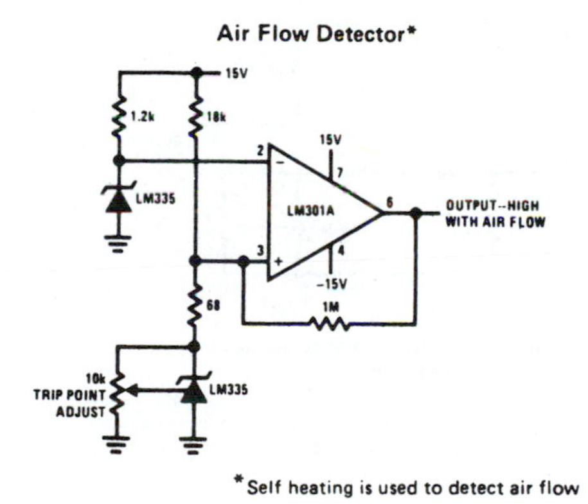


*Self heating is used to detect air flow

## Isolated Temperature Sensor

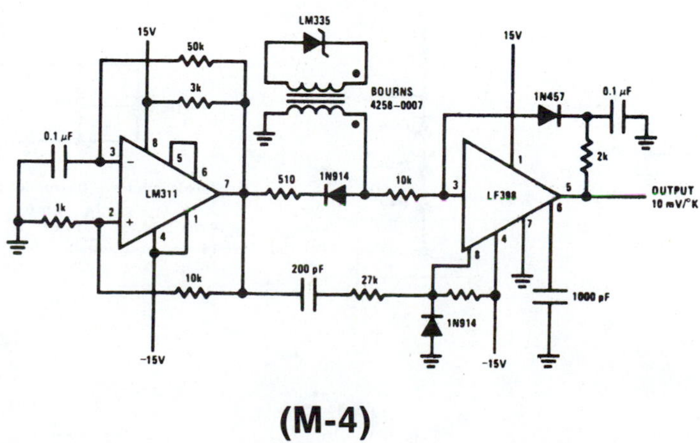

### Ground Referred Fahrenheit Thermometer

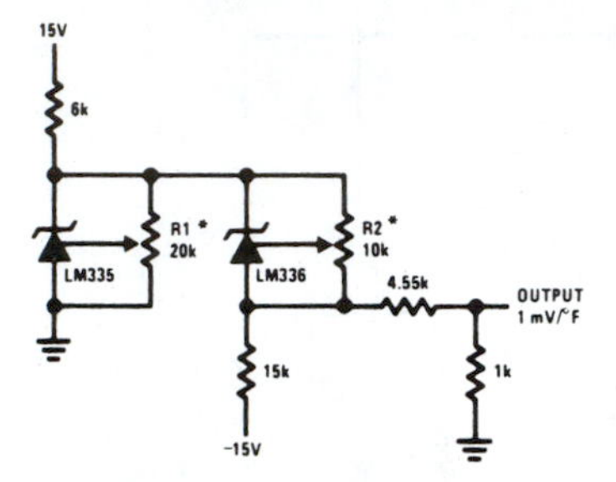


*Adjust R2 for 2.554V across LM336.
Adjust R1 for correct output.

### Centigrade Thermometer

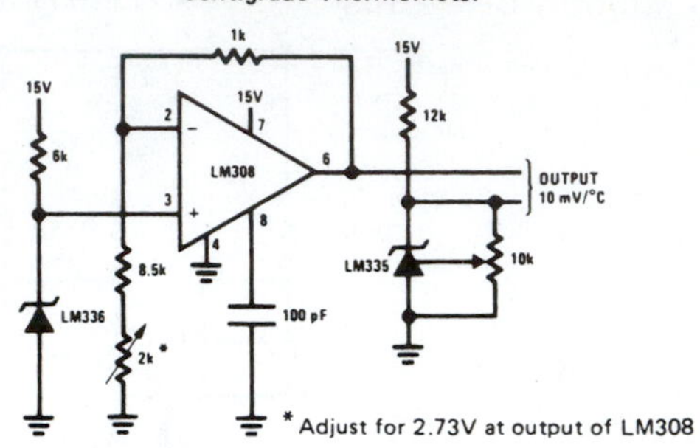


*Adjust for 2.73V at output of LM308

### Fahrenheit Thermometer

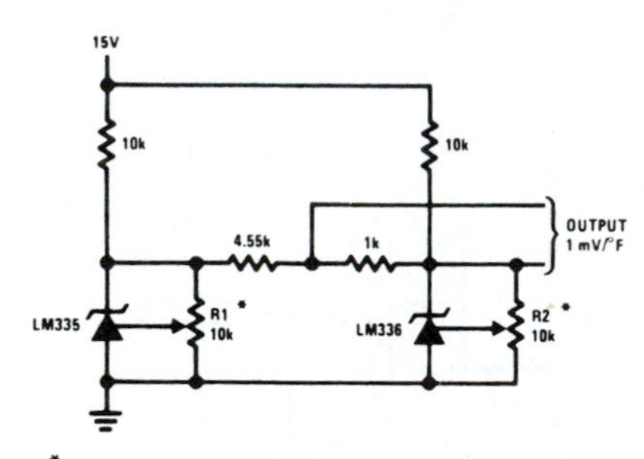


*To calibrate adjust R2 for 2.554V across LM336.
Adjust R1 for correct output.

### Minimum Temperature Sensing

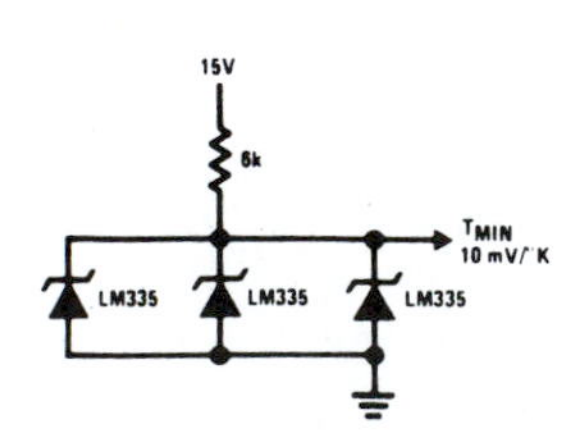


### Average Temperature Sensing

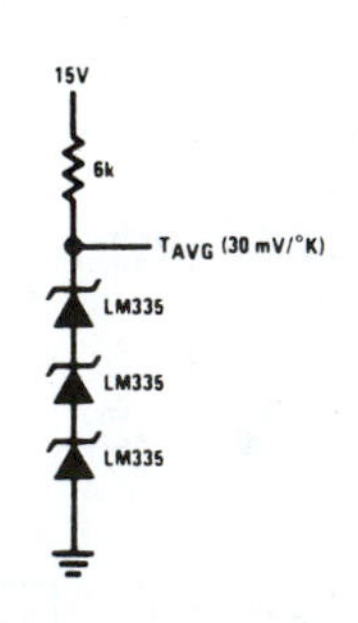


### Remote Temperature Sensing

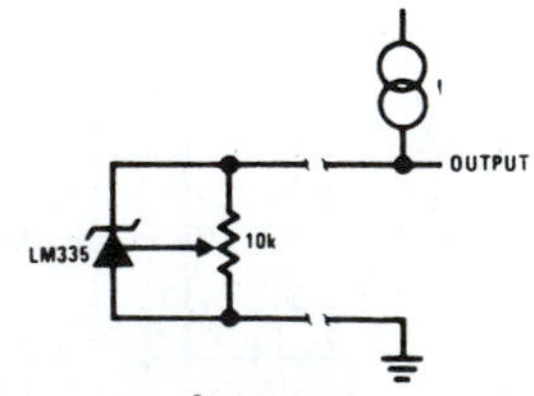


Wire length for 1°C error due to wire drop

| AWG | $I_R$ = 1 mA FEET | $I_R$ = 0.5 mA FEET |
|---|---|---|
| 14 | 4000 | 8000 |
| 16 | 2500 | 5000 |
| 18 | 1600 | 3200 |
| 20 | 1000 | 2000 |
| 22 | 625 | 1250 |
| 24 | 400 | 800 |

## Definition of Terms

**Operating Output Voltage:** The voltage appearing across the positive and negative terminals of the device at specified conditions of operating temperature and current.

**Uncalibrated Temperature Error:** The error between the operating output voltage at 10 mV/°K and case temperature at specified conditions of current and case temperature.

**Calibrated Temperature Error:** The error between operating output voltage and case temperature at 10 mV/°K over a temperature range at a specified operating current with the 25°C error adjusted to zero.

## Connection Diagrams

**TO-92**
**Plastic Package**

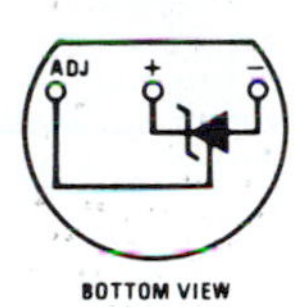


BOTTOM VIEW

**TO-46**
**Metal Can Package***

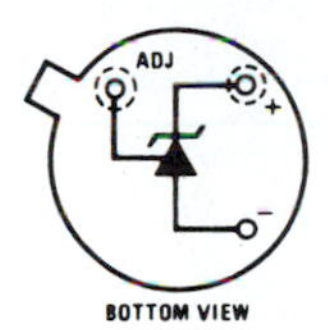


BOTTOM VIEW

*Case is connected to negative pin

## Physical Dimensions inches (millimeters)

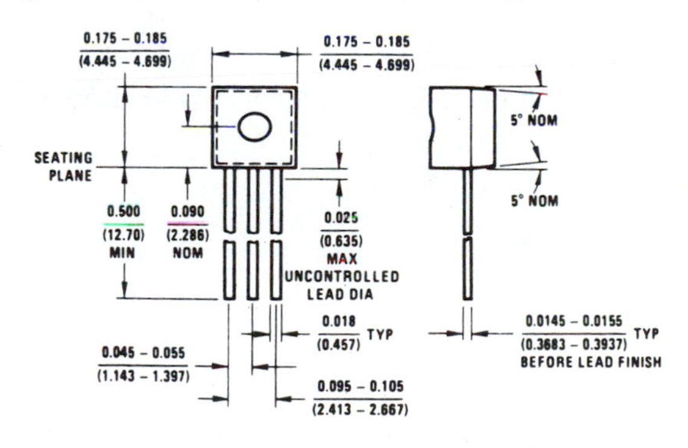


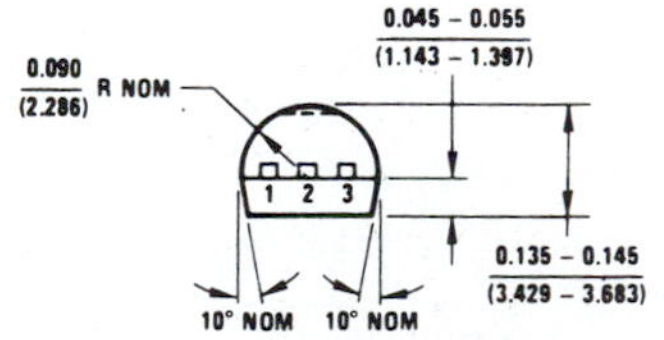


**TO-92 Plastic Package (Z)**
**Order Number LM335Z**
**or LM335AZ**
**NS Package Number Z03A**

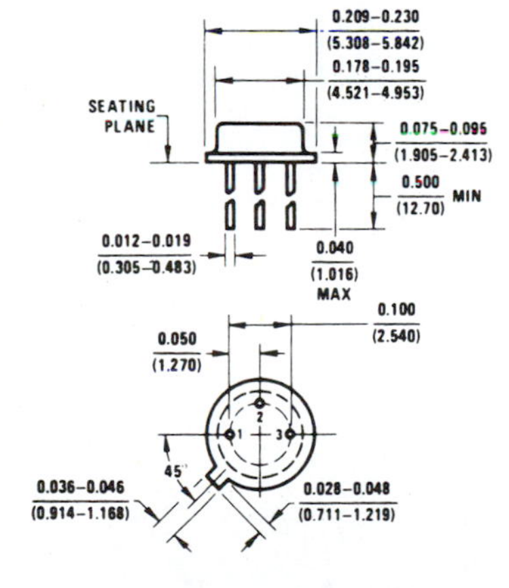


**TO-46 Metal Can Package (H)**
**Order Number LM135H,**
**LM235H, LM335H, LM135AH,**
**LM235AH or LM335AH**
**NS Package Number H03H**

**National Semiconductor Corporation**
2900 Semiconductor Drive
Santa Clara, California 95051
Tel.: (408)737-5000
TWX: (910)339-9240

**National Semiconductor GmbH**
Eisenheimerstrasse 61/II
8000 Munchen 21
West Germany
Tel.: (089)576091
Telex: 05-22772

**NS International Inc., Japan**
Miyake Building
1-9 Yotsuya, Shinjuku-ku 160
Tokyo, Japan
Tel.: (03)355-3711
TWX: 232-2015 NSCJ-J

**National Semiconductor (Hong Kong) Ltd.**
8th Floor,
Cheung Kong Electronic Bldg.
4 Hing Yip Street
Kwun Tong
Kowloon, Hong Kong
Tel.: 3-899235
Telex: 43866 NSEHK HX
Cable: NATSEMI

**NS Electronics Do Brasil**
Avda Brigadeiro Faria Lima 844
11 Andar Conjunto 1104
Jardim Paulistano
Sao Paulo, Brasil
Telex: 1121008 CABINE SAO PAULO

**NS Electronics Pty. Ltd.**
Cnr. Stud Rd. & Mtn. Highway
Bayswater, Victoria 3153
Australia
Tel.: 03-729-6333
Telex: 32096

National does not assume any responsibility for use of any circuitry described, no circuit patent licenses are implied, and National reserves the right, at any time without notice, to change said circuitry.

# Answers to Selected Questions and Problems

Questions and Problems requiring essay answers are not included in this answer section.

## CHAPTER 1

1.1. +5 V
0
−2 V

## CHAPTER 2

2.5. a. −40 to +85°C; b. 7.5 mV; c. 70 dB; d. ±30 mV
2.7. Equal to supply voltage
2.11. 10 kHz
2.13. 4 mV
2.15. 74 dB
2.17. 100.8 mW
2.19. 25 mA

## CHAPTER 3

3.1. (1) b,c,h; (2) i; (3) c; (4) b,g,h,i,k,l; (5) g,k,l; (6) a,c,f; (7) b,h; (8) e; (9) j; (10) d
3.3. 24 V
3.5. a. 20 mV; b. 4.4 mV; c. 48.3 mV

## CHAPTER 4

4.5. 0.56 μs
4.11. $V_{uth}$ = 2.45 V; $V_{lth}$ = 1.55 V
4.13. $V_H$ = 0.90 V
4.15. $C$ = 25 μF

## CHAPTER 5

5.11. 25%
5.19. 15 V

## CHAPTER 6

6.5. (1) a,c,e; (2) d,f,g; (3) b; (4) a; (5) b,c; (6) f; (7) d; (8) g; (9) b
6.15. 60 MHz
6.33. a. 2.838 MΩ; b. 660 μs; c. 1515 Hz
6.35. 10 s
6.37. 10 MΩ
6.39. a. 30.914 ms; b. 34 ms
6.41. 0.167 μF
6.43. 3.14 V
6.45. a. 762 nF; b. 12.5 kΩ

## CHAPTER 7

7.3. select $C = 1\ \mu F$, $R_2 = 10\ k\Omega$, $R_3 = 20\ k\Omega$; then $R_1 = 1600\ \Omega$; $A_{dB} = 9.5$ dB; breakpoint = 6.5 dB

7.5. a. scaling: select $C_1 = 0.05\ \mu F$; then $R_1 = R_2 = 22.5\ k\Omega$; $C_2 = 0.1\ \mu F$; $A_{dB} = 0$ dB; breakpoint = -3 dB
b. equal components: select $C_1 = C_2 = 0.05\ \mu F$; then $R_1 = R_2 = 31.8\ k\Omega$; $A_{dB} = 4$ dB; breakpoint = 1 dB

7.7. a. wideband filter; b. assume equal components, $A_v = 1.59$; c. $A_{dB} = 4$ dB; d. $BW = 2700$ Hz; e. second order

7.9. $A_v = 4.22$; $A_{dB} = 12.5$ dB

7.11. $f_{high} = 2200$ Hz; $Q = 5$

7.13. $f_{low} = 53$ Hz; $f_{high} = 67$ Hz

7.17. 1

## CHAPTER 8

8.21. a. 3.14; b. 2.14

8.23. a. -15.468 V; b. -15.418 V; c. 0.05 V

8.25. a. 2.0833 kΩ; b. 1.0204 kΩ; c. 251.26 Ω; d. 143.27 Ω; e. 66.756 Ω

8.27. a. 10 kHz; b. 20 kHz; c. 90 kHz

8.29. a. 0.002 V/μs; b. 0.005 V/μs; c. 0.01 V/μs

## CHAPTER 9

9.7. 1024

9.9. a. 2.5 V; b. 1.75 V; c. 3.75 V; d. 4.5 V; e. 63.75 V

9.11. 0.25 V

9.15. a. 0101010110; b. 3.42 ms; c. 0.098%

9.17. 100 μs

## CHAPTER 10

10.5. refer to Figures 10–6 and 10–12

10.7. refer to Figure 10–20

10.9. 30 MHz

10.11. 6 mV

10.13. 100%

## CHAPTER 11

11.13. 242 μs

11.15. 27.3 μF

11.19. 110 ms

11.21. a. 1.04 ms; b. 0.3475 ms; c. 1.3875 ms; d. 720 Hz

11.23. a. 34000 s; b. 17000 s; c. 51000 s

11.25. select $C = 470\ \mu F$; then $R = 394.7$ MΩ

11.27. a. 4.54 ms, 92%; b. 4.54 ms, 8%

## CHAPTER 12

12.5. Channel 3 = 61.25 MHz; Channel 4 = 67.25 MHz; $L_1 = 0.14\ \mu H$ (tunable); $C_2 = 39$ pF; $C_7 = 12$ pF (tunable)

12.15. 909 nF

12.17. 220 ms

# Index